The SAGE
Handbook of

Environmental Change

Volume 2

SAGE has been part of the global academic community since 1965, supporting high quality research and learning that transforms society and our understanding of individuals, groups, and cultures. SAGE is the independent, innovative, natural home for authors, editors and societies who share our commitment and passion for the social sciences.

Find out more at: **www.sagepublications.com**

The SAGE
Handbook of
Environmental Change

Volume 2

Edited by
John A. Matthews, Editor-in-Chief
Patrick J. Bartlein, Associate Editor
Keith R. Briffa, Associate Editor
Alastair G. Dawson, Associate Editor
Anne De Vernal, Associate Editor
Tim Denham, Associate Editor
Sherilyn C. Fritz, Associate Editor
Frank Oldfield, Associate Editor

Los Angeles | London | New Delhi
Singapore | Washington DC

First published 2012

SAGE Publications Ltd
1 Oliver's Yard
55 City Road
London EC1Y 1SP

SAGE Publications Inc.
2455 Teller Road
Thousand Oaks, California 91320

SAGE Publications India Pvt Ltd
B 1/I 1 Mohan Cooperative Industrial Area
Mathura Road, Post Bag 7
New Delhi 110 044

SAGE Publications Asia-Pacific Pte Ltd
33 Pekin Street #02-01
Far East Square
Singapore 048763

Library of Congress Control Number: 2011923232

British Library Cataloguing in Publication data

A catalogue record for this book is available from the British Library

ISBN 978-0-85702-360-5

Typeset by Cenveo Publisher Services
Printed in India at Replika Press Pvt Ltd
Printed on paper from sustainable resources

Contents

VOLUME 2   HUMAN IMPACTS AND RESPONSES

List of Figures   ix
List of Tables   xiii

**SECTION IV   HUMAN-INDUCED ENVIRONMENTAL CHANGES
AND THEIR IMPACTS ON GEO-ECOSYSTEMS**   1

25   Monitoring Global Land Cover   3
     Sietse O. Los and Jamie Williams

26   Human Impacts on Terrestrial Biota and Ecosystems   25
     Craig Miller and Iain Gordon

27   Human Impacts on Lacustrine Ecosystems   47
     Richard W. Battarbee, Helen Bennion, Peter Gell and Neil Rose

28   Human Impacts on Coastal and Marine Geo-Ecosystems   71
     Ben Daley

29   Human Impacts on the Atmosphere   95
     Kevin J. Noone

**SECTION V   PATTERNS, PROCESSES AND IMPACTS OF
ENVIRONMENTAL CHANGE AT THE REGIONAL SCALE**   111

30   Environmental Change in the Humid Tropics and Monsoonal Regions   113
     Mark B. Bush and William D. Gosling

31   Environmental Change in the Arid and Semi-Arid Regions   141
     Xiaoping Yang

32   Environmental Change in the Mediterranean Region   163
     Miryam Bar-Matthews

33   Environmental Change in the Temperate Forested Regions   188
     Matt McGlone, Jamie Wood and Patrick J. Bartlein

34   Environmental Change in the Temperate Grasslands and Steppe   215
     Pavel E. Tarasov, John W. Williams, Jed O. Kaplan, Hermann Österle,
     Tatiana V. Kuznetsova and Mayke Wagner

35   Environmental Change in the Arctic and Antarctic   245
     Marianne S. V. Douglas

36   Environmental Change in Mountain Regions   262
     Martin Beniston

37   Environmental Change in Coastal Areas and Islands   282
     Patrick Nunn

SECTION VI   PAST, PRESENT AND FUTURE RESPONSES OF PEOPLE
**   TO ENVIRONMENTAL CHANGE**   **299**

38   Testing the Role of Climate Change in Human Evolution   301
     Simon P. E. Blockley, Ian Candy and Stella M. Blockley

39   The Origins and Spread of Early Agriculture and Domestication:
     Environmental and Cultural Considerations   328
     Deborah M. Pearsall and Peter W. Stahl

40   Complexity, Causality and Collapse: Social Discontinuity in History
     and Prehistory   355
     Georgina Endfield

41   Vulnerabilities and the Resilience of Contemporary Societies to
     Environmental Change   374
     Donald R. Nelson

42   Disease, Human and Animal Health and Environmental Change   387
     Matthew Baylis and Andrew P. Morse

43   Policy and Management Options for the Mitigation of Environmental Change   406
     Katie Moon and Chris Cocklin

44   Socioeconomic Adaptation to Environmental Change: Towards Sustainable
     Development   426
     Chris J. Barrow

     Index   447

VOLUME 1   APPROACHES, EVIDENCE AND CAUSES

List of Figures   ix
List of Tables   xv
Notes on Contributors   xvii

Introduction

1   Background to the Science of Environmental Change   1
     John A. Matthews, Patrick J. Bartlein, Keith R. Briffa, Alastair G. Dawson,
     Anne De Vernal, Tim Denham, Sherilyn C. Fritz and Frank Oldfield

SECTION I   APPROACHES TO UNDERSTANDING ENVIRONMENTAL
**   CHANGE**   **35**

2   Philosophical and Methodological Perspectives on the Science of
     Environmental Change   37
     Stephan Harrison

3   Direct Observation and Monitoring of Climate and Related Environmental Change   53
     Keith Alverson

4   Reconstructing and Inferring Past Environmental Change   67
     Frank M. Chambers

5   Dating Environmental Change and Constructing Chronologies   92
     Mike Walker

6   Modelling Environmental Change and Developing Future Projections   116
     Reto Knutti

7   Approaches to Understanding Long-term Human–Environment Interactions:
     Past, Present and Future   134
     John A. Dearing

SECTION II   EVIDENCE OF ENVIRONMENTAL CHANGE AND THE
**   GEO-ECOLOGICAL RESPONSE**   **163**

8   Environmental Change in the Geological Record   165
     Jane Francis, Alan M. Haywood, Daniel Hill, Paul Markwick
     and Claire McDonald

9   Evidence of Environmental Change from the Marine Realm   181
     Ian D. Goodwin and William R. Howard

10   Evidence of Environmental Change from the Cryosphere   211
     Shawn Marshall

11   Evidence of Environmental Change from Terrestrial Palaeohydrology   239
      Wim Z. Hoek

12   Evidence of Environmental Change from Terrestrial and Freshwater
      Palaeoecology   254
      Alison J. Smith

13   Evidence of Environmental Change from Aeolian and Hillslope Sediments
      and Other Terrestrial Sources   284
      Joseph A. Mason

14   Environmental Change and Archaeological Evidence   305
      Tim Denham

15   Evidence of Environmental Change from Annually Resolved Proxies
      with Particular Reference to Dendrochronology and the Last Millennium   320
      Eugene R. Wahl and David Frank

16   Early-Instrumental and Documentary Evidence of Environmental Change   345
      Cary J. Mock

**SECTION III   CAUSES, MECHANISMS AND DYNAMICS OF
                ENVIRONMENTAL CHANGE   361**

17   Plate Tectonics, Continental Drift, Vulcanism and Mountain Building   363
      Paul Bishop

18   Extraterrestrial Causes of Environmental Catastrophes   384
      Elisabetta Pierazzo and H. Jay Melosh

19   Astronomical Theory and Orbital Forcing   405
      André Berger and Qiuzhen Yin

20   Millennial-Scale Climatic Events During the Last Glacial Episode   426
      Siwan M. Davies and Anders Svensson

21   Solar and Volcanic Forcing of Decadal- to Millennial-scale Climatic Variations 444
      Raimund Muscheler and Erich Fischer

22   Ocean–Atmosphere Interactions on Interannual to Decadal Time Scales   471
      Mathias Vuille and René D. Garreaud

23   Responses of Biogeochemical Cycles in the Sea to Environmental Change   497
      Thomas F. Pedersen and Rainer Zahn

24   Anthropogenic Drivers of Environmental Change   517
      Jemma L. Gornall, Andrew J. Wiltshire and Richard A. Betts

      Index   537

List of Figures

(* colour figures)

25.1   Comparison of global albedo derived from the Matthews and
       Wilson–Henderson-Sellers land-cover classifications   7
25.2* Mean monthly NDVI for December and June   9
25.3* Deviations in NDVI from the mean Austral summer NDVI (DJF) during
       warm El Niño Southern Oscillation (ENSO) events (1982, 1986, 1987
       and 1997)   10
25.4* (a) Landsat TM subscene of an area in Rondonia, Brazil in 1986. (b) Mean
       annual NDVI for the region with 200–400 mm mean annual rainfall   18
26.1   (a) The estimated proportion of all genera extinct at the boundary of the
       five big extinction periods. (b) The extinction intensity of marine mammals 28
26.2   (a) The saline and arid landscape of Minqin County, China.
       (b) A farmer confronting salinisation, New South Wales, Australia   30
26.3   Introduced brushtail possum (*Trichosurus vulpecula*) and ship rat (*Rattus rattus*)
       eating song thrush nestlings (*Turdus philomelos* – also introduced)   31
26.4   Conceptual stock-and-flow model depicting a three-level predator–prey system 38
26.5   Graphical representation of the behaviour of the predator–prey system model 39
27.1   Summary diatom diagram of Lake of Menteith   51
27.2   Comparison of pH reconstruction outputs and annual measured pH   55
27.3   Declining representation of thalassic diatoms in the Ramsar wetlands
       on the lower River Murray   60
27.4   Temporal trends of Pb and Hg concentrations, and ΣDDT and ΣPBDE fluxes
       from the sediment record of Lochnagar, Scotland   63
28.1   Dugongs caught in Hervey Bay, Queensland, Australia, *c.*1937   77
28.2   Corals taken from the Great Barrier Reef, Australia, *c.*1940   80
29.1   Monthly mean CO_2 concentration measured at Mauna Loa observatory
       in Hawaii, USA   98
29.2   Mauna Loa CO_2 data together with data from Lüthi, D. et al. (2008)
       EPICA Dome C Ice Core 800 ka carbon dioxide data   99
29.3   Mean ozone levels at Halley station for the months of February and October 100
29.4   Ozone concentrations above the NOAA South Pole station for 22 August
       and 25 October in 2008   101
29.5* Spatial distributions of (a) NO_2, (b) non-sea-salt sulfate, (c) methane and
       (d) carbon dioxide   103
29.6* Annual average pH of rainwater at the global scale   105
30.1   (a) Exposed continental area during the Last Glacial Maximum.
       (b) Eustatic sea-level change during the transition from glacial to
       interglacial conditions   116

30.2 Tropical climate systems for: (a) January and (b) July 119
30.3 Block diagram showing inferred lake level in the southern hemisphere
tropical Andes and western Amazonia 120
30.4 Early-Holocene antiphasing of selected paleoclimatic records
from South America 124
30.5 Early-Holocene antiphasing of selected paleoclimatic records
from Africa and Asia 125
31.1 Global key regions of drylands and savannas 143
31.2 On-site evidence of formerly wetter environments in the present-day
hyper-arid zones 144
31.3 Synthesis of late Quaternary palaeoclimatic changes in the
southwestern Kalahari, Namibia 145
31.4 Schematic picture of morphogenetic stories in northern (a) and
southern (b) Africa caused by regional climates 146
31.5 Alternating climatic conditions in the Kalahari and Sahara due to
changes in solar radiation in each hemisphere 147
31.6 Wetter climate (recognized from lacustrine processes) in the Taklamakan
Desert of China and colder times (recognized from permafrost processes) 149
31.7 Simplified history of pluvial Lake Mojave fluctuations and aeolian
depositional periods in the Mojave Desert 151
32.1* Proxy records of the last two interglacial–glacial cycles 167
32.2 $\delta^{18}O_{G.ruber}$ from marine cores (a) 9501 SE of Cyprus and (b) 9509 off the
Nile plum vs. age superimposed on Soreq Cave speleothems record 168
32.3 $\delta^{18}O_{G.ruber}$ of the last 250 ka from marine core MD84651 169
32.4 $\delta^{18}O$ record of Soreq Cave speleothems for the last 180 ka 170
32.5 Alkenones-based paleo SST from (a) various marine cores in the EM and
(b) speleothems-based paleotemperatures 171
32.6 Mid-Holocene cultural changes superimposed on the smooth curve of
the $\delta^{13}C$ record of Soreq Cave speleothems 177
33.1* Megabiomes 190
34.1 (a) The distribution of temperate and tropical grasslands. (b) Mean monthly
values of precipitation and (c) mean monthly temperature 216
34.2 Reconstructed shifts in the prairie-forest ecotone during the Holocene 225
34.3 Selected time series of environmental indicators from the
grassland-dominated North American mid-continent 226
34.4 The major pollen types in the Late Glacial–Holocene records from the
northern Great Plains 228
34.5 Selected pollen types in the Late Glacial–Holocene records from the
grassland-dominated regions of East Europe 230
34.6 Selected Holocene palaeoclimate records from northern Eurasia 232
34.7 Changes in the total area of crop and pasture (a–c) and in the
total area of natural grasslands (d–f) reconstructed for the three regions 235
34.8 Temperate grasslands and graminoid and forb tundra simulated by the
BIOME4 model for four time slices 237
35.1 Polar projection showing various boundaries used to demarcate the Arctic 247
35.2 Polar projection showing various boundaries used to demarcate the Antarctic 248
36.1* Mountains and upland regions 263
36.2* Changes in the average mass balance of mountain glaciers of
different continents 266

36.3* Snow-cover duration as a function of winter (DJF) minimum temperature
and precipitation for two Swiss locations under current climate and for the
last three decades of the twenty-first century (IPCC A2 emissions scenario) 273
37.1 (a) Late Quaternary sea-level changes. (b) Changes in the geography of
southwest Pacific islands 284
37.2 Tossed as though it were a pebble, this giant piece of reefrock
(named Kasakanja) was thrown 30 m up onto the clifftop of
southern Okinawa Island, Japan 287
37.3 Examples of submarine landslides resulting from island-flank collapses 288
37.4 The coast of Niue Island (central Pacific Ocean) exhibits evidence for
recent uplift associated with movement up the flank of a lithospheric flexure 290
37.5 Eroding shoreline of Luamotu Island, Funafuti Atoll, Tuvalu 291
38.1 Key events in human evolution and dispersal over the last glacial cycle for
Africa and Eurasia plotted against the NGRIP and GRIP ice core record 309
38.2 The middle to upper Palaeolithic transition in Europe plotted against the
NGRIP ice core record 322
39.1* Phenotypes of some crops and their progenitors illustrating changes in
fruit form, seed dispersal, size, and plant architecture 330
39.2 Overview of the geography of plant domestication 332
39.3 The spread of the West Asian founder crops and sites discussed in the text 333
39.4 The spread of rice agriculture in East Asia and sites discussed in the text 336
39.5 The spread of maize and manioc in the Americas and sites
discussed in the text 340
40.1 Hypothetical relationships between climatic events, vulnerability and societal
adaptation or collapse 360
42.1 A schematic framework of the effects of climate change on diseases of
humans and animals 391
42.2 Projection of the effect of climate change on the future risk of transmission
of bluetongue in northern Europe 397
43.1 The four main interrelationships between mitigation and adaptation action 409
43.2 Estimated sectoral economic potential for global mitigation for different
regions as a function of carbon price in 2030 from bottom-up studies 419
44.1 Adaptation to environmental change and variability 427
44.2 Linkages for climate change policies responding to climate change and
seeking to support sustainable development 440
44.3 Different approaches to linking adaptation and development 441

List of Tables

25.1   Confusion matrix for five most frequent classes in the Matthews (1983, 1984) classification   6
25.2   Some commonly adopted classification schemes   14
25.3   Comparison of thematic agreement (in %) at nominal 0.5° resolution grid of remotely sensed derived DISCover   15
28.1   Summary of human impacts on coastal and marine geo-ecosystems   72
29.1   Source strength, mass loading, lifetime and optical depth for five major aerosol types   104
30.1   Palaeoenvironmental records   121
31.1   Arid and semi-arid areas of the world ($km^2 \times 10^6$)   142
33.1   Climatic descriptors for moist forest types   189
33.2   North America and Europe. Vegetation history from selected sites and regions   196
33.3   Asia and southern South America. Vegetation history from selected sites and areas   197
33.4   Australasia. Vegetation history for selected sites and areas   198
33.5   Late Quaternary extinctions of megafauna from temperate regions   203
34.1   Linear trends of the annual temperature (TANN) and precipitation (PANN) in the selected regions   223
34.2   Linear trends in northern hemisphere winter (December–February), spring (March–May), summer (June–August) and autumn (September–November) temperature (°C) calculated in seven bioclimatic regions   235
34.3   Contribution of different seasons to the temperature rise between 1936 and 2007 in selected climatic regions   236
34.4   Area of temperate grassland simulated by the BIOME4 model for four time slices in the Last Glacial and Holocene   236
38.1   General timeline of the main species in hominid evolution   304
41.1   Priorities for environmental change policy: Comparing three response typologies   382
42.1   The major diseases transmitted by arthropod vectors to humans and livestock   393
43.1   Similarities and differences between mitigation and adaptation action   410
43.2   Summary of three decision-making cultures   414
43.3   Summary of key greenhouse gas mitigation policy instruments   415
43.4   Key mitigation technologies and practices by sector   418
44.1   Seven Asian rivers, which are highly dependent on glaciers/snowmelt to maintain summer flows, with estimated basin populations   431
44.2   Some reasons for failure to adapt to environmental and other challenges   435

Human-induced Environmental Changes and their Impacts on Geo-ecosystems

Monitoring Global Land Cover

Sietse O. Los and Jamie Williams

1 INTRODUCTION

Monitoring of land-surface vegetation is important for several reasons. It provides information to study biodiversity and how it may change over time; it provides information on the impact of climate events, such as droughts or warmer springs, on vegetation; and it provides data for land-surface and ecological models. These models are used to understand the global carbon cycle, water cycle and energy budget. The present chapter provides an overview of the literature on measurement techniques and methods to assemble land-surface vegetation data.

Land-surface vegetation is an important component of the climate system. Vegetation affects the land-surface albedo and thereby affects the amount of solar radiation absorbed and reflected by the land surface. The process of photosynthesis by vegetation has a predominant role in the carbon cycle, involving the direct storage of carbon in live vegetative material. On the global scale, an even larger store of carbon is formed by soil organic matter that is maintained by an influx of dead plant material. Respiration of vegetation and decomposition of soil organic matter release carbon dioxide (CO_2) back into the atmosphere. For the globe, the CO_2 fluxes from the land to the atmosphere and back are almost in balance on an annual basis; the difference between them is approximately 2–5 per cent. Photosynthesis regulates the release of H_2O to the atmosphere (referred to as transpiration), and affects the water cycle. Vegetation thereby regulates the partitioning of the latent heat flux (the energy used for transpiration and evaporation) and sensible heat flux from the land surface. The litter layer aids infiltration of precipitation and protects the soil from erosion. The height and shape of vegetation affect the wind profile and the atmospheric turbulence near the land surface. Vegetation, through the release of carbonic and humic acids at the roots, enhances weathering and the uptake of nutrients from the soil. Information of the state of vegetation is required at frequent intervals to model and understand the global mass and energy budgets. It is impractical to rely only on field measurements, hence satellites, which provide global coverage at frequent intervals, play an important role in collecting land-surface vegetation data.

There are various approaches to model the global water, energy and carbon cycles. However, the discussion in the present chapter is limited to information on vegetation data required for land-surface parameterizations and ecological models. Land-surface parameterizations are used either as standalone models or as models coupled to general circulation models of the atmosphere. Land-surface parameterizations simulate the exchange of water, energy and momentum between the land and atmosphere at half-hourly to hourly time steps. Examples of land-surface parameterizations are the Biosphere–Atmosphere Transfer Scheme (BATS; Dickinson et al., 1986), the Simple Biosphere model (SiB; Baker et al., 2008; Sellers et al., 1996a), the Integrated Biosphere Simulator Model (IBIS; Foley et al., 1996, 2005), and the Joint UK Land Environment Simulator (JULES) derived from the UK Met Office Surface Exchange Scheme (MOSES; Essery et al., 2001). Land-surface parameterizations require biophysical parameters; that is, quantitative measures of vegetation, to calculate photosynthesis and the energy and mass budgets. Biophysical parameters used in land-surface parameterizations include:

1 Leaf area index (LAI), defined as the one-sided leaf area per unit surface area. LAI is used to calculate albedo and interception of precipitation.
2 The fraction of photosynthetically active radiation (PAR) region absorbed by the green parts of the vegetation canopy (fAPAR or FPAR). fAPAR is used to calculate the photosynthetic rate and the stomatal conductance.

The objective of ecological models is to simulate aspects of the biosphere such as competition between species or nutrient cycles regulated by growth and decomposition of vegetation at daily to annual time scales. Examples of ecological models are CENTURY (Parton et al., 1987; see http://www.nrel.colostate.edu/projects/century/), a mapped atmosphere–plant–soil system (MAPPS; Neilson, 1995), the Carnegie Ames Stanford Approach (CASA; Potter et al., 1993), the BIOME biogeochemical model (BIOME-BGC; Running and Hunt Jr., 1993; Thornton et al., 2002), the BIOME model (Haxeltine and Prentice, 1996) and the Lund–Potsdam–Jena Dynamic Global Vegetation Model (LPJ; Sitch et al., 2003). Some of these ecological models calculate biophysical parameters and use observed biophysical

parameters for validation; others require biophysical parameters as input. Similar parameters are used in ecological models and land-surface parameterizations, although ecological models tend not to use parameters linked to the energy budget (albedo) and exchange of momentum (surface roughness) and tend to calculate net primary production over periods of a month, rather than calculate photosynthesis at half-hourly time steps. The net primary production of vegetation is the difference between the amount of CO_2 absorbed by photosynthesis and the amount of CO_2 respired by vegetation for maintenance. Other examples of parameters used in either ecological models, land-surface parameterizations or both are the vegetation cover fraction, biomass, canopy shape, vegetation height, leaf dimensions, leaf optical properties, leaf orientation and leaf density distribution.

Prior to the 1980s, vegetation parameters were obtained from global land-cover classification maps that contain the global distribution of land-cover types such as needle-leaf forests, agricultural lands, tundra, tropical forests and grasslands. These classifications were combined with look-up tables that provided information on the seasonal variation of leaf area index, fAPAR, biomass, and so on. The look-up tables were assembled from site measurements and literature surveys. Since the 1990s biophysical parameters are also derived from satellite data. Satellite data provide effective wall-to-wall coverage of the globe at weekly to monthly intervals. With satellite data it is possible to obtain information on deforestation rates, year-to-year variations in vegetation greenness of desert margins, and the response of vegetation to climate change. Because of the short time over which satellite data are available there is a strong focus of the present chapter on the recent past. Long-term changes in the environment are discussed elsewhere in the Handbook.

The remainder of the chapter is organized as follows. In section 2, a discussion is provided of global land-cover classifications assembled from inventories. Land cover is used in a broader sense by the land-cover change community and includes vegetation, as well as surface water, rock and bare soil. In section 3, the monitoring of land cover and estimation of biophysical parameters from satellite is discussed, and later sections assess several land-cover classification products obtained from satellite data. The purpose of the present chapter is to provide a flavour of recent developments in this area and to give an introduction to some of the key literature. Owing to the plethora of new material that has appeared over the past decades, the discussion is by no means complete or exhaustive.

2 GLOBAL LAND-COVER CLASSIFICATIONS BASED ON INVENTORIES

During the 1980s the availability of more powerful computers allowed models to perform simulations for the global scale. A need for global vegetation data emerged and several authors assembled global land-cover maps to estimate biophysical parameters. A significant amount of work was put into these inventories; material was collected from various archives and sources and interpretation of this material assembled for different time periods was an enormous challenge. The assemblage of a consistent global product required interpretation and judgment of the interpreter at several stages.

Global ecological models and land-surface parameterizations from the 1970s and 1980s combine look-up tables of biophysical parameters with a land-cover classification map to obtain estimates of their spatial distribution (Dorman and Sellers, 1989). The look-up tables are obtained from literature surveys and site measurements and provide typical values of biophysical parameters for a particular biome. These biophysical parameters (e.g. albedo, leaf area index or fAPAR) are then assigned to each class. A criticism of this method is that 'typical' values for a

biome may reflect optimum rather than average conditions. For example, leaf area index measurement for a boreal forest may be measured for areas with dense tree cover and the measurements are therefore not representative for average conditions where tree cover is lower. Another criticism of land-cover classifications assembled from inventories is that they do not provide information on the spatial variation of biophysical parameters within a class and they do not provide information on interannual variations.

The following issues need to be addressed to obtain a consistent global land-cover map.

- Definitions of source material must be translated to a common set of classes. A frequently used classification system is the UNESCO (1973) land classification scheme.
- Different collection protocols for source material need to be translated to one standard.
- Differences in age of source material, and sometimes uncertainty in the age of source material, needs to be taken into account. In some cases newer information may have been copied from previous work without proper acknowledgement.
- Sometimes the documentation is poor as to how the data were obtained or the source material was interpreted.
- Judgement calls must be made in cases where there is contradictory information in the source material. The judgement calls depend on the expertise of analysts and knowledge of a particular area. Because there are many judgement calls made, they may not always be documented.
- The compilations are made for different purposes (land inventory, ecological function, calculation of carbon stocks or calculation of albedo) and different classifications may not be directly comparable as a result.

Three examples of compilations of these data are the Olson global land-cover classification (Olson et al., 1985), the Matthews global land-cover classification (Matthews, 1983, 1984) and the Wilson and Henderson-Sellers global land-cover classification (Wilson and Henderson-Sellers, 1985). Each global compilation represents a large body of work, in some cases as much as 20 years of research.

The Olson global land-cover classification (Olson et al., 1985; see http://cdiac.ornl.gov/) was developed to model the global carbon cycle. It was derived from the land-cover data set by Hummel and Reck (1979) which was used to calculate global albedo fields. The Olson et al. (1985) data set aims to be an inventory of carbon stocks (biomass) for the 1980s and provides information on carbon stocks for potential vegetation (defined as vegetation prior to the Iron Age) as well. It provides estimates where vegetation is destroyed and where it has increased and translates this information to comparable changes in carbon stocks. Ranges of uncertainty are also provided. For some areas estimates of potential vegetation are derived from climate data (Olson et al., 1985).

The Matthews global land-cover classification (Matthews, 1983, 1984; see http://data.giss.nasa.gov) was assembled for use in climate models. Its primary purpose is to derive land-surface biophysical properties such as albedo for the Goddard Institute of Space Studies (GISS) general circulation model (GCM) of the atmosphere (Hansen et al., 1983;

Matthews, 1984). The Matthews land-cover classification is based on the UNESCO classification, and can be used for multiple purposes. The classification in 32 classes allows a large degree of flexibility. The data also contain information on land use.

The Wilson–Henderson-Sellers (see http://dss.ucar.edu) land-cover data was intended for use in climate models and to derive land-surface biophysical properties for the National Center for Atmospheric Research (NCAR) GCM in the mid 1980s (Wilson and Henderson-Sellers, 1985).

A test of accuracy known as a confusion matrix provides an indication of the variation that results from differences in source material, interpretation of source material and the classification scheme adopted. This test of accuracy was applied to the three most frequent classes in the Olson, Mathews and Wilson land-cover classes (Table 25.1). Tundra (class 22, Matthews; class 61, Wilson-Henderson-Sellers; class 53, Olson) agrees well between the three land-cover classifications. Tropical evergreen rain forest (M 1; W 50) agrees well for the Matthews and

Table 25.1 Confusion matrix for 5 most frequent classes in the Matthews (1983, 1984) classification (M), the Wilson & Henderson-Sellers (1985) classification (W) and the Olson et al. (1985) land-cover classification (O). Antarctic ice was excluded. Numbers indicate the numbers of 0.5° × 0.5°cells. M/W confusion matrix of Matthews vs. Wilson; M/O Matthews vs. Olson, W/O Wilson vs. Olson. The Matthews and Wilson-Henderson Sellers data were sampled to 0.5° × 0.5° resolution by repeating the land-cover types of the 1° × 1° resolution

M\W	31	40	50	61	71	M\O	21	31	41	51	53
1		7	3086			1			115	204		6
8	307	216		78		8	1611	43	11	2	274
22	64	12		3526		22	201	4	2		3303
25	783	277	4		485	25		312	1104	584	16
30	296	28		447	2595	30		23	830	2273	659
						W\O	21	31	41	51	53
						31	186	206	676	76	460
						40	3	1170	204	36	5
						50		94	180	2	5
						61	61				3806
						71		11	832	2147	29

Legend for Matthews classes shown: 1: Tropical evergreen rain forest, mangrove forest. 8: temperate/sub polar evergreen needle leaf forest 22 (M): arctic - alpine tundra, mossy bog. 25 (P): tall/medium/short grassland with shrub cover. 30 (U): desert. Legend for Wilson and Henderson Sellers classes: 31: Temperate rough grazing. 40: Arable cropland. 50: Equatorial rain forest 61: Tundra 71: Scrub desert and semi desert. Legend for Olson classes 21: Main Taiga. 31: Other crop, settlements and marginal lands, warm or hot farms, towns. 41: Grass and shrub complexes, main grassland or shrub land, warm or hot shrub and grassland. 51: non-polar desert or semi-desert, other (not cool or sand desert) desert and semi-desert. 53: Tundra. Empty cells have zero values.

Wilson-Henderson-Sellers classifications, but differs for the Olson classification. Desert agrees well between the three classification schemes. Some of the mismatch between classes occurs between classes with similar features; for example, between tundra and grassland. A more important question in terms of using the data as input to models is how the disagreement between classification schemes translates to differences in biophysical parameters. An example of this is shown in Figure 25.1, where the albedo for the Matthews and Wilson data is calculated; the average values look similar, but several outliers can be noticed from the 1:1 line.

The compilation of global land-cover maps from inventories represented important progress, both in terms of the information made available as well as the recognition that information on the land is crucially important for the modelling of biogeochemical cycles and climate.

3 LAND-COVER MONITORING FROM SATELLITE

Wall-to-wall global monitoring of land-surface vegetation became possible with the launch of the NOAA satellites; the NOAA series started with the launch of NOAA-7 in August 1981 and continues to the present day (Kidwell, 1998; Rodwell, 2009). The NOAA satellites are polar-orbiting, operational satellites; that is, their operation is deemed essential and after failure every attempt is made to replace a satellite. The NOAA satellites carry the advanced very-high-resolution radiometer (AVHRR). The AVHRR collects data over a swath width of about 2,500 km and at a nadir resolution of 1.1×1.1 km; the resolution becomes coarser with increasing viewing angle. Usually only data from afternoon satellites are used for vegetation monitoring. The equatorial local overpass time for the afternoon satellites varies between 1pm

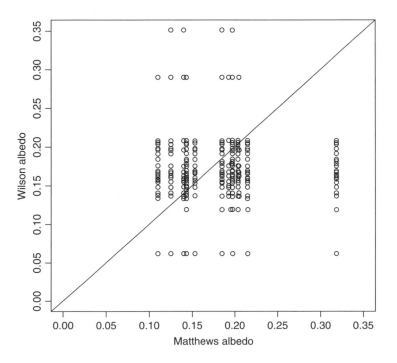

Figure 25.1 Comparison of global albedo derived from the Matthews and Wilson–Henderson-Sellers land-cover classifications. A soil albedo of 0.35 for the Wilson–Henderson-Sellers data is assumed.

just after launch to 4 or 5pm after 4–6 years of the operation of a satellite. Data are collected by the AVHRR at 1.1 × 1.1 km resolution at nadir and are sampled on board to a 5.5 × 3.3 km resolution; only four 1.1 km pixels of one out of three lines represents a 5.5 × 3.3 km pixel. The global area coverage GAC data (3.3 × 5.5 km resolution at nadir) are collected for most of the globe twice daily, once during the night and once during the day. Data at 1.1 km resolution are available for selected areas and times during the 1980s, the coverage improves from the 1990s onwards when more receiving stations across the world started collecting AVHRR data on a routine basis. The AVHRR was not designed for observation of global land cover, but was nevertheless very successful for this purpose. Its success justified the launch of various subsequent missions such as the Satellite Pour l'Observation de la Terre (SPOT) Vegetation, Sea-viewing Wide Field-of-view Sensor (SeaWiFS), the Advanced Along Track Scanning Radiometer (AATSR), the Moderate-resolution Imaging Spectroradiometer (MODIS) and the Medium Resolution Imaging Spectrometer (MERIS).

The AVHRR detects radiation reflected from or emitted by the land surface in five bands; for vegetation monitoring, the red and near-infrared bands are the most important. Detection of vegetation from satellite is based on its unique spectral properties – vegetation strongly absorbs radiation in the visible wavelengths and strongly reflects radiation in the near infrared bands. Other observable entities from space; for example, bare soil, clouds, water and snow, tend to reflect radiation in the red and near-infrared bands in roughly similar amounts. A vegetation index; for example, the normalized difference vegetation index (NDVI; see Figure 25.2) exploits the unique spectral property of vegetation (Tucker, 1979). NDVI calculates the difference between the near infrared and red reflectance and divides this by the sum:

$$V = \frac{\rho_2 - \rho_1}{\rho_2 + \rho_1} \qquad (25.1)$$

where V = NDVI, ρ_1 = the visible reflectance, and ρ_2 = the near-infrared reflectance. NDVI values of dense vegetation tend to be around 0.7 or higher, the NDVI of bare soil tends to be slightly above zero and the NDVI of snow tends to be slightly below zero.

AVHRR NDVI data collected from August 1981 until the present provide information on the seasonality and the spatial distribution of vegetation (Figure 25.2) and on the interannual variations in vegetation (Figure 25.3). Seasonal variations in vegetation at the global scale as detected by the AVHRR are linked to seasonal variations in drawdown of atmospheric CO_2 (Tucker et al., 1986). Interannual variations in NDVI are used to detect variations in vegetation in desert margins (Prince et al., 2007; Tucker et al., 1991); the same observations are also used to detect the impacts of droughts leading to below-normal crop production (Hutchinson, 1991), or unusually wet conditions in desert margins leading to locusts outbreaks (Hielkema and Snijders, 1994). Early warning systems monitor these conditions on a regular basis, although this information is not always put to use by the authorties (Ceccato et al., 2007).

The NDVI is linked to the fraction of photosynthetically active radiation absorbed by vegetation (fAPAR or FPAR; see Asrar et al., 1986; Myneni et al., 1995; Tucker and Sellers, 1986), and this link has been exploited to derive biophysical parameters for land-surface models and ecosystem models (Los et al., 2000; Potter et al., 1993; Sellers et al., 1996b). Some of the commonly available AVHRR NDVI data are the global inventory, monitoring and modelling studies (GIMMS) NDVI data for August 1981–December 2006 (Tucker et al., 2005; available from the global land cover facility at http://www.landcover.org/), and the FASIR NDVI data for 1982–1988 (Los et al., 2000, 2005; available from International Satellite Land-Surface Climatology Project (ISLSCP) Initiative II hosted by the Oak Ridge National Laboratory distributed active archive center: http://daac.ornl.gov.).

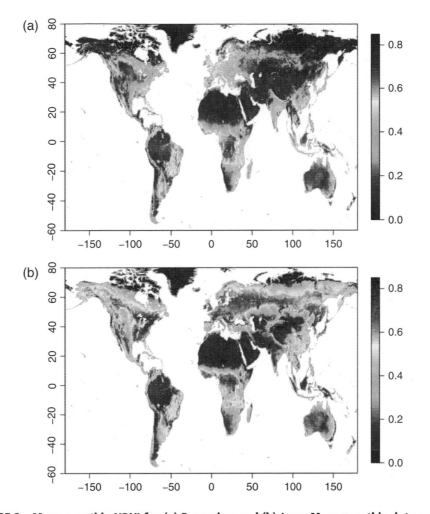

Figure 25.2 Mean monthly NDVI for (a) December and (b) June. Mean monthly data are obtained by averaging for all months of December or June over the period 1982–1999. The data are corrected for sensor degradation, atmospheric effects and view angle and illumination effects. Missing data caused by cloud effects have been filled using spatial and temporal interpolation. The corrections are collectively referred to as FASIR corrections (James and Kalluri, 1994; Sellers et al., 1996b; Los, 1998; Los et al., 2000, 2005).

AVHRR data have several limitations for vegetation monitoring. The first is a lack of onboard calibration of the sensor. Measurements drift over time and are not comparable between sensors. Corrections of AVHRR data for sensor degradation and lack of inter-calibration are based on analysis of signals from stable targets such as deserts, clouds, ice and measurements of deep space (Kidwell, 1998; Los, 1998; Rodwell, 2009). The relative accuracy that can be obtained is

around 1–2 per cent reflectance; with the absolute accuracy probably not smaller than 5 per cent reflectance. The second source of error in the data is linked to the orbital drift of the NOAA afternoon satellite. Directly after launch, the AVHRR crosses the equator in the early afternoon; the equatorial crossing time shifts by more than 20 minutes every year to later times of the afternoon. Signals from one year to the next are not directly comparable as a result and need to be

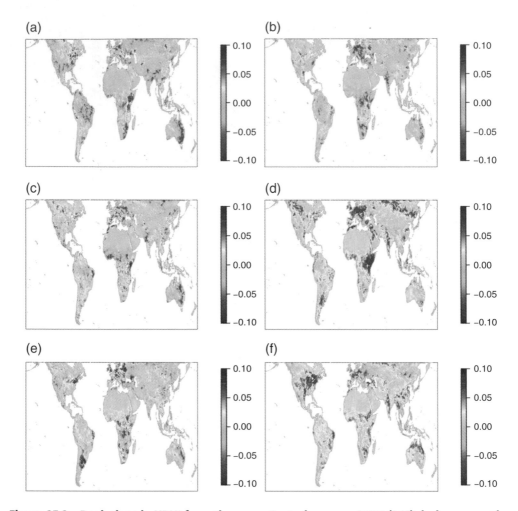

Figure 25.3   Deviations in NDVI from the mean Austral summer NDVI (DJF) during warm El Niño southern oscillation (ENSO) events in (a) 1982, (b) 1986, (c) 1987 and (d) 1997) and cold ENSO events in (d) 1988 and (e)1998. The year refers to the year of the December month. Drought during a warm ENSO event tends to occur in eastern Brazil, Southern Africa and parts of Australia. The reverse tends to occur during a cold ENSO event, although each ENSO event has unique properties. The NDVI in high northern latitudes can be affected by variations in snow cover. The difference in NDVI is shown over the range –0.1 to 0.1.

corrected for variations in view angle and illumination angle (Gutman, 1991; Los et al., 2005; Wanner et al., 1995).

Other sources of error occur in all satellite-derived vegetation data and are linked to temporal variations in atmospheric composition (atmospheric water vapour content, stratospheric ozone, tropospheric and stratospheric volcanic aerosols; Holben et al., 1992; Vermote and El Saleous, 1995), seasonal

variations in illumination angles and cloud effects (Gutman, 1991; Holben, 1986; Los et al., 2000; Sellers et al., 1996b).

Modern satellite sensors (e.g., SPOT Vegetation, SeaWiFS, AATSR, MODIS, MISR and MERIS) detect radiation in bands that are more suitable for vegetation monitoring, and in some cases obtain these measurement for multiple viewing angles and illumination angles. The sensor calibration is much improved in these sensors as well. Correction algorithms for atmospheric scattering and absorption of radiation have been applied (Holben et al., 1992; Vermote et al., 2001). Biophysical parameters are estimated by numerical inversion of models or by inversion of look-up tables generated by models (Myneni et al., 2003). Numerical estimation of biophysical parameters from satellite data is hampered by the fact that there are more knowns than unknowns; therefore no unique solution to the inversion exists. The solution to the inversion is usually constrained by using *a priori* values for some of the unknown parameters (e.g., leaf angle distribution, leaf optical properties and soil background reflectance); these parameters are usually derived from land-cover classification maps. Other biophysical parameters, leaf area index, fAPAR and vegetation cover fraction are subsequently estimated. Examples of biophysical parameter data derived from satellite data are the FASIR-AVHRR biophysical parameters, fAPAR, vegetation cover fraction and LAI (see http://daac.ornl.gov/; Sellers et al., 1996a; Los et al., 2000, 2005), the MODIS biophysical parameters (http://modis-land.gsfc.nasa.gov/; Myneni et al., 2003), the CYCLOPES LAI and fAPAR products (http://postel.mediasfrance.org/; Baret et al., 2007) and the JRC fAPAR products (http://fapar.jrc.ec.europa.eu/; Gobron et al., 2005). At present, differences between these data exist, reflecting uncertainties in absolute values; however, the seasonal variations and variations between years in these data are more similar (Weiss et al., 2007).

4 BIOMASS ESTIMATION FROM SATELLITE DATA

Monitoring of vegetation using passive optical sensors (section 25.3) is based on the unique spectral properties of vegetation. The information obtained from these sensors is closely linked to the leaf area of vegetation and the amount of PAR absorbed for photosynthesis. Variables linked to the height of vegetation, canopy shape and biomass can sometimes be inferred from spectral signatures (e.g., Steiniger, 2000, linked optical remote sensing data to biomass in the Amazon with varying degrees of success). Land-cover maps can be used to obtain an estimate of biomass, vegetation height or canopy structure (Dorman and Sellers, 1989); this results in a spatial map representing average, time-invariant estimates of a particular parameter for a particular biome. Biomass, vegetation height or canopy structure can also be estimated from models. An example is provided by van der Werf et al. (2006), who used the CASA model to calculate biomass from the long-term accumulation of net primary production and used estimates of biomass to calculate the spatial distribution of fuel load and biomass consumed by fires.

Direct estimates of biomass can be obtained from synthetic aperture radar (SAR) measurements (Waring et al., 1995). Synthetic aperture radar is a method to extend the length of a radar antenna virtually along the flight path of the aircraft or spacecraft. Because of the greater length of the virtual antenna, data can be collected by an SAR satellite at an effective spatial resolution of around 30 m. The atmosphere and clouds are largely transparent to microwave radiation, whereas microwave radiation is affected by water in vegetation and in soils and by the structure of vegetation. SAR bands at shorter wavelengths, the X-band and C-band, provide information on leaves and small branches, whereas SAR bands at longer wavelengths (the L and P bands) penetrate deeper into the vegetation and provide information on larger elements such as branches and stems (Waring et al., 1995). The ability to retrieve biomass from single waveband SAR is limited to about 100–150 t/ha; SAR becomes saturated with higher biomass values (Waring et al., 1995). An example of the retrieval of biomass for Siberia is provided by Wagner et al. (2003).

Light Detection and Ranging (LiDAR) measurements have shown potential for measuring biomass and vegetation structure (Harding et al., 2001; Lefsky et al., 2002). A study of forests in in Oregon found that the LiDAR did not show saturation to biomass values of at least 1300 t/ha and may saturate at even higher values (Means et al., 1999). Spaceborn LiDAR is available from the ICESat GLAS mission (Lefsky et al., 2005; Zwally et al., 2002); GLAS data have been used to obtain estimates of biomass over large regions (Nelson et al., 2009; Rosette et al., 2008).

5 LAND-COVER CLASSIFICATIONS FROM SATELLITE OBSERVATIONS

Satellite data collected in visible and near-infrared wave bands (Section 3) are also used to derive land-cover classifications. Land cover type cannot be measured directly from satellite; the radiation reflected from the land surface in visible and near-infrared spectral bands and the seasonal variations in these spectral signatures are instead linked to land-cover type. Monthly data to describe the temporal variation in vegetation can be substituted by metrics describing the seasonal cycle; for example, the seasonal amplitude, the time-of-start of the growing season and the length of the growing season.

Two approaches to classification in general can be distinguished: unsupervised classification where the data are grouped according to their spectral signatures, and supervised classification where the (time-variant) spectral properties of the land are compared to the spectral properties of a known area of land-cover, usually determined from the ground (see, e.g., Di Gregorio and Jansen, 2005).

The use of satellite data to obtain global land-cover products to classify land-cover has some advantages when compared to products compiled from inventories (Section 2). The data for the classification are obtained from one type of instrument or series of instruments for one particular period. The classification

itself is carried out by one investigator or one group of investigators and this can reduce variations caused by differences in conventions. Validation and testing can be designed according to a single protocol, but can also be carried out by independent teams of scientists (Loveland et al., 1999; Scepan, 1999). The satellite-derived land-cover data sets can be used in land-surface and ecosystem models similar to the conventional land-cover inventories. It is, however, possible to derive biophysical parameters such as land-surface albedo, leaf area index and fAPAR, directly from satellite data (Section 3). Thus, the use of land-cover classifications as an intermediate appears circumspect when information on vegetation seasonality is directly available from satellite data. There are, however, some advantages. The first is that the land-cover classifications can be used directly in models without adapting these models first; simulations with conventional inventories and satellite-derived inventories can therefore be directly compared. The second use of satellite derived land-cover classifications is that the solutions of the algorithms to infer biophysical parameters from satellite data are under-determined (Section 3). Thus, land-cover classifications can be used to obtain reasonable values of parameters needed to obtain a unique solution to the inversion; the advantage of this approach is that the land-cover classification is obtained from similar data as the biophysical parameters.

The accuracy assessment of global land-cover classifications can be approached in different ways and depends to a large extent on the application for which it is used. At its most basic level, one would like to obtain a high degree of agreement between land-cover classes on the ground and those estimated by satellite. Foody (2002) provides a detailed discussion of various issues related to classification accuracy assessment. In general, four levels of quality assessment are distinguished (Congalton, 1994). The first level of quality assessment is visual appraisal. This assessment has met with severe criticism, the main objection being that it is too subjective and dependent on the skill of the interpreter. Although subjective, the human eye is an incredibly powerful tool and is much more capable of interpreting spatial and temporal patterns than any software. Severe problems with a classification algorithm will be identified quite easily. Visual appraisal is thus a necessary stage in quality control but needs to be backed up with appropriate statistical measures. The second level of quality assessment is to see if the percentage cover of the classification matches the percentage cover of the ground data. Although a necessary quantitative assessment, it does not tell us anything about land-cover types matching in the right locations. The third is a comparison of classes for the same locations. At the heart of this comparison is the confusion matrix (see, e.g., Table 25.1). The fourth level is an expansion of the third level where additional statistics are used to provide quantitative measures of the degree of success of the classification.

For global land-cover classifications, confusion occurs frequently between 'similar' land-cover types (e.g. between deserts and shrub lands with bare soil; shrub lands with dense cover and woodlands; grassland and shrub lands; and in temperate regions broadleaf deciduous land cover and agricultural land). In some cases, a poor map in terms of accuracy in its description of biome type may still yield good results if the biophysical parameters derived from it are similar for the confused biome types (DeFries and Los, 1999). For use in models, a confusion matrix and measures of (dis)similarity of land-cover type can be replaced by assessment of the prediction of the biophysical parameter considered (see, e.g., Figure 25.1).

A relatively new approach to classification of the land into cover types is the use of continuous fields. Rather than assigning one type of cover to a cell, a percentage cover for each type is provided (Hansen and DeFries, 2004). This classification approach links in well with developments in land-surface modelling where a mosaic of land cover types is used to calculate fluxes for the grid

cell rather than one dominant vegetation type (Essery et al., 2003; Koster and Suarez, 1992). A limitation is that the spatial distribution of land-cover types is unknown within a grid cell. Thus, models will produce the same results for cells with equal percentages of the same land-cover type regardless of whether classes are spread evenly throughout the box or are grouped next to each other. Phenomena such as edge effects (e.g., accumulation of snow along a forest boundary) are ignored as a result.

Recent efforts are directed towards the harmonization of classification systems to obtain results that are both spatially and temporally consistent (Herold et al., 2008; Neumann et al., 2007). Given uncertainties in global land-cover maps, the reported agreement between different global land-cover products is between 70 and 80 per cent (Loveland et al., 1999), there is clearly a need for harmonization since real-world changes in land cover are likely smaller than differences between adopted classification schemes. Perhaps the biggest advantage of a consistent system is that it allows change detection; for example, a recent analysis of aerial photographs of Europe provided consistent estimates with a range of uncertainty of increases in urbanization and changes in agricultural land cover. Land-use change, in particular of intensification of land use for a similar land-cover type, was more difficult to detect (Gerard et al., 2010).

6 LAND-COVER CLASSIFICATION SCHEMES

Table 25.2 shows four frequently used classification schemes. There are variations in philosophy towards the classification approach in each of these and as a result classes may not be directly comparable. For example, the spatial distribution of wooded grassland in the SiB classification is limited to the tropics, whereas the spatial distribution of the wooded grassland in the

Table 25.2   Some commonly adopted classification schemes

	USGS/IGBP 1992–1993	*UMD 1992–1993*	*SiB*	*MODIS*
1	Evergreen needle leaf	Broad leaf evergreen	Evergreen broad leaf	Evergreen needle leaf
2	Evergreen broad leaf	Coniferous leaf evergreen	Deciduous broad leaf	Evergreen broad leaf
3	Deciduous needle leaf	High lat. deciduous leaf	Needle and broad leaf	Deciduous needle leaf
4	Deciduous broad leaf	Tundra	Evergreen needle leaf	Deciduous broad leaf
5	Mixed forest	Mixed deciduous and evergreen	Deciduous needle leaf	Shrub
6	Closed shrublands	Broad leaf deciduous	Wooded grassland	Grass
7	Open shrublands	Wooded grassland	Grassland	Cereal crop
8	Woody savannah	Grassland	Shrubs	Broad leaf crop
9	Savannas	Shrubs and bare ground	Shrubs and bare soil	Urban and built-up
10	Grasslands	Bare ground	Tundra	Snow and ice
11	Persistent wetlands	cultivated crops	Bare soil	Barren or sparse cover
12	Croplands		Agriculture	Water
13	Urban and built-up			
14	Cropland/other			
15	Snow and ice			
16	Mostly barren			
17	Water			

UMD classification extends to areas outside the tropics.

One of the earliest satellite-derived global land-cover classifications is the University of Maryland (UMD) land-cover classification (DeFries and Townshend, 1994). This data set is derived from one year of $1° \times 1°$ NDVI data (Los et al., 1994). The UMD classification builds on an early land-cover classification based on NDVI data of Africa by Tucker et al. (1985). The first UMD classification uses the SiB classification scheme. The division in classes is based on differences in the seasonal NDVI cycle and takes into account hemispheric differences in seasonality. Training data are obtained from an analysis of the Wilson, Olson and Matthews land-cover classifications (Section 2). These land-cover classifications are translated to the SiB classification scheme. Training and validation sites are identified from cells that have the same SiB classification in all three data sets. The seasonal profile was obtained for each of the training sites and were split by hemisphere and continent where necessary; for example, cultivated crops appeared to have very different temporal profiles between continents. The separability of the temporal signatures of the training data was assessed and groups were merged where no distinction

between spectral profiles could be made. Some distinct groups (e.g. grasslands and cultivated crops) were difficult to distinguish because they had similar temporal profiles whereas other, similar groups had distinct temporal profiles (e.g. rain forests in Africa have a greater degree of seasonality in vegetation greenness than rain forests in Asia and South America).

The USGS/IGBP data set (Loveland et al., 2001) was derived from a 1×1 km global AVHRR data set for the period April 1992 to March 1993. The global 1×1 km data set was assembled as a result of a collaboration between several research groups operating AVHRR receiving stations around the world (Eidenshink and Faundeen, 1994). An unsupervised classification scheme was adopted that distinguished 961 seasonal global land-cover types based on metrics such as the seasonal amplitude, inferred start of the growing season, and length of the growing season. Look-up tables were developed to translate the 961 classes into other classification schemes such as the Olson classification, the SiB classification and the USGS/IGBP classification. An extensive validation of the USGS land-cover map was carried out based on the interpretation by local experts of 25–60 m resolution Landsat thematic mapper and multi spectral scanner data (Loveland et al., 1999; Scepan, 1999). The agreement between the USGS 1 km classification and the interpretation of Landsat data was around 60 per cent; the agreement improved if the 1 km data were grouped into fewer classes. A similar degree of agreement was found between the USGS 1 km land-cover classification and an updated UMD land-cover classification that was derived from 1 km data using a decision tree classification protocol (Hansen and Reed, 2000; Hansen et al., 2000). The agreement ranged from about 70 per cent when the classifications was based on five major land-cover types to about 85 per cent when only two classes, short and tall vegetation, were considered (Table 25.3).

More recent land-cover classification schemes use improved data from more modern sensors (SPOT Vegetation, MODIS, MERIS). These sensors have better calibration, measure in more and narrower spectral bands, and have higher spatial resolution. A key objective of these newer land-cover classification schemes is to obtain a greater degree of consistency both in the classification algorithms as well as in the definition of classes adopted. Moreover, with the availability of longer times series, attempts are made to obtain estimates of

land-cover change over the past two decades (Hansen and DeFries, 2004). Examples of improved or higher resolution land-cover classifications products are the GLOBCOVER product. Version 1 is based on 1.1 km SPOT Vegetation data and version 2 on 300 m MERIS data, both with a reported accuracy around 70 per cent (Arino et al., 2008). The GLC 2000 land-cover classification is based on 300 m MERIS data (Barttholomé and Belward, 2005) and has a reported classification accuracy of about 69 per cent (Mayaux et al., 2006) and the MODIS collection 5 global land-cover product at 500 m resolution and has a a reported classification accuracy of about 75 per cent.

Regional land-cover classifications for particular continents are developed from higher resolution data; for example, Landsat data or aerial photography. These regional classifications are based on schemes used for global land-cover classifications. For example, the Coordination of Information on the Environment (CORINE) program is to provide comparable information on land cover in digital form for Europe and Northern Africa. Land-cover types are derived from Landsat images analyzed by country at a scale of 1:100,000. Each country uses the same CORINE nomenclature and interpretation

Table 25.3 Comparison of thematic agreement (in %) at nominal 0.5° resolution grid of remotely sensed derived DISCover and UMD maps and ground-based 'Global Distribution of Vegetation at 1° × 1°, compiled by Matthews (1983) and 'Carbon in Live Vegetation of Major World Ecosystems' by Olson et al. (1983) (analysis from Hansen and Reed 2000, values rounded to nearest %)

	USGS/UMD			
	Forest/woodland	Grass/shrubs	Crops	Bare ground
Forest/woodland	89	9	2	0
Grass/shrubs	16	69	10	5
Crops	9	12	79	0
Bare ground	0	9	0	91
	Olson/Matthews			
	Forest/woodland	Grass/shrubs	Crops	Bare ground
Forest/woodland	70	22	7	1
Grass/shrubs	15	60	7	19
Crops	16	30	51	3
Bare ground	1	15	1	84

methods to ensure comparability between countries. The regional product (Europe and North Africa) is produced by merging the national products (Neumann et al., 2007; Perdigao and Annoni, 1997). The CORINE land-cover classification has been carried out for 1990 and 2000. When the 2000 classification was carried out, the 1990 classification was updated simultaneously to enhance consistency between the two data sets and allow the data to be used for change detection (Perdigao and Annoni, 1997).

The BIOPRESS project uses the CORINE classification scheme to obtain simultaneously detailed land-cover classifications of selected sites throughout Europe for 1950, 1990 and 2000. The classification is carried out simultaneously on aerial photographs for 1950, 1990 and 2000 and Landsat data for 1990 and 2000. The joint analysis allows for more accurate change detection even in the case of inaccuracies in establishing the land-cover type. The analysis of one interpreter is verified by an independent expert for the selected areas. The consistency of the interpretation for the selected sites was around 94 per cent. Although full details of land-cover change throughout Europe were not obtained, BIOPRESS data provide information on changes in forest management in the Boreal and Alpine regions, indications of abandonment and intensification of agriculture in the Mediterranean, and conversion of agricultural land into urban areas in the continental and Atlantic regions of Europe. Changes in land use such as a change in the intensity of agricultural practices cannot usually be detected (Gerard et al., 2010).

7 LAND-COVER CHANGE DATABASES

Some of the land-cover classifications discussed in previous sections contain information on land-cover change. The Matthews (1983) database and Olson et al. (1985) database contain estimates of potential vegetation, and current vegetation or agricultural

land use. The BIOPRESS data (Gerard et al., 2010) provide estimates of land-cover type in Europe for 1950, 1990 and 2000 and pay particular attention to the consistency of the classification between these years. The record of 20–30 years of satellite data currently available provides information on seasonal and interannual variations in global vegetation greenness or biophysical parameters (Myneni et al., 1997; Los et al., 2001; Slayback et al., 2003; Tucker et al., 2001). These data do not provide direct information on changes in land-cover type, however, although some classification schemes have been applied to these data to obtain estimates of variations in tree cover (Hansen and DeFries, 2004).

Two global data land-cover change databases exist that provide the spatial distribution of land cover for the entire land surface from 1700 to 1990; these databases are the historical croplands data set (http://www.sage.wisc.edu/) assembled by Ramankutty and Foley (1999) and the history database of the global environment (HYDE; http://www.pbl.nl/en/themasites/hyde/index.html) assembled by Klein Goldewijk (2001). Both data sets provide and estimate of land-cover types between 1700 and 1950 every 50 years and from 1950 until 1990 every 10 years. Both use the IGBP global land-cover classification data of 1992–1993 (Loveland et al., 2001) as their starting point and work back in time to reconstruct the land-cover of the past. These land-cover change data sets base their estimates on crop statistics obtained from the Food and Agriculture Organization (FAO) (Food and Agricultural Organization, 1996) and country statistics. Missing data are estimated by modelling and interpolation. A map of potential vegetation (i.e. vegetation not disturbed by land use) is obtained from the BIOME 3 model (Haxeltine and Prentice, 1996). Differences between the potential vegetation map and the land-cover map indicate the amount of land conversion. There are some differences between the two historic land-cover data sets; the Ramankutty and Foley (1999) data provide percent cover of crop land for each cell, whereas the HYDE data only provide an estimate of the dominant cover type. The HYDE database provides estimates of both crop land and pasture; the Ramankutty and Foley (1999) data provide only crop land, although an updated version is in preparation that contains pasture as well. The HYDE data model fills any gaps in the historic land-use data based on population per country and the Ramankutty and Foley (1999) base the interpolation on remote sensing data. Differences between the two data sets increase when going back in time and are most pronounced for South America, Africa and Oceania and can be explained by poor data coverage for these continents (Klein Goldewijk and Ramankutty, 2004).

8 LAND-COVER CHANGE DETECTION FROM SATELLITE

Several examples of studies exist where remote sensing data are used to monitor changes in land cover that occurred over the past three decades or so. Remote sensing data are a unique source of information, in particular, for areas where information cannot be obtained in any other way.

The first example of land-cover change studies from satellite is the monitoring of deforestation in South America, Africa and Asia. Tropical forests are important because of their high biodiversity; and they play an important role in the global carbon cycle, the water cycle and energy cycle. Changes in vegetation cover in tropical forests can be monitored with optical satellite sensors with a spatial resolution of around 50 m or better. SAR methods to monitor deforestation would be more suitable because of the ability of SAR to look through clouds (Kuplich, 2006), but data coverage is less frequent than for optical systems. Landsat multi-spectral scanner (MSS) data, later replaced by Landsat thematic mapper (TM) data, have been available since 1973. The data coverage is sufficient to obtain updates of land cover every

10 years (Skole and Tucker, 1993). Since the 1980s more satellites are available to monitor deforestation, such as SPOT and the recently launched disaster-monitoring constellation (DMC). The MERIS and MODIS sensors with 300 m and 250 m resolution, respectively, are also useful for this purpose and provide frequent coverage. Deforestation rates estimated from satellites indicate that between 1990 and 1997 the highest rate of deforestation occurred in Asia and South America (2.5×10^6 ha per year); in Africa the annual deforestation was about 0.9×10^6 ha. As a percentage of remaining tropical forest cover, the deforestation in Asia is the highest and in South America the lowest (Achard et al., 2002). The area disturbed by deforestation is higher because the construction of roads allows easier access to the remaining forest (Skole and Tucker, 1993).

The second example where satellite observations provided useful information is the study of desertification. Desertification is thought to result in the (permanent) loss of biological potential of the land. Causes of desertification are mostly attributed to poor management of the land; for example, overgrazing of range lands, use of wood lands for fuel or frequent burning of vegetation in savannas, but desertification has also been attributed to changes in climate. There are potential interactions between the land surface and atmosphere, where a reduction in vegetation can lead to a reduction of precipitation usually as a result of decreased evapotranspiration or convection (Charney et al., 1977). The Sahel, the southern border of the Sahara, experienced a marked decrease in precipitation from the 1950s to the early 1980s. The rapid decrease in rainfall and associated decline in vegetation, as well as evidence of local, human-induced reductions in vegetation, led to concerns of continued, irreversible degradation of land. Satellite-based studies indicated large year-to-year variations in vegetation cover, largely driven by year-to-year variations in rain.

Figure 25.4 shows the maximum extent of the drought was observed during 1984 after

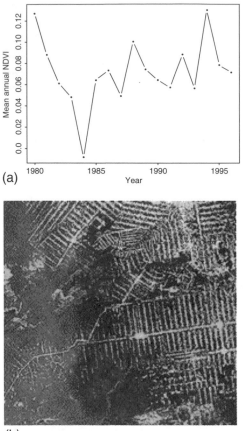

(a)

(b)

Figure 25.4   (a) Landsat TM subscene of an area in Rondonia, Brazil in 1986; red colours indicate vigorous vegetation; bright green colours indicate cleared land (based on Remote Sensing Tutorial, NASA Goddard Space Flight Centre). (b) Mean annual NDVI for the region with 200–400 mm mean annual rainfall south of the Sahel (data courtesy of Compton Tucker, NASA GSFC).

which a (partial) recovery of vegetation occurred (Prince et al., 2007; Tucker et al., 1991). This dispelled the notion of desertification as an irreversible process. Model studies indicate that the land surface does play a role in sustaining droughts (Koster et al., 2006); and analysis of observations shows that 8–15 per cent of the variance in rainfall of a particular month can be explained by variations in vegetation greenness of the previous month (Los et al., 2006).

A third example of land-cover change detection from satellite is the detection of urban sprawl; that is, the rapid conversion of agricultural land and natural areas into urban areas. Urban areas can be detected from the analysis of (low) night-time light levels detected by the Defence Meteorological Satellite Program (DMSP) Operational Linescan System (OLS) (Imhoff et al., 1997; Milesi et al., 2003). The OLS is sensitive to visible and near-infrared light emitted at a low-level, such as city lights, fires, volcanoes and lightning. The OLS data are analyzed to identify light from stable sources that can be attributed to built-up areas. The light data are calibrated with measures such as the number of houses per unit area or population density (Imhoff et al., 1997). The comparison of multiple years of data provides information on changes in urban areas; a major proportion of recent urban sprawl in the US occurs in those areas that are the most suitable for agriculture (Milesi et al., 2003).

Fire is an important disturbance of vegetation and, at the global scale, accounts for a large amount of vegetation destroyed on an annual basis (Crutzen and Andreae, 1990). Thermal emissions from wild fires are at a maximum in the mid-infrared wavelengths (2–4 μm); the optimum wavelength varies with the temperature of the fire (Dozier, 1981). Fire detection algorithms need to filter out data from cloud and water. Tests are applied on near-infrared, mid-infrared and thermal infrared channels, and data from surrounding pixels are also considered in order to obtain an estimate of the background temperature for nonfire areas (Giglio et al., 2003). Polar orbiting satellites such as MODIS Aqua and Terra or satellites with a circular orbit such as the Tropical Rainfall Monitoring Mission (TRMM) make, at most, two measurements per day. Instruments aboard geostationary satellites such as the Meteosat Spinning Enhanced Visible and Infrared Imager (SEVERI) can obtain measurements every 15 minutes. The total sum of all energy released by a fire over its duration, the fire radiative energy, can be linked to the total amount of biomass burned

(Wooster et al., 2005). The area burned after a fire can be detected from satellite as well (Justice et al., 2002). Satellite-based estimates of biomass burning indicate that the largest amount of biomass burning occurs in Africa (about 270 Mha annually) followed by Australia (54 Mha) and South America (24 Mha; see Giglio et al., 2010). The area burned annually in Australia and equatorial Asia shows a correlation with the occurrence of an El Niño (Giglio et al., 2010). Emissions by biomass burning are highest in Asia, predominantly in the equatorial regions, and in Central and South America (van der Werf et al., 2006).

CONCLUSION

The purpose of the present chapter is to highlight some of the research in land-cover monitoring. The availability of global satellite data from various instruments measuring radiation in different spectral bands at various spatial, spectral and temporal resolutions has clearly made a significant impact on the progress in this area and the literature reviewed is only a proportion of what is available. Both qualitative indicators of land-cover vegetation (classification of the land in biomes, plant functional types or classes) and quantitative indicators (vegetation indices, biophysical parameters, biomass, disturbance) have been derived from satellites and in several cases satellite data have been used to detect changes in these indicators over time.

There is clearly a need for greater consistency in satellite derived data. Sources of inconsistencies are linked to lack of stability of the instruments, variability in the atmosphere and viewing and illumination angles, and changes in the spectral characteristics and specifications of sensors. Compatibility is usually not an objective of the design of new sensors, hence research is needed to translate measurements of first generation sensors such as the AVHRR to measurements of more modern sensors such as AATSR, SeaWiFS, MODIS and MERIS.

The availability of new information globally on land-surface vegetation has led to the development and improvement of ecological models and of land-surface parameterizations in climate models. Satellite data provide realistic estimates of disturbance by fires and interannual variations in vegetation, and this allows the study of the sensitivity of the carbon cycle, hydrological cycle and energy budget to these changes.

Future missions are expected to provide additional data on land vegetation. At present, most variables measured from space are linked to photosynthesis or leaf area. Data from the ICESat GLAS instrument provide measurements of vegetation height, vegetation structure and biomass. Missions designed specifically for measuring vegetation structure (e.g. the vegetation canopy LiDAR (VCL) and the Deformation, Ecosystem Structure and Dynamics of Ice (DESDynI; http://desdyni.jpl.nasa.gov/) mission (Bergen et al., 2009) have been cancelled due to budget cuts.

There is still much that is unknown about land vegetation. There are indications that land surface vegetation is important for the prediction of rainfall patterns and that a disturbance in land vegetation can lead to their disruption. At present, the strength of this interaction is largely unknown. Another major issue is how land vegetation will change in response to a warmer climate and increased atmospheric CO_2 levels. Some models indicate die back of the Amazon if global temperatures continue to increase and an accelerated release of CO_2 from the land into the atmosphere as a result. Synthesis between vegetation data and atmospheric measurements as well as improved simulations of climate models may provide answers to these questions.

REFERENCES

Achard F., Eva H. D., Stibig H. J., Mayaux P., Gallego J., Richards T. and Malingreau J. P. 2002. Determination of deforestation rates of the world's humid tropical forests. *Science*: 297(999–1002).

Arino O., Bicheron P., Achard F., Latham J., Witt R. and Weber J. L. 2008. GLOBCOVER — The most detailed portrait of Earth. *ESA Bulletin — European Space Agency* 136: 24–31.

Asrar G., Kanemasu E. T., Miller G. P. and Weiser R. L. 1986. Light interception and leaf-area estimates from measurements of grass canopy reflectance. *IEEE Transactions on Geoscience and Remote Sensing* 24: 76–82.

Baker I. T., Prihodko L., Denning A. S., Goulden M., Miller S. and da Rocha H. R. 2008. Seasonal drought stress in the Amazon: Reconciling models and observations. *Journal of Geophysical Research* 113, G00B01.

Baret F., Hagolle O., Geiger B., Bicheron P., Miras B., Huc M. et al. 2007. LAI, FAPAR, and FCover CYCLOPES global products derived from Vegetation. Part 1: principles of the algorithm. *Remote Sensing of Environment* 110: 305–316.

Barttholomé E. and Belward A. S. 2005. GLC2000: a new approach to global land cover mapping from earth observation data. *International Journal of Remote Sensing* 26(9): 1959–1977.

Bergen K. M., Goetz S. J., Dubayah R. O., Henebry G. M., Hunsaker C. T., Imhoff M. L. et al. 2009. Remote sensing of vegetation 3-D structure for biodiversity and habitat: Review and implications for LiDAR and RADAR spaceborne missions. *Journal of Geophysical Research-Biogeosciences* 114: G00E06.

Ceccato P., Cressman K., Giannini A. and Trzraska S. 2007. The desert locust upsurge in West Africa (2003–2005): Information on the desert locust early warning system and the prospects for seasonal climate forecasting. *International Journal of Pest Management* 53: 7–13.

Charney J., Quirk W. J., Chow S. H. and Kornfield J. 1977. Comparative study of effects of albedo change on drought in semi-arid regions. *Journal of the Atmospheric Sciences* 34(9): 1366–1385.

Congalton R. G. 1994. Accuracy assessment of remotely sensed data: future needs and directions. *Proceedings of Pecora 12 Land Information From Space-Based Systems.* Bethesda: ASPRS, pp. 383–388.

Crutzen P. J. and Andreae M. O. 1990. Biomass burning in the tropics: Impact on atmospheric chemistry and biogeochemical cycles. *Science* 250: 1669–1677.

DeFries R. S. and Los S. O. 1999. Implications of land-cover misclassification for parameter estimates in global land-surface models: an example from the simple biosphere model (SiB2). *Photogrammetric Engineering and Remote Sensing* 65: 1083–1088.

DeFries R. S. and Townshend J. R. G. 1994. NDVI-derived land cover classifications at a global scale. *International Journal of Remote Sensing* 15: 3567–3586.

Di Gregorio A. and Jansen L. J. M. 2005. *Land Cover Classification System. Classification Concepts and User Manual. Software version (2).* Environment and Natural Resources Series, vol. 8. Rome: Food and Agriculture Organization of the United Nations.

Dickinson R. E., Henderson-Sellers A., Kennedy P. J. and Wilson M. F. 1986. *Biosphere Atmosphere Transfer Scheme (BATS) for the NCAR Community Climate Model.* Technical Report Note NCAR/TN–387+STR. Boulder: NCAR.

Dorman J. L. and Sellers P. J. 1989. A global climatology of albedo, roughness length, and stomatal resistance for atmospheric general circulation models as represented by the Simple Biosphere Model (SiB). *Journal of Applied Meteorology* 28: 833–855.

Dozier J. 1981. A method for satellite identification of surface temperature fields of subpixel resolution. *Remote Sensing of Environment* 11: 221–229.

Eidenshink J. C. and Faundeen J. L. 1994. The 1 km AVHRR global land data set – 1st stages in implementation. *International Journal of Remote Sensing* 15: 3443–3462.

Essery R. L. H., Best M. J. and Cox P. M. 2001. *MOSES 2.2 Technical Documentation.* Exeter: Met Office.

Essery R. L. H., Best M. J., Betts R. A., Cox P. M. and Taylor C. M. 2003. Explicit representation of subgrid heterogeneity in a GCM land-surface scheme. *Journal of Hydrometeorology* 4(3): 530–543.

Foley J. A., Prentice I. C., Ramankutty N., Levis S., Pollard D., Sitch S. and Haxeltine A. 1996. An integrated biosphere model of land surface processes, terrestrial carbon balance, and vegetation dynamics. *Global Biogeochemical Cycles* 10: 603–628.

Foley J. A., Kucharik C. J. and Polzin D. 2005. *Integrated Biosphere Simulator Model (IBIS), Version 2.5. Model product.* Oak Ridge: Oak Ridge National Laboratory Distributed Active Archive Center. Available at http://daac.ornl.gov

Food and Agricultural Organization. 1996. *FAOSTAT.* Rome: Food and Agricultural Organization of the UN. Available at: http://apps.fao.org/

Foody G. M. 2002. Status of land cover classification accuracy assessment. *Remote Sensing of Environment* 80: 185–201.

Gerard F., Petit S., Smith G., Thomson A., Brown N., Manchester S. et al. 2010. Land-cover change in Europe between 1950 and 2000. determined employing aerial photography. *Progress in Physical Geography* 34: 183–205.

Giglio L., Kendall J. D. and Mack R. 2003. A multi-year active fire dataset for the tropics derived from the TRMM VIRS. *International Journal of Remote Sensing* 24: 4505–4525.

Giglio L., Randerson J. T., van der Werf G. R., Kasibhatla P. S., Collatz G. J., Morton D. C. and DeFries R. S. 2010. Assessing variability and long-term trends in burned area by merging multiple satellite fire products. *Biogeosciences* 7: 1171–1186.

Gobron N., Pinty B., Taberner M., Mélin F., Verstraete M. M. and Widlowski J.-L. 2005. Monitoring the photosynthetic activity vegetation from remote sensing data. *Advances in Space Research*, 38, 2196–2202.

Gutman G. G. 1991. Vegetation indexes from AVHRR — an update and future prospects. *Remote Sensing of Environment* 35: 121–136.

Hansen J., Russell G., Rind D., Stone P., Lacis A., Lebedeff S. et al. 1983. Efficient three-dimensional global models for climate studies: Models I and II. *Monthly Weather Review* 111: 609–662.

Hansen M. C. and DeFries R. S. 2004. Detecting long-term global forest change using continuous fields of tree-cover maps from 8-km advanced very high resolution radiometer (AVHRR) data for the years 1982–99. *Ecosystems* 7: 695–716.

Hansen M. C., DeFries R. S., Townshend J. R. G. and Sohlberg R. 2000. Global land cover classification at 1 km spatial resolution using a classification tree approach. *International Journal of Remote Sensing* 21: 1331–1364.

Hansen M. C. and Reed B. 2000. A comparison of the IGBP DISCover and University of Maryland 1 km global land cover products. *International Journal of Remote Sensing* 21(6 and 7): 1365–1373.

Harding D. J., Lefsky M. A., Parker G. G. and Blair J. B. 2001. Laser altimeter canopy height profiles — Methods and validation for closed-canopy, broadleaf forests. *Remote Sensing of Environment* 76: 283–297.

Haxeltine A. and Prentice I. C. 1996. BIOME3: An equilibrium terrestrial biosphere model based on ecophysiological constraints, resource availability, and competition among plant functional types. *Global Biogeochemical Cycles* 10: 693–709.

Herold M., Mayaux P., Woodcock C. E., Baccini A. and Schmullius C. 2008. Some challenges in global land cover mapping: An assessment of agreement and accuracy in existing 1 km datasets. *Remote Sensing of Environment* 112: 2538–2556.

Hielkema J. U. and Snijders F. L. 1994. Operational use of environmental satellite remote-sensing and satellite-communications technology for global food security and locust control by FAO — The ARTEMIS and DIANA systems. *Acta Astronautica* 32: 603–616.

Holben B. N. 1986. Characteristics of maximum-value composite images from temporal AVHRR data. *International Journal of Remote Sensing* 7: 1417–1434.

Holben B. N., Vermote E. F., Kaufman Y. J., Tanré D. and Kalb V. 1992. Aerosol retrieval over land from AVHRR data — application for atmospheric correction. *IEEE Transactions on Geoscience and Remote Sensing* 30: 212–222.

Hummel J. and Reck R. 1979. A global surface albedo model. *Journal of Applied Meteorology* 18: 239–253.

Hutchinson, C. F. 1991. Uses of satellite data for famine early warning in sub-Saharan Africa. *International Journal of Remote Sensing* 12: 1405–1421.

Imhoff M. L., Lawrence W. T., Elvidge C. D., Paul T., Levin E., Privaslky M. V. and Brown V. 1997. Using nighttime DMSP/OLS images of city lights to estimate their impact on urban land-use on soil resource in the United States. *Remote Sensing of Environment* 59: 105–117.

James M. E. and Kalluri S. N. V. 1994. The Pathfinder AVHRR land data set – An improved coarse resolution data set for terrestrial monitoring. *International Journal of Remote Sensing* 15(17): 3347–3363.

Justice C. O., Giglio L., Korontzi S., Owens J., Morisette J. T., Roy D. et al. 2002. The MODIS fire products. *Remote Sensing of Environment* 83: 244–262.

Kidwell K. B. 1998. *NOAA Polar Obiter Data User's Guide*. Suitland: NOAA NESDIS NCDC. Available at http://www.ncdc.noaa.gov/

Klein Goldewijk K. 2001. Estimating global land use change over the past 300 years: The HYDE database. *Global Biogeochemical Cycles* 15(2): 417–433.

Klein Goldewijk K. and Ramankutty N. 2004. Land cover change over the last three centuries due to human activities: The availability of new global data sets. *GeoJournal* 61: 335–344.

Koster, Randal D. and Suarez M. J. 1992. Modeling the land-surface boundary in climate models as a composite of independent vegetation stands. *Journal of Geophysical Research-Atmospheres* 97: 2697–2715.

Koster R. D., Guo Z, Dirmeyer P. A., Bonan G., Chan E., Cox P., et al. 2006. GLACE: The Global land-atmosphere coupling experiment. Part I: Overview. *Journal of Hydrometeorology* 7(4): 590–610.

Kuplich T. M. 2006. Classifying regenerating forest stages in Amazônia using remotely sensed images and a neural network. *Forest Ecology and Management* 234: 1–9.

Lefsky M., Harding D., Keller M., Cohen W., Carabajal C., Del Bom Espirito-Santo F. et al. 2005. Estimates of forest canopy height and aboveground biomass using ICESat. *Geophysical Research Letters* 32: L22S02.

Lefsky M. A., Cohen W. B., Parker G. G. and Harding D. J. 2002. LiDAR remote sensing for ecosystem studies. *BioScience* 52: 19–30.

Los S. O. 1998. Estimation of the ratio of sensor degradation between NOAA AVHRR channels 1 and 2 from monthly NDVI composites. *IEEE Transactions on Geoscience and Remote Sensing* 36(1): 206–213.

Los S. O., Justice C. O. and Tucker C. J. 1994. A global 1-degrees-by-1-degrees NDVI data set for climate studies derived from the GIMMS continental NDVI data. *International Journal of Remote Sensing* 15: 3493–3518.

Los S. O., Collatz G. J., Sellers P. J., Malmstrom C. M., Pollack N. H., DeFries R. S. et al. 2000. A global 9-yr biophysical land surface dataset from NOAA AVHRR data. *Journal of Hydrometeorology* 1(2): 183–199.

Los S. O., Collatz G. J., Bounoua L., Sellers P. J. and Tucker C. J. 2001. Global interannual variations in sea surface temperature and land surface vegetation, air temperature, and precipitation. *Journal of Climate* 14: 1535–1549.

Los S. O., North P. R. J., Grey W. M. F. and Barnsley M. J. 2005. A method to convert AVHRR Normalized Difference Vegetation Index time series to a standard viewing and illumination geometry. *Remote Sensing of Environment* 99(4): 400–411.

Los S. O., Weedon G. P., North P. R. J., Kaduk J. D., Taylor C. M. and Cox P. M. 2006. An observation-based estimate of the strength of rainfall-vegetation interactions in the Sahel. *Geophysical Research Letters* 33: L16402.

Loveland T. R., Estes J. E. and Scepan J. 1999. Introduction to the PE&RS Special Issue on global land-cover mapping and validation. *Photogrammetric Engineering and Remote Sensing* 65: 1011–1012.

Loveland T. R., Reed B. C., Brown J. F., Ohlen D. O., Zhu J., Yang L. and Merchant J. W. 2001. Development of a global land cover characteristics database and IGBP DISCover from 1-km AVHRR Data. *International Journal of Remote Sensing* 21: 1303–1330.

Matthews E. 1983. Global vegetation and land use: New high-resolution data bases for climate studies. *Journal of Climate and Applied Meteorology* 22: 474–487.

Matthews E. 1984. *Prescription of land-surface boundary conditions in GISS GCM II: A simple method based on fine-resolution data bases.* NASA Tech. Memo. 86096. NASA GISS. New York: NASA Goddard Institute for Space Studies.

Mayaux P., Strahler A., Eva H., Herold M., Shefali A., Naumov S. et al. 2006. Validation of the global land cover 2000 map. *IEEE Transactions on Geoscience and Remote Sensing* 44(7): 1728–1739.

Means J. E., Acker S. A., Harding D. J., Blair J. B., Lefsky M. A., Cohen W. B. et al. 1999. Use of large-footprint scanning airborn LiDAR to estimate forest stand characteristics in the Western Cascades of Oregon. *Remote Sensing of Environment* 67: 298–308.

Milesi C., Elvidge C. D., Nemani R. R. and Running S. W. 2003. Assessing the impact of urban land development on net primary productivity in the southeastern United States. *Remote Sensing of Environment* 86: 401–410.

Myneni R., Knyazikhin Y., Glassy J., Votava P. and Shabanov N. 2003. *User's Guide FPAR LAI (ESDT: MOD15A2) 8-day Composite NASA MODIS Land Algorithm.* Terra MODIS Land Team.

Myneni R. B., Hall F. G., Sellers P. J. and Marshak A. L. 1995. The interpretation of specral vegetation indexes. *IEEE Transactions on Geoscience and Remote Sensing* 33: 481–486.

Myneni R. B., Keeling C. D., Tucker C. J., Asrar G. and Nemani R. R. 1997. Increased plant growth in the northern high latitudes from 1981 to 1991. *Nature* 386: 698–702.

Neilson R. P. 1995. A model for predicting continental-scale vegetation distribution and water-balance. *Ecological Applications* 5: 362–385.

Nelson R., Ranson K. J., Sun G., Kimes D. S., Kharuk V. and Montesano P. 2009. Estimating Siberian timber volume using MODIS and ICESat/GLAS. *Remote Sensing of Environment*, 113: 691–701.

Neumann K., Herold M., Hartley A. and Schmullius C. 2007. Comparative assessment of CORINE2000 and GLC2000: Spatial analysis of land cover data for Europe. *International Journal of Applied Earth Observation and Geoinformation* 9: 425–437.

Olson J. S., Watts J. A. and Allison L. J. 1985. *Major World Ecosystem Complexes Ranked by Carbon in Live Vegetation: A Database.* NDP–017. Oak Ridge: Oak Ridge National Laboratory.

Parton W. J., Schimel D. S., Cole C. V. and Ojima D. S. 1987. Analysis of factors controlling soil organic matter levels in Great Plains grasslands. *Soil Science Society of America Journal*, 51: 1173–1179.

Perdigao V. and Annoni A. 1997. *Technical and Methodological Guide for Updating CORINE Land-cover Data Base.* Brussels: Joint Research Centre, European Commission.

Potter C. S., Randerson J. T., Field C. B., Matson P. A., Vitousek P. M., Mooney H. A. and Klooster S. A. 1993. Terrestrial ecosystem production: A process model based on global satellite and surface data. *Global Biogeochemical Cycles* 7: 811–841.

Prince S. D., Wessels K. J., Tucker C. J. and Nicholson S. 2007. Desertification in the Sahel: a reinterpretation of a reinterpretation. *Global Change Biology* 13: 1308–1313.

Ramankutty N. and Foley J. A. 1999. Estimating historical changes in global land cover: croplands from 1700 to 1992. *Global Biogeochemical Cycles* 13: 997–1028.

Rodwell J. 2009. *NOAA KLM User's Guide*. NOAA NESDIS NCDC. Available at: http://www.ncdc.noaa.gov/

Rosette J. A. B., North P. R. J. and Suarez J. C. 2008. Vegetation height estimates for a mixed temperate forest using satellite laser altimetry. *International Journal of Remote Sensing* 29(5): 1475–1493.

Running S. W. and Hunt Jr. E. R. 1993. *Generalization of a Forest Ecosystem Process Model for other Biomes, BIOME-BGC, and an Application for Global-scale Models. Vol. Scaling Physiological Processes: Leaf to Globe*. San Diego: Academic Press, pp. 141–158.

Scepan J. 1999. Thematic validation of high-resolution global land-cover data sets. *Photogrammetric Engineering and Remote Sensing* 65(9): 1051–1060.

Sellers P. J., Collatz G. J. and Randall D. A. 1996a. A revised land-surface parameterization (SiB2) for atmospheric GCMs. Part 1: Model formulation. *Journal of Climate* 9: 676–705.

Sellers P. J., Los S. O., Tucker C. J., Justice C. O., Dazlich D. A., Collatz G. J. and Randall D. A. 1996b. A revised land-surface parameterization (SiB2) for atmospheric GCMs. Part 2: The generation of global fields of terrestrial biophysical parameters from satellite data. *Journal of Climate* 9: 706–737.

Sitch S., Smith B., Prentice I. C., Arneth A., Bondeau A., Cramer W. et al. 2003. Evaluation of ecosystem dynamics, plant geography and terrestrial carbon cycling in the LPJ dynamic global vegetation model. *Global Change Biology* 9: 161–185.

Skole D. and Tucker C. J. 1993. Tropical deforestation and habitat fragmentation in the Amazon – Satellite data from 1978–1988. *Science* 260: 1905–1910.

Slayback D. A., Pinzon J. E., Los S. O. and Tucker C. J. 2003. Northern hemisphere photosynthetic trends 1982–99. *Global Change Biology* 9(1): 1–15.

Steiniger M. K. 2000. Satellite estimation of tropical secondary forest above-ground biomass: data from Brazil and Bolivia. *International Journal of Remote Sensing* 21: 1139–1157.

Thornton P. E., Law B. E., Gholz H. L., Clark K. L., Falge E., Ellsworth D. S. et al. 2002. Modeling and measuring the effects of disturbance history and climate on carbon and water budgets in evergreen needle-leaf forests. *Agricultural and Forest Meteorology* 113: 185–222.

Tucker C. J. 1979. Red and photographic infrared linear combinations for monitoring vegetation. *Remote Sensing of Environment* 8: 127–150.

Tucker C. J., Dregne H. E. and Newcomb W. W. 1991. Expansion and contraction of the Sahara desert from 1980 to 1990. *Science* 253: 299–301.

Tucker C. J., Fung I. Y., Keeling C. D. and Gammon R. H. 1986. Relationship between atmospheric variations and a satellite-derived vegetation index. *Nature* 319: 195–199.

Tucker C. J., Pinzon J. E., Brown M. E., Slayback D. A., Pak E. W., Mahoney R. et al. 2005. An extended AVHRR 8-km NDVI Data Set Compatible with MODIS and SPOT Vegetation NDVI Data. *International Journal of Remote Sensing* 26: 4485–5598.

Tucker C. J. and Sellers P. J. 1986. Satellite remote sensing of primary production. *International Journal of Remote Sensing* 7: 1395–1416.

Tucker C. J., Slayback D. A., Pinzon J. E., Los S. O., Myneni R. B. and Taylor M. G. 2001. Higher northern latitude normalized difference vegetation index and growing season trends from 1982 to 1999. *International Journal of Biometeorology* 45: 184–190.

Tucker C. J., Townshend J. R. G. and Goff T. E. 1985. African land-cover classification using satellite data. *Science* 227: 369–375.

UNESCO. 1973. *International Classification and Mapping of Vegetation*. Paris: UNESCO.

van der Werf G. R., Randerson J. T., Giglio L., Collatz G. J., Kasibhatla P. S. and Arellano Jr A. F. 2006. Interannual variability in global biomass burning emissions from 1997. to 2004. *Atmospheric Chemistry and Physics* 6: 3423–3441.

Vermote E. F. and El Saleous N. 1995. Stratospheric aerosol perturbing effect on the remote sesning of vegetation – operational method for the correction of AVHRR composite NDVI. *Proceedings of the Society of Photo-Optical Instrumentation Engineers (SPIE)*, 2311: 19–29.

Vermote E. F., Justice C. O., Descloitres J., El Saleous N., Roy D. P., Ray J., Margerin B. and Gonzalez L. 2001. A SeaWiFS global monthly coarse-resolution reflectance dataset. *International Journal of Remote Sensing* 22(6): 1151–1158.

Wagner W., Luckman A., Vietmeier J., Tansey K., Balzter H., Schmullius C. et al. 2003. Large-scale mapping of boreal forest in SIBERIA using ERS tandem coherence and JERS backscatter data. *Remote Sensing of Environment* 85: 125–144.

Wanner W., Li X. and Strahler A. H. 1995. On the derivation of kernels for kernel-drive models of bidirectional reflectance. *Journal of Geophysical Research-Atmospheres* 100: 21077–21089.

Waring R. H., Way J., Hunt Jr. E. R., Morrissey L., Ranson K. J., Weishampel J. F. et al. 1995. Imaging RADAR for ecosystem studies. *BioScience* 45: 715–723.

Weiss M., Baret F., Garrigues S., Lacaze R. and Bicheron P. 2007. LAI, fAPAR and fCover CYCLOPES global products derived from VEGETATION. Part 2: Validation and comparison with MODIS Collection 4 products. *Remote Sensing of Environment* 110: 317–331.

Wilson M. F. and Henderson-Sellers A. 1985. A global archive of land cover and soils data for use in general circulation climate models. *Journal of Climatology* 5: 119–143.

Wooster M. J., Roberts G., Perry G. L. and Kaufman Y. J. 2005. Retrieval of biomass combustion rates and totals from fire radiative power observations: FRP derivation and calibration relationships between biomass consumption and fire radiative energy release. *Journal of Geophysical Research-Atmospheres* 110.

Zwally H. J., Schutz R., Abdalati W., Abshire J., Bentley C., Brenner A. et al. 2002. ICESat's laser measurements of polar ice, atmosphere, ocean, and land. *Journal of Geodynamics* 34(3–4): 405–445.

Human Impacts on Terrestrial Biota and Ecosystems

Craig Miller and Iain Gordon

1 INTRODUCTION

> When human beings evolved, the challenge was survival in a world dominated by systems we could barely influence but that determined how we lived and died. Today the challenges we face are the result of systems that we have created (Sterman, 2002).

In the 3.7-billion-year history of life on Earth only one species, *Homo sapiens*, has been able to singlehandedly change ecosystem structure, function and condition at a global scale. From humble beginnings in Africa, modern humans have spread to occupy most terrestrial regions of the Earth and, enabled by agriculture, have developed the capacity to move beyond subsistence living to extracting personal wealth from ecosystem services and natural resources. Many human societies now view themselves as entities separate and above the ecosystems of which they are a part and from which they derive the necessities of life.

In the past, when the human population was small, sparsely distributed, and did not have the technological capacities of today, the impact of humans on ecosystem resources would have been localised and of no more or less importance to the ecosystem than any other species of animal. As human populations began to grow the impacts would have increased at the local scale. These impacts would have been both direct; for example, on prey or forage species; or indirect, through changes in fire frequency and intensity. The first evidence of moderate-scale deforestation through the purposeful use of fire occurs in the Mesolithic period (e.g., Brown, 1997; Carcaillet, 1998; Innes and Blackford, 2003). At this stage the local impacts are likely to have been transitory because humans would have exploited an area until the resources dwindled below a certain level and then moved on to new areas. However, with the development of agriculture about 10,000 years ago, humans began to settle in defined locations and develop complex social, political and economic systems. The global human population is estimated to have been 1–3 million at the time. With settlement the indirect impacts of human activities on local species and ecosystems would have increased, especially with the removal of trees for agriculture and construction, and the advent of livestock domestication.

The Industrial Revolution of the late eighteenth century led to a major change in the capacity of humans to convert natural resources into wealth. The rate and scale of deforestation and the conversion of ecosystems to agro-ecosystems accelerated in order to meet human needs and wants, with the system feedbacks resulting in the human population increasing from an estimated 791 million to over 6 billion currently (US Census Bureau, 2009; United Nations, 2009), and with an increasing rate and scale of ecosystem change. The unintended consequences of industrialisation, particularly since World War II, include the local, regional or global extinction of plant and animal species, wholesale loss of distinct biological communities and ecosystems, and changes in the capacity of ecosystems to provide the services that sustain human wealth and welfare (Millennium Ecosystem Assessment, 2005a, 2005b).

In this chapter we consider the impact of humans on terrestrial ecosystems through a social–ecological system lens, acknowledging that the Millennium Ecosystem Assessment (MEA 2005a), provides the most comprehensive review to date of the global condition of terrestrial biodiversity and ecosystems. We seek to advance the MEA's premise that human-modified ecosystems are losing or have lost the capacity to provide essential ecosystem services and we identify a potential path forwards.

The pressure to remediate past degradation and avoid future degradation is more pressing than ever, as the global population and its demand for ecosystem services grows. We remind ecologists and environmental scientists that the unmet challenge is to move beyond a research agenda highlighting the problems (e.g., Carson, 1962) and writing obituaries for extinct species or the living dead (e.g., Janzen, 1986; Norton, 1991) to seeking and providing the solutions (e.g., Kremen and Ostfeld, 2005; Norton and Miller, 2000). This challenge requires interdisciplinary research into the relationship between people and the environment in order to develop the biophysical, social, economic and institutional means for restoring and sustaining the ecosystem services provided by

nature (e.g., Carpenter et al., 2009; Committee on Facilitating Interdisciplinary Research, 2004; Holling, 2001; White et al., 2009).

2 APPLYING A SOCIAL–ECOLOGICAL SYSTEMS LENS

Throughout the history of life on Earth, organisms have interacted with their environment, self-organising to form complex and dynamic communities and ecosystems in the absence of a conscious guiding force (Dawkins, 2009). Indeed, the gradual accumulation of free oxygen in the atmosphere as a by-product of photosynthesis by prokaryotic and then eukaryotic organisms led to the Great Oxidation Event some 2.4 billion years ago (Anbar et al., 2007; Frei et al., 2009), without which life on Earth as we know it would not have evolved. With the evolution of human consciousness came the capacity to create cultural or economic systems and political institutions to proactively derive greater benefit from, or have greater control over the use of, ecosystems. As such, humans have moved from being a minor variable operating within an independently existing ecosystem, to creating new systems that intentionally or unintentionally drive many of the ecological patterns and processes of ecosystems (Westley et al., 2002).

The problems of biodiversity loss and environmental degradation do not persist because anyone wants them to, but because they are intrinsically outcomes of complex system dynamics. The undesirable patterns and behaviours of species-loss and ecosystem-degradation are emergent nonlinear behaviours that arise from system feedback structures (Meadows, 2008). Humans have a tendency to look for, or blame external causes of problematic system behaviour, when it is the internal dynamics of the system, created by humans, which causes the undesirable behaviour. We forget that social and economic systems are created structures, becoming immersed in them and losing sight of the fact that they can be changed (Westley et al., 2002). People may see biodiversity loss and perceive it as the result of simple trade-offs with economic development, but our demonstrated inability to deal with those trade-offs is because the proximal drivers of the trade-off are a product of the whole system, therefore not amenable to isolated intervention. Ultimately the solutions to loss of biodiversity and ecosystem services lie in changing the structure of those systems (Meadows, 2008).

3 TERRESTRIAL BIODIVERSITY AND ECOSYSTEMS: STATUS, CONDITIONS AND TRENDS

The recent MEA provides the most comprehensive review of the state and trends of the globe's ecosystems and the drivers of these conditions (MEA, 2005a, 2005b). We do not attempt to replicate this here, but will draw attention to key issues, including the fact that the globe's ecosystems have been changed more extensively by humans since World War II than at any other time in human history. The global population has doubled in the past 40 years and will reach 9 billion by the middle of this century if present trends continue (United Nations, 2009). Global economic activity has increased by a factor of seven in the same time period, although there are significant disparities between and within wealthy nations, developing nations and nations experiencing extreme poverty despite being rich in natural resources. Most of the changes observed in ecosystems during this time have been due to the dramatically increased demand for ecosystem services such as food, water and timber. This demand is met by consuming an increasing and unsustainable fraction of the available supply; in effect, killing the goose that lays the golden eggs.

Recently it has been suggested that environmental degradation will decrease as human wealth grows because it is hypothesised that incomes grow as a country develops economically after which social awareness turns to the environment. This hypothesised relationship is known as the Environmental Kuznets Curve (EKC) (Kuznets, 1955; Stern, 2004). However, this parabolic relationship between economic development and environmental degradation has been demonstrated to be an artefact of inadequate statistical analysis of empirical data, and there is not a simple parabolic relationship between biodiversity or ecosystem condition, or other environmental issues such as pollution, and income (Perman and Stern, 2003; Stern, 2004). Similarly, ecological footprint analyses suggest that higher incomes or economic activity are tied to the increased use of natural resources and degradation of ecosystems and ecosystem services, not necessarily a reduction in use or degradation (Foran et al., 2005; Wackernagel et al., 2004).

Five periods of mass extinction are recognised over the last half-billion years (Figure 26.1a) although extinctions have occurred continuously over this time (e.g., Figure 26.1b), and the current complement of Earth's species represents only about 2–4 per cent of all those that have ever lived (May et al., 2002). The current surge of extinctions due to human impact is considered by many to be the sixth extinction crisis (Leakey and Lewin, 1995; May et al., 2002). Notably, species are going extinct thousands of times faster than they have in the recent past with, for example, 0.5–2 per cent of forest-ecosystem-based species being lost each year (Nott et al., 1995; Pimm et al., 1995; Rosenzweig, 2003). Approximately 150 mammal species have gone extinct in the last 300 years, and more than one-quarter of the 5,487 known mammal species are currently threatened with extinction. The IUCN Red Data Book lists some 17,291 of 47,677 assessed species that are threatened with extinction, including 70 per cent of plants, 30 per cent of known amphibians, 28 per cent of reptiles, 21 per cent of known mammals, and 12 per cent of all known birds (Vie et al., 2009). While there are many factors associated with species extinction, human population growth alone is

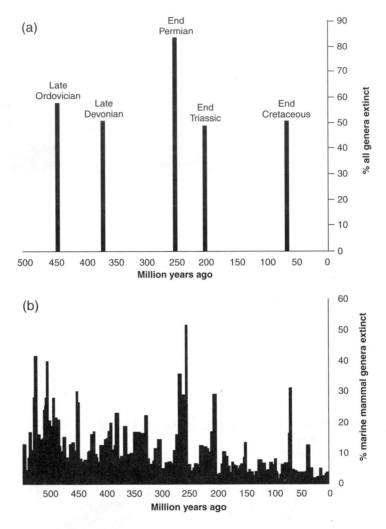

Figure 26.1   (a) The estimated proportion of all genera extinct at the boundary of the five big extinction periods (Sepkoski, 1989, 1990). (b) The extinction intensity of marine mammals (i.e., the proportion of genera present during each interval of time but that do not exist in the following interval) during the Phanerozoic eon, but excluding the Holocene. The fossil record of marine mammals is more complete than for the terrestrial environment, so is used here for illustrative purposes.

Source: Rohde and Muller (2005).

predicted to increase the number of threatened species in an average nation by 7 per cent by 2020 and 14 per cent by 2050 (McKee et al., 2004).

Of the 13 terrestrial biomes assessed by the MEA, 6 have lost more than 40 per cent of their area in historical times and another 5 will have lost over 25 per cent of their area by 2050. These have been converted largely to agricultural ecosystems. Only the boreal forests and tundra have remained relatively untouched. However, this is likely to change over the next century with the effects of climate change on temperature and rainfall patterns (Vince, 2009) and subsequent agricultural expansion. In the past hundred years, the global extent of forests has shrunk by approximately 40 per cent. Historically, temperate forests have been the most affected by clearance and fragmentation, but tropical forests have experienced greater rates of deforestation in the last 100 years, increasing rapidly in the last few decades. Southeast Asia is currently experiencing deforestation rates more than double that of other tropical regions, and this is likely to lead to mass extinctions of tropical species (Brook et al., 2003). One-third of countries have completely lost their forest cover, while another 25 countries, such as the Philippines, have lost more than 90 per cent of their forest cover (FAO, 2001). Over 40 per cent of the world's grasslands and more than 50 per cent of the world's wetlands have also been lost. A significant proportion of the wetland loss occurred in the northern hemisphere from the mid-1900s, again associated with agricultural development. The significant increase in human population size and density in the tropics and subtropics have also led to massive coastal development, with some countries losing over 80 per cent of their mangroves due to conversion to aquaculture, overexploitation for timber, or storm damage (Sodhi et al., 2006). This has knock-on impacts on coastal fisheries affecting the livelihoods and food security of millions of people.

The outlook is for even higher rates of extinction and increased degradation of ecosystems with the impact of climate change and increases in human population pressure (exacerbated by the associated expectations of affluence and consumptive parity). Indications are that this will happen unless there are fundamental systemic changes in human behaviour and consumption.

4 HUMAN-INDUCED PRESSURES ON TERRESTRIAL ECOSYSTEMS

The structure and functioning of the world's ecosystems have been affected by many factors associated with human activities.

The mass clearance of forests and drainage of wetlands to create new ecosystem structures, that is, agro-ecosystems, is the most overt example of human-induced pressures on ecosystems. Irrigated agriculture, thought to have first developed in the valley of the Nile and Mesopotamia some 6,000 years ago, has enabled agricultural production and settlement in areas with poor or spatially limited natural water supplies (Perry, 1986). It has also led to desertification and the inevitable local or regional collapse of the civilisations relying on the agriculture. The most prominent root causes of this are the feedbacks between climatic factors, institutional structures and policies, population growth and remote economic influences, that have led to local cropland expansion, overgrazing and infrastructure extension (Geist, 2005). For example, a system of oases in the Minqin Basin, on the historic Silk Road, led to various historic dynasties expanding irrigated agriculture. They did this through policies that included the development of infrastructure and irrigation technology, and tax relief (Xie et al., 2009). Problems with salinisation and alkalisation emerged, and increasing desertification in the area corresponded with a climate shift from warm-humid to arid over the intervening 2,000 years (Figure 26.2a). Subsequently, wind-driven sand and dune coverage made it almost impossible to restore the oases and led to the loss of regional towns (Xie et al., 2009). Salinity due to inappropriate land management is an issue in other parts of the world too, including Australia (Figure 26.2b) (Anderies et al., 2006).

Similarly, and more recently, an increase in crop production and the loss of natural vegetation has led to soil crusting and desertification throughout Sahelian regions of Africa, despite increases in rainfall, with feedbacks between population growth, reduced soil productivity and increased wood fuel harvest exacerbating the problem (Descroix et al., 2009). The risks of desertification are not just restricted to poor equatorial countries or arid zones; similar problems

Figure 26.2   (a) The saline and arid landscape of Minqin County, China. (Photograph courtesy of Yaowen Xie, Lanzhou University). (b) A farmer confronting salinisation, New South Wales, Australia. (Photograph courtesy of Willem van Aken, CSIRO).

have emerged in countries such as Romania (Nicolaescu et al., 2009).

Other systemic pressures have been created by the human-assisted movement of species from where they evolved to new countries and ecosystems, and in the process altering food webs, affecting population dynamics (including the extinction rates of native species), and changing ecosystem processes. Rats (*Rattus norvegicus*, *R. rattus*, and *R. exulans*) and mice (*Mus musculus*) have perhaps been the most widely spread animal, being commensal with humans. Their effect on local biota and ecosystems has differed, depending on factors such as whether similar taxa exist in the ecological niche and the life history strategies of potential prey species. The fauna of isolated islands, including New Zealand, have proven particularly vulnerable to rodent invasion (e.g., Angel et al., 2009; Harris, 2009; Mulder et al., 2009) (Figure 26.3). Some species, such as the cane toad (*Bufo marinus*), have been purposely introduced to new environments to act as biocontrol agents but have become major problems in their own right. The cane toad was introduced to Australia in 1935 to control a beetle affecting sugar cane production, but it was unable to interact with the beetle as they occupied different strata within cane

Figure 26.3 Introduced brushtail possum (*Trichosurus vulpecula*) and ship rat (*Rattus rattus*) eating song thrush nestlings (*Turdus philomelos* – also introduced) from nest, New Zealand. (Photograph courtesy of Nga Manu Images).

fields. Cane toads have since spread, causing the decline of many populations of native frog-eating species because of their toxicity (Doody et al., 2009). Thus the 'solution' to a problem became a new problem.

Exotic plant species have been found to affect the structure of native Hawaiian rainforest and colonising vegetation on young lava flows through competitive exclusion or by changing nitrogen dynamics and decomposition rates (Asner et al., 2008; Kurten et al., 2008; Rothstein et al., 2004). Tuttle et al. (2009) compared the relative impact of an introduced invasive tree (*Falcataria moluccana*) and an introduced predatory frog (*Eleutherodactylus coqui*) on the soil invertebrate community of an Hawaiian rainforest, and found that litter from the tree had a greater role in restructuring the community than did predation. Invasive, non-native, woody species are a major problem for conservation management in the South African fynbos, changing ecosystem structure and function, and affecting the provision of ecosystem services (Roura-Pascual et al., 2009). Introduced plants can also affect soil microorganisms. For example, in Britain, *Rhododendron ponticum* is considered to have an allelopathic effect on soil amoeba (Sutton and Wilkinson, 2007). Similarly, many weed species can have an allelopathic effect on agricultural crops (Malkomes, 2006). Non-native plant species may also affect pollinator population and community dynamics by changing the availability of preferred or required nectar sources, particularly when they occur at high density (Munoz and Cavieres, 2008; Nienhuis et al., 2009; Stout and Morales, 2009).

Introduced invertebrate species are also known to affect community structure and ecological processes. Argentine ants (*Linepithema humile*) are one of the most widespread invasive ant species. In Australia they displace other ant species and consequently may affect key ecological processes such as seed dispersal (Rowles and O'Dowd, 2009; Walters, 2006). The endemic blue-tailed day-gecko (*Phelsuma cepediana*), endemic to the island of Mauritius, is the sole pollinator and seed disperser of the critically endangered *Roussea simplex*. The introduced ant *Technomyrmex albipes* has been observed to monopolise and exclude the gecko from the flowers and fruits, and has been shown to reduce pollination and seed dispersal of this endangered plant (Hansen and Muller, 2009).

The yellow crazy ant (*Anoplolepis gracilipes*), listed as one of the world's worst 100 invasive species (Lowe et al., 2000), has been present on Christmas Island since the early 1900s. In the early to mid 1990s, populations of this species exploded, forming super colonies in a number of places. These supercolonies killed endemic red crabs (*Gecarcoidea natalis*) wherever they encountered them, triggering an 'invasional meltdown' (Abbott, 2006; O'Dowd et al., 2003). The red crabs are a keystone species in the forest ecosystem, eating leaves and seedlings on the forest floor and helping in litter breakdown, and their elimination has led to a change in forest structure and composition (O'Dowd et al., 2003). It has also been suggested that these changes provided favourable conditions for secondary invasions, cementing the ants' new dominance in these forest ecosystems (Green et al., 2001).

But it is not all one way. The invasion of new species may provide alternate resources for, or be a driver of adaptation and evolution in, existing populations of native species. For example, recent research suggests that cane toads may be more susceptible to predation by Australian 'meat ants' (*Iridomyrmex reburrus*) than the native frogs (Ward-Fear et al., 2009), while native phytophagous insects, such as North American and Australian soapberry bugs (e.g., *Leptocoris tagalicus*), have evolved recent structural adaptations to introduced plants such as balloon vine (*Cardiospermum grandiflorum*) (Carroll, 2007b).

The overexploitation and collapse of marine fisheries is well known (and continuing) (e.g., Gjoster et al., 2009; Grafton et al., 2009; Hutchinson, 2008; Liu and de Mitcheson, 2008), but this is also occurring in terrestrial

environments, particularly with wildlife harvesting for food, the traditional medicine market, or the pet trade (Bradshaw et al., 2009; Cheung and Dudgeon, 2006; Waite, 2007). For example, the combination of wildlife harvesting and habitat loss has led to a critical reduction in populations of orangutan (*Pongo abelii* – critically endangered; *P. pygmaeus* – endangered.) in Borneo and Sumatra, and this is unlikely to change in the near future (Marshall et al., 2006; Nantha and Tisdell, 2009). Initially this overexploitation would have been local, limited by transport options and tribal or political boundaries, but increased global transport and recent détente has led to the expansion of existing markets and the opening up of new ones.

Pollution from industrial and urban wastes is also a significant disruptor of ecosystem and species health, but the most widespread distribution of pollutants has come through the excessive application of nutrients, herbicides and pesticides to increase agricultural production (Bennett et al., 2001; Lew et al., 2009; Parton et al., 2005; Vitousek et al., 1997, 2009). The effect of nutrient pollution from the land is most evident in freshwater ecosystems, with eutrophication and hypoxic zones being the most obvious impacts. However, excess nutrients can also change the dynamics of plant communities, including increasing weed invasion (e.g., De Cauwer et al., 2008; Pfeifer-Meister et al., 2008; Prober et al., 2005; Sala et al., 2007), and residual pesticides can change the dynamics of soil micro-organisms (Lew et al., 2009). Prober et al. (2005) found that it was possible to restore the structure and function of Australian temperate grassy woodlands, including the reduction of annual weed species, by actively manipulating soil nutrient status.

At a global scale, humans have changed the rate of carbon cycling through the biosphere, atmosphere, hydrosphere and geosphere, leading to a climate system response that includes warming in global, regional and local annual temperatures, changed precipitation patterns, and changed patterns of extreme events. The Intergovernmental Panel on Climate Change (IPCC) has reviewed the likely impacts of climate change and research is continuing in this area. It is predicted that climate change will result in extensive changes in the distribution and population dynamics of species and the patterns and processes occurring in ecosystems (IPCC, 2007). For example, Australia's entire alpine ecosystem is likely to be lost if there is a change in global temperatures that exceeds 1.4°C (Hughes, 2003; Pickering et al., 2008).

The reality is that human impacts on ecosystems, whether they are through vegetation clearance, invasive species or climate change, do not act in isolation, and detrimental impacts on terrestrial biodiversity may be compounded by their interaction (Asner et al., 2006; Brook, 2008; Carroll, 2007a; Dunlop and Brown, 2008; Henry et al., 2006; McGeoch et al., 2008). Walker et al. (2006a) note that environmental or social pathologies occur when the management of ecological systems for economic or social goals leads to a loss of system resilience. These pathologies are exacerbated by a general lack of understanding of the roles of accumulation and feedback interactions in complex social–ecological systems and the impact of the consequent time lags, and nonlinear cause-and-effect relationships (Forrester, 1961; Geist, 2005; Meadows, 2008).

5 ECOSYSTEM SERVICES

Ecosystems, whether they are largely natural or modified, provide people with goods and services such as food, freshwater, timber, climate regulation, nutrient cycling, protection from disease and natural hazards, and recreation and cultural activities (Diaz et al., 2006; Tallis et al., 2008). Collectively these goods and services are referred to as ecosystem services. Ecosystem services can be broken down into four different types: (1) provisioning services, which provide water,

food, fuel; (2) regulatory services, which control the climate, flooding, water quality and disease; (3) cultural services, such as human's relationship with nature, be it spiritual, educational, aesthetic and (4) supporting services, which provide the input to primary production.

These services support human livelihoods and wellbeing through the provision of the basic materials of life such as health, security and freedom of choice and action, but the quality of these services is threatened by overconsumption and ecosystem degradation (MEA, 2005a). For example, habitat provision, water quality improvement, flood abatement, and carbon sequestration are key functions that are impaired when wetlands are lost or degraded (Zedler and Kercher, 2005); pollinators are being isolated from natural and agricultural ecosystems due to land use change and vegetation clearance (Kremen et al., 2007); and the loss of Amazonian rainforest is affecting the modulation of regional climate patterns and the spread of infectious diseases (Foley et al., 2007). Similarly, the capacity of tropical forests to store carbon into the future will be affected by their composition (Bunker et al., 2005). The decrease in terrestrial food chain length, with species from upper trophic levels disappearing, parallels the results of overexploitation observed in the oceans (Dobson et al., 2006). There are likely to be a number of surprising consequences of the change in trophic structure on the provision of ecosystem services, including such things as an increased risk of Lyme disease in humans due to a loss of vertebrate hosts (Ostfeld and LoGiudice, 2003).

The MEA (2005a, 2005b) reported that biodiversity loss is contributing to food and energy insecurity and poorer health, increasing vulnerability to natural disasters such as floods or tropical storms, reducing availability and quality of water, and eroding cultural heritage. The scientific literature abounds with the problems of trying to directly attribute ecosystem services to 'biodiversity' (Balvanera et al., 2006; Chapin et al., 2000;

Hooper et al., 2005; Kremen, 2005). We suspect that this may partially be due to the fact that biodiversity is a collective term for a wide range of ecological attributes. The accepted official definition of biodiversity is:

> The variability among living organisms from all sources, including, inter alia, terrestrial, marine and other aquatic ecosystems and the ecological complexes of which they are a part; this includes diversity within species, between species and of ecosystems (Convention on Biological Diversity, 1992).

However, there are many different applications of the term (Faith, 2003). For example, there has been an uneasy dichotomy in the ecological and conservation literature between an inventory and/or preservation focus on individual species or collections of species (often referred to as stamp collecting) and a focus on ecosystem processes (Norton, 2001). Norton (2001) considers that the process perspective will replace the object perspective, allowing a focus on the processes that created and sustained those objects or elements, rather than the elements themselves.

We argue for a whole-systems approach, contending that the either–or dichotomy of states and processes is inadequate and hinders progress in understanding and protecting biodiversity and ecosystem services. Humans will always value and quite naturally focus on 'things', hence the conservation of biodiversity will always be about the protection of species, communities or biomes, rather than carbon cycles. Therefore, there is a need to be able to understand and communicate the consequences of interactions between system components without losing sight of elements or processes. We suggest that the perspective and language of system dynamics offers a way forward for ecologists and environmental scientists, whereby species and ecosystem services are viewed as stocks; that is, components of the system that can be measured at any point in time, and that can accrue or be depleted, and that ecological processes are viewed as the flows of material or information that add to or deplete the stocks (Sterman, 2000).

6 AGRICULTURE AND THE FUTURE

Over the past 50 years, the greatest impacts on terrestrial ecosystems and biodiversity have come about as a consequence of the conversion of natural ecosystems into agricultural ecosystems. For example, cultivated cropping land now occupies more than 25 per cent of the world's surface and this is likely to expand as global demand for food increases. Associated with this has been the intensification of agricultural activities, with increases in the use of fertilisers, herbicides and pesticides (e.g., Vitousek et al., 1997) and the introduction of new crop and livestock species around the globe. Unintended or unexpected consequences arising from agriculture and agricultural development on terrestrial biodiversity and ecosystems are rife, and many of these have been identified in previous sections. Further examples include the selection of the best breeds for agriculture paradoxically leading to a loss of genetic diversity and/or loss of populations of wild stock (e.g., Borner, 2008; Carvalho et al., 2009; Mariante et al., 2009; Peter-Schmid et al., 2008), the emergence of salinity as a major issue reducing the productivity of landscapes (Anderies et al., 2006), and proposed biocontrol agents, such as cane toads and stoats (*Mustela erminea*), becoming pests (Doody et al., 2009; Hagman et al., 2009; White and King, 2006).

It has been estimated that food production will need to double by 2030 in order to feed the projected human population (Bruinsma, 2003). Arguments that rising agricultural productivity will allow us to meet our food needs are fraught for a number of reasons including the various issues caused by high-input agriculture (e.g., Vitousek et al., 2009), the loss of productive agricultural land to urbanisation (Foley et al., 2005), and the degradation of existing agricultural land (e.g., Anderies et al., 2006; Barbier, 2000). Recent analyses of regional and global ecological footprints suggest that the planet's human carrying capacity of 1.91 global hectares per capita was reached in 1978 (Monfreda et al., 2004; Wackernagel et al., 2002, 2004), and that it now effectively takes 1.2 years for the Earth to recover the natural capital that was used in 1 year (Wackernagel et al., 2002). This over-shoot of a sustainable harvesting rate of natu-ral resources has been fuelled by population and economic growth, increases in per capita consumption, and urbanisation (Geist and Lambin, 2002; York et al., 2003). Enduring and undesirable features of system over-shoots are the subsequent downturns or system crashes (Meadows, 2008). Con-sequently, the need to protect natural capital and the capacity of species' populations and ecosystems to renew or regenerate them-selves is now greater than ever (MEA, 2005a, 2005b; Monfreda et al., 2004).

7 SYSTEMS THINKING OUR WAY TO SOLUTIONS

The environmental and associated social problems that we currently face are an unin-tended consequence of the social, economic, religious and political systems that we have created to build livelihoods and wealth through the use of natural resources (Meadows, 2008; Sterman, 2002; Westley et al., 2002). These problems persist despite our best efforts to remedy them because we tend to apply linear thinking to a nonlinear world and oper-ate from a position of individual or collective bounded rationality (Allison and Hobbs, 2006; Meadows, 2008; Simon, 1972). As ecologists and environmental scientists, we have the opportunity and moral obligation to seek to understand how the systems we care about function, and to find the appropriate leverage points to ensure that they can be changed in a positive direction.

A relevant system archetype is that social–ecological systems have a natural resilience whereby the desired effect of any policy intervention is often muted or defeated by the response of the system to the intervention itself (Meadows, 1982). This is referred to as policy resistance and comes about, from a human perspective, because we continuously compare the state of the physical or social environment with our desired goals and act according to the difference between our desired situation and our perceived situa-tion (Sterman, 2000). But of course we are not the only ones assessing the state of the environment against personal or institutional goals. As we alter the state of the environ-ment, other people then react in order to bring it back towards their desired state, lead-ing in effect to systems that appear to be stuck in undesirable patterns of behaviour, such as the ongoing loss of plant and animal species.

The options to overcome policy resistance are through wielding power or letting go of ineffective policies, seeking to understand the feedbacks and bounded rationality behind them, and finding creative ways to meet the goals of the people engaged in the system while collectively moving the state of the system in a better direction (Meadows, 2008; Sterman, 2000). Letting go of 'control' in natural resource management is difficult (e.g., McDermott, 2001; Selfa and Endter-Wada, 2008), but as Meadows (2008) notes, a significant amount of the effort expended in policy implementation is often to 'correct' the response of others to the policy. Harmonisation of individual goals is not always possible or desirable, but it is possible to work towards common higher-order goals that engage the community of interest and provide for sustainable human livelihoods.

The work by the IPCC (2007) and the MEA (2005a, 2005b) has engaged govern-ments and communities across the world to focus on issues of common good, and is driv-ing change by highlighting the role that wild-life and ecosystems play in national and global economies (Balmford et al., 2002; Costanza et al., 1998; Farber et al., 2002, 2006) and general wellbeing (Diaz et al., 2006).

However, the majority of people still rank environmental issues as low on their own personal agendas, especially in developing countries where per capita income is low and people struggle to meet their basic needs, creating a feedback loop that perpetuates poverty, nonsustainable natural resource use and vulnerability (e.g., Dasgupta et al., 2005; Rayner and Malone, 2001).

Traditionally, the protection of biodiversity and ecosystems has been focused within protected areas; there are over 100,000 areas, or roughly 10 per cent of the global land surface, protected for conservation purposes. However, this is demonstrably inadequate (MEA, 2005a, 2005b) and the land area is unlikely to be increased significantly in the current global environmental paradigm, given that conservation is often viewed as a separate and competing land use to agriculture or natural resource use. Many ecologists and environmental scientists are actively promoting the integration of conservation and productive land uses (e.g., Banks, 2004; Brown, 2003; Craig et al., 2000; Harvey et al., 2008; Norton, 2000; Norton and Miller, 2000) and we view this as a step towards restoring natural capital and truly managing and valuing the provision of ecosystem services from our landscapes.

As ecologists and environmental scientists, we have a responsibility to provide plausible options for the restoration and sustainable use of natural resources, based on a better understanding of the interactions between people, their goals and aspirations, and other components of social–ecological systems (Carpenter et al., 2009; Kremen and Ostfeld, 2005; Loreau et al., 2001). We can contribute to the redesign of agricultural production systems to address issues such as degradation (e.g., salinity; Anderies et al., 2006), biodiversity conservation (Craig et al., 2000), and the provision of ecosystem services, including food production (Bennett and Balvanera, 2007) in the face of climate change. This may include the restoration of ecosystem processes that support more sustainable agricultural practices. One innovative approach

to driving the restoration of degraded landscapes is through payments for the provision of ecosystem services. For example, carbon trading schemes could enable governments to fulfil their Kyoto commitments, and communities to derive income from their natural resources, by restoring land cleared for agricultural production, providing a carbon sink (Harper et al., 2007) and benefits for biodiversity (Woinarski et al., 2007). This needs to be coupled with other social and economic mechanisms in developing nations to address poverty and vulnerability due to nonsustainable land and resource use (Dellink and Ruijs, 2008). Similarly the redesign of agricultural landscapes is unlikely to be taken up unless it is framed with an understanding of commodity system dynamics (Sawin et al., 2003).

Carpenter et al. (2009) note that biodiversity research in isolation of the social–ecological context is incomplete and, therefore, inadequate for addressing the sustainable management of ecosystem services. For example, it is just as important to understand the effect of institutional arrangements and governance structures on ecosystem services as it is to understand the biophysical effect of climate change. Similarly, it is important to recognise that stakeholders at different scales have different interests in ecosystem services, therefore one-size solutions are counterproductive (Hein et al., 2006).

But it is necessary to go beyond putting economic values on ecosystem services (Costanza et al., 1998) to actually creating stewardship incentives or markets for ecosystem services that provide a financial incentive for sustainable harvest (Niesten et al., 2002; Whitten et al., 2009) and encourage the restoration of natural capital. Markets have increasingly developed for ecotourism, although their contribution to actual conservation has been questioned (Gössling, 1999; Isaacs, 2000; Kiss, 2004), and sustainable wildlife hunting practices that do not result in short- or long-term economic hardship are being encouraged in many places (e.g., Bodmar and Lozano, 2001).

8 LEARNING TO AVOID SOLUTIONS THAT FAIL

We live in a world of dynamic complexity, whereby the simplest of interactions between agents can create the most complex or counterintuitive system behaviours (Senge, 1991). Research has demonstrated that even the most technically competent and educated people have difficulty understanding the lowest levels of dynamic complexity, yet we are expected to successfully manage systems with high-level complexity everyday (Booth Sweeney and Sterman, 2000; Cronin et al., 2009). Adaptive management has been proposed as a way of institutional learning for managing natural resources and ecosystems in the face of uncertainty (Holling, 1978; Walters, 1986; Walters and Holling, 1990) although its application has been more conceptual than applied, and is often used to rebadge conventional monitoring without feedback to policy refinement (Lee, 1999; Stem et al., 2005). We, like Forrester (1991) and Sterman (2000, 2002, 2008), recommend using the systems dynamics modelling process to make explicit the mental models of cause and effect that underpin policy, test them against real system behaviours, and test potential policy options. Forrester (1961) initially developed the field and language of systems dynamics to understand how industrial systems respond to human decision making, and this remains the most accessible, yet probably least used, approach to complex systems analysis available at present.

Systems dynamics is underpinned by the concepts of accumulation and feedback, and an understanding of these concepts is essential for effective policy development in complex systems (Sterman, 2008). Accumulation drives the dynamics of social, economic, biological and physical systems (Cronin et al., 2009). Accumulations are the state variables of a system and these are added to or depleted by state-change processes (Forrester, 1961). State-change processes operate at finite rates of flow, thereby causing delayed system response to natural changes and human actions. Such delays are the source of 'inertia' in complex systems. They inhibit experiential learning and introduce uncertainty into the design and implementation of policies. This is one reason why effective policy making in complex systems requires a good understanding of the dynamical effects caused by accumulations and the related state-change processes.

The interconnections and relationships between stocks (accumulations) and flows (state-change processes) lead to overall system behaviour. Simple or complex feedback loops; that is, the process by which changes in the value of the stock affect the flows into or out of the stock, are the mechanisms that create consistent system behaviour, leading to growth or decline, resilience or collapse. Information delays, in conjunction with feedbacks, can also lead to oscillatory or unstable system behaviours. Perturbations external to the system of interest may cause temporary fluctuations in system behaviour, with negative feedback loops acting to bring the system behaviour back into the typical pattern; that is, system resilience. The capacity of the system to recover may be affected if the perturbation is too large, positive feedback loops have become dominant, and/or if essential stocks or flows have been changed, potentially leading to system collapse and the development of a new system and new system behaviour (Folke et al., 2004; Kinzig et al., 2006; Walker et al., 2004; Walker et al., 2006b).

We present a simple conceptual stock-and-flow model (Figure 26.4) which has the well-known feedback structure of a predator–prey system to illustrate the capacity of a simple system to generate complex behaviour and to depict how systems dynamics modelling could be used to test policies or hypotheses *a priori*. In the model, the birth and death rates of each species depends on the abundance (stock) of the species and on the interactions between the species. Thus, the carnivore birth rate depends on

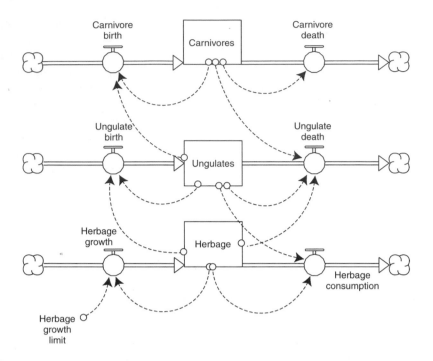

Figure 26.4   Conceptual stock-and-flow model depicting a three-level predator–prey system. In this diagram the rectangles represent accumulations (stocks), the double-lined arrows represent processes (flows), and the single-lined arrows represent feedbacks and influence links. The 'tap' symbol represents process rates and the 'cloud' symbol represents sources or sinks that lie outside the boundaries of the model. Such models can be used to demonstrate the complex behaviour generated by the feedback relations between the stocks and flows of a system and the nonlinear, and often counterintuitive, effects of management on the behaviour of these systems.

the availability of ungulates, the ungulate birth rate depends on the availability of herbage, the ungulate death rate depends on the abundance of carnivores and herbage, and the herbage consumption rate depends on the number of ungulates.

The behaviour of this model is illustrated in Figure 26.5. The model parameters have been set so that the system is initially in equilibrium. That is, the stocks of carnivores, ungulates, and herbage are constant over time. Then, at time $t = 100$ a program of culling of carnivores is initiated, thereby increasing their death rate. As carnivore numbers decrease, so the number of ungulates increases. The increased grazing pressure drives down the stocks of herbage until it cannot survive. When the herbage dies out

there is mass starvation of ungulates and their numbers collapse. The oscillations seen in the time series represent the natural feedback dynamics of a predator–prey system. This type of behaviour was observed, for example, when conservationists sought to protect the Kaibab mule deer (*Odocoiles hemionus crooki*) by removing predators such as mountain lions (*Puma concolor*) in the Kaibab Plateau region on the North Rim of the Grand Canyon, Arizona, in 1906 (McCulloch, 1986; Young, 2002). A system dynamics approach that takes into account the stock and flow structure of the social–ecological systems offers one way to minimise the probability that policy initiatives will produce such unexpected and unwelcome outcomes (Sterman, 2000).

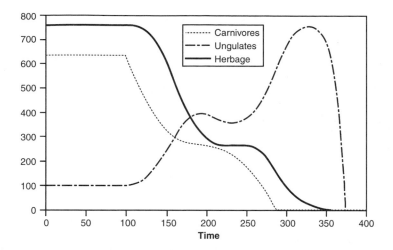

Figure 26.5   Graphical representation of the behaviour of the predator–prey system model from Figure 26.4. Note that (a) the scale of the *y*-axis is arbitrary, and that the behaviour of each organism should be compared to itself over time; that is, in reality, carnivore population numbers are not higher than herbivore population numbers, and (b) the model is at equilibrium until time *t* = 100, after which culling of carnivores is initiated.

9 OUTLOOK

The outlook for terrestrial biodiversity and the provision of ecosystem services as we know them is not good under the current regional and global trajectories (Czech, 2008; MEA, 2005a, 2005b). This is likely to be exacerbated by climate change and population growth (IPCC, 2007; McKee et al., 2004) and requires concerted global action, a change in the way we perceive and utilise natural resources, and a change in many of our economic and social systems. Some of these changes may be forced upon us as critical environmental thresholds are exceeded, but we do have the opportunity and incentive to act now to protect, rehabilitate or restore degraded or vulnerable biodiversity, and ensure the future provision of essential ecosystem services. The issues are a consequence of the systems that we have created and the solutions will be found in restructuring those social–ecological systems.

It is a human imperative to find the solutions (MEA, 2005a, 2005b) as the problems we face are irrelevant to the Earth which will continue to exist long after we are gone. The current suite of extinctions follow five other mass extinction periods in the Earth's long history (Leakey and Lewin, 1995) and new species will evolve and disperse over the next million years. New communities and ecosystems will self-organise from the suite of invasive and agricultural species we have moved around the world over the next millennia. Climate change will drive global ecology and evolution as it always has done but the quality of human life is affected by the loss of biodiversity and ecosystem services now.

REFERENCES

Abbott K. L. 2006. Spatial dynamics of supercolonies of the invasive yellow crazy ant, *Anoplolepis gracilipes*, on Christmas Island, Indian Ocean. *Diversity and Distributions* 12: 101–110.

Allison H. and Hobbs R. 2006. *Science and Policy in Natural Resource Management. Understanding*

System Complexity. Cambridge: Cambridge University Press.

Anbar A. D., Duan Y., Lyons T. W., Arnold G. L., Kendall B., Creaser R. A. et al. 2007. A whiff of oxygen before the great oxidation event? *Science* 317: 1903–1906.

Anderies J. M., Ryan P. and Walker B. H. 2006. Loss of resilience, crisis, and institutional change: Lessons from an intensive agricultural system in southeastern Australia. *Ecosystems* 9: 865–878.

Angel A., Wanless R. M. and Cooper J. 2009. Review of impacts of the introduced house mouse on islands in the Southern Ocean: are mice equivalent to rats? *Biological Invasions* 11: 1743–1754.

Asner G. P., Hughes R. F., Vitousek P. M., Knapp D. E., Kennedy-Bowdoin T. et al. 2008. Invasive plants transform the three-dimensional structure of rain forests. *Proceedings of the National Academy of Sciences of the United States of America* 105: 4519–4523.

Asner G. P., Martin R. E., Carlson K. M., Rascher U. and Vitousek P. M. 2006. Vegetation-climate interactions among native and invasive species in Hawaiian rainforest. *Ecosystems* 9: 1106–1117.

Balmford A., Bruner A., Cooper P., Costanza R., Farber S., Green R. E. et al. 2002. Economic reasons for conserving wild nature. *Science* 297: 950–953.

Balvanera B., Pfisterer A. B., Buchmann N., He J.-S., Nakashizuka T., Raffaelli D. and Schmid B. 2006. Quantifying the evidence for biodiversity effects on ecosystem functioning and services. *Ecology Letters* 9: 1146–1156.

Banks J. E. 2004. Divided culture: integrating agriculture and conservation biology. *Frontiers in Ecology and the Environment* 2: 537–545.

Barbier E. B. 2000. The economic linkages between rural poverty and land degradation: some evidence from Africa. *Agriculture, Ecosystems & Environment* 82: 355–370.

Bennett E. M. and Balvanera P. 2007. The future of production systems in a globalized world. *Frontiers in Ecology and the Environment* 5: 191–198.

Bennett E. M., Carpenter S. R. and Caraco N. F. 2001. Human impact on erodable phosphorus and eutrophication: a global perspective. *BioScience* 51: 227–234.

Bodmar R. E. and Lozano E. P. 2001. Rural development and sustainable wildlife use in Peru. *Conservation Biology* 15: 1163–1170.

Booth Sweeney L. and Sterman J. D. 2000. Bathtub dynamics: initial results of a systems thinking inventory. *System Dynamics Review* 16: 249–286.

Borner A. 2008. Plant genetic resources for future breeding. *Proceedings of the 18th EUCARPIA General Congress on Modern Variety Breeding for Present and Future Needs, Valencia, Spain, 9–12 September*, pp. 37–48.

Bradshaw C. J. A., Sodhi N. S. and Brook B. W. 2009. Tropical turmoil: a biodiversity tragedy in progress. *Frontiers in Ecology and the Environment* 7: 79–87.

Brook B. W. 2008. Synergies between climate change, extinctions and invasive vertebrates. *Wildlife Research* 35: 249–252.

Brook B. W., Sodhi N. S. and Ng P. K. L. 2003. Catastrophic extinctions follow deforestation in Singapore. *Nature* 424: 420–423.

Brown K. 2003. Integrating conservation and development: a case of institutional misfit. *Frontiers in Ecology and the Environment* 1: 479–487.

Brown T. 1997. Clearances and clearings: deforestation in Mesolithic/Neolithic Britain. *Oxford Journal of Archaeology* 16: 133–146.

Bruinsma J. (ed.) 2003. *World Agriculture: Towards 2015/2030.* Rome: Food and Agricultural Organization of the United Nations.

Bunker D. E., DeClerck F., Bradford J. C., Colwell R. K., Perfecto I., Phillips O. L. et al. 2005. Species loss and aboveground carbon storage in a tropical forest. *Science* 310: 1029–1031.

Carcaillet C. 1998. A spatially precise study of Holocene fire history, climate and human impact within the Maurienne Valley, North French Alps. *Journal of Ecology* 86: 384–396.

Carpenter S. R., Mooney H. A., Agard J., Capistrano D., DeFries R. S., Diaz S. et al. 2009. Science for managing ecosystem services: Beyond the Millennium Ecosystem Assessment. *Proceedings of the National Academy of Sciences of the United States of America* 106: 1305–1312.

Carroll C. 2007a. Interacting effects of climate change, landscape conversion, and harvest on carnivore populations at the range margin: Marten and Lynx in the northern Appalachians. *Conservation Biology* 21: 1092–1104.

Carroll S. P. 2007b. *Facing Change: Forms and Foundations of Contemporary Adaptation to Biotic Invasions.* Los Angeles: Blackwell Publishing, pp. 361–372.

Carson R. 1962. *Silent Spring.* Boston: Houghton Mifflin.

Carvalho A., Lima-Brito J., Macas B. and Guedes-Pinto H. 2009. Genetic diversity and variation among botanical varieties of old Portuguese wheat cultivars revealed by ISSR assays. *Biochemical Genetics* 47: 276–294.

Chapin F. S., Zavaleta E. S., Eviner V. T., Naylor R. L., Vitousek P. M., Reynolds H. L. et al. 2000. Consequences of changing biodiversity. *Nature* 405: 234–242.

Cheung S. M. and Dudgeon D. 2006. Quantifying the Asian turtle crisis: market surveys in southern China, 2000–2003. *Aquatic Conservation-Marine and Freshwater Ecosystems* 16: 751–770.

Committee on Facilitating Interdisciplinary Research. 2004. *Facilitating Interdisciplinary Research.* Washington DC: The National Academies Press.

Convention on Biological Diversity. 1992. Convention on Biological Diversity.

Costanza R., d'Arge R., de Groot R., Farber S., Grasso M., Hannon B. et al. 1998. The value of the world's ecosystem services and natural capital. *Ecological Economics* 25: 3–15.

Craig J. L., Mitchell N. and Daunders D. A. (eds) 2000. *Conservation in Production Environments: Managing the Matrix.* Chipping Norton: Surrey Beatty & Sons.

Cronin M. A., Gionzalez C. and Sterman J. D. 2009. Why don't well-educated adults understand accumulation? A challenge to researchers, educators and citizens. *Organizational Behaviour and Human Decision Processes* 108: 116–130.

Czech B. 2008. Prospects for reconciling the conflict between economic growth and biodiversity conservation with technological progress. *Conservation Biology* 22: 1389–1398.

Dasgupta S., Deichmann U., Meisner C. and Wheeler D. 2005. Where is the poverty-environment nexus? Evidence from Cambodia, Lao PDR, and Vietnam. *World Development* 33: 617–638.

Dawkins R. 2009. *The Greatest Show on Earth.* London: Bantam Press.

De Cauwer B., Reheul D., Nijs I. and Milbau A. 2008. Management of newly established field margins on nutrient-rich soil to reduce weed spread and seed rain into adjacent crops. *Weed Research* 48: 102–112.

Dellink R. B. and Ruijs A. (eds) 2008. *Economics of Poverty, Environment and Natural-resource Use.* Doetinchem: Springer.

Descroix L., Mahe G., Lebel T., Favreau G., Galle S., Gautier E. et al. 2009. Spatio-temporal variability of hydrological regimes around the boundaries between Sahelian and Sudanian areas of West Africa: a synthesis. *Journal of Hydrology (Amsterdam)* 375: 90–102.

Diaz S., Fargione J., Chapin F. S., III and Tilman D. 2006. Biodiversity loss threatens human well-being. *PLoS Biol* 4: e277.

Dobson A., Lodge D., Alder J., Cumming G. S., Keymer J., McGlade J. et al. 2006. Habitat loss, trophic collapse, and the decline of ecosystem services. *Ecology* 87: 1915–1924.

Doody J. S., Green B., Rhind D., Castellano C. M., Sims R. and Robinson T. 2009. Population-level declines in Australian predators caused by an invasive species. *Animal Conservation* 12: 46–53.

Dunlop M. and Brown P. R. 2008. *Implications of Climate Change for Australia's National Reserve System: A Preliminary Assessment.* Report to the Australian Department of Climate Change. Canberra: Department of Climate Change.

Faith D. P. 2003. Biodiversity, in Zalta E. N. (ed.) *The Stanford Encyclopedia of Philosophy.* Stanford: Stanford University.

FAO 2001. State of World's forests. Rome: Food and Agriculture Organization of the United Nations.

Farber S., Costanza R., Childers D. L., Erickson J., Gross K., Grove M. et al. 2006. Linking ecology and economics for ecosystem management. *BioScience* 56: 121–133.

Farber S. C., Costanza R. and Wilson M. A. 2002. Economic and ecological concepts for valuing ecosystem services. *Ecological Economics* 41: 375–392.

Foley J. A., Asner G. P., Costa M. H., Coe M. T., DeFries R., Gibbs H. K. et al. 2007. Amazonia revealed: forest degradation and loss of ecosystem goods and services in the Amazon Basin. *Frontiers in Ecology and the Environment* 5: 25–32.

Foley J. A., DeFries R., Asner G. P., Barford C., Bonan G., Carpenter S. R. et al. 2005. Global consequences of land use. *Science* 309: 570–574.

Folke C., Carpenter S., Walker B., Scheffer M., Elmqvist T., Gunderson L. and Holling C. S. 2004. Regime shifts, resilience, and biodiversity in ecosystem management. *Annual Review of Ecology Evolution and Systematics* 35: 557–581.

Foran B. D., Lenzen M. and Dey C. 2005. *Balancing Act: A Triple Bottom Line Analysis of the Australian Economy.* Canberra: University of Sydney and CSIRO.

Forrester J. W. 1961. *Industrial Dynamics.* Cambridge: MIT Press.

Forrester J. W. 1991. System dynamics and the lessons of 35 years, in De Greene K. B. (ed.) *The Systemic Basis of Policy Making in the 1990s.* Cambridge, MA: MIT Press, pp. 5–34.

Frei R., Gaucher C., Poulton S. W. and Canfield D. E. 2009. Fluctuations in Precambrian atmospheric oxygenation recorded by chromium isotopes. *Nature* 461: 250–253.

Geist H. 2005. *The Causes and Progression of Desertification.* Aldershot: Ashgate Publishing.

Geist H. J. and Lambin E. F. 2002. Proximate causes and underlying driving forces of tropical deforestation. *BioScience* 52: 143–150.

Gjoster H., Bogstad B. and Tjelmeland S. 2009. Ecosystem effects of the three capelin stock collapses in the Barents Sea. *Marine Biology Research* 5: 40–53.

Gössling S. 1999. Ecotourism: a means to safeguard biodiversity and ecosystem functions? *Ecological Economics* 29: 303–320.

Grafton R. Q., Kompas T. and Pham V. 2009. Cod today and none tomorrow: the economic value of a marine reserve. *Land Economics* 85: 454–469.

Green P. T., O'Dowd D. J. and Lake P. S. 2001. From resistance to meldown: secondary invasion of an island rain forest, in Ganeshaiah K. N., Uma Shaanker R. and Bawa K. S. (eds) *Proceedings of the International Conference on Tropical Ecosystems, New Delhi, India, 15–18 July.* Oxford: Oxford & IBH, pp. 451–455.

Hagman M., Phillips B. L. and Shine R. 2009. Fatal attraction: adaptations to prey on native frogs imperil snakes after invasion of toxic toads. *Proceedings of the Royal Society B – Biological Sciences* 276: 2813–2818.

Hansen D. M. and Muller C. B. 2009. Invasive ants disrupt gecko pollination and seed dispersal of the endangered plant Roussea simplex in Mauritius. *Biotropica* 41: 202–206.

Harper R. J., Beck A. C., Ritson P., Hill M. J., Mitchell C. D., Barrett D. J. et al. 2007. The potential of greenhouse sinks to underwrite improved land management. *Ecological Engineering* 29: 329–341.

Harris D. B. 2009. Review of negative effects of introduced rodents on small mammals on islands. *Biological Invasions* 11: 1611–1630.

Harvey C. A., Komar O., Chazdon R., Ferguson B. G., Finegan B., Griffith D. M. et al. 2008. Integrating agricultural landscapes with biodiversity conservation in the Mesoamerican hotspot. *Conservation Biology* 22: 8–15.

Hein L., van Koppen K., de Groot R. S. and van Ierland E. C. 2006. Spatial scales, stakeholders and the valuation of ecosystem services. *Ecological Economics* 57: 209–228.

Henry H. A. L., Chiariello N. R., Vitousek P. M., Mooney H. A. and Field C. B. 2006. Interactive effects of fire, elevated carbon dioxide, nitrogen deposition, and precipitation on a California annual grassland. *Ecosystems* 9: 1066–1075.

Holling C. S. (ed.) 1978. *Adaptive Environmental Assessment and Management.* New York: John Wiley & Sons.

Holling C. S. 2001. Understanding the complexity of economic, ecological, and social systems. *Ecosystems* 4: 390–405.

Hooper D. U., Chapin F. S., Ewel J. J., Hector A., Inchausti P., Lavorel S. et al. 2005. Effects of biodiversity on ecosystem functioning: A consensus of current knowledge. *Ecological Monographs* 75: 3–35.

Hughes L. 2003. Climate change and Australia: Trends, projections and impacts. *Austral Ecology* 28: 423–443.

Hutchinson W. F. 2008. The dangers of ignoring stock complexity in fishery management: the case of the North Sea cod. *Biology Letters* 4: 693–695.

Innes J. B. and Blackford J. J. 2003. The ecology of Late Mesolithic woodland disturbances: Model testing with fungal spore assemblage data. *Journal of Archaeological Science* 30: 185–194.

IPCC. 2007. Climate Change 2007: Synthesis Report, in Pachauri R. K. and Reisinger A. (eds) *Contribution of Working Groups I, II and III to the Fourth Assessment Report of the Intergovernmental Panel on Climate Change.* Geneva: Intergovernmental Panel on Climate Change, p. 104.

Isaacs J. C. 2000. The limited potential of ecotourism to contribute to wildlife conservation. *Wildlife Society Bulletin* 28: 61–69.

Janzen D. A. 1986. Blurry catastrophes. *Oikos* 47: 1–2.

Kinzig A. P., Ryan P., Etienne M., Allison H., Elmqvist T. and Walker B. H. 2006. Resilience and regime shifts: Assessing cascading effects. *Ecology and Society* 11(1): 20.

Kiss A. 2004. Is community-based ecotourism a good use of biodiversity conservation funds? *Trends in Ecology & Evolution* 19: 232–237.

Kremen C. 2005. Managing ecosystem services: what do we need to know about their ecology? *Ecology Letters* 8: 468–479.

Kremen C. and Ostfeld R. S. 2005. A call to ecologists: measuring, analyzing, and managing ecosystem services. *Frontiers in Ecology and the Environment* 3: 540–548.

Kremen C., Williams N. M., Aizen M. A., Gemmill-Herren B., LeBuhn G., Minckley R. et al. 2007. Pollination and other ecosystem services produced by mobile organisms: a conceptual framework for the effects of land-use change. *Ecology Letters* 10: 299–314.

Kurten E. L., Snyder C. P., Iwata T. and Vitousek P. M. 2008. Morella cerifera invasion and nitrogen cycling on a lowland Hawaiian lava flow. *Biological Invasions* 10: 19–24.

Kuznets S. 1955. Economic growth and income inequality. *American Economic Review* 451: 1–28.

Leakey R. and Lewin R. 1995. *The Sixth Extinction: Biodiversity and its Survival.* New York: Doubleday.

Lee K. N. 1999. Appraising adaptive management. *Conservation Ecology* 3(2): 3.

Lew S., Lew M., Szarek J. and Mieszczynski T. 2009. Effect of pesticides on soil and aquatic environmental microorganisms – a short review. *Fresenius Environmental Bulletin* 18: 1390–1395.

Liu M. and de Mitcheson Y. S. 2008. Profile of a fishery collapse: why mariculture failed to save the large yellow croaker. *Fish and Fisheries* 9: 219–242.

Loreau M., Naeem S., Inchausti P., Bengtsson J., Grime J. P., Hector A. et al. 2001. Biodiversity and ecosystem functioning: current knowledge and future challenges. *Science* 294: 804–808.

Lowe S., Browne M., Boudjelas S. and De Poorter M. 2000. *100 of the World's Worst Invasive Alien Species. A Selection From the Global Invasive Species Database.* Auckland: The Invasive Species Specialist Group.

Malkomes H. P. 2006. *Allelopathy of Middle European Agricultural Weeds – An Overview.* Stuttgart: Eugen Ulmer Gmbh, pp. 435–445.

Mariante A. D. S., Albuquerque M. D. S. M., Egito A. A., McManus C., Lopes M. A. and Paiva S. R. 2009. Present status of the conservation of livestock genetic resources in Brazil. *Livestock Science* 120: 204–212.

Marshall A. J., Nardiyono, Engstrom L. M., Pamungkas B., Palapa J., Meijaard E. and Stanley S. A. 2006. The blowgun is mightier than the chainsaw in determining population density of Bornean orangutans (Pongo pygmaeus morio) in the forests of East Kalimantan. *Biological Conservation* 129: 566–578.

May R. M., Lawton J. H. and Stork N. E. 2002. Assessing extinction rates, in Lawton J. H. and May R. M. (eds) *Extinction Rates.* Oxford: Oxford University Press.

McCulloch C. Y. 1986. A. history of predator control and deer productivity in northern Arizona. *The Southwestern Naturalist* 31: 215–220.

McDermott M. 2001. Invoking community: indigenous people and ancestral domain in Palawan, Philippines, in Agarwal A. and Gibson C. (eds) *Communities and the Environment: Ethnicity, Gender and the State in Community-based Conservation.* New Brunswick: Rutgers University Press, pp. 32–62.

McGeoch L., Gordon I. and Schmitt J. 2008. Impacts of land use, anthropogenic disturbance, and harvesting on an African medicinal liana. *Biological Conservation* 141: 2218–2229.

McKee J. K., Sciulli P. W., Fooce C. D. and Waite T. A. 2004. Forecasting global biodiversity threats associated with human population growth. *Biological Conservation* 115: 161–164.

Meadows D. H. 1982. Whole earth models and systems. *CoEvolution Quarterly* Summer, 98–108.

Meadows D. H. 2008. *Thinking in Systems.* White River Junction: Chelsea Green Publishing.

Millennium Ecosystem Assessment. 2005a. *Ecosystems and Human Well-being: Current State and Trends.* Washington DC: Island Press.

Millennium Ecosystem Assessment. 2005b. Ecosystems and Human Well-being: Synthesis. Washington DC: Island Press.

Monfreda C., Wackernagel M. and Deumling D. 2004. Establishing national natural capital accounts based on detailed Ecological Footprint and biological capacity assessments. *Land Use Policy* 21: 231–246.

Mulder C. P. H., Grant-Hoffman M. N., Towns D. R., Bellingham P. J., Wardle D. A., Durrett M. S. et al. 2009. Direct and indirect effects of rats: does rat eradication restore ecosystem functioning of New Zealand seabird islands? *Biological Invasions* 11: 1671–1688.

Munoz A. A. and Cavieres L. A. 2008. The presence of a showy invasive plant disrupts pollinator service and reproductive output in native alpine species only at high densities. *Journal of Ecology* 96: 459–467.

Nantha H. S. and Tisdell C. 2009. The orangutan-oil palm conflict: economic constraints and opportunities for conservation. *Biodiversity and Conservation* 18: 487–502.

Nicolaescu M., Lupascu N. and Chirila E. 2009. *Land Degradation and Desertification Risk in Dobrogea Region.* Brasov: Gh Asachi Technical Univ Iasi, pp. 911–914.

Nienhuis C. M., Dietzsch A. C. and Stout J. C. 2009. The impacts of an invasive alien plant and its removal on native bees. *Apidologie* 40: 450–463.

Niesten E., Frumhoff P. C., Manion M. and Hardner J. J. 2002. Designing a carbon market that protects forests in developing countries. *Philosophical Transactions: Mathematical, Physical and Engineering Sciences* 360: 1875–1888.

Norton B. G. 2001. Conservation biology and environmental values: can there be a universal earth ethic?, in Potvin C., Kraenzel K. and Seutin G. (eds) *Protecting Biological Diversity: Roles and Responsibilities.* Montreal: McGill-Queen's University Press.

Norton D. A. 1991. *Trilepidea adamsii:* An obituary for a species. *Conservation Biology* 5: 52–57.

Norton D. A. 2000. Conservation biology and private land: shifting the focus. *Conservation Biology* 14: 1221–1223.

Norton D. A. and Miller C. J. 2000. Conservation of native biodiversity in rural New Zealand. *Ecological Management & Restoration* 1: 26–34.

Nott M. P., Rogers E. and Pimm S. 1995. Extinction rates – modern extinctions in the kilo-death range. *Current Biology* 5: 14–17.

O'Dowd D. J., Green P. T. and Lake P. S. 2003. Invasional 'meltdown' on an oceanic island. *Ecology Letters* 6: 812–817.

Ostfeld R. S. and LoGiudice K. 2003. Community disassembly, biodiversity loss, and the erosion of an ecosystem servive. *Ecology* 84: 1421–1427.

Parton W. J., Neff J. and Vitousek P. M. 2005. Modelling phosphorus, carbon and nitrogen dynamics in terrestrial ecosystems. *Organic Phosphorus in the Environment,* 325–347.

Perman R. and Stern D. I. 2003. Evidence from panel unit root and cointegration tests that the Environmental Kuznets Curve does not exist. *The Australian Journal of Agricultural and Resource Economics* 47: 325–347.

Perry R. A. 1986. Desertification processes and impacts in arid regions. *Climatic Change* 9, 43–47.

Peter-Schmid M. K. I., Kolliker R. and Boller B. 2008. Value of permanent grassland habitats as reservoirs of Festuca pratensis Huds. and Lolium multiflorum Lam. populations for breeding and conservation. *Euphytica* 164: 239–253.

Pfeifer-Meister L., Cole E. M., Roy B. A. and Bridgham S. D. 2008. Abiotic constraints on the competitive ability of exotic and native grasses in a Pacific Northwest prairie. *Oecologia* 155: 357–366.

Pickering C., Hill W. and Green K. 2008. Vascular plant diversity and climate change in the alpine zone of the Snowy Mountains, Australia. *Biodiversity and Conservation* 17: 1627–1644.

Pimm S. L., Russell G. J., Gittleman J. L. and Brooks T. M. 1995. The future of biodiversity. *Science* 269: 347–350.

Prober S. M., Thiele K. R., Lunt I. D. and Koen T. B. 2005. Restoring ecological function in temperate grassy woodlands: manipulating soil nutrients, exotic annuals and native perennial grasses through carbon supplements and spring burns. *Journal of Applied Ecology* 42: 1073–1085.

Rayner S. and Malone E. L. 2001. Climate change, poverty, and intragenerational equity: the national level. *International Journal of Global Environmental Issues* 1: 175–202.

Rohde R. A. and Muller R. A. 2005. Cycles in fossil diversity. *Nature* 434: 208–210.

Rosenzweig M. L. 2003. Reconciliation ecology and the future of species diversity. *Oryx* 37: 194–205.

Rothstein D. E., Vitousek P. M. and Simmons B. L. 2004. An exotic tree alters decomposition and nutrient cycling in a Hawaiian montane forest. *Ecosystems* 7: 805–814.

Roura-Pascual N., Richardson D. M., Krug R. M., Brown A., Chapman R. A., Forsyth G. G. et al. 2009. Ecology and management of alien plant invasions in South African fynbos: Accommodating key complexities in objective decision making. *Biological Conservation* 142: 1595–1604.

Rowles A. D. and O'Dowd D. J. 2009. Impacts of the invasive Argentine ant on native ants and other invertebrates in coastal scrub in south-eastern Australia. *Austral Ecology* 34: 239–248.

Sala A., Verdaguer D. and Vila M. 2007. Sensitivity of the invasive geophyte Oxalis pes-caprae to nutrient availability and competition. *Annals of Botany* 99: 637–645.

Sawin B., Hamilton H., Jones A., Rice P., Seville D., Sweitzer S. and Wright D. 2003. *Commodity Systems Challenges. Moving Sustainability into the Mainstream of Natural Resource Economies.* Hartland: Sustainability Institute.

Selfa T. and Endter-Wada J. 2008. The politics of community-based conservation in natural resource management: a focus for international comparative analysis. *Environment and Planning A* 40: 948–965.

Senge P. M. 1991. *The Fifth Discipline: The Art and Practice of the Learning Organization.* New York: Doubleday.

Sepkoski J. J., Jr. 1989. Periodicity in extinction and the problem of catastrophism in the history of life. *Journal of the Geological Society of London* 146: 7–19.

Sepkoski J. J., Jr. 1990. The taxonomic structure of periodic extinction, in Sharpton L. and Ward P. D. (eds) *Global Catastrophes in Earth History,* Boulder: Geological Society of America.

Simon H. 1972. Theories of bounded rationality, in Radner R. and McGuire C. B. (eds) *Decision and Organisation.* Amsterdam: North Holland Publishing Company.

Sodhi N. S., Brooks T. M., Koh L. P., Acciaioli G., Erb M., Tan A. K. J. et al. 2006. Biodiversity and human livelihood crises in the Malay archipelago. *Conservation Biology* 20: 1811–1813.

Stem C., Margoluis R., Salafsky N. and Brown M. 2005. Monitoring and evaluation in conservation: a review of trends and approaches. *Conservation Biology* 19: 295–309.

Sterman J. D. 2000. *Business Dynamics. Systems Thinking and Modeling for a Complex World*. Boston: Irwin McGraw-Hill.

Sterman J. D. 2002. All models are wrong: reflections on becoming a systems scientist. *System Dynamics Review* 18: 501–531.

Sterman J. D. 2008. Risk communication on climate: mental models and mass balance. *Science* 322: 532–533.

Stern D. I. 2004. The rise and fall of the Environmental Kuznets Curve. *World Development* 32: 1419–1439.

Stout J. C. and Morales C. L. 2009. Ecological impacts of invasive alien species on bees. *Apidologie* 40: 388–409.

Sutton C. A. and Wilkinson D. M. 2007. The effects of Rhododendron on testate amoebae communities in woodland soils in north west England. *Acta Protozoologica* 46: 333–338.

Tallis H., Kareiva P., Marvier M. and Chang A. 2008. An ecosystem services framework to support both practical conservation and economic development. *Proceedings of the National Academy of Sciences* 105: 9457–9464.

Tuttle N. C., Beard K. H. and Pitt W. C. 2009. Invasive litter, not an invasive insectivore, determines invertebrate communities in Hawaiian forests. *Biological Invasions* 11: 845–855.

US Census Bureau. 2009. *U.S. & World population clocks*. New York: US Census Bureau.

United Nations. 2009. *World Population Prospects. The 2008 Revision*. New York: United Nations.

Vie J.-C., Hilton-Taylor C. and Stuart S. N. 2009. *Wildlife in a Changing World – An Analysis of the 2008. IUCN Red List of Threatened Species*. Gland: IUCN.

Vince G. 2009. Surviving in a warmer world. *New Scientist* 2697: 28–33.

Vitousek P. M., Aber J. D., Howarth R. W., Likens G. E., Matson P. A., Schindler D. W. et al. 1997. Human alteration of the global nitrogen cycle: Sources and consequences. *Ecological Applications* 7: 737–750.

Vitousek P. M., Naylor R., Crews T., David M. B., Drinkwater L. E., Holland E. et al. 2009. Nutrient imbalances in agricultural development. *Science* 324: 1519–1520.

Wackernagel M., Monfreda C., Schulz N. B., Erb K.-H., Haberl H. and Krausmann F. 2004. Calculating national and global ecological footprint time series: resolving conceptual challenges. *Land Use Policy* 21: 271–278.

Wackernagel M., Schulz N. B., Deumling D., Linares A. C., Jenkins M., Kapos V. et al. 2002. Tracking the

ecological overshoot of the human economy. *Proceedings of the National Academy of Sciences of the United States of America* 99: 9266–9271.

Waite T. A. 2007. Revisiting evidence for sustainability of bushmeat hunting in West Africa. *Environmental Management* 40: 476–480.

Walker B., Gunderson L., Kinzig A., Folke C., Carpenter S. and Schultz L. 2006a. A handful of heuristics and some propositions for understanding resilience in social-ecological systems. *Ecology and Society* 11(1): 13.

Walker B., Hollin C. S., Carpenter S. R. and Kinzig A. 2004. Resilience, adaptability and transformability in social-ecological systems. *Ecology and Society* 9(2): 5.

Walker B. H., Anderies J. M., Kinzig A. P. and Ryan P. 2006b. Exploring resilience in social-ecological systems through comparative studies and theory development: Introduction to the special issue. *Ecology and Society* 11(1): 12.

Walters A. C. 2006. Invasion of Argentine ants (Hymenoptera: Formicidae) in South Australia: Impacts on community composition and abundance of invertebrates in urban parklands. *Austral Ecology* 31: 567–576.

Walters C. 1986. *Adaptive Management of Renewable Resources*. New York: Macmillan.

Walters C. J. and Holling C. S. 1990. Large-scale management experiments and learning by doing. *Ecology* 71: 2060–2068.

Ward-Fear G., Brown G. P., Greenlees M. J. and Shine R. 2009. Maladaptive traits in invasive species: in Australia, cane toads are more vulnerable to predatory ants than are native frogs. *Functional Ecology* 23: 559–568.

Westley F., Carpenter S. R., Brock W. A., Holling C. S. and Gunderson L. H. 2002. Why systems of people and nature are not just social and ecological systems, in Gunderson L. H. and Holling C. S. (eds) *Panarchy: Understanding Transformations in Human and Natural Systems*. Washington DC: Island Press, pp. 103–120.

White P. C. L. and King C. M. 2006. Predation on native birds in New Zealand beech forests: the role of functional relationships between Stoats Mustela erminea and rodents. *Ibis* 148: 765–771.

White P. C. L., Taylor A. C., Boutin S., Myers C. and Krebs C. J. 2009. Wildlife research in a changing world. *Wildlife Research* 36: 275–278.

Whitten S., Coggan A. and Shelton D. 2009. Markets for ecosystem services in Australia: Practical design and a case study, in Bhatbagar M. (ed.) *Environmental*

Service Markets: Global Scenario. Hyderabad: Icfai Books.

Woinarski J., Mackey B., Nix H. and Traill B. 2007. *The Nature of Northern Australia: Its Natural Values, Ecological Processes and Future Prospects.* Canberra: Australian National University Press.

Xie Y., Chen F. and Qi J. G. 2009. Past desertification processes of Minqin Oasis in arid China. *International Journal of Sustainable Development and World Ecology* 16: 260–269.

York R., Rosa E. A. and Dietz T. 2003. Footprints on the earth: the environmental consequences of modernity. *American Sociological Review* 68: 279–300.

Young C. C. 2002. *In the Absence of Predators: Conservation and Controversy on the Kaibab Plateau.* Lincoln: University of Nebraska Press.

Zedler J. B. and Kercher S. 2005. Wetland resources: Status, trends, ecosystem services, and restorability. *Annual Review of Environment and Resources* 30: 39–74.

27

Human Impacts on Lacustrine Ecosystems

Richard W. Battarbee, Helen Bennion,
Peter Gell and Neil Rose

1 INTRODUCTION

Lacustrine ecosystems vary across the world in their form, origin, age and biodiversity ranging from ancient natural lakes of Tertiary age, such as Lake Baikal and Tanganyika, to newly created artificial lakes and impoundments constructed for human use. Most lakes throughout the world have been modified, intentionally or unintentionally, by human activity. Many are used as a source of water for irrigation, drinking water supply and hydro power. Large lakes especially have been hydrologically controlled to prevent flooding or modified to create reservoirs for potable water supply causing habitat instability, particularly in the littoral zones. And water abstraction in dry and subhumid regions of the world have caused severe problems for flow and water-level causing, in some cases, salinisation or complete desiccation.

In addition to such hydromorphological modifications, lake ecosystems have been significantly changed over time by accelerated infilling rates caused by soil erosion, by excessive nutrient and organic matter loading, by acid deposition from fossil fuel combustion and by the release of toxic substances directly and indirectly, through air pollution, to rivers and lakes. Many have been altered by the impact of invasive species and there are concerns over the impact of climate change and long-distance atmospherically transported pollutants, especially nitrogen.

In developed countries in particular, programmes of pollution control and habitat restoration are in place aiming to improve the health of freshwater ecosystems (e.g., US Clean Water Act, EU Water Framework Directive, Australian National Water Initiative), but their success is as yet limited and their future is uncertain. Lake ecosystems typically are affected by many stresses acting together, ecological resilience has been reduced or lost and new threats, especially from climate change, may impair restoration strategies.

Identifying and disentangling the relative roles of different kinds of human activity on lake ecosystems is challenging. It requires an understanding of processes that operate on interannual and decadal time scales based on the examination and analysis of historical records. These include both long-term data sets from monitoring programmes and palaeo data sets from sediment records. Long-term observational records for lakes are becoming increasingly long and most lakes contain conformably accumulating sediments that can be used to reconstruct lake history in detail (Smol, 2008). In some instances there are now substantial time overlaps between observational time-series and the record of change in recent sediments enabling the former to be used to verify the latter (Battarbee et al., 2005a).

In this chapter, we review evidence for the impact of human activity on lakes based on evidence from their sediments, and where appropriate, from long-term time series. The quality of the evidence from sediment records varies, for example, in terms of the preservation of biological remains and the temporal resolution of the sediment record, but palaeoecologists have an increasingly large range of analytical and dating techniques to call upon (e.g., Smol et al., 2001) and in most cases analyses can be extended back beyond the range of historical observations providing unique insights into the history of lake ecosystems.

We consider here how human activity has altered lake ecosystems with respect to the principal stresses they face. These include eutrophication, acidification, salinisation and contamination by toxic substances, and we compare their status at the present day with conditions prior to the main impact of human activity, often referred to as reference conditions (cf. Bennion and Battarbee, 2007). In each case, we briefly outline the key processes that are operating, we illustrate evidence for their impact using examples from lake sediment records and consider problems for the future, especially with regard to the potential impact of climate change. The examples used are mainly from sites in temperate latitudes.

2 NUTRIENTS

The two key nutrients that are essential for life in freshwater ecosystems are phosphorus (P) and nitrogen (N). Phosphorus compounds are usually measured as total phosphorus (TP) and soluble reactive phosphorus (SRP) and nitrogen is usually measured as total nitrogen (TN), ammonia (NH_3^+), nitrate (NO_3^-) and nitrite (NO_2^-). Phosphorus is commonly the growth-limiting nutrient in freshwaters, exerting a strong control on species composition and primary productivity. Nitrogen can also be a limiting or co-limiting nutrient with phosphorus. In particular, nitrogen can be important such that blue-green algae blooms occur in low N:P waters and there is evidence for nitrogen limitation in upland waters (Maberly et al., 2002). The general pattern in temperate, stratifying lakes is for phosphorus concentrations to be at their highest during the winter and their lowest in spring when physical conditions limit algal growth, and in summer when nutrient uptake is at a maximum. In shallow, enriched lakes, internal cycling of phosphorus can result in highly variable phosphorus concentrations, often with summer peaks when phosphorus is released from the sediments under anoxic conditions. Nitrogen concentrations in temperate lakes typically experience winter maxima, followed by a period of low concentrations through summer and autumn, a pattern usually attributed to increased rates of NO_3–N assimilation by algae throughout the growing season.

2.1 The problem of cultural eutrophication

The process of nutrient enrichment of waters is known as eutrophication. Early limnologists considered eutrophication to occur naturally and slowly over thousands of years. However, human activity has accelerated this process, leading to enrichment over time scales of decades. It has been recognised as a global problem since the 1960s, the chief cause being excess nutrient input from agricultural runoff and urban wastewater. The consequent high algal biomass leads to filtration problems for the water industry, oxygen depletion, recreational impairment, loss of biodiversity, fish mortality, and decline or loss of submerged plants. During the 1970s, attention was particularly drawn to the problem of phosphorus enrichment, and research focused on the relationship between phosphorus loading and algal biomass to predict the impact of nutrient levels on water quality (Vollenweider, 1968). Cultural eutrophication remains one of the foremost environmental issues threatening the quality of surface waters.

2.2 Identifying evidence in sediment records

Palaeolimnological studies provide the opportunity to assess the onset, extent and causes of lake eutrophication. Until the 1990s, studies focused largely on changes in diatom species composition in sediment cores in response to lake enrichment. By the late 1990s, transfer functions were being generated to reconstruct past lake TP concentrations from changes in diatom assemblages allowing baseline TP concentrations and the degree of enrichment to be estimated (Hall and Smol, 2010).

A range of other fossil groups have been employed to assess lake response to eutrophication. Cladocera (microscopic crustaceans) can be used to infer changes in fish population density and habitat shifts, particularly in shallow lakes (e.g., Jeppesen et al., 2001) and chironomids (Diptera), representing the benthic-epiphytic invertebrate community, have been used to reconstruct lake trophic status (e.g., Brooks et al., 2001) and oxygen conditions (e.g., Quinlan and Smol, 2001). Plant macrofossils can be employed to determine the past composition, structure and dynamics of in-lake macrophyte communities, particularly the fate of submerged vegetation as lakes become increasingly productive (e.g., Davidson et al., 2005). Algal composition and past primary production can be inferred using fossil pigments of photosynthetic organisms (e.g., Leavitt et al., 1989). By using several fossil indicators from a range of food-web components, the palaeorecord

potentially allows the structural and functional characteristics of pre-enrichment aquatic ecosystems to be reconstructed (e.g., Sayer et al., 1999).

Geochemical records are also valuable for recording eutrophication histories although interpreting sedimentary phosphorus profiles is difficult due to variable retention and post-depositional diagenesis. Nevertheless, there are some sites in which an accurate, relatively undisturbed record of epilimnetic phosphorus concentrations is preserved. Jordan et al. (2001) employed diatom-phosphorus transfer functions and geochemical-phosphorus measurements in combination with mass balance equations to provide a novel approach to modelling diffuse phosphorus loads to lakes. The sediment record can be used in combination with other methods such as export coefficient modelling to provide multiple yet independent lines of evidence for eutrophication (e.g., Bennion et al., 2005).

2.3 Examples of early eutrophication in agriculturally rich lowland catchments

Some lakes have been subject to nutrient loading over relatively long time scales of hundreds or thousands of years. The study of Dallund Sø, Denmark (Bradshaw et al., 2005) provided evidence of early enrichment during the forest clearances of the Late Bronze Age (1000–500 BC) and the expansion of arable agriculture during the Iron Age (500 BC to AD 1050). More marked eutrophication in the medieval period (c.AD 1050–1550) was observed as a result of further agricultural intensification and the use of the lake for retting. Early land-use influences associated with Neolithic and Bronze Age forest clearance and cultivation were also found in a study of Diss Mere, England (Birks et al., 1995). However, the most marked increases in diatom-inferred phosphorus were seen in the upper parts of the core, highlighting the heavy human impact over the last 150 years.

Palaeostudies of this kind are valuable for placing enrichment patterns observed in modern times into a longer temporal context.

2.4 Wastewater, detergents and fertilisers

Increased nutrient loading to freshwaters has been brought about by intensified use of lowland landscapes for human settlement, industry and agriculture. The most significant sources of phosphorus pollution are point sources in the form of wastewaters and detergents, whilst the main sources of nitrogen pollution are diffuse run-off of nitrate and ammonia-based fertilisers and animal waste from agricultural land. There are numerous examples worldwide of lakes that have experienced enrichment in recent decades from one or more sources of nutrient pollution. One of the best documented cases is Lake Windermere, England, where long-term monitoring data track the increase in phosphorus concentrations associated with the expansion of tourism in the region and the installation of major sewage treatment works in 1964 (Talling and Heaney, 1988).

The value of the sediment record for assessing recent eutrophication is illustrated by the diatom study of Lake of Menteith, Scotland (Bennion et al., unpublished) (Figure 27.1). The first major change in the diatom assemblages occurred in c.1920 (Zone 2/Zone 3) with the appearance of *Stephanodiscus hantzschii*, a taxon generally considered an indicator of high nutrient concentrations, and a number of other planktonic diatoms. This is reflected in the diatom-inferred TP values which rose gradually between 1920 and the mid-1960s and then increased more sharply after c.1965. Inferred TP values were approximately double those estimated for the pre-1920 period. From c.1980, the plankton-dominated community increased markedly at the expense of the low nutrient taxa (Zone 3/Zone 4). These findings are consistent with the first reports of algal blooms on the lake in the early 1980s.

Concerns have been raised about the impact of a commercial trout hatchery which has operated at the site for the last few decades (Marsden et al., 1995). However, the sediment record indicates that whilst the lake has experienced further enrichment since c.1980 that there was evidence of eutrophication prior to this time with potential nutrient sources including agricultural activities, afforestation, and increased inputs of sewage and phosphate detergents. The relative stability of the diatom community prior to c.1920 suggests that the inferred TP of c.10 µg L^{-1} for this period can be used as a target value for nutrient reduction.

2.5 Restoration measures and their effectiveness

Efforts to restore enriched systems have increased over the last few decades and there are now numerous examples of lakes in recovery. Point-source control at sewage treatment works has been particularly effective at reducing external nutrient loads. Nutrient pollution from diffuse agricultural sources has proved more difficult to control as it is dispersed over large areas. Nevertheless, restoration schemes that promote use of buffer strips, good agricultural practice and wetland regeneration, plus legislation have all contributed to the reduction of nutrient loading from agricultural sources (Sharpley et al., 2000). In deep, well-flushed lakes, eutrophication is often reversed by the reduction in phosphorus inputs alone. In the classic case of Lake Washington, phosphorus concentrations fell dramatically, phytoplankton biomass declined and there were sustained increases in transparency following effluent diversion and treatment (Edmondson and Lehman, 1981). However, in shallow lakes internal phosphorus loading can delay recovery and external phosphorus reduction is often combined with other management measures such as dredging or biomanipulation (Søndergaard et al., 2007). In an analysis of long-term datasets from 35 restored lakes,

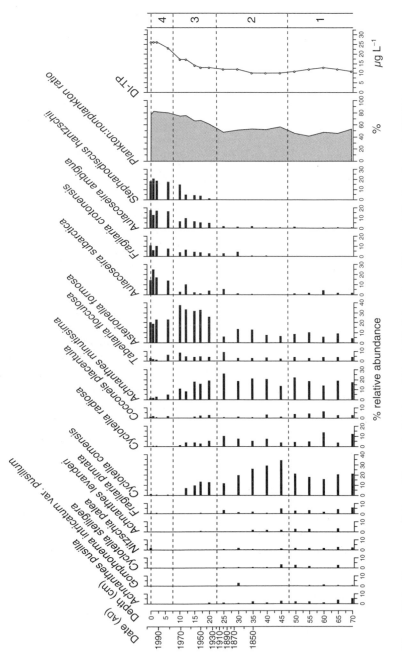

Figure 27.1 Summary diatom diagram of Lake of Menteith (showing all taxa >3% abundance); DI-TP is diatom-inferred total phosphorus (μg L⁻¹)

Jeppesen et al. (2005) showed that internal loading delayed recovery, but in most lakes a new equilibrium for TP was reached after 10–15 years.

The advent of the Water Framework Directive (WFD) with its aim of restoring European waters to good status or better, has increased the need for effective restoration programmes for all lakes. Within the WFD, ecological status is based on the degree to which present day conditions deviate from those expected in the absence of significant anthropogenic influence, termed 'reference conditions'. Consequently, there has been a wave of research aimed at defining reference conditions and development of tools for estimating deviation from them. Palaeoecological methods have played a key role, particularly in determining pre-enrichment reference conditions and degree of eutrophication (Bennion and Battarbee, 2007). The strengths and weaknesses of using a palaeoecological approach to the definition of reference conditions has been reviewed by Bennion et al. (2011). As many restoration programmes progress, there is great potential to employ a combination of limnological and sediment records to track recovery using the pre-eutrophication baseline as a benchmark (Battarbee et al., 2005a).

2.6 Forward look and the problem of climate change

In recent decades, diffuse nutrient sources have become relatively more significant than urban wastewater pollution and losses from agricultural land are now the biggest challenge. There has been a growing literature on the need to reduce nitrogen loads as well as phosphorus in order to reverse eutrophication, particularly in shallow lakes with moderate phosphorus levels where high summer nitrogen concentrations stimulate algal growth and cause loss of submerged plants (e.g., Jeppesen et al., 2007). However, in a 37-year fertilisation experiment on Lake 227, Canada, results suggest that controlling nitrogen inputs could aggravate the dominance of nitrogen-fixing cyanobacteria and therefore the need for reduction of phosphorus inputs is argued (Schindler et al., 2008). No doubt the debate over the relative importance of nitrogen and phosphorus will continue (Smith and Schindler, 2009). In the emerging economies of countries such as India and China, nutrient pollution control is only in its infancy and algal blooms arising from wastewater discharges are widespread (e.g., Shen et al., 2003). On a global scale, therefore, it is very likely that eutrophication will continue to be a growing problem.

Climate change is predicted to result in higher water temperatures, shorter periods of ice cover and longer summer stratification. Will these changes exacerbate the symptoms of eutrophication and confound recovery efforts? Models suggest that lakes with long residence times may experience higher phosphorus levels in the future under warmer temperatures (Malmaeus et al., 2006) and shallow lakes may be particularly susceptible. Ecological consequences might include earlier appearance of spring blooming phytoplankton and increased proportions of cyanobacteria. In some systems, negative effects may be compensated by greater predation pressure by zooplankton which is known to be positively temperature dependent. However, fish activity may also increase in warmer temperatures thereby reducing zooplankton populations through increased predation (Moss et al., 2003). In addition, changes in mixing may influence the availability of nutrients in the photic zone and higher temperatures may enhance sediment-phosphorus release, whilst higher winter precipitation is likely to enhance nutrient loss from cultivated fields (see Battarbee et al., 2008b for a review). Models that predict likely outcomes of climate change on nutrient regimes will play a vital role in improving our understanding of future lake response and in guiding management decisions (e.g., Whitehead et al., 2006). Whilst sediment records cannot be used in a predictive capacity, they provide an opportunity to validate hindcasts derived from dynamic models (Anderson et al., 2006). They should, therefore, play an increasingly important role in assessing uncertainty associated with future predictions.

3 ACIDITY

Lake water pH varies enormously from exceptionally acidic volcanic crater lakes with pH values as low as 2 through to soda lakes with pH values above 10 (cf. Wetzel, 1975). Most lakes, however, have pH values between about 5.5 and 8.5, regulated by the carbon dioxide–bicarbonate–carbonate buffering mechanism that mainly reflects the geochemistry of catchment rocks and soils. However, pH varies seasonally and intra-annually depending on weather patterns and on in-lake variation in lake metabolism. On longer time scales mean pH can change in response to overall shifts in catchment biogeochemistry related, for example, to long-term base depletion of soils (cf. Renberg, 1990; Whitehead et al., 1989), changes in catchment vegetation (e.g., Nygaard, 1956), agriculture (e.g., Renberg et al., 1993a) or acid deposition (e.g., Battarbee et al., 1990).

3.1 The acid rain debate

Acidification of surface waters as a matter of environmental concern came to attention in the late 1960s and early 1970s when Scandinavian scientists claimed that 'acid rain' was the principal reason why fish populations had declined dramatically in Swedish, Norwegian and Canadian lakes (Almer et al., 1974; Beamish and Harvey, 1972). The cause of the problem was attributed to the deposition of sulphur and nitrogen compounds from long-distance transported air pollutants generated by fossil fuel combustion, principally from coal- and oil-fired power stations. Alternative explanations were also advanced, principally that acidification was the result of

land-use change or due to natural processes (e.g., Pennington, 1984; Rosenquist, 1978). The debate that followed drew heavily on the use of palaeolimnological research that was designed to evaluate these different hypotheses (e.g., Battarbee et al., 1990; Charles and Whitehead, 1986). Diatom analysis was the principal tool used to reconstruct pH history of lakes (cf. Birks et al., 1990) and the various research programmes sought to compare lakes with different land-use histories, differing sensitivity to acid deposition and receiving differing amounts of acid deposition (cf. Battarbee, 1990; Charles, 1990).

In the UK one of the first lakes to be studied was the Round Loch of Glenhead in Galloway, Scotland, a lake with a moorland catchment and granitic bedrock. The diatom diagram for a ^{210}Pb-dated sediment core (Flower and Battarbee, 1983) showed that diatoms characteristic of circumneutral water such as *Brachysira vitrea* began to decline in the latter half of the nineteenth century and be replaced by more acidophilous species. Acidification continued through the twentieth century and by the early 1980s the diatom flora of the lake was dominated only by acid-tolerant taxa such as *Tabellaria quadriseptata* and *T. binalis*. The changes indicated that pH had declined in the loch from around pH 5.5 in 1850 to 4.7 by 1980. As there had been no change in catchment land-use over this time period and the sediment contained high concentrations of spheroidal carbonaceous particles, derived from fossil fuel combustion, the core data provided strong evidence in support of the 'acid rain' hypothesis.

Equally clear changes were observed in the uppermost sediments of lakes in similarly sensitive regions throughout Europe and North America (Battarbee and Charles, 1986; Cumming et al., 1992; Renberg et al., 1993b). As more sites have been studied and data have accumulated (cf. Battarbee et al., 2011), it can now be demonstrated that the spatial pattern of acidification matches the regional pattern of sulphur deposition, both in Europe and in North America. Strongly acidified sites are only found in areas of high sulphur deposition. In areas of very low sulphur deposition acidification on a regional scale only occurs in the most sensitive areas. At the most remote sites, sensitive sites show no evidence for recent acidification (e.g., Cameron et al., 2002).

3.2 Reference conditions, restoration targets and models

The recognition of the 'acid rain' problem as a major environmental issue in the 1980s led to the introduction of legislation both in North America and Europe to limit the emissions of sulphur and nitrogen gases into the atmosphere from fossil fuel burning. Over the last few decades, there has been a very significant reduction in emissions, especially in sulphur emissions, and a number of monitoring schemes have been set up to assess the chemical and biological response of lakes and streams to the reductions that have taken place. The key question now is the extent to which acidified lakes and streams can be restored to 'good ecological health' defined in Europe as a state that is minimally impacted by human activity, also referred to as the reference condition (see previously). While this can be defined using comparisons to the condition of low-alkalinity lakes and streams in nonpolluted regions in the present day, it can also be defined by comparison with the past using palaeolimnological techniques (cf. Bennion and Battarbee, 2007), as for eutrophication problems. Diatom analysis, in particular, can be used to establish both the diatom assemblages that characterised the pre-acidification status of acidified lakes or to reconstruct preacidification lake pH values from diatom-pH transfer functions (Battarbee et al., 2008a). Although there are uncertainties in the transfer function methodology, target pH values can be defined in this way and used as a measure of restoration success. In the case of the Round Loch of Glenhead, for example, diatom analysis of three sediment cores and three different diatom-pH

training sets indicated that the pre-acidification pH of the lake was between pH 5.5 and 5.8 (Figure 27.2). Although there has been a slight increase in pH over the last 20 years to approximately pH 5.0 (Figure 27.2), *Tabellaria quadriseptata* remains one of the dominant diatoms and *Brachysira vitrea* is still rare, indicating that only a slight recovery has taken place so far.

Diatom-based pH reconstruction also has an important role in helping to improve models that are needed to assess the effectiveness of emissions policies by comparing model output with palaeodata (e.g., Battarbee et al., 2005b). In the Round Loch of Glenhead example, the diatom-inferred pH data were compared with output from the MAGIC model (Cosby et al., 1985). The results showed that the model estimates for pre-acidification pH yielded higher values than the diatom-based reconstructions (Figure 27.2; Battarbee et al., 2005b). The mismatch was attributed to a weakness in the model with respect to its treatment of dissolved organic carbon (DOC) and its failure to allow DOC values to be higher under natural conditions in the past (cf. Monteith et al., 2007).

3.3 Influences on restoration success

Although there is evidence that acidified lakes are now beginning to recover as a result of the major reductions in emissions that have taken place (Battarbee et al., 2008a; Dixit et al., 2002; Majewski and Cumming, 1999), it is not clear whether the recovery process will continue and eventually lead to the re-establishment of pre-acidification conditions or whether fundamentally new systems will be created. It might be expected that not all taxa lost as a result of acidification will successfully recolonise as alkalinity increases, especially if the ultimate alkalinity increase falls short of the reference target, and water quality improvement might be limited by a failure to reduce nitrate concentrations. Despite considerable reduction in

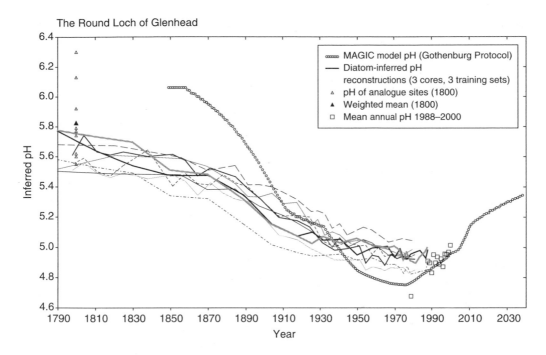

Figure 27.2   Comparison of pH reconstruction outputs and annual measured pH. Chronology of diatom-inferred pH according to SWAP, UK and EDDI models (fine lines) for ^{210}Pb-dated samples from three sediment cores (RLGH 81, RLGH 3 and K05). The RMSEP of the SWAP, UK and EDDI training sets are 0.38, 0.31, and 0.25 pH units, respectively. Modern annual pH of nine lakes providing the strongest biological analogues for a pre-acidification (c.1800) sediment sample (open triangles) and the weighted average of these (filled triangle). MAGIC model pH reconstruction (open circles) and mean annual average pH for the period 1988–2000 and the year 1979 (open squares).

Source: Battarbee et al. (2005b). Reproduced with permission. © 2005, Elsevier.

nitrogen deposition in Europe, nitrate concentrations in some surface waters remain relatively high, replacing sulphate as the dominant acid anion (Curtis et al., 2005) and potentially causing eutrophication.

Climate change may also have an influence on recovery, although in certain cases this may be positive. Psenner and Schmidt (1992) showed that the diatom-inferred pH from two cores in the Austrian Tyrol corresponded closely to instrumental temperature records during the nineteenth century before acid deposition became the dominant influence on pH. Sommaruga-Wögrath et al. (1997), using the same approach but at a higher alkalinity site insensitive to acid deposition, showed that pH increased towards the present day in line with global warming. As sulphur deposition continues to decrease, temperature change may emerge as the strongest driver of lake acidity at such sites and pH values may reach levels only previously experienced in the warmer early Holocene (e.g., Larsen et al., 2006), potentially higher than nineteenth-century pre-acidification levels. On the other hand, where climate change is characterised by an increase in winter precipitation, as is projected for northwest Europe, pH levels may be depressed, limiting the success of restoration measures.

4 SALINITY

Lake salinity strongly influences lake eco-systems not least because of the strength of osmotic pressure on the function of key organisms. As fewer species are adapted to saline waters, diversity declines as salinity increases. In freshwater systems salinity has less influence on ecosystem function relative to changes in light, pH and nutrients. While some organisms respond to changes in salinity at low concentrations, several thresholds of salinity tolerance exist in salinity levels between 1 and 10 gL^{-1}. In particular, the sensitivity of aquatic macrophytes to salinity produces structurally simple lake ecosystems at higher levels. Tolerance to salinity may vary with changing light conditions and with changing brine type with, for example, evaporation series groups defining ostracod faunas (Radke et al., 2002).

4.1 Salinity systems throughout the world: coastal versus inland waters

Cyclic salts bring considerable loads onto land over long time periods but concentrations in rain reduce with distance from oceans, therefore lake systems at altitude tend to be fresh to oligosaline. Notable exceptions are the endoreic lakes of the Andean altiplano that have inherited salts through epeirogenesis and have retained salts due to their closed catchment (Sylvestre, 2002). Lowland systems are more saline as they accumulate salts shed from catchments, retaining the legacy of past marine incursions, therefore both ground and surface waters are influenced by connate salts. Lakes strongly influenced by both cyclic and connate salts usually retain brine characteristics of the ocean and are often sodium and chloride dominated (Williams, 1981). Where cyclic salts are derived from large inland lakes, such as in eastern Africa, downwind systems inherit this ionic makeup and may be carbonate or sulphate rich. The chemistry of lakes in volcanic contexts receive geologic salts from their catchment as well as metals or other compounds such as phosphates that may generate naturally eutrophic saline systems (Hammer, 1986).

4.2 Salt lake types

Large playa lakes, such as Lake Eyre and Great Salt Lake, accumulate salts owing to the internal drainage of their catchments. These fill readily because of the amplifying effect of the size of their catchment, yet often exist in arid zones with prolonged water deficit and so are often dry for long periods. Saline lakes also exist in smaller catchments with internal drainage. These include groundwater-window lakes that are formed in depressions developed through interaction with surface saline water tables. Crater lakes, owing to their closed nature and small catchment-to-lake area ratio and relative freedom from the influence of groundwater, are valuable recorders of past climate. These lakes retain salts and their record of lake salinity is an indirect measure of effective rainfall. Lastly, salt lakes include those at the coast that were once open estuaries, but have evolved geomorphically into either closed or intermittently open coastal lakes, or remain as open estuaries regularly flushed by tides. Depending on the nature of the sea connection they can be largely thalassic ecosystems, or protected from tides and so mostly athalassic, yet retain some of the ionic ratio features of a tidal history.

Rapid filling in intermittent or ephemeral systems with large catchments allows the lake to remain fresh until the salt crust dissolves elevating salt concentrations. These events can be recognised from beaches of sand formed from the heightened wave energy of the deeper water. Overflow allows for the loss of saline water from the system representing a net loss of salt load. On recession, salts evapo-concentrate and the lake salinity increases with lake depth. As ions have different solubility limits, evapo-concentration leads to the precipitation of the salts from least to most soluble. A simple model would follow the progressive precipitation of $CaCO_3$, $MgSO_4$, and lastly $NaCl$ leading, in turn, to the deposition of calcite, magnesium calcite, dolomite, magnesite and halite. As ions precipitate, the lake is relatively enriched in the remaining more soluble ions, generating a correlation between lake salinity and brine type. As the lake level falls, there is a strong interaction with the groundwater allowing a salt crust to form. Salts may be deflated from the lake surface and accumulate downwind on source bordering dunes or lunettes. If the water table falls, the salt crust may leach from the surface leaving a sodic clay surface that is also readily deflated downwind.

4.3 Closed lakes as archives of effective rainfall

In endoreic systems with large catchments, there is a considerable amplification effect with rain falling within the watershed funnelling through drainage lines to the lake. In these circumstances lake level can increase rapidly. However, interpreting past rainfall from records is complicated as the area collecting the water volume is unknown. Further, large systems may be strongly groundwater controlled complicating the interpretation of changes of effective rainfall and changing catchment water yield efficiency, which may arise through natural vegetation responses or direct modification by people. These represent complex systems that render the relative influence of climate, vegetation cover and direct anthropogenic catchment change difficult to interpret from the palaeorecord. Some wetlands, for example, may retain surface water due to low evaporation rates even when rainfall is reduced. Also, lakes in high latitudes or altitudes may freshen with increased temperatures owing to increased catchment input through glacial melting (Boomer et al., 2009; Zhang et al., 2004). Lakes with low catchment to lake area ratios are more responsive to changes to effective

rainfall and, in many instances, crater lakes largely removed from regional groundwater, and with a well-defined catchment boundary, act as rain-gauges. Such sites exist in many of the volcanic provinces of the world and long-term climatic variations together with hydrological modelling can be used to infer recent hydrological balance. When compared to previous cycles of change, it can be used to estimate the relative contributions of anthropogenic climate change to changes in the lake system (Jones et al., 2001). Care needs to be exercised in regionally extrapolating results from individual sites as the response may be partly specific to the lake's topographic and groundwater context. Nearby lakes have been shown to operate independently (Fritz, 2008).

4.4 Evidence for changing lake salinity in the sediment record

Analysis of past lake salinity complements other geomorphic evidence for lake level change. Beach ridges, past shorelines and strandlines are used to develop a portfolio of evidence to reconstruct lake level through time (e.g., Magee et al., 1995). The main features of the sediments analysed are grain size and geochemistry, particularly stable isotopes of inorganic and biogenic carbonates. The ^{18}O contents reflect the hydrological balance while that of ^{13}C reflects water mixing and residence time (e.g., Gasse et al., 1987). Geochemical analyses for elements such as Fe, Mn, Sr, Mg, Ca and Zn can reflect water and salinity levels, and the relative contributions of ground and surface water. The main biological indicators used have been mostly ostracods and diatoms for which detailed salinity tolerances of common taxa are determined by sampling modern lakes of varying salinity. Halobian systems for classifying diatom species were established as early as 1927 (by Kolbe) and these were widely used owing to the largely cosmopolitan distribution of diatom taxa. Ostracod species–salinity relations were established on a more regional basis due to their more restricted distributions globally. Because of their abundance and species diversity, diatom-salinity training sets were developed (e.g., Fritz et al., 1991) to reconstruct salinity through time from fossil assemblages subsampled from sediment cores extracted from lakes, using statistical methods similar to those for reconstructing TP and pH. Stable isotope and trace element chemistry of ostracod shells is also now widely used based on temperature and salinity controls on their uptake into the shell (Holmes, 2001). As in sediment-based studies of eutrophication, reconstructions of past salinity have been made more robust by the use of multiple indicators from the same sediment samples (e.g., Gasse et al., 1987).

4.5 Natural variability of saline lake systems through the Holocene

One of the challenges for understanding the impact of humans on saline lake systems is to distinguish direct catchment change from natural climate variability. Even more challenging is to separate out the impact of people on climate change and variability. In both instances records of past lake salinity provide an understanding of historic variability from which recent change can be qualified. Palaeosalinity records attesting to past water balance through the Holocene are now widespread, particularly in the climatically sensitive, semi-arid parts of the world. Where many sites have been analysed across a region, large-scale interpretations can be made that overcome the specific geomorphic and hydrologic conditions at a particular site. In northern Africa, multiple records demonstrate a regional pattern of increasing humidity inducing lake salinity minima by the mid-Holocene (Gasse, 2002), but rapid drying from 6 to 4 ka BP. These general trends are interrupted by periods of increased lake salinity inferring regional drying phases at 8.2, 7.2 and 4.2 ka BP. Similar patterns of change are documented from saline systems

globally. In several instances, drying phases have impacted upon human societies such as the collapse of Neolithic civilisations in northern Africa after 5 ka BP. More recent climatic fluctuations reveal regional variations in lake salinity and lake level with out-of-phase responses in central African lakes (Fritz, 2008) to the Medieval Climate Anomaly drying, and Little Ice Age wetting, experienced in east African lakes (Verschuren et al., 2000). Further, decadal changes in salinity reflect recent changes in water balance and reveal the relative magnitude of past and modern droughts (e.g., Verschuren et al., 1999). These recent changes in climate-driven water availability can be out of phase globally. Since the La Niña event of the mid-1970s, southeast Australian lakes have fallen and increased in salinity yet, at the same time, the level of Laguna mar Chiquita (Piovano et al., 2004) in the Pampean Plains of Argentina rose generating the lowest salinity concentrations for the last 230 years.

4.6 Catchment diversions

The transfer of freshwater from uplands through river systems is the principal means by which floodplain and coastal lakes and estuaries are diluted. Estuaries in particular may be strongly influenced by tidal influx of seawater and the variability in salinity is dependent, spatial and temporally, on the yield of freshwater from the catchment. Coastal lakes, and those in saline floodplains, can be freshened by river input yet may evapo-concentrate salts when the river is disconnected under low flows. In these lakes the natural state and regime of variability is intrinsically linked, not only to the natural variability in river flow, but also the influence of regulation, diversion and abstraction of that flow. Changes to flow by human activity began at a small scale with modification of channels to trap fish, but expanded greatly with the emergence of agriculture, most notably in Mesopotamia and the

Indus valley. Reduced river-based freshening leaves tidal and evapo-concentration forces to dominate lake condition (MacGregor et al., 2005). In closed systems, diversion and abstraction raise lake salinity independent of the rainfall balance as witnessed in east Africa (Verschuren et al., 1999) and the Aral Sea (Boomer et al., 2009). River impoundments can also impact on floodplain lake systems by raising the lake level rendering hydrologically connected lakes permanent and stable, acting as sediment sinks rather than throughflow systems. Along the Murray River this has accelerated lake sediment accumulation rates but has also induced the net accumulation of sulphurous salts under anoxic conditions. Drought has recently exposed sulphidic sediments and several wetlands have rapidly become both saline and acidic (Gell et al., 2006).

4.7 Problems of salinisation

While there are many instances where river regulation was developed for navigation and domestic water supply, it has widely evolved hand-in-hand with the development of irrigated agriculture, focused in valleys and lowlands to exploit enriched floodplain soils. The application of irrigation water contributes surface water to the ground. The water table rises through saline substrata, dissolving salts en route, bringing saline groundwater within the capillary zone of plants and generating saline groundwater seeps and streams. These problems emerged early in the development of irrigated agriculture quickly impacting on agricultural productivity in the Tigris and Euphrates, and the Indus. Secondary salinisation occurs in dryland circumstances where loss of vegetation cover, particularly in recharge zones, can lead to net increases in the percolation of rainwater into aquifers. In small catchments the distance between recharge and discharge zone is short and the response of the lake or stream can be rapid. One floodplain wetland in the Murray Darling Basin showed signs of salinisation in

the late eighteenth century soon after the onset of pastoralism (Gell et al., 2005). In large systems there can be a considerable lag in the response but a long and sustained rise in water tables can lead to the upslope migration of the hinge line below which regional groundwaters discharge (Macumber, 1991). Where this water is saline surface seeps can reactivate and saline drainage can reach channels and be transferred to rivers and lakes. Several wetlands have received saline groundwater disposal under short-term management measures while others denied irrigation return water have rapidly salinised (Gell et al., 2002).

4.8 The application of palaeoecology in assessing natural ecological conditions

The impact of people on the salinisation of freshwater systems is widely recognised and for some decades now, societies have endeavoured to mitigate impacts and restore, or at least rehabilitate, wetlands. This is enshrined in legislation such as the EU Water Framework Directive but can be driven by measures at a range of political and spatial scales. The utility of palaeoecological approaches to identifying 'good ecological condition' under the WFD is detailed by Bennion and Battarbee (2007). It is applied mainly to lakes but it can also be used to review the ecological character of wetlands including those listed under the Ramsar protocol. In the case of the estuary of the River Murray palaeoecological evidence has shown that the ecological character description at the time of listing under Ramsar reflects recent disturbed states rather than the long-term historic condition (Fluin et al., 2007). Regulation, diversion and abstraction have caused a coastal lagoon to become hypersaline and the lower lakes to become largely fresh, at variance from their naturally tidal state (Figure 27.3). In both instances their natural ecological character was wrongly inferred, complicating efforts to remedy for recent degradation.

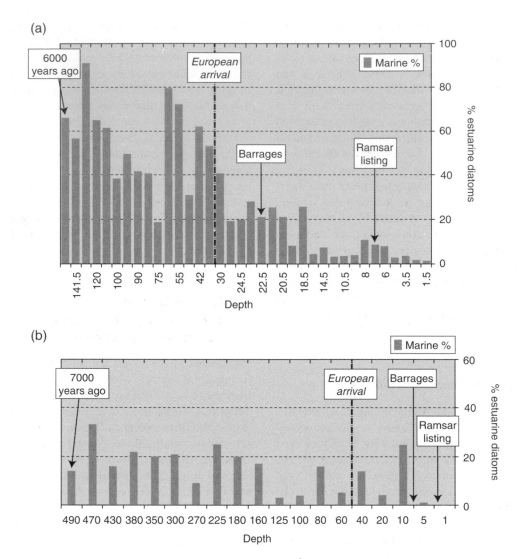

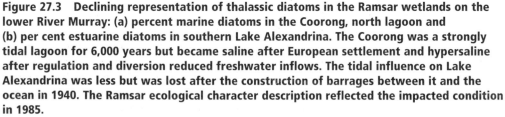

Figure 27.3 Declining representation of thalassic diatoms in the Ramsar wetlands on the lower River Murray: (a) percent marine diatoms in the Coorong, north lagoon and (b) per cent estuarine diatoms in southern Lake Alexandrina. The Coorong was a strongly tidal lagoon for 6,000 years but became saline after European settlement and hypersaline after regulation and diversion reduced freshwater inflows. The tidal influence on Lake Alexandrina was less but was lost after the construction of barrages between it and the ocean in 1940. The Ramsar ecological character description reflected the impacted condition in 1985.

Source: Fluin et al. (2007).

4.9 Forward look and the role of climate change

The influence of the surface and groundwater budget on the salinity of lakes and wetlands ensures that accelerated climate change will inevitably bring further changes in the future. Some lakes are likely to freshen and rise due to increased effective rainfall and others will fall. Strengthening or weakening of monsoons will be important in tropical and subtropical regions. Devils Lake in North Dakota, a lake used to reconstruct past climates by Fritz et al. (1991), is now overflowing and threatening the local township. The rain-gauge crater lakes in western Victoria (Jones et al., 2001) continue to increase in salinity owing to a decreasing trend in effective rainfall under warming climates. Across China some of the Tibetan Plateau lakes (Zhang et al., 2004) are also falling, yet others, in glaciated catchments, are rising rapidly through increased supply of meltwater. Rapid climate change will inevitably move some wetlands outside their historic range of variability, and represent a considerable additive effect on salinity impacts brought on by more direct catchment modification.

5 TOXIC SUBSTANCES

Toxic substances in freshwater ecosystems fall into three main classes: trace metals, persistent organic pollutants (POPs), and organometallic compounds. The trace metals of most concern are mercury (Hg), lead (Pb) and cadmium (Cd), although nickel (Ni), zinc (Zn) and copper (Cu) are often included. The EU Water Framework Directive (WFD) names Pb and Ni and their compounds as 'priority substances' while Hg and Cd and their compounds are listed as 'priority hazardous substances' and hence are considered a greater threat. For an organic compound to be classed as a POP, it must fulfil criteria for persistence, bioaccumulation and toxicity (the PBT criteria). Examples include polychlorinated biphenyls (PCBs), polychlorinated dibenzo-p-dioxins (PCDDs) and polychlorinated dibenzofurans (PCDFs), many organochlorine pesticides (e.g., DDTs, toxaphene) and brominated flame retardants (BFRs, including polybrominated diphenyl ethers (PBDEs)). Many of these compounds are present on the WFD priority lists and comprise the initial 12 compounds of the Stockholm Convention. The main

organometallic compounds of concern in freshwaters are methylmercury (MeHg), the biologically available form of Hg and hence the most toxic form in the environment, and tributyl tin (TBT) a powerful anti-fouling biocide widely used since the 1960s (Leung et al., 2007) and linked to adverse effects in many organisms. TBT and its degradation products are also listed as priority hazardous substances by the WFD.

5.1 Background concentrations and evidence of early contamination

As all trace elements have natural mineral origins it is essential to determine the level and range of natural inputs to a system in order that the true scale of anthropogenic contamination can be identified. For freshwater ecosystems, a background concentration or input flux can be identified from the lake sediment record. In Europe, the anthropogenic lead signal, as determined from an identifiable increase in sediment lead concentration and concomitant decline in $^{206}Pb/^{207}Pb$ isotope ratio, may be first detected after c.2000 BC (e.g., Brännvall et al., 2001) and a peak in contamination may be observed in the last centuries BC and early centuries AD as a result of Greek and Roman silver and lead production. The use of lead isotopes in lake sediment studies is particularly useful in identifying early human influence. Natural ratios of the isotopes ^{206}Pb:^{207}Pb are generally higher than those of 'pollution lead' and hence a decline in this ratio, simultaneous with an increase in sediment lead concentration, provides strong evidence of anthropogenic influence.

While the earliest records for organic pollutants in the sediment record are generally limited by when they were first produced, there are examples of natural and early human production prior to the industrial period. Polycyclic aromatic hydrocarbons (PAHs) are produced from the combustion of organic matter and so have a long-term record from events such as forest fires. The sediment records of particular PAHs indicative of wood combustion such as retene and the ratio between 1,7-dimethylphenanthrene and 2,6-dimethylphenanthrene (Fernàndez et al., 2000) can therefore be used as wood combustion indicators. Natural sources of halogenated POPs are more scarce although natural biochlorination in peat bogs (Silk et al., 1997), forest fires (Harrad, 2001) and domestic combustion of coastal peats (Meharg and Killham, 2003) may all provide pre-industrial signals of PCDD/Fs in lake sediment cores albeit that the scale is much lower than any industrial contamination.

5.2 Post-1800 contamination and food-chain uptake

The Industrial Revolution gave rise to an enormous, and initially uncontrolled, increase in pollutant emissions to the atmosphere and signalled the onset of major contamination by trace metals in surface waters. Starting in the latter part of the eighteenth century, the introduction of steam power fuelled primarily by coal, led to the widespread use of powered machinery, and resulted in vastly elevated pollution emissions and the widespread contamination of the environment. The lake sediment record of contaminants in the form of trace metals (e.g., Farmer et al., 1999; Gallon et al., 2005; Yang and Rose, 2003) and fly-ash particles (Griffin and Goldberg, 1981; Rose, 2001) reflect this historical contamination with a marked increase from the early- to mid-nineteenth century.

In the middle decades of the twentieth century two significant developments resulted in a further increase in surface water contamination in many areas. The introduction of tall chimneys at industrial installations from the 1930s led to greatly increased pollutant dispersal and transport and was followed a decade or two later by a major increase in contamination as a result of the introduction of cheap fuel-oil and an increased demand for electricity following the

Second World War. In many lake sediment studies a major increase in contamination is observed from this time. However, since the 1970s, there has been a trend towards increased pollution legislation and control, greater fuel efficiency, changes in heavy industrial practices and the widespread introduction of natural gas as a fuel. From the 1970s therefore, trace metal contamination has been observed to decline, particularly in Europe and North America, often by up to 80–90 per cent. In sediment cores these temporal trends are observed as a decline in mercury and lead concentrations (Figure 27.4) and a reversal of lead-isotope ratios towards more 'natural' values. In developing countries, industrial expansion follows different temporal patterns. In China, for example, increases in emissions and freshwater contamination are observed more recently and increase through to the present (e.g., Rose et al., 2004).

Although some organochlorine compounds were first synthesised in the late-nineteenth century (e.g., DDT) the contamination of freshwater ecosystems by POPs is essentially a twentieth and twenty-first century phenomenon and this is reflected in the lake sediment record. Temporal trends observed in sediments reflect the sequential nature of synthesis, increased usage and production and then decline following a ban and, lastly, the legacy of persistence. Sediment concentrations of DDT, for example, peak in the 1950s and 1960s when it became central to the World Health Organisation antimalarial programme but declines after this due to concerns over environmental effects and emerging resistance within the insect population. Although no longer widely used, DDT and its degradation products remain ubiquitous within the aquatic environment and the same is true of many other organochlorine compounds. The commercial production of brominated compounds such as PBDEs began in the 1970s but penta- and octa-BDE products were banned in the EU from 2004 and a few years later in the US.

Where lakes receive contamination only from atmospheric sources, their sediment

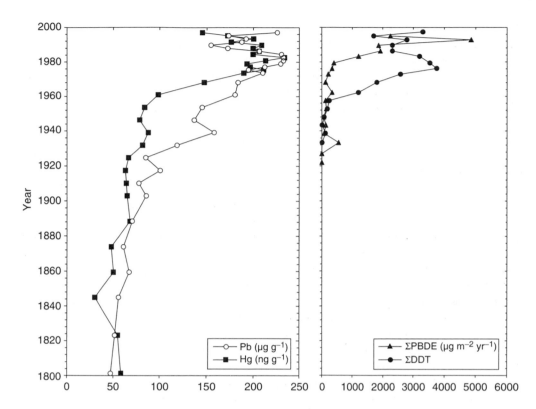

Figure 27.4   Temporal trends of Pb (O) and Hg (■) concentrations, and ΣDDT (●) and ΣPBDE (▲) fluxes from the sediment record of Lochnagar, Scotland showing difference in time trends and the persistence of DDT. x-axis units for Hg are ng g^{-1}; for Pb μg g^{-1}; for ΣDDT and ΣPBDE units are ng m^{-2} year^{-1}. y-axis is date (AD). Hg and Pb data are from Yang et al. (2002); dates from spheroidal carbonaceous particle analysis (Rose unpublished); ΣDDT and ΣPBDE data are from Muir and Rose (2007).

archives faithfully record these sequences even when the site is remote from sites of production and application, emphasising the long-range and often intercontinental nature of pollutant transport. Figure 27.4 shows the temporal trends of mercury, lead, ΣDDT and ΣPBDE for Lochnagar, a remote lake in Scotland (Rose, 2007). It shows that the historical record for the two trace metals is similar, reflecting the dominance of coal as a major source over much of the period, while the sediment record of the selected POPs shows the sequence of production and usage as described above. Where comparable studies exist a similar succession of contaminant input is also observed (e.g., Pearson et al., 1997; Song et al., 2005).

However, it is the effects of these pollutants on aquatic biota and aquatic food-webs that are the main concern. While all trace metals may bioconcentrate, only mercury, as methylmercury, biomagnifies in the aquatic food chain. Hence, although present as a small fraction (0.1–5.0 per cent) of total mercury in air and water, it represents 90–100 per cent in invertebrates and fish. Furthermore, a piscivorous fish may contain a mercury concentration two to three times higher than a nonpiscivorous specimen of the same age (Rosseland et al., 2007). Adverse effects of trace metal contamination to aquatic biota includes impacts on growth, development and, in extreme cases, deformity, impacts on reproduction such as reduced brood sizes and the number of reproducing adults (de Schamphelaere et al., 2004), inhibition of egg development and abnormal embryonic development (Leung et al., 2007), reduced hatching success and transgenerational effects (e.g., deformity in second generation after mercury exposure to the parent).

Many POPs accumulate in the tissues of biota due to their lipophilicity and resistance to degradation. POPs may be transported over long distances by repeated volatilisation and condensation cycles (the 'grasshopper' effect) (Wania and Mackay, 1996) and retention of these compounds in freshwater ecosystems is therefore dependent on water temperature. Some less volatile compounds will preferentially remain and accumulate in waters that are colder by virtue of their latitude or altitude, and altitudinal gradients in the concentrations of organochlorines have been observed for both sediments and fish that reflect these processes (Vives et al., 2004a). However, altitudinal gradients have not yet been observed for PBDEs (Vives et al., 2004b). Adverse effects of POPs include impacts on all stages of growth and development including reproductive effects, teratogenicity (embryonic effects), neurotoxicity and carcinogenicity (Harrad, 2001). Further, some POPs including pesticides, PCBs and PAHs may mimic oestrogens or counteract natural hormones and are known as endocrine disruptors (e.g., Matthiessen and Johnson, 2007). Hormones often act in pg L^{-1} or ng L^{-1} levels and therefore these chemical mimics may have an impact even at the remotest sites (Garcia-Reyero et al., 2005). Their primary action is often on embryos, foetuses or larvae although the full effect may not be expressed until adulthood; for example, the feminisation of male fish.

5.3 Forward look and climate change

While there have been significant reductions in the emissions and deposition of trace metals and some POPs over the last 30 years, predicted trends for the future under a changing climate are very uncertain. There are three main concerns. First, insecticides and pesticide usage may increase as a result of the increased need to control invasive species and agricultural pests. Second, in a warmer atmosphere those compounds which volatilise readily will be able to travel greater distances prior to deposition. Increasing lake-water temperatures will also reduce the retention of volatile POPs and as a consequence there may be a shift in the altitudinal (or latitudinal) gradients of POPs accumulation (e.g., Grimalt et al., 2001, Vives et al., 2004a). Third, recent studies have shown that remobilisation of trace metals from catchment soils as a result of soil erosion and, to a lesser extent leaching of metals bound to dissolved organic carbon (DOC), may impact on metal inputs to lakes (Rose et al., 2009) and rivers (Rothwell et al., 2007). These processes are exacerbated by increased winter rainfall, decreased summer rainfall and increased frequency of high intensity rain events. Hence, predicted climatic changes will elevate pollutant transfer from catchments to surface waters. The catchment storage of deposited pollutants is a massive potential reservoir of contamination that may keep contaminant fluxes elevated for many decades and maintain and elevate exposure of aquatic biota to these contaminants.

New substances are continually being developed and employed both industrially and domestically. 'Emerging' chemicals of concern include pharmaceuticals and persistent polar pollutants (P^3) such as perfluorinated acids (PFAs) including perfluorooctanoate (PFOA) and perfluorooctyl sulphonate (PFOS) used in fire-fighting foams and in oil and water resistant coatings for textiles, paper and so on. These substances, being polar, are more soluble in water but are toxic to aquatic organisms and accumulate in fish thereby fulfilling the PBT criteria. Their presence has already been reported in many European rivers (Loos et al., 2009).

6 CONCLUSIONS

Lake sediment analysis provides unique insights into the history of freshwater ecosystems giving evidence for the nature and timing of ecosystem change, enabling in many cases the causes of change to be identified and providing a record of human impact that can be indispensable in developing strategies for ecosystem management in the present day. Although the most intense impacts of human activity on freshwater can be shown to have occurred over the last 100 to 200 years, during the Anthropocene, there is clear evidence of some impact as early as the Bronze Age. For most freshwater, the impact has been cumulative with stresses multiplying, intensifying and interacting over time. In the most extreme cases freshwaters have suffered from the combined effects of hydromorphological alteration, eutrophication, long-range transported air pollutants and invasive species. Even freshwater ecosystems in remote regions are not beyond the reach of air pollutants and now global warming has the potential to alter the nature of all systems.

The extent to which freshwaters have been degraded by human activity has been recognised in many regions of the world and legislative programmes are now in force to reduce pollution and restore aquatic ecosystems to good health. However, although there are many individual success stories, especially in reducing the pressures on ecosystems, there is considerable uncertainty about whether restoration targets can be easily achieved. Recovery may be a slow process as ecosystems take time to readjust to reduced stresses, and new pressures, especially from future global warming, may counter restoration strategies, especially those designed to reverse problems of eutrophication and salinisation and reduce the remobilisation of toxic substances. The future shape of freshwater ecosystems is difficult to predict or model. Continued research is needed to understand how these different stresses might interact in future under a range of climate scenarios. High-quality, site-specific monitoring networks will need to be maintained or newly created to constrain model simulations and identify future threats.

ACKNOWLEDGEMENTS

We should like to thank Cath d'Alton for preparing the figures.

REFERENCES

Almer B., Dickson W., Ekström C., Hörnström E. and Miller U. 1974. Effects of acidification on Swedish lakes. *Ambio* 3: 30–36.

Anderson N. J., Bugmann H., Dearing J. A. and Gaillard M. J. 2006. Linking palaeoenvironmental data and models to understand the past and to predict the future. *Trends in Ecology and Evolution* 21: 696–704.

Battarbee R. W. 1990. The causes of lake acidification, with special reference to the role of acid deposition. *Philosophical Transactions of the Royal Society London B* 327: 339–347.

Battarbee R. W., Anderson N. J., Jeppesen E. and Leavitt P. R. 2005a. Combining palaeolimnological and limnological approaches in assessing lake ecosystem response to nutrient reduction. *Freshwater Biology* 50: 1772–1780.

Battarbee R. W. and Charles D. F. 1986. Diatom-based pH reconstruction studies of acid lakes in Europe and North America: A synthesis. *Water Air and Soil Pollution* 31: 347–354.

Battarbee R. W. Kernan M. Livingstone D. Nickus U. Verdonschot P. Hering D. et al. 2008b. Freshwater ecosystem responses to climate change: the Eurolimpacs project, in Quevauviller P., Borchers U., Thompson C. and Simonart T. (eds) *The Water Framework Directive – Ecological and Chemical Status Monitoring (Water Quality Measurements)*. London: John Wiley and Sons.

Battarbee R. W., Mason J., Renberg I. and Talling J. F. (eds) 1990. *Palaeolimnology and Lake Acidification*. London: The Royal Society.

Battarbee R. W., Monteith D. T., Juggins S., Evans C. D., Jenkins A. and Simpson G. L. 2005b. Reconstructing pre-acidification pH for an acidified Scottish loch: a comparison of palaeolimnological and modelling approaches. *Environmental Pollution* 137: 135–149.

Battarbee R. W., Monteith D. T., Juggins S., Simpson G. L., Shilland E. M., Flower R. J. and Kreiser A. M. 2008a. Assessing the accuracy of diatom-based transfer functions in defining reference pH conditions for acidified lakes in the United Kingdom. *The Holocene* 8: 57–67.

Battarbee R. W., Morley D., Bennion H., Simpson G. L., Hughes M. and Bauere V. 2011. A palaeolimnological meta-database for assessing the ecological status of lakes. *Journal of Paleolimnology* 45: 405–414.

Beamish J. and Harvey H. H. 1972. Acidification of La Cloche Mountain lakes, Ontario, and resulting fish mortalities. *Journal of the Fisheries Research Board of Canada* 29: 1131–1143.

Bennion H., Johnes P., Ferrier R., Phillips G. and Haworth E. 2005. A comparison of diatom phosphorus transfer functions and export coefficient models as tools for reconstructing lake nutrient histories. *Freshwater Biology* 50: 1651–1670.

Bennion H. and Battarbee R. 2007. The European Union Water Framework Directive: opportunities for palaeolimnology. *Journal of Paleolimnology* 38: 285–295.

Bennion H., Battarbee R. W., Sayer C. D., Simpson G. L., Davidson T. A. 2011. Defining reference conditions and restoration targets for lake ecosystems using palaeolimnology: a synthesis. *Journal of Paleolimnology* 45: 533–544.

Birks H. J. B., Anderson N. J. and Fritz S. C. 1995. Post-glacial changes in total phosphorus at Diss Mere, Norfolk inferred from fossil diatom assemblages, in Patrick S. T. and Anderson N. J. (eds) *Ecology and Palaeoecology of Lake Eutrophication, an Informal Workshop*. Copenhagen: Geological Survey of Denmark and Greenland, pp. 48–49.

Boomer I., Wünneman B., Mackay A. W., Austin P., Sorrel P., Reinhardt C. et al. 2009. Advances in understanding the late Holocene history of the Aral Sea region. *Quaternary International* 194: 79–90.

Bradshaw E. G., Rasmussen P. and Odgaard B. V. 2005. Mid- to late-Holocene change and lake development at Dallund Sø, Denmark: synthesis of multi-proxy data, linking land and lake. *The Holocene* 15: 1152–1162.

Bränvall M. L., Bindler R., Emteryd O. and Renberg I. 2001. Four thousand years of atmospheric lead pollution in northern Europe: a summary from Swedish lake sediments. *Journal of Paleolimnology* 25: 421–435.

Brooks S. J., Bennion H. and Birks H. J. B. 2001. Tracing lake trophic history with a chironomid-total phosphorus inference model. *Freshwater Biology* 46: 511–532.

Cameron N. G., Schnell O. A., Rautio M. L., Lami A., Livingstone D. M., Appleby P. G. et al. 2002. High-resolution analyses of recent sediments from a Norwegian mountain lake and comparison with instrumental records of climate. *Journal of Paleolimnology* 28: 79–93.

Charles D. F. 1990. Effects of acidic deposition on North American lakes: paleolimnological evidence from diatoms and chrysophytes. *Philosophical Transactions of the Royal Society of London B* 327: 403–412.

Charles D. F. and Whitehead D. R. 1986. The PIRLA project: Paleoecological Investigation of Recent Lake Acidification. *Hydrobiologia* 143: 13–20.

Cosby B. J., Ferrier R. C., Jenkins A. and Wright R. F. 2001. Modelling the effects of acid deposition: refinements, adjustments and inclusion of nitrogen dynamics in the MAGIC model. *Hydrology and Earth System Sciences* 5: 499–517.

Cumming B. F., Smol J. P., Kingston J. C., Charles D. F., Birks H. J. B., Camburn K. E. et al. 1992. How much acidification has occurred in Adirondack region lakes (New York, USA) since preindustrial times? *Canadian Journal of Fisheries and Aquatic Sciences* 49: 128–141.

Curtis C. J., Evans C. D., Helliwell R. C. and Monteith D. T. 2005. Nitrate leaching as a confounding factor in chemical recovery from acidification in UK upland waters. *Environmental Pollution* 137: 73–82.

Davidson T., Sayer C., Bennion H., David C., Rose N. and Wade M. 2005. A 250 year comparison of historical, macrofossil and pollen records of aquatic plants in a shallow lake. *Freshwater Biology* 50: 1671–1686.

de Schamphelaere K. A. C., Canli M., Van Lierde V., Forrez I., Vanhaecke F. and Janssen C. R. 2004. Reproductive toxicity of dietry zinc to *Daphnia magna*. *Aquatic Toxicology* 70: 233–244.

Dixit S. S., Dixit A. S. and Smol J. P. 2002. Diatom and chrysophyte transfer functions and inferences of post-industrial inferences of post-industrial acidification and recent recovery trends in Killarney lakes (Ontario, Canada). *Journal of Paleolimnology* 27: 79–96.

Edmondson W. T. and Lehman J. T. 1981. The effect of changes in the nutrient income on the condition of Lake Washington. *Limnology and Oceanography* 26: 1–29.

Farmer J. G., Eades L. J. and Graham M. C. 1999. The lead content and isotopic composition of British coals and their implications for past and present releases of lead to the UK environment. *Environmental Geochemistry and Health* 21: 257–272.

Fernández P., Vilanova R. M., Martinez C., Appleby P. and Grimalt J. O. 2000. The historical record of atmospheric pyrolytic pollution over Europe registered in the sedimentary PAH from remote mountain lakes. *Environmental Science & Technology* 34: 1906–1913.

Flower R. J. and Battarbee R. W. 1983. Diatom evidence for recent acidification of two Scottish lochs. *Nature* 20: 130–133.

Fluin J., Gell P., Haynes D. and Tibby J. 2007. Paleolimnological evidence for the independent evolution of neighbouring terminal lakes, the Murray Darling Basin, Australia. *Hydrobiologia* 591: 117–134.

Fritz S. C. 2008. Deciphering climatic history from lake sediments. *Journal of Paleolimnology* 39: 5–16.

Fritz S. C., Juggins S., Battarbee R. W. and Engstrom D. R. 1991. Reconstruction of past changes in salinity and climate using a diatom-based transfer function. *Nature* 352: 706–708.

Gallon C. L., Tessier A., Gobeil C. and Beaudin L. 2005. Sources and chronology of atmospheric lead deposition to a Canadian Shield lake: Inferences from Pb isotopes and PAH profiles. *Geochimica et Cosmochimica Acta* 69: 3199–3210.

Garcia-Reyero N., Piña B., Grimalt J. O., Fernández P., Fonts R., Polvillo O. and Martrat B. 2005. Estrogenic activity in sediments from European mountain lakes. *Environmental Science & Technology* 39: 1427–1435.

Gasse F. 2002. Diatom-inferred salinity and carbonate oxygen isotopes in Holocene waterbodies of the western Sahara and Sahel (Africa). *Quaternary Science Reviews* 21: 737–767.

Gasse F., Fontes J-Ch., Plaziat J. C., Carbonel P., Kaczmarska I., De Deckker P. et al. 1987. Biological remains, geochemistry and stable isotopes for the reconstruction of environmental and hydrological changes in the Holocene lakes from North Sahara. *Palaeogeography Palaeoclimatology Palaeoecology* 60: 1–46.

Gell P., Bulpin S., Wallbrink P., Bickford S. and Hancock G. 2005. Tareena Billabong – A palaeolimnological history of an everchanging wetland, Chowilla Floodplain, lower Murray-Darling Basin. *Marine and Freshwater Research* 56: 441–456.

Gell P. A., Sluiter I. R. and Fluin J. 2002. Seasonal and inter-annual variations in diatom assemblages in Murray River-connected wetlands in northwest Victoria, Australia. *Marine & Freshwater Research* 53: 981–992.

Gell P.. Fluin J., Tibby J., Haynes D., Khanum S., Walsh B. et al. 2006. Changing fluxes of sediments and salts as recorded in lower River Murray wetlands, Australia. *International Association of Hydrological Sciences* 306: 416–424.

Griffin J. J. and Goldberg E. D. 1981. Sphericity as a characteristic of solids from fossil-fuel burning in a Lake Michigan sediment. *Geochimica et Cosmochimica Acta* 45: 763–769.

Grimalt J. O., Fernández P., Berdie L., Vilanova R. M., Catalan J., Psenner R. et al. 2001. Selective trapping of organochlorine compounds in mountain lakes of temperate areas. *Environmental Science & Technology* 35: 2690–2697.

Hall R. I. and Smol J. P. 2010. Diatoms as indicators of lake eutrophication, in Stoermer E. F. and Smol J. P. (eds) *The Diatoms: Applications for the Environmental and Earth Sciences*, 2nd edition. Cambridge: Cambridge University Press, pp. 122–151.

Hammer U. T. 1986. *Saline Lake Ecosystems of the World*. Dordrecht: Junk.

Harrad S. 2001. The environmental behaviour of persistent organic pollutants, in Harrison R. M. (ed.) *Pollution: Causes, Effects and Control*, 4th edition. Cambridge: Royal Society of Chemistry, pp. 445–473.

Holmes J. A. 2001. Ostracoda, in Smol J. P., Birks H. J. B. and Last W. M. (eds) *Tracking Environmental Change Using Lake Sediments, Volume 4: Zoological Indicators* Dordrecht: Kluwer, pp. 125–151.

Jeppesen E., Leavitt P., De Meester L. and Jensen J. P. 2001. Functional ecology and palaeolimnology: using cladoceran remains to reconstruct anthropogenic impact. *Trends in Ecology & Evolution* 16: 191–198.

Jeppesen E., Søndergaard M., Meerhoff M., Lauridsen T. L. and Jensen J. P. 2007. Shallow lake restoration by nutrient loading reduction—some recent findings and challenges ahead. *Hydrobiologia* 584: 239–252.

Jeppesen E., Søndergaard M., Jensen J. P., Havens K., Anneville O., Carvalho L. et al. 2005. Lake responses to reduced nutrient loading – an analysis of contemporary long-term data from 35 case studies. *Freshwater Biology* 50: 1747–1771.

Jones R. N., McMahon T. A. and Bowler J. M. 2001. Modelling historical lake levels and recent climate change at three closed lakes, Western Victoria, Australia (c.1840–1990). *Journal of Hydrology* 246: 159–180.

Jordan P., Rippey B. and Anderson N. J. 2001. Modeling diffuse phosphorus loads from land to freshwater using the sedimentary record. *Environmental Science and Technology* 35: 815–819.

Kolbe R. W. 1927. Zur ökologie, morphologie, und systematik der brackwasser-Diatomeen. *Pflanzenforschung* 7: 1–146.

Larsen J., Jones V. J. and Eide W. 2006. Climatically driven pH changes in two Norwegian alpine lakes. *Journal of Paleolimnology* 36: 57–69.

Leavitt P. R., Carpenter S. R. and Kitchell J. F. 1989. Whole-lake experiments: the annual record of fossil pigments and zooplankton. *Limnology and Oceanography* 34: 700–717.

Leung K. M. Y., Grist E. P. M., Morley N. J., Morritt D. and Crane M. 2007. Chronic toxicity of tributyltin to development and reproduction of the European freshwater snail *Lymnaea stagnalis* (L.). *Chemosphere* 66: 1358–1366.

Loos R., Gawlik B. M., Locoro G., Rimaviciute E., Contini S. and Bidoglio G. 2009. EU-wide survey of polar organic persistent pollutants in European river waters. *Environmental Pollution* 157: 561–568.

Maberly S. C., King L., Dent M. M., Jones R. I. and Gibson C. E. 2002. Nutrient limitation of phytoplankton and periphyton growth in upland lakes. *Freshwater Biology* 47: 2136–2152.

MacGregor A. J., Gell P. A., Wallbrink P. J. and Hancock G. 2005. Natural and post-disturbance variability in water quality of the lower Snowy River floodplain, Eastern Victoria, Australia. *River Research and Applications* 21: 201–213.

Macumber P. G. 1991. *Interaction Between Groundwater and Surface Systems in Northern Victoria.* Melbourne: Department of Conservation and Environment.

Magee J. W., Bowler J. M., Miller G. H. and Williams D. L. G. 1995. Stratigraphy, sedimentology, chronology and palaeohydrology of Quaternary lacustrine deposits at Madigan Gulf, Lake Eyre, South Australia. *Palaeogeography Palaeoclimatology Palaeoecology* 113: 3–42.

Majewski S. P. and Cumming B. F. 1999. Paleolimnological investigation of the effects of post-1970 reductions of acidic deposition on an acidified Adirondack lake. *Journal of Paleolimnology* 21: 207–213.

Malmaeus J. M., Blenckner T., Markensten H. and Persson I. 2006. Lake phosphorus dynamics and climate warming: A mechanistic model approach. *Ecological Modelling* 190: 1–14.

Marsden M. W., Fozzard I. R., Clark D. McLean N. and Smith M. R. 1995. Control of phosphorus inputs to a freshwater lake: a case study. *Aquaculture Research* 26: 527–538.

Matthiessen P. and Johnson I. 2007. Implications of research on endocrine disruption for the environmental risk assessment, regulation and monitoring of chemicals in the European Union. *Environmental Pollution* 146: 9–18.

Meharg A. A. and Killham K. 2003. A pre-industrial source of dioxins and furans. *Nature* 421: 909–910.

Monteith D. T., Stoddard J. L., Evans C. D., de Wit H. A., Forsius M., Høgåsen T. et al. 2007. Dissolved organic carbon trends resulting from changes in atmospheric deposition chemistry. *Nature* 450: 537–540.

Moss B, McKee D., Atkinson D., Collings S. E., Eaton J. W., Gill A. B. et al. 2003. How important is climate? Effects of warming, nutrient addition and fish on phytoplankton in shallow lake microcosms. *Journal of Applied Ecology* 40: 782–792.

Muir D. C. G. and Rose N. L. 2007. Persistent organic pollutants in the sediments of Lochnagar, in Rose N. L. (ed.) *Lochnagar: The Natural History of a Mountain Lake.* Dordrecht: Springer, pp. 375–402.

Nygaard G. 1956. Ancient and recent flora of diatoms and chrysophyceae in Lake Gribsö. Studies on the humic acid lake Gribsö. *Folia Limnologica Scandinavica* 8: 32–94.

Pearson R. F., Swackhammer D. L., Eisenreich S. J. and Long D. T. 1997. Concentrations, accumulations and inventories of toxaphene in sediments of the Great Lakes. *Environmental Science and Technology* 31: 3523–3529.

Pennington W. 1984. Long-term natural acidification of upland sites in Cumbria: evidence from post-glacial lake sediments. *Freshwater Biological Association Annual Report* 52: 28–46.

Piovano E. L., Ariztegui D., Bernasconi S. M. and MacKenzie J. A. 2004. Stable isotopic record of hydrological changes in subtropical Laguna Mar Chiquita (Argentina) over the last 230 years. *The Holocene* 14: 525–535.

Psenner R. and Schmidt R. 1992. Climate-driven pH control of remote alpine lakes and effects of acid deposition. *Nature* 356: 781–783.

Quinlan R. and Smol J. P. 2001. Chironomid-based inference models for estimating end-of-summer hypolimnetic oxygen from south-central Ontario shield lakes. *Freshwater Biology* 46: 1529–1551.

Radke L. C., Juggins S., Halse S. A., DeDeckker P. and Finston T. 2003. Chemical diversity in south eastern Australian saline lakes II. Biotic implications. *Marine and Freshwater Research* 54: 895–912.

Renberg I. 1990. A 12 600 year perspective of the acidification of Lilla Öresjön, southwest Sweden. *Philosophical Transactions of the Royal Society London B* 327: 357–361.

Renberg I., Korsman T. and Birks H. J. B. 1993a. Prehistoric increases in the pH of acid-sensitive

Swedish lakes caused by land-use changes. *Nature* 362: 824–826.

Renberg I., Korsman T. and Anderson N. J. 1993b. A temporal perspective of lake acidification in Sweden. *Ambio* 22: 264–271.

Rose N. L. 2001. Fly-ash particles, in Last W. M. and Smol J. P. (eds) *Tracking Environmental Change Using Lake Sediments: Volume 2. Physical and Chemical Techniques.* Dordrecht: Kluwer Academic Publishers, pp. 319–349.

Rose N. L. (ed.) 2007. *Lochnagar: The Natural History of a Mountain Lake.* Dordrecht: Springer.

Rose N. L., Boyle J. F., Du Y., Yi C., Dai X., Appleby P. G., Bennion H., Cai S. and Yu L. 2004. Sedimentary evidence for changes in the pollution status of Taihu in the Jiangsu region of eastern China. *Journal of Paleolimnology* 32: 41–51.

Rose N. L., Yang H. and Turner S. D. 2009. *An assessment of the mechanisms for the transfer of lead and mercury from atmospherically contaminated organic soils to lake sediments with particular reference to Scotland, UK.* Deliverable No. 379 Report from EU-FP6 Project Euro-limpacs. (Integrated project to evaluate the impacts of global change on European Freshwater Ecosystems; Project No GOCE-CT–2003–505540.)

Rosenqvist I. T. 1978. Alternative sources for acidification of river water in Norway. *The Science of the Total Environment* 10: 39–49.

Rosseland B. O. Rognerud S. Collen P. Grimalt J. O. Vives I. Massabuau J.-C. et al. 2007. Brown trout in Lochnagar: Population and contamination by metals and organic micropollutants, in Rose N. L. (ed.) *Lochnagar: The Natural History of a Mountain Lake.* Dordrecht: Springer, pp. 253–285.

Rothwell J. J., Evans M. G. and Allott T. E. H. 2007. Lead contamination of fluvial sediments in an eroding blanket peat catchment. *Applied Geochemistry* 22: 446–459.

Sayer C., Roberts N., Sadler J., David C. and Wade P. M. 1999. Biodiversity changes in a shallow lake ecosystem: a multi-proxy palaeolimnological analysis. *Journal of Biogeography* 26: 97–114.

Schindler D. W. Hecky R. E. Findlay D. L. Stainton M. P. Parker B. R. Paterson M. J. et al. 2008. Eutrophication of lakes cannot be controlled by reducing nitrogen input: Results of a 37-year whole-ecosystem experiment. *Proceedings of the National Academy of Sciences of the USA* 105: 11254–11258.

Sharpley A., Foy B. and Withers P. 2000. Practical and innovative measures for the control of agricultural phosphorus losses to water: an overview. *Journal of Environmental Quality* 29: 1–9.

Shen P. P., Shi Q., Hua Z. C., Kong F. X. and Wang Z. G. 2003. Analysis of microcystins in cyanobacteria blooms and surface water samples from Meiliang Bay, Taihu Lake, China. *Environment International* 29: 641–647.

Silk P. J., Lonergan G. C., Arsenault T. L. and Boyle C. D. 1997. Evidence of natural organochlorine formation in peat bogs. *Chemosphere* 35: 2865–2880.

Smith V. H. and Schindler D. W. 2009. Eutrophication science: where do we go from here? *Trends in Ecology and Evolution* 24: 201–207.

Smol J. P. 2008. *Pollution of Lakes and Rivers: A Paleoenvironmental Perspective.* Malden, MA: Blackwell.

Smol J. P. Birks H. J. B. and Last W. M. (eds) 2001. *Tracking environmental change using lake sediments Volume 3: Terrestrial, algal and siliceous indicators.* Dordrecht: Kluwer.

Sommaruga-Wögrath S., Koinig K. A., Schmidt R., Sommaruga R., Tessadri R. and Psenner R. 1997. Temperature effects on the acidity of remote alpine lakes. *Nature* 387: 64–67.

Song W. L., Ford J. C., Li, A., Sturchio N. C., Rockne K. J., Buckley D. R. and Mills W. J. 2005. Polybrominated diphenyl ethers in the sediments of the Great Lakes. 3. Lakes Ontario and Erie. *Environmental Science and Technology* 39: 5600–5605.

Søndergaard M., Jeppesen E., Lauridsen T., Van Nes S. C. H., Roijackers R., Lammens E. and Portielje R. 2007. Lake restoration: successes, failures and long-term effects. *Journal of Applied Ecology* 44: 1095–1105.

Sylvestre F. 2002. A high resolution diatom reconstruction between 21,000 and 17,000 ^{14}C yr BP from the southern Bolivian Altiplano (18–23°S). *Journal of Paleolimnology* 27: 45–57.

Talling J. F. and Heaney S. I. 1988. Long-term changes in some English (Cumbrian) lakes subjected to increased nutrient inputs, in Round F. (ed.) *Algae and the Aquatic Environment.* Bristol: Biopress, pp. 1–29.

Verschuren D., Laird K. R. and Cumming B. F. 2000. Rainfall and drought in equatorial east Africa during the past 1,100 years. *Nature* 403: 410–414.

Verschuren D., Tibby J., Leavitt P. R. and Roberts C. N. 1999. The environmental history of a climate-sensitive lake in the former 'White Highlands' of Central Kenya. *Ambio* 28: 494–501.

Vives I., Grimalt J. O., Catalan J., Rosseland B. O. and Battarbee R. W. 2004a. Influence of altitude and age in the accumulation of organochlorine compounds in fish from high mountain lakes. *Environmental Science and Technology* 38: 690–698.

Vives I., Grimalt J. O., Lacorte S., Guillamón M., Barceló D. and Rosseland B. O. 2004b. Polybromodiphenyl ether flame retardants in fish from lakes in European high mountains and Greenland. *Environmental Science and Technology* 38: 2338–2344.

Vollenweider R. A. 1968. *Scientific Fundamentals of the Eutrophication of Lakes and Flowing Waters, with Particular Reference to Nitrogen and Phosphorus as Factors in Eutrophication.* Technical Report DAS/CSI/68, 27. Paris: OECD.

Wania F. and Mackay D. 1996. Tracking the distribution of persistent organic pollutants. *Environmental Science and Technology* 30: 390A–396A.

Wetzel R. G. 1975. *Limnology.* Philadelphia: W.B. Saunders.

Whitehead D. R., Charles D. F., Jackson S. T., Smol J. P. and Engstrom D. R. 1989. The developmental history of Adirondack (N.Y.) lakes. *Journal of Paleolimnology* 2: 185–206.

Whitehead P. G., Wilby R. L., Butterfield D. and Wade A. J. 2006. Impacts of climate change on nitrogen in lowland chalk streams: adaptation strategies to minimise impacts. *Science of the Total Environment* 365: 260–273.

Williams W. D. 1981. Inland salt lakes: an introduction. *Hyrobiologia* 81: 1–14.

Yang H. and Rose N. L. 2003. Distribution of mercury in six lake sediment cores across the UK. *Science of the Total Environment* 304: 391–404.

Zhang M., Qi W., Jiang X., Zhao Y. and Li M. 2004. Trend of salt lake changes in the background of global warming and tactics for adaptation to the changes. *Acta Geologica Sinica* 78: 795–807.

28

Human Impacts on Coastal and Marine Geo-Ecosystems

Ben Daley

1 INTRODUCTION

Human impacts on many – if not all – parts of the Earth system have grown, accelerated and intensified over the last three centuries, and particularly since around 1950 (Crutzen and Stoermer, 2001; Turner et al., 1990). Coastal and marine environments and ecosystems have not escaped those influences; indeed, many profound impacts have been sustained by coastal and marine geo-ecosystems, and transformations of those systems continue to occur at increasing rates and intensities (Halpern et al., 2008; Table 28.1). Yet, in contrast to their terrestrial counterparts, coastal and marine themes have been comparatively neglected in many accounts of environmental change. Analysis of the coverage of environmental history studies, for instance, reveals a relative paucity of coastal and marine themes. As Roberts (1998: 7) has acknowledged, 'climate, forests and rivers have their histories' – but so too do oceans, seas, coastlines, rocky shores, beaches, deltas, estuaries, mangrove ecosystems, coral reefs, islands and many other coastal and marine habitats, together with their myriad associated species. Those histories are no less important or interesting than are the accounts of terrestrial environmental change. Such an imbalance is perhaps surprising, given the magnitude of the global ocean, the extent of the coastal zone and the fact that the hydrosphere is now widely acknowledged to play a critical role in the functioning of the Earth system in relation to biogeochemical cycling, as a key component of the global climate system and as a major sink for pollutants. The comparative neglect of coastal and marine geo-ecosystems in accounts of environmental change may reflect the (historical) relative scarcity of data for many aspects of those environments and ecosystems, as well as the difficulty of reconstructing environmental changes in what are often vast, complex and inaccessible habitats. Nevertheless, attempts are increasingly being made to redress the relative deficit in the scientific understanding of coastal and marine environmental change (Holm et al., 2001).

This chapter presents an overview of the main human impacts on coastal and marine geo-ecosystems, based on a review of relevant scientific and environmental history studies. The aim of this chapter is to illustrate the broad scope of human impacts on coastal and marine environments and ecosystems; yet, whilst those impacts are diverse, some

Table 28.1 Summary of human impacts on coastal and marine geo-ecosystems

Human activity	Environmental impacts
Exploitation of coastal and marine resources	
Hunting of whales, seals and other large marine animals	Depletion of populations
	Extinction of species
	Reduction of biodiversity
	Release of carbon to the atmosphere
Industrial and artisanal fishing	Changes in the species composition of communities
	Changes in the size structure of fish populations
	Reduction of biodiversity
	Discarded by-catch
	Destruction of bottom habitat by benthic trawling
	Destruction of coral reefs by dynamite and cyanide
	Increase in populations of scavenging species
	Changes in coral reef ecology
Exploitation of other coastal, marine and island resources	Depletion of populations
	Extinction of species
	Reduction of biodiversity
	Transformation of islands
	Transformation of coral reefs
Changes in island land-use	Depletion of populations
	Extinction of species
	Reduction of biodiversity
	Transformation of islands
Pollution of coastal and marine environments	
Increased sediment input	Sedimentation and siltation of coastal habitats
	Transport of reactive chemical species to the coastal zone
	Changes in coastal geomorphology
	Ecological effects on coastal biota
Increased nutrient input and eutrophication	Nutrient enhancement of coastal waters
	Degradation of water quality
	Noxious and toxic algal blooms
	Increased turbidity
	Loss of submerged aquatic vegetation
	Oxygen deficiency and the formation of 'dead zones'
	Disruption of ecosystem functioning
	Loss of habitat
	Loss of biodiversity
	Changes in food webs
	Loss of harvestable fisheries
Other forms of coastal and marine pollution	Toxicity due to contamination by oil and other substances
	Toxicity due to the use of chemical dispersants
	Accumulation of drill cuttings
	Degradation of sea floor habitat
	Bioaccumulation of toxic substances (e.g., methyl mercury)
	Accumulation of synthetic polymer (plastic) debris
	Harmful effects on biota due to entanglement and ingestion
	Other ecological effects (e.g., interference with reproduction)
Effects of climate change on coasts and seas	
Emission of greenhouse gases	Increased mean sea surface temperature (SST)
	Changes in global mean and local sea levels
	Changes in sea surface salinity (SSS)
	Changes in ocean currents
	Ocean acidification
	Ecological effects on coastal and marine biota
	Loss of coastal habitats

Table 28.1   Cont'd

Human activity	Environmental impacts
Effects of ozone-depleting substances on coasts and seas	
Emission of ozone-depleting substances	Increased ultraviolet radiation at the Earth's surface
	Ecological effects on coastal and marine biota
Changes in coastlines	
Coastal development, coastal engineering, land reclamation, dam construction, dredging and coastal management schemes	Changes in the position and extent of coastlines
	Changes in sediment and nutrient fluxes
	Coastal erosion
	Coastal pollution
	Introduction of exotic species
	Disruption of the patterns of fish migration
	Changes in salinity
	Changes in food webs
	Habitat fragmentation and loss
	Thermal pollution

recurring themes are prominent. In particular, several key problems are highlighted: the unsustainable exploitation of coastal and marine resources; coastal and marine pollution; the effects of climate change on coasts and seas; and the effects of coastline change. Whilst this chapter draws on studies found within the literatures of environmental science and environmental history, it is worth acknowledging that those disciplinary areas remain largely separate, although some efforts have been made, recently, to adopt more integrated, multidisciplinary approaches. This chapter also points to the urgent need to enact effective conservation measures for coastal and marine environments, including a global network of marine protected areas that is adequate to the challenge of marine conservation – and that is based on the principles of adaptive management. The account presented in this chapter covers several key types of impact: resource exploitation, pollution, climate change and changes in coastlines. Each of those types of impact represents a major threat to the integrity of coastal and marine systems, for various reasons: due to the alarming rate of decline of marine biodiversity; due to the creation of 'dead zones' in coastal and marine waters as a result of eutrophication; and due to the changes in sea-surface temperature (SST), sea-surface salinity (SSS), sea levels,

ocean currents, ocean acidity and ecosystems that are occurring as a result of climate change. Yet human impacts on coasts and seas do not occur in isolation; and, increasingly, studies are focusing on the effects of multiple impacts on coastal and marine environments and ecosystems and the need to devise management approaches and frameworks that can cope with complexity in, and uncertainty about, ecosystem responses. This chapter concludes by emphasising that, given that profound impacts are being sustained by coastal and marine environments and ecosystems, and that climate change presents an increasing threat, the need for effective protection and management of those environments and ecosystems is already urgent.

2 EXPLOITATION OF COASTAL AND MARINE RESOURCES

One broad category of human impact on coastal and marine geo-ecosystems is the exploitation of resources, which has occurred since early human populations first commenced hunting, fishing and gathering – a mode of production which supported the majority of the world's human population until around 5,000 years ago and which remains vital for the livelihoods of many coastal communities today. The earliest archaeological evidence of the consumption of seafood products by humans appears – in the form of shell middens – at around the onset of the Holocene. Later, during the Mesolithic, seafood products assumed a major role in some site economies, and a variety of methods was used to exploit both coastal and deep-sea fish (Rick and Erlandson, 2008; Shackleton and van Andel, 1980).

During the early Holocene, human impacts on coasts and seas were limited to the exploitation of coastal and marine species in order to support subsistence activities and the small-scale collection and trade of resources between clan groups. Thus, those impacts were generally small in their magnitude, localised in their geographical extent and brief in their duration. In many cases, human impacts due to subsistence activities – especially fishing – were transient, periodic or seasonal and were associated with clan group migrations; in other cases, camps or settlements were established on a more permanent basis. For some groups – notably for island populations – coastal and marine resources formed the basis of entire economies; however, even in most of those instances, impacts on the resource base were probably relatively small and were, in any case, confined to the vicinity of the islands. As a result, hunting, fishing and gathering activities have typically been regarded as being a mode of production which uses natural ecosystems without transforming them, and in many cases the impacts of pre-modern human populations on coastal and marine resources were limited.

With time, however, and with the growth of human populations, the direct exploitation of coastal and marine resources increased in its spatial extent, in its intensity and in the number of species that were used. The range of coastal and marine resources used by humans came to include many coastal and island mineral, plant and animal products such as sand, salt, kelp, coral, shells, shellfish, sea cucumbers (bêche-de-mer or trepang), fish, seabirds, island wildlife species (such as the moa), marine turtles and marine mammals (such as seals, dugongs, manatees and whales). Indeed, the traditional use of such resources persists today in some coastal indigenous groups that undertake hunting, gathering and fishing activities; in many of those societies, considerable economic, social and cultural significance is still attached to particular coastal and marine resources (such as dugongs and marine turtles; Marsh et al., 1997). The progressive increase of human impacts on coastal and marine resources meant that, in some cases – especially in relation to islands – the cumulative, localised effects even of hunting, fishing and gathering activities could become severe. The overexploitation of seal and bird populations in the Pacific region by Polynesian settlers has been well documented; in conjunction with other environmental impacts (deforestation and soil degradation), those changes ultimately resulted in the extinction of the moas of New Zealand and in the collapse of the Easter Island economy, society and environment. Such examples illustrate that, in some cases, even premodern human impacts on island environments and ecosystems could be severe at the local scale. Nevertheless, overall, the impacts of hunting, fishing and gathering were small in comparison with the changes that occurred later – and which have become prominent during the Anthropocene.

During the course of the Anthropocene, human demands on coastal and marine environments and ecosystems have increased substantially, and many human activities have contributed to the unsustainable exploitation of coastal and marine resources. Those activities include the operation of commercial and artisanal fisheries based on large marine animals (such as whales, seals, dugongs, manatees and marine turtles), fish and other seafood products (such as bêche-de-mer and oysters); the hunting of seabirds and island bird species; and the extraction of other coastal and island resources (such as coral, guano and rock phosphate). As a result of such activities, some species have been driven to extinction whilst the populations of others have been depleted with varying severity. The unsustainable nature of many fisheries is arguably one of the most significant impacts within this category, given the importance of fisheries for food security in many human societies. As a result of unsustainable fishing, at the global scale, around 5 per cent of fish species are now threatened with extinction – and that proportion is increasing (Pimm et al., 1995). In addition, some coastal and island habitats have been profoundly transformed by industrial activities, and habitat degradation has had feedback effects on the populations of many coastal and marine species. Various human impacts that may be regarded as forms of unsustainable exploitation of coastal and marine resources are now discussed in turn.

2.1 Impacts on large marine animals

Whale populations have sustained profound impacts due to exploitation by humans, and that exploitation is now acknowledged to have occurred much earlier, on a far greater scale and more intensively than was previously thought (Holm et al., 2001). Many whale populations have consequently diminished to a tiny proportion of their former sizes. Those declines have been due largely to commercial whaling, an activity that reached its peak in the interwar period, yet which continues today. Accurate historical reconstruction of the population dynamics of whales before, during and after their exploitation is vital for marine ecological restoration and to inform debates about future commercial whaling. Yet it is difficult to estimate pre-exploitation whale stocks and hence to quantify the magnitude of the impacts that have been sustained by whale populations, partly because population dynamic models require accurate information about historical catches, as well as estimates of intrinsic rates of increase and current abundance, the details of which remain uncertain. Nevertheless, some recent progress in improving historical population reconstructions has been made,

using population genetic techniques (although those studies also contain large uncertainties, especially about rates of mutational substitution and gene flow). Several studies suggest that historical whaling – especially during the sixteenth and seventeenth centuries – could have been responsible for very high whale mortality rates, with the implication that current population sizes represent only a very small fraction of their past numbers (Baker and Clapham, 2004; Jackson et al., 2008; Palumbi and Cipriano, 1998; Roman and Palumbi, 2003). Further research is required in order to reconcile conflicting estimates of pre-exploitation abundance of whales (which currently differ by at least an order of magnitude) and to develop more robust baselines against which the recovery of whale populations may be assessed.

Individual whale populations have been affected to varying degrees by historical whaling. Populations of bowhead, humpback, fin and minke whales sustained particularly heavy impacts due to commercial exploitation. For instance, the population of humpback whales (*Megaptera novaeanglia*) in east Australian waters was severely depleted (from around 10,000 animals to less than 500 over the course of a decade) as a result of commercial whaling between 1952 and 1962. This activity was based at the Tangalooma whaling station on Moreton Island, despite the fact that concurrent regulation and scientific monitoring of that industry occurred (Jones, 2002; Paterson et al., 1994). At the global scale, although whaling bans have been introduced and some whale populations are now showing signs of recovery, the status of some baleen whale populations continues to be a cause of concern. Phylogenetic analysis has shown that the unauthorised trade of whale products occurs, involving meat from three whale species (fin, sei and Antarctic minke whales) that are currently protected from international trade yet which have been killed as part of Japan's scientific whaling programme (Baker and Palumbi, 1994). The persistence of such illegal trade illustrates the need for independent, transparent and effective monitoring of scientific research programmes into whale populations, as well as of any future commercial whaling. Furthermore, the effects of whaling are compounded by a range of other human impacts on whales: boat strikes; ingestion of pollutants and debris; and acoustic disturbance due to seismic exploration, the operation of motor vessels and the use of sonar equipment.

Recent research has indicated that the environmental impact of whaling extended beyond the depletion of whale populations; the destruction of whales also contributed to anthropogenic climate change through the release of additional carbon to the atmosphere. Pershing (2010) has argued that whales play an important role in the global carbon cycle by storing carbon within their bodies, much of which may be released to the atmosphere when they are killed. When whales die naturally, their bodies sink and transport carbon to the ocean floor. In contrast, and especially during the early period of commercial whaling when whales were captured and used to manufacture products such as lamp oil, which was subsequently burned, the carbon stored in the animals was instead released to the atmosphere. By calculating the annual carbon-storing capacity of whales during their growth, Pershing (2010) estimated that historical whaling was responsible for releases of carbon to the atmosphere totalling 9 million tonnes. Thus, through the conservation and restoration of whale populations, substantial amounts of carbon could potentially be sequestered. Further research is required to determine the quantity of carbon that could be sequestered in a fully populated stock of whales (or of other large marine animals, such as bluefin tuna or white sharks) and to explore ways in which such biota might be protected in exchange for marine carbon credits (Pershing, 2010).

Besides whales, other large marine animals have been exploited unsustainably by humans. Seals, such as the southern fur seal (*Arctocephalus gazella*), have been hunted commercially since the eighteenth century

and their populations have been significantly reduced. By the late eighteenth century, over 10 million southern fur seal skins had been taken, and by the early nineteenth century the species had been hunted almost to extinction and the industry had collapsed. The Caribbean monk seal (*Monachus tropicalis*) had been hunted to extinction by 1952, whilst the Mediterranean monk seal (*Monachus monachus*) was also overexploited and remains critically endangered (Engelman et al., 2008: 12). The Steller's sea cow (*Hydrodamalis gigas*) had been hunted to extinction by 1768, and the Gulf porpoise (*Phocoena sinus*) was also severely depleted and remains a critically endangered species (Engelman et al., 2008: 12). Conservation concerns about the three extant species of manatee have been documented; impacts on those species include poaching and incidental drowning in fishing nets, compounded in some places by the effects of boat strikes, pollution and habitat loss (de Thoisy et al., 2003; Jiménez, 2002; Marsh and Lefebvre, 1994). The dugong (*Dugong dugon*) has sustained severe impacts in Australian waters as a result of commercial dugong hunting (for dugong oil, meat, hides and bones) between 1847 and 1969; those impacts occurred in conjunction with an increasingly wide range of other effects, such as seagrass degradation, boat strikes, entanglement in nets set for bather protection and hunting (Chilvers et al., 2005; Marsh et al., 2004, 2005; Preen and Marsh, 1995; Preen et al., 1995; Sheppard et al., 2006, 2007; Figure 28.1).

The conservation status of marine turtles has been the focus of considerable concern as a result of various human impacts. Those impacts include the unsustainable harvest of marine turtles and their eggs, the accidental by-catch of the animals in trawl nets, boat strikes, ingestion of litter and the degradation of nesting habitats. Consequently, worldwide, populations of marine turtles declined during the twentieth century, with some species becoming 'endangered' or 'critically endangered', according to the criteria used by the International Union for Conservation

Figure 28.1   Dugongs caught in Hervey Bay, Queensland, Australia, *c.*1937.

Source: Mrs Anita Jensen, courtesy of Professor Helene Marsh, used with permission.

of Nature (Dobbs, 2001). Marine turtles are particularly vulnerable to human impacts as a result of their life history, which involves a very high natural mortality of hatchlings and of small juvenile turtles and, at the same time, a very low mortality rate of large juveniles and adults, a limited number of nesting beaches, high fidelity to nesting sites and feeding grounds, limited interaction between genetic stocks, and long maturation periods. Thus, marine turtles are long-lived, slow-maturing animals whose populations are rapidly depleted following unsustainable exploitation. Marine turtles also undertake long migrations, and monitoring their abundance presents considerable challenges to researchers. Consequently, the true magnitude of anthropogenic impacts on marine turtles is not immediately apparent but instead emerges over decadal timescales. In some parts of their ranges, marine turtle populations have been severely depleted due to commercial over-exploitation, as occurred in the tortoise-shell industry and in the commercial turtle fisheries of the nineteenth and twentieth centuries, and also in the turtle farming enterprise that operated in the Torres

Strait between 1970 and 1979. At times, such forms of unsustainable exploitation occurred despite the fact that regulatory frameworks and scientific monitoring programmes were in place. Many instances have been documented in which the overexploitation of marine turtles led to the localised decline of turtle stocks and to the collapse of the industries they had previously supported. In addition to the depletion of marine turtles, other marine reptiles have suffered unsustainable exploitation; for instance, saltwater crocodile numbers have been dramatically reduced.

2.2 Impacts of fishing

One of the most widespread and extensive forms of unsustainable exploitation of coastal and marine environments and ecosystems is associated with fisheries. Marine fish have been exploited by humans since antiquity but, with time, that exploitation has intensified, expanded and affected an increasing number of species (Bolster, 2008; Engelman et al., 2008; Kowaleski, 2000;

Pauly et al., 1998, 2002, 2003). Globally, annual fish catches have increased more than five-fold since 1950; in some cases, the unsustainable nature of that trend has resulted in the precipitous decline and collapse of major fisheries (Bellwood et al., 2004; Gowers, 2008; Hughes et al., 2003, 2005, 2007; Jackson, 1997; Jackson et al., 2001; Koslow et al., 1988; Munro, 1998; Myers and Worm, 2003; Pandolfi et al., 2003, 2005). Today, marine fisheries may be categorised as large-scale (industrial) and small-scale (artisanal) fisheries; together, those fisheries yield catches in excess of 100 million tonnes, both for human consumption and for the production of agricultural and aquaculture feed (Pauly et al., 1998, 2002, 2003). Whilst excessive catches of targeted species are now occurring routinely, concerns have also been raised about the problem of by-catch: discarded fish that have not been targeted. By-catch occurs, in particular, as a result of the use of nonselective fishing methods, such as trawlers, long-lines and driftnets. A further problem, as Pauly et al. (2002) have acknowledged, is that the global fishing fleet is characterised by substantial over-capacity due to the persistence of government subsidies designed to support commercial fisheries despite their diminishing resource bases. As a result, many fish populations have exhibited severe declines worldwide (Jackson et al., 2001; Myers and Worm, 2003; Watson and Pauly, 2001). Consequently, the capacity of fisheries to contribute to food security, as well as the integrity of coastal and marine ecosystems and biodiversity, has become an urgent cause for concern (Folke et al., 1998; Pauly et al., 2002; Worm et al., 2006).

Pauly et al. (1998, 2002, 2003) have argued that, in response to the depletion of coastal fish stocks, commercial fisheries have expanded into deeper waters (further offshore), and into new fishing grounds, and that they have increasingly targeted smaller fish occupying the lower trophic levels in marine food webs. This phenomenon may be explained as follows: (1) fishing deeper and further offshore has been facilitated by the development of larger vessels and of equipment capable of exploiting deeper waters, notably the factory trawler, which has been responsible for immense ecological damage to biogenic structure and to sessile and benthic organisms; (2) substantial geographical expansion of commercial fisheries has occurred, particularly since the 1960s, with the result that fisheries have been exploited along the coasts of developing countries and southward into Antarctic waters; and (3) as numbers of large fish have declined, fishers have successively 'fished down' marine food webs by exploiting smaller fish that occupy lower trophic levels in those food webs (Kaczynski and Fluharty, 2002; Pauly et al., 1998, 2002, 2003; Watson and Pauly, 2001). This process of fishing down marine food webs has resulted in the progressive removal of large organisms, the decline of species diversity and the gradual replacement of more recent evolutionary groups (such as marine mammals and bony fish) with more primitive ones (such as invertebrates – especially jellyfish – and bacteria).

At the global scale, the activities of fisheries have resulted in a historical sequence that has progressed through three general phases, as described by Engelman et al. (2008) and Pauly et al. (1998, 2002, 2003). In the first, '*pristine*' phase, before human activities exerted a significant impact on marine ecosystems, abundances of fish (and of other marine animals) were greater than today by several orders of magnitude, with a much larger and more diverse group of top predators (Jackson et al., 2001). The second, 'exploited' phase – which represents the current situation in many parts of the world – is characterised by the decreasing biomass of large fish, and by declines in the sizes, diversity and trophic levels of fish catches (the fishing-down phenomenon), as well as by a variety of other, interrelated ecological effects (Pauly et al., 1998, 2002, 2003). The third, 'fully degraded' phase is expected to occur if present trends continue; however, in some places, such as Chesapeake Bay, many of the features of the fully degraded state are already apparent. In that state, overfishing has elimi-nated virtually all large animals together with benthic filter-feeders, resulting ultimately in the creation of 'dead zones' in which oxygen levels are low, nutrient levels are raised, chronic pollution occurs and algal blooms are regularly observed (Jackson et al., 2001; see also Chapter 23). Pauly et al. (1998, 2002, 2003) have argued that an increasing number of seasonal dead zones are found throughout the world.

Unless an effective global network of marine reserves is urgently created, the out-look for fisheries is bleak, with the likelihood that many pelagic and demersal species will be fished to extinction over a decadal times-cale. Consequently, many authors have called for the creation of no-take marine reserves, and for the wider adoption of responsible fishing practices (Amaral and Jablonski, 2005; Dulvy et al., 2003; Pauly et al., 1998, 2002, 2003; Sadovy, 2005). A wide range of economic, social and cultural issues sur-rounds the introduction and adoption of such practices (Cinner, 2007; Cinner and Aswani, 2007; Cinner et al., 2005, 2008).

Recent research has emphasised the point that unsustainable fishing practices have important, interrelated effects, not only on the viability of particular fisheries (such as the bluefin tuna fishery), but also on the quality of coastal and marine habitats as well as on ecosystem structure and function, especially in shallow marginal seas and in many coral reef ecosystems where the overexploitation of herbivorous fish has prompted the rapid development of algal blooms and has led to potentially irreversible changes ('phase shifts') in coral reef composition (Hughes, 1994; Hughes et al., 2003, 2005, 2007). Blaber et al. (2000) have documented the global effects of fisheries on estuarine and nearshore environments, arguing that fishing has exerted clear impacts on the structure and functioning of many ecosystems; those impacts may be categorised as effects on target organisms, nontarget organisms, nurs-ery functions, trophic effects, habitat change, reduced water quality, the human environ-ment, and the potential for local extinctions (Blaber et al., 2000). Jennings and Kaiser (1998) have argued that fishing has caused significant direct and indirect impacts on ben-thic habitats as well as on the structure, pro-ductivity and diversity of benthic communities; thus the introduction of fishing in previously unexploited ecosystems has led to dramatic changes in fish community structure, such as pronounced declines in the abundance of many forage fishes, with knock-on effects for the reproductive success and abundance of birds and marine mammals.

Jennings and Kaiser (1998) have argued that fishing has caused some coral reef eco-systems to undergo phase shifts to alternate stable states, due to the local removal of populations of predatory fishes – an effect that is compounded by the tight predator–prey coupling that exists between invertebrate feeding fishes and sea urchins in coral reef ecosystems. Jennings and Kaiser (1998) have also acknowledged that fisheries have caused increases in the populations of scavenging animals, particularly seabirds, especially due to the disposal of by-catch. Caddy (2000) has argued that, in semi-enclosed seas, the effects of fisheries are compounded by those of another anthropogenic impact – that of enhanced nutrient runoff – which may trigger ecological changes and the dominance of exotic species. Caddy (2000) has argued that, since eutrophic processes make demersal eco-systems particularly sensitive to the distur-bance of bottom habitats, the effects of fishing with bottom gear may be particularly acute under such conditions (see also Jones,1992).

Many studies have made suggestions and recommendations for the regulation and management of fisheries. Charles (1994) proposed the use of an integrated sustainabil-ity assessment framework for fisheries man-agement, involving the evaluation of ecological, socioeconomic, community and institutional sustainability; he argued that such a framework should incorporate several key principles and approaches: adaptive management principles, integrated strategies to address resource system complexity, the enhancement of local control and decision making, the establishment of appropriate property rights systems, and the combination of comprehensive planning with economic diversification. Murray et al. (1999) have argued that improved management approaches are required in order to reduce the rate of depletion of exploited marine populations, including the establishment and expansion of networks of no-take marine reserves. Such no-take reserve networks could potentially reduce economic and social risks by providing insurance for fishery managers against the overexploitation of individual fish populations; however, those reserve networks should be based on adaptive management principles so that their boundaries and regulations may be modified in the light of changing conservation priorities. Blaber et al. (2000) have argued that the effects of the economically important fisheries of many estuaries and coastal waters, combined with various other anthropogenic impacts in those areas, create multiple interactions and require an integrated approach to coastal zone management. Jacquet and Pauly (2007) have investigated the role of seafood awareness campaigns in promoting the preservation of fisheries, arguing that governments, rather than consumers, should take the initiative in protecting fisheries – for instance, by introducing legislation to restrict the amount of seafood used in animal feed.

2.3 Exploitation of other coastal, marine and island resources

Many other coastal and marine resources have also been exploited, including bêche-de-mer (sea cucumbers), pearl-shell, trochus and oysters, which were harvested by commercial fisheries during the nineteenth and twentieth centuries, as well as by artisanal fishers over a much longer period. Coral, shells, guano, rock phosphate and vegetation, have also been extracted from coastal and marine environments, often at the cost of profound environmental impacts – including,

in some cases, the transformation of entire islands or coral reefs (Bruckner, 2001; Charlier, 2002; Dulvy et al., 2003; Gomez, 1983; Guzman et al., 2003; Huber, 1994; Wong, 2003; Figure 28.2). Impacts have also been sustained by bird populations, both of seabirds and of island species. Some bird species were driven to extinction as a result of unsustainable exploitation, although that exploitation has often occurred in conjunction with various other impacts. Island bird populations were rapidly destroyed due to hunting and following the introduction of cats, dogs and rats by mariners and settlers, with the result that some species were driven to extinction, including the dodo (*Raphus cucullatus*), the endemic flightless

Figure 28.2   Corals taken from the Great Barrier Reef, Australia, c.1940.

Source: Cairns Historical Society, used with permission.

rail (*Rallus wakensis*) of Wake Island, and the flightless rail of Laysan (*Porzanula palmeri*). Populations of seabirds such as terns, noddies and boobies have been drastically reduced as a result of the clearance of island vegetation, and also due to the introduction of cats and rats. Some bird species, such as albatross, have been deliberately culled in order to reduce the hazard they present to military aircraft operations at island bases; the breeding activities of others, such as the little tern (*Sterna albifrons*), have been disrupted by the activities of fishers and bathers on nesting beaches. Bird populations have also been affected by coastal and marine pollution, as in the case of seabirds whose breeding has been disrupted by the effects of DDT on eggshell thickness.

3 POLLUTION OF COASTAL AND MARINE ENVIRONMENTS

One of the most significant human impacts on coastal and marine geo-ecosystems has occurred as a result of the multiple, and often interrelated, effects of various types of pollution. In coastal waters, dramatically enhanced inputs of sediment and nutrients (especially nitrogen and phosphorus) due to human activities have led, in many places, to cultural eutrophication, which in turn has caused the formation of algal blooms and the creation of hypoxic conditions and 'dead zones'. Those increased inputs of sediment and nutrients have occurred due to land clearance, agriculture, pastoralism, mining, soil erosion, coastal development, fertiliser runoff and discharges of sewage and industrial waste. Furthermore, both coastal waters and the open ocean – particularly those areas in the vicinity of shipping lanes – have been affected by oil pollution and by the accumulation of litter and nondegradable plastic waste; in addition, marine environments have been degraded due to the introduction of a variety of toxic substances such as lead, mercury, synthetic organic compounds and radionuclides.

3.1 Increased sediment input

In many parts of the world, coastal waters have experienced substantially increased sediment inputs from adjacent terrestrial sources. Increased sediment fluxes have occurred as a result of land clearance, agricultural and pastoral practices, mining operations, accelerated soil erosion and coastal development in coastal catchments. Examples from several parts of the world illustrate this trend. Geochemical investigations of long-lived corals (*Porites* spp.) in the Great Barrier Reef of Australia have indicated that, since European settlement commenced in that region, in around 1870, the sediment load discharged by the Burdekin River has increased by a factor of five to ten (Lough and Barnes, 1997; McCulloch et al., 2003). Estimates of the total sediment runoff to the Great Barrier Reef World Heritage Area from terrestrial catchments suggest that a three- to four-fold increase has occurred since around 1800 (Commonwealth of Australia Productivity Commission, 2003; Furnas, 2003; Williams, 2001; Williams et al., 2002). Sediment accumulation has also been documented in Tillamook Bay, Oregon; in Chesapeake Bay; and in the San Francisco Bay (Komar et al., 2004; Willard et al., 2003). Sediment flux to the Atlantic seaboard of the US has increased four- or five-fold over pre-European settlement values. In turn, those increased inputs have led to the sedimentation of coastal habitats, including bays, estuaries and coral reefs. Increased sediment fluxes to the coastal zone from adjacent terrestrial environments have exerted profound effects on coastal geomorphology and on some ecosystems, partly due to the direct effects of the sedimentation and siltation of nearshore habitats (such as coral reefs) and partly due to the indirect effects of reactive chemical pollutants that are transported into coastal waters with the sediment load. McLaughlin et al. (2003) have investigated anthropogenic influences on sediment flows along hydrological pathways at the global scale, arguing that terrigenous sediment input is likely to be a major factor influencing the distribution of coral reef ecosystems, although they also argued that the analysis should be carried out at higher resolution and with a greater number of specific variables.

3.2 Increased nutrient input and eutrophication

Increased sediment flux to the coastal zone has typically been accompanied by another deleterious effect: that of the increased input of nutrients as a result of the enhanced nutrient runoff to coastal waters from adjacent coastal land. In particular, during the Anthropocene, the concentrations of nitrogen and phosphorus in coastal waters have dramatically increased as a result of the runoff of agricultural fertilisers, together with other nitrogenous compounds including some of the organic waste products of coastal populations. Human activities have altered the nitrogen cycle – significantly accelerating the natural rate of nitrogen fixation – as a result of very large increases in the rates of production and application of synthetic nitrogenous fertilisers, together with pollution of coastal waters with sewage. Additionally, the anthropogenic burning of fossil fuels has increased the nitrogen content of oceanic surface waters. The most significant of these effects is that of fertiliser application, which continues to increase at the global scale (Mayewski et al., 1990; Peierls et al., 1991; Seitzinger et al., 2002).

Nutrient enhancement of coastal waters also occurs as a result of the increased flux of phosphorous to the coastal zone, due to the application of agricultural fertiliser (which increased significantly in the latter half of the twentieth century), to deforestation and to urban and industrial waste disposal. In particular, increases in nutrient concentrations have been observed in enclosed (or partially enclosed) seas such as the Adriatic, Baltic, Black, Mediterranean and North Seas, Chesapeake Bay and the Gulf of Mexico, where pollutants are relatively

concentrated and diluting capacity is limited (Mee, 1992; Rabalais, 2002); however, they have also been documented in continental shelf ecosystems such as the Great Barrier Reef (Furnas, 2003; Williams, 2001; Williams et al., 2002). Rabalais (2002) has highlighted the nonlinear, positive relationships existing among nitrogen and phosphorus flux, phytoplankton primary production and fisheries yield; when critical thresholds are crossed and the load of nutrients to estuarine, coastal and marine systems exceeds the capacity for assimilation of nutrients, then degradation of water quality occurs. As Rabalais (2002) has acknowledged, the effects include noxious and toxic algal blooms, increased turbidity with a subsequent loss of submerged aquatic vegetation, oxygen deficiency, disruption of ecosystem functioning, loss of habitat, loss of biodiversity, shifts in food webs and loss of harvestable fisheries.

Nitrogen and phosphorus enrichment due to fertiliser runoff and sewage and industrial waste discharge have significantly degraded nearshore waters worldwide. However, whilst scientific understanding of freshwater eutrophication and its effects on algal-related water quality is already detailed and is developing rapidly, important research gaps exist in the scientific understanding of the effects of eutrophication on estuarine and coastal marine ecosystems (Smith, 2003). The main effect of nutrient enhancement is to cause eutrophication which, in turn, encourages excessive algal growth and the formation of algal blooms, and which thereby creates hypoxic (or even anoxic) conditions that may have harmful or lethal ecological effects (Heathwaite et al., 1996; Rabalais, 2002).

Marine eutrophication has been observed in coastal areas since the nineteenth century, and the biomass of marine phytoplankton is known to respond sensitively and predictably to changes in the external supplies of nitrogen and phosphorus (Smith, 2003; Smith et al., 1999). This effect occurs because, in aquatic ecosystems, nitrogen and phosphate are generally the critical limiting nutrients that restrict primary productivity, and many coastal and marine organisms are adapted to live in oligotrophic (nutrient-poor) conditions; thus any increase of nitrogen and/or phosphorus concentrations tends to create conditions in which primary productivity can rapidly increase, causing the severe depletion of oxygen in coastal waters and leading to the formation of 'dead zones' that are characterised by episodes of faunal out-migration, mortality and population declines. Smith (2003) has highlighted the need for further research to develop models that quantitatively link ecosystem-level responses to nutrient loading in both freshwater and marine systems. In particular, there is an urgent need for research into the linkages between nutrient sources, transport and impacts.

3.3 Other forms of coastal and marine pollution

Other forms of coastal and marine pollution include significant contamination of coastal and marine waters by oil pollution, as a result of spillages from oil tankers (including spills following major shipping accidents, such as the Amoco Cadiz, Torrey Canyon, Exxon Valdez and Braer disasters), seepage or larger-scale industrial accidents associated with offshore oil installations (most notably the 2010 Deepwater Horizon disaster, which affected the coastal zone of the Gulf of Mexico) and the discharge of ballast water from tanker holds (although some natural seepage of oil also occurs). Some technological and operational progress has been made in reducing the last of those effects through the use of 'load on top' systems and of segregated/clean ballast systems. Oil pollution has major ecological effects on coastal and marine fauna and flora; grazing molluscs and seabirds, in particular, are vulnerable to the effects of oil spills. Ecological damage is caused not only by the oil slicks themselves but also by the chemical dispersants that are typically used to clean up the slick, as in the case of the 10,000-plus litres of detergent used to disperse the oil discharged from the Torrey Canyon, and by the physical cleaning processes that may involve the use of high water temperatures and pressures.

A range of other environmental impacts associated with oil and gas exploration and extraction have also been documented, such as the accumulation of drill cuttings around drilling platforms. The environmental impacts of the extraction of energy resources from marine environments are likely to increase with the development of methane hydrate exploitation. Other forms of deep-sea mining – such as manganese nodule mining and polymetallic sulphide mining – represent important potential threats to marine environments which could affect considerable areas of sea floor as these forms of resource exploitation are developed.

Other forms of pollution entering coastal and marine environments are diverse; they include the presence of discarded solid structures (such as wrecked vessels and oil- and gas-storage structures), munitions (both chemical and conventional), low- and intermediate-level radioactive wastes, sewage sludge and dredge spoils, halogenated hydrocarbons (such as polychlorinated biphenyls (PCBs), dichlorodiphenyltrichloroethane (DDT) and related pesticides), trace metals (such as lead, mercury and cadmium) and radionuclides. As Smith et al. (2008: 344) have acknowledged:

> The deep sea is often considered one of the most pristine environments on Earth, relatively unaffected by anthropogenic pollutants because of its distance from pollution sources, the slow rates of physical exchange between near-surface and deep-water masses, and its vast diluents capacity. In fact, the deep sea and deep-sea organisms are an important sink and often the ultimate repository for many of the most persistent and toxic of human pollutants.

Many of these pollutants become concentrated in animal tissues by the process of bioaccumulation. Thus, in large, long-lived, deepwater fish, concentrations of substances such as methyl mercury sometimes exceed maximum permissible levels for human consumption, and the bioaccumulation of DDT has been shown to be responsible for the breeding failure of some seabirds.

Whilst discharges of some well-known pollutants, such as DDT, are now controlled in some parts of the world, they continue elsewhere. The problem of the high toxicity of such pollutants also persists due to the long lifetimes of substances such as organochlorine compounds (Haynes et al., 2000). Furthermore, the processes of cycling of such pollutants in the deep ocean are poorly understood, with the implication that the full effects of pollutants on deep-sea biota cannot yet be evaluated or effectively managed (Smith et al., 2008). Sonak et al. (2009) have acknowledged the effects of organotin-based antifouling paints, including tributyltin (TBT), which has been described as the most toxic substance ever introduced into the marine environment.

In addition to the problem of highly toxic marine pollutants, human activity has been responsible for the accelerating accumulation of decay-resistant synthetic polymer (plastic) debris, including litter, fishing nets and the residues of industrial processes. Derraik (2000) has provided a review of the impacts of plastic debris on the marine environment, indicating that many marine species are harmed or killed by plastic debris, especially by entanglement in and ingestion of plastic litter, although other effects of plastic debris include its role in vectoring invading species and the absorption into animal tissues of polychlorinated biphenyls from ingested plastics. Cadée (2002) has investigated the impact of floating plastic debris on seabirds, especially the harm caused to marine animals by ingestion of plastics, and Mallory (2008) has documented the effects of marine plastic debris on northern fulmars from the Canadian High Arctic. Moore (2008) has argued that plastic debris in the marine environment represents a rapidly increasing long-term threat, with plastics now forming one of the most widespread, persistent pollutants in oceanic waters and on beaches worldwide. The harmful impacts of plastic on marine biota extend to at least 267 species of marine organisms, including albatross, fulmars, shearwaters, petrels, marine turtles and cetaceans, with millions of animals affected each year (Moore, 2008). Corcoran et al. (2009) have investigated the effects of oceanic plastic debris following its deposition on beaches, where it is particularly susceptible to both chemical and mechanical breakdown.

4 EFFECTS OF CLIMATE CHANGE ON COASTS AND SEAS

Historical and present-day human impacts on coastal and marine environments and ecosystems are increasingly being reframed in the context of anthropogenic climate change. The increase in the global mean surface temperature, the thermal expansion of the global ocean and the resulting changes in global mean and local sea levels are projected to have profound consequences for many aspects of coastal and marine geo-ecosystems. Those consequences include alterations in the position of the global coastline; in mean SST and SSS; in the pattern and strength of ocean currents; in ocean acidification; and in ecological interactions and species distributions. At the local scale, coasts and estuaries will probably be affected by climate change in highly diverse ways. However, estuaries, coastal marshes and coral reefs are amongst the habitats most seriously threatened by climate change since they are some of the environmental systems that are least able to keep pace with their projected rates of inundation and of warming.

Considerable efforts are being made to determine the ways in which coastal and marine environments and ecosystems are likely to change in response to climate change. In the case of coral reefs, for instance, very large declines in coral species abundance are projected to occur as a result of the increased occurrence of coral bleaching events and diseases (Carpenter et al., 2008; Hoegh-Guldberg, 1999). As Oldfield (2005) acknowledged, the implications of such changes for human vulnerability – and for the sustainability of human economies and societies – is a complex function of resilience, vulnerability, risk, sustainability and adaptive capacity, at all spatial and organisational scales. Given the fact that large human populations inhabit coastal areas, and that many human economies and societies are highly dependent on coastal and marine resources, it is clear that the need to devise effective, co-ordinated strategies to adapt to, and to mitigate, climate change is already urgent (Carpenter et al., 2008).

Scavia et al. (2002) have summarised the potential impacts of climate change on US coastal and marine geo-ecosystems, including effects on shorelines, estuaries, coastal wetlands, coral reefs and ocean margin ecosystems. Those authors argued that, over the decadal timescale, accelerating sea-level rise and the greater intensity and frequency of coastal hurricanes and storms will present greater threats to shorelines, wetlands and coastal developments. Scavia et al. (2002) found that estuarine productivity is likely to change due to variations in the magnitude and timing of freshwater, nutrient and sediment fluxes. In addition, increasing SST and changes in freshwater delivery are expected to cause chances in estuarine stratification, residence time and eutrophication. Increasing SST is also expected to trigger coral bleaching events; and increasing ocean acidification could reduce coral calcification, with the result that coral reef ecosystems will be more vulnerable to other environmental changes. Furthermore, increasing SST is projected to cause poleward shifts in the ranges of many coastal and marine organisms, together with secondary ecological effects.

Scavia et al. (2002) have emphasised that, whilst the potential impacts of climate change will vary between environments and ecosystems, those impacts will nevertheless be superimposed upon – and may intensify – other environmental impacts, especially pollution, resource exploitation, habitat destruction, the spread of invasive species and the effects of extreme natural events. The cumulative effects of those environmental impacts may be severe. Similar conclusions were reached in a study of the effects of climate change on the Mediterranean Sea by Lejeusne et al. (2010), who acknowledged the fact that little is known about the extent of climate change impacts on marine ecosystems. Lejeusne et al. (2010) documented a variety of environmental changes and climate change impacts on the Mediterranean Sea: steadily increasing SST, more frequent extreme climatic events and related disease outbreaks, shifts in faunal populations and distributions, and the spread of invasive species. Lejeusne et al. (2010) argued that the Mediterranean Sea represents a mesocosm of the global ocean. Thus, they highlighted the research opportunity presented by that ecosystem to investigate the various sources of disturbances that interact synergistically in the marine realm. Those authors acknowledged that critical questions are faced by researchers: in particular, how resilient are marine ecosystems, and how will their current functioning be modified under various climate change scenarios?

Increasing SST is a particular concern in relation to coral reef ecosystems, since it is likely to mean that the thermal tolerances of reef-building corals will be exceeded on an annual basis over the decadal timescale. In a study of mass coral bleaching, Hoegh-Guldberg (1999) has argued that the capacity of corals to adjust to rising SST has already been exceeded; consequently, at the global scale, their adaptation will be too slow to avert a decline in the health of coral reef ecosystems. Hoegh-Guldberg (1999) concluded that the projected rate of change of SST will present a major threat to tropical marine ecosystems, and that unrestrained warming will inevitably mean the destruction and degradation of coral reefs at the global scale. Other studies of the effects of increasing SST on coral reef ecosystems have indicated that warmer oceanic waters will preclude the formation of dimethyl sulphide (DMS) by algae living in coral tissue, with the result that cloud formation over coral reefs may be reduced. Consequently, coral reefs would be exposed to greater localised heating – an effect that would compound the original perturbation as well as having secondary effects on adjacent coastal ecosystems, such as the rainforests of coastal north Queensland (Jones and Ristovski, 2010; Jones et al., 2007).

A major topic of concern in relation to the effects of climate change on marine geo-ecosystems is that of ocean acidification, which is occurring as the oceans absorb carbon dioxide – released by human activities – which in turn forms carbonic acid. It is now acknowledged that the acidity of the global ocean is increasing more rapidly than at any time during the last 55 million years – with potentially severe consequences for many marine species, habitats and ecosystems, especially given that the rate of change of ocean acidity may be too rapid for many species to adapt to their new conditions (Caldeira and Wickett, 2003; Orr et al., 2005).

Ocean acidification can affect the migration and communication patterns of some marine animals. In particular, acidification allows the transmission of sound over greater distances – including the noise from drilling, sonar and vessel engines – with adverse effects on whale and dolphin species. Whales are also likely to be affected by ocean acidification through the depletion of their summer feeding grounds in the North Atlantic, North Pacific and Arctic waters – areas in which acidification is likely to be most pronounced. By causing a significant decrease in concentrations of aragonite (a form of calcium carbonate which is particularly susceptible to carbonic acid), ocean acidification affects the skeletons of many marine organisms and the carbonate structures of coral reefs. Thus it may affect the abundance of algae and plankton, with secondary effects on marine food chains. Under more acidic conditions, brittle stars (*Ophiothrix fragilis*) are expected to produce fewer larvae as they will be forced to expend more energy maintaining their skeletons in more acid waters, with implications for the herring populations which prey on them. Some algal species, such as *Calcidiscus leptoporus* (upon which salmon feed) may be unable to survive in more acidic conditions, with potential consequences for their predators. Juvenile clownfish (*Amphiprion percula*) have displayed direct changes in behaviour in response to increasing acidity, and are expected to become unable to detect their anemone habitat due to the effects of acidification in disrupting the olfactory cues received by the fish (Munday et al., 2009a, 2009b; Wilson et al., 2010).

Particular concerns about ocean acidification focus on the effects of increased acidity on coral reef ecosystems, with recent research indicating that, by 2050, 98 per cent of the world's reef habitats are likely to have become too acidic for corals to grow. Increasing acidity causes damage to corals by causing their aragonite skeletons to become brittle and unable to withstand the damage caused by bioerosion and other pressures. Such a loss of corals and their associated species would have a major effect on food security for human populations, as well as other effects in terms of reducing tourism revenue and leaving coastlines unprotected against storm surges. Levels of aragonite in the oceans have been modelled for the period from preindustrial times until the end of the twenty-first century, under three emission scenarios, with the conclusion that, even if atmospheric carbon dioxide concentrations are stabilised at 550 ppm – a scenario that would require concerted international action – no existing coral reef will survive in such conditions (Anthony et al., 2008; Carpenter et al., 2008; De'ath et al., 2009; Hoegh-Guldberg et al., 2007). Ocean acidification – and the effects of climate change in general – are therefore regarded as a conservation challenge of unprecedented scale, one that highlights the urgent need for effective marine management and protection.

The impacts of climate change on coastal and marine geo-ecosystems will not occur in isolation but will involve multiple, interrelated and complex effects. A number of studies have investigated responses to multiple impacts. Jackson et al. (1997) have argued that Caribbean coastal ecosystems have long been severely degraded due to a combination of effects, although they argue that overfishing has been a prime impact on those ecosystems. Jackson et al. (1997) pointed to the drastic changes that accompany the loss of an entire trophic level, and the synergistic interactions of those changes with other environmental impacts such as pollution, degradation of water quality and anthropogenic climate change.

Pandolfi et al. (2003) have demonstrated that the degradation of coral reef ecosystems commenced centuries ago, although those authors acknowledged that no global-scale account of the magnitude of change yet existed. Consequently, Pandolfi et al. (2003) analysed records of the status of a wide range of species from 14 regions; that analysis suggested that large animals declined before small animals and architectural species, and that Atlantic coral reefs declined before those in the Red Sea and in Australian waters. However, the authors emphasised that the trajectories of coral reef decline have been markedly similar worldwide (see also Pandolfi et al., 2005). Pandolfi et al. (2003) argued that, globally, coral reefs were significantly degraded due to human activities long before recent threats – such as coral bleaching and outbreaks of disease – had become prominent concerns. Those authors argued that coral reefs require immediate protection from human exploitation over large areas in order to ensure their survival.

In another study, Hughes et al. (2003) investigated the relations among climate change, human impacts and the resilience of coral reefs, arguing that the projected increases in carbon dioxide emissions and temperature over a 50-year period will exceed the conditions in which coral reefs have flourished over the last 500,000 years. They argued that, rather than disappearing entirely,

coral reefs will change, since some species have much greater tolerance to increasing SST and to coral bleaching than others. Hence, Hughes et al. (2003) have called for the international integration of management strategies in order to promote coral reef resilience, in addition to urgent action to mitigate the effects of climate change. That study was followed by another in which the interactions of phase shifts, herbivory and the resilience of coral reefs to climate change were investigated (Hughes et al., 2007). These authors pointed to the important role played by large herbivorous fish in promoting the fecundity, recruitment and survival of corals (through keeping the abundance of macroalgae in check), with the implication that the effective management of fish stocks is a vital component in preventing phase shifts in, and improving the resilience of, coral reef ecosystems. Such an approach could improve the ability of coral reefs to withstand the effects of coral bleaching events which are impractical to manage directly. The relationship between fish abundance and coral reef health has also been studied by Feary et al. (2007), who showed that the global degradation of coral reefs is profoundly affecting the structure and species diversity of associated fish assemblages, not least because coral mortality alters the process of replenishment of coral reef fish communities.

Another study, by Carpenter et al. (2008), focused on the interrelated effects of climate change and local impacts on coral reefs. They demonstrated that 32.8 per cent of reef-building corals are at an increased risk of extinction due to those combined effects. Using the IUCN Red List criteria, Carpenter et al. (2008) assessed the conservation status of 845 zooxanthellate reef-building coral species, and showed that declines in coral abundance are associated with coral bleaching and diseases due to increased SST, with the extinction risk further increased by localised anthropogenic disturbances. They argued, moreover, that the proportion of coral species at risk of extinction has increased substantially recently and now exceeds that of most terrestrial groups. Furthermore, they stated that the Caribbean has the largest proportion of corals in high extinction risk categories, whilst coral reefs in the western Pacific have the highest proportion of species in all categories of elevated extinction risk. In common with other researchers, Carpenter et al. (2008) have therefore emphasised the widespread threats to coral reefs and the urgent need to implement effective conservation measures. Similar conclusions have been reached by Alvarez-Filip et al. (2009), who demonstrated that the combination of white-band disease and increasing SST has degraded Caribbean coral reefs over a 40-year period, with the result that those reefs have lost the branching elkhorn and staghorn coral species that provided habitats for reef fish and other biota. Their study, which was based on data from 500 surveys of 200 Caribbean coral reefs, identified two phases of degradation which led to the dominance of flat coral species in the region. Alvarez-Filip et al. (2009) acknowledged the significant difficulties involved in enacting effective coral reef management in the context of multiple anthropogenic pressures (see also Mora, 2008).

5 CHANGES IN COASTLINES

Changes in coastlines also occur as a result of the direct and indirect effects of land reclamation, quarrying, coastal engineering schemes, changing sediment fluxes, coastal erosion, coastal development, subsidence (e.g., following the removal of groundwater or oil), the clearance of natural vegetation (such as mangroves), the introduction of exotic species, aquaculture and coastal protection and management schemes. These human activities have become an increasingly dominant cause of changes in the position and morphology of coastlines during the Anthropocene, although their influence is often entangled with the effects of concurrent natural processes.

Land reclamation results in dramatic changes to the position and extent of coastlines, as exemplified by the changing coastlines of Hong Kong and Singapore. The quarrying of rock and sediment results in changes to sediment budgets in the coastal zone, as has occurred at Hallsands, in the UK, where coastal erosion resulted from the dredging of offshore sediment. Sediment budgets are also modified by the dredging of ports and shipping lanes, as well as by the dumping of the dredge spoil offshore. The construction of dams has had a major impact on sediment supply to the coastal zone, in some places, with the result that river flows have been disrupted, coastal and marine habitats have received reduced sediment loads, and ecological impacts have occurred. Vegetation change in the coastal zone has involved both vegetation removal (as where mangrove species have been removed, mangrove areas have been dramatically reduced, and mangrove swamps have been converted to agricultural and aquaculture uses) as well as the introduction of new species (as where the marsh grass species, *Spartina anglica*, has spread in salt marshes in the UK).

Extensive coastal development – characterised by a rapid increase in the rate of construction of infrastructure – has also modified many coastal areas, worldwide, particularly where coastal cities, together with their associated port and industrial facilities, have expanded. In many cases, the effects of infrastructure development have been spatially concentrated in estuaries, particularly during the nineteenth and twentieth centuries, resulting in substantial modification of estuarine environments and ecosystems. Those effects include the broad range of impacts that have been sustained by coastal areas more generally: changes in sediment and nutrient fluxes; other forms of pollution; the introduction of exotic species; disruption of the patterns of fish migration; changes in salinity; changes in food webs; land reclamation; changes in shoreline position and extent; the construction of infrastructure (such as causeways and bridges); habitat fragmentation and loss; thermal pollution; and also some positive effects (such as the rehabilitation of damaged habitats and the creation of recreational beaches using synthetic materials).

6 CONCLUSION

This chapter has provided an overview of the main human impacts on coastal and marine geo-ecosystems. It illustrates that a broad range of such impacts have occurred and that in many cases those impacts have occurred for longer, in more places and more intensively than has previously been acknowledged (Halpern et al., 2008). Yet this account has highlighted some dominant human impacts: in particular, the effects of the unsustainable exploitation of coastal and marine resources, of coastal and marine pollution, of climate change, and of changes in coastlines. Climate change now presents a significant threat to coastal and marine geo-ecosystems: changes in global mean SST and SSS, global mean sea level (and local sea levels), ocean currents, oceanic acidity and ecology are projected to have profound implications for the integrity – and even the survival – of many coastal and marine ecosystems and biota, hence the urgent need to promote improved forms of environmental protection and management.

At present, coastal and marine management initiatives are sharply focused on the management of coastal erosion. This issue significantly affects large stretches of the global coastline and presents a direct hazard to human populations; it also generates enormous costs due to the construction and maintenance of coastal defences (Wong, 2003). The effects of sea-level rise are very likely to increase the need for effective coastal defences. Thus, coastal protection and management will become an increasingly urgent challenge for many societies. However, coastal and marine protection and management must increasingly adopt an integrated

approach to a broader range of coastal (and other environmental) issues, including the unsustainable exploitation of resources, pollution, coastal erosion – and, increasingly, climate change. Such an approach has already emerged in the form of adaptive management (Folke et al., 2002; Hughes et al., 2007; Murray et al., 2009; Olsson et al., 2004).

Overall, coastal and estuarine management now focuses on the need to sustain the progress that has been made to improve water quality and to build up resource bases that have been depleted or degraded by pollution, unsustainable exploitation and infrastructure development. Such progress needs to be extended more widely to the marine realm: in particular, through the creation of a global network of marine protected areas that is adequate to the challenge of marine conservation. Yet the task of promoting coastal and marine conservation is a formidable one. Recent studies have demonstrated that, in coastal and marine systems, the 'cycle' of degradation and recovery is not necessarily straightforward (Hughes et al., 2003, 2005, 2007). Instead, coastal and marine environmental changes may involve high degrees of hysteresis, where recovery may occur over much longer periods than degradation and where a return to the original ('predisturbance') state may be impossible.

In this respect, Oldfield (2005: 174–5) has identified various critical questions:

- Which ecosystems have been irreparably damaged, when and how?
- What were the degradation trajectories?
- How have human impacts interacted with climate variability and ecosystem processes to generation degradation trajectories of different kinds?
- Where 'restoration' or 'conservation' are the main goals, what state is to be restored, or conserved, and what are the key processes that should ensure this in the long term?

As yet, most of those questions lack convincing answers, partly because of the discontinuities between the coverage of 'contemporary' and 'historical' studies, and partly because of an incomplete understanding of the interactions between environmental and societal processes. Such observations suggest the need for research that bridges the continuing divide between the scientific and environmental history literatures, and that is based on more integrated, multidisciplinary approaches. In the context of coastal and marine geo-ecosystems, the nature and extent of human impacts, combined with the increasing problem of climate change, means that the need for further research, linked to greater protection and more effective management, has already become urgent.

REFERENCES

Alvarez-Filip L., Dulvy N. K., Gill J. A., Cote I. M. and Watkinson A. R. 2009. Flattening of Caribbean coral reefs: region-wide declines in architectural complexity. *Proceedings of the Royal Society of London B* 276: 3019–3025.

Amaral A. C. Z. and Jablonski S. 2005. Conservation of marine and coastal biodiversity in Brazil. *Conservation Biology* 19: 625–631.

Anthony K. R. N., Kline D. I., Diaz-Pulido G., Dove S. and Hoegh-Guldberg O. 2008. Ocean acidification causes bleaching and productivity loss in coral reef builders. *PNAS* 105: 17442–17446.

Baker C. S. and Clapham P. J. 2004. Modeling the past and future of whales and whaling. *Trends in Ecology and Evolution* 19(7): 365–371.

Baker C. S. and Palumbi S. R. 1994. Which whales are hunted? A molecular genetic approach to monitoring whaling. *Science* 265(5178): 1538–1539.

Bellwood D. R., Hughes T. P., Folke C. and Nyström M. 2004. Confronting the coral reef crisis. *Nature* 429: 827–833.

Blaber S. J. M., Cyrus D. P., Albaret J.-J., Ching C. V., Day J. W., Elliott M. et al. 2000. Effects of fishing on the structure and functioning of estuarine and nearshore ecosystems. *ICES Journal of Marine Science* 57: 590–602.

Bolster J. 2008. Putting the ocean in Atlantic history: maritime communities and marine ecology in the northwest Atlantic, 1500–1800. *American Historical Review* 113(1): 19–47.

Bruckner A. W. 2001. Tracking the trade in ornamental coral reef organisms: the importance of CITES and its limitations. *Aquarium Sciences and Conservation* 3: 79–94.

Caddy J. F. 2000. Marine catchment basin effects versus impacts of fisheries on semi-enclosed seas. *ICES Journal of Marine Science* 57: 628–640.

Cadée G. C. 2002. Seabirds and floating plastic debris. *Marine Pollution Bulletin* 44(11): 1294–1295.

Caldeira K. and Wickett M. E. 2003. Anthropogenic carbon and ocean pH. *Nature* 425: 365.

Carpenter K. E., Abrar M., Aeby G., Aronson R. B., Banks S., Bruckner A. et al. 2008. One-third of reef-building corals face elevated extinction risk from climate change and local impacts. *Science* 321: 560–563.

Charles A. T. 1994. Towards sustainability: the fishery experience. *Ecological Economics* 11(3): 201–211.

Charlier R. H. 2002. Impact on the coastal environment of marine aggregates mining, *International Journal of Environmental Studies* 59(3): 297–322.

Chilvers B. L., Lawler I. R., MacKnight F., Marsh H., Noad M. and Paterson R. 2005. Moreton Bay, Queensland, Australia: an example of the co-existence of significant marine mammal populations and large-scale coastal development. *Biological Conservation* 122: 559–571.

Cinner J. E. 2007. Designing marine reserves to reflect local socioeconomic conditions: lessons from long-enduring customary management systems. *Coral Reefs* 26: 1035–1045.

Cinner J. E. and Aswani S. 2007. Integrating customary management into marine conservation. *Biological Conservation* 140: 201–216.

Cinner J. E., Daw T. and McClanahan T. R. 2008. Socioeconomic factors that affect artisanal fishers readiness to exit a declining fishery. *Conservation Biology* 23(1): 124–130.

Cinner J. E., Marnane M. J., McClanahan T. R., Clark T. H. and Ben J. 2005. Trade, tenure, and tradition: influence of sociocultural factors on resource use in Melanesia. *Conservation Biology* 19(5): 1469–1477.

Commonwealth of Australia Productivity Commission. 2003. *Industries, Land Use and Water Quality in the Great Barrier Reef Catchment*. Research Report. Canberra: Commonwealth of Australia Productivity Commission.

Corcoran P. L., Biesinger M. C. and Grifi M. 2009. Plastics and beaches: a degrading relationship. *Marine Pollution Bulletin* 58(1): 80–84.

Crutzen P. J. and Stoermer E. 2001. The 'Anthropocene'. *International Geosphere Biosphere Programme Global Change Newsletter* 14: 12–13.

De'ath G., Lough J. M. and Fabricius K. E. 2009. Declining coral calcification on the Great Barrier Reef. *Science* 323: 116–119.

Derraik J. G. B. 2002. The pollution of the marine environment by plastic debris: a review. *Marine Pollution Bulletin* 44(9): 842–852.

de Thoisy B., Spiegelberger T., Rousseau S., Talvy G., Vogel I. and Vié J.-C. 2003. Distribution, habitat, and conservation status of the West Indian manatee *Trichechus manatus* in French Guiana. *Oryx* 37(4): 431–436.

Dobbs K. 2001. *Marine Turtles in the GBRWHA: A Compendium of Information and Basis for the Development of Policies and Strategies for the Conservation of Marine Turtles.* Townsville: GBRMPA.

Dulvy N. K., Sadovy Y. and Reynolds J. D. 2003. Extinction vulnerability in marine popultions. *Fish and Fisheries* 4: 25–64.

Engelman R., Pauly D., Zeller D., Prinn R. G., Pinnegar J. K. and Polunin N. V. C. 2008. Introduction: climate, people, fisheries and aquatic ecosystems, in Polunin N. V. C. (ed.) *Aquatic Ecosystems: Trends and Global Prospects.* Cambridge: Cambridge University Press, pp. 1–18.

Feary D. A., Almany G. R., Jones G. P. and McCormick M. I. 2007. Coral degradation and the structure of tropical reef fish communities. *Marine Ecology Progress Series* 333: 243–248.

Folke C., Carpenter S., Elmqvist T., Gunderson L., Holling C. S., Walker B et al. 2002. *Resilence and Sustainable Development: Building Adaptive Capacity in a World of Transformations.* Paris: International Council for Science.

Folke C., Kautsky N., Berg H., Jansson, Å. and Troell M. 1998. The ecological footprint concept for sustainable seafood production: a review, *Ecological Applications* 8(Suppl. 1): S63–S71.

Furnas M. 2003. *Catchments and Corals: Terrestrial Runoff to the Great Barrier Reef.* Townsville: AIMS.

Gomez E. D. 1983. Perspectives on coral reef research and management in the Pacific. *Ocean Management* 8: 281–295.

Gowers R. J. 2008. Selling the 'untold wealth' in the seas: a social and cultural history of the south-east Australian shelf trawling industry, 1915–1961. *Environment and History* 14(2): 265–287.

Guzmán H. M., Guevara C. and Castillo A. 2003. Natural disturbances and mining of Panamanian coral reefs by indigenous people. *Conservation Biology* 17(5): 1396–1401.

Halpern B. S., Walbridge S., Selkoe K. A., Kappel C. V., Micheli F., D'Agrosa C. et al. 2008. A global map of human impact on marine ecosystems. *Science* 319(5865): 948–952.

Haynes D., Müller J. and Carter S. 2000. Sources, fates and consequences of pollutants in the Great Barrier Reef. *Marine Pollution Bulletin* 41: 279–287.

Heathwaite A. L., Johnes P. J. and Peters N. E. 1996. Trends in nutrients. *Hydrological Processes* 10: 263–293.

Hoegh-Guldberg O. 1999. Climate change, coral bleaching, and the future of the world's coral reefs. *Marine and Freshwater Research* 50: 839–866.

Hoegh-Guldberg O., Mumby P. J., Hooten A. J., Steneck R. S., Greenfield P., Gomez E. et al. 2007. Coral reefs under rapid climate change and ocean acidification, *Science* 318(5857): 1737–1742.

Holm P., Smith T. D. and Starkey D. J. (eds) 2001. *The Exploited Seas: New Directions for Marine Environmental History.* Research in Maritime History, 21. St Johns: International Maritime Economic History Association.

Huber M. E. 1994. An assessment of the status of the coral reefs of Papua New Guinea. *Marine Pollution Bulletin* 29: 69–73.

Hughes T. P. 1994. Catastrophes, phase shifts and large-scale degradation of a Caribbean reef. *Science* 265: 1547–1551.

Hughes T. P., Baird A. H., Bellwood D. R., Card M., Connolly S. R., Folke C. et al. 2003. Climate change, human impacts, and the resilience of coral reefs. *Science* 301(5635): 929–933.

Hughes T. P., Bellwood D. R., Folke C., Steneck R. and Wilson J. 2005. New paradigms for supporting the resilience of marine ecosystems. *Trends in Ecology and Evolution* 20(7): 380–386.

Hughes T. P., Rodrigues M. J., Bellwood D. R., Ceccarelli D., Hoegh-Guldberg O., McCook L. et al. 2007. Phase shifts, herbivory, and the resilience of coral reefs to climate change. *Current Biology* 17: 360–365.

Jackson J., Patenaude N., Carroll E. and Baker C. S. 2008. How few whales were there after whaling? Inference from contemporary mtDNA diversity. *Molecular Ecology* 17: 236–251.

Jackson J. B. C. 1997. Reefs since Columbus. *Coral Reefs* 16(S): S23–S32.

Jackson J. B. C., Kirby M. X., Berger W. H., Bjorndal K. A., Botsford L. W., Bourque B. J. et al. 2001. Historical overfishing and the recent collapse of coastal ecosystems. *Science* 293: 629–638.

Jacquet J. L. and Pauly D. 2007. The rise of seafood awareness campaigns in an era of collapsing fisheries. *Marine Policy* 31: 308–313.

Jennings S. and Kaiser M. J. 1998. The effects of fishing on marine ecosystems. *Advances in Marine Biology* 34: 201–212.

Jiménez I. 2002. Heavy poaching in prime habitat: the conservation status of the West Indian manatee in Nicaragua. *Oryx* 36(3): 272–278.

Jones D. 2002. The whalers of Tangalooma 1952–1962, in M. Johnson (ed.) *Brisbane: Moreton Bay Matters.* Brisbane History Group Papers 19. Brisbane: Brisbane History Group, pp. 87–94.

Jones G. B., Curran M., Broadbent A. D., King S., Fischer E. and Jones R. J. 2007. Factors affecting the cycling of dimethylsulfide and dimethylsulfoniopropionate in coral reef waters of the Great Barrier Reef. *Environmental Chemistry* 4(5): 310–322.

Jones G. B. and Ristovski Z. 2010. Reef emissions affect climate. *Australasian Science* 31(5): 26–28.

Jones J. B. 1992. Environmental impact of trawling on the seabed: a review. *New Zealand Journal of Marine and Freshwater Research* 26: 59–67.

Kaczynski V. M. and Fluharty D. L. 2002. European policies in West Africa: who benefits from fisheries agreements? *Marine Policy* 26(2): 75–93.

Komar P. D., McManus J. and Styllas M. 2004. Sediment accumulation in Tillamook Bay, Oregon: natural processes versus human impacts. *Journal of Geology* 112(4): 455–469.

Koslow J. A., Hanley F. and Wicklund R. 1988. Effects of fishing on reef fish communities at Pedro Bank and Port Royal Clays, Jamaica. *Marine Ecology Progress Series* 43: 201–212.

Kowaleski M. 2000. The expansion of the south-western fisheries in late medieval England. *Economic History Review* 53(3): 429–454.

Lejeusne C., Chevaldonné P., Pergent-Martini C., Boudouresque C. F. and Pérez T. 2010. Climate change effects on a miniature ocean: the highly diverse, highly impacted Mediterranean Sea. *Trends in Ecology and Evolution* 25(4): 250–260.

Lough J. M. and Barnes D. J. 1997. Centuries-long records of coral growth on the Great Barrier Reef, in D. R. Wachenfeld et al. (eds) *State of the Great Barrier Reef World Heritage Area Workshop: Proceedings of a Technical Workshop held in Townsville, Queensland, Australia, 27–29 November 1995.* Townsville: GBRMPA., pp. 149–157.

Mallory M. L. 2008. Marine plastic debris in northern fulmars from the Canadian high Arctic. *Marine Pollution Bulletin* 56(8): 1501–1504.

Marsh H., Death G., Gribble N. and Lane B. 2005. Historical marine population estimates: triggers or targets for conservation? The dugong case study. *Ecological Applications* 15: 481–492.

Marsh H., Harris A. N. M. and Lawler I. R. 1997. The sustainability of the indigenous dugong fishery in Torres Strait, Australia/Papua New Guinea. *Conservation Biology* 11(6): 1375–1386.

Marsh H., Lawler I., Kwan D., Delean S., Pollock K. and Alldredge M. 2004. Aerial surveys and the potential biological removal technique indicate that the Torres Strait dugong fishery is unsustainable. *Animal Conservation* 7: 435–443.

Marsh H. and Lefebvre L. W. 1994. Sirenian status and conservation efforts. *Aquatic Mammals* 20(3): 155–170.

Mayewski P. A., Lyons W.B., Spencer M. J., Twickler M. S., Buck C. F. and Whitlow, S.1990. An ice-core record of atmospheric response to anthropogenic sulphate and nitrate. *Nature* 346: 554–556.

McCulloch M., Fallon S., Wyndham T., Hendy E., Lough J. and Barnes D. 2003. Coral record of increased sediment flux to the inner Great Barrier Reef since European settlement. *Nature* 421: 727–730.

McLaughlin C. J., Smith C. A., Buddemeier R. W., Bartley J. D. and Maxwell B. A. 2003. The supply of flux of sediment along hydrological pathways: anthropogenic influences at the global scale. *Global and Planetary Change* 39(1–2): 191–199.

Mee L. D. 1992. The Black Sea in crisis: a need for concerted international action. *Ambio* 21: 278–286.

Moore C. J. 2008. Synthetic polymers in the marine environment: A rapidly increasing, long-term threat, *Environmental Research* 108(2): 131–139.

Mora C. 2008. A clear human footprint in the coral reefs of the Caribbean. *Proceedings of the Royal Society of London B* 275: 767–773.

Munday P. L., Dixson D. L., Donelson J. M., Jones G. P., Pratchett M. S., Devitsina D. V. and Døving K. B. 2009a. Ocean acidification impairs olfactory discrimination and homing ability of a marine fish. *PNAS* 106(6): 1848–1852.

Munday P. L., Donelson J. M., Dixson D. L. and Endo G. G. K. 2009b. Effects of ocean acidification on the early life history of a tropical marine fish. *Proceedings of the Royal Society of London B* 276: 3275–3283.

Munro J. L. 1998. Comment on the paper by J. B. C. Jackson: Reefs since Columbus (*Coral Reefs,* Supplement to 16: S23–S32). *Coral Reefs* 17: 191–192.

Murray S. N., Ambrose R. F., Bohnsack J. A., Botsford L. W., Carr M. H., Davis G. E. et al. 1999. No-take reserve networks: sustaining fishery populations and marine ecosystems. *Fisheries* 24: 11–25.

Myers R. A. and Worm B. 2003. Rapid worldwide depletion of predatory fish communities. *Nature* 423: 280–283.

Oldfield F. 2005. *Environmental Change: Key Issues and Alternative Approaches.* Cambridge: Cambridge University Press.

Olsson P., Folke C. and Berkers F. 2004. Adaptive comanagement for building resilience in social-ecological systems. *Environmental Management* 34: 75–90.

Orr J. C., Fabry V. J., Aumont O., Bopp L., Doney S. C., Feely R. A. et al. 2005. Anthropogenic ocean acidification over the twenty-first century and its impact on calcifying organisms. *Nature* 437: 681–686.

Palumbi S. R. and Cipriano F. 1998. Species identification using genetic tools: the value of nuclear and mitochondrial gene sequences in whale conservation. *Journal of Heredity* 89(5): 459–464.

Pandolfi J. M., Bradbury R., Sala E., Hughes T. P., Bjorndal K. A., Cooke R. G. et al. 2003. Global trajectories of the long-term decline of coral reef ecosystems. *Science* 301: 957.

Pandolfi J. M., Jackson J. B. C., Baron N., Bradbury R., Guzman H. M., Hughes T. P. et al. 2005. Are U.S. coral reefs on the slippery slope to slime? *Science* 307(5716): 1725–1726.

Paterson R., Paterson P. and Cato D. 1994. The status of Humpback whales Megaptera novaeanglia in east Australia thirty years after whaling. *Biological Conservation* 70: 135–142.

Pauly D., Christensen V., Guénette S., Pitcher T. J., Sumaila U. R., Walters C. J., Watson R. and Zeller D. 2002. Towards sustainability in world fisheries. *Nature* 418: 689–695.

Pauly D., Alder J., Bennett E., Christensen V., Tyedmers P. and Watson R. 2003. The future for fisheries. *Science* 302: 1359–1361.

Pauly D., Christensen V., Dalsgaard J., Froese R. and Torres Jr F. 1998. Fishing down marine food webs. *Science* 279: 860–863.

Peierls B. L., Caraco N. F., Pace M. L. and Cole J. J. 1991. Human influence on river nitrogen. *Nature* 350: 386.

Pershing A. J. 2010. Climate impacts on whales (and vice versa). Presentation IT44D–01 at the American Geophysical Unions Ocean Sciences Meeting, 22–26 February 2010, Portland, Oregon.

Pimm S. L., Russell G. J., Gittleman J. L. and Brooks T. M. 1995. The future of biodiversity, *Science* 269: 347–350.

Preen A. R., Lee Long W. J. and Coles R. G. 1995. Flood and cyclone related loss, and partial recovery, of more than 1000 km^2 of seagrass in Hervey Bay, Queensland, Australia. *Aquatic Botany* 52(1–2): 3–17.

Preen A. and Marsh H. 1995. Response of dugongs to large-scale loss of seagrass from Hervey Bay, Queensland, Australia. *Wildlife Research* 22: 507–519.

Rabalais N. N. 2002. Nitrogen in aquatic ecosystems. *Ambio* 31(2): 102–112.

Rick T. C. and Erlandson J. M. (eds) 2008. *Human Impacts on Ancient Marine Ecosystems: A Global Perspective.* Berkeley: University of California Press.

Roberts N. 1998. *The Holocene: An Environmental History,* 2nd edition. Malden: Blackwell.

Roman J. and Palumbi S. R. 2003. Whales before whaling in the North Atlantic. *Science* 301: 508–510.

Sadovy Y. 2005. Trouble on the reef: the imperative for managing vulnerable and valuable fisheries. *Fish and Fisheries* 6: 167–185.

Scavia D., Field J. C., Boesch D. F., Buddemeier R. W., Burkett V., Cayan D. R. et al. 2002. Climate change impacts on US coastal and marine ecosystems. *Estuaries and Coasts,* 25(2): 149–164.

Seitzinger S. P., Kroeze C., Bouwman A. F., Caraco N., Dentener F. and Styles R. V. 2002. Global patterns of dissolved inorganic and particulate nitrogen inputs to coastal systems: recent conditions and future projections. *Estuaries* 25: 640–655.

Shackleton J. C. and van Andel T. H. 1980. Prehistoric shell assemblages from Franchthi cave and evolution of the adjacent coastal zone. *Nature* 288: 357–359.

Sheppard J. K., Lawler I. R. and Marsh H. 2007. Seagrass as pasture for seacows: landscape-level dugong habitat evolution. *Estuarine, Coastal and Shelf Science* 71: 117–132.

Sheppard J. K., Preen A. R., Marsh H., Lawler I. R., Whitling A. D. and Jones R. E. 2006. Movement heterogeneity of dugongs, *Dugong dugon* (Müller), over large spatial scales. *Journal of Experimental Marine Biology and Ecology* 334: 64–83.

Smith C. R., Levin L. A., Koslow A., Tyler P. A. and Glover A. G. 2008. The near future of the deep-sea floor ecosystems, in Polunin N. V. C. (ed.) *Aquatic Ecosystems: Trends and Global Prospects.* Cambridge: Cambridge University Press, pp. 334–349.

Smith V. H. 2003. Eutrophication of freshwater and coastal marine ecosystems: a global problem. *Environmental Science and Pollution Research* 10(2): 126–139.

Smith V. H., Tilman G. D. and Nekola J. C. 1999. Eutrophication: impacts of excess nutrient inputs on freshwater, marine, and terrestrial ecosystems. *Environmental Pollution* 100: 179–196.

Sonak S., Pangam P., Giriyan A. and Hawaldar K. 2009. Implications of the ban on organotins for protection of global coastal and marine ecology. *Journal of Environmental Management* 90: S96–S108.

Turner B. L. II., Kasperson R. E., Meyer W. B., Dow K. M., Golding D., Kasperson J. X. et al. 1990. Two types of global environmental change. *Global Environmental Change* 4: 15–22.

Watson R. and Pauly D. 2001. Systematic distortions in world fisheries catch trends. *Nature* 414: 534–536.

Willard D. A., Cronin T. M. and Verardo S. 2003. Late-Holocene climate and ecosystem history from Chesapeake Bay sediment cores, USA. *The Holocene* 13: 201–214.

Williams D. McB. 2001. *Impacts of Terrestrial Run-Off on the Great Barrier Reef World Heritage Area.* Report to CRC Reef. Townsville: CRC Reef Research Centre.

Williams D. McB., Roth C. H., Reichelt R., Ridd P., Rayment G. E., Larcombe P. et al. 2002. The current level of scientific understanding on impacts of terrestrial run-off on the Great Barrier Reef World Heritage Area. Consensus Statement. Townsville: CRC Reef Research Centre.

Wilson S. K., Adjeroud M., Bellwood D. R., Berumen M. L., Booth D., Bozec Y.-M. et al. 2010. Crucial knowledge gaps in current understanding of climate change impacts on coral reef fishes. *Journal of Experimental Biology* 213(6): 894–900.

Wong P. P. 2003. Where have all the beaches gone? Coastal erosion in the tropics. *Singapore Journal of Tropical Geography* 24(1): 111–132.

Worm B., Barbier E. B., Beaumont N., Duffy J. E., Folke C., Halpern B. S. et al. 2006. Impacts of biodiversity loss on ocean ecosystem services. *Science* 314: 787–790.

29

Human Impacts on the Atmosphere

Kevin J. Noone

1 INTRODUCTION

The Earth has entered the Anthropocene – the current epoch in which human activities have come to rival and even dominate many of the global biogeochemical cycles in the Earth system (Crutzen, 2002; Steffen et al., 2007). Prior to the Anthropocene, the environmental impacts of human activities were mostly limited to local and sometimes regional scales. Now, however, the human impact has reached the planetary scale, and this impact is clearly reflected in changes in the chemistry and physics of the atmosphere.

The aim of this chapter is to use a number of atmospheric constituents to illustrate some of the main human impacts on the atmosphere, beginning with a short historical perspective on air pollution. A limited number of atmospheric constituents are then used to map out the spatial scales of the human impact, as well as look at how this impact has changed over time. The idea is not to be exhaustive in the treatment of every chemical species emitted by human activities, but rather to use example compounds to show the different spatial and temporal characteristics of human influence on the atmosphere.

Finally, a prospective look is taken into what the atmosphere may look like at the end of the century.

2 A SHORT HISTORY OF AIR POLLUTION

The human impact on the atmosphere has been present for a long time. Anyone familiar with sitting around a campfire inhaling wood smoke whenever the wind moves to an unfavorable direction knows about the local impact of biofuel combustion. These local effects range from minor irritations (having to move to another spot around the campfire) to severe health issues caused by indoor heating and cooking using wood and dung. Certainly early human settlements would have been subject to the effects of indoor air pollution, but human impacts on the atmosphere as a whole were not recorded until after the development of cities.

The history of air pollution is closely related to energy production and fuel use. As transition between major fuel sources (e.g., wood, coal, oil) were crossed and as the human population grew, impacts on the atmosphere changed from local to regional and finally to global scales, and the nature of the impacts changed from local health effects through crop damage and visibility degradation to changes in global climate. These changes in the scale and nature of the human impacts on the atmosphere can be illustrated with four snapshots along the trajectory of past societal development. An excellent, more detailed account can be found in Brimblecombe (2008).

2.1 Kos, Greece, c.400 BC

Hippocrates of Kos describes the relationship between the seasons, local climate, air quality and human health. In his treatise 'On Airs, Waters and Places', Hippocrates explains how 'hot winds' and 'cold winds' cause very different effects in the inhabitants of cities exposed to these winds.[1] The conditions do not sound particularly appealing. For city dwellers exposed to hot winds, 'the heads of the inhabitants are of a humid and pituitous constitution, and their bellies subject to frequent disorders, owing to the phlegm running down from the head; the forms of their bodies, for the most part, are rather flabby; they do not eat nor drink much'. On the other hand, those living in cities exposed to cold winds, '[t]heir heads are sound and hard, and they are liable to burstings (of vessels?) for the most part. The diseases which prevail epidemically with them, are pleurisies, and those which are called acute diseases.' While ascribing detailed health issues to the effects of hot and cold winds seems a bit of a stretch today, it does reflect a solid combination of observation and logic more than 2,400 years ago.

2.2 Tel Aviv, Israel, c. AD 100–200

The Mishnah is a collection of oral interpretations of Jewish religious and cultural tradition (Mamane, 1987). Several zoning laws regulate the establishment and location of different commercial activities. For instance,

in the section on civil laws and court procedures there is the clause, 'Carcasses, cemeteries and tanneries must be removed from the town to a distance of fifty cubits. A tannery must not be established except on the east side of town.' The prevailing winds in the region are from the northwest, meaning that such activities should not be located upwind of major settlements. A weakness in the zoning law is that the distance of 50 cubits (approximately 30 m) would not be sufficient to mitigate the pungent smell of a tannery if the winds were light or from the east.

2.3 Constantinople, AD 533

Flavius Anicius Iustinianus (Roman Emperor Justinian) brought order to a confused state of Roman law with the publication of the *Institutes of Justinian* in AD 533. The *Institutes* cover a number of different topics, from marriage and paternal power to 'The Different Kinds of Things'. In this article, Justinian defines a number of resources common to all: 'Thus, the following things are by natural law common to all – the air, running water, the sea, and consequently the sea shore.' While the *Institutes* do not refer to air pollution, they are a very early example of the recognition that the atmosphere is a common resource, and must be recognized as having a special character in terms of jurisprudence. We still struggle today with similar issues, as exemplified by conventions and treaties of varying success on ozone-depleting compounds, transboundary air pollution and greenhouse gases.

2.4 London, England, AD 1661

The full title of John Evelyn's treatise gives a good flavor of the nature and impact of air pollution in seventeenth century London. *FUMIFUGIUM: or The Inconveniencie of the AER and SMOAK of LONDON Dissipated together with some Remedies* begins with the observation that smoke coming from tunnels not far from Scotland Yard filled the royal court to such an extent that people in a crowd could hardly discern one another, and were caused 'manifest Inconveniency'. Evelyn wisely chose to avoid the kind of association of health effects with the physical state of the air such as taken up by Hippocrates. He does, however, identify industrial coal burning (and not domestic cooking fires) as the major culprit for the noxious air, and for somewhat diminishing the town's attractiveness:

> Whilst there are belching it forth their sooty jaws, the City of London resembles the face rather of Mount Ætna, the Court of Vulcan, Stromboli, or the Suburbs of Hell, then an Assembly of Rational Creatures, and the Imperial seat of our incomparable Monarch.

Evelyn goes on to connect atmospheric pollution to impurities in precipitation, anticipating the acid rain issues of a period more than three centuries after his own:

> It is this horrid Smoake which obscures our Churches, and makes our Palaces look old, which fouls our Clothes, and corrupts the waters, so as the very Rain, and refreshing Dews which fall in the several Seasons, precipitate this impure vapour, which, with its black and tenacious quality, spots and contaminates whatsoever is expos'd to it.

He also notes a correlation between atmospheric pollution and adverse human health effects, although then (as now) the exact causal links remain to be completely established.

These four examples serve to illustrate two main points about the human impact on the atmosphere in the preindustrial era: (1) humans have had impacts on the atmosphere for much of recorded history; (2) air pollution has long been connected with other adverse effects on humans and ecosystems. There is one more remark to be made before moving on to quantifying some of the atmospheric changes due to human activities. As mentioned in the introduction, we now are well into the Anthropocene – an era in which human activities rival or even dominate many natural biogeochemical cycles. Nobel laureate Paul Crutzen (who originally coined the term 'Anthropocene') estimated that this epoch started around the latter part of the eighteenth century (Crutzen, 2002). However, William Ruddiman advanced the hypothesis that humans began to have a planetary-scale impact much earlier (Ruddiman, 2003, 2007). Ruddiman postulates that human activities caused increases in atmospheric carbon dioxide (CO_2) and methane (CH_4) more than 5,000 years ago – increases sufficiently large to prevent a new glaciation. Regardless of exactly when the Anthropocene started, humans have had an impact on the atmosphere on scales from local to global from very early on in our history.

3 MAPPING THE HUMAN IMPACT ON THE ATMOSPHERE

Assessing the human impact on the atmosphere requires comparisons of the state of the atmosphere at times and locations in which human activities have and have not had discernable influences. The fact that different species emitted to the atmosphere due to human activities can have very different life cycles and residence times must also be accounted for. In the following sections, a small number of example species are used to map out some of the important spatial and temporal characteristics of this impact.

3.1 Changes with time: carbon dioxide and ozone

The human imprint on the atmosphere can be seen by following concentration changes with time of many different compounds and elements. In this section we will focus on the temporal evolution of two compounds as surrogates for all of the others: carbon dioxide and ozone.

Perhaps the most iconic graph depicting the human impact on the atmosphere is the time series of carbon dioxide measurements from Mauna Loa in Hawaii. The measurements were started by Charles Keeling of the Scripps Institution of Oceanography in March 1958, and (with the exception of a short period in 1964 when funding cutbacks prevented staff from maintaining the equipment) have been ongoing ever since. Figure 29.1 shows monthly mean CO_2 concentrations from March 1958 to February 2009. The rapid increase from 316 ppmv in 1958 to 387 ppmv in 2009 is striking. The oscillations in the signal represent the 'breathing' of the terrestrial biosphere of the northern hemisphere, which dominates the seasonal signal at this site. Carbon dioxide concentrations decrease in the northern hemisphere spring and summer as the terrestrial biosphere absorbs CO_2, and increase in the autumn and winter as respiration puts CO_2 back into the atmosphere.

The rate of increase during this period has not been constant. The average annual increase in atmospheric CO_2 over the entire period is 1.4 ppmv yr^{-1}, increasing from less than 1 ppmv yr^{-1} in the early 1960s to around 2 ppmv yr^{-1} after 2005. There are, however, variations around these values, with a primary period of about 3.6 years. Atmospheric CO_2 concentrations reflect the balance between sources, such as fossil fuel combustion, cement production and land-use change, and sinks, such as uptake by the oceans and the terrestrial biosphere. These sources and sinks vary with time, leading to the observed atmospheric concentrations of CO_2 (Canadell et al., 2007).

The human impact on atmospheric CO_2 concentrations becomes even more striking if the last five decades of atmospheric measurements are plotted together with CO_2 measured in Antarctic ice cores. Figure 29.2 shows the data from Mauna Loa presented in Figure 29.1 plotted together with the composite CO_2 concentrations extracted from several Antarctic ice cores (Lüthi et al., 2008).

In Figure 29.2, the datasets have been combined on a common time scale, plotted as years before 2009. Care needs to be taken when combining datasets from different sources; however, on the time scales shown in Figure 29.2, any differences due to interhemispheric transport are not important. The most recent measurement from the composite ice core is from 138 years before 2009,

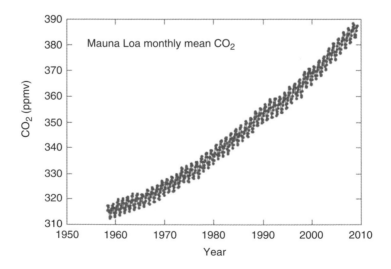

Figure 29.1   Monthly mean CO_2 concentration measured at Mauna Loa observatory in Hawaii, US. Data available online from Dr Pieter Tans, NOAA/ESRL at: www.esrl.noaa.gov/ gmd/ccgg/trends/

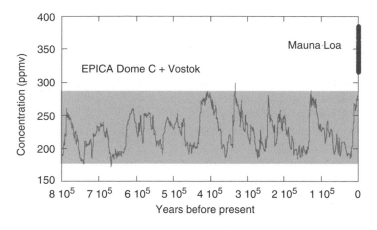

Figure 29.2 Mauna Loa CO$_2$ data (see Figure 29.1) together with data from Lüthi, D. et al. (2008), EPICA Dome C ice core 800 Ka carbon dioxide data, IGBP PAGES/World Data Center for Paleoclimatology Data Contribution Series # 2008-055, NOAA/NCDC Paleoclimatology Program.

and the oldest is dated at approximately 798,510 years before 2009. During this entire period – spanning nine glacial/interglacial cycles – atmospheric CO$_2$ varied from between about 180 ppmv in glacial periods to about 290–300 ppmv in previous interglacial periods (shaded light gray in Figure 29.2). The value of 386 ppmv for 2008 is clearly far outside the range of natural variability over the last 800,000 years, and the rate of increase is also unprecedented in this period. The human impact on the atmosphere in terms of carbon dioxide concentrations is both clear and profound.

The human impact is even greater for atmospheric methane. Atmospheric CH$_4$ varied between about 350 to 700 ppbv over the previous 650,000 years (Spahni et al., 2005). The current atmospheric methane concentration is about 1790 ppbv. Unlike CO$_2$, however, annual average atmospheric methane concentrations have been relatively stable for the last decade – but still more than double the maximum levels observed in the six glacial/interglacial cycles preceding the Anthropocene.

Humanity's move into the Anthropocene is evident in the atmospheric concentration of these two important greenhouse gases.

Human impact on the atmosphere is clearly discernable from natural variability over at least the last six glacial/interglacial cycles.

Another iconic image of the human impact on the atmosphere was the appearance of the 'ozone hole' over Antarctica in the 1980s. In contrast to the situation for CO$_2$ in which the Mauna Loa time series starting in the late 1950s provided an ongoing picture of our impact on the atmosphere (coupled with the recognition since Arrhenius' time that burning fossil fuels could elevate CO$_2$ concentrations in the atmosphere), the Antarctic ozone hole came as a complete surprise to the scientific community. Scientists at the British Antarctic Survey had been measuring total ozone concentrations at Halley station, Antarctica (76°S, 27°W) since 1957 – about the same time that the Mauna Loa measurements started (corresponding to the International Geophysical Year of 1957–1958). However, it was not until the pioneering work of J. C. Farman, B. G. Gardiner and J. D. Shanklin (Farman et al., 1985) that careful observations showed that total ozone concentrations at the site in the month of October had decreased by more than 100 Dobson units (DU)– from an average slightly above 300 DU in the period 1957–1973 to less than 200 DU in 1983. Total ozone values at the site returned to 'normal' levels by Austral summer at roughly the beginning of December.

Figure 29.3 shows the average concentrations of ozone over Halley station for the months of October and February from 1956 to 2008. Column ozone concentrations at the site were essentially equal prior to the mid 1970s. By 1980, a clear decrease was observed, which continued for about the next decade. The median February and October ozone levels at the site for the 20-year period 1956–1975 were 298 and 295 DU respectively, with no statistically significant difference between the months. For the 20-year period between 1989 and 2008, the February mean was 267 DU while the October mean was 150 DU, with a probability of less than 10^{-4} of this difference occurring by chance.

The observations by Farman and colleagues started a tremendous amount of research and debate about the causes for the decrease. A summary of model estimates of stratospheric ozone depletion published in the mid-1980s (Brasseur and Solomon, 1984) predicted decreases in stratospheric ozone of less than about 10 per cent over the coming 50–100 years. The initial debate about the unexpected decrease in Antarctic stratospheric ozone was about whether chemical reactions or atmospheric circulations were the main cause of the decrease in springtime ozone in the Antarctic. Through a concerted effort of field campaigns, satellite observations and model investigations, the scientific community was able to show that heterogeneous chemical reactions on a previously unknown kind of stratospheric cloud produced chlorine compounds which, when sunlight returned to the Antarctic stratosphere in the Austral spring, led to an effective elimination of ozone between about 15–20 km (Solomon, 1988, 1999).

A snapshot example of an ozone 'hole' can be seen in Figure 29.4. It shows ozone concentrations measured from sondes launched at the US National Oceanographic and Atmospheric Administration (NOAA) South Pole station on 22 August (solid line) and 25 October (dots), 2008. The vertical profiles show very low ozone concentrations in the altitude range between about 14 and 21 km.

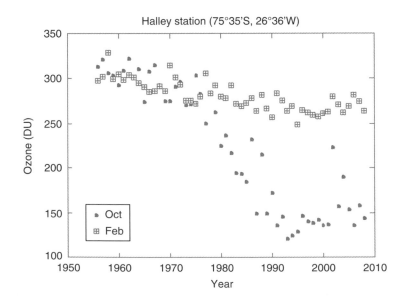

Figure 29.3   Mean ozone levels at Halley station for the months of February and October. Data courtesy of J. D. Shanklin, British Antarctic Survey, Cambridge CB3 0ET, UK.

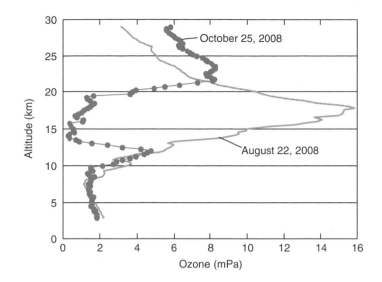

Figure 29.4   Ozone concentrations above the NOAA South Pole station for 22 August (solid line) and 25 October (dots) in 2008. Data from the NOAA Earth System Research Laboratory Global Monitoring Division archive.

A very abbreviated explanation of the causes of the ozone hole can serve as a nice illustration of the ways in which the human impact on the atmosphere can manifest itself in complex and unexpected ways. The story starts with emissions of man-made chlorofluorocarbon (CFC) compounds. These compounds were invented in the 1920s, and were used as heat exchange fluids in, for example, refrigerators and air conditioners, and as aerosol propellants. They were viewed prior to the 1970s as wonderful compounds; they were relatively inert and nontoxic and, for the uses to which they were put, much better than the compounds they replaced. Their chemical stability meant that once they were released into the atmosphere, they would remain there for decades to centuries – more than enough time to be transported into the stratosphere. Once in the stratosphere, Molina and Rowland (1974) showed that they could be photolyzed by radiation in the far ultraviolet – radiation that does not reach the Earth's surface. For instance, CFC-11 ($CFCl_3$) can be photolyzed to form a chlorine atom (Cl) and a radical ($CFCl_2$).

These species could then catalytically destroy ozone in a similar fashion to the cycle for nitrogen oxides (Crutzen, 1970). Further refinement of these early findings – particularly the identification of reactions that terminate the catalytic destruction of ozone and produce stable 'reservoir' compounds like hydrochloric (HCl) and nitric (HNO_3) acids, and chlorine nitrate ($ClONO_2$) – led to the expectation in the early 1980s that stratospheric ozone would decrease slightly, relatively evenly across the globe. In this context, the Antarctic ozone hole appeared out of nowhere. A combination of theoretical advances and concerted observational expeditions eventually showed that clouds containing crystals made up of different combinations of water, nitric acid and sulfuric acid were forming under the special thermodynamic conditions present in the polar lower stratosphere. These polar stratospheric clouds (PSCs) provided the substrate on which heterogeneous chemical reactions converted reservoir compounds into reactive ones. The following set of reactions shows how reactions on the surface of a PSC crystal convert HCl and $ClONO_2$ into reactive chlorine (*s* refers to the solid phase):

$$HCl(s) + ClONO_2 \rightarrow Cl_2 + HNO_3(s)$$
$$Cl_2 + h\upsilon \rightarrow 2Cl$$
$$Cl + O_3 \rightarrow ClO + O_2$$
$$ClO + NO_2 + M \rightarrow ClONO_2 + M$$
$$Net: HCl(s) + NO_2 + 2O_3$$
$$\rightarrow ClO + HNO_3(s) + O_2$$

The net reaction in this chain destroys two molecules of ozone, and forms chlorine nitrate that if in sufficient concentrations can itself participate in a separate catalytic cycle destroying even more ozone. The two key elements of this reaction scheme that explain the ozone hole phenomenon in the Austral spring are the reactions on the surfaces of solid PSC crystals to convert reservoir compounds into reactive chlorine (which can occur as long as the clouds are present), and the photolysis of Cl_2 to form chlorine atoms (which occurs when sunlight returns to the Antarctic stratosphere in the spring).

The ozone hole is a classic example of a truly surprising human impact on the atmosphere. It shows that human emissions of materials that we expect to be benign can nonetheless have large-scale, unexpected impacts.

3.2 Spatial imprints: trace gases and aerosols

The two greenhouse gases described in the preceding sections have relatively long atmospheric residence times: carbon dioxide has an atmospheric lifetime of about a century, while methane's lifetime is on the order of a decade. Because of their long residence times, these greenhouse gases are relatively evenly distributed around the globe, and thus human emissions of these gases will have global-scale effects.

Many other species – such as oxides of nitrogen (NO_x) and aerosols – have much shorter atmospheric residence times, and the human impact of these species on the atmosphere will be more heterogeneous. However, human impacts can and do aggregate in such a way that even our emissions of short-lived species can have global effects.

The spatial distribution of trace species in the atmosphere will be determined by the nature of the sources (e.g., point or area sources), the atmospheric lifetime of the substance in question, and the atmospheric transport time.

Figure 29.5 shows the spatial distributions of four different species with different atmospheric lifetimes. Panel A shows the average tropospheric NO_2 slant column density (units: 10^{15} molecules cm^{-2}) derived from SCHIMACHY satellite observations between 2003 and 2006 (Wagner et al., 2008). Panel B shows the column density of non-sea salt sulfate (units: mg S m^{-2}) for the year 2000 derived from the AEROCOM (aerosol comparisons between observations and models) model database.[2] Panels C and D depict CH_4 (Schneising et al., 2009) and CO_2 concentrations (Schneising et al., 2008). The panels are ordered by increasing atmospheric residence time: NO_2 has a residence time of approximately one day, aerosols of a few days to one week, the lifetime of CH_4 is about a decade, and the lifetime of CO_2 is about a century.

Because of its short lifetime, high NO_2 concentrations are found close to the sources, the largest of which are fossil fuel combustion, release from soils, and biomass burning. The spatial distribution of NO_2 shows considerable sub-regional heterogeneity, with the highest concentrations primarily in areas of high fossil fuel combustion. Richter et al. (2005) use satellite observations to show that relative to 1996, NO_2 concentrations over the continental US have been steady or decreasing, concentrations over Europe decreasing, and concentrations over eastern China and Hong Kong increasing. NO_2 is a key component in determining surface ozone concentrations, together with volatile organic compounds. Typical daily maximum ozone

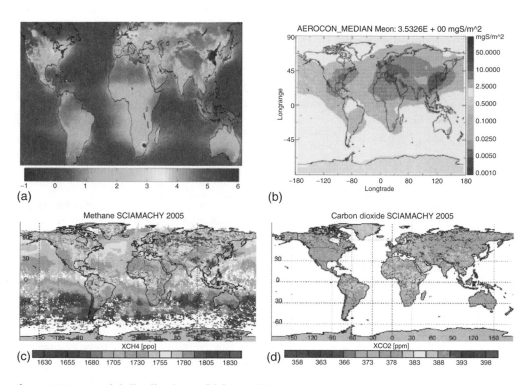

Figure 29.5   Spatial distributions of (a) NO$_2$, (b) non-sea-salt sulfate, (c) CH$_4$ and (d) CO$_2$.

concentrations in background locations are between 20 and 40 ppbv, 50 and 100 ppbv in rural locations, and 100 and 400 ppbv in urban and suburban locations (Seinfeld and Pandis, 1998). Thus, human emissions of oxidized nitrogen compounds impact the atmosphere both directly (Figure 29.5a) and indirectly through the formation of surface ozone.

Intercontinental transport of trace substances in the atmosphere normally occurs on time scales of days to a week or more. Compounds with short lifetimes like NO$_2$ are therefore generally removed within the same region in which they are emitted. On some occasions, however, extremely rapid intercontinental transport can occur. Stohl et al., (2003) show a case in which NO$_x$ from the eastern seaboard of the US was transported very rapidly across the North Atlantic and reached parts of northern Europe. While relatively rare, Stohl et al. (2003) estimate that the meteorological 'bomb' conditions that

make very rapid intercontinental transport of air masses possible could account for about two-thirds of the intercontinental atmospheric transport of short-lived substances.

The longer atmospheric lifetime of aerosols is reflected in a more heterogeneous distribution within a given region, as compared with NO$_2$. The maxima in non-sea-salt sulfate column densities shown in Figure 29.5b over the eastern US, central Europe and eastern Asia and Japan are much more regionally homogeneous than the short-lived NO$_2$. An added complexity with atmospheric aerosols compared with gas-phase substances is the fact that they are composed of a number of different compounds and elements, and range in size from a few nanometers to millimeters in diameter. Table 29.1 shows the source strengths, mass loading, lifetime and resultant optical depth of five major aerosol types. The data (other than for nitrate) are based on results from 16 models from the Aerosol Comparisons between

Table 29.1   Source strength, mass loading, lifetime and optical depth for five major aerosol types, given as median values and ranges.

Aerosol type	Source (Tg a^{-1})	Mass loading (Tg)	Lifetime (days)	Optical depth (550nm)	Anthrop. fraction[5] (%)
Sulfate[1]	190 (100–230)	2.0 (0.9–2.7)	4.1 (2.6–5.4)	0.034 (0.015–0.051)	61
Black Carbon[1]	11 (8–20)	0.2 (0.05–0.5)	6.5 (5.3–15)	0.004 (0.002–0.009)	n/a
Organics[1]	100 (50–140)	1.8 (0.5–2.6)	6.2 (4.3–11)	0.019 (0.006–0.030)	51
Dust[1]	1600 (700–4000)	20 (5–30)	4.0 (1.3–7)	0.032 (0.012–0.054)	22
Sea salt[1]	6000 (2000–120000)	6 (3–13)	0.4 (0.03–1.1)	0.030 (0.020–0.067)	0
Nitrate (non-dust)[2,3]	21.3[4]	0.3 (0.04–0.63)	4.5	n/a	n/a

Sources: (1) Chin et al. (2009); (2) (Tsigaridis et al. (2006); (3) Liao et al. (2004); (4) gas-to particle conversion from nitric acid; and (5) anthropogenic fraction estimated from supplementary material from Ramanathan et al. (2001).

Observations and Models (AeroCom) project (Kinne et al., 2006; Textor et al., 2006).

Sea salt has the largest source strength of all of the aerosol categories, but since it is rapidly removed from the atmosphere its lifetime is less than a day and its mass loading is small relative to its source strength. The lifetimes of the other aerosol components are comparable, all being in the range of a few days to about a week. Soil dust has the largest source strength after sea salt, and has the highest median mass loading of the major aerosol components. Human activities account for nearly two-thirds of the sulfate source to the atmosphere, and about half of the particulate organic matter. The anthropogenic fraction of the black carbon source is not estimated due to the difficulty of distinguishing natural from human-induced forest and bush fires; however, the fraction is expected to be large since black carbon originates from incomplete combustion processes. Human activities also contribute to the dust source via industrial processes and land use change.

Nitrogen compounds such as nitrate and ammonium are more difficult to estimate, since they cycle much more readily between the gas and condensed phases during their lifetime in the atmosphere. While the anthropogenic fraction of nitrate is not estimated here, Galloway et al. (2004) estimate that natural processes created 233 Tg a^{-1} of reactive nitrogen in the 1990s, while human activities created 389 Tg a^{-1} in the same period; humans accounted for 63 per cent of the total reactive nitrogen production. While this fraction cannot be translated directly to atmospheric nitrate, we can expect the anthropogenic fraction for nitrate to be comparable to sulfate and organics. Interestingly, Liao et al. (2006) predict significant changes in aerosol nitrate concentrations in the future. As global annual average temperatures increase due to the effects of anthropogenic greenhouse gases, more of the nitrogen remains in the gas phase as nitric acid, instead of being in the form of aerosol nitrate.

Aerosol sulfate, nitrate and ammonium were the subject of considerable research and attention in the 1980s and 1990s, when the issue of acid deposition was at a peak. Strictly speaking, acid deposition is not a human impact on the atmosphere, since the ultimate impact is primarily on terrestrial ecosystems. On the other hand, since the acid–base state of precipitation is the result of a balance between acidifying compounds such as sulfate, nitrate and organic acids and alkaline compounds such as ammonium and alkaline material in soil dust – all of which have anthropogenic as well as natural sources – acid deposition is worth at least a mention in this chapter. Rodhe et al. (2002) present an analysis of the global distribution of acidifying wet deposition. Figure 29.6 shows a map of the annual average pH of rainwater. With the exception of central and

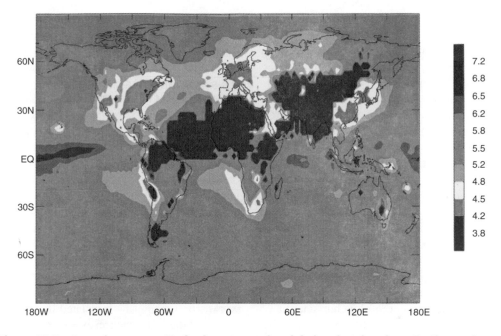

Figure 29.6   Annual average pH of rainwater at the global scale taken from Rodhe et al. (2002). Reprinted with permission of *Environmental Science and Technology*.

south Asia and the Saharan and eastern central Africa, the map of rainwater pH closely resembles the sulfate aerosol distribution shown in Figure 29.5b, with low pH values over the eastern seaboard of North America, central Europe and east Asia. This is not surprising, since the acidifying compounds have significant sources in these regions, and their lifetimes prevent them from being transported long distances before they are removed from the atmosphere. In fact, wet deposition (as reflected in the low pH of rainwater) is the major removal pathway for these compounds.

Some additional aspects of the human impact on the atmosphere in terms of aerosols are discussed in the following 'Connections' section. It is worth noting here, however, that the direct radiative effects of aerosols in the atmosphere are to scatter and absorb incoming solar radiation. These effects on the radiative balance of the Earth are expressed in the optical depth; a measure of the attenuation of radiation passing through

a scattering or absorbing medium. For comparison with the values in Table 29.1, the optical depth of a particle-free (Rayleigh) atmosphere at sea level and a latitude of $45°$ (for a wavelength of 550 nm) is 0.097 (Bodhaine et al., 1999). The presence of particles at least doubles the total optical depth of the atmosphere; in polluted locations and in areas of high relative humidity the aerosol optical depth can be much more than ten times Rayleigh extinction.

Along with the development of reasonably accurate models and observations of global aerosol distributions came the realization that particles could have an important climatic effect. R. J. Charlson and collaborators showed in the early 1990s that sulfate aerosols could have a cooling effect on the Earth's climate that was on the same order of magnitude as the warming effect of anthropogenic greenhouse gases (Charlson et al., 1990, 1991, 1992; Charlson and Wigley, 1994). However, the spatial and temporal distributions of the climate forcing due to aerosols and greenhouse gases are very different, due to their different lifetimes and spatial distributions. Greenhouse gas radiative forcing occurs day and night, in clear and cloudy conditions. On the other hand, direct aerosol radiative forcing happens during daytime, is strongest in cloud-free areas and is concentrated in regions of high aerosol loading. The regional cooling effects due to aerosols, combined with the spatial distribution of maxima in warming due to greenhouse gases actually makes the spatial distribution of the anthropogenic radiative forcing even more complex (Charlson and Wigley, 1994). The notion that anthropogenic aerosols could be used as a method of offsetting greenhouse gas warming – even ignoring all the other adverse effects of aerosols – turned out to be a chimera.

Moving toward the long-lived species, the global distributions of CH_4 and CO_2 do show a good deal of spatial variability; however, the relative spatial changes in concentration are far lower than for NO_2 and aerosols. While NO_2 and aerosol concentrations vary by factors of ten or more, CH_4 and CO_2 concentrations typically vary by about ten percent or less. In addition to the seasonal variability shown in Figure 29.1 for CO_2, there is a gradient between the northern and southern hemispheres for both CO_2 and CH_4, with higher concentrations in the north. Wetlands in Siberia are a conspicuous source of CH_4. However, since panels (c) and (d) are shown in Figure 29.5 as yearly averages, the seasonal variability is smoothed out, and the relationship between specific sources and atmospheric concentrations is obscured.

4 CONNECTIONS

While the main focus of this chapter is the human impact on the atmosphere, an even more interesting aspect of anthropogenically driven changes in atmospheric composition is how these changes are connected to a range of different effects. Acid deposition

was mentioned earlier. Here I will use ocean acidification and aerosols as cases in point to illustrate how human influences on the atmosphere connect to many other phenomena (see also Chapter 23).

Roughly half of all the CO_2 emitted to the atmosphere through human activities stays there. Of the 50 percent that disappears, roughly half is taken up by terrestrial ecosystems, and the other half ends up in the oceans. This latter removal process includes both dissolution of CO_2 into seawater, and the uptake of carbon by marine organisms – and the uptake is not evenly distributed (Sabine et al., 2004). In this sense, nature provides an extremely valuable service by removing CO_2 from the atmosphere – a service that may well change in a warmer world.

One of the consequences of CO_2 being dissolved into the oceans is that it causes a decrease of the pH of the surface ocean water. Many marine organisms – especially those that use carbonate dissolved in the ocean to manufacture solid calcium carbonate shells – are very sensitive to changes in these parameters. This issue (ocean acidification) has been the focus of intense research efforts in recent years. Guinotte et al. (2003) showed that ocean pH has changed a great deal since preindustrial times, and how it may change in the future as the result of human emissions of CO_2. One of the results of this acidification is that coral reefs – and the extensive, biodiverse and very important ecosystems that are associated with them – are in danger of being exposed to conditions that are marginal or dangerous by as early as 2050. Perhaps even more importantly, if the pH of the oceans decreases sufficiently, calcium carbonate (in two main forms: aragonite and calcite) produced by marine organisms to make their solid shells becomes soluble; the aragonite or calcite saturation ratio for the ocean water goes below unity (Feely et al., 2004). Orr et al. (2005) showed that parts of the southern ocean could reach this state by as early as 2050–2060. Aragonite saturation is generally reached before calcite, meaning that organisms forming aragonite will be affected first.

The consequences stemming from significant parts of the oceans becoming corrosive to aragonite is the subject of intense research. While the details of what may happen cannot be predicted in detail, it is safe to conclude that very substantial changes in ecosystem composition and dynamics would result of going below aragonite saturation (Riebesell et al., 2000). In addition to the effects on coral communities, any marine biological communities that contain significant numbers of organisms that form aragonite would certainly be severely impacted, presumably with ripple effects up the food chain.

Aerosols have long been known to directly influence the Earth's radiation balance directly by scattering incoming radiation back to space (Charlson et al., 1992) or indirectly by influencing cloud reflectivity and lifetime (Albrecht, 1989; Twomey, 1977). Aerosols can also influence the hydrological cycle through altering precipitation formation mechanisms in clouds (Ferek et al., 2000; Rosenfeld, 2000). Recent investigations have shown that aerosols may have a substantial influence on the Asian monsoon circulation (Lau and Kim, 2006; Lau et al., 2008; Ramanathan et al., 2005). Lau and Kim (2006) showed that absorbing aerosols over the Indo-Gangetic plain near the foothills of the Himalayas act as an extra heat source aloft, enhancing the incipient monsoon circulation. The same aerosols lead to a surface cooling over central India, shifting rainfall to the Himalayan region. This 'elevated heat pump' effect leads to the monsoon rain beginning earlier in May–June in northern India and the southern Tibetan plateau, and increase in monsoon rainfall over all of India in July–August, and a corresponding reduction in rainfall over the Indian Ocean. While it is generally accepted that aerosols influence the Asian monsoon, there is still a great deal of uncertainty in the physical processes underlying the effects, and in their interactions.

From the aspect of human health effects, Cohen et al. (2005) examined the relationship between fine particulate air pollution (PM2.5) and mortality for three outcomes: cardiopulmonary causes in adults, lung cancer, and acute respiratory infections in children up to 5 years of age. They estimated that outdoor particulate air pollution was responsible for about 3 per cent of adult cardiopulmonary disease mortality, about 5 per cent of tracheal, bronchial and lung cancer mortality, and about 1 percent mortality from acute respiratory infection in children in urban areas worldwide. These effects convert to about 800,000 premature deaths and 6.4 million lost life years. Additionally, they found that the burden of disease due to air pollution is predominately in developing countries, particularly in Asia. Ezzati et al. (2002) ranked the major risk factors for global and regional disease burden. They arrived at very similar estimates for mortality to urban air pollution as did Cohen et al. (2005). In addition, they found that mortality due to exposure to indoor smoke from solid fuels to be about double that of urban air pollution (roughly 1.6 million deaths), and that exposure to occupational airborne particulates accounted for roughly 300,000 deaths. In both cases, developing countries were the most affected.

The same components (such as particulates, tropospheric ozone, oxides of sulfur and nitrogen) lead to other deleterious effects. Crop damage from exposure to ozone, forest degradation due to acidic precipitation, changes in global precipitation patterns and energy balance due to particles are examples of such indirect effects of air pollution on humans. Clearly the human impact on the atmosphere is tightly coupled with impacts in many other areas.

5 A LOOK FORWARD

This chapter began with a historical perspective on the human impacts on the atmosphere. While the Ruddiman (2003) conjecture is still being debated, we can without doubt say that the human impact on the atmosphere became truly global as we entered into the Anthropocene within the last two centuries. Human activities have come to dominate many of the major biogeochemical cycles, all of which impact on atmospheric composition. What does the future have in store? Perhaps a rephrasing of the question is even more germane: what does the human species have in store for atmospheric composition in the future?

Greenhouse gases will continue to be emitted. Raupach et al. (2007) have shown that current anthropogenic emissions of CO_2 are actually above the most pessimistic of the scenarios of the Intergovernmental Panel on Climate Change (IPCC). Unless very significant progress is made in limiting greenhouse gas emissions in the near future, even more significant changes in atmospheric composition will be faced, with resultant effects on the coupled human–environmental Earth system. Given the correlation between aerosol loading and adverse human health effects, it looks increasingly likely that coordinated legislation will be developed to further reduce the anthropogenic emissions that lead to aerosol formation. Acting wisely to reduce adverse human health effects and problems associated with acid deposition could actually exacerbate anthropogenic greenhouse warming if the aerosols that act to scatter incoming solar radiation back to space are removed faster than greenhouse gases are reduced (Andreae et al., 2005). Unfortunately, current international conventions and negotiations on air pollution and climate are proceeding on parallel tracks. The United Nations Framework Convention on Climate Change (UNFCCC) and the United Nations Economic Commission for Europe (UNECE) convention on Long-Range Transboundary Air Pollution (LRTAP) are both developing suggestions for international agreements on similar substances, but are doing so in effective separation. Black carbon is another example of the human impact on the atmosphere that falls somewhere between these two conventions. As with other aerosol species, black carbon has both a climatic effect

(in this case warming the atmosphere by absorbing shortwave radiation instead of cooling by scattering radiation back to space), as well as human health effects. There is increasing realization that targeting black carbon for reductions could serve both to slow global warming as well as alleviate the adverse health effects of the particles (Gustafsson et al., 2009).

The world has made significant progress on reducing some pollutant emissions to the atmosphere. As one example, emissions of sulfur and nitrogen oxides have decreased in Europe and North America after legislation to reduce regional air pollution. Given the growing realization of the global character and impact of anthropogenic emissions, and the fact that changes in the atmosphere are tightly connected to changes elsewhere in the Earth system, it is incumbent on the present generation to pursue policies on a global scale that will reduce the human impact of future generations on the atmosphere.

NOTES

1 Available at: http://ebooks.adelaide.edu.au/h/hippocrates/airs/
2 Available at: http://dataipsl.ipsl.jussieu.fr/AEROCOM/aerocomhome.html

REFERENCES

Albrecht B. A. 1989. Aerosols, cloud microphysics, and fractional cloudiness. *Science* 245: 1227–1230.
Andreae M. O. et al. 2005. Strong present-day aerosol cooling implies a hot future. *Nature* 435: 1187–1190.
Bodhaine B. A. et al. 1999. On Rayleigh optical depth calculations. *Journal of Atmosphere and Ocean Technology* 16: 1854–1861.
Brasseur G., and S. Solomon 1984. *Aeronomy of the Middle Atmosphere*. Hingham, MA: D. Reidel.
Brimblecombe P. 2008. Air pollution history, in Sokhi E. (ed.) *World Atlas of Atmospheric Pollution*. London: Anthem Press, pp. 7–18.

Canadell J. G. et al. 2007. Contributions to accelerating atmospheric CO_2 growth from economic activity, carbon intensity, and efficiency of natural sinks. *Proceedings of the National Academy of Sciences* 104(47): 18866–18870.

Charlson R. J. et al. 1990. Sulphate aerosol and climate. *Nature* 348: 22.

Charlson R. J. et al. 1991. Perturbation of the Northern Hemisphere radiative balance by backscattering of anthropogenic sulfate aerosols. *Tellus* 43AB: 152–163.

Charlson R. J. et al. 1992. Climate forcing by anthropogenic aerosols. *Science* 255: 423–430.

Charlson R. J. and Wigley T. M. L. 1994. Sulfate aerosol and climate change. *Scientific American* 270: 48–57.

Chin M. et al. 2009. *CCSP 2009: Atmospheric Aerosol Properties and Climate Impacts A. Report by the U.S. Climate Change Science Program and the Subcommittee on Global Change Research.* Washington DC: National Aeronautics and Space Administration,

Cohen A. J., et al. 2005. The global burden of disease due to outdoor air pollution, *Journal of Toxicology and Environmental Health, Part A* 68: 1301–1307.

Crutzen P. 1970. The influence of nitrogen oxides on the atmospheric ozone content, *Quartery Journal of the Royal Meteorological Society* 96: 320–325.

Crutzen P. J. 2002. Geology of mankind. *Nature* 415: 23.

Ezzati M. et al. 2002. Selected major risk factors and global and regional burden of disease. *The Lancet* 360: 1347–1360.

Farman J. C. et al. 1985. Large losses of total ozone in Antarctica reveal seasonal ClOx/NOx interaction. *Nature* 315: 207–210.

Feely R. A. et al. 2004. Impact of anthropogenic CO_2 on the $CaCO_3$ system in the oceans. *Science* 305: 362–366.

Ferek R. J. et al. 2000. Drizzle suppression in ship tracks. *Jornal of Atmospheric Science* 57: 2707–2728.

Galloway J. N. et al. 2004. Nitrogen cycles: past, present, and future. *Biogeochemistry* 70: 153–226.

Guinotte J. M. et al. 2003. Future coral reef habitat marginality: temporal and spatial effects of climate change in the Pacific basin. *Coral Reefs* 22: 551–558.

Gustafsson O. et al. 2009. Brown clouds over South Asia: Biomass or fossil fuel combustion? *Science* 323: 495–498.

Kinne S. et al. 2006. An AeroCom initial assessment – optical properties in aerosol component modules of global models. *Atmospheric Chemistry and Physics* 6: 1815–1834.

Lau K. M. and Kim K. M. 2006. Obervational relationships between aerosol and Asian monsoon rainfall, and circulation. *Geophysics Research Letters* 33: L21810.

Lau K. M. et al. 2008. The joint aerosol-monsoon experiment: A new challenge for monsoon climate research. *Bulletin of the American Meteorological Society*.

Liao H. et al. 2004. Global radiative forcing of coupled tropospheric ozone and aerosols in a unified general circulation model. *Journal of Geophysical Research* 109.

Liao H. et al. 2006. Role of climate change in global predictions of future tropospheric ozone and aerosols, *Journal of Geophysical Research* 111.

Lüthi D. et al. 2008. High-resolution carbon dioxide concentration record 650,000–800,000 years before present. *Nature* 453: 379–382.

Mamane Y. 1987. Air pollution control in Israel during the first and second century. *Atmospheric Environment* 21: 1861–1863.

Molina M. J., and F. S. Rowland 1974. Stratospheric sink for chlorofluoromethanes: chlorine atomcatalysed destruction of ozone. *Nature* 249: 810–812.

Orr J. C. et al. 2005. Anthropogenic ocean acidification over the twenty-first century and its impact on calcifying organisms. *Nature* 437: 681–686.

Ramanathan V. et al. 2001. Aerosols, climate and the hydrological cycle. *Science* 294: 2119–2124.

Ramanathan V. et al. 2005. Atmospheric brown clouds: impacts on South Asian climate and hydrological cycle. *Procedings of the National Academy of Sciences USA* 102: 5326.

Raupach M. R. et al. 2007. Global and regional drivers of accelerating CO_2 emissions. *Proceedings of the National Academy of Sciences* 104: 10288–10293.

Richter A. et al. 2005. Increase in tropospheric nitrogen dioxide over China observed from space. *Nature* 437: 129–132.

Riebesell U. et al. 2000. Reduced calcification of marine plankton in response to increased atmospheric CO_2. *Nature* 407: 364–367.

Rodhe H. et al. 2002. The global distribution of acidifying wet deposition. *Environmental Science and Technology* 36: 4382–4388.

Rosenfeld D. 2000. Suppression of rain and snow by urban and industrial air pollution. *Science* 287: 1793–1796.

Ruddiman W. F. 2003. The Anthropogenic greenhouse era began thousands of years ago. *Climatic Change* 61: 261–293.

Ruddiman W. F. 2007. The early anthropogenic hypothesis: Challenges and responses. *Reviews of Geophysics* 45: RG4001.

Sabine C. L. et al. 2004. The oceanic sink for Anthropogenic CO_2. *Science* 305: 367–371.

Schneising O. et al. 2008. Three years of greenhouse gas column-averaged dry air mole fractions retrieved from satellite – Part 1: Carbon dioxide. *Atmospheric Chemistry and Physics* 8: 3827–3853.

Schneising O. et al. 2009. Three years of greenhouse gas column-averaged dry air mole fractions retrieved from satellite – Part 2: Methane. *Atmospheric Chemistry and Physics* 9: 443–465.

Seinfeld J. H. and Pandis S. N. 1998. *Atmospheric Chemistry and Physics: From Air Pollution to Climate Change.* New York: John Wiley & Sons.

Solomon S. 1988. The mystery of the Antarctic ozone 'hole'. *Reviews of Geophysics* 26: 131–148.

Solomon S. 1999. Stratospheric ozone depletion: A review of concepts and history. *Reviews of Geophysics* 37: 275–316.

Spahni R. et al. 2005. Atmospheric methane and nitrous oxide of the late Pleistocene from Antarctic ice cores. *Science* 310: 1317.

Steffen W. et al. 2007. The Anthropocene: Are humans now overwhelming the great forces of nature. *AMBIO: A. Journal of the Human Environment* 36: 614–621.

Stohl A. et al. 2003. Rapid intercontinental air pollution transport associated with a meteorological bomb. *Atmospheric Chemistry and Physics* 3: 969–985.

Textor C. et al. 2006. Analysis and quantification of the diversities of aerosol life cycles within AeroCom. *Atmospheric Chemistry and Physics* 6: 1777–1813.

Tsigaridis K. et al. 2006. Change in global aerosol composition since preindustrial times. *Atmospheric Chemistry and Physics* 6: 5143–5162.

Twomey S. 1977. The influence of pollution on the shortwave albedo of clouds. *Journal of Atmospheric Science* 34: 1149–1152.

Wagner T. et al. 2008. Monitoring of atmospheric trace gases, clouds, aerosols and surface properties from UV/vis/NIR satellite instruments. *Journal of Optics A: Pure and Applied Optics* 10: 104019.

Patterns, Processes and Impacts of Environmental Change at the Regional Scale

Environmental Change in the Humid Tropics and Monsoonal Regions

Mark B. Bush and William D. Gosling

1 INTRODUCTION

Research over the past few decades has illuminated the dynamic nature of tropical palaeoclimates and the important role that the tropics played in global climate patterns. Increasingly, the realization has grown that climate changes at high latitudes cannot simply be extrapolated into tropical regions. Rather, the tropics have sometimes led the changes or have responded differently when compared with their high-latitude counterparts. Similarly, the tropics, spanning 46° of latitude in two hemispheres on three continents, do not respond uniformly to a given forcing. In this chapter we review evidence from the last glacial maximum (defined in the high latitude northern hemisphere) to modern, and consider features of environmental change ranging from climatic variability to the extinction of megafauna.

2 THE LAST DEGLACIATION

Global glaciation had two obvious physical impacts on tropical regions. First, at high altitudes, glaciers formed and advanced causing major perturbation of ecosystems, hydrological systems and landforms. Second, sea level descended exposing continental shelf and resulting in the expansion of low-lying coastal areas and the interconnection of island chains. The impact of generally warming global climates after the global Last Glacial Maximum (LGM; 21 ± 2 thousand years ago [ka]) did not have a geographically even impact. For example, major mountain chains (the Andes and central Asia) saw landmasses appear from beneath glaciers, while in Southeast Asia a continental-sized area was submerged by rising sea levels and transformed into an island chain. In the non-coastal tropical lowlands, the magnitude, timing, and rate of climatic changes between 21 and 11 ka initiated some of the most profound environmental alteration experienced during the last glacial cycle. However, the timing of these events was not necessarily coincident with those of the North Atlantic region.

2.1 Tropical glaciers

The establishment of ice caps and glaciers at low latitudes was restricted to high altitudes during the LGM. Today, the most extensive tropical glaciers occur in the Andes and on the Tibetan Plateau, as well as on isolated peaks in East Africa and New Guinea (Thompson et al., 2005). During the last glacial period glaciers formed on the tropical mountains of the Andes (Clapperton, 1972; Smith et al., 2008), Central America (Lachniet, 2004; Weyl, 1956), the Tibetan Plateau (Herzschuh, 2006; Lehmkuhl and Owen, 2005), East Africa (Coetzee, 1964; Mark and Osmaston, 2008; Osmaston and Harrison, 2005), New Guinea (Peterson et al., 2004; Prentice et al., 2005; Walker and Flenley, 1979) and even Hawaii (Pigati et al., 2008; Porter, 2001). However, the timing of glacial expansion was not uniform.

Montane glaciers needed both cold temperatures and high precipitation to expand. All neotropical pollen records of glacial age indicated a cooling sometime between 30 and 21 ka. However, the timing and extent of glacial advance was heterogeneous (Clark et al., 2009). Tropical glacial advance was probably broadly tied to monsoonal activity that, in turn, was influenced by orbital precession (Clement et al., 2004). Peaks in seasonality resulting from the precessional ($c.21$ ka) cycle occurred in the southern hemisphere at $c.20$ ka (15°S) and in the northern hemisphere at $c.10$ ka (15°N) (Berger and Loutre, 1991). A key influence other than precession, may have been the southerly displacement of the Inter Tropical Convergence Zone (ITCZ) during cold events (Haug et al., 2001; Newell, 1973). The spatially variable timing of tropical glacial maxima relative both to other tropical sites and the northern hemispheric LGM conditions suggested a climate system in which precipitation played at least as large a role as temperature.

The thermal conditions existed for glacier expansion in many tropical regions between $c.60$ and 19 ka. Temperatures in tropical South America and Southeast Asia are thought to have generally been between 4 and 10°C lower than the modern day during the global LGM, with the possibility of slightly greater cooling in mountain regions (Bush et al., 2007a; Groot et al. 2011; Kershaw et al., 2007; Punyasena et al., 2008; van der Hammen, 1974; van der Hammen and Gonzalez; 1959; Walker and Flenley, 1979), while in Africa the cooling was possibly slightly less, $c.2–5°C$ (Bonnefille, 2007; Livingstone, 1967; Mark and Osmaston, 2008; Powers et al., 2005).

The maximum extent of tropical glacial advance in the southern hemisphere is thought to predate the LGM ($c.44–34$ ka; Clapperton, 1987; Smith et al., 2005); while in the northern hemisphere, Himalayan advances occurred during the equivalent of Marine Isotope Stage (MIS) 3 (24–59 ka) and early Holocene (Lehmkuhl and Owen, 2005; Martinson et al., 1987). Thompson et al. (2005) noted that the oldest ice present today in southern hemisphere tropical glaciers, represented by the Sajama ice cap, Bolivia, formed $c.25$ ka. In contrast, in the northern hemisphere the oldest tropical glaciers were Holocene in origin (e.g., Dasuopu $c.6$ ka, Himalayas). This difference probably reflected the time since last major drought rather than changes in temperature. The timing for tropical glacial inception and advance were therefore neither synchronous between hemispheres nor necessarily coincident with the LGM; the period of maximum global ice volume and inferred lowest global temperatures.

Most tropical records show ice retreat beginning between 22 and 19 ka (Hodell et al., 2008; Seltzer et al., 2002), which at least in the southern tropical Andes was a wet period (Baker et al., 2001a; Hillyer et al., 2009). These data argue that the tropics warmed earlier than the high northern latitudes. The process of deglaciation at high northern latitudes began with the Bølling/Allerød (B/A) warming (14.7–14.0 ka) which was punctuated by the Younger Dryas (YD) cold event (12.9–11.7 ka) prior to persistent warming into the Holocene (Rasmussen et al., 2006; Björck et al., 1998). While the driving forces behind these changes have attracted much attention (e.g., Bradley and England, 2008; Firestone et al., 2007) they probably originated in the northern hemisphere. Regardless of the trigger mechanism, the YD cooling induced a reduction in the Atlantic Meridional Overturning Circulation (AMOC) of the North Atlantic (Broecker, 1998). A weakened AMOC pooled heat in the tropics and led to a southward displacement of the ITCZ. Although potentially a global event, whether the YD was truly manifested in the tropics has been debated (Bennett et al., 2000). In the northernmost tropics of Mexico and Central America, the YD was a dry event caused by the reduced transport of tropical moisture because the ITCZ failed to migrate as far north during the boreal summer. The southern displacement of the ITCZ influenced its northern limit, but because of a substantial overlap in the preceding and YD ranges of the ITCZ there was relatively little detectable impact for much of tropical South America.

The most highly resolved records for deglacial times and the Holocene from tropical South America come from pollen, diatom, and speleothem records (Ekdahl *et al.*, 2008; Valencia *et al.*, 2010; Wang et al. 2008). Most of the southern tropical records reveal that the changes in AMOC between 60 and 30 ka induced larger changes than those of the YD (e.g. Cruz et al. 2005) (Baker et al., 2001b, Fritz et al., 2010) (see discussion later in this chapter), but there is debate regarding whether the YD is represented. The Sajama ice-core record reveals evidence of a dry event that was suggested to align with the YD (Thompson et al., 1998); in addition the Pacucha pollen record demonstrates a reversal during this period of dry conditions that limited forest expansion (Valencia et al., 2010).

While the continental records of deglaciation have been tentatively linked to northern hemisphere events, evidence from the mountains of Hawaii, in the tropical Pacific, suggests a possible disconnection between high and low latitude climate mechanisms during both the last glacial maximum period and the subsequent period of deglaciation (Porter, 2001). The largest glacial advances on Mauna Kea (Hawaii) occurred at 23 and

13 ka (Pigati et al., 2008). The younger of these expansions occurred just prior to the initiation of the YD cooling, and probably ended while the northern hemispheric ice masses were still expanding. Pigati et al. suggest that this finding indicates that low latitude climate mechanisms, such as Walker circulation, insolation, seasonality, and/or El Niño southern oscillation (ENSO)-type variations were a stronger influence on tropical glacial extent than northern hemispheric ice volume.

What is clear from these data is that moisture availability plays a proportionally more important role (than temperature) in the expansion of tropical glaciers compared with high latitudes. In recent decades, a clear and unusually uniform trend is apparent, that of a worldwide retreat of tropical ice masses. A profound realization follows: the tropical glacial system has changed from one dominated by precipitation to one dominated by rising global temperatures. A chilling prediction is that, within a few decades, for the first time in at least 25,000 years there will be no tropical ice caps (Thompson et al., 2005).

2.2 Sea level

Global sea level is estimated to have been 121 ± 5 meters below the current mean level during the LGM (Clark et al., 2009; Fairbanks, 1989). In South America and Africa, lowered sea levels at the LGM resulted in the flooding of a relatively small area of continental shelf during deglaciation, but may have had a profound influence on local hydrological regimes. For example, in South America the 121 m lower than modern sea level at the LGM would have, on average, almost doubled the gradient of the Amazon River from the western tributaries to its mouth. The western river channels, however, were so remote from the ocean that the majority of the downcutting would have been in the modern tidal section of the river; that is, the eastern half of the basin. Here the river would have been constrained within an incised channel and vast areas of modern flood forest (locally termed várzea) would have been above the annual flood stages. As sea level and the river rose, the great várzea landscapes re-established about 5 ka (Behling, 2002).

In Southeast Asia the change in sea level had stronger biogeographic implications than in other tropical regions. The lowering of sea level had created dry land between many of the islands on the Sunda Shelf and the mainland. Deep channels such as that separating Bali from Lombok were not bridged by sea-level change during the Quaternary (last c. 2.6 million years) and consequently maintained strong biogeographic divides, i.e. Wallace's line, (Wallace, 1876). However, the lowered sea levels allowed migrations and genetic reassortment of many other populations emanating from the Malay Peninsula to the west and Australasia to the east.

During deglaciation, sea-level rise tracked the global ice sheet/temperature patterns. Consequently, there were two global phases of rapid sea-level rise of c.15 m per millennium: between c.16.0 and 12.5 ka, and again between c.11.5 and 8.0 ka (Lambeck et al., 2002). The peak rate of sea-level rise appears to have been c.14.7–14.4 ka when local sea levels rose by c.16 m in just 300 years (Hanebuth et al., 2000). Overall, the change in sea level reduced the bottomlands of Southeast Asia by more than 50 per cent in the space of a few thousand years (Figure 30.1).

The ecological impacts of this dramatic flooding were spatially uneven. Flood forests developed along low-lying rivers, the most notable of which was the development of Amazonian várzeas, and the migration of coastal/littoral zones. However, the impacts on Africa and South America pale in significance compared with the inundation of the Sunda Shelf. In two events of just a few hundred years each, almost 50 per cent of what had been lowland forests at the LGM were flooded by rising seawater. The consequent fragmentation vastly reduced available habitat area for many thousands of species (Bird et al., 2005) and possibly resulted in a ~10 per cent extinction or, at least, the

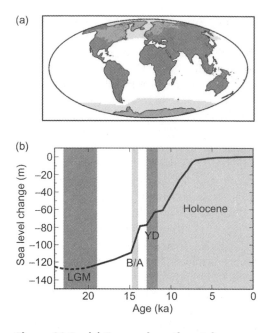

(a)

(b)

Figure 30.1   (a) Exposed continental area during the Last Glacial Maximum (c.21 ka): dark grey = exposed continental landmass, mid grey = terrestrial ice sheet, light-grey = sea ice (CLIMAP data); thin grey line indicates modern coast line. (b) Eustatic sea-level change during the transition from glacial to interglacial conditions: dark grey vertical bars indicate cool events noted in text (LGM = Last Glacial Maximum, YD = Younger Dryas); light grey vertical bar indicates warming (B/A = Bølling/Allerød and Holocene). Both images based on http://www.globalwarmingart.com.

establishment of an extinction debt (*sensu* Tilman et al., 1994) as newly formed islands were initially supersaturated with species (Diamond, 2002).

2.3 Lowland tropical continental interiors

The impact of high latitude glaciation on tropical ecosystems and their response to deglaciation has been long debated. There is not space to discuss in detail these debates here; instead a brief synopsis and citations for reviews are provided.

In the Neotropics, discussion has focused on the extent to which the Amazon Basin during the last glacial period was occupied by savannah (Colinvaux et al., 2000; Haffer, 1969; Haffer and Prance, 2001) or seasonally dry forest (Mayle, 2004; Pennington et al., 2000; Prado and Gibbs, 1993) instead of the evergreen moist forest found there today. A consensus has emerged from the increasing number of palaeoenvironmental records being published (Bush et al., 2007a). Amazonian forest area was probably largely unchanged during the glacials with some movement along ecotonal boundaries, (e.g. Absy et al., 1991; Mayle et al., 2000), but little overall change in biome type. A corridor of low modern rainfall passes north to south through eastern Amazonia. Within this area it is entirely possible that an intermittent dry forest became established during the driest times of glacial and interglacial periods (Bush, 1994). However, for the great majority of Amazonia, there is no evidence to suggest a break in forest cover. Species compositions were different from those at present, reflecting cooler conditions. Consequently, the changes from the LGM into the Holocene were relatively subtle and included: (1) the taxonomic composition of the moist evergreen forest gradually became more similar to the forests we see today, (2) moist evergreen forest expanded at the dry margins of Amazonia, and (3) dry, and cold, forest taxa were outcompeted within the Amazon Basin and migrated towards the modern dry forest regions or uplands. New data from older, middle Pleistocene, sediments on the eastern flank of the Andes suggests that forest ecosystems also existed during earlier glacial and interglacial periods but, that reassortment of forest taxa occurred in response to climate change (Cárdenas et al., 2011).

During the last glacial period, much of lowland Africa was cooler and drier than today resulting in the replacement of forest with savannah (Coetzee, 1964, 1967; Lahr and Foley, 1998; Shanahan et al., 2006; van Zinderen Barker, 1964). However, palaeoenvironmental data (Bonnefille, 2007) and vegetation modelling (Cowling et al., 2008) suggest that the tropical African lowlands during the LGM were, like their South American counterparts, predominantly forested notwithstanding alterations to composition and structure. Altered sea-surface temperatures in the Atlantic and Indian Oceans that resulted in migration of the ITCZ and varying strength of the Indian monsoon induced strong climate changes. Transition from the LGM into the Holocene in Africa can be characterized, on a continental scale, by an expansion of broadleaf forest north of the equator and a reduction of such forests in southern Africa (Cowling et al., 2008; Tierney and Russell, 2007; Tierney et al., 2008).

In Southeast Asia, few lowland lake records extend to the LGM, and the majority of understanding is based on low-resolution marine sediment records. These records generally show an expansion of grasslands at the expense of forest at the LGM, suggesting dry conditions (reviewed in Kershaw et al., 2007).

2.4 Overview: deglaciation in the tropics

Global temperatures responding to the $c.100$ ka eccentricity cycle, provided an underlying framework for glacial advance and sea-level change. Contrastingly, precipitation and its close correlate soil moisture availability, were modulated by changes to monsoonal climate systems driven by the $c.21$ ka precessional cycle. Moisture availability provided local variability in the timing of maximum glacier extent and change in the composition and structure of lowland tropical ecosystems.

During deglaciation the relative impact of warming and precipitation change mechanisms across the tropics was dependent on the geographic situation of each region. The large, flat, continental areas of South America and Africa had relatively narrow continental shelves resulting in a limited impact of post-LGM sea-level rise. In contrast, the mountainous, volcanic landscapes of Southeast Asia suffered a $c.50$ per cent loss of lowland forest area as sea level rose, potentially resulting in a 10 per cent reduction in species. Although tropical South America is basically a flat landscape, the presence and orientation of the north–south Andean cordilleras were vital corridors for species migration. The Andean mountain chain provided habitat-space variability both in elevation and latitude for taxa to occupy when confronted with changing climates. In comparison west African taxa faced with changing climates had only isolated peaks, which could not offer habitat continuity for population retreat or expansion across much of the continent. It is probably no coincidence that the highest biodiversity in Africa is associated with lands backed by the eastern arc mountains (Lovett et al., 2007).

3 THE HOLOCENE

Glacials and interglacials are the product of a continuum of climatic forcing and response. Consequently the same eccentricity and precessional influences structure both late Pleistocene and Holocene (11.7 ka to present) climate (Denton and Karlen, 1973). Holocene climate changes are best viewed on a number of different timescales to help tease apart the mechanisms involved (Mayewski et al., 2004). Superimposed on the orbital patterns, variation in the AMOC, and monsoonal activity produced millennial-scale changes in climate. Shorter quasi-cyclic events, e.g., ENSO and the Atlantic multi-decadal oscillation, induced climate change on centennial to annual time scales.

3.1 Climate structure and millennial-scale change in the Neotropics, the Palaeotropics and tropical Australasia

As in the deglacial period, one of the manifestations of changing ocean circulation was the migration of the ITCZ. The ITCZ is a maritime phenomenon that forms over the region of warmest surface waters and therefore migrates seasonally, tracking changes in sea-surface temperature (Hastenrath, 1997). The latitudinal limits are controlled by the differential seasonal warming of surface waters and consequently the southerly limit of the ITCZ occurs in January, and the northerly limit in July (Figure 30.2). Because the intensity of regional rainfall is so strongly associated the convection of the ITCZ, its arrival directly controls the timing and quantity of precipitation for much of the tropics and, in addition, affects regional monsoon systems.

Climate models (e.g., Liu et al., 2009) can produce outputs that match empirical data for both sea-surface temperature and precipitation with remarkable accuracy. A simulated cooling of sea-surface temperatures in the northern extra-tropics and a concomitant warming in the southern extra-tropics has been shown to result in a southerly shift in the position of the ITCZ and alterations in the associated wind fields and Hadley cell (Broccoli et al., 2006). Evidence from records of past environmental change in the tropics indicates that the position of the ITCZ has fluctuated during the Holocene (Haug et al., 2001; Liu et al., 2009; Peterson et al., 2000).

Neotropics

The climate of tropical South America is principally influenced by three key features: (1) the ITCZ, (2) the South American Summer Monsoon (SASM) (Vera et al., 2006; Zhou and Lau, 1998), and (3) the South American low level jet (SALLJ) (Marengo et al., 2004). Interaction of these systems leads to seasonal variations in climate. The following description is summarised from the detailed review by (Garreaud et al., 2009). During the austral winter (June, July, August) maximum continental precipitation is north of the equator, roughly in line with the position of the ITCZ, and southern Amazonia (the continental interior) is dry. During the austral summer (December, January, February) the southward migration of the ITCZ and a monsoon-like

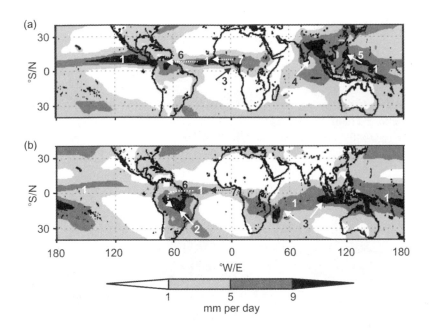

Figure 30.2   Tropical climate systems for: (a) January and (b) July. Base map is monthly mean precipitation rate (mm/day) average 1979–2008; data from the Global Precipitation Climatology Project (http://cics.umd.edu/~yin/GPCP/). Numbers indicate climate systems associated with precipitation: 1 = Inter-Tropical Convergence Zone (ITCZ) - solid arrows indicate monsoonal systems; 2 = South American Summer Monsoon (SASM); 3 = West African Monsoon (following path of low-level westerly jet); 4= Indian Ocean Monsoon; 5 = Asian Monsoon – dashed arrows indicate direction of jet streams; 6 = South American Low Level Jet (SALLJ); 7 = African Easterly Jet.

system bring copious precipitation to the continental interior including the eastern Andes and southern Amazon Basin. Additional summer moisture transport occurs across the Amazon Basin (east to west) during the summer because the warming causes an intensification of the SALLJ (~1,000 m above sea level) (Marengo et al., 2004).

The major feature of climate change in the southern hemisphere tropical Andes during the Holocene is a spatially time transgressive dry event (Figure 30.3). The Andes has one of the highest densities of palaeoenvironmental records anywhere in the tropics. However, the picture of environmental change is not clear cut and two fundamental questions remain about the nature of the moisture balance fluctuations: (1) what were

the climatic factors that were altered (i.e. decrease in precipitation, increased evaporation or a combination of the two); and (2) are the same mechanisms responsible for all the changes? First, we examine the palaeoenvironmental evidence for the moisture balance changes in the southern hemisphere tropical Andes and then discuss the possible mechanisms behind these.

Sites closest to the equator became dry during the early Holocene (*c*.11 ka). Abundant pollen taxa indicative of high Andean grassland (Puna) at Laguna Surucucho indicate dry conditions (Colinvaux et al., 1997), while the pollen spectra from Laguna Pallcacocha indicate the start of a transition from Puna (dry) to mountain woodland (moist) vegetation (Hansen et al., 2003). Higher moisture availability is indicated at these sites

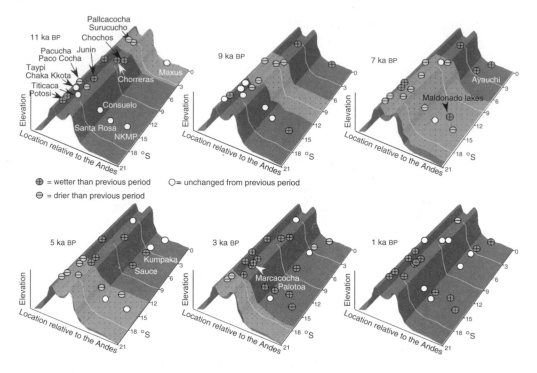

Figure 30.3 **Block diagram showing inferred lake level in the southern hemisphere tropical Andes and western Amazonia. Lighter band indicates time-transgressive dry zone.**

(abundance of woodland taxa) by 9 ka. These lower latitude lakes then maintained a positive moisture balance through the majority of the Holocene (Figure 30.3).

A few hundred kilometres towards the south, at Laguna Chochos, a lake low-stand occurred between c.9.5 and 7.3 ka; as evident from an increased abundance of shallow water aquatics (Bush et al., 2005). Further south still, a peak of aridity is indicated at c.10 ka by positive $\delta^{18}O$ values and abundant grassland pollen taxa at Junin (Hansen et al., 1984; Seltzer et al., 2000) and abundant benthic diatoms at Pacucha (Hillyer et al., 2009). At both lakes the peak in aridity is followed by a period of highly variable moisture balance. While Junin appears to become wetter after c.7.3 ka, Pacucha enters a second dry phase, coincident with those documented on the Altiplano (e.g., Baker et al., 2001b; Mourguiart et al., 1998; Paduano et al., 2003), which ends around 4.4 ka.

Palaeoenvironmental records from the most southerly section of the southern hemisphere tropical Andes suggest relatively wet conditions during the early Holocene (Figure 30.3). On the eastern Andean flank, pollen data from Lake Consuelo suggest the driest period here occurs between 9 and 5 ka (Bush et al., 2004; Urrego et al., 2010). Slightly further south and in the high Andes a dry event is indicated between 8.0 and 4.3 ka by arid pollen taxa within sediment from the main sub-basin of Lake Titicaca (Lago Grande) (Hanselman et al., 2005; Paduano et al., 2003). In the adjacent, Huiñaimarca (or Wiñaymarka) sub-basin water levels did not rise high enough to connect with Lago Grande until c.3.5 ka (Abbott et al., 1997a).

Modern precipitation in the southern tropical Andes is principally governed by the SASM and therefore peaks in the Andes during the austral summer (DJF) (Garreaud et al., 2003; Vuille et al., 2000). Temperature is

Table 30.1   Palaeoenvironmental records marked in Figure 30.3

Site name	Location	Duration (ka)	Reference
Maxus	0° 27'S, 76° 37'W; 220 masl	c.9.5	Athens and Ward (1999); Weng et al. (2002)
Ayauch[i]	2°05'S, 78° 1'W; 500 masl	c.7	Bush and Colinvaux (1988)
Chorreras	2°44'S, 79°10'W; 3700 masl	c.17	Hansen et al. (2003)
Pallcacocha	2°46'S, 79°14'W; 4200 masl	c.14	Hansen et al. (2003); Moy et al. (2002)
Kumpak[a]	3° 02'S, 77° 49'W; 250 masl		Liu and Colinvaux (1988)
Surucucho	3° 03'S, 79° 09'W; 3180 masl	c.13	Colinvaux et al. (1997)
Sauce	6°42'S 76°15'W; 600 masl	c.8	Correa-Metrio et al. (n.d.)
Chochos	7° 38'S, 77° 28'W; 3285 masl	c.17	Bush et al. (2005)
Junin	11° 00'S, 76° 08'W; 4100 masl	>43	Hansen et al. (1984); Seltzer et al. (2000)
Maldonado lakes	11° 44'–12° 08'S, 68° 52'–69° 14'W; 246–302 masl	c.8.2	Bush et al. (2007c); Bush et al. (2007b)
Marcacocha	13° 13'S, 72° 12'W; 3355 masl	c.4	Chepstow-Lusty et al. (1998); Chepstow-Lusty et al. (2003); Chepstow-Lusty et al. (2007), Sterken et al. (2006)
Pacucha	13° 36'S, 73°29W; 3095 masl	c.25	Valencia (2010); Hillyer et al. (2009)
Paco Cocha	13° 54'S, 71° 52'W; 5580 masl	c.15	Abbott et al. (2003)
Consuelo	13° 57'S, 68° 59'W; 1360 masl	c.48	Bush et al. (2004); Urrego et al. (2005); Urrego et al. (2010)
Santa Rosa	14° 28'S 67 ° 52'W 352 masl		Urrego (2006) 350 mas
Palotoa	12 ° 28'S 71 ° 32W 1360 masl	3.3	Bush (n.d.)
Taypi Chaka Kkota	16° 12'S, 68° 21W'; 4300 masl	c.14	Abbott et al. (1997b), Abbott et al. (2000), Abbott et al. (2003); Wolfe et al. (2001)
Titicaca	16.0-17.5°S, 68.5–70.0°W; 3810 masl	c.370	Abbott et al. (1997a), Gosling et al. (2008); Hanselman et al. (2005); Paduano et al. (2003); Gosling et al. (2009); Hanselman et al. (2011)
Potosi	19° 38'S, 65° 41'W; 5025 masl	c.12	Hansen et al. (2003)

primarily a function of altitude. Changes in either of these factors would result in an alteration of the moisture balance. Throughout the Holocene the variations in the precessional cycle have altered the amount and seasonality of solar radiation (Berger and Loutre, 1991). Precessional forcing has been linked with changes to moisture availability in the high Andes by comparing proxies for past environmental change with predicted orbital patterns (Hillyer et al., 2009; Seltzer et al., 2000). However, the evidence from the palaeoenvironmental records does not show a direct (linear) response between the peak in insolation and change in moisture balance (Figure 30.3). This complexity is because of the many contributing factors governing regional moisture availability. For example, at some locations (e.g., high Andes) the principal impact of higher solar radiation might be increased evaporation and consequently results in a decrease in moisture availability (Seltzer et al., 2000). Alternatively, at lower elevations, increased heating may lead to increased moisture availability through a strengthening of the monsoonal system (Baker et al., 2001a) or altering the strength of the SALLJ (Bush, 2005). Interactions between these effects could also explain abrupt changes evident in the palaeoenvironmental records.

In summary, precessional changes in insolation seem to prescribe the broad structure of changes in moisture balance in the southern hemisphere tropical Andes during the Holocene. However, the direction and degree of change is highly variable on a site-specific basis with local factors, such as elevation and topography, having significant influence.

Palaeotropics

Holocene climate in Africa can be split into two sections, the early Holocene (roughly 11.7–5.5 ka) and late Holocene (c.5.5 ka onwards). The early Holocene is often referred to as the 'African Humid Period' (Ritchie et al., 1985). However, this term refers principally to northern Africa. Wetter-than-modern conditions between 14.8 and 5.5 ka in northern Africa are attributed to increased moisture availability caused by the northerly movement of the west African monsoon, and shifts in two associated climate features: (1) the low-level westerly jet (10–12.5°N), and (2) the African easterly jet (15°N) (Liu et al., 2007). The strength and location of the monsoon is driven by summer insolation heating. Precessional orbital cycles (c.21 ka) principally control the latitudinal position of the perihelion; that is, latitudinal position of maximum summer insolation. If precessional variations in seasonality are the key drivers of moisture balance change, and therefore vegetation change, in Africa we would anticipate an antiphasing of wet/dry events north/south of the equator.

In northern Africa, during the early Holocene, lake levels were higher than they are today (Gasse, 2002; Shanahan et al., 2006), and the Sahara seems likely to have

been almost completely vegetated (Claussen, 2009; Claussen and Gayler, 1997). These data are supported by vegetation models for 6 ka (deMenocal et al., 2000), which based on multiple pollen and macro-fossil records indicate the presence of vegetation that would have required wetter than modern conditions to survive in northern Africa (Hoelzmann et al., 1998; Jolly et al., 1998). For the same period in southern Africa, palaeoclimatic inferences derived from vegetation models point towards drier-than-modern conditions, revealing that environmental change was not uniform across the continent during the Holocene (Jolly et al., 1998). Evidence from δ^{13}C in sediment from Lake Malawi (10° 15′S, 34° 19′E) through the Holocene indicates that southeastern Africa was drier (more C4 grassland vegetation) during the early Holocene (11.0–7.7 ka) and wettest (more C3 woodland vegetation) 7.7–2.0 ka (Castañeda et al., 2007, 2009; de Busk, 1998; Tierney et al., 2008). It should be noted, however, that the north–south split does not run directly along the equator, isotopic data from Lake Tanganyika at 6°S has a 'northern' African pattern, while Lake Malawi at c.14°S has a 'southern' pattern. Tierney and Russell (2007) suggest that rather than being driven solely by migration of the ITCZ, these East African lake records are responding to different seasons of the Indian monsoon as the ITCZ migrates across them.

Supporting these findings, pollen extracted from pack rat middens in Namibia (western southern Africa, c.20°S) indicates higher-than-modern moisture availability between c.6 and 1 ka (Gil-Romera et al., 2006; Scott, 1996). In addition, an abrupt decline in woodland taxa pollen, and a concomitant increase in aquatic spores in sediments off the western coast of Angola (12°S) at c.4 ka (Dupont et al., 2007), could also be interpreted as indicative of increased moisture availability. The suggested explanation for this pattern is that established savannah trees are more resilient to drought conditions, compared with grasses and herbs, due to more extensive rooting systems (Gil-Romera et al., 2007).

While the overriding orbital mechanism for moisture balance change in Africa is reasonably well established, the nature, rate, and mechanisms behind the transitions remain controversial. The termination of the humid period in northern Africa, c.5.5 ka, is particularly hotly debated. deMenocal et al. (2000) interpreted abrupt changes in the dust record from marine sediments off the cost of Mauritania (20° 45´N, 18° 35´W) as evidence of rapid (10–100 year) change, which could only be modelled by allowing biogeophysical feedbacks to amplify the change driven by orbital forcing. In contrast, Kröpelin et al. (2008b) infer a gradual drying of the northern African climate from fossil pollen, chironomid, diatom and sedimentary data from Lake Yao, northern Chad (19.03°N, 20.31°E) through the last 6 ka. Kröpelin et al. suggest the progressive drying is a direct result of insolation forcing (as opposed to rapid change effected by biogeophysical feebacks). However, the exact nature of any feedbacks and regional applicability of both these records remains unclear (Brovkin and Claussen, 2008; Kröpelin et al., 2008a, 2008b).

Various coupled climate models have been used to explore the mechanisms behind the wet–dry transition in northern Africa. Models have been able to simulate two steady states in northern African climate (vegetated and nonvegetated) but there is, as yet, no consensus on the mechanisms responsible for 'flipping' the Sahara between these two states. For example, a coupled atmosphere–ocean–terrestrial ecosystem model for the Holocene indicates that in northern Africa the relationship between vegetation and precipitation is nonlinear (Liu et al., 2007). Simulations show a rapid collapse of vegetation in northern Africa c.5 ka while precipitation continues to decrease gradually. Liu et al. (2007) interpret that this vegetation collapse is due to the crossing of an intrinsic bioclimatic vegetation threshold rather than an abrupt climate shift. In contrast, the complex interaction between long-term (orbitally forced) and rapid (feedback driven) climate change

highlights the importance of understanding global climate change on a variety of temporal scales.

Tropical Australasia

The record from Lynch's Crater from Queensland, Australia, reveals the replacement of sclerophyll forest with mesic tropical forest at the start of the Holocene, suggesting a transition from relatively dry to wetter conditions. This pattern is reproduced in almost all lowland Southeast Asian records (e.g., Hope et al., 2004). The Lynch's Crater record also demonstrates the influence of precession on this system and the suggested presence of millennial-scale variability between El Niño-dominant and La Niña dominant phases throughout the last glacial period (Turney et al., 2004). High-quality data sets providing Holocene palaeoclimatic histories of Southeast Asia are still sparse. Data from Thailand (Kealhofer and Penny, 1998; White et al., 2004), Java (van der Kaars and Dam, 1997; van der Kaars et al., 2001), New Guinea (Hope, 2009) and offshore cores from the Aru and Banda seas provide some of the best palaeovegetation records for this region (van der Kaars et al., 2000).

Climate change within the Holocene is driven by the monsoonal system. Maxwell (2001) produced an estimate of monsoon activity from Lake Kara, Cambodia, based on pollen and changes in organic/inorganic carbon. Most of the lowland sites appear to have human occupation leading to elevated levels of burning. Although fire histories associated with human occupation date to the last glacial maximum in New Guinea, there appears to be an intensification of human burning of these landscapes after c.7 ka (Hope and Golson, 1995).

Overview: tropical climate patterns

Through the Holocene orbital forcing produced long-term gradual change in climate that was broadly antiphased between the tropical northern and southern hemispheres (Figures 30.4 and 30.5). However, feedback mechanisms that accelerated or weakened

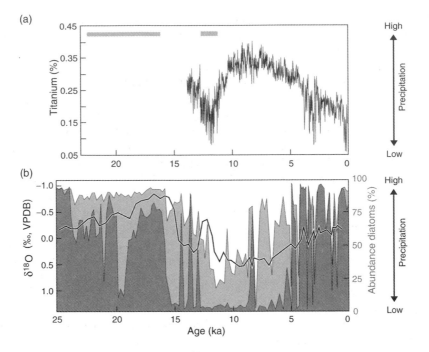

Figure 30.4   Early-Holocene (11-6 ka) antiphasing of selected paleoclimatic records from South America. (a) Black line = Cariaco Basin (Venezuela), titanium (9°N) (Haug et al., 2001) and grey bars = northeastern Brazil; (10°S) pluvial phases indicated by travertine deposits (Wang et al., 2004). (b) dark gray = planktonic-Lake Pacucha (Peru) diatom abundance (13°S), dark gray = planktonic, mid gray = shallow water, white = benthic species (Hillyer et al., 2009) and black line = Botuverá Cave δ¹⁸O southern Brazil (27°S) (Cruz et al., 2005; redated by (Wang et al., 2008).

AMOC, such as rapid melting of ice sheets and freshwater pulses, induced sudden shifts in the position of the ITCZ and strengths of the monsoon circulation. Consequently, while local climatic variability is present in all records, the signatures of these larger trends is usually apparent. At tropical latitudes, where climate is dominated by the migration of the ITCZ, the broad pattern of change will be consistently wetter (drier) as the influence of the ITCZ increases (decreases). As would be predicted, broad antiphasing between wet/dry events in northern and southern Africa through the Holocene suggests that orbital precessional control on seasonality is an important driver of moisture availability which in turn is the major driver of vegetation change (Castañeda et al., 2007). While antiphasing on a millennial scale is evident between lakes Malawi and Tanganyika in the east African

Rift, it is very noticeable that Andean lakes are similarly antiphased; that is, the Fuquene record of Colombia 9°N and the Titicaca record at 16°S across the hemispheres.

The coherence of the antiphasing also holds up as one compares records on different continents. Though this relationship has been suggested for some time (e.g., Baker et al., 2001a; Marchant and Hooghiemstra, 2004; Thompson et al., 2005), it is only as higher-resolution data sets have become available that the degree of (mis)match is evident.

The δD values derived from leaf waxes for Lake Tanganyika closely parallel the isotopic data from subtropical cave records in eastern China (both 'northern' signatures), while the Neotropical records have a 'southern' signature. Note that in the neotropics there is antiphasing between northern (Cariaco

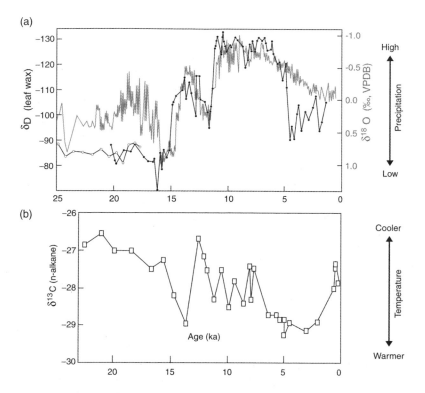

Figure 30.5   Early-Holocene (11–6 ka) antiphasing of selected paleoclimatic records from Africa and Asia. (a) Grey line = Hulu Cave, eastern China, $\delta^{18}O$ (32°N) (Wang et al., 2001) and black line = Lake Tanganyika, δD leaf wax (6°S) (Tierney et al., 2008). (b) Lake Malawi, $\delta 13C$ (10°S) (Castañeda et al., 2009).

Basin) and southern locations (Pacucha and southeastern Brazil). High levels of Titanium in the Cariaco record indicate high stream discharge, at a time when the southern records are experiencing lake lowstands. While we can speculate on the teleconnections that could explain intercontinental linkages, it is hoped that the next generation of climate models will make clear the causality underlying these empirical observations.

3.2 Submillennial scale oscillations

Millennial-to-decadal cycles evident in ocean records, have been (controversially) attributed to variance in solar activity and insolation (Bond et al., 2001). While the true mechanism of millennial-scale variability is

unknown, Dansgaard–Oeschger oscillations and related changes in the AMOC that dominated glacial climate change at high latitudes probably continued, albeit with moderated intensity, into the Holocene.

Bond et al. (2001) suggested that a quasi-periodic cycle of c.1500 years was a persistent feature of both Holocene and full glacial climates. Bond et al. observed that Arctic sea ice expanded rapidly eight times in the last 11.5 ka, in events that became known as 'Bond Cycles'. The ice-expansion events appeared to follow a c.1500 year rhythm, although Bond et al. were careful to call this quasi-cyclic and not periodic. The timing of these drift-ice extensions in the North Atlantic mirrored records of [10]Be and ^{14}C that were taken to be proxies for solar activity (high [10]Be concentrations corresponding to reduced solar output). With some data smoothing, periods of reduced solar output appeared to match stacked records of sea-ice expansion. Bond et al. suggested a causal mechanism of cascading effects that originated with insolation minima reducing the formation of stratospheric ozone that, in turn, initiated a cooling of the northern hemispheric polar airmasses. A consequence of that new temperature gradient would be the southward displacement of the northern subtropical jet that led to weakened northward migration of the Hadley cell. Bond et al. (2001) predicted that the effects of these cycles should be global rather than local, and suggested that one mechanism for transference to low latitudes would be through a weakening of the ITCZ in the northern hemisphere leading to drought.

As noted above, the long-term (precessional) changes in insolation induced a mid-Holocene hypsithermal in many locations that drove drying in South America and wetter conditions in Saharan Africa. However, superimposed on this trend were the shorter oscillations for which the Greenland ice-core records have become the *de facto* benchmark. As Bond et al. noted, changes in millennial-scale solar activity in the tropics would be expected to be manifested in precipitation rather than temperature changes. Speleothem, ocean sediment, and lake proxy data all suggest a strong linkage between Greenland ice core data (and by inference variation in the circulation of the North Atlantic) and monsoonal activity in Africa, Asia and Central America.

Indian and Asian monsoon systems

Morrill et al. (2003) synthesized 36 records of Asian monsoon activity and found a general pattern of strengthening at 11.5 ka, with a peak perhaps around 7 ka and then an abrupt weakening between 5 and 4.5 ka and again at *c*.1.3 ka. While concordant in general pattern with Morrill et al.'s findings, a high-resolution speleothem record from Dongge Cave, China, adds considerable detail (Wang et al., 2005). Four dry events

punctuated the period of intense monsoon activity between 11.5 and 8 ka, all of which appear to be coeval with events in Greenland. High-resolution data for the period between 8.11 and 8.25 ka revealed significant peaks of wet and dry episodes with periodicities of 208 and 86 years. The similarity of this pattern to the 88-year and 208-year cycles of solar activity (Wang et al., 2001) suggested a solar-forcing component of these decadal-centennial scale events in China.

The Indian Ocean monsoon was also found to respond synchronously with the Greenland (GRIP) δO^{18} record. Data from southern Oman documented the strength of the Indian monsoon (Fleitmann et al., 2003) and from 10.3 to 8 ka the two records were strikingly similar. However, after 8 ka the relationship with Greenland became less certain. Fleitmann et al. suggested that the presence of northern hemispheric ice masses may have provided causal linkage between these records, but as ice cover waned, other factors played an increasing role in shaping Indian Ocean monsoon circulation. The issue of cause in what appear to be teleconnected records has yet to be fully resolved.

The climate of East Africa and the Indian subcontinent is strongly influenced by the Indian Ocean monsoon. Data from the Arabian Sea in the glacial show a strong relationship to events in the North Atlantic. In the Holocene, off the coast of Oman (ODP Hole 723A) this relationship is still evident with seven events of weakened monsoons that correlate with Bond events (Gupta et al., 2003). The proxy in this case is the percentage occurrence of *Globigerina bulloides*, a foraminifera species that requires upwelling, and hence is most frequent during strong monsoon episodes. It is important to note that the amplitudes of change for the proxy of Holocene monsoon oscillations are markedly smaller than those of the glacial but, even so, the impacts on the terrestrial system are very strong.

Further analysis of the same core revealed a strong response to the 8.2 ka event, that is generally associated with meltwater discharge to the Atlantic, and both the Medieval Warm Period (MWP; wet) and the Little Ice Age (LIA; dry) (Gupta et al., 2003). Moving from correlation to causation in palaeoclimatology is often problematic, but Gupta et al. (2005) suggest that the Indian Ocean monsoon responds directly to solar forcing of the convection associated with the ITCZ. Simply put, strong solar activity translates to strong convection and a strong monsoon, with failures of rains associated with weaker solar activity.

Intertropical Convergence Zone

Very similar patterns were evident in Central America. A high-resolution record from Lake Petén Itzá in Guatemala, provided data for the last 85 ka (Bush et al., 2009; Hodell et al., 2008). Although data from this record for a variety of proxies are preliminary, magnetic susceptibility, δO^{18} on fossil ostracods, fossil pollen and charcoal, portray a detailed record of climate change. The Heinrich Events of the last glacial period are represented in the fossil pollen record by unusual assemblages dominated by taxa indicative of drier than modern conditions but without modern analogue (Correa-Metrio et al. n.d.). A surprising finding of these analyses was the relatively moist conditions of the last glacial maximum. Hodell et al. (2008) suggested that a southward displacement of the ITCZ, leading to a reduction in wet season rainfall, and of the polar jet stream increasing penetration by cold, wet winter airmasses, increased rainfall during the winter (dry) season. The driest time of the last 85 ka was the deglacial period from c.19 to 14.7 ka, when the cool glacial-age forest dominated by *Quercus* and *Pinus* was replaced by *Acacia* woodland. The Petén Itzá record faithfully reflects the deglacial oscillations found in the North Atlantic, including Heinrich Event 1 (16.8 ka), the 14 ka and 8.2 ka meltwater events, and the YD. Mid- and late-Holocene changes in this record are probably masked by human activity in the watershed.

The Cariaco Basin off the coast of Venezuela provided an exceptional palaeoclimate record of the last 50 ka (Hughen et al., 2000; Peterson et al., 2000). The glacial climate of this coast was dry, and conditions became progressively wetter during deglaciation. A major reduction in stream discharge leaving the continent was reflected in decreased titanium and iron transport during the Younger Dryas as dry conditions returned. The early Holocene was wet with the maximum discharge of Fe and Ti between 10.5 and 5.4 ka. A steady trend toward drier conditions in modern times was punctuated by an episode of strong droughts between 3.8 and 2.8 ka, with minima of Fe and Ti deposition at 3.8, 3.4, 3.0 and 2.8 ka. Another low in offshore transport suggests the LIA (550–200 years BP) to have been dry. These data for cool periods equating to low precipitation, suggest that the ITCZ failed to bring rain to this latitude (10°N) (Haug et al., 2001; Hughen et al., 2000; Peterson et al., 2000). Within the Holocene, large-scale changes associated with the mid-Holocene warm event suggest profound ecosystem changes between c.11 and 4 ka, according to location.

Solar cycles

Quasi-cyclic climatic activity inferred from Andean records probably reflect variations in solar output. At Laguna Chochos, Peru, a 1500-year cycle is evident between 17 and 8 ka; however, it is also evident that this signal could be a multiple of shorter wavelength cycles of c.210 years (Bush et al., 2005) that, within margins of dating error, would correspond to the 208-year solar cycle detected in the ^{14}C record (Peristykh and Damon, 2003).

In Central America, a 208-year periodicity from Lake Chichancanab, Guatemala, of recurrent drought in the last 1,500 years was reflected in $CaCO_3$ concentrations and δO^{18} signatures forming ~50 year cycles (Hodell et al., 2001). The driest times coincided with the times of solar minima suggesting that the northern limb of the ITCZ was weakened during solar minima. The Asian monsoon is also forced at similar intervals with data from the Arabian Sea off Oman (Anderson and

Prell, 1993), and Dongge Cave, China, that reveal an 11, 86, and 208-year periodicity (Wang et al., 2005; Yuan et al., 2004).

The decadal to centennial-scale changes in solar output are much less than the $c.8$ per cent change in insolation induced via precessional changes. Solar cycles of 11, 88, and 208 years are known to occur (Peristykh and Damon, 2003), and long records of sunspot cycles demonstrate solar output minima occurred at AD 1790–1820 (Dalton Minimum); AD 1645–1715 (Maunder Minimum); AD 1420–1570 (Spörer Minimum); AD 1280–1340 (Wolf Minimum); AD 1010–1050 (Oort Minimum). The Spörer and Maunder Minima are often suggested to relate to the early and late phase cooling of the Little Ice Age. However, solar output during the regular 11-year cycle fluctuates by only about 0.1 per cent, whereas, the Maunder Minimum was estimated to reduce solar output by $c.0.24$ per cent (Lean et al., 1995). The 0.24 per cent reduction suggested for the Maunder Minimum would translate to a cooling of 0.2–0.6°C, compared with the observed global cooling of $c.1$°C. If these subtle differences in solar output induced climate change an associated amplifying factor or positive feedback mechanism would be needed.

In Amazonia, 228 dated samples of soil charcoal were found to cluster around three solar minima between $c.$ AD 800 and 1500, but that two later minima (Spörer and Maunder) showed much lower fire frequency (Bush et al., 2008). If it is accepted that solar output minima can influence climate, the data are consistent with human activity providing sources of ignition that, when coupled with solar-forced drought, spawned widespread wildfires. That later minima did not repeat this pattern probably reflected the human population crash of $c.90$ per cent associated with conquest and the arrival of European diseases (Black, 1992).

Overview: submillennial scale oscillations
Through the Holocene submillennial scale oscillations have perturbed global climates. Across the tropics rapid climate change

events are generally best recorded as fluctuations in moisture availability. Variation to the climate systems which distribute moisture at tropical latitudes, such as monsoons and the ITCZ, are probably the mechanisms that underlie the oscillations. The driving force behind these wet/dry events remains ambiguous. Changes in solar irradiance and links with high latitude ice-sheet expansion and collapse may be causally related to tropical moisture patterns. Further well-dated, high-resolution records from the tropics are required to determine the magnitude and extent of rapid climate change within tropical environments during the Holocene.

4 HUMAN IMPACTS

4.1 Fire

Human impacts on landscapes in the tropics are as old as human occupation of the tropics, and hence geographically very variable. Arguably, the greatest changes wrought by pre-historic peoples were through introducing fire to systems that were not previously fire prone perhaps and causing animal extinctions. In Africa, the long history of Plio-Pleistocene human evolution is intimately tied to the gradual drying of continental Africa. Fire is probably natural within African ecosystems; nevertheless, archaeologists report human-induced fires. Dynamic global vegetation models suggest that East Africa would support woodland rather than savanna in the absence of fires (Bond et al., 2005). The extent to which human-induced fires played a role in the positive feedback loop that led to savanna expansion is unclear.

In the other dry tropical continent, Australia, human arrival at $c.50$ ka (Bowler et al., 2003) is closely associated with extinction of megafauna (Miller et al., 2005 though see Wroe and Field (2006) for a different viewpoint), and a shift to increasing fire-adaptedness and sclerophylly between 35 and 45 ka (Singh and Geissler, 1985). The period of human arrival, MIS 3, was generally taken to be climatically stable, and consequently increased fire frequency in Australia was suggested to be attributable to human activity. However, data from marine cores, ODP Site 820 and MD 98-2167 (Kershaw et al., 1993; Moss and Kershaw, 2000), suggest that this was a time of relatively strong climate changes, perhaps ENSO related, but also of floristic changes without parallel in the previous glacial cycle. Evidence of vanishing mega-lakes in the interior of Australia at the time of the megafaunal extinctions suggests strong climatic changes could be implicated (Cohen et al., 2011). Vine thickets rich in *Olea* that had been a characteristic component of the Queensland landscape for 250 ka were abruptly replaced by *Eucalyptus* forest *c.*45 ka (Kershaw et al., 2007). That fire was involved in these transitions is suggested by the presence of charcoal in the soils of Queensland rainforest near its modern southern limit. Those charcoal fragments generally date between 30 and 8 ka (Hopkins et al., 1990). In Australia, a combination of climate change promoting flammability and the presence of humans to provide the ignition for the fire seems plausible (Hope et al., 2004); a synergy that was to be repeated in the Americas.

4.2 Megafaunal extinctions

The arrival of humans and the loss of megafauna (animals weighing >44 kg) throughout the tropics, with the notable exception of Africa, is a recurrent theme. The extinction of megafauna in South America was even more severe than in North America with at least 37 genera and 85 species being lost (Barnosky, 2004, Barnosky et al., 2008). Steadman et al. (2005) put most of these extinctions between 12 ka and 11 ka, but it is apparent that some taxa survived into the early Holocene, probably disappearing between 9 and 8 ka (Hubbé et al., 2007; Barnosky and Lindsey, 2010). Humans have been in the South American tropics since at least 15 ka (Dillehay, 2009) and in Amazonia

since at least 13 ka (Roosevelt et al., 1996). Consequently, it is apparent that people and megafauna coexisted in South America for as much as 7 ka – much longer than the generally accepted 1–2 ka in North America.

The cause of megafaunal decline has become polarized with advocates of a human-induced 'overkill hypothesis' *versus* those attributing the event to climate change (e.g., Grayson and Meltzer, 2003; Martin and Klein, 1984). Both sides end up with positions that are very difficult to defend in their pure form. The original concept of the overkill was applied to the peopling of the Americas by hunters using Clovis technology (Martin, 1973). Clovis hunters had a distinctive arsenal of stabbing spears and tools clearly adapted to hunting and butchering large game. The Clovis hunters were thought to have crossed Beringia when sea levels were low enough to allow them to walk across from Siberia. Fanning out across the North American continent, the Clovis exploited naïve prey that had never encountered a two-legged hunter. Within *c.*1,000 years, herbivores such as wild horses, mammoth, glyptodonts, ground-sloth and camelids; predators such as American lion and short-faced bear; and scavengers such as the giant condor, were extinct. The best evidence of all for the overkill was that megafaunal populations collapsed at different times around the world, and that each time it appeared to coincide with the arrival of humans. While the timing of disappearance was enough to convince many of the importance of human activity, and Clovis points were found at some mammoth and ground sloth kill sites, much of the rest of the argument was speculative. Models of human population growth and food consumption demonstrated that given basic assumptions it would have been possible for a front of hunters to exterminate big game animals in North America (Alroy, 2001). However, the underlying assumptions of those models about human reproductive success, demand for meat (as opposed to fish, shellfish and other popular subsistence prey), and rates of spread, can certainly be questioned. So, in short, people were involved in the demise of some megafauna, but were they *causal* in these population collapses?

The other side of the debate argued that the dating of the extinctions was fuzzy, and that many species may have gone extinct before humans arrived, while others persisted for some thousands of years alongside humans (Wroe et al., 2006; Barnosky and Lindsey, 2010). They argued that the Clovis period itself was now redefined to represent a very brief window of just 500 years (Waters and Stafford 2007), probably not long enough to extirpate the megafauna. While it was certainly plausible that humans might have been effective hunters of big game in the early Holocene grasslands of North America, it was harder to see how they could be as effective in the rainforests of Central and South America, or even the forests and swamps of the eastern US. A similar argument was made for megafaunal loss in New Guinea. Diamond (1989) invoked the terms *Blitzkrieg* (lightning war) extinction to represent the overkill as suggested by Martin, while a modified form of extinction he labelled as *Sitzkrieg* (phoney war). Diamond's *Sitzkrieg* refers to the indirect changes that humans can engender in addition to hunting pressure; for example, fire, habitat loss/modification, introduction of non-native predators, competitors or pathogens, or initiating ecological cascades that lead to extinction in non-prey organisms. Not satisfied with inclusion of humans as the causal agents of megafaunal loss, Grayson and Meltzer (e.g., Grayson and Meltzer, 2003, 2004) have mounted a steady opposition to the overkill hypothesis and suggest that the rapid climatic oscillations of the last 30 ka caused, for species-specific reasons, the demise of the megafauna.

The greatest argument against the climate-based extinction is why it would occur in this ice-age transition and not one of the previous 20+ glaciations of the Quaternary. Although there is evidence of Gymnosperm extinctions in Tasmania and Southern Australia at the Plio-Pleistocene boundary, it is generally assumed that the glaciations that followed the 41 ka obliquity cycle (MIS 17–104) were milder than those of the last 500 ka. Thus a wave of extinction might have been expected with the first of these major glaciations, but none is identified there.

Lowland tropical forests may have been less susceptible to drought than savanna and montane settings. The great majority of sites yielding megafaunal fossils in the lowland Neotropics have been from relatively dry settings (Borrero, 2009). In these settings the landscape is drier and more seasonal than Amazonia. Indeed, it has become widely accepted that the megafauna were creatures of grasslands and woodlands rather than forests. This preference for grassy settings is supported by the high-crowned dentition (hypsodonty) common to many of the megafauna that suggests a grazing diet. Perhaps the ecology of the megaherbivores led them to exist in the habitats most vulnerable to climate change. Further research is needed to establish the extent to which megafauna used Amazonian settings.

If climate caused the collapse of megafauna, it is very unlikely to have been a synchronous event across the neotropics. In North America, the last gasp of the ice age, experienced as a sudden chilling between *c.*12.9 and 11.7ka in many areas, termed the Younger Dryas, was followed by a rapid warming. This rapid climatic oscillation has often been invoked as a source of stress for megafauna, though Gill et al, (2009) used the dung-fungus *Sporormiella* to demonstrate megafaunal population declines beginning about 15–14ka in Indiana. Those same oscillations are all recorded in the fossil sedimentary sequence from Lake Petén-Itzà, Guatemala; however, the variable that was changing was primarily precipitation not temperature (Hodell et al., 2008). The tropical glacial had been cooler (probably 5–7 °C) than the Holocene, but the deglaciation began around 19 ka and produced a nonanalogue flora. Oak, myrtle, and pine died out locally, a local extinction, but close analogues to this glacial age woodland are today's forest at *c.*1,500 m elevation in Mexico, suggesting that species ranges contracted upslope. In place of the oak–pine–myrtle forest that surrounded the glacial age lake (or the seasonal tropical rainforest found there today), an *Acacia*-scrub woodland developed, in which fire was frequent and productivity low (Bush et al., 2009). This period of scrubby vegetation from *c.*19 to 14.7 ka was the driest of the last 85 ka, with near-modern conditions established abruptly around 11 ka. These changes in vegetation were in many ways more profound than those at higher latitude during the Younger Dryas event. Could this regional aridity and frequent fire have contributed to the loss of megafauna?

In Amazonia, there is a somewhat different story, though parallels to the Central American one can be found. The warming associated with the termination of the last ice age was gradual in Amazonia and the adjacent Andes. Indeed *the rate* of climate change measured in 200-year time slices, at least in terms of temperature, did not differ between 45 and 10 ka (Bush et al., 2004; Urrego et al., 2005). A gradual warming begins at *c.*19 ka and continued until 11 ka at a rate of about 0.6°C per millennium.

Deglacial precipitation and effective moisture availability was highly temporally and spatially heterogeneous. In the wet region of northern Amazonia, the driest time of the last 180 ka varied was between 33 and 27 ka, whereas the period from 33 to 15 ka was the driest in sections of eastern Amazonia, whereas 25 to 15 ka was dry in southeastern Amazonia (Bush et al., 2008). All of these events coincided with relatively cool climates, perhaps reducing the stress on plants and megafauna; certainly they do not seem to coincide with extinctions. Indeed the period of northern hemispheric deglaciation was generally wet in Amazonia, though the early Holocene was a time of intense droughts. Is it chance that charcoal first appears in Amazonian lake records at c. 9 ka, coinciding with the extinction of the last megafauna? Again, is it coincidence that the later megafaunal decline in Amazonia than in Central America, also coincides with a period of

increased fire frequency and strong drought? While the overkill concept in its pure form seems untenable, a satisfactory explanation for the timing, of these extinctions has yet to be identified. It appears very likely that people exterminated the last of the megafauna, but these mammal populations may have been at a natural low ebb due to climate change, rendering them especially vulnerable to hunting pressure (Cione et al., 2003; 2009).

4.3 Agricultural impacts

Independent inventions of agriculture occurred in Africa, Southeast Asia, and the Americas in the early Holocene. While grasses formed an important component in all regions (rice in Asia, maize in Central America, and millet in Africa), legumes and root crops were also an essential source of carbohydrate for tropical populations (see Diamond, 2002). Rather than covering this vast topic we concentrate on one facet: what was the scale of human impact prior to European arrival?

Humans spread rapidly through many tropical landscapes, with many records exhibiting human influence. However, because of the inherent bias that most palaeoecological data are drawn from lakes and people like living beside lakes and rivers, the density of settlement in the drier hinterlands is largely unknown. Where large human populations are known to have existed; for example, in Central America, Java, Sri Lanka, and Vietnam, the impact on landscapes appears to have been profound, with deforestation and fire greatly reducing forest area. In other areas (e.g., Amazonia, and the Sahel), the influence of human activity is far from certain. Archaeological evidence clearly points to distinct cultural evolution and the establishment of substantial settlements (Heckenberger et al., 2003, 2008), but whether these spectacular archaeological sites are truly characteristic of the broader landscape awaits investigation. Certainly, not all researchers are convinced that Amazonia was extensively modified by human activity prior to the arrival of Columbus. Analysis of soil charcoal as a proxy for fire, in areas that do not burn naturally, suggest that occupation was widespread but its effects were highly localized (McMichael et al. in press).

If Amazonia is shrouded in mystery, the history of Africa is even more poorly known. The Sahara was undoubtedly occupied when it was a fertile landscape, but whether the loss of vegetation due to climate change was abetted by overgrazing needs to be tested.

A pattern that has been suggested in the Americas is that human occupation and impacts achieved their peak intensity in moderately seasonal areas (3–5 month dry season). Areas that were everwet were unpleasant places to live, while areas that were very strongly seasonal also have the most erratic climates and are probably difficult places to base a settlement (Bush et al., 2008).

4.4 Overview: human impact

Hominids evolved in Africa, and arrived in tropical Australasia and America at *c.*50 and 16 ka respectively. The principal way in which early humans modified their environment was through the use of fire. Peaks in charcoal records are generally associated with the arrival of humans. It is possible that the introduction by humans of fire into ecosystems that did not previously burn could have altered the vegetative make up and broader functioning. The coincidence of human dispersal patterns and megafaunal extinctions underlies a compelling case for a hunting driven demise for many creatures. However, the low density of human populations at this time means it is unlikely that hunting was the sole driver in the demise of large mammals. More likely is that the additional pressure placed on the large mammals by humans combined with environmental stresses, massive rapid global climate oscillations and habitat loss (particularly in Southeast Asia), to drive populations below

the point of sustainability. Human occupation of the tropics appears to have been widespread, with high intensity exploitation and impacts probably restricted to areas of optimal seasonality and fertility.

5 CONCLUSIONS

The last deglaciation had a spatially variable impact on tropical ecosystems largely dependent on topography. In Southeast Asia around 50% of the forested area disappeared beneath rising seas. While in South America and Africa the orientation and location of the mountain chains determined the opportunity for species to migrate in response to global climate change events. The Holocene has been a time of intense environmental change in the tropics, with sharp changes in precipitation as the most significant drivers of climate variability. Some of these events have followed long-term orbital patterns, while others have been millennial-scale quasi-cyclic oscillations. As a result of these changes, major ecotonal boundaries have drifted. The results of these events were often broadly opposite in the northern *versus* southern hemisphere. However, other factors, such as the loss of the megafauna and the spread of human populations, have ensured that the Holocene was utterly unlike any previous interglacial period. While lags in the influence of humans may be determined by geography, unlike climatic change, the outcome is uniform: a reduction in landscape and biotic diversity.

ACKNOWLEDGMENTS

This work was funded by grants from NSF and the Gordon and Betty Moore Foundation. This is publication number 44 of the Florida Institute of Technology's Institute for Research on Global Climate Change. Gosling was supported by a Research Councils UK Academic Fellowship at The Open University.

REFERENCES

Abbott M. B., Binford M. W., Brenner M. and Kelts K. 1997a. A 3500 ^{14}C high-resolution record of water-level changes in Lake Titicaca, Bolivia/Peru. *Quaternary Research* 47: 169–180.

Abbott M. B., Seltzer G. O., Kelts K. R. and Southon J. 1997b. Holocene paleohydrology of the tropical Andes from lake records. *Quaternary Research* 47: 70–80.

Abbott M. B., Wolfe B. B., Aravena R., Wolfe A. P. and Seltzer G. O. 2000. Holocene hydrological reconstructions from stable isotopes and paleolimnology, Cordillera Real, Bolivia. *Quaternary Science Reviews* 19: 1801–1820.

Abbott M. B., Wolfe B. B., Aravena R., Wolfe A. P., Seltzer G. O., Mark B. G. et al. 2003. Holocene paleohydrology and glacial history of the central Andes using multiproxy lake sediment studies. *Palaeogeography Palaeoclimatology Palaeoecology* 194: 123–138.

Absy M. L., Cleef A., Fornier M., Servant M., Siffedine A., Da Silva M. F. et al. 1991. Mise en evidence de quatre phases d'ouverture de la foret dense dans le sud-est de l'Amazonie au cours des 60 000 dernieres annees. *Comptes Rendues Academie des Sciences Paris* 313: 673–678.

Alroy, J. 2001. A multispecies overkill simulation of the end-Pleistocene megafaunal mass extinction. *Science* 292: 1893–1896.

Anderson D. M. and Prell W. L. 1993. A 300 kyr record of upwelling off Oman during the late Quaternary: evidence of the Asian southwest monsoon. *Paleoceanography* 8: 193–208.

Athens J. S. and Ward J. V. 1999. The late Quaternary of the western Amazon: climate, vegetation and humans. *Antiquity* 73: 287–302.

Baker P. A., Rigsby C. A., Seltzer G. O., Fritz S. C., Lowenstein T. K., Bacher N. P. and Veliz C. 2001a. Tropical climate changes at millennial and orbital timescales on the Bolivian Altiplano. *Nature* 409: 698–701.

Baker P. A., Seltzer G. O., Fritz S. C., Dunbar R. B., Grove M. J., Tapia P. M. et al. 2001b. The history of South American tropical precipitation for the past 25,000 years. *Science* 291: 640–643.

Barnosky A. D. 2008. Megafauna biomass tradeoff as a driver of Quaternary and future extinctions. *Proceedings of the National Academy of Sciences USA* 105: 11543–11548.

Barnosky A. D., Koch P. L., Feranec R. S., Wing S. L. and Shabel A. B. 2004. Assessing the causes of late Pleistocene extinctions on the continents. *Science* 306: 70–75.

Barnosky A. D. & Lindsey E. L. (2010) Timing of Quaternary megafaunal extinction in South America in relation to human arrival and climate change. *Quaternary International* 217: 10–29.

Behling H. 2002. Impact of the Holocene sea-level changes in coastal, eastern, and central Amazonia. *Amazoniana* 17: 41–52.

Bennett K. D., Haberle S. G. and Lumley S. H. 2000. The Last Glacial–Holocene Transition in Southern Chile. *Science* 290: 325–328.

Berger A. and Loutre M. F. 1991. Insolation values for the climate of the last 10 million years. *Quaternary Science Reviews* 10: 297–317.

Bird M. I., Taylor D. and Hunt C. 2005. Palaeoenvironments of insular Southeast Asia during the Last Glacial Period: A savanna corridor in Sundaland? *Quaternary Science Reviews* 24: 2228–2242.

Björck S., Walker M. J. C., Cwynar L. C., Johnsen S., Knudsen K.-L., Lowe J. J. and Wohlfarth B. 1998. An event stratigraphy for the last termination in the North Atlantic region based on the Greenland ice-core record: a proposal by the INTIMATE group. *Journal of Quaternary Science* 13: 283–292.

Black F. L. 1992. Why did they die? *Science* 258: 1739–1740.

Bond G., Kromer B., Beer J., Muscheler R., Evans M. N., Showers W. et al. 2001. Persistent solar influence on north atlantic climate during the Holocene. *Science* 294: 2130–2136.

Bond W. J., Woodward F. I. and Midgley G. F. 2005. The global distribution of ecosystems in a world without fire. *New Phytologist* 165: 525–538.

Bonnefille R. 2007. Rainforest responses to past climatic changes in tropical Africa, in Bush M. B. and Flenley. J. R. (eds) *Tropical Rainforest Response to Climatic Change.* Chichester: Springer, pp. 117–170.

Borrero L. A. (2009) The elusive evidence: The archaeological record of the South American extinct megafauna, in Haynes, G. (ed.) *American megafaunal extinctions at the end of the Pleistocene* Springer, pp. 145–168.

Bowler J. M., Johnston H., Olley J. M., Prescott J. R., Roberts G., Shawcross W. and Spooner N. A. 2003. New ages for human occupation and climatic change at Lake Mungo, Australia. *Nature* 421: 837–840.

Bradley R. S. and England J. H. 2008. The Younger Dryas and the Sea of Ancient Ice. *Quaternary Research* 70: 1–10.

Broccoli A. J., Dahl K. A. and Stouffer R. J. 2006. Response of the ITCZ to Northern Hemisphere cooling. *Geophysical Research Letters* 33: L01702.

Broecker W. S. 1998. Paleocean circulation during the last deglaciation: a bipolar seesaw? *Paleoceanography* 13: 119–121.

Brovkin V. and Claussen M. 2008. Comment on 'Climate-driven ecosystem succession in the Sahara: The past 6000 years'. *Science* 322: 1326–1327.

Bush M. B. 1994. Amazonian speciation: A necessarily complex model. *Journal of Biogeography* 21: 5–18.

Bush M. B. 2005. Of orogeny, precipitation, precession and parrots. *Journal of Biogeography* 32: 1301–1302.

Bush M. B. and Colinvaux P. A. 1988. A 7000-year pollen record from the Amazon lowlands, Ecuador. *Vegetatio* 76: 141–154.

Bush M. B., Correa-Metrio A., Hodell D. A., Brenner M., Ariztegui D., Anselmetti F. S. et al. 2009. Re-evaluation of climate change in lowland Central America during the Last Glacial Maximum using new sediment cores from Lake Peten Itza, Guatemala, in Vimeaux. F., Sylvestre. F. and Khodri. M. (eds) *Past Climate Variability From the Last Glacial Maximum to the Holocene in South America and Surrounding Regions.* Chichester: Springer/Praxis, pp. 113–128.

Bush. M. B., Gosling. W. D. and Colinvaux. P. A. 2007a. Climate change in the lowlands of the Amazon Basin, in Bush M. B. and Flenley J. R. (eds) *Tropical Rainforest Responses to Climatic Change.* Chichester: Springer/Praxis, pp. 55-76.

Bush M. B., Hansen B. C. S., Rodbell D. T., Seltzer G. O., Young K. R., Leon B. et al. 2005. A 17,000-year history of Andean climate and vegetation change from Laguna de Chochos, Peru. *Journal of Quaternary Science* 20: 703–714.

Bush M. B., Silman M. R., De Toledo M. B., Listopad C., Gosling W. D., Williams C., et al. 2007b. Holocene fire and occupation in Amazonia: records from two lake districts. *Philosophical Transactions of the Royal Society of London B* 362: 209–218.

Bush M. B., Silman M. R. and Listopad C. M. C. S. 2007c. A regional study of Holocene climate change and human occupation in Peruvian Amazonia. *Journal of Biogeography* 34: 1342–1356.

Bush M. B., Silman M. R., Mcmichael C. and Saatchi S. 2008. Fire, climate change and biodiversity in Amazonia: A Late-Holocene perspective. *Philosophical Transactions of the Royal Society B: Biological Sciences* 363: 1795–1802.

Bush M. B., Silman M. R. and Urrego D. H. 2004. 48,000 years of climate and forest change in a biodiversity hotspot. *Science* 303: 827–829.

Cárdenas M. L., Gosling W. D., Sherlock S. C., Poole I., Pennington R. T. and Mothes P. 2011. The response of vegetation on the Andean flank in western Amazonia to Pleistocene climate change. *Science* 331: 1055–1058.

Castañeda I. S., Werne J. P. and Johnson T. C. 2007. Wet and arid phases in the southeat African tropics since the Last Glacial Maximum. *Geology* 35: 823–826.

Castañeda. I. S., Werne. J. P., Johnson. T. C. and Filley. T. R. 2009. Late Quaternary vegetation historu of southeast Africa: The molecular isotopic record from Lake Malawi. *Paleogeography Paleoclimatology Paleoecology* 275: 100–112.

Chepstow-Lusty A., Frogley M. R., Bauer B. S., Bush M. B. and Herrera A. T. 2003. A late Holocene record of arid events from the Cuzco region, Peru. *Journal of Quaternary Science* 18: 491–502.

Chepstow-Lusty A. J., Bennett K. D., Fjeldsa J., Kendall A., Galiano W. and Tupayachi Herrera A. 1998. Tracing 4000 years of environmental history in the Cuzco area, Peru, from the pollen record. *Mountain Research and Development* 18: 159–172.

Chepstow-Lusty A. J., Frogley M. R., Bauer B. S., Leng M. J., Cundy A. B., Boessenkool K. P. and Gioda A. 2007. Evaluating socio-economic change in the Andes using oribatid mite abundances as indicators of domestic animal densities. *Journal of Archaeological Science* 34: 1178–1186.

Cione A. L., Tonni E. P. and Soibelzon L. (2009) Did humans cause the late Pleistocene-early Holocene mammalian extinctions in South America in a context of shrinking open areas? *American megafaunal extinctions at the end of the Pleistocene* (ed. by G. Haynes), pp. 125–144. Springer.

Cione A. L., Tonni E. P. and Soibelzon L. (2003) The broken zig-zag: Late Cenozoic large mammal and tortoise extinctions in South America. *Revista Museo de Historia Natural Bernardino Rivadavia*, 5, 1–19.

Clapperton C. M. 1972. The Pleistocene moraine stages of west-central Peru. *Journal of Glaciology* 11: 255–263.

Clapperton C. M. 1987. Maximal extent of the late Wisconsin glaciation in the Ecuadorian Andes, in Rabassa. J. (ed.) *Quaternary of South America and Antarctic Peninsula 5.* Rotterdam: A.A. Balkema, pp. 165–179.

Clark P. U., Dyke A. S., Shakun J. D., Carlson A. E., Clark J., Wohlfarth B. et al. 2009. The Last Glacial Maximum. *Science* 325: 710–714.

Claussen M. 2009. Late Quaternary vegetation-climate feedbacks. *Climate of the Past* 5: 203–216.

Claussen M. and Gayler V. 1997. The greening of the Sahara during the mid-Holocene: Results of an interactive atmosphere-biome model. *Global Ecology and Biogeography Letters* 6: 369–377.

Clement A. C., Hall A. and Broccoli A. J. 2004. The importance of precessional signals in the tropical climate. *Climate Dynamics* 22: 327–341.

Coetzee J. A. 1964. Evidence for a considerable depression of vegetation belts during the upper Pleistocene on the east African mountains. *Nature* 204: 564–566.

Coetzee J. A. 1967. Pollen analytical studies in East and Southern Africa. *Palaeoecology of Africa* 3: 1–146.

Cohen, T.J., Nanson, G.C., Jansen, J.D., Jones, B.G., Jacobs, Z., Treble, P. et al. 2011. Continental aridification and the vanishing of Australia's megalakes. *Geology* 39: 167–170.

Colinvaux P. A., Bush M. B., Steinitz-Kannan M. and Miller M. C. 1997. Glacial and postglacial pollen records from the Ecuadorian Andes and Amazon. *Quaternary Research* 48: 69–78.

Colinvaux P. A., De Oliveira P. E. and Bush M. B. 2000. Amazonian and neotropical plant communities on glacial time-scales: The failure of the aridity and refuge hypotheses. *Quaternary Science Reviews* 19: 141–169.

Cowling S. A., Cox P. M., Jones C. D., Maslin M. A., Peros M. and Spall S. A. 2008. Simulated glacial and interglacial vegetation across Africa: Implications for species phylogenies and trans-African migration of plants and animals. *Global Change Biology* 14: 827–840.

Cruz Jr F. W., Burns S. J., Vuille M., Karmann I., Viana Jr O., Sharp W. D. et al. 2005. Insolation-driven changes in atmospheric circulation over the past 116,000 years in subtropical Brazil. *Nature* 434: 63–66.

de Busk G. H., Jr .1998. A 37,500 year pollen record from Lake Malawi and implications for the biogeography of Afromontane forests. *Journal of Biogeography* 25: 479–500.

deMenocal P., Ortiz J., Guilderson T., Adkins J., Sarnthein M., Baker L. and Yarusinsky M. 2000. Abrupt onset and termination of the African Humid Period: rapid climate responses to gradual insolation forcing. *Quaternary Science Reviews* 19: 347–361.

Denton G. H. and Karlen W. 1973. Holocene climatic variations – their pattern and possible cause. *Quaternary Research* 3: 155–174.

Diamond J. 1989. Quaternary megafaunal extinctions: Variations on a theme by Paganini. *Journal of Archaeological Science* 16: 167–175.

Diamond J. 2002. Evolution, consequences and future of plant and animal domestication. *Nature* 418: 700–707.

Dillehay, T.D. 2009. Probing deeper into first American studies *Proceedings of the National Academy of Sciences* 106: 971–978.

Dupont L. M., Behling H., Jahns S., Marret F. and Kim J.-H. 2007. Variability in glacial and Holocene marine pollen records offshore from west southern Africa. *Vegetation History and Archaeobotany* 16: 87–100.

Ekdahl, E.J., Fritz, S.C., Baker, P.A., Rigsby, C.A. and Coley, K. 2008. Holocene multidecadal- to millennial-scale hydrologic variability on the South American Altiplano. *The Holocene* 18: 867–876.

Fairbanks R. G. 1989. A 17,000 year glacio-eustatic sea level record: influence of glacial melting rates on Younger Dryas event and deep ocean circulation. *Nature* 342: 637–642.

Fiedel S. and Haynes G. 2004. A premature burial: Comments on Grayson and Meltzer's 'Requiem for overkill'. *Journal of Archaeological Science* 31: 121–131.

Firestone R. B., West A., Kennett J. P., Becker L., Bunch T. E., Revay Z. S. et al. 2007. Evidence for an extra-terrestrial impact 12,900 years ago that contributed to the megafaunal extinctions and the Younger Dryas cooling. *Proceedings of the National Academy of Sciences USA* 104: 16016–16021.

Fleitmann D., Burns S. J., Mudelsee M., Neff U., Kramers J., Mangini A. and Matter A. 2003. Holocene forcing of the Indian monsoon recorded in a stalagmite from Southern Oman. *Science* 300: 1737–1739.

Fritz, S.C., Baker, P.A., Ekdahl, E., Seltzer, G.O. and Stevens, L.R. 2010. Millennial-scale climate variability during the last glacial period in the tropical Andes. *Quaternary Science Reviews*, 29, 1017–1024.

Garreaud R., Vuille M. and Clement A. C. 2003. The climate of the Altiplano: Observed current conditions and mechanisms of past changes. *Palaeogeography Palaeoclimatology Palaeoecology* 194: 5–22.

Garreaud R. D., Vuille M., Compagnucci R. and Marengo J. 2009. Present-day South American climate. *Palaeogeography Palaeoclimatology Palaeoecology* 281: 180–195.

Gasse F. 2002. Diatom-inferred salinity and carbonate oxygen isotopes in Holocene waterbodies of the western Sahara and Sahel (Africa). *Quaternary Science Reviews* 21: 737–767.

Gill, J. L., Williams, J. W., Jackson, S. T., Lininger, K. B. and Robinson, G. S. Pleistocene megafaunal collapse, novel plant communities, and enhanced fire regimes in North America. *Science* 326: 1100–1103.

Gil-Romera G., Scott L., Marais E. and Brook G. A. 2006. Middle- to late-Holocene moisture changes in the desert of northwest Namibia derived from fossil hyrax dung pollen. *Holocene* 16: 1073–1084.

Gil-Romera G., Scott L., Marais E. and Brook G. A. 2007. Late Holocene environmental change in the northwestern Namib Desert margin: New fossil pollen evidence from hyrax middens. *Palaeo-geography, Palaeoclimatology, Palaeoecology* 249: 1–17.

Gosling W. D., Bush M. B., Hanselman J. A. and Chepstow-Lusty A. 2008. Glacial-Interglacial changes in moisture balance and the impact on vegetation in the southern hemisphere tropical Andes (Bolivia/Peru). *Palaeogeography Palaeo-climatology Palaeoecology* 259: 35–50.

Gosling W. D., Hanselman J. A., Knox C., Valencia B. G. and Bush M. B. 2009. Long term drivers of change in Polylepis woodland distribution in the central Andes. *Journal of Vegetation Science* 20: 1041–1052.

Grayson D. K. and Meltzer D. J. 2003. A requiem for North American overkill. *Journal of Archaeological Science* 30: 585–593.

Grayson D. K. and Meltzer D. J. 2004. North American overkill continued? *Journal of Archaeological Science* 31: 133–136.

Groot M. H. M., Bogotá R. G., Lourens L. J., Hooghiemstra H., Vriend, M., Berrio, J. C., et al. 2011. Ultra-high resolution pollen record from the northern Andes reveals rapid shifts in montane climates within the last two glacial cycle. *Climate of the Past* 7: 299–316.

Gupta A. K., Anderson D. M. and Overpeck J. T. 2003. Abrupt changes in the Asian southwest monsoon during the Holocene and their links to the North Alantic Ocean. *Nature* 421: 354–357.

Gupta A. K., Das M. and Anderson D. M. 2005. Solar influence on the Indian summer monsson during the Holocene. *Geophysical Research Letters* 32: 1–4.

Haffer J. 1969. Speciation in Amazonian forest birds. *Science* 165: 131–137.

Haffer J. and Prance G. T. 2001. Climate forcing of evolution in Amazonia during the Cenozoic: On the refuge theory of biotic differentiation. *Amazoniana* XVI: 579–607.

Hanebuth T., Stattegger K. and Grootes P. M. 2000. Rapid flooding of the Sunda Shelf: A late-glacial sea-level record. *Science* 288: 1033–1035.

Hanselman, J.A., Bush, M.B., Gosling, W.D., Collins, A., Knox, C., Baker, P.A. and Fritz, S.C. 2011. A 370,000-year record of vegetation and fire history around Lake Titicaca (Bolivia/Peru). *Palaeogeography Palaeoclimatology Palaeoecology* 305: 201–214.

Hanselman J. A., Gosling W. D., Ralph G. M. and Bush M. B. 2005. Contrasting histories of MIS 5e and the Holocene from Lake Titicaca (Bolivia/Peru). *Journal of Quaternary Science* 20: 663–670.

Hansen B. C., Rodbell D. T., Seltzer G. O., Leon B., Young K. R. and Abbott M. 2003. Late-glacial and Holocene vegetational history from two sites in the western Cordillera of southwestern Ecuador. *Palaeogeography Palaeoclimatology Palaeoecology* 194: 79–108.

Hansen B. C. S., Wright H. E., Jr and Bradbury J. P. 1984. Pollen studies in the Junin area, Central Peruvian Andes. *Geological Society of America Bulletin*, 1454–1465.

Hastenrath S. 1997. Annual cycle of upper air circulation and convective activity over the tropical Americas. *Journal of Geophysical Research* 102: 4267–4274.

Haug G. H., Hughen K. A., Sigman D. M., Peterson L. C. and Rohl U. 2001. Southward migration of the Intertropical Convergence Zone through the Holocene. *Science* 293: 1304–1308.

Heckenberger M. J., Kuikuro A., Kuikuro U. T., Russell J. C., Schmidt M., Fausto C. and Franchetto B. 2003. Amazonia 1492: Pristine forest or cultural parkland? *Science* 301: 1710–1714.

Heckenberger M. J., Russell J. C., Fausto C., Toney J. R., Schmidt M. J., Pereira E. et al. 2008. Pre-Columbian urbanism, anthropogenic landscapes, and the future of the Amazon. *Science* 321: 1214–1217.

Herzschuh U. 2006. Palaeo-moisture evolution in monsoonal Central Asia during the last 50,000 years. *Quaternary Science Reviews* 25: 163–178.

Hillyer R., Valencia B. G., Bush M. B., Silman M. R. and Steinitz-Kannan M. 2009. A 24,700-yr paleolimnological history from the Peruvian Andes. *Quaternary Research* 71: 71–82.

Hodell D. A., Anselmetti F. S., Ariztegui D., Brenner M., Curtis J. H., Gilli A. et al. 2008. An 85-ka record of climate change in lowland Central America. *Quaternary Science Reviews* 27: 1152–1165.

Hodell D. A., Brenner M., Curtis J. H. and Guilderson T. 2001. Solar forcing of drought frequency in the Maya lowlands. *Science* 291: 1367–1370.

Hoelzmann P., Jolly D., Harrison S. P., Laarif F., Bonnefille R. and Pachur H.-J. 1998. Mid-Holocene land-surface conditions in northern Africa and the Arabian peninsula: A data set for the analysis of biogeophysical feedbacks in the climate system. *Global Biogeochemical Cycles* 12: 35–51.

Hope G. 2009. Environmental change and fire in the Owen Stanley Ranges, Papua New Guinea. *Quaternary Science Reviews* 28: 2261–2276.

Hope G. and Golson J. 1995. Late Quaternary change in the mountains of New Guinea. *Antiquity* 69: 818–830.

Hope G., Kershaw A. P., Van Der Kaars S., Xiangjun S., Liew P.-M., Heusser L. E. et al. 2004. History of vegetation and habitat change in the Austral-Asian region. *Quaternary International* 118–119: 103–126.

Hopkins M. S., Graham A. W., Hewett R., Ash J. and Head J. 1990. Evidence of late Pleistocene fires and eucalypt forest from a North Queensland humid tropical rainforest site. *Austral Ecology* 15: 345–347.

Hubbé A., Hubbé M. and Neves W. 2007. Early Holocene survival of megafauna in South America. *Journal of Biogeography* 34: 1642–1646.

Hughen K. A., Southon J. R., Lehman S. J. and Overpeck J. T. 2000. Synchronous Radiocarbon and Climate Shifts During the Last Deglaciation. *Science* 290: 1951–1954.

Jolly D., Prentice I. C., Bonnefille R., Ballouche A., Bengo M., Brenac P. et al. 1998. Biome reconstruction from pollen and plant macrofossil data for Africa and the Arabian peninsula at 0 and 6000 years. *Journal of Biogeography* 25: 1007–1027.

Kealhofer L. and Penny D. 1998. A combined pollen and phytolith record for fourteen thousand years of vegetation change in northeastern Thailand. *Review of Palaeobotany and Palynology* 103: 83–93.

Kershaw A. P., McKenzie G. M. and McMinn A. 1993. A Quaternary vegetation history of northeastern Queensland from pollen analysis of ODP Site 820. *Proc., scientific results, ODP., Leg 133, northeast Australian margin*, 107–114.

Kershaw A. P., Van Der Kaars S. and Flenley J. R. 2007. The Quaternary history of far eastern rainforests, in Bush M. B. and Flenley J. R. (eds) *Tropical Rainforest Responses to Climatic Change*. Chichester: Springer/Praxis, pp. 77–116.

Kröpelin S., Verschuren D. and Lézine A.-M. 2008a. Response to comment on 'Climate-driven ecosystem succession in the Sahara: The past 6000 years'. *Science* 322: 1326.

Kröpelin S., Verschuren D., Lézine A.-M., Eggermont H., Cocquyt C., Francus P. et al. 2008b. Climate-driven ecosystem succession in the Sahara: The past 6000 years. *Science* 320: 765–768.

Lachniet M. S. 2004. Late Quaternary glaciations of Costa Rica and Guatemala, in Ehlers J. and Gibbard P. L. (eds) *Quaternary Glaciations – Extent and Chronology, Part III*. Amsterdam: Elsevier Science, pp. 135–138.

Lahr M. M. and Foley R. A. 1998. Towards a theory of modern human origins: Geography, demography, and diversity in recent human evolution. *American Journal of Physical Anthropology* 107: 137–176.

Lambeck K., Yokoyama Y. and Purcell T. 2002. Into and out of the last glacial maximum: Sea-level change during oxygen isotope stages 3 and 2. *Quaternary Science Reviews* 21: 343–360.

Lean, J., Beer, J. and Bradley, R. 1995. Reconstruction of solar irradiance since 1610: Implications for climate change. *Geophysical Research Letters* 22: 3195–3198.

Lehmkuhl F. and Owen L. A. 2005. Late Quaternary glaciation of Tibet and the bordering mountains: A review. *Boreas* 34: 87–100.

Liu K.-B. and Colinvaux P. A. 1988. A 5200-year history of Amazon Rain Forest. *Journal of Biogeography* 15: 231–248.

Liu Z., Otto-Bliesner B. L., He F., Brady E. C., Tomas R., Clark P. U. et al. 2009. Transient simulation of last deglaciation with a new mechanism for bolling-allerod warming. *Science* 325: 310–314.

Liu Z., Wang Y., Gallimore R., Gasse F., Johnson T., Demenocal P. et al. 2007. Simulating the transient evolution and abrupt change of Northern Africa atmosphere-ocean-terrestrial ecosystem in the Holocene. *Quaternary Science Reviews* 26: 1818–1837.

Livingstone D. A. 1967. Postglacial vegetation of the Ruwenzori Mountains in equatorial Africa. *Ecological Monographs* 37: 25–52.

Lovett J. C., Marchant R., Marshall A. R. and Barber J. 2007. Tropical moist forests. *Issues in Environmental Science and Technology* 25: 161–192.

Marchant R. and Hooghiemstra H. 2004. Rapid environmental change in African and South American tropics around 4000 years before present: A review. *Earth-Science Reviews* 66: 217–260.

Marengo J. A., Soares W. R., Saulo C. and Nicolini M. 2004. Climatology of the low-level jet east of the Andes as derived from the NCEP-NCAR reanalyses: Characteristics and temporal variability. *Journal of Climate* 17: 2261–2280.

Mark B. G. and Osmaston H. A. 2008. Quaternary glaciation in Africa: Key chronologies and climatic implications. *Journal of Quaternary Science* 23: 589–608.

Martin P. S. 1973. The discovery of America. *Science* 179: 969–974.

Martin P. S. and Klein R. G. 1984. *Quaternary Extinctions: A Prehistoric Revolution*. Tucson: University of Arizona Press.

Martinson D. G., Pisias N. G., Hays J. D., Imbrie J., Moore T. C., Jr and Shackleton N. J. 1987. Age dating and the orbital theory of the ice ages: Development of a high-resolution 0 to 300,000-year chronostratigraphy. *Quaternary Research* 27: 1–29.

Maxwell A. L. 2001. Holocene monsoon changes inferred from lake sediment pollen and carbonate records, northeastern Cambodia. *Quaternary Research* 56: 390–400.

Mayewski P. A., Rohling E. E., Stager J. C., Karle N. W., Maasch K. A., Meeker L. D. et al. 2004.

Holocene climate variability. *Quaternary Research* 62: 243–255.

Mayle F. E. 2004. Assessment of the Neotropical dry forest refugia hypothesis in the light of palaeoecological data and vegetation model simulations. *Journal of Quaternary Science* 19: 713–720.

Mayle F. E., Burbridge R. and Killeen T. J. 2000. Millennial-scale dynamics of southern Amazonian rain forests. *Science* 290: 2291–2294.

McMichael C., Bush M. B., Piperno D. R., Silman M., Zimmerman A. and Anderson C. (In press) Spatial and temporal scales of pre-columbian disturbance associated with western Amazonian lakes. *The Holocene.*

Miller G. H., Fogel M. L., Magee J. W., Gagan M. K., Clarke S. J. and Johnson B. J. 2005. Ecosystem collapse in Pleistocene Australia and a human role in megafaunal extinction. *Science* 309: 287–290.

Morrill C., Overpeck J. T. and Cole J. E. 2003. A. synthesis of abrupt changes in the Asian summer monsoon since the last deglaciation. *The Holocene* 13: 465–476.

Moss P. T. and Kershaw A. P. 2000. The last glacial cycle from the humid tropics of northeastern Australia: Comparison of a terrestrial and a marine record. *Palaeogeography Palaeoclimatology Palaeoecology* 155: 155–176.

Mourguiart P., Correge T., Wirrmann D., Argollo J., Montenegro M. E., Pourchet M. and Carbonel P. 1998. Holocene palaeohydrology of Lake Titicaca estimated from an ostrocod-based transfer function. *Palaeogeography Palaeoclimatology Palaeoecology* 143: 51–72.

Moy C. M., Seltzer G. O., Rodbell D. T. and Anderson D. M. 2002. Variability of El Nino/Southern Oscillation activity at millenial timescales during the Holocene epoch. *Nature* 420: 162–165.

Newell R. E. 1973. Climate and the Galapagos Islands. *Nature* 245: 91–92.

Osmaston H. A. and Harrison S. P. 2005. The Late Quaternary glaciation of Africa: A regional synthesis. *Quaternary International* 138–139: 32–54.

Paduano G. M., Bush M. B., Baker P. A., Fritz S. C. and Seltzer G. O. 2003. A. vegetation and fire history of Lake Titicaca since the Last Glacial Maximum. *Palaeogeography Palaeoclimatology Palaeoecology* 194: 259–279.

Pennington R. T., Prado D. E. and Pendry C. A. 2000. Neotropical seasonally dry forests and Quaternary vegetation changes. *Journal of Biogeography* 27: 261–273.

Peristykh A. N. and Damon P. E. 2003. Persistence of the Gleissberg 88-year solar cycle over the last

12,000 years: Evidence from cosmogenic isotopes. *Journal of Geophysical Research A: Space Physics* 108: A1 1003.

Peterson J. A., Chandra S. and Lundberg C. 2004. Landforms from Quaternary glaciation of Papua New Guinea: An overview of ice extent during the Last Glacial Maximum, in Ehlers J. and Gibbard P. L. (eds) *Quaternary Glaciations – Extent and Chronology, Part III.* Amsterdam: Elsevier, pp. 313–319.

Peterson L. C., Haug G. H., Hughen K. A. and Rohl U. 2000. Rapid changes in the Hydrologic cycle of the Tropical Atlantic during the Last Glacial. *Science* 290: 1947–1951.

Pigati J. S., Zreda M., Zweck C., Almasi P. F., Elmore D. and Sharp W. D. 2008. Ages and inferred causes of Late Pleistocene glaciations on Mauna Kea, Hawai'i. *Journal of Quaternary Science* 23: 683–702.

Porter S. C. 2001. Snowline depression in the tropics during the Last Glaciation. *Quaternary Science Reviews* 20: 1067–1091.

Powers L. A., Johnson T. C., Werne J. P., Castan Eda I. S., Hopmans E. C., Sinninghe Damste J. S. and Schouten S. 2005. Large temperature variability in the southern African tropics since the Last Glacial Maximum. *Geophysical Research Letters* 32: 1–4.

Prado D. E. and Gibbs P. E. 1993. Patterns of species distributions in the seasonal dry forests of South America. *Annals of the Missouri Botanical Garden* 80: 902–927.

Prentice M. L., Hope G. S., Maryunani K. and Peterson J. A. 2005. An evaluation of snowline data across New Guinea during the last major glaciation, and area-based glacier snowlines in the Mt Jaya region of Papua, Indonesia, during the Last Glacial Maximum. *Quaternary International* 138–139: 93–117.

Punyasena. S. W., Mayle. F. E. and McElwain. J. C. 2008. Quantitative estimates of glacial and Holocene temperature and precipitation change in lowland Amazonian Bolivia. *Geology* 36: 667–670.

Rasmussen S. O., Andersen K. K., Svensson A. M., Steffensen J. P., Vinther B. M., Clausen H. B. et al. 2006. A new Greenland ice core chronology for the last glacial termination. *Journal of Geophysical Research D: Atmospheres* 111: D06102.

Ritchie J. C., Eyles C. H. and Haynes C. V. 1985. Sediment and pollen evidence for an early to mid-Holocene humid period in the eastern Sahara. *Nature* 314: 352–355.

Roosevelt A. C., Lima Da Costa M., Lopes Machado C., Michab M., Mercier N., Valladas H. et al. 1996. Paleoindian cave dwellers in the Amazon: The peopling of the Americas. *Science* 272: 373–384.

Scott L. 1996. Palynology of hyrax middens: 2000 years of palaeoenvironmental history in Namibia. *Quaternary International* 33: 73–79.

Seltzer G. O., Rodbell D. T., Baker P. A., Fritz S. C., Tapia P. M., Rowe H. D. and Dunbar R. B. 2002. Early warming of tropical South America at the last glacial-interglacial transition. *Science* 296: 1685–1686.

Seltzer G. O., Rodbell D. T. and Burns S. 2000. Isotopic Evidence For Late Quaternary climatic change in tropical South America. *Geology* 28: 35–38.

Shanahan T. M., Overpeck J. T., Wheeler C. W., Beck J. W., Pigati J. S., Talbot M. R. et al. 2006. Paleoclimatic variations in West Africa from a record of late Pleistocene and Holocene lake level stands of Lake Bosumtwi, Ghana. *Palaeogeography Palaeoclimatology Palaeoecology* 242: 287–302.

Singh G. and Geissler E. A. 1985. Late Cainozoic history of fire, lake levels, and climate at Lake George, New South Wales, Australia. *Philosophical Transactions of the Royal Society of London* 311: 379–447.

Smith J. A., Mark B. G. and Rodbell D. T. 2008. The timing and magnitude of mountain glaciation in the tropical Andes. *Journal of Quaternary Science* 23: 609–634.

Smith J. A., Seltzer G. O., Farber D. L., Rodbell D. and Finkel R. C. 2005. Early local last glacial maximum in the tropical Andes. *Science* 308: 687–681.

Steadman D.W., Martin P. S., MacPhee R. D. E., Jull A. J. T., McDonald H. G., Woods, C. A. et al. 2005. Asynchronous extinction of late Quaternary sloths on continents and islands. *Proceedings of the National Academy of Sciences USA* 102: 11763–11768.

Sterken M., Sabbe K., Chepstow-Lusty A., Frogley M., Vanhoutte K., Verleyen E. et al. 2006. Hydrological and land-use changes in the Cuzco region (Cordillera Oriental, South East Peru) during the last 1200 years: A diatom-based reconstruction. *Archiv fur Hydrobiologie* 165: 289–312.

Thompson L. G., Davis M. E., Mosley-Thompson E., Lin P.-N., Henderson K. A. and Mashiotta T. A. 2005. Tropical ice core records: Evidence for asynchronous glaciation on Milankovitch timescales. *Journal of Quaternary Science* 20: 723–733.

Thompson L. G., Davis M. E., Mosley-Thompson E., Sowers T. A., Henderson K. A., Zagorodnov V. S., Lin P.-N., Mikhalenko V. N., Campen R. K., Bolzan J. F., Cole-Dai J. and Francou B. 1998. A 25,000-year tropical climate history from Bolivian ice cores. *Science* 282: 1858–1864.

Tierney J. E. and Russell J. M. 2007. Abrupt climate change in southeast tropical Africa influenced by Indian monsoon variability and ITCZ migration. *Geophysical Research Letters* 34: L15709.

Tierney J. E., Russell J. M., Huang Y., Sinninghe Damst J. S., Hopmans E. C. and Cohen A. S. 2008. Northern hemisphere controls on tropical southeast African climate during the past 60,000 years. *Science* 322: 252–255.

Tilman D., May R. M., Lehman C. L. and Nowak M. A. 1994. Habitat destruction and the extinction debt. *Nature* 371: 65–66.

Turney C. S. M., Kershaw A. P., Clemens S. C., Branch N., Moss P. T. and Fifield L. K. 2004. Millennial and orbital variations of El Niño/Southern Oscillation and high-latitude climate in the last glacial period. *Nature* 428: 306–310.

Urrego D. H. 2006. Long-term vegetation and climate change in western Amazonia. Department of Biological Sciences, Florida Institute of Technology. Unpublished PhD thesis: 278.

Urrego D. H., Silman M. R. and Bush M. B. 2005. The Last Glacial Maximum: Stability and change in a western Amazonian cloud forest. *Journal of Quaternary Science* 20: 693–701.

Urrego D. H., Bush M. B. and Silman M. R. 2010. A long history of cloud and forest migration from Lake Consuelo, Peru. *Quaternary Research* 73: 364–373.

Valencia B. G., Urrego D. H., Silman M. R. and Bush M. B. 2010. From Ice Age to modern: A record of landscape change in an Andean cloud forest. *Journal of Biogeography* 37: 1637–1647.

van Der Hammen T. 1974. The Pleistocene changes of vegetation and climate in tropical South America. *Journal of Biogeography* 1: 3–26.

van Der Hammen T. and Gonzalez E. 1959. Historia de clima y vegetacion del Pleistocene Superior y del Holoceno de la Sabana de Bogota y alrededeores. Dept Cundinamarca Official Report of the Servicio Geologico Nacional.

van Der Kaars S. and Dam R. 1997. Vegetation and climate change in West-Java, Indonesia during the last 135,000 years. *Quaternary International* 37: 67–71.

van Der Kaars S., Penny D., Tibby J., Fluin J., Dam R. A. C. and Suparan P. 2001. Late quaternary palaeoecology, palynology and palaeolimnology of a tropical lowland swamp: Rawa Danau, West-Java, Indonesia. *Palaeogeography Palaeoclimatology Palaeoecology* 171: 185–212.

van Der Kaars S., Wang X., Kershaw P., Guichard F. and Setiabudi D. A. 2000. A Late Quaternary palaeoecological record from the Banda Sea, Indonesia: Patterns of vegetation, climate and biomass burning in Indonesia and northern Australia. *Palaeo-geography Palaeoclimatology Palaeoecology* 155: 135–153.

van Zinderen Barker E. M. 1964. A late-glacial and post-glacial climatic correlation between East Africa and Europe. *Nature* 194: 201–203.

Vera C., Higgins W., Amador J., Ambrizzi T., Garreaud R., Gochis D. et al. 2006. Toward a unified view of the American monsoon systems. *Journal of Climate* 19: 4977–5000.

Vuille M., Bradley R. S. and Keimig F. 2000. Interannual climate variability in the Central Andes and its relation to tropical Pacific and Atlantic forcing. *Journal of Geophysical Research D: Atmospheres* 105: 12447–12460.

Walker D. and Flenley J. R. 1979. Late Quaternary vegetational history of the Enga province of upland Papua New Guinea. *Philosophical Transactions of the Royal Society B: Biological Sciences* 286: 265–344.

Wallace A. R. 1876. *The Geographical Distribution of Animals.* London: Macmilian.

Wang X., Auler A. S., Edwards R. L., Cheng H., Cristalli P. S., Smart P. L. et al. 2004. Wet periods in northeastern Brazil over the past 210 kyr linked to distant climate anomalies. *Nature* 432: 740–743.

Wang X., Cruz Jr F. W., Auler A. S., Cheng H. and Edwards R. L. 2008. Millennial-scale climate variability recorded in Brazilian speleothems. *PAGES News* 16: 31–32.

Wang Y., Cheng H., Edwards R. L., He Y., Kong X., An Z. et al. 2005. The Holocene Asian monsoon: Links to solar changes and North Atlantic climate. *Science* 308: 854–857.

Wang Y. J., Cheng H., Edwards R. L., An Z. S., Wu J. Y., Shen C. C. and Dorale J. A. 2001. A. high-resolution absolute-dated Late Pleistocene monsoon record from Hulu Cave, China. *Science* 294: 2345–2348.

Waters M. R. and Stafford Jr. T. W. 2007. Redefining the age of Clovis: implications for the peopling of the Americas. *Science* 315: 1122–1126.

Weng C., Bush M. B. and Athens J. S. 2002. Holocene climate change and hydrarch succession in lowland Amazonian Ecuador. *Review of Palaeobotany and Palynology* 120: 73–90.

Weyl R. 1956. Eiszeitliche gletscherspuren in Costa Rica (Mittelamerika). *Zeitschrift für Gletscherkunde und Glazialgeologie* 3: 317–325.

White J. C., Penny D., Kealhofer L. and Maloney B. 2004. Vegetation changes from the late Pleistocene through the Holocene from three areas of archaeological significance in Thailand. *Quaternary International* 113: 111–132.

Wolfe B. B. R. A., Abbott M. B., Seltzer G. O. and Gibsond J. J. 2001. Reconstruction of paleohydrology and paleohumidity from oxygen isotope records in the Bolivian Andes. *Palaeogeography Palaeoclimatology Palaeoecology* 176: 177–192.

Wroe S. and Field J. 2006. A review of the evidence for a human role in the extinction of australian megafauna and an alternative interpretation. *Quaternary Science Reviews*, 25: 2692–2703.

Wroe S., Field J. and Grayson D. K. 2006. Megafaunal extinctions: climate, humans and assumptions. *Trends in Ecology and Evolution* 21: 61–62.

Yuan D., Cheng H., Edwards R. L., Dykoski C. A., Kelly M. J., Zhang M. et al. 2004. Timing, duration, and transitions of the Last Interglacial Asian Monsoon. *Science* 304: 575–578.

Zhou J. and Lau K.-M. 1998. Does a monsoon climate exist over South America? *Journal of Climate* 11: 1020–1040.

31

Environmental Change in the Arid and Semi-Arid Regions

Xiaoping Yang

1 INTRODUCTION

Arid and semi-arid regions are of great significance for understanding the Earth system because they occupy a large portion of its surface. In addition, desertification is one of the major environmental issues threatening many dryland regions of the world. Thus, research on past environmental change in these areas may provide critical insights into the causes of past ecosystem fluctuations, which may aid in the interpretation of contemporary processes. The recognition of climate change in arid regions of western China (Richthofen, 1877) was one of the earliest reports on Quaternary climatic instability; it predates even the landmark publications on Quaternary glaciation in the Alps (Penck and Brückner, 1909). Indeed, arid regions potentially offer great insights into the science of environmental change, because some imprints of climate change in arid zones are large-scale and easily visible (e.g., the desiccation of lakes and rivers, shifting of the sand sea boundary, and stabilization or reworking of dunes). But over a century later, compared with the development of knowledge about global change in other regions of Earth, what is known about changes in arid and semi-arid regions is characterized by a variety of inconsistencies and gaps in knowledge. Investigating the age and nature of environmental change in arid and semi-arid regions remains a particularly challenging task, due to uncertainties in both dating and interpretation of proxy records.

The availability of palaeoenvironmental data varies greatly from continent to continent. High time-resolution is rarely achievable in the available records from drylands (e.g., Williams et al., 1998), and biotic remains are generally rare in open-air sites in desert regions due to low primary production, generally oxygen-rich conditions, and the lack of long-lasting anoxic lakes and swamps. In addition, the shortage of groundwater also limits the development of useful nonorganic proxies of past climate change, such as stalagmites (e.g., Scott, 2003). The rapid improvement of luminescence dating technology (see Chapter 5), including thermoluminescence (TL) dating and, more recently, optically stimulated luminescence (OSL), has enabled pure quartz and feldspar grains to be dated and enabled the aeolian records to be directly used for palaeoclimatic reconstructions. However, it should be borne in mind that a comprehensive understanding of the individual depositional system is required beforehand to interpret the aeolian records correctly, because sand deposition, for example, may represent aridity in one location but humidity or wind strength in another (Busacca et al., 2004; Chase and Thomas, 2007; Wells et al., 2003).

Based on the publications about palaeoenvironmental change in drylands of Africa, Asia, the Americas, and Australia, this chapter aims to provide a global view and to develop a coherent hypothesis about environmental changes that have occurred in the major arid and semi-arid regions since the late Pleistocene. Greater attention is given to evidence directly from arid regions, including geomorphologic, sedimentologic, hydrologic, geochemical and biological records. In the latter part of the chapter, desertification and the global impacts of current environmental changes in drylands are briefly discussed. The arid regions of Antarctic are discussed in Chapter 35, and the semi-arid regions in temperate zones in Chapter 34. The background information in individual research methodologies and the potential mechanisms that trigger environmental changes are only briefly touched upon here, as they are the focus of other thematic chapters of this Handbook.

2 PRESENT-DAY DISTRIBUTION OF ARID AND SEMI-ARID REGIONS

In general, regions with a mean annual precipitation below 200 mm can be defined as arid, and those with a mean annual precipitation between 200 and 500 mm as semi-arid, although aridity does not depend solely on the amount of precipitation, and evapotranspiration rates and temperature must also be taken into account. Arid and semi-arid regions, usually described as drylands, occupy about one-third of the global continents' surface, occurring in all continents (Table 31.1), including Antarctica. There are two major areas of drylands (Figure 31.1).

Table 31.1   Arid and semi-arid areas of the world ($km^2 \times 10^6$)

Continent	Arid area	Semi-arid area	Sum
Africa	11.862	6.081	17.943
Asia	8.96	7.516	16.476
North America	1.31	2.657	3.967
South America	1.388	1.626	3.014
Australia	3.864	2.517	6.381
Europe	0.171	0.844	1.015
Sum	27.555	21.241	48.796

One, directly related to the subtropical high-pressure cells, is centered on the Tropics of Cancer and Capricorn and extends 10° to 15° poleward and equatorward from there. The main reason for the formation of arid climate in these low latitudes is subsidence in the subtropical high-pressure systems. The other dryland area is the continental interiors, particularly of the northern hemisphere. The aridity in the heart of landmasses is linked to the remoteness from the oceanic moisture supply or/and the blockage of moisture pathways by high mountains and plateaus. Along western coasts in the subtropical zone, ocean currents reinforce the effect of the subtropical anticyclones in suppressing precipitation. On the eastern sides of the oceanic anticyclones or the western coasts of continents, cold water from the polar seas moves equatorward. Air moving onshore from the west is cooled from below by these cold currents, which increases airmass stability and consequently causes greater aridity.

Arid climate zones correspond to desert, and semi-arid zones to steppe. One often associates deserts with sand seas, but the land surface in arid regions is covered to a larger degree by bare exposures of regoliths, alluvium and bedrock. For example, active sand seas cover between 15 and 30 percent of the arid areas in the Sahara, Arabian Peninsula, Australia, and Southern Africa. Active and stable dunes occupy as much as 45 percent of deserts in China. In contrast,

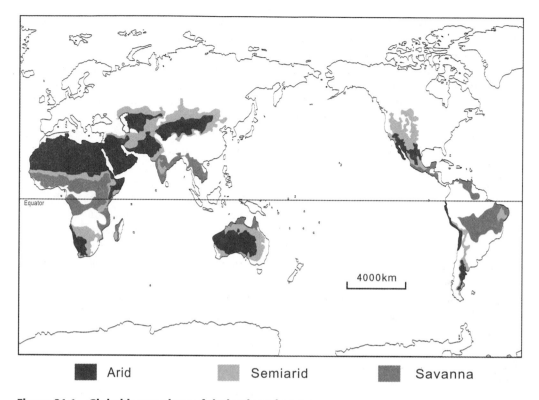

Figure 31.1   Global key regions of drylands and savannas.

aeolian sand covers only less than one per-cent of the arid zone in the Americas (Yang and Goudie, 2007).

In the transitional zone between the sub-tropical high pressure cells and the Inter-tropical Convergence Zone (ITCZ), a tropical savanna climate occurs on either side of the equator, extending over large areas particu-larly in Africa, Central and South America (Figure 31.1). As the latitudinal and pressure belts shift seasonally, savanna areas are under the influence of the rainy ITCZ for part of the year and the dry subtropical highs for the other part. Situated between the rainforest climate on one side and steppe climate on the other, the savanna has some of the features of both climate zones (i.e. a grassland with scat-tered small tress and shrubs). Savanna located close to rainforest may have rain during every month, and total annual precipitation may reach 1,800 mm. In contrast, the dry margins of savanna can have long periods of drought and low annual rainfalls, less than 1,000 mm (see Chapter 30).

3 ENVIRONMENTAL CHANGES IN THE MAJOR ARID AND SEMI-ARID REGIONS SINCE THE LATE PLEISTOCENE

3.1 Africa

Arid and semi-arid landscapes account for ~60 percent of the African continent. In the Sahara, which covers about one-third of Africa and has 27 great sand seas, the high level of aridity became established probably at around 2–3 Ma BP, more or less contempo-raneous with the time of onset of mid-latitude glaciation (Goudie, 2002). The earliest dune activities in the Western Desert of Egypt started at around 700 ka BP and ended at about 200 ka BP according to archaeological evidence in the inter-dune sites and sedimentologic evidence found in the pre-Nile sediments in the Nile Valley (Embabi, 2004). Palaeolacustrine sediments indicate that pluvial conditions took place during the last interglacial in the hyper-arid central Sahara (Petit-Maire et al., 1994), confirming the large amplitude of Quaternary climatic variations (Figure 31.2a).

Figure 31.2   On-site evidence of formerly wetter environments in the present-day hyper-arid zones. (a) Yardangs developed in Pleistocene lacustrine deposits in the Western Desert, Egypt, showing change from wetland to aeolian environments. (b) Mid-Holocene shorelines (marked by the outer boundary of the vegetation, dark-colored) in the interdune depression between megadunes in the Badain Jaran Desert, western China. The white patches are the relics of a formerly large single lake.

Linear dunes of the western Sahara extended onto the continental shelf during the Pleistocene glacial periods and were truncated by the subsequent sea level rise (Sarnthein and Diester-Haas, 1977). Many studies suggest that it was particularly arid in Sahara during the last glacial maximum, accompanied by shrinkage of lakes, disintegration of drainage system, increase in dust loading in the atmosphere (see Section 5 of this chapter) and southward migration of dunes (Goudie, 2002, and references therein). At this time the extension of dunes occurred across a broad front from Senegal in the west to Kordofan of Sudan in the east (Grove and Warren, 1968), and the extent of Guinean forest in West Africa contracted 10° of latitude (Lezine, 1989). The initial phase of dune formation in Mauritania was OSL dated to 25–15 ka BP (Lancaster et al., 2002), consistent with the marine proxy record that shows a high dust input from 25 to 14.8 ka BP (deMenocal et al., 2000). Lake Chad experienced desiccation and had an area of 7 percent of its present extension at the Last Glacial Maximum (LGM; Adams and Tetzlaff, 1985), and at this time the dunes were mobilized in the basin. During moister phases, areas that are now hydrologically inactive contributed to the inflow to the former Chad basin. The Jebel Marra crater lake of the southern Sahara experienced its lowest level at *c*.20.2 ka BP (Williams et al., 1980).

The date of the initial aridity in the Namib of southern Africa has been suggested as early as 80 Ma BP (Ward et al., 1983). Other authors, such as Tankard and Rogers (1978), however, conclude that the aridity in the Namib was initiated in the late Tertiary, but only fully developed in the Quaternary. It appears that there have been relatively frequent changes between dry and moist conditions since the late Pleistocene in the southern part of Africa (Figure 31.3). A large number of publications show evidence of increased aridity and decreased temperature during the last glacial maximum (Blümel et al., 1998; Heine, 2002, and references therein), while others conclude that the LGM was wet and cold

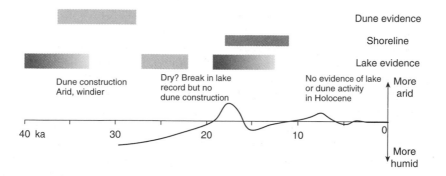

Figure 31.3   Synthesis of late Quaternary palaeoclimatic changes in the southwestern Kalahari, Namibia (Blümel et al., 1998) and temporal dynamics of Lake Tsodilo in the northwestern Kalahari, Botswana (Thomas et al., 2003), both indicating long-lasting wetter conditions before and after the short drier LGM (c.18 ka BP).

(Chase and Meadows, 2007). Luminescence ages of cover sands of the west coast show that most phases of aeolian activity occurred during glacial periods, with sediments dating to 4–5, 16–24, 30–33, 41–49 and 63–73 ka BP. These aeolian deposits reflect sediment supply from invigorated fluvial sources, caused by a potential increase in humidity in the region (Chase and Thomas, 2007). It is possible that the stronger aeolian activity along the west coast of South Africa during glacial periods was triggered by variation in windiness and the increase in fluvial sediment supply. Based on the composite distribution of ^{14}C dates of scattered and diverse archives, Lancaster (2002a) attributed relatively wet conditions to the periods at 30–24 ka BP, and 14–10 ka BP, respectively.

Marine cores taken off the Namibian coast suggest periods of aridity 21–17.5 ka BP, 14.3 –12.6 ka BP and 11–8.9 ka BP (Shi et al., 2000). Sections from fluvial channels in the northern Namib show, however, fairly continuous aggradation from 27.7 ± 3.6 ka BP to 15.6 ± 2.7 ka BP on the basis of OSL ages (Srivastava et al., 2004). Fluvial aggradation was assumed to be related with higher annual rainfall and more intense storms. Srivastava et al. (2004) suggested that the difference in interpretation might originate from chronological uncertainty, as the ages of these fluvial sediments would be between 17 and

13 ka BP if two sigma errors were considered for these OSL ages and all potential error terms were subtracted. Lake stands in the northwestern Kalahari indicate that wetter regional conditions than present occurred at 40–32 ka BP, with a possible drying out at 22–19 ka BP (Figure 31.3; Thomas et al., 2003). East of Kalahari, a lacustrine record indicates that precipitation during the LGM was reduced by 15–20 percent (Partridge et al., 1997).

In the sense of climatic geomorphology, establishment of new landscape types would normally need a time span of about 10^4 years, but signals of millennial-scale climatic variations can be recognized in areas with petrologically sensitive basement. Forms of fluvial erosion and accumulation, preserved at elevations from 1,000 to 2,000 m in the Tibesti Mountains of the central Sahara, suggest that frequent transitions between hyper-arid and hyper-humid conditions have taken place during the last 15,000 years. Three phases of valley erosion were recognized on the basis of radiocarbon ages of fossils and carbonates found in the fluvial terraces that occur in almost all valleys of western Tibetsi: prior to 15 ka BP, 14–6 ka BP and at c.2 ka BP (Hövermann, 1972). The erosion prior to 15 ka BP seems to be confirmed in a broad sense by a subsequent study that reconstructed lacustrine and fluvial formation in the Tibesti at the Glacial Maximum rather than the early

Holocene (Maley, 2000). Busche (1998) reported four different terraces in palaeo-rivers of Central Sahara and suggested that these terraces are the remnants of four periods of fluvial activity; the precise dating of these terraces should be a future research focus. It is likely that there are more periods of strong fluvial processes in the central Sahara during the Quaternary, but the remnants of earlier processes may have been destroyed or erased by later geomorphologic processes. Based on the climatic-geomorphologic types (Figure 31.4) and the development trends of

geomorphologic processes, Hövermann (1988) suggests that Sahara and Kalahari/Namib may have experienced opposite humidity trends during the last 50 ka BP due to changes in solar radiation received in each hemisphere, where high radiation causes an increase in humidity and temperature. This concept appears rational provided that southern Africa is generally less arid than northern Africa at the present time (Figure 31.5).

At long time scales, orbitally induced changes in solar radiation do account for a large part of climatic changes in the subtropical

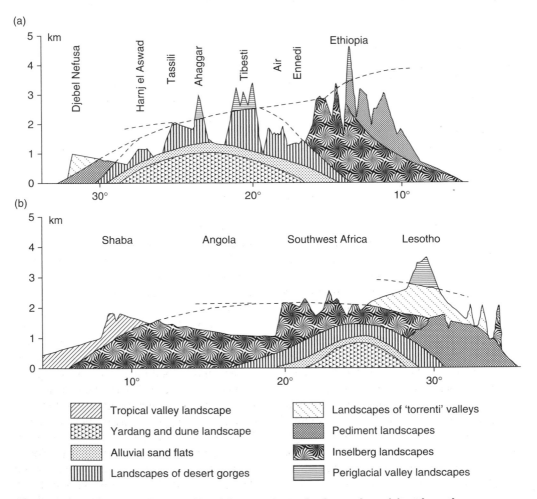

Figure 31.4 Schematic picture of morphogenetic stories in northern (a) and southern (b) Africa caused by regional climates (Hövermann, 1985), indicating more extensive arid areas (both in terms of vertical upper boundary and horizontal extension of desert landscapes such as Yardangs, dunes, alluvial sand flats and desert gorges) in northern than in southern Africa.

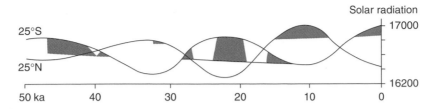

Figure 31.5   Alternating climatic conditions in the Kalahari and Sahara due to changes in solar radiation in each hemisphere. The wetter periods (dark) linked to the canonical summer half-year radiation at latitude 25° during the last 50 ka, radiation units. Meinardus (1944) and Hövermann (1988).

drylands of Africa. The generally drier conditions during or around LGM must be associated with lower land- and sea-surface temperatures and a weakened hydrological cycle. Based on the studies of the dinoflagellate cyst and pollen records from marine sediments off the southwestern African coast, Shi et al. (2000) concluded that the aridification recorded in these marine cores correspond to enhanced oceanic upwelling off Namibia and cooler temperature in Antarctica and possibly to oceanic thermohaline circulation. From the studies of stalagmites in the Makapansgat Valley of northern South Africa, Holmgren et al. (2003) postulated that the trigger for the millennial- and centennial-scale changes deciphered in this record should be atmospheric circulation changes associated with a change in the southern hemisphere circumpolar wind vortex.

The environment of the Sahara was quite green from the early- to mid-Holocene (*c*.10–6 ka BP) due to an increase in precipitation, and many studies deal with widespread sedimentary and cultural evidence of moist conditions during this period (Bubenzer et al., 2007; Scott, 2003). The northern Sahelian vegetation boundary was shifted by at least 500 km northward from the east to the west (Scott, 2003, and the references therein), and the Sahara might not have existed at that time due to the increase in moisture (Roberts, 1989). This increase in precipitation seems to have been regulated by orbital precessional forcing that increases monsoon strength

(Street-Perrott and Perrott, 1993). Several lacustrine episodes are identified in the Lake Chad Basin during the Quaternary (Goudie, 2002). Most recent studies of geomorphologic and lake deposits reconfirm the occurrence of long-lasting Lake Mega-Chad (LMC) covering more than 350,000 km². The latest LMC episode started after 7 ka BP (Gumnior and Thiemeyer, 2003) and lasted until *c*.5.3–4.4 ka BP (Schuster et al., 2005).

Various records indicate that it was relatively dry in the early Holocene in the northern Namib. A marine core off the west coast of Namibia shows a very dry interval from 10.9 to 9.3 ka BP and the warmest and wettest Holocene epoch between 6.3 and 4.8 ka BP (Shi et al., 1998). A synthesis of fluvial, lacustrine, speleothems, groundwater recharge and palynological records in Namibia and Botswana indicate three well-defined intervals of wetness during the Holocene at 7–5, 2.5–1.7 and *c*. 1.0 ka BP, and dryness prior to *c*.8 ka BP (Brook et al., 2007 and reference therein). The northwestern Kalahari was geomorphologically stable during the Holocene, with no evidence of hydrological variations sufficient to generate either water bodies or sand dune activities (Thomas et al., 2003).

Data from stalagmites in the Makapansgat Valley of northern South Africa document millennial- and centennial-scale climate changes in the savannas of northern South Africa over the past 25 ka BP. Drier and cooler conditions with sparser grass cover were suggested for c. 23–21, 19.5–17.5 and 15.5–13.5 ka BP (Holmgren et al., 2003). The early Holocene (between c. 12 and 10 ka BP) was characterized by warm and evaporative conditions with fewer C_4 grasses (Scott, 1982). The carbon isotopic data from the stalagmites show a rapid decease after 10.2 ka BP with its lowest values at 9–8.4 ka BP, indicating vegetation-poor, considerably drier environments. A rapid increase in vegetation cover and rise in soil CO_2 production, as well as enhanced bedrock weathering, marked the period between 10 and 6 ka BP, and then a gradual cooling trend occurred in the mid- and late Holocene, except for a warm, wet epoch from 1.2–0.6 ka BP (Holmgren et al., 2003).

3.2 Asia

Located in the interior of the largest continent, Eurasia, the dry lands in Asia are distributed in a wide range of geomorphologic and tectonic settings, from 155 m below sea level to altitudes of more than 5,000 m above sea level. Environmental changes of these arid and semi-arid regions are related to changes of three different climate systems (i.e., the East Asian monsoon, Indian Ocean monsoon and Northern Hemisphere westerlies). Therefore, evidence of changes of Asian arid and semi-arid areas is often used to interpret the histories and development of these three climate systems. Numerical simulation shows that the formation of the Siberian–Mongolian High, which is the most crucial cause of the arid climate in China and Central Asia, was caused by the uplifting of the Tibetan Plateau (Manabe and Terpstra, 1974; see Chapter 17). The initial age of the desert formation in China has been controversially debated, while the estimates for the uplifting of the Tibetan Plateau range from the Pliocene (~2–4 Ma BP) to the early Cenozoic (Tooth, 2008). Much of the earlier understandings about the climatic change in deserts of China and Central Asia have been inferred from loess-palaeosol sequences in

the Chinese Loess Plateau, as the loess is brought from the deserts by aeolian processes. Denying lacustrine or fluvial origins, Richthofen (1877) firstly attributed the loess in China to aeolian process based on two features: (1) the appearance of loess on all kinds of landforms (even on ridges of mountains); and (2) the occurrence of land snails and plant remains in the loess. The gradual changes of loess grain size in the Loess Plateau of China reconfirm that loess in China was sourced from the deserts located in the northwestern (Liu, 1985). The Miocene loess-soil sequences in the western Loess Plateau suggested that deserts existed in the Asian interior at latest 22 Ma ago (Guo, 2002). The establishment of magnetostratigraphical chronology of the loess deposits in the Loess Plateau of China (Heller and Liu, 1982) has been a key step for recognizing loess-palaeosol sequences as the longest terrestrial records of palaeoclimate. The grain size changes in the loess–red clay sequence on the northern margin of the Loess Plateau indicate advances and retreats of the desert in northern China in response to global glacial–interglacial climate changes during the last 3.5 Ma. The distinct increase in sand fractions indicates that significant southward migration occurred four times during the last 3.5 Ma, roughly at 2.6, 1.2, 0.7 and 0.2 Ma, caused presumably by a stepwise weakening of the East-Asian summer monsoon in relation to the changes in global ice volume (Ding et al., 2005). Although the Asian loess-paleosol sequences are generally not well dated by physical dating methods, changes in their grain size and magnetic features may also match climatic oscillations discovered in and around the North Atlantic. By comparing grain size curves of loess-paleosol sequences with oxygen isotope records of Greenland NGRIP ice core, Vandenberghe et al. (2006) concluded that the Atlantic signals, characterized by short-term oscillations, penetrated eastwards to the semi-arid northeastern Tibetan Plateau and the western part of the Loess Plateau during times of full glacial conditions. In contrast, the monsoonal effects, marked by changes at orbital scales, were found to be predominant in these regions during interglacials (Vandenberghe et al., 2006). More details about environmental changes inferred from loess deposits can be found in Chapter 13.

Although the loess is generally sourced from desert regions, there are areas where loess originates from local sources. The fluvial deposits of the Yellow River have been a significant component of the loess in the southeast margin of the Loess Plateau since marine isotope stage six (Jiang et al., 2007).

In contrast to the conclusions inferred from loess-palaeosol sections, desert geomorphologists often propose a much younger age for the deserts, based on investigations within the deserts. Among various deserts, great attention has been given to the studies of the Taklamakan Desert, the largest sand sea in China. Norin (1932) reported that there was a fresh lake with an extensive area in the southern part of the Taklamakan during the Quaternary, meaning that the desert is of young age. Zhu et al. (1980) concluded that the Taklamakan was formed from fluvial and alluvial sediments during the middle Pleistocene, because aeolian loess had been deposited since that time on the north slope of Kunlun Mountains. Jäkel (1991) suggested a Quaternary mega lake Lop Nuor that covered much of the eastern Taklamakan.

Geomorphologic and sedimentologic evidence in the large deserts of western China suggests distinct environmental changes during the late Pleistocene and Holocene, although a precise correlation between these records is still impossible due to the low temporal resolution. In general, it was wetter in the deserts of western China at around 30 ka BP and during the middle Holocene (Yang et al., 2004). Fluvial terraces and lacustrine sediments occur in the dune fields of the Taklamakan Desert, indicating an increase in water availability and a potentially moist climate. The highest terrace of the Keriya River was dated to $28,740 \pm 1,500$ [14]C BP (Yang et al., 2002), and three layers of lacustrine sediments in the interior of the Taklamakan Desert were OSL dated to $39,800 \pm 2,900$, $29,200 \pm 2,600$ and $28,000 \pm 2,300$ BP, respectively (Yang et al., 2006; Figure 31.6). The highest lake terrace of the former large lake at the terminus of the Black River northwest of the Badain Jaran Desert was radiocarbon dated to $33,700 \pm 1,300$ BP (Norin, 1980), while the calcareous cementation indicating pedogenic process was developed at around $31,750 \pm 485$ [14]C BP and $19,100 \pm 770$ [14]C BP, respectively (Yang et al., 2003).

Lacustrine records in the terminal lakes of dryland rivers in China have enabled various reconstructions of Holocene climate changes but have caused some inconsistencies in the interpretation of environmental histories, because changes in the lake systems of western China can be caused by variation in the surface runoff directly linked to melting of mountain glaciers or by alteration of river courses. The evidence in the desert interiors of western China, probably reflecting the climates onsite, suggests a drier earlier Holocene, followed by a warmer/wetter mid Holocene and a return to a drier environment

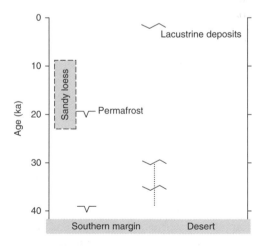

Figure 31.6   Wetter climate (recognized from lacustrine processes) in the Taklamakan Desert of China and colder times (recognized from permafrost processes) on its south margin.

Source: Yang et al. (2006).

in the Late Holocene (Yang et al., 2004; Figure 31.2b). The climate was arid in the Mongolia during the LGM, because almost all dune fields were active, and the aridity lasted probably until 14 ka BP. High lake levels at numerous sites of western and central Mongolia and pollen assemblages suggest a pluvial period between 8 and 4 ka BP, consistent with the general trend in western China (Grunert et al., 2000; Grunert and Dasch, 2004).

One third of the Arabian Peninsula is covered by the Rub Al Khali sand sea, which at 660,000 km^2 in area is the largest sandy desert in the world. It was essentially a large drainage basin during the Early Pleistocene, where the Quaternary dunes probably appeared for the first time at about 920–880 ka BP (Edgell, 2006). Multiple wetter periods have occurred since then, and the latest major wetter episodes took place at about 35–25 ka BP and 10–6 ka BP, respectively. Between 35 and ~25 ka BP the Rub Al Khali desert was under moister grassland to semi-desert conditions, while lacustrine sediments were deposited in the interdunes of its southwestern part, with fluvial deposits in its northern region (Bray and Stokes, 2004; Edgell, 2006; Schultz and Whitney, 1986). During the early Holocene, it was relatively cold and arid, and linear dune activities in the northeast were OSL dated to the period between 13.5 ± 0.7 and 9.1 ± 0.3 ka BP (Goudie et al., 2000). The late Early-Holocene to Mid-Holocene wetter interval was identified in marine sediment cores and in various terrestrial geoarchives, although reasonable differences in timing of this moister period exist between various records (Edgell, 2006).

The arid areas of the Indian subcontinent are neither particularly dry nor windy, as both the westerlies and summer southwest monsoon bring some rainfall to this region. Most of the sand dunes in India were relict landforms and were of late Pleistocene age (Goudie, 2002). The two most recent major phases of aeolian activity occurred at around 11–13 ka and at *c*.5.2 ka BP, respectively (Thomas et al., 1999).

3.3 The Americas

The arid regions in southern North America owe their aridity to the influence of a subtropical high-pressure cell, while orographical barriers are especially important in the northern region and in parts of California. In South America, the continent's plate tectonic history and the development of the Andean cordillera are crucial to the establishment of arid environments. The impact of aeolian processes in the currently relatively humid Argentinean Pampas demonstrates that arid conditions at different times in the Quaternary have been more extensive than they are today (Goudie, 2002; Muhs and Zárate, 2001).

During the last four decades, significant progresses have been achieved in deciphering palaeoclimatic histories from the lacustrine records in the drylands of the western US (e.g., Gillespie et al., 2004). The Owens Lake diatom record for the last 200 ka indicates that this lake responded mainly to climate change rather than tectonic or volcanic perturbations of its watershed (Bradbury, 1997). It was found that late Pleistocene water budgets of Great Basin lakes might have been controlled by the size and shape of the northern hemisphere ice sheets, which determined the mean position of storm tracks (Oviatt, 1997). The changes of *Quercus* pollen abundance in the lacustrine records from northern California indicate that relatively warm and dry climates occurred frequently during MIS 5 when continental-ice volumes were minimal (Adam, 1988). The pollen records from the continental margin of northern and southern California also suggest the frequent occurrence of a Mediterranean type of climate during the MIS 5 (Heusser et al., 2000). Millennial-scale-resolution records from the lakes of the western US provide support to the opinion that solar insolation changes affect northern hemisphere climate. Correlation of a centennial-scale record from the Summer Lake with the Greenland Ice Core Record GISP2 shows that the northern Great Basin, the coolest of the North American deserts, was anomalously wet during Dansgaard–Oescher interstades. A centennial-scale climate record from the Mono Lake and lake-level changes in Lake Bonneville indicate that some lakes in the western US may have experienced lower levels at times of Heinrich events (Benson, 2004; Oviatt, 1997).

Radiocarbon-dated lacustrine sediments and palaeo-shorelines from the Mojave Desert of southern California suggest that two major high and persistent lake stands occurred in the desert between *c.*18.4 and 16.6 ka BP (Lake Mojave I) and 13.7 and 11.4 ka BP (Lake Mojave II; Figure 31.7), caused by significantly increased precipitation and annual large-scale foods (Wells et al., 2003). Stratigraphic correlations and luminescence dating confirm that major alluvial fan deposition occurred between 14 and 9.4 ka BP, largely due to an increase in sediment yields caused possibly by extreme storm events, and an increase in tropical cyclones (McDonald et al., 2003). In the Mojave Desert, the rates and styles of episodes of aeolian construction have largely been determined by sediment supply from the Mojave River. In the Mojave Desert where source of sand is a limiting factor for sand dune formation, increased aeolian processes are likely associated with storm and flooding events caused by climate change (Figure 31.7). Reconstruction of the Holocene records for the deserts in the southwestern US indicate that conditions have been close to the threshold for aeolian processes through much of the Holocene (Busacca et al., 2004).

Palaeolimnological, mineralogical and geochemical data indicate that repeated cycles of reactivation of dunes have taken place in the Nebraska Sand Hills over long periods of time, and stratigraphic and radiocarbon data show that this sand sea, the largest dune field in the western hemisphere, has everywhere been active at least once in the past few thousand years, although not necessarily at the same time (Muhs et al., 1997; Jacobs et al., 2007). Several droughts in past two millennia were more severe than the 1930s Dust Bowl drought according to lacustrine and tree-ring records, and were dry enough to remove the vegetation cover on dunes (Muhs et al., 1997). (See Chapter 13 for an extensive discussion of the Quaternary history of this region.)

Pollen and packrat middens data suggest wetter conditions in the southwestern US and the Rocky Mountains at *c.*10,000 BP, and by 6,800 BP, the environment in the southwest became more like present conditions (Thompson and Anderson, 2000). In south central New Mexico and northern Mexico a gradual increase in the dominance of desert plants is recorded after *c.*5,700 BP (Betancourt et al., 2001). By studying the transects of

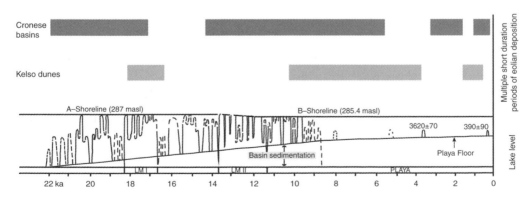

Figure 31.7   Simplified history of pluvial Lake Mojave fluctuations (decrease in lake level from stage LM I to LM II) and aeolian depositional periods in the Mojave Desert.

Source: Wells et al. (2003) and Busacca et al. (2004).

sediment cores and subsurface geophysical profiles from three lakes near the Continental Divide in Montana and Colorado, Shuman et al. (2009) found out that a severe and widespread aridity occurred between 7,000 and 4,500 BP, consistent with the extent of insolation-induced aridity in regional climate model simulations. For millennial-scale moisture variation it was suggested that an abruptly initiated and terminated wet period occurred in Colorado from 4,400 to 3,700 BP. In addition, they came to the conclusion that a wide range of settings in the region were much drier than today before 3,000–2,000 years ago (Shuman et al., 2009).

From a well-dated series of hydrologic variation inferred from a 220-m-long core in Salar de Uyuni in tropical South America, the world's largest salt flat, alterations between wet and dry periods during the last 170,000 years were reconstructed (Fritz et al., 2004). The stratigraphy of this core is characterized by alternating mud and salt units, and the mud layers are longer in duration in the upper part of the sequence than below. The dominance of salt between 170,000 and 140,000 years ago shows that much of the penultimate glacial was relatively dry, in contrast to the wet conditions in the LGM. The wetter periods of the past 70,000 years occurred during times of cold North Atlantic temperatures, but prior to that time the relationship is not obvious. Consequently, it was assumed that the relative influence of insolation forcing on regional moisture budgets may have become stronger during the past 50,000 years than in earlier times (Fritz et al., 2004).

Fossil rodent middens and wetland deposits from the central Atacama Desert (22–24°S) suggest a distinct increase in summer precipitation and grass cover, as well as groundwater levels from 16.2 to 10.5 ka BP, and a less pronounced episode of higher groundwater levels from 8 to 3 ka BP (Betancourt et al., 2000). Potentially high aridity was suggested for the period 35 to 16.2 ka BP, because of the paucity of middens and wetland sediments. Changes of plant species

indicate that the annual precipitation increased to ~70 to 100 mm between 11.8 and 10.5 ka BP at a site that receives ~20 mm at present, probably due to strengthening Pacific trade winds (easterlies) and persistent La Niña-like conditions (Betancourt et al., 2000). Occurrence of fossil mollusks and loess deposits in the hyper-arid coastal desert of Peru (14.5°S) indicates a semi-desert environment between 13.5 and 10.5 ka BP, an open grassland from 10.5 to 4.5 ka BP, and returning to desert afterwards (Eitel et al., 2005). However, it was predominantly arid between 5,000 and 4,000 years in the western and southern fringe of the Pampean plains according to aeolian, pollen and vertebrate fossils records (Muhs and Zárate, 2001). Further west, the record from the Laguna Miscanti, a lake located at 4,140 m above sea level in the Atacama of northern Chile, suggests more arid conditions between *c.*8,000 and 3,600 ^{14}C years (Valero-Garcés et al., 1996; Grosjean et al., 2001). The collapse of Paracas and Nasca cultures existing mainly along rivers in the northern Atacama in Peru between ~900 BC and AD 700, was probably caused by an increase in aridity at ~AD 600, limited reoccupation occurred during the late Intermediate Period (~AD 1200–1400) due to increase in rainfall, but fully abandoned since then because of renewed increase in aridity (Unkel et al., 2007).

3.4 Australia

Australia is dominated by an arid zone (Figure 31.1) but is, however, only moderately dry and generally vegetated. Even the driest areas of the continent, in the vicinity of Lake Eyre, are able to support a low level of commercial cattle grazing at the present time. On long time scales, shifts in latitude have triggered significant climate changes in the past 50 million years while Australia moved northward from a near-polar climatic zone through a middle-latitude humid zone into a zone of tropical and subtropical climates (Gale, 1992). A progressive increase in aridity during the middle to late Quaternary has been recognized in Australia from a variety of proxy climatic data and is particularly recognizable over the past 350 ka (Hesse, 1994; Kershaw et al., 2003; Nanson et al., 2008). By the early Pleistocene, the modern climatic regime had been set up, but precipitation was higher than today (Martin, 2006). According to $^{10}Be/^{21}Ne$ dating, the Australian stony deserts developed during the global cooling and onset of the Quaternary ice ages (Fujioka et al., 2005). Cosmogenic nuclides ^{10}Be and ^{26}Al together with quartz OSL in dune sand indicate that the aeolian dune building in central Australia commenced at $c.1$ Ma (i.e., the time of the shifting of global climate from 41 ka glacial cycles to 100 ka cycles with larger amplitude). This intensified aridity, which began earlier in the Australian center than at the periphery (Fujioka et al., 2009, and the references therein).

On short time scales, aeolian records show that the cold epochs of the late Pleistocene glacial cycles were dry in all parts of Australia, although the nature of aridity varies from north to south. In northern Australia rainfall has been positively correlated with sea level and global temperature, and therefore warm, wet interglacials alternate with cooler, drier glacials. In southern Australia the relationship is much less clear, because very low runoff occurred during the peak interglacials. During glacials it was much colder with much sparser vegetation and greater aeolian activity, but runoff may have been much higher, at least seasonally (Hesse et al., 2004). At $c.15–13$ ka BP there was considerable climatic amelioration across much of Australia.

Luminescence dating has identified the latter part of MIS 5 and MIS 3 as periods of source-bordering dune formation on Cooper Creek, central Australia. These dunes as a whole have been remarkably stable without migration since MIS 3 although aeolian reworking of the upper parts of some dunes has continued up to the present (Maroulis et al., 2007). In the Strezlecki and Tirari

deserts, stronger dune activities took places in the episodes around 73–66, 35–32, 22–18 and 14–10 ka BP according to OSL ages from 26 sites in these two deserts. Intermittent partial mobilization persisted at other times and an intensification of dune activity appears to have occurred during the late Holocene (Fitzsimmons et al., 2008), consistent with the geological and botanical evidence, suggesting that the present interglacial period is drier than the previous interglacial (Martin, 2006). Widespread dune activation in the Strezlecki and Tirari deserts seems to occur during the cooler periods, because its timing coincided with glaciation in southeastern Australia, along with colder temperatures in the adjacent oceans and Antarctica (Fitzsimmons et al., 2008).

4 LAND-USE AND DESERTIFICATION

Land degradation refers to the detriment of all aspects of the biophysical environment by human activities (Conacher and Conacher, 2000). Land degradation in arid, semi-arid and dry subhumid zones is defined as desertification by the United Nations Convention to Combat Desertification (UNCCD, 1999). According to this definition, the cause of desertification is closely related to unsustainable land use practices. Thus, the giant deserts formed by purely natural factors, such as subtropical air pressures or rain shadows in the lee of mountain ranges, should not be considered as cases of desertification (Mensching, 1990; Yang et al., 2007), although desertification is also a serious problem in hyperarid regions (e.g., Yang et al., 2006a). At the UN Conference on Environment and Development in Rio de Janeiro in 1992, climatic variation was accepted as a factor triggering desertification (Williams and Balling, 1995). Within the context of desertification, climatic variation may refer only to the short time scale anomalies, such as the droughts in the Sahel in the 1970s, but not the long time scale changes like glacial–interglacial alternations.

Some scientists suggested that 'desertification' was a politically charged term and should be substituted with land degradation or landscape degradation (e.g., Seuffert, 2001).

The human activities causing desertification include direct land use practices such as overgrazing, agricultural reclamation in areas not suitable for cultivation, mismanagement of irrigated lands and deforestation for fuel and fodder collection, as well as indirect government policies such as failed population planning policies and war (see Chapter 24). Indeed, there is a very long history of human activities in arid and semi-arid regions. For example, large areas of arid climate in Australia were occupied by aboriginal people throughout the last 30,000 years, including the LGM (Ross et al., 1992). The semi-arid desert fringes were the basis of some of the world's oldest civilizations; for instance, the Fertile Crescent between the Rivers Tigris and Euphrates is the original area of domestication of wheat and barley (see Chapter 39). Rain-fed agriculture and transhuman herding prevailed for over ten millennia in Southwest Asia, whereas the problem of sustainability has developed only over the last 70 years due to the dynamic interplay between various natural and social factors (Hole, 2009). Indeed, humans have been becoming an increasingly heavy trigger for the environmental changes on Earth. Due to the human-induced changes of atmospheric composition, Crutzen (2002) suggested that the time since the beginning of the nineteenth century should be termed the 'Anthropocene'. It has also been argued that the rice farming in China more than 5,000 years ago had probably already caused an increase in the greenhouse gas (methane) concentrations, in turn triggering an increase in temperature (Ruddiman et al., 2008). Global warming might become a key factor causing the reworking of the stable desert dunes in the near future, as the case of the Kalahari in southern Africa suggests. All stable dunes in Kalahari are likely to be reactivated by 2099, because the ratio between precipitation and potential evaporation will decline and mean wind velocity will increase according to the general climate model simulations (Thomas et al., 2005).

On a regional scale, desertification-induced hazards mainly include sand encroachment, soil deflation, salinization and alkalinization, and disappearance of wetlands. One of the dramatic cases of desertification is the desiccation of the Aral Sea, the formerly fourth largest inland water body of the world. The Aral Sea is the endorheic lake of two major Central Asian rivers, Amudarya and Syrdarya. Irrigation has been expended greatly along the two rivers during the last five decades, causing a shrinkage of approximately 75 percent of the lake surface (Mensching, 1990; Zavialov, 2005). In this region, industrial pollutants such as heavy metals and large quantities of pesticides used to control parasites and weeds have accumulated not only in water but also in soil and have been deposited over large areas by aeolian transport (O'Hara et al., 2000).

The issue of desertification has been given great attention by individual scientists and various organizations. In 1996, the UNCCD legally entered into force following its fiftieth ratification, and 2006 was the United Nations' International Year of Deserts and Desertification. The UNCCD (1999) reported that 70 per cent of global drylands are affected by desertification that threatens the livelihood of around one billion people. In contrast, it was commented that none of the available global-scale assessments on the extent of desertification can be considered a reliable source of information (Safriel, 2007). The UN's estimate is also questionable, because it is mainly derived from data of varying authenticity and consistency (Thomas and Middleton, 1994). Based on regional monitoring and assessment of annually integrated NDVI and annual rainfall, Helldén and Tottrup (2008) came to the conclusion that a systematic desertification trend at the regional–global level did not exist, and a 'greening-up' seemed to occur over large regions. Verstraete et al. (2008) comment that unraveling the tangled syndrome of desertification has remained an elusive goal, although a large number of publications on this subject appear each year.

Over the years, good concepts and efficient technologies have been developed to combat desertification, while many countries have established research and demonstration stations to improve and disseminate knowledge about desertification control. Details on land use changes and interactions between climate and human impacts can be found in Chapters 7, 24 and 25.

5 GLOBAL IMPACTS OF THE ENVIRONMENTAL CHANGES IN ARID AND SEMI-ARID REGIONS

Various studies have indicated that environmental change in arid and semi-arid regions has a significant global impact (e.g., Goudie, 2009; Tooth, 2008). Remote sensing and geochemical data have demonstrated that many of the world's major dust source regions are parts of arid zones that have a mean annual rainfall of less than 100 mm (Goudie and Middleton, 2006; Yang et al., 2007a). The environmental conditions of these areas have a crucial role in generating dust storms. The relationships between dust storms and the Earth system have been intensively explored from various perspectives in recent years. Dust loading may influence air temperature both at regional and global scales through absorption and scattering of solar radiation, affect sulfur dioxide levels in the atmosphere by physical absorption or by heterogeneous reactions, and have an impact on marine primary productivity by fertilization. Consequently the content of carbon dioxide in the atmosphere is related to the global dust cycle via the delivery of iron and nutrients to the oceans (Goudie, 2009, and references therein). Simulations indicate that strong Asian dust emissions during glacials may have prevented the occurrence of permanent snow cover in Northern Asia and may even have been decisive in determining the position and extent of the last giant ice sheets (Krinner et al., 2006).

The late Quaternary dust records reconstructed from various ice cores reveal that each climatic fluctuation towards more negative $\delta^{18}O$ values (glacial) is matched by an abrupt increase in dust concentrations. The glacials are characterized not only by a higher dust content, but also by larger dust particles, indicating that strong winds lifted and transported the dust from continental regions when climate was colder and drier and vegetation in the source areas was less abundant. In the Vostok ice core of Antarctica, the dust concentrations rise from about 50 ng g^{-1} during interglacials to 1,000–2,000 ng g^{-1} during cold stages 2, 4, 6, 8 and 10 (Petit et al., 1999). In the Greenland ice cores, dust concentrations generally show levels up to 70 times higher during the last glaciation than during the present interglaciation (Dansgaard et al., 1984). The colder intervals also contain larger amounts of sea-salt aerosol entrained from ocean surface, because strong winds over the oceans plucked more of these ions out of the salty water spray tossed up above the rough sea surface and carried them to the ice. Geochemical analysis of the dust deposited in Greenland ice during the millennial oscillations indicates that the main source region is not in nearby North America but in more distant eastern Asian deserts (Svensson et al., 2000). The predominant sources of windborne sediments during the LGM in the southern latitudes are from Patagonia in Argentina (Gaiero et al., 2004). On orbital time scales, the changes in dust concentrations reconstructed from ice cores are supportive of the histories of environmental changes in arid and semi-arid regions, as demonstrated earlier in this chapter. More detailed information from the arid and semi-arid zones is needed in order to compare the millennial changes between the data from drylands and from ice cores. More evidence of environmental change obtained from ice cores is presented in Chapter 10.

In addition, dust emission and transportation play an important role in global biogeochemical cycles. Desertification and deterioration of the ecosystem, as well as biomass burning in drylands, can transfer carbon currently stored in soils and in plants into the atmosphere, contributing to the increase in greenhouse gases. However, a systematic study in a sandy area of northern China suggests that the earlier estimates (e.g., Duan et al., 2001) of the carbon dioxide already released from desertified lands in China should be treated with caution, because the important portion of carbon lost on site was redeposited in downwind regions via deflation and redeposition. Therefore, the amount of carbon initially contained in the deflated surface layer should not be evaluated as exchange between atmosphere and desertified lands (Yang et al., 2008).

Slaymaker (2008) has emphasized that changes in relief should be given equal treatment as climate in global change studies in order to fully understand Earth system change, as land or landscape can record natural and anthropogenic impacts more directly than the atmosphere. In this regard, the global impact of environmental changes in arid and semi-arid regions would be easy to follow, because dust originating from desert regions can be transported over huge distances. For instance, dust from the Sahara has remarkably influenced the nature of soils in Barbados, the Bahamas and Florida (Muhs et al., 2007), and the Andes (Boy and Wilcke, 2008).

6 CONCLUSIONS

Knowledge about the environmental change in arid and semi-arid regions is of great importance for a comprehensive understanding of the Earth system, because one-third of the land surface on Earth is under arid and semi-arid conditions today, and the proportion was considerably larger during some periods of past glacials. Compared with many other regions of the world, information on environmental changes in arid and semi-arid regions is still poor mainly due to two factors: (1) the complicated nature of the dryland system; and (2) the lack of long-lasting, high-resolution records, although arid and semi-arid regions react very sensitively to global climate change because of their fragile constitutions. Each system in arid and semi-arid regions needs to be individually studied in order to correctly interpret the meaning of its aeolian, lacustrine and fluvial records, because sand deposition, for example, may represent aridity in one case but humidity in another. From a global perspective on climate and biogeochemical cycles, environmental changes occurring in arid and semi-arid regions may influence the nature of the Earth system by generating dust storms and impacting dust cycles, and quantification of these impacts needs to be confirmed by future research. The environmental conditions in drylands have been facing great challenges and worsening due to the continuous increase in land use pressure, particularly in more recent decades.

The date of the initial onset of aridity varies greatly from continent to continent. Multiple Quaternary periods of both increase and decrease in aridity have been recognized in all arid and semi-arid regions of the world. The considerable climatic variability that exists across the arid and semi-arid regions of the world causes some confusion in the reconstruction of late-Pleistocene and Holocene climates. However, when laid out geographically, the data suggests generally dry conditions during the last glacial maximum and expanding of arid zones during the glacials in the most arid and semi-arid regions of the globe. During the Pleistocene, glacial-period linear dunes of the Western Sahara extended onto the continental shelf, accompanied by shrinkage of lakes in the region, southward migration of dunes and large-scale contraction of Guinean forest in West Africa. Increased aridity occurred in South Africa and Australia as well as in parts of the Americas due to a decrease in precipitation. However, the glacial increase in aridity was much less clear and inconsistent in the drylands of Asia. Lacustrine records from western China indicate an increase in effective moisture around 20 to 18 ka BP was likely due to both an intensification of westerly winds and an expansion of glaciers in the surrounding mountain ranges. Wetter epochs also occurred during the MIS 3 in western China. Between 35 and ~25 ka BP the world's largest sand sea, Rub Al Khali, was under moister grassland to semi-desert conditions. Similarly, the LGM was characterized by lacustrine and fluvial formations in the great mountains of Sahara like Tibesti.

At least one distinctly wetter period has occurred in every part of arid and semi-arid regions of the world during the Holocene, although the timing of this epoch may vary. From the early to mid-Holocene (c.10–6 ka BP), a green Sahara was suggested, when the last appearance of the Lake Mega-Chad occurred from 7 ka BP to c.5 ka BP. In arid and semi-arid regions of China and Mongolia, it was probably dry in the early Holocene, considerably warmer and wetter at mid-Holocene, and drier again during the late Holocene, accompanied by distinct lake-level changes in the arid regions and dunes' activation and stabilization in semi-arid regions. In today's hyper-arid coast desert of Peru, there was an open grassland from c.10 ka BP to 4.5 ka BP. Persistent high lake stands occurred earlier in Mojave Desert of the US (i.e., between 18.4 and 16.6 ka BP, and from 13.7 to 11.4 ka BP). A gradual increase in aridity after the mid-Holocene is also recorded in the southwestern US. In Australia, the Holocene appears to be much more stable with just a minor intensification of dune activity, although an increase in aridity occurs on glacial–interglacial time scales, and today's interglacial is drier than the previous one.

REFERENCES

Adams D. 1988 Correlations of the Clear Lake, California, core CL–73–4 pollen sequence with other long climate records. *Geological Society of America Special Paper* 214: 81–95.

Adams L. and Tetzlaff G. 1985. The extension of Lake Chad at about 18,000 yr BP. *Zeitschrift für Gletscherkunde und Glazialgeologie* 21: 115–123.

Benson L. 2004. Western lakes, in Gillespie A., Porter S. and Atwater B. (eds) *The Quaternary Period in the United States.* Amsterdam: Elsevier, pp. 185–204.

Betancourt J., Latorrre C., Rech J., Quade J. and Rylander K. 2000. A 22,000-year record of monsoonal precipitation from northern Chile's Atacama Desert. *Science* 289: 1542–1546.

Betancourt J., Rylander K., Penalba C. and McVickar J. 2001. Late Quaternary vegetation history of Rough Canyon, south-central New Mexico, USA. *Palaeogeography, Palaeoclimatology, Palaeoecology* 165: 71–95.

Blümel W., Eitel B. and Lang A. 1998. Dunes in southeastern Namibia: evidence for Holocene environmental changes in the southwestern Kalahari based on thermoluminescence data. *Palaeogeography, Palaeoclimatology, Palaeoecology* 138: 139–149.

Boy J. and Wilcke W. 2008. Tropical Andean forest derived calcium and magnesium from Saharan dust. *Global Biogeochemical Cycles* 22: GB1027.

Bradbury J. 1997. A diatom record of climate and hydrology for the past 200 ka from Owens Lake, California with comparison to other Great Basin records. *Quaternary Science Reviews* 16: 203–219.

Bray H. and Stokes S. 2004. Temporal patterns of arid-humid transitions in the south-eastern Arabian Peninsula based on optical dating. *Geomorphology* 59: 271–280.

Brook G., Marais E., Srivastava P. and Jordan T. 2007. Timing of lake-level changes in Etosha Pan, Namibia, since the middle Holocene from OSL ages of relict shorelines in the Okondeka region. *Quaternary International* 175: 29–40.

Bubenzer O., Besler H. and Hilgers A. 2007. OSL data expanding ^{14}C chronologies of late Quaternary environmental change in the Libyan Desert. *Quaternary International* 175: 41–52.

Busacca A., Beget J., Markewich H., Muhs D., Lancaster N. and Sweeney M. 2004. Eolian sediments, in Gillespie A., Porter S. and Atwater B. (eds) *The Quaternary period in the United States.* Amsterdam: Elsevier, pp. 275–310.

Busche D. 1998. *Die Zentrale Sahara.* Gotha: Justus Perthes.

Chase B. and Meadows M. 2007. Late Quaternary dynamics of southern Africa's winter rainfall zone. *Earth-Science Reviews* 84: 103–138.

Chase B. and Thomas D. 2007. Multiphase late Quaternary aeolian sediment accumulation in western South Africa: Timing and relationship to palaeoclimatic changes inferred from the marine record. *Quaternary International* 166: 29–41.

Conacher J. and Conacher A. 2000. Policy responses to land degradation in Australia in A. Conacher (ed.) *Land Degradation.* Dordrecht: Kluwer Academic Publishers, pp. 363–385.

Crutzen P. 2002. Atmospheric chemistry in the 'Anthropocene', in Steffen W., Jäger J., Carson D. and Bradshaw C. (eds) *Challenges of a Changing Earth: Proceedings of the Global Change Open Science Conference, Amsterdam, The Netherlands, 2001.* Berlin: Springer, pp. 45–48.

Dansgaard W., Johnsen S., Clausen H., Dahl-Jensen D., Gundestrup N., Hemmer C. and Oeschger H. 1984. North Atlantic climatic oscillations revealed by deep Greenland ice cores, in Hansen J. and Takahashi T. (eds) *Climate Processes and Climate Sensitivity.* Washington DC: AGU, pp. 288–298.

deMenocal P., Ortiz J., Guliderson T., Adkin J., Sarnthein M., Baker L. and Yarusinsky M. 2000. Abrupt onset and termination of the African Humid Period: Rapid climate responses to gradual insolation forcing. *Quaternary Science Reviews* 19: 347–361.

Ding Z., Derbyshire E., Yang S., Sun J. and Liu T. 2005. Stepwise expansion of desert environment across northern China in the past 3.5 Ma and implication for monsoon evolution. *Earth and Planetary Science Letters* 237: 45–55.

Duan Z., Xiao H., Dong Z., He X. and Wang G. 2001. Estimate of total CO_2 output from desertified sandy land in China. *Atmospheric Environment* 35: 5915–5921.

Edgell H. 2006. *Arabian Deserts: Nature, Origin, and Evolution.* Dordrecht: Springer.

Eitel B., Hecht S., Mächtle B., Schukraft G., Kadereit A., Wagner G., Kromer B., Unkel I. and Reindel M. 2005. Geoarchaeological evidence from desert Loess in the Nazca-Palpa Region, Southern Peru: Palaeo-environmental changes and their impact on pre-Columbian cultures. *Archaeometry* 47: 137–158.

Embabi S. 2004. *The Geomorphology of Egypt: Landforms and Evolution, Volume I. The Nile Valley and the Western Desert.* Cairo: The Egyptian Geographical Society,

Fitzsimmons K., Rhodes E., Magee J. and Barrows T. 2007. The timing of linear dune activity in the Strzelecki and Tirari Deserts, Australia. *Quaternary Science Reviews* 26: 2598–2616.

Fritz S., Baker P., Lowenstein T., Seltze G., Rigsby C., Dwyer G. et al. 2004. Hydrologic variation during the last 170,000 years in the southern hemisphere tropics of South America. *Quaternary Research* 61: 95– 104.

Fujioka T., Chappell J., Honda M., Yatsevich I., Fifield L. and Fabel D. 2005. Global cooling initiated stony deserts in central Australia 2–4 Ma, dated by cosmogenic $^{21}Ne–^{10}Be.$ *Geology* 33: 993–996.

Fujioka T., Chappell J., Fifield L. and Rhodes E. 2009. Australia desert dune fields initiated with Pliocene-Pleistocene global climatic shift. *Geology* 37: 51–54.

Gaiero D., Depetris P., Probst J., Bidart S. and Leleyter L. 2004. The signature of river- and wind-borne materials exported from Patagonia to the southern latitudes: a view from REEs and implications for paleoclimatic interpretations. *Earth and Planetary Science Letters* 219: 357–376.

Gale S. 1992. Long-term landscape evolution in Australia. *Earth Surface Processes and Landforms* 17: 323–343.

Gillespie A., Porter S. and Atwater B. (eds) 2004. *The Quaternary Period in the United States.* Amsterdam: Elsevier.

Goudie A. 2002. *Great Warm Deserts of the World: Landscapes and Evolution.* Oxford: Oxford University Press.

Goudie A. 2009. Dust storms: Recent developments. *Journal of Environmental Management* 90: 89–94.

Goudie A. and Middleton N. 2006. *Desert Dust in the Global System.* Berlin: Springer.

Goudie A., Colls A., Stokes S., Parker A., White and Al-Farraj A. 2000. Latest Pleistocene and Holocene dune construction at the north-eastern edge of the Rub'Al Khali, United Arab Emirates. *Sedimentology* 47: 1011–1021.

Grosjean M., van Leeuwen F., van der Knaap W., Geyh M., Ammann B., Tanner W. et al. 2001. A. 22,000 ^{14}C year BP sediment and pollen record of climate change from Laguna Miscanti (23°S), northern Chile. *Global and Planetary Change* 28: 35–51.

Grove A. and Warren A. 1968 Quaternary landforms and climate on the south side of the Sahara. *Geographical Journal* 134: 194–208.

Grunert J. and Dasch D. 2004. Dynamics and evolution of dune fields on the northern rim of the Gobi Desert (Mongolia). *Zeitschrift für Geomorphologie N.F.* 133(Suppl.-Bd): 81–106.

Grunert J., Lehmkuhl F. and Walther M. 2000. Palaeoclimatic evolution of the Uvs Nuur Basin and adjacent areas (Western Mongolia). *Quaternary International* 65/66: 171–192.

Gumnior M. and Thiemeyer H. 2003. Holocene fluvial dynamics in the NE Nigerian Savanna: some preliminary interpretations. *Quaternary International* 111: 51–58.

Guo Z., Ruddiman W., Hao Q., Wu H., Qiao Y., Zhu R. et al. 2002. Onset of Asian desertification by 22 myr ago inferred from loess deposits in China. *Nature* 416: 159–163.

Heine K. 2002. Sahara and Namib/Kalahari during the late Quaternary – inter-hemispheric contrasts and comparison. *Zeitschrift für Geomorphologie N.F.* 126(Suppl.-Bd): 1–29.

Helldén U. and Tottrup C. 2008. Regional desertification: A global synthesis. *Global and Planetary Change* 64: 169–176.

Heller F. and Liu T. 1982. Magnetostratigraphic dating of loess deposits in China. *Nature* 300: 431–433.

Hesse P. 1994. The record of continental dust from Australia in Tasman Sea sediments. *Quaternary Science Reviews* 13: 257–272.

Hesse P., Magee J. and van der Kaars S. 2004. Late Quaternary climates of Australian arid zone: a review. *Quaternary International* 118/119: 87–102.

Heusser L., Lyle M. and Mix A. 2000. Vegetation and climate of the northwest coast of North America during the last 500K Y: high-resolution pollen evidence from the Northern California margin. *Proceedings of the Ocean Drilling Program Scientific Results* 167: 217–226.

Hole F. 2009. Drivers of unsustainable land use in the semi-arid Khabur River basin, Syria. *Geographical Research* 47: 4–14.

Holmgren K., Lee-Thorp J., Cooper G., Lundblad K., Partridge T., Scott L. et al. 2003. Persistent millennial-scale climatic variability over the past 25,000 years in Southern Africa. *Quaternary Science Reviews* 22: 2311–2326.

Hövermann J. 1972. Die periglaziale Region des Tibesti und ihr Verhältnis zu angrenzenden Formungsregion. *Göttinger Geographische Abhandlungen* 60: 235–260.

Hövermann J. 1985 Das System der klimatischen Geomorphologie auf landschaftskundlicher Grundlage. *Zeitschrift für Geomorphologie N. F.* 56(Suppl.-Bd.): 143–153.

Hövermann J. 1988. The Sahara, Kalahari and Namib deserts: a geomorphological comparison, in Dardis G. and Moon B. (eds) *Geomorphological Studies in Southern Africa.* Totterdam: Balkema, pp. 71–83.

Jacobs K., Fritz S. and Swinehart J. 2007. Lacustrine evidence for moisture changes in the Nebraska Sand Hills during Marine Isotope Stage 3. *Quaternary Research* 67: 246–254.

Jäkel D. 1991. The evolution of dune fields in the Taklimakan Desert since the Late Pleistocene – Notes on the 1:2 500 000 map of dune evolution in the Taklimakan. *Die Erde Erg.-H* 6: 191–198.

Jiang F., Fu J., Wang S., Sun D. and Zhao Z. 2007. Formation of the Yellow River, inferred from loess palaeosol sequence in Mangshan and lacustrine sediments in Sanmen Gorge, China. *Quaternary International* 175: 62–70.

Kershaw P., van der Kaars S. and Moss P. 2003. Late Quaternary Milankovitch-scale climate change and variability and its impact on monsoonal Australia. *Marine Geology* 201: 81–95.

Krinner G., Boucher O. and Balkanski Y. 2006. Ice-free glacial northern Asia due to dust deposition on snow. *Climate Dynamics* 27: 613–625.

Lancaster N., Kocurek G., Singhvi A., Pandey V., Deynoux M., Ghienne J. and Lô K. 2002. Late Pleistocene and Holocene dune activity and wind regimes in the western Sahara Desert of Mauritania. *Geology* 30: 991–994.

Lancaster N. 2002a How dry was dry? Late Pleistocene palaeoclimates in the Namib Desert. *Quaternary Science Reviews* 21: 769–782.

Lezine A. 1989. Late Quaternary vegetation and climate of the Sahel. *Quaternary Research* 32: 317–334.

Liu T. 1985 *Loess and the Environment China.* Beijing: Ocean Press

Maley J. 2000. Last Glacial Maximum lacustrine and fluviatile formations in the Tibesti and other Sahara mountains, and large-scale climatic teleconnections linked to the activity of the Subtropical Jet Stream. *Global and Planetary Change* 26: 121–136.

Manabe S. and Terpstra T. 1974. The effects of mountains on the general circulation of the atmosphere as identified by numerical experiences. *Journal of Atmospheric Sciences* 31: 3–42.

Martin H. 2006. Cenozoic climatic change and the development of the arid vegetation in Australia. *Journal of Arid Environments* 66: 533–563.

Maroulis J., Nanson G., Price D. and Pietsch T. 2007. Aeolian-fluvial interaction and climate change: source-bordering dune development over the past ~100 ka on Cooper Creek, central Australia. *Quaternary Science Reviews* 26: 386–404.

Meinardus W. 1944 Zum Kanon der Erdbestrahlung. *Geologisches Rundschau* 34: 748–762.

Mensching H. 1990. *Desertifikation: ein weltweites Problem der ökologischen Verwüstung in den Trockengebieten der Erde.* Darmstadt: Wissenschaftliche Buchgesellschaft.

McDonald E., McFadden L. and Wells S. 2003. Regional response of alluvial fans to the Pleistocene-Holocene climatic transition, Mojave Desert, California, in

Enzel Y., Wells S. and Lancaster N. (eds) *Paleoenvironments and Paleohydrology of the Mojave and Southern Great Basin Deserts.* Boulder: Geological Society of America, pp. 189–205.

Muhs D. and Zárate M. 2001. Late Quaternary eolian records of the Americas and their paleoclimatic significance, in Markgraf V. (ed.) *Interhemispheric Climate Linkages.* San Diego: Academic Press, pp. 183–216.

Muhs D., Stafford T., Swinehart J., Cowherd S., Mahan S., Bush C. et al. 1997. Late Holocene eolian activity in the mineralogically mature Nebraska Sand Hills. *Quaternary Research* 48: 162–176.

Muhs D., Budahn J., Prospero J. and Carey S. 2007. Geochemical evidence for African dust inputs to soils of western Atlantic islands: Barbados, the Bahamas and Florida. *Journal of Geophysical Research* 112: F02009.

Nanson G., Price D., Jones B., Maroulis J., Coleman M., Bowman H. et al. 2008. Alluvial evidence for major climate and flow regime changes during the middle and late Quaternary in eastern central Australia. *Geomorphology* 101: 109–129.

Norin E. 1932. Quaternary climatic changes within the Tarim Basin. *Geographical Review* 22: 591–598.

Norin E. 1980 *Sven Hedin central Asia atlas, Memoir on maps. Vol.III.* Stockholm: Statens Etnografiska Museum.

O'Hara S., Wiggs G., Mamedov B., Davidson G. and Hubbard R. 2000. Exposure to airborne dust contaminated with pesticide in the Aral Sea region. *Lancet* 355: 627–628.

Oviatt C. 1997. Lake Bonneville fluctuations and global climate change. *Geology* 25: 155–158.

Partridge T., Demenocal P., Lorentz S., Paiker M. and Vogel J. 1997. Orbital forcing of climate over South Africa: a 200 000-year rainfall record from the Pretoria Saltpan. *Quaternary Science Reviews* 16: 1–9.

Penck A. and Brückner E. 1909 *Die Alpen im Eiszeitalter 3 Bde.* Leipzig: Tauchnitz.

Petit J., Jouzel J., Raynaud D., Barkov N., Barnola J., Basile I. et al. 1999. Climate and atmospheric history of the past 420,000 years from the Vostok ice core, Antarctica. *Nature* 399: 429–436.

Petit-Maire N., Reyss J. and Fabre J. 1994. Un paléolac du dernier interglaciare dans une zone hyperaride du Sahará Malien (23°N). *Comptes Rendus Académie des Sciences* 319: 805–809.

Richthofen F. 1877 *China: Ergebnisse eigener Reisen und darauf gegründeter Studien Bd 1.* Berlin: Dietrich Reimer.

Roberts N. 1989. *The Holocene.* Oxford: Blackwell.

Ross A., Donnelly T. and Wasson R. 1992. The peopling of the arid zone: human-environment interactions, in Dodson J. (ed.) *The Naive Lands: Prehistory and Environmental Change in Australia and the Southwest Pacific.* Melbourne: Longman Cheshire, pp. 76–114.

Ruddiman W., Guo Z., Zhou X., Wu H. and Yu Y. 2008. Early rice farming and anomalous methane trends. *Quaternary Science Reviews* 27: 1291–1295.

Sarnthein M. and Diester-Hass. 1977. Eolian sand turbidities. *Journal of Sedimentary Petrology* 47: 868–890.

Safriel U. 2007. The assessment of global trends in land degradation, in Savakumar M. and Ndiangui N. (eds) *Climate and Land Degradation.* Berlin-Heidelberg: Springer, pp. 1–38.

Schuster M., Roquin C., Duringer P., Brunet M., Caugy M., Fontugne M. et al. 2005. Holocene Lake Maga-Chad palaeoshorelines from space. *Quaternary Science Reviews* 24: 1821–1827.

Schulz E. and Whitney J. 1986. Upper Pleistocene and Holocene lakes in the An Nafud, Saudi Arabia. *Hydrobiologia* 143: 175–190.

Scott L. 1982. A Late Quaternary pollen record from the Transvaal bushveld, South Africa. *Quaternary Research* 17: 339–370.

Scott L. 2003. The Holocene and middle latitude arid areas, in Mackay A., Battarbee R., Birks J. and Oldfield F. (eds) *Global Change in the Holocene.* London: Arnold, pp. 396–405.

Seuffert O. 2001. Landschafts(zer)störung – Ursachen, Prozesse, Produkte, Definition & Perspektiven. *Geo-Öko* 22: 91–102.

Shi N., Duppont L., Beug H. and Schneider R. 1998. Vegetation and climate changes during the last 21000 years in S.W. Africa based on marine pollen record. *Vegetation History and Archaeobotany* 7: 127–140.

Shi N., Duppont L., Beug H. and Schneider R. 2000. Correlation between vegetation in southwestern Africa and oceanic upwelling in the past 21,000 years. *Quaternary Research* 54: 72–80.

Shuman B., Henderson A., Colman S., Stone J., Fritz S., Stevens L. et al. 2009. Holocene lake-level trends in the Rocky Mountains U.S.A. *Quaternary Science Reviews* 28: 1861–1879.

Slaymaker O. 2008. Why geomorphology matters in global environmental change? *Abstract CD of the 33rd International Geological Congress* Organizing Committee of 33rd IGC, Oslo.

Srivastava P., Brook G., Marais E. 2004. A record of fluvial aggradation in the northern Namib Desert during the Late Quaternary. *Zeitschrift für Geomorphologie N.F.* 133(Suppl-Bd.): 1–18.

Street-Perrott A. and Perrott R. 1993. Holocene vegetation, lake-levels, and climate of Africa, in Wright H., Kutzbach J., Webb T., Ruddiman W., Street-Perrott A. and Bartlein P. (eds) *Global Climates Since the Last Glacial Maximum.* Minneapolis: University of Minnesota Press, pp. 318–356.

Svensson A., Biscaye P. and Grousset F. 2000. Characterization of late glacial continental dust in the Greenland Ice Core Project ice core. *Journal of Geophysical Research* 105(D4): 4637–4656.

Tankard A. and Rogers J. 1978 Late Cenozoic palaeoenvironments on the west coast of southern Africa. *Journal of Biogeography* 5: 319–337.

Thomas D. and Middleton N. 1994. *Desertification: Exploding the Myth.* Chichester: John Wiley & Sons.

Thomas D., Knight M. and Wiggs G. 2005. Remobilization of southern African desert dune systems by twenty-first century global warming. *Nature* 435: 1218–1221.

Thomas D., Brook G., Shaw P., Bateman M., Haberyan K., Appleton C. et al. 2003. Late Pleistocene wetting and drying in the NW Kalahari: an integrated study from the Tsodilo Hills, Botswana. *Quaternary International* 104: 53–67.

Thomas J., Kar A., Kailath A., Juyal N., Rajagurus S. and Singhvi A. 1999. Late Pleistocene-Holocene history of aeolian accumulation in the Thar Desert, India. *Zeitschrift für Geomorphologie N. F.* 116(Suppl.-Bd.): 181–194.

Thompson R. and Anderson K. 2000. Biomes of western North America at 18,000, 6000 and 0 ^{14}C yr BP reconstructed from pollen and packrat midden data. *Journal of Biography* 27: 555–584.

Tooth S. 2008. Arid geomorphology: recent progress from an Earth System Science perspective. *Progress in Physical Geography* 31: 81–101.

UNCCD. 1999. *United Nations Convention to Combat Desertification in Those Countries Experiencing Serious Drought and/or Desertification, Particularly in Africa: Text with Annexes.* Bonn: Secretariat of the Convention to Combat Desertification.

Unkel I., Kadereit A., Mächtle B., Eitel B., Kromer B., Wagner G. and Wacker L. 2007. Dating methods and geomorphic evidence of palaeo-environmental changes at the eastern margin of the South Peruvian coastal desert (14°30′ S) before and during the Little Ice Age. *Quaternary International* 175: 3–28.

Valero-Garcés B., Grosjean M., Schwalb A., Geyh M., Messerli B. and Kelts K. 1996. Limnogeology of Laguna Miscanti: evidence for mid to late Holocene moisture changes in the Atacama Altiplano (Northern Chile). *Journal of Paleolimnology* 16: 1–21.

Vandenberghe J., Renssen H., van Huissteden K., Nugteren G., Konert M., Lu H. et al. 2006. Penetration of Atlantic westerly winds into Central and East Asia. *Quaternary Science Reviews* 25: 2380–2389.

Verstraete M., Brink A., Scholes R., Beniston M. and Smith M. 2008. Climate change and desertification: Where do we stand, where should we go? *Global and Planetary Change* 64: 105–110.

Ward J., Seely M. and Lancaster N. 1983 On the antiquity of the Namib. *South African Journal of Science* 79: 175–183.

Wells S., Brown W., Enzel Y., Anderson R. and McFadden L. 2003. Late Quaternary geology and paleohydrology of pluvial Lake Mojave, southern California, in Enzel Y., Wells S. and Lancaster N. (eds) *Paleoenvironments and Paleohydrology of the Mojave and Southern Great Basin Deserts.* Boulder: Geological Society of America, pp. 79–114.

Williams M. and Balling R. 1995. *Interactions of Desertification and Climate.* London: Edward Arnold.

Williams M., Adamson D., Williams F., Morton W. and Parry D. 1980. Jebel Marra Volcano: a link between the Nile Valley, the Sahara and Central Africa, in Williams M. and Faure H. (eds) *The Sahara and the Nile.* Rotterdam: Balkema, pp. 305–337.

Williams M., Dunkerley D., De Deckker P., Kershaw P. and Chappell J. 1998. *Quaternary Environments*, 2nd edition. London: Arnold.

Yang X. and Goudie A. 2007. Geomorphic processes and palaeoclimatology in deserts. *Quaternary International* 175: 1–2.

Yang X., Zhu Z., Jaekel D., Owen L. and Han J. 2002. Late Quaternary palaeoenvironment change and landscape evolution along the Keriya River, Xinjiang, China: the relationship between high mountain glaciation and landscape evolution in foreland desert regions. *Quaternary International* 97: 155–166.

Yang X., Liu T. and Xiao H. 2003. Evolution of megadunes and lakes in the Badain Jaran Desert, Inner Mongolia, China during the last 31000 years. *Quaternary International* 104: 99–112.

Yang X., Rost K., Lehmkuhl F., Zhu Z. and Dodson J. 2004. The evolution of dry lands in northern China and in the Republic of Mongolia since the Last Glacial Maximum. *Quaternary International* 118–119: 69–85.

Yang X., Preusser F. and Radtke U. 2006. Late Quaternary environmental changes in the Taklamakan Desert, western China, inferred from OSL-dated lacustrine and aeolian deposits. *Quaternary Sciences Reviews* 25: 923–932.

Yang X., Liu Z., Zhang F., White P. and Wang X. 2006a. Hydrological changes and land degradation in the southern and eastern Tarim Basin, Xinjiang, China. *Land Degradation & Development* 17: 381–392.

Yang X., Ding Z., Fan X., Zhou Z. and Ma N. 2007. Processes and mechanisms of desertification in northern China during the last 30 years, with a special reference to the Hunshandake Sandy Land, eastern Inner Mongolia. *Catena* 71: 2–12.

Yang X., Liu Y., Li C., Song Y., Zhu H. and Jin X. 2007a. Rare earth elements of aeolian deposits in Northern China and their implications for determining the provenance of dust storms in Beijing. *Geomorphology* 87: 365–377.

Yang X., Zhu B., Wang X., Li C., Zhou Z., Chen J., Wang X., Yin J. and Lu Y. 2008. Late Quaternary environmental changes and organic carbon density in the Hunshandake Sandy Land, eastern Inner Mongolia, China. *Global and Planetary Change* 61: 70–78.

Zavialov P. 2005. *Physical Oceanography of the Dying Aral Sea.* Chichester: Springer/Praxis.

Zhu Z., Wu Z., Liu S. and Di X. 1980. *An Outline of Chinese Deserts.* Beijing: Science Press.

32

Environmental Change in the Mediterranean Region

Miryam Bar-Matthews

1 CHARACTERIZATION AND PRESENT-DAY DISTRIBUTION OF THE MEDITERRANEAN-TYPE REGIONS

Mediterranean climate zones are associated with the five largest subtropical high pressure cells of the oceans. Geographically, the connection leads to the duality of deserts adjacent to Mediterranean-type climate regions. The Azores High is associated with the Sahara Desert and the Mediterranean climate which covers most of the area surrounding the Mediterranean Basin. The South Atlantic High is associated with the Namib Desert and the climate of southwestern South Africa. The North Pacific High is related to the Sonoran Desert and California's climate. The South Pacific High is related to the Atacama Desert and the climate of central Chile which covers parts of central Chile and west-central Argentina. The Indian Ocean High is related to the deserts of western and south-central Australia. These high-pressure cells shift polarward in the summer and towards the equator in winter and play a major role in the formation of the world's tropical deserts.

During summer, regions with Mediterranean climate are dominated by subtropical high pressure cells, making rainfall almost impossible. Summer temperatures are usually hot but vary depending on the region. In winter, the polar jet stream and associated periodic storms penetrate into the lower latitudes of the Mediterranean zones, bringing rain at lower elevations and snow at higher elevations. As a result, areas with Mediterranean climate receive almost all of their yearly precipitation during the winter season. Because regions with a Mediterranean climate are always near large bodies of water, temperatures are generally moderate but variable. However, precipitation and temperature vary considerably in relation to the topography and distance from the sea. At sea level, temperatures only occasionally reach freezing, but in the surrounding mountains, freezing and snow can occur during winters.

The Mediterranean Basin, lying between 30° and 46°N, is the largest area of the world to experience Mediterranean type climate, with summer drought, winter rain of cyclonic origin and a mean annual temperature of $15 \pm 5°C$.

2 THE MEDITERRANEAN REGION AS A BRIDGE BETWEEN THE ATLANTIC OCEAN AND THE MEDITERRANEAN SEA

The Mediterranean Sea (MS) is connected to the Atlantic Ocean via the Gibraltar Straights in the west, and is almost completely enclosed by land: in the north by Anatolia and Europe, in the south by Africa, and in the east by the Levant. The coastline extends for 46,000 km. The average depth of the MS is 1,500 m. The basin is divided into two main sub-basins, the Western Mediterranean (WM) and Eastern Mediterranean (EM), which are separated by a shallow submarine ridge (the Strait of Sicily) between the island of Sicily and the coast of Tunisia. The western basin covers an area of 0.85 million km², the larger eastern basin is ~1.65 million km². The MS is further subdivided into number of smaller seas: the Alboran Sea between Spain and Morocco; the Balearic Sea between mainland Spain and the Balearic Islands; the Ligurian Sea between Corsica and Liguria (Italy); the Adriatic Sea between Italy and Slovenia, Croatia, Bosnia and Herzegovina, Montenegro and Albania; the Ionian Sea between Italy and Albania and Greece; the Aegean Sea between Greece and Turkey and the Sea of Marmara between the Aegean Sea and Black Sea.

The sea surface conditions are diverse. Evaporation greatly exceeds precipitation and river runoff (Bethoux, 1979), a fact that is central to the water circulation within the basin. A marked contrast between surface and deep waters is due to an anti-estuarine exchange between the MS and the Atlantic Ocean (Bormans et al., 1986; Thunell and Williams, 1989). At the Straits of Gibraltar, which have a depth of 284 m, there is also an interfacial layer between inflowing Atlantic waters and out flowing Mediterranean waters (Baringer and Price, 1999; Bryden and Kinder, 1991; Fiúza et al., 1998). The Atlantic surface waters are relatively warm and light, and after some lateral and vertical mixing

processes, become Modified Atlantic Waters (MAW) that flow to the EM sub-basins whilst simultaneously increasing in salinity. Complementary to the MAW is the dense, deep outflow of Mediterranean, Mediterranean Outflow Water (MOW) (García-Lafuente et al., 1998; Pierre, 1999; Tintore et al., 1988; Vazquez-Cuervo et al., 1996; Viudez et al., 1998). The penetration of cold dry Arctic air to the EM region brings about an increase in surface water density, leading to the formation of the Levantine Intermediate Water (LIW) off the island of Rhodes (Malanotte-Rizzoli et al., 1999; Pinardi and Masetti, 2000). This air pressure gradient pushes relatively cool, low-salinity water from the Atlantic across the WM basin; it warms and becomes saltier as it travels east, then sinks in the region of the Levant and circulates westward, to spill over the Straits of Gibraltar. Sea-surface salinity (SSS) and sea-surface temperature (SST) increase eastward associated with an increase in sea-surface oxygen isotopic composition ($\delta^{18}O$) values by up to ~1 per mille relative to the WM (Pierre, 1999).

During severe winters, northwesterly winds blowing over platform zones cause surface waters to downwell. As noted above, LIW are formed at continental shelves of the EM and the WMDW originated by the same mechanism in the Gulf of Lions and the Tyrrhenian Sea (Leaman and Schott, 1991; Pinardi and Masetti, 2000; Schott and Leaman, 1991). In the EM, deepwater formation sites are close to the shelves of the Adriatic and Aegean seas, forming the EM Deep Water (EMDW) (Lascaratos et al., 1999). This process has long been dominated by the sinking and spreading of cool and salty dense water masses that form in the Adriatic Sea (Malanotte-Rizzoli and Hecht 1988). During the late 1980s and early 1990s an abrupt shift to the dominance of Aegean Deep water in the EM deep sea ventilation occurred, known as the EM Transient (EMT) (Malanotte-Rizzoli et al., 1999; Roether et al., 1996) which eventually influenced the density of the Mediterranean outflow into the

North Atlantic (Millot et al., 2006). This profound reorganization in the EM Sea was preconditioned by a long-term increase in the net evaporation from the EM Sea (Bethoux et al., 1998; Boscolo and Bryden, 2001).

The EM–Levant region has the most variable and extreme climatic conditions in the entire Mediterranean region. The region is located at the meeting of the Eurasian continent, the Saharan–Arabian desert and the MS and at the boundary between the high-to-mid latitude and tropical–subtropical climate systems. The MS moderates the climate, and without its moderating affects the Levant would have been part of the larger northern, low, mid-latitude, warm, dry desert extending from North Africa into Western Asia.

The synoptic-scale system in the EM is associated with an extratropical cyclone – the Cyprus Low – with rainfall fronts that originate in the northeast Atlantic Ocean that pass over Europe and the MS and are responsible for most of the annual rainfall (Rindsberger et al., 1983; Sharon and Kutiel, 1986). Precipitation is generally higher on the northern shores of the Mediterranean and on westward-facing mountains. The influence of the Saharan arid zone to the south is felt primarily during summer, when anticyclonic conditions migrate north to create seasonal drought.

The air masses bringing moisture are rapidly influenced by the warm surface of the MS with the result that isotopic characteristics of rainfall in the EM region are unique and differ from the global isotopic characteristics as defined by the relationships between deuterium and oxygen ($\delta^{18}O$–δD relationships). $\delta^{18}O$–δD relationships define a local Mediterranean Meteoric Water Line (MMWL) with a deuterium excess of 20–30 per mille compared to the global value of 10 per mille (Ayalon et al., 1998; Gat, 1996; McGarry et al., 2004; Rindsberger et al., 1983; Rozanski et al., 1993) arising from the addition of water vapor from the warm MS into the relatively cold, dry air from the North Atlantic sourced air masses (McGarry et al., 2004). Occasionally, rainfall is associated

with subtropical warm air passing along the southern Mediterranean (Dayan, 1986). IAEA monitoring data (IAEA/WMO, 2001), shows that mean weighted value of $\delta^{18}O$ of precipitation decreases eastward and northward from ~4 per mille in North Africa to ~10 per mille in the uplands of eastern Turkey. This gradient fits within global patterns of $\delta^{18}O$ in precipitation linked to rain-out effects from the tropics northwards (e.g., Bowen and Wilkinson, 2002) and is locally associated with increasing elevation. Locally, changes in monthly to yearly values of $\delta^{18}O$ of precipitation are dependent on changes in climatic factors such as temperature, precipitation amount and air mass trajectory (Bar-Matthews et al., 1997, 2003; Rindsberger et al., 1983).

Winter rains brought by the mid-latitude westerlies are common to the coastal zones. The marine influence diminishes sharply eastwards due to the orographic effect of the mountain ridge running parallel to the EM Sea coast 'rain shadow' desert towards Jordan and Syria, and southward towards the Israeli Negev Desert, where the coastline change its direction from north–south to east–west (Develle et al., 2010; Enzel et al., 2008; Vaks et al., 2003). The transition from a humid coastal zone to the desert occurs in less than 100 km from west to east and from north to south. Mean annual precipitation ranges between 400 and 800 mm along the coast to ~2,000 mm in the high mountains in the north and drops to less than 100 mm southward and eastward. As a whole, the region exhibits steep north–south and east–west precipitation gradients with high interannual to decadal rainfall variability.

3 PALEOCLIMATE OF THE MEDITERRANEAN SEA

The region contains one of the longest histories of the human civilization. It has been suggested that there are causal links between major northern hemisphere and Mediterranean climatic events and the evolution of human civilization in the region (e.g., Bar-Matthews et al., 1998; Bar-Yosef, 1998; deMenocal, 2001; Issar and Zohar, 2004; Migowski et al., 2006; Stein et al., 2010; Vaks et al., 2007).

The wide coverage of marine core studies from the MS is shown in Hayes et al. (2005: figure 4). Information on the paleoclimate and environmental conditions is based on the wide use of proxy records. The most commonly is the oxygen isotopic record ($\delta^{18}O$) of the surface-dwelling planktonic foraminifera *Globigerinoides ruber* in the EM basin and *Globigerinoides bulloides* in the WM. This proxy is used widely for recording the geological time framework, ice volume, temperature and salinity. Other commonly used proxies are foraminiferal assemblages, sediment characteristics such as organic compounds (U^{K}_{37} alkenone method) and their content, color and geochemical composition (e.g., Almogi-Labin et al., 2009; Brassell et al., 1986; Chapman et al., 1996; Emeis et al., 2000; Fontugne and Calvert, 1992; Prahl et al., 1988; Volkman et al., 1980).

3.1 Western-Mediterranean marine records

The paleoclimate of the WM is largely based on marine cores ODP-977A (Martrat et al., 2004, 2007) and MD-95-2043 (Cacho et al., 1999a, 1999b) which were retrieved from the eastern sub-basin of the Alboran Sea. The $\delta^{18}O$ record of *G. bulloides* displays a well-defined orbital modulation of glacial and interglacial marine isotope stages (Martinson et al., 1987) and also correlation with the Dansgaard–Oeschger (DO) cycles observed in the Greenland $\delta^{18}O$ ice core record (Dansgaard et al., 1993; Grootes et al., 1993; NGRIP members, 2004). Alkenone-based SSTs display sequences of rapid warming and cooling events, occurring during mainly glacials and contemporaneous with the variations in the $\delta^{18}O$ record of *G. bulloides* (Cacho et al., 1999a, 1999b; Martrat et al.,

2004, 2007). These events are known as Alboran interstadials (AI-1 to AI-26 and AI-1' to AI-15' for the last and penultimate climate cycles, respectively) and stadials (AS-1 to AS-26 and AS-1' to AS-15', respectively) (Martrat et al., 2004, 2007; Walker et al., 1999) (Figure 32.1).

Among the lowest SST interstadials, six minima AS-2a, AS-2c, AS-5, AS-9, AS-13 and AS-18 (Figure 32.1) correlate with the Heinrich Event time periods when melting of massive icebergs resulted in the deposition of detritic particles larger than 150 μm, known as ice rafted detritus (IRD) in the open north Atlantic (Bond et al., 1993; Heinrich, 1988), and the western Iberian margin (Cayre et al., 1999). Although IRD were not identified in the MS, their impact on SST of the MS is prominent.

Together with the drops in SST, there were a coeval increases in the relative proportion of the polar planktonic foraminifera *Neogloboquadrina pachyderma* (sinistral) (Cacho et al., 1999b; Perez-Folgado et al., 2003), and with organic markers reflecting the influx of low-salinity water masses (Bard et al., 2000) (Figure 32.1). Not all the minimum SSTs are correlated with Heinrich Events; some are correlated with minimum temperatures in the northern hemisphere (minimum $\delta^{18}O$ in GRIP ice core), pointing to a teleconnection of the northern hemisphere climate and with the WM.

Interestingly, during relatively warm and largely ice-free interglacials there are large SST fluctuations that are often comparable to the glacial–interglacial amplitude described by McManus et al. (1999) as a key aspect of the Pleistocene climate. However, unlike glacials, interglacial SST fluctuations last only a few centuries and their reasons are not fully understood. Martrat et al. (2004) suggest that the long, warm, stable periods result from the more pronounced eccentricity of the Earth's orbit and seasonal contrast, and strengthening of the North Atlantic Deep Water (NADW) formation which buffers the climate. SST changes are also coeval with major changes in vegetation around the

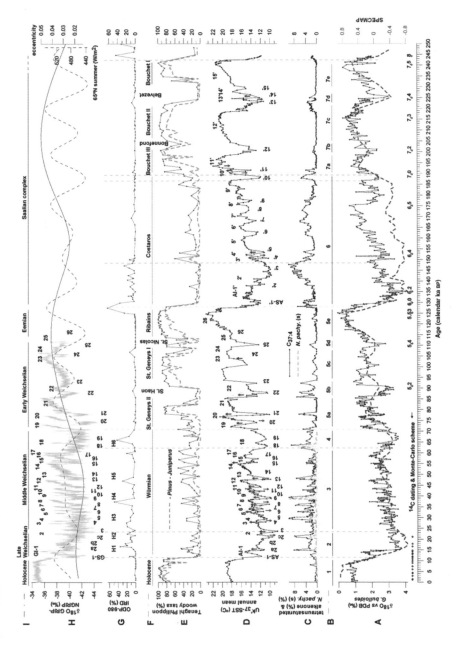

Figure 32.1 Proxy records of the last two interglacial–glacial cycles (based on Martrat et al., 2004). (a) $\delta^{18}O_{G.bulloides}$ from core ODP-977A compared with the SPECMAP isotope curve (dotted line). (b) Marine Isotope Stages (MIS) 7e-1. (c) The relative proportion of heptatriatereaenone to total alkenone together with percentage of *N. pachyderma* sinistral as a tracer for cold and/or low salinity. (d) Sea-surface temperature (SST) record showing six interstadials, AI-1 to 26, and stadials, AS-1 to 26 during glacial. During interglacials 15 warm phases, AI-1 to 15, and 15 cold phases, AS-1 to 15, are shown. Red arrows indicate abrupt warming, blue arrows indicate abrupt cooling. (e) The pollen record from Tenaghi Philippon. Woody taxa are shown by the green line, dotted green line mark the temperate pollen curve. (f) The pollen stratigraphy. (g) The relative proportion of ice-rafted debris (IRD) in North Atlantic sediments from in 55°N. (H) $\delta^{18}O$ in GRIP and NGRIP. Greenland interstadials are marked as GI, and stadials as GS. Eccentricity at 65°N during summer is also shown.

Mediterranean (Tzedakis et al., 2003; Figure 32.1). When SSTs are at their highest, there was a much higher percentage of woody taxa.

3.2 The Eastern Mediterranean Sea

As with the WM, $\delta^{18}O$ record of *G. ruber* displays a well-defined orbital modulation of glacial and interglacial marine isotope stages (Fontugne and Calvert, 1992) and correlation with DO cycles and Heinrich Events (Almogi-Labin et al., 2009). Although the EM Sea record follows the global record it is characterized by larger-amplitude $\delta^{18}O$ variations (Fontugne and Calvert, 1992; Kallel et al., 1997; Vergnaud-Grazzini et al., 1986). The difference in $\delta^{18}O_{G.ruber}$ values between the last glacial maximum and the beginning of the interglacials is ~5 per mille compared with 3–4 per mille in the WM, and there is ~3.5 per mille amplitude in the EM during glacials compared with ~2 per mille in the

WM (Figures 32.1 and 32.2). The large glacial–interglacial amplitude not only reflects the ice-volume effect, but also to a considerable extent the semi-enclosed continental conditions of the region whereby climate change in the EM Sea is amplified. This large isotopic shift is also recorded on land, as evident from the speleothem $\delta^{18}O$ records of the EM caves (e.g., Bar-Matthews et al., 2000, 2003).

3.3 West–east transect

The sedimentary record from the EM differs from that of the WM not only because of the amplifying effect of the EM but also because of its unique sequence composed of alternating light and dark sediments showing distinct cyclicity (Fontugne and Calvert, 1992). The dark sediments are composed of organic rich layers known as sapropels which developed in suboxic to anoxic conditions. Sapropels are finely laminated, devoid of benthic fauna,

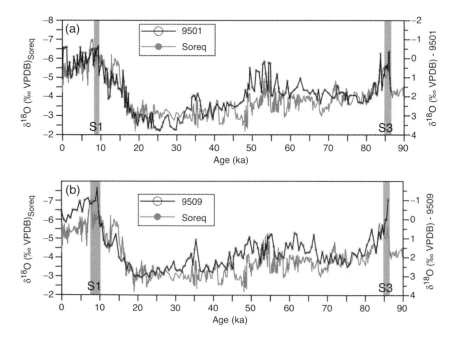

Figure 32.2 $\delta^{18}O_{G.ruber}$ **from marine cores (a) 9501 SE of Cyprus and (b) 9509 off the Nile plum *versus* age superimposed on Soreq Cave speleothems record (based on Almogi-Labin et al., 2009). The vertical gray bars represent the timing of sapropels.**

rich in sulfides and organic matter (Almogi-Labin et al., 2009; Bethoux, 1993; Cita et al., 1977; Kallel et al., 2000) and were mainly deposited during interglacial stages with a few exceptional glacial sapropels. Sapropel formation occurs mainly in the EM (Emeis et al., 2000; Murat and Got, 2000; Rossignol-Strick et al., 1982, 1998), in contrast to the weaker episodes known as sapropelites from the WM (deKaenel et al., 1999; Martinez-Ruiz et al., 2003; Weldeab et al., 2003). Their formation is associated with a weakening in deep and intermediate water formation due to strong intermediate to surface water stratification (Jorissen, 1999) caused by pronounced weakening of density differences between Atlantic and Mediterranean waters (Bethoux and Pierre, 1999). All sapropel events coincide with astronomically driven maximum summer insolation at 65°N latitude and maxima in radiation at the 19–23 ka cycles of orbital precession (e.g., Haynes, 1987; Rossignol-Strick, 1985; Rossignol-Strick et al., 1982), and are associated with intensification of the Africa monsoon system influencing the hydrography of the region, leading to an increase of the freshwater River Nile

input to the Levantine Basin (e.g., Calvert and Fontugne, 2001; Rohling, 1994; Fontugne et al., 1994). Additional sources of freshwater into the EM are the rivers emitting from the Tibesti Mountains in Libya, as evident from the radiogenic Nd-isotopic values in planktonic foraminifera at ODP Site 967 (Scrivner et al., 2004) which reflect a northward shift of the monsoonal belt (Osborne et al., 2008; Rohling et al., 2002). Coincident with strengthening of the monsoonal system, rainfall increased from Atlantic–Mediterranean sources throughout the entire MS (Bar-Matthews et al., 2000; Rohling et al., 2002; Vaks et al., 2007).

Nine sapropel layers are recognized in the sedimentary record during the last 250 ka (Fontugne and Calvert, 1992) (Figure 32.3). Low foraminiferal $\delta^{18}O$ values reflect the combined effects of sea-surface warming and enhanced fresh water supply. Bar-Matthews et al. (2000) dated the duration of the sapropel events according to the contemporaneous low $\delta^{18}O$ values of land speleothems: S5: 124–119 ka BP; S4: 108–100 ka BP; S3: 85–79 ka BP; missing glacial sapropel S2: 55–52 ka BP, the Holocene sapropel S1: 8.5–7.0 ka BP

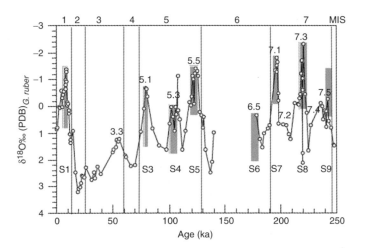

Figure 32.3 $\delta^{18}O_{G.ruber}$ **of the last 250 ka from marine core MD84651 located at 33°02′N, 32°38′E (based on Fontugne and Calvert, 1992). Marine Isotopic Stages 7 to 1 and substages are shown. The vertical grey bars represent the timing of EM sapropels S9-S1 (Bar-Matthews et al., 2000).**

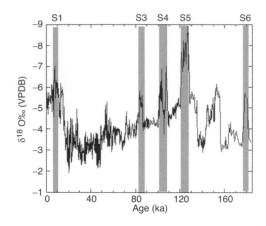

Figure 32.4 $\delta^{18}O$ **record of Soreq Cave speleothems for the last 180 ka. The gray bars show periods with minimum $\delta^{18}O$ values coinciding with deposition of sapropels S6–S1 (based on Bar-Matthews et al., 2000).**

(Figures 32.2 and 32.4). These events are associated with high Nile River discharge as evident from increasing total organic carbon (TOC) to several percent (Almogi-Labin et al., 2009), and minimum Ti/Al, Si/Al and K/Al (Calvert and Fontugne, 2001). Pollen assemblages from the southeastern Mediterranean basin show a marked increase in deciduous oak, also supporting wet conditions (Cheddadi and Rossignol-Strick, 1995; Langgut, 2008; Rohling and Hilgen, 1991; Rossignol-Strick, 1985). Mediterranean fronts probably penetrated farther south over northeast Africa relative to their present-day track (Arz et al., 2003), while the African monsoonal system shifted to its maximal northern position (cf. Haynes et al., 1987; Rohling et al., 2002; Street-Perrott and Perrott, 1993). The simultaneous occurrence of humid climate conditions in areas influenced by the low-latitude African monsoonal system together with areas affected by the high-latitude northeast Atlantic system are cave deposits in the Levant and in the northeastern parts of the Sahara Desert (Bar-Matthews et al., 2000; Vaks, 2008; Vaks et al., 2007) suggest a common cause (at least in a broader sense). The deeper southward

penetration of rainfall associated with the northeast Atlantic–Mediterranean area into the present-day arid and hyper-arid Negev Desert, and northward migration of the low-latitude African monsoonal system during sapropel events resulted in narrowing of the Sahara Desert, have opened climatic 'windows' for dispersal of hominids and animals out of the African continent (Vaks et al., 2007).

Studies of marine cores from the central (SL21) and south-eastern (LC21) Aegean Sea (Marino et al., 2009), show a marked ~1.3 per mille negative shift in $\delta^{18}O$ between 10.7 and 9.7 ka BP, which was taken to suggest influx of fresh water from the Black Sea.

Although the unique sedimentary records of sapropels are most strongly evident in the EM basin, speleothems on land from the WM and the EM show coeval decrease in their $\delta^{18}O$ record (Ayalon et al., 2002; Bard et al., 2002; Bar-Matthews et al., 2000; Verheyden et al., 2008; Zanchetta et al., 2007) consistent with the timing of sapropel deposition reflecting large-scale hydrological activity over the entire Mediterranean. Indeed, SSS points to freshening of the entire Mediterranean basin (Kallel et al., 1997).

Several short, dry intervals interrupted the wet sapropel periods and are contemporary with global cooling and the development of arid conditions in higher latitudes. This suggests that the 8.2 ka BP cooling event during the middle part of S1 sapropel and at ~122 ka BP, during the middle part of the S5 sapropel, were due to periodic strengthening of higher latitude circulation and the temporary retreat of the monsoonal system to lower latitudes (e.g., Rohling et al., 2002).

Interglacial substages between wet sapropel events (e.g., MIS 5.2, 5.4, 7.2, 7.4, etc.) were drier. $\delta^{18}O$ values of foraminifera (and EM speleothems) are 1–2 per mille higher than during the sapropel intervals (Figures 32.2–32.4). Increase in Ti/Al, Si/Al and K/Al ratios to values similar to those of glacial periods point to increasing aridity and dust supply from eolian sources (Almogi-Labin et al., 2004; Calvert and Fontugne, 2001). These climate and environmental changes follow the southward retreat of the African monsoonal system from its maximum position during sapropel periods (Haynes, 1987), and the shift northward of the Mediterranean fronts to their present-day position; both changes leading to the expansion to their current dimensions of the Sahara and Arabian deserts.

The clear west–east climate gradient is also evident from pollen records and alkenone-based paleotemperatures. In the EM basin the temperatures rose by ~6°C to ~10°C from the Last Glacial Maximum (LGM) to the Holocene (Figure 32.5). Alkenone-based paleotemperatures between 30 and 15 ka BP were ~6°C colder in the WM

Sea compared to the EM Sea (Figure 32.5). This difference is significantly larger than the 2–3°C difference typical of the Holocene and the present day (Figure 32.5). Essallami et al. (2007) present a comprehensive basin-scale view of hydrological changes, expressed as the differences in water $\delta^{18}O$ ($\Delta\delta_w$) between the WM and EM during the LGM. Their reconstruction is based on west–east transects from the Atlantic Ocean off Portugal to the EM basin (core MD84-632), and reveals a progressive isotopic enrichment of the surface water from the North Atlantic to the Eastern basin. The gradient at the LGM is steeper than today and indicates a drastic change from the central part of the Mediterranean to the east owing to a substantial increase of the evaporation–precipitation

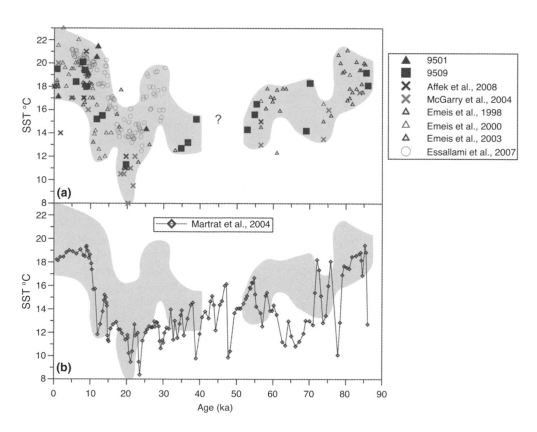

Figure 32.5   Alkenones-based paleo SST (taken from Almogi-Labin et al., 2009) from (a) various marine cores in the EM and (b) speleothems-based paleotemperatures (Affek et al., 2008; McGarry et al., 2004) and alkenone-based paleo SST from the WM (Martrat et al., 2004).

balance in the eastern basin, and a decrease in precipitation in the southern and eastern Levantine Basin. Even though colder temperatures prevailed during the LGM, evaporation still exceeded precipitation. Cold events coincide with a decreased δ_w/salinity in the central Mediterranean, similar to the North Atlantic, whereas in the EM basin cold events are associated with increase δ_w/salinity due to a different response to heat transport. During cold events (Heinrich Events and the Younger Dryas (YD)), reduction of the Atlantic heat transport to the high latitudes affected the European continent. Depleted δ_w values in the central MS result from the southward migration of the polar front of the North Atlantic allowing less saline waters to enter the MS. At the same time the EM Sea underwent higher evaporation over precipitation compared to today.

Lower-resolution alkenone-based temperature studies from the EM core off the Israeli coastline (cores 9509 and 9501) suggest that the difference between the two basins during the glacial is only in the order of ~2–3°C (Figure 32.5), which is confirmed also by land temperatures revealed from speleothems from the EM region (Affek et al., 2008; McGarry et al., 2004).

The overall temperature difference between the LGM and the present is in the order of ~6–10°C. This temperature difference is also confirmed by calculated EM paleotemperature on land (Figure 32.5). The cold LGM is also associated with severe aridity (Cheddadi and Rossignol-Strick, 1995; Ganopolski et al., 1998; Horowitz, 1979; Langgut, 2008). Calvert and Fontugne (2001) suggested that intensified wind and dust transport over the Levantine Basin could have contributed to glacial aridity. The dramatic warming and increase in moisture from the LGM to the Holocene was associated with a major change in vegetation and the spread of Mediterranean forest combined with increasing moisture (Baruch and Bottema, 1999; Bottema and van Zeist, 1981; Martrat et al., 2007; Langgut, 2008; Weinstein-Evron, 1990).

3.4 North–south transect of the Eastern Mediterranean

To understand the role of the Sahara Desert and the influence of the monsoonal *versus* the Atlantic system on paleoenvironmental conditions of the EM, two marine cores retrieved under the influence of the Nile plume (core 9509) and south of Cyprus (core 9591) were compared in high resolution (Almogi-Labin et al., 2009). This work clearly demonstrates the greater influence of the River Nile on the southern Levant Basin. Sedimentation rates are nearly four times higher than that in the northern core due to the higher sediment load carried seasonally during the flooding period of the Nile, which is also consistent with other isotopic records from the EM, such as $^{87}Sr/^{86}Sr$ and $^{143}Nd/^{144}Nd$ ratios, which indicate a limited River Nile contribution further north (e.g., Warning and Brumsack, 2000; Weldeab et al., 2002). During sapropel deposition, the north–south gradient diminishes, reflecting the increased hydrological activity over the entire Mediterranean basin, leading to a more homogeneous SST and SSS throughout the basin (Emeis et al.,1998; Kallel et al., 1997). Almogi-Labin et al. (2009) showed that when the Holocene sapropel S1 was interrupted during the 8.2 ka BP cooling event there was a decrease in River Nile influence in both the northern and southern basins. However, whereas in the southern basin the sapropel period continued 200 years after this cooling event, in the northeast Mediterranean the termination of S1 sapropel occurred immediately after 8.2 ka BP. They also showed that the 1–2°C difference between the present-day cooler and wetter northern basin compared with the warmer southern Levantine Basin (Marullo et al., 1999) remained unchanged during most of the last 90 ka. An exception was the period between 58 ka BP and 49 ka BP coinciding with DO interstadial 14 and maximal insolation at 65°N, when the northern basin was warmer than the south. This interval was also the warmer in the WM (Figure 32.5).

4 MEDITERRANEAN LAKES

Our knowledge of the paleoclimate, environment and hydrology of the Mediterranean region is also based on land records, which include mainly the isotopic composition of lake-level records and pollen assemblages (e.g., Bartov et al., 2002, 2003; Bookman et al., 2006; Enzel et al., 2003; Haase-Schramm et al., 2004; Hazan et al., 2005; Jones et al., 2007; Katz et al., 1977; Kolodny et al., 2005; Lisker et al., 2009; Migowski et al., 2006; Sadori et al., 2008; Stein et al., 2010; Stevens et al., 2006; Torfstein et al., 2009). Roberts et al. (2008) summarized the paleohydrology of the Mediterranean region primarily based on oxygen isotopic composition of lake carbonates from eastern, western and central Anatolia, the Levant, Greece, Italy, Spain and Morocco, the Alpine region and most of Central Europe. Mediterranean lakes reflect local, catchment-specific processes resulting in a wide range of $\delta^{18}O$ values from isotopically depleted fresh waters to enriched saline waters sometimes even in the same region. Despite the heterogeneity, almost all EM lakes shifted to more depleted $\delta^{18}O$ values during the transition from Late Glacial to Holocene climate, known as 'Mediterranean-type' shift, at the start of the Holocene (Finsinger et al., 2008). A similar shift to isotopically light interglacial composition took place during Marine Isotope Stage (MIS) 5e (Eemian) (Frogley et al., 1999; Torfstein et al., 2009). Thus, there have been clear, hydrologically coherent, regional isotopic trends within the Mediterranean basin since the LGM (Roberts et al., 2008). The low carbonates $\delta^{18}O$ occur during the Bølling–Allerød (BA) and high lake carbonates $\delta^{18}O$ values occur during Heinrich and YD events, in Mediterranean lakes compared to the Holocene partly reflect source area effects and partly greater aridity during Heinrich and YD events. Almost all EM lake records shifted to more depleted $\delta^{18}O$ after the Last Glacial, far more (up to ~6 per mille) than can be attributed to source effect alone, which points to a change in the hydrological balance, most probably a marked increase in precipitation.

The marked depletion of lakes $\delta^{18}O$ around the northern rim of the EM associated with marine sapropel formation cannot be explained by freshwater input through alone. Jones et al. (2007) calculated that rainfall was higher by ~20 per cent in the early Holocene in the northern rim of the EM, similar to Bar-Matthews et al.'s (2003) calculation for the southern Levant based on speleothem records. This is in contrast to WM lakes and speleothems (McDemott, 2004) which show no such pronounced $\delta^{18}O$ difference, suggesting a northwest–southeast contrast as was seen in the marine record.

Contradictory paleoclimate interpretations come from reconstruction of lake-level fluctuations of Lake Lisan, the late Pleistocene precursor of the Dead Sea that existed from ~70 ka BP to 15 ka BP in the Jordan Valley–Dead Sea basin. The lake evolved through frequent water-level fluctuations between ~340 and 160 m below mean sea level (msl) (Bartov et al., 2002, 2003; Enzel et al., 2008). The lake-level changes in a gauge for wetter/dryer climate, and the existence of the larger Lake Lisan during glacial compared with the Holocene Dead Sea, suggest a much wetter glacial in the Levant (Enzel et al., 2008). A lake-level reconstruction based on the position of paleoshores suggests that the lake reached its maximum elevation of about 164 m below msl between 26 ka BP and 23 ka BP, and that major drops in lake level during the last glacial occurred at ~16 ka BP, ~24 ka BP, ~30 ka BP, ~38 ka BP and at 45-47 ka BP, coinciding with H1–H5 events, respectively (Bartov et al., 2003) confirming the climatic linkage between the EM and the North Atlantic. However, such a linkage does not explain why the lake reached its highest stand between 26 ka BP and 23 ka BP. It is argued that local changes in rainfall amount are not the only control on the evaporative Lake Lisan levels (Roberts et al., 2008), and that changes in temperature, effective precipitation, precipitation/evaporation ratio (P/E) and distal supply of water from the surrounding mountains in the east and north and from Judean Mountains in the west, must be also taken into account. In a recent study based on the U-Th dating of algae stromatolites in caves situated at the lower part of Dead Sea Fault Escarpment, Lisker et al. (2009) showed that the lake already reached its highest level at ~41.5 ka BP and remained in high stand until ~17.5 ka BP suggesting that the climate in the Dead Sea region has generally been moister than interglacial climate.

Do such high lake-level stands require much wetter conditions relative to the Holocene? If so, what are the unique conditions that enabled much wetter conditions in the Dead Sea Basin when the marine record, other EM lakes, pollen, tufa and speleothem records suggest that the last glacial was mostly drier? Enzel et al. (2008) suggest that unlike today, EM cyclones during the glacial were forced by the ice and snow-covered Europe and Turkey, the lower sea level and the Sahara to be funneled along the Mediterranean directly east into the Levant leading to higher precipitation, but determining how much wetter the climate was in terms of rainfall is also a difficult. This suggestion does not explain why lake levels had already dropped during the LGM.

One has to be cautious in taking the present Dead Sea as an analog for more/less rainfall. Lake Lisan existed when temperatures were much colder (6–8°C) (Affek et al., 2008; McGarry et al., 2004), with different relative humidity (Gat, 1995) and probably a higher P/E ratio (Lisker et al., 2010; Vaks et al., 2003). Deposition of speleothems in caves situated in the lower part of Dead Sea Fault Escarpment during glacials compared to their scarcity during interglacials (Lisker et al., 2010) indicate that the Dead Sea area received more water during cool glacial intervals. Increased P/E ratio during glacials was the mechanism proposed by Vaks et al. (2003) to explain the deposition of speleothems in the Jordan Valley region, whereby high evaporation during interglacial periods prevented rainfall recharging the area. This apparent contradiction between a 'wet' Dead Sea region during most of the glacial on one hand and 'drier Mediterranean' on the other hand show the different response of various proxies to the combinations of various factors such as temperature, precipitation, evaporation, and so on. Kolodny et al. (2005) argue that the fluctuations in $\delta^{18}O$ values of the Lake Lisan carbonate are good recorders of the long-term changes in the isotopic composition of the source (MS) and override the effects of temperature changes and regional hydrological balance, and that the Lake Lisan level is indeed the best recorder for more/less rainfall.

Since it is difficult to extract paleoclimate interpretation from the carbonate oxygen isotope records in a region where the climate gradient is sharp and several factors control the terrestrial $\delta^{18}O$ signals, Develle et al. (2010) used a different approach. They studied the $\delta^{18}O$ values of ostracod valves as an indicator of paleohydrologic regime in a small karstic, hydrologically open freshwater lake basin from Lebanon (Yammouˆneh Basin), where water flow is rapid and evaporative conditions, which might affect the isotopic composition, is small or negligible. Their study addressed the question of what the dominant factors controlling the $\delta^{18}O$ values of lake water and carbonates in this open lake during the LGM and the Holocene are and how the Yammouˆneh record can be compared with other carbonate $\delta^{18}O$ records from the region. They found a negative shift in the isotopic composition of the lake water at the beginning of the Holocene, which is especially marked during the period of sapropel S1 in the EM basin, and interpreted it to reflect increasing precipitation during this period. Increasing precipitation in Lebanon at this time is supported also by the low speleothem $\delta^{13}C$ values of the Jeita cave, indicative of increased soil moisture and C3- type vegetation cover, suggesting conditions wetter than those of the present day (Verheyden et al., 2008). Pollen data also support a wetter, warm early Holocene. Deciduous oak forests reflecting wet, warm conditions developed from ~10.0 ka BP to 6 ka BP around the Aammiq marsh in the Beqaa valley (Hajar et al., 2008).

The debate of paleoclimate interpretation based on lake level and carbonate isotopic composition also holds for the Dead Sea. On one hand, the Dead Sea level reconstruction indicates wet conditions relative to the modern era during the early Holocene (Frumkin, 1997) and agrees with the other proxies from the EM. However, during the BA and the YD the Dead Sea record differs from other EM lakes, and from marine, speleothem and pollen records. Pollen records show an increase in Chenopodiaceae and Artemisia suggesting a sharp drop in rainfall (Langgut, 2008). Increased aridity during the YD is also supported by the speleothem record (Bar-Matthews et al., 2003). Cool SSTs (~13°C) are recorded in the marine record (Almogi-Labin et al., 2009; Emeis et al., 2000, 2003). In the WM the records show colder, dryer conditions and higher salinities (e.g., Cacho et al., 2001; Kallel et al., 1997). On the other hand, Torfstein et al. (2009) and Stein et al. (2010) argue, based on Dead Sea levels, that in the Levant the BA was the most arid period of the late Pleistocene and that the YD was wet.

Thus, the paleoclimate record derived for the entire MS record, and speleothem, lake and pollen records, usually contradict the paleoclimate interpretation of the Levant deduced from the terminal lakes, Lake Lisan and the Dead Sea, leaving the debate over wetter/drier glacial in the Levant a controversial one.

5 THE SPELEOTHEM RECORD WITH EMPHASIS ON THE HOLOCENE

Intensive speleothem-based research in the EM and the WM regions in Israel, Italy, and Lebanon during the last decade has explored various aspects of their climate-related proxy records: growth periods and duration, petrography, stable oxygen and carbon isotopes,

Sr isotopic composition and trace elements, fluid inclusions, and most recently paleotemperature estimation using the mass 47 anomaly ('clumped isotopes') (Affek et al., 2008; Ayalon et al., 1999; Bar-Matthews et al., 1997, 2000; Frumkin and Stein, 2004; McDermott, 2004; McGarry et al., 2004; Vaks et al., 2006, 2007) . The data from these studies enables reconstruction of the paleoclimate conditions of central and northern Israel during the last 250 ka (Bar-Matthews et al., 2003; Frumkin et al., 1999); central Lebanon from ~12 ka to 1 ka ago (Verheyden et al., 2008) and central and western Italy during the Holocene and a large part of MIS 5 (Drysdale et al., 2006, 2007; Zanchetta et al., 2007). The speleothem record shows that the moisture source (sea surface isotopic composition) is the dominant control of oxygen isotopic composition, but other parameters, mainly the amount of rainfall, atmospheric conditions and vegetation type, are also crucial in understanding the paleoclimate using speleothems.

5.1 Eastern Mediterranean speleothems

The most accurately dated high-resolution long-term climate record comes from the EM, mainly from central Israel, Soreq Cave speleothems (Figures 32.2 and 32.4). $\delta^{18}O$ values show similar isotopic trends and amplitude as the EM Sea record of planktonic foraminifera (Almogi-Labin et al., 2009; Bar-Matthews et al., 2000, 2003), thus implying that the speleothems mainly record the Atlantic-Mediterranean cyclonic systems. This conclusion is supported by $\delta^{18}O$–δD relationships of the fluid inclusions trapped within the speleothems, which follow the Mediterranean Meteoric Water Line (Matthews et al., 2000; McGarry et al., 2004).

A major advantage of the central Israel speleothem record for paleoclimate reconstruction is their continuous growth through glacial and interglacial intervals

(Figures 32.2 and 32.4), indicating that this region did not experience severe aridification (or freezing) during several glacial–interglacial cycles. This compares with the 'rain-shadow' Jordan Valley region, or 100 km further south in the Negev Desert where deposition of speleothems was discontinuous and occurred only when enough water reached the unsaturated zone (Vaks et al., 2003, 2006, 2007). The continuous record of speleothems from central Israel was used to precisely date periods of increased rainfall in the EM, such as the sapropel events (Figure 32.4) and other short duration climate changes such as Heinrich Events (Bar-Matthews et al., 1999), warming events during the glacial (Almogi-Labin et al., 2009) and glacial–interglacial paleotemperature changes. These studies demonstrate that rainfall in the region both during glacials and interglacials was associated with Atlantic–Mediterranean systems and that glacial temperatures were colder by 4°C during most of the glacial and up to 10°C colder during extreme cold intervals such as the LGM, compared to the present (Affek et al., 2008; McGarry et al., 2004).

Based on the speleothem record, the wettest intervals in the EM region, where the speleothem record shows the lowest $\delta^{18}O$ values, are those that coincide with the accumulation of sapropels (Figure 32.4). The petrography of speleothems from these high rainfall periods, their chemical and isotopic composition, and much higher paleo-pool levels in the cave are indicative of formation from fast dripping water, resulting in a large input of detritus and oxides and a larger input of the dolomitic host-rock due to enhanced weathering (Ayalon et al., 1999; Bar-Matthews et al., 2003; Kaufman et al., 1998). At the same time, speleothems were deposited in the present-day arid and hyperarid region of the Negev Desert, Israel at the northeast margins of the Sahara Desert. Deposition of speleothems in such an arid environment requires a major increase in precipitation. Petrographic, isotopic and geochemical evidence indicate that this speleothem formation

is associated with the southward migration of Atlantic Mediterranean fronts (Vaks, 2008; Vaks et al., 2007) as was previously suggested by Arz et al. (2003) from studies of the Red Sea marine cores.

The climate changes during the mid to late Holocene were less severe than those occurring during the LGM, the LGM–Holocene transition and the transition from early to mid Holocene. Sea level only changed slightly (Wanner et al., 2008), and temperatures remained almost constant (Almogi-Labin et al., 2009; Martrat et al., 2004). In order to illustrate climate changes for this period, Bar-Matthews and Ayalon (2011) discussed high-resolution isotopic records (in the order of single years or even seasonal records) from the Soreq Cave (Figure 32.6). The present-day calibration between rainfall amount and its weighted mean annual $\delta^{18}O$ values in the cave area is used to determine the correlation of the isotopic changes to rainfall amount and vegetation. The present-day calibration shows that increase in rainfall amount correlates negatively with $\delta^{18}O$ of precipitation, and that a decrease of 1 per mille in the annual weighted $\delta^{18}O$ value of precipitation is equivalent to an increase of ~250 mm in annual rainfall. Although other factors potentially can affect the rainfall $\delta^{18}O$ values, in view of the similar conditions to the present day (i.e., isotopic composition of sea water, sea level, temperatures) the present-day relationships between rainfall amount and isotopic composition can be applied to the mid to late Holocene and that the lowest $\delta^{18}O$ events revealed in the Soreq Cave isotopic record reflect an increase in annual rainfall and vice versa.

Variations in the $\delta^{13}C$ values of calcite speleothems mainly reflect changes of the vegetation type in the vicinity of the cave and arise because of the differences in the photosynthetic pathways between C3- and C4-type vegetation. Other factors controlling the carbon isotopic composition are the presence or absence of soil cover, intensity of carbonate host-rock dissolution and water-rock interactions in the unsaturated zone

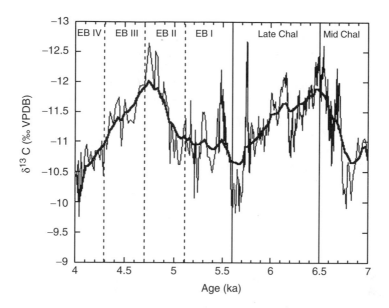

Figure 32.6 Mid-Holocene cultural changes superimposed on the smooth curve of the $\delta^{13}C$ record of Soreq Cave speleothems (Bar-Matthews and Ayalon, 2011). The vertical lines separate between archeological periods. Mid-Chal – middle Chalcolithic; Late Chal – late Chalcolithis; EB I – early Bronze I; EB II – early Bronze II; EB III – early Bronze III; EB IV – early Bronze IV.

(Bar-Matthews et al., 1996; Genty et al., 2001). During the mid to late Holocene, $\delta^{13}C$ values of Soreq Cave speleothems range between ~13 and ~9.5 per mille, indicating a dominance of C3-type vegetation. There is no evidence for vegetation change during this period, thus we suggest that the oscilations in $\delta^{13}C$ seen in Figure 32.6 must be climate related, because they closely follow the $\delta^{18}O$ variations (Bar-Matthews and Ayalon, 2011). The correlation between low $\delta^{18}O$ values (indicative of increased rainfall) and low $\delta^{13}C$ values suggests that lower $\delta^{13}C$ values most probably reflect changes in the composition of the soil organic matter, probably due to different proportions of organic compounds that are formed during winter/summer months (Orland et al., 2009).

The high-resolution $\delta^{13}C$ speleothem record (Figure 32.6) shows that the mid Holocene in the EM was characterized by sinusoidal cycles alternating between wetter and drier climates, each cycle lasting ~1,500 years. Several short-lasting (20–100 years)

wet and dry events can also be identified. The longer, 1,500-year cycles match the Bond events, which are related to the North Atlantic climate fluctuations (Bond et al., 1997, 2001), and with the microcharcoal content from two crater lake basins in central Turkey (Turner et al., 2008). Since rainfall in the EM is associated with Atlantic–Mediterranean fronts, it is highly probable that Bond cycles are also recorded in the EM speleothem profile. Mayewski et al. (2004) suggested that periods of significant rapid climate change (RCC) during the mid and late Holocene are influenced by polar cooling, tropical aridity and major atmospheric circulation changes.

5.2 Archeological change and the EM speleothem-climate record

Changes in the isotopic composition of carbon and oxygen are shown to correspond to cultural changes and the climatic cyclicity is also evident in the archeological cultural

record. When comparing archeological periods, as summarized by Bar-Yosef and
Garfinkel (2008), with the isotopic record
presented here (Figure 32.6) several observations can be made. The transitions between
the main three archeological periods, mid
Chalcolithic to late Chalcolithic, late
Chalcolithic to early Bronze, and early
Bronze II to early Bronze III, match the
peaks and troughs of the major sinusoidal
isotopic cycles. The transition from mid
Chalcolithic to late Chalcolithic periods and
from the early Bronze II to early Bronze III
occurs at the peaks of the wet climate conditions that prevailed during the mid Holocene
(6,550–6,450 BP and 4,800–4,700 BP respectively), when annual precipitation most probably exceeded ~700 mm. The end of the
Chalcolithic period coincides with the end of
a long, dry period of the Mid Holocene
(5,700–5,600 BP), with an estimated annual
rainfall of ~350 mm. This is also the case for
the transition from early Bronze IV to the
mid Bronze, which occurred at the end of a
long dry trend that reach its peak at 4,200–
4,050 BP. Mayewski et al. (2004) identified
six major widespread arid RCCs, two of
them occurring during the mid Holocene, at
6,000–5,000 BP and at 4,200–3,800 BP. They
suggested that these mid Holocene rapid climate change events were caused by high-
latitude cooling and low-latitude aridity.

The 6,000–5,000 BP event was considered
by Weninger et al. (2009) to last between
6,000 and 5,200 BP. They suggested that this
was a widespread event through Europe and
its rapid climate change triggered social
change in southeast Europe, the collapse of
the southeast European Copper Age, and the
onset of the southeast European early Bronze
Age at the end of the event. Based on Soreq
Cave records, the peak of this event occurred
in two main phases, the first at 5,700–5,600
BP, a period that was the driest throughout the
mid Holocene, and the second, a shorter dry
event at 5,250–5,170 BP. This second dry
event is superimposed on the general trend
towards a wetter climate. The first event was
also observed by Verheyden et al. (2008) at

between 6,500 and 5,800 BP based on the
speleothem record from Jeita Cave in
Lebanon. They attribute the increase in $\delta^{18}O$
in the speleothems, and a decrease in their
growth rate and diameter, to dry climate. The
second, shorter, dry event is not always evident as a distinct event in other records.
Staubwasser and Weiss (2006) related the dry
event at ~5,200 BP to a drought that caused
societal change in which the late Uruk period
society in southern and northern Mesopotamia
collapsed (Postgate, 1986). Based on lacustrine sediments retrieved from Lake Zazari,
Northern Greece, Cavallari and Rosenmeier
(2007) found that the major peaks of dry
events occurred at 6,250–5,950 BP, 5,320–
4,750 BP, and 4,050–3,500 BP. These dry
periods partly overlap the dry periods noted
above in the speleothem studies.

Two other dry events at the beginning of
mid Holocene, at 6,650–6,600 BP and 6,250–
6,180 BP, may be included as part of a longer
dry period evident in Lebanon (Verheyden
et al., 2008). The second event is described
by Cavallari and Rosenmeier (2007) at
6,250–5,950 BP as a dry event in northern
Greece.

With respect to the important rapid climate
change event that occurred at the end of the
mid Holocene, between 4,200 and 3,800 BP,
Mayewski et al. (2004) suggested that of all
potential climate forcing mechanisms, the
most probable was solar variability superimposed on long-term changes in insolation.
This event in the Soreq Cave speleothem
record occurs at the end of longer 1,500-year
cyclicity, terminating a trend towards aridity
that started at 4,700 BP and reached its maximum at ~4,200–4,050 BP. The end of this dry
event marks the transition to the mid Bronze
period. Although, based on the $\delta^{18}O$ record of
Soreq Cave speleothems, this event does not
indicate severe aridity as in the dry event
between 5,700 and 5,600 BP, the $\delta^{13}C$ record
does suggest climate change of the same
order of magnitude. It is possible that the
$\delta^{18}O$ values of Soreq Cave speleothems do
not fully record the most extreme years of
aridity during the 4,200–4,050 BP event due to

the lower resolution (~20 years) compared with the very high-resolution (~3.5 years) during the 5,700–5,600 BP event. During the 4,200–3,800 BP event, drought was felt in northern Mesopotamia (Weiss, 2000; Weiss et al., 1993). Staubwasser and Weiss (2006) refer to this event as a 'collapse as adaptation to rapid climate change'. They propose that synchronous changes among early Bronze Age societies indicate a causal link with climate. This causal link is that between reduced precipitation and the collapse of the politic-oeconomic superstructures that depended on cereal agriculture. This event is observed on global scale as evident from speleothems in Italy (Drysdale et al., 2006), brine sediments from the northern Red Sea (Arz et al., 2006), deep sea sediments from core of the Gulf of Oman (Cullen et al., 2000), and ice cores from tropical Africa (Thompson et al., 2002). Most RCC events tend to show a rapid change towards a wetter climate, but slow and gradual change to drier conditions (Figure 32.6).

Many RCC studies tend to emphasize dry events and their connection to social collapse. However, it is important to note that several wet periods also characterize Mid Holocene cultural changes. The most important ones occurred at the transition from the mid to late Chalcolitic period at 6,550–6,450 BP, and at the transition from early Bronze II to early Bronze III at 4800–4700 BP (Figure 32.6), as well as shorter wet events lasting 20–70 years at 6700–6680 BP, 6170–6100 BP, 5760–5740 BP, and 5500–5450 BP. It is not been defined how and if these shorter-scale events are linked to cultural changes.

The climate event marked by the ~4.2 ka BP is of a large regional scale. This event has been recorded in Egypt (Hassan, 1996) and Mesopotamia (Weiss et al., 1993), and was claimed to cause the end of the Old Kingdom in Egypt and the collapse of northern Mesopotamian civilization. Taken together, the evidence indicates that the increasing aridity at ~4.2 ka BP affected a vast area including the Middle East as well as northeastern and eastern equatorial Africa (e.g., Thompson et al., 2002).

5.3 Western Mediterranean speleothems

Paleoclimate records based on WM speleothems come mainly from the Buca della Renella and Antro del Corchia caves located in the Alpi Apuane karst of northwestern Italy (southwestern MS) on speleothems that grew between ~7.0 and 1.0 ka BP, between 14.0 ka BP and ~1.0 ka BP and ~ 120 ka BP and 90 ka BP, respectively (Drysdale et al., 2006, 2007; Zanchetta et al., 2007). Their study demonstrates that between ~112.0 ka BP and ~109.0 ka BP, and between ~105.0 and 102.6 ka BP the timing of climate change is in broad agreement with Greenland stadials recorded in the NGRIP ice core $\delta^{18}O$ data and with cold events in North Atlantic marine cores. They noticed apparent disagreement over ages assigned to these events in the various proxy records, highlighting the difficulty of obtaining precise age estimates from ice and marine archives, thus proposing that the speleothem record from the WM can constitute new tuning points for the NGRIP record.

In a study of Corchia Cave, Zanchetta et al. (2007) show that the speleothem record supports enhanced rainfall during the early Holocene in the WM as was evident also from the EM speleothems record. Drysdale et al. (2006) show that using various geochemical proxies combined with carbon and oxygen isotopic record in speleothems they were able to identify extreme values at ~4.1–3.8 ka BP. They suggest that during this interval there was significant reduced moisture reaching the cave interior similar to Soreq Cave speleothems, confirming the link of increased aridity recorded in Egypt, Mesopotamia and southwestern Asia with that of the EM.

The drop in rainfall amount during the later part of the Holocene agrees with the general trend toward aridity in North Africa and the Middle East, a result of the intensification of the African Monsoon (deMenocal et al., 2000). It is associated with drops in rainfall and lake level and in the deterioration of vegetation (e.g., Bar-Matthews et al., 1998; Rossignol-Strick, 1985; Schilman et al., 2001; Street and Grove, 1976). The reduction in the regional biosphere during the aridification process released large amounts of carbon to the atmosphere causing a gradual increase in atmospheric CO_2 from 7000 years ago (Indermuhle et al., 1999).

ACKNOWLEDGMENTS

I would like to thank A. Matthews, A. Ayalon and A. Almogi-Labin for fruitful discussions and collaboration over the years.

REFERENCES

Affek H. P., Bar-Matthews M., Ayalon A., Matthews A. and Eiler J. M. 2008. Glacial/interglacial temperature variations in Soreq cave speleothems as recorded by 'clumped isotope' thermometry. *Geochimica et Cosmochimica Acta* 72: 5351–5360.

Almogi-Labin A., Bar-Matthews M. and Ayalon A. 2004. Climate variability in the Levant and northeast Africa during the Late Quaternary based on marine and land records, in Goren-Inbar N.and Speth J. D. (eds) *Human Paleoecology in the Levantine Corridor.* Oxford: Oxbow Press, pp. 117–134.

Almogi-Labin A., Bar-Matthews M., Shriki D., Kolosovsky E., Paterne M., Schilman B. et al. 2009. Climatic variability during the last ~90 ka of the southern and northern Levantine basin as evident from marine records and speleothems. *Quaternary Science Reviews* 28: 2882–2896.

Arz H. W., Lamy F., Pätzold J., Müller P. J. and Prins M. 2003. Mediterranean moisture source for an early-Holocene humid period in the northern Red Sea. *Science* 300: 118–121.

Ayalon A., Bar-Matthews M. and Kaufman A. 1999: Petrography, strontium, barium, and uranium concentrations, and strontium and uranium isotope ratios in speleothems as paleoclimatic proxies: Soreq Cave, Israel. *The Holocene* 9, 715–722.

Ayalon A., Bar-Matthews M. and Kaufman A. 2002. Climatic conditions during marine isotopic stage 6 in the Eastern Mediterranean region as evident from the isotopic composition of speleothems: Soreq Cave, Israel. *Geology* 30: 303–306.

Ayalon A., Bar-Matthews M. and Sass E. 1998. Rainfall-recharge relationships within a karstic terrain in the Eastern Mediterranean semi-arid region, Israel: $\delta^{18}O$ and δD characteristics. *Journal of Hydrology* 207: 18–31.

Bar-Matthews M. and Ayalon A. 2011. Mid-Holocene climate variations revealed by high-resolution speleothems record of Soreq Cave, Israel and their correlation with cultural changes. *The Holocene* 21: 163–171.

Bar-Matthews, M., Ayalon, A., Matthews, A., Sass, E. and Halicz, L., 1996. Carbon and oxygen isotope study of the active water-carbonate system in a karstic Mediterranean cave: implications for paleoclimate research in semiarid regions. Geochim. Cosmochim. Acta 60: 337–347.

Bar-Matthews M., Ayalon A. and Kaufman A. 1997. Late Quaternary paleoclimate in the eastern Mediterranean region from stable isotope analysis of speleothems at Soreq Cave, Israel. *Quaternary Research* 47: 155–68.

Bar-Matthews M., Ayalon A. and Kaufman A. 1998. Middle to late Holocene (6500 years period) paleoclimate in the Eastern Mediterranean region from stable isotopic composition of speleothems from Soreq Cave, Israel, in Issar A.S. and Brown N. (eds) *Water, Environment and Society in Times of Climate Change.* Dordrecht: Kluwer, pp. 203–214.

Bar-Matthews M., Ayalon A. and Kaufman A. 2000. Timing and hydrological conditions of Sapropel events in the Eastern Mediterranean, as evident from speleothems, Soreq cave, Israel. *Chemical Geology* 169: 145–156.

Bar-Matthews M., Ayalon A., Gilmour M., Matthews M. and Hawkesworth C. 2003. Sea-land isotopic relationships from planktonic foraminifera and speleothems in the Eastern Mediterranean region and their implications for paleorainfall during interglacial intervals. *Geochimica et Cosmochimica Acta* 67: 3181–99.

Bar-Matthews M., Ayalon A., Kaufman A. and Wasserburg G. J. 1999. The Eastern Mediterranean paleoclimate as a reflection of regional events: Soreq Cave, Israel. *Earth and Planetary Science Letters* 166: 85–95.

Bar-Yosef O. 1998. On the nature of transition: The Middle to Upper Palaeolithic and the Neolithic revolution. *Cambridge Archaeological Journal* 8(2): 63–141.

Bar-Yosef O. and Garfinkel Y. 2008. *The Prehistory of Israel, Human Cultures Before Writing.* Jerusalem: Ariel Publishing.

Bard E., Rostek F., Turon J.-L. and Gendreau S. 2000. Hydrological impact of Heinrich events in the subtropical northeast Atlantic. *Science* 289: 1321–1324.

Bard E., Delaygue G., Rostek F., Antonioli F., Silenzi S., Schrag D. 2002. Hydrological conditions in the western Mediterranean basin during the deposition of Sapropel 6 (ca 175 kyr). *Earth and Planetary Science Letters* 202: 481–494.

Baringer M. O. N. and Price J. F. 1999. A review of the physical oceanography of the Mediterranean outflow. *Marine Geology* 155: 63–82.

Bartov Y., Stein M., Enzel Y., Agnon A. and Reches Z. 2002. Lake levels and sequence stratigraphy of Lake Lisan, the late Pleistocene precursor of the Dead Sea. *Quaternary Research* 57: 9–21.

Bartov Y., Goldstein S. L., Stein M. and Enzel Y. 2003. Catastrophic arid episodes in the Eastern Mediterranean linked with the North Atlantic Heinrich events. *Geology* 31: 439–442.

Baruch U. and Bottema S. 1999 A new Pollen diagram from Lake Hula, in Kawanabe H., Coulter G. W. and Roosevelt A. C. (eds) *Ancient Lakes: Their Cultural and Biological Diversity*. Kenobi Productions, Belgium pp. 75–86.

Bethoux J. P. 1979. Budgets of the Mediterranean Sea. Their dependence on the local climate and on characteristics of the Atlantic waters. *Oceanologica Acta* 2: 157–163.

Bethoux J. P. 1993. Mediterranean sapropel formation, dynamic and climatic viewpoints. *Oceanologica Acta* 16: 127–133.

Bethoux J. P., Pierre C. 1999. Mediterranean functioning and sapropel formation: respective influences of climate and hydrological changes in the Atlantic and the Mediterranean. *Marine Geology* 153: 29–39.

Bethoux J.P., Gentili B. and Tailliez D. 1998. Warming and freshwater budget change in the Mediterranean since the 1940s, their possible relation to the greenhouse effect. *Geophysical Research Letters* 25: 1023–1026.

Bond G., Broecker W., Johnsen S., McManus J., Labeyrie L., Jouzel J. and Bonani G. 1993. Correlations between climate records from North Atlantic sediments and Greenland. *Nature* 365: 143–147.

Bond G., Showers W., Cheseby M., Lotti R., Almasi P., deMenocal P. et al. 1997. A pervasive millennial-scale cycle in North Atlantic Holocene and Glacial climates. *Science* 278: 1257–1266.

Bond G., Kromer B., Beer J., Muscheler R., Evens M.N., Showers W. et al. 2001. Persistent solar influence on North Atlantic climate during the Holocene. *Science* 294: 2130–2136.

Bookman R., Bartov Y., Enzel Y. and Stein M. 2006. The levels of the Quaternary lakes in the Dead Sea

basin: two centuries of research, in Enzel Y., Agnon A. Stein M. (eds) *New Frontiers in Dead Sea Paleoenvironmental Research*. Geological Society of America Special Paper 401, pp. 155–170.

Bormans M., Garret C. and Thompson K. R. 1986. Seasonal variability of the surface inflow through the strait of Gibraltar. *Oceanologica Acta* 9: 403–414.

Boscolo R. and Bryden H. 2001. Causes of long-term changes in Aegean sea deep water. *Oceanologica Acta* 24(6): 519–527.

Bottema S. and Van Zeist W. 1981. Palynological evidence for climatic history of the Near East, 50,000–6,000 B.P., in Cauvin J. and Sanlaville, P. (eds) *Préhistoire Du Levant*. Paris: CNRS. pp. 111–132.

Bowen G. J. and Wilkinson B. 2002. Spatial distribution of d18O in meteoric precipitation. *Geology* 30(4): 315–318.

Brassell S.C., Eglinton G., Marlowe I.T., Pflaumann U. and Sarnthein M. 1986. Molecular stratigraphy: a new tool for climate assessment. *Nature* 320: 129–133.

Bryden H. L., Kinder T. H. 1991. Steady two-layer exchange through the strait of Gibraltar. *Deep-Sea Research* 38: S445–S463.

Cacho I., Grimalt J. O., Canals M., Sbaffi L., Shackleton N. J., Schönfeld J., and Zahn R. 2001. Variability of the western Mediterranean Sea surface temperature during the last 25,000 years and its connection with the Northern Hemisphere climatic changes. *Paleoceanography* 16: 40-52.

Cacho I., Grimalt J. O., Pelejero C., Canals M., Sierro F. J., Flores J. A. and Shackleton N. 1999a. Dansgaard-Oeschger and Heinrich event imprints in Alboran Sea paleotemperatures. *Paleoceanography* 14: 698–705.

Cacho I., Pelejero C., Grimalt J. O., Calafat A., Canals M. 1999b. C37 alkenone measurements of sea surface temperature in the gulf of Lions (NW Mediterranean). *Organic Geochemistry* 30: 557–566.

Calvert S. E. and Fontugne M. R. 2001. On the late Pleistocene-Holocene sapropel record of climatic and oceanographic variability in the eastern Mediterranean. *Paleoceanography* 16: 78–94.

Cavallari B. J. and Rosenmeier M. F. 2007. A Multi-Proxy Paleoclimate Record of Rapid Holocene Climate Variability From Northern Greece/Greek Macedonia. American Geophysical Union, Fall Meeting 2007, abstract #PP31A–0180.

Cayre O., Lancelot Y., Vicent E. and Hall M. A. 1999. Paleocenoagraphic reconstructions from planktonic foraminifera off the Iberian margin: temperature, salinity and Heinrich events. *Paleoceanography* 14: 384–396.

Chapman M. R., Shackleton N. J., Zhao M., Eglinton, G. 1996. Faunal and alkenone reconstructions of subtropical North Atlantic surface hydrography and paleotemperature over the last 28 kyr. *Paleoceanography* 11: 343–357.

Cheddadi R. and Rossignol-Strick M. 1995. Eastern Mediterranean Quaternary palaeoclimates from pollen and isotope records of marine cores in the Nile cone area. *Paleoceanography* 10: 291–300.

Cita M. B., Vergnaud-Grazzini C., Robert C., Chamley H., Ciaranfi N. and d'Onofrio S. 1977. Paleoclimatic record of a long deep sea core from the Eastern Mediterranean. *Quaternary Research* 8: 205–235.

Cullen H.M., deMenocal P.B., Hemming S., Hemming G., Brown F.H., Guilderson T. and Sirocko F. 2000. Climate change and the collapse of the Akkadian empire: evidence from the deep sea. *Geology* 28: 379–382.

Dansgaard W., Johnsen S. J., Clausen H. B., Dahl-Jensen D., Gundestrup N., Hammer C. U. et al. 1993. Evidence for general instability of past climate from a 250-kyr ice-core record. *Nature* 364: 218–220.

Dayan U. 1986. Climatology of back trajectories from Israel based on synoptic analysis. *Journal of Climate and Applied Meteorology* 25: 591–595.

deKaenel E., Siesser W. G. and Murat A. 1999. Pleistocene calcareous nannofossil biostratigraphy and the western Mediterranean sapropels, sites 974 to 977 and 979, in Zahn R., Comas M. C. and Klaus A. (eds) *Proceedings of the Ocean Drilling Program, Scientific Results, College Station, TX (Ocean Drilling Program)* 161: 159–183.

deMenocal P. B. 2001. Cultural responses to climate change during the Late Holocene. *Science* 292: 667–673.

deMenocal P. B., Ortiz J., Guilderson T. and Sarnthein M. 2000. Coherent high- and low-latitude climate variability during the Holocene warm period. *Science* 288: 2198–2202.

Develle, A-L., Herreros J., Vidal L., Sursock A., Gasse F. 2010. Controlling factors on a paleo-lake oxygen isotope record (Yammoûneh, Lebanon) since the Last Glacial Maximum. *Quaternary Science Reviews* 29: 865–886.

Drysdale R., Zanchetta G., Hellstrom J., Maas R., Fallick A., Pickett M. et al. 2006. Late Holocene drought responsible for the collapse of Old World civilizations is recorded in an Italian cave flowstone. *Geology* 34: 101–104.

Drysdale R.N., Zanchetta G., Hellstrom J.C., Fallick A.E., McDonald J. and Cartwright I. 2007. Stalagmite evidence for the precise timing of North Atlantic cold events during the early last glacial. *Geology* 35: 77–80.

Emeis K. C., Sakamoto T., Wehausen R. and Brumsack H.-J. 2000. The sapropel record of the eastern Mediterranean Sea – results of Ocean Drilling Program Leg 160. *Palaeogeography, Palaeoclimatology, Palaeoecology* 158: 371–395.

Emeis K. C., Schulz H., Struck U., Rossignol-Strick M., Erlenkeuser H., Howell M. W. et al. 2003. Eastern Mediterranean surface water temperatures and 18O composition during deposition of sapropels in the late Quaternary. *Paleoceanography* 18: 1005–1029.

Emeis K.C., Schulz H.M., Struck U., Sakamoto T., Doose H., Erlenkeuser H., Howell M., Kroon D. and Paterne M. 1998. Stable isotope and alkenone temperature records of sapropels from sites 964 and 967: constraining the physical environment of sapropel formation in the eastern Mediterranean Sea, in Robertson et al. (eds) *Proceedings of the Ocean Drilling Program* 160: 309–331.

Enzel Y., Bookman (Ken-Tor) R., Sharon D., Stein M., Gvirtzman H. and Dayan U. 2003. Dead Sea lake level variations and Holocene climates in the Near East: implications to historical responses and modern water resources. *Quaternary Research* 60: 263–273.

Enzel Y., Amit R., Dayan U., Crouvi O., Kahana R., Ziv B., Sharon D. 2008. The climatic and physiographic controls of the eastern Mediterranean over the late Pleistocene climates in the southern Levant and its neighboring deserts. *Global and Planetary Change* 60: 165–192.

Essallami L., Sicre M.A., Kallel N., Labeyrie L. and Siani G. 2007. Hydrological changes in the Mediterranean Sea over the last 30,000 years. *Geochemistry Geophysics Geosystems* 8Q07002. doi:10.1029/2007GC001587

Finsinger W., Belis C., Blockley S. P. E., Eicher U., Leuenberger M., Lotter A. F. and Ammann B. 2008. Temporal patterns in lacustrine stable isotopes as evidence for climate change during the late glacial in the Southern European Alps. *Journal of Paleolimnology* 40: 885–895.

Fiúza A. F. G., Hamann M., Ambar I., Diaz del Rio G., Gonzalez N. and Cabanas J. M. 1998. Water masses and their circulation off western Iberia during May 1993. *Deep Sea Research I* 45: 1127–1160.

Fontugne M. R. and Calvert S. E. 1992. Late Pleistocene variability of the carbon isotopic composition of organic matter in the Eastern Mediterranean: monitor of changes in carbon sources and atmospheric CO_2 concentrations. *Paleoceanography* 7: 1–20.

Fontugne M. R., Arnold M., Labeyrie L., Paterne M., Calvert S. E. and Duplessy J.-C., 1994. Palaeo-environment, sapropel chronology and Nile River discharge during the last 20,000 years as indicated by deep sea sediment records in the Eastern Mediterranean, in O. Bar-Yosef, and R.S. Kra (eds) *Late Quaternary Chronology and Paleoclimates of the Eastern Mediterranean* . *Radiocarbon*: 75–88.

Frogley M. R., Tzedakis P. C., Heaton T. H. E. 1999. Climate variability in Northwest Greece during the last interglacial. *Science* 285: 1886–1889.

Frumkin A. 1997. The Holocene history of the Dead Sea levels, in Niemi T. M., Ben-Avraham Z. and Gat Y. (eds) *The Dead Sea – The Lake and its Settings*. Oxford: Oxford University Press, pp. 237–248.

Frumkin A., Ford D. C. and Schwarcz H. P. 1999. Continental oxygen isotopic record of the last 170,000 years in Jerusalem. *Quaternary Research* 51: 317–327.

Frumkin A. and Stein M. 2004. The Sahara-East Mediterranean dust and climate connection revealed by strontium and uranium isotopes in a Jerusalem speleothem. *Earth and Planetary Science Letters* 217: 451–464.

Ganopolski A., Kubatzki C., Claussen M., Brovkin V. and Petoukhov V. 1998. The influence of vegeta-tion–atmosphere–ocean interaction on climate during the mid-Holocene. *Science* 280: 1916–1919.

Garcia-Lafuente J., Cano N., Vargas M., Rubín J. P. and Hernández-Guerra A. 1998. Evolution of the Alboran sea hygrographic structures during July 1993. *Deep-Sea Research* 45: 39–65.

Gat J. R. 1995. Stable isotopes of fresh and saline lakes, in Lerman A., Imboden D. and Gat J.R. (eds) *Physics and Chemistry of Lakes*. Berlin: Springer, pp. 139–166.

Gat J. R. 1996. Oxygen and hydrogen isotopes in the hydrologic cycle. *Annual Reviews Earth Planetary Science* 24: 225–262.

Genty G., Baker A., Massault M., Proctor C., Gilmour M., Pons-Branchu E. and Hamelin B. 2001. Dead carbon in stalagmites: carbonate bedrock paleodissolution vs. ageing of soil organic matter. Implications for ^{13}C variations in speleothems. *Geochimica et Cosmochimica Acta* 65: 3443–3457.

Grootes P. M., Stuiver M., White J. W. C., Johnsen S. J. and Jouzel J. 1993. Comparison of oxygen isotope records from the GISP2 and GRIP Greenland ice cores. *Nature* 366: 552–554.

Hajar L., Khater C. and Cheddadi R. 2008. Vegetation changes during the late Pleistocene and Holocene in Lebanon: a pollen record from the Beqaa Valley. *The Holocene* 18: 1089–1099.

Haase-Schramm A., Golstein S. L. and Stein M. 2004. U-Th dating of Lake Lisan (Late Pleistocene Dead Sea) aragonite and implications for glacial East Mediterranean climate change. *Geochimica et Cosmochimica Acta* 68: 985–1005.

Hassan F. A. 1996. Nile floods and political disorder in early Egypt, in Dalfes H. N. et al. (eds) *Third Millennium B.C. Climate Change and Old World Collapse*, NATO ASI Series I., 49. Berlin/Heidelberg: Springer-Verlag, pp. 39–66.

Hayes A., Kucerab M., Kallel N., Sbaffid L. and Rohling E. J. 2005. Glacial Mediterranean sea sur-face temperatures based on planktonic foraminiferal assemblages. *Quaternary Science Reviews* 24: 999–1016.

Haynes C. V. Jr. 1987. Holocene migration rates of the Sudano-Sahelian wetting front, Arba'in Desert, Eastern Sahara, in Close, A. (ed.) *Prehistory of Arid North Africa*. Texas: Southern Methodist University Press, pp. 69–84.

Hazan N., Stein M., Agnon A., Marco S., Nadel D., Negendank J. F. W., Schwab M. J., Neev D. 2005. The late Quaternary limnological history of Lake Kinneret (Sea of Galilee), Israel. *Quaternary Research* 63: 60–77.

Heinrich H. 1988. Origin and consequences of cyclic ice rafting in the Northeast Atlantic Ocean during the past 130,000 years. *Quaternary Research* 29: 142–152. Horowitz A. 1979. *The Quaternary of Israel*. New York: Academic Press.

IAEA/WMO. 2001. Global network of isotopes in pre-cipitation. The GNIP Database. Available at: http://www.isohis.iaea.org.

Indermuhle A., Stocker T.F., Joos F., Fisher H., Smith H.J., Wahlen M. et al. 1999. Holocene carboncycle dynamics based on CO2 trapped in ice at Taylor Dome, Antarctica. *Nature* 398: 121–126.

Issar A. S. and Zohar M. 2004. *Climate Change: Environment and Civilization in the Middle East*. Berlin: Springer-Verlag.

Jones M.D., Roberts C.N. and Leng M.J. 2007. Quantifying climatic change through the last glacial–interglacial transition based on lake isotope palaeo-hydrology from central Turkey. *Quaternary Research* 67: 463–473.

Jorissen F. 1999. Benthic foraminiferal successions across Late Quaternary Mediterranean sapropels. *Marine Geology* 153: 91–101.

Kallel N., Duplessy J.-C., Labeyrie L., Fontugne M., Paterne M. and Montacer M. 2000. Mediterranean pluvial periods and sapropel formation during the last 200,000 years. *Palaeogeography, Palaeoclimatology, Palaeoecology* 157: 45–58.

Kallel N., Paterne M., Duplessy J.-C., Vergnaud-Grazzini C., Pujol C., Labeyrie L. et al. 1997. Enhanced rainfall in the Mediterranean region during the last sapropel event. *Oceanologica Acta* 20: 697–712.

Katz A., Kolodny Y. and Nissenbaum A. 1977. The geochemical evolution of the Pleistocene Lake Lisan–Dead Sea system. *Geochimica Cosmochimica Acta* 41: 1609–1626.

Kaufman A., Wasserburg G. J., Porcelli D., Bar-Matthews M., Ayalon A. and Halicz L. 1998. U-Th isotope systematics from the Soreq Cave Israel and climatic correlations. *Earth and Planetary Science Letters* 156: 141–155.

Kolodny Y., Stein M. and Machlus M. 2005. Sea–rain–lake relation in the Last Glacial East Mediterranean revealed by $\delta^{18}O/\delta^{13}C$ in Lake Lisan aragonites. *Geochimica Cosmochimica Acta* 16: 4045–4060.

Langgut D. 2008. Late Quaternary palynological sequences from the Eastern Mediterranean Sea. *Israel Geological Survey Report* GSI/16/08, Jerusalem.

Lascaratos A., Roether W., Nittis K. and Klein B. 1999. Recent changes in deep water formation and spreading in the eastern Mediterranean sea: a review. *Progress in Oceanography* 44: 5–36.

Leaman K.D. and Schott F. 1991. Hydrographic structure of the convection regime in the gulf of Lions: winter 1987. *Journal Physical Oceanography* 21: 575–598.

Lisker S., Vaks A., Bar-Matthews M., Porat R. and Frumkin A. 2009. Stromatolites in caves of the Dead Sea Fault Escarpment: Implications to latest Pleistocene lake levels and tectonic subsidence. *Quaternary Science Review* 28: 80–92.

Lisker S., Vaks A., Bar-Matthews M., Porat R. and Frumkin A. 2010. A Late Pleistocene palaeoclimatic and palaeoenvironmental reconstruction of the Dead Sea area (Israel), based on speleothems and cave stromatolites. *Quaternary Science Review* 29: 1201–1211.

Malanotte-Rizzoli P. and Hecht A. 1988. Large-scale properties of the Eastern Mediterranean: A review. *Oceanologica Acta* 11/4: 323–335.

Malanotte-Rizzoli P., Manca B. B., d'Alcala M. R., Theocharis A., Brenner S., Budillon G., Ozsoy E. 1999. The Eastern Mediterranean in the 80s and in the 90s: the big transition in the intermediate and deep circulations. *Dynamics of Atmospheric and Oceans* 29: 365–395.

Marino G., Rohling E. J., Sangiorgi F., Hayes A., Casford J. L., Lotter A. F. et al. 2009. Early and middle Holocene in the Aegean Sea: interplay between high and low latitude climate variability. *Quaternary Science Reviews* 28: 3246–3262.

Martinez-Ruiz F., Paytan A., Kastner M., Gonzalez-Donoso J. M., Linares D., Bernasconi S. M. and Jimenez-Espejo F. J. 2003. A comparative study of the geochemical and mineralogical characteristics of the S1 sapropel in the western and eastern Mediterranean. *Palaeogeography, Palaeoclimatology, Palaeoecology* 190: 23–37.

Martinson D. G., Pisias N., Hays J. D., Imbrie J., Moore T. C. J. and Schackleton N. J. 1987. Age dating and the orbital theory of the ice ages: development of the high-resolution 0–300,000 year chronostratigraphy. *Quaternary Research* 27: 1–29.

Martrat B., Grimalt J. O., Lopez-Martinez C., Cacho I., Sierro F. J., Flores J. A. et al. 2004. Abrupt temperature changes in the Western Mediterranean over the past 250,000 years. *Science* 306: 1762–1765.

Martrat B., Grimalt J. O., Shackleton N. J., de Abreu L., Hutterli M. A. and Stocker T. F. 2007. Four climate cycles of recurring deep and surface water destabilizations on the Iberian margin, *Science* 317: 502–507.

Marullo S., Santoleri R., Malanotte-Rizzoli P. and Bergamasco A. 1999. The sea surface temperature field in the Eastern Mediterranean from advanced very high resolution radiometer (AVHRR) data. Part I. Seasonal variability. *Journal of Marine Systems* 20: 63–81.

Matthews A., Ayalon A. and Bar-Matthews M. 2000. D/H ratios of fluid inclusions of Soreq Cave (Israel) speleothems as a guide to the Eastern Mediterranean Meteoric Line relationships in the last 120 ky. *Chemical Geology* 166: 183–91.

Mayewski P. A., Rohling E. E., Stager J. C., Karlen W., Maascha K. A., Meekler L. D. et al. 2004. Holocene climate variability. *Quaternary Research* 62: 243–255.

McDermott F. 2004. Palaeo-climate reconstruction from stable isotope variations in speleothems: a review. *Quaternary Science Reviews* 23: 901–918.

McGarry S., Bar-Matthews M., Matthews A., Vaks A., Schilman B. and Ayalon A. 2004. Constraints on hydrological and paleotemperature variations in the Eastern Mediterranean region in the last 140ka given by the δD values of speleothem fluid inclusions. *Quaternary Science Reviews* 23: 919–934.

McManus J. F., Oppo D. W. and Cullen J. L. 1999. A 0.5 Million-year record of millennial-scale climate variability in the north Atlantic. *Science* 283: 971–975.

Migowski C., Stein M., Prasad S., Negendank J. F. W. and Agnon A. 2006. Holocene climate variability and cultural evolution in the Near East from the

Dead Sea sedimentary record. *Quaternary Research* 66: 421–431.

Millot C., Candela J., Fuda J. L. and Tber Y. 2006. Large warming and salinification of the Mediterranean outflow due to changes in its composition. *Deep-Sea Research* 53: 656–666.

Murat A. and Got H. 2000. Organic carbon variations of the eastern Mediterranean Holocene sapropel: A key for understanding formation processes. *Palaeogeography, Palaeoclimatology, Palaeoecology* 158: 241–257.

NGRIP (North Greenland Ice Core Project) members. 2004. High-resolution record of Northern Hemisphere climate extending into the last interglacial period. *Nature* 431, 147–151.

Osborne A. H., Vance D., Rohling E. J., Barton N., Rogerson M., and Fello N. 2008. A humid corridor across the Sahara for the migration of early modern humans out of Africa 120,000 years ago. *PNAS* 105: 16444–16447.

Orland I. J., Bar-Matthews M., Kita N. T., Ayalon A., Matthews A. and Valley J. W. 2009. Climate deterioration in the eastern Mediterranean as revealed by ion microprobe analysis of a speleothem that grew from 2.2 to 0.9 ka in Soreq Cave Israel. *Quaternary Research* 71: 27–35.

Perez-Folgado M., Sierro F. J., Flores J. A., Cacho I., Grimalt J. O., Zahn R. and Shackleton N. 2003. Western Mediterranean planktonic foraminifera events and millennial climatic variability during the last 70 kyr. *Marine Micropaleontology* 48: 49–70.

Pierre C. 1999. The oxygen and carbon isotope distribution in the Mediterranean water masses. *Marine Geology* 153: 51–55.

Pinardi N. and Masetti E. 2000. Variability of the large scale general circulation of the Mediterranean Sea from observation and modelling: a review. *Palaeogeography, Palaeoclimatology, Palaeoecology* 158: 153–173.

Postgate N. 1986. The transition from Uruk to Early Dynastic: continuities and discontinuities in the record of settlement, in Finkbeiner U. and Rollig W. (eds) *Ðamdat Nasr: Period or Regional Style?* Reichert: Wiesbaden, pp. 90–106.

Prahl F. G., Muehlhausen L. A. and Zahnle D. L. 1988. Further evaluation of long-chain alkenones as indicators of paleocenographic conditions. *Geochimica et Cosmochimica Acta* 52: 2303–2310.

Rindsberger M., Magaritz M., Carmi I. and Gilad D. 1983. The relation between air mass trajectories and the water isotope composition of rain in the Mediterranean Sea area. *Geophysical Research Letters* 10: 43–46.

Roberts N., Jones M. D., Benkaddour A., Eastwood W. J., Filippi M. L., Frogley M. R. et al. 2008. Stable isotope records of Late Quaternary climate and hydrology from Mediterranean lakes: the ISOMED synthesis. *Quaternary Science Reviews* 27: 2426–2441.

Roether W. H., Manca B. B., Klein B., Bregant D., Georgopoulos D., Beitzel V. et al. 1996. Recent changes in eastern Mediterranean deep waters. *Science* 271: 333–335.

Rohling E. J. 1994. Review and new aspects concerning the formation of Mediterranean sapropels. *Marine Geology* 122: 1–28.

Rohling E. J. and Hilgen F. J. 1991. The eastern Mediterranean climate at times of sapropel formation, a review. *Geologie en Mijnbouw* 70: 253–264.

Rohling E. J., Cane T. R., Cooke S., Sprovieri M., Bouloubassi I., Emeis K.-C. et al. 2002. African monsoon variability during the previous interglacial maximum. *Earth and Planetary Science Letters* 202: 61–75.

Rossignol-Strick M. 1985. Mediterranean Quaternary sapropels, an immediate response of the African monsoon to variation of insolation. *Palaeogeography, Palaeoclimatology, Palaeoecology* 49: 237–263.

Rossignol-Strick M., Nesteroff W., Olive P. and Vergnaud-Grazzini C. 1982. After the deluge: Mediterranean stagnation and sapropel formation. *Nature* 15: 105–110.

Rossignol-Strick M., Paterne M., Bassinot F. C., Emeis K.-C. and De Lange G. J. 1998. An unusual mid-Pleistocene monsoon period over Africa and Asia. *Nature* 392: 269–272.

Rozanski K., Araguas-Araguas L. and Gonfiantini R. 1993. Isotopic patterns in modern global precipitation, in Swart et al. (eds) *Climate Change in Continental Isotopic Record*. Geophys. Monograph 78. Washington DC: American Geophysical Union, pp. 1–37.

Sadori L., Zanchetta G. and Giardini M. 2008. Last glacial to Holocene palaeoenvironmental evolution at Lago di Pergusa (Sicily, southern Italy) as inferred by pollen, microcharcoal and stable isotopes. *Quaternary International* 181: 4–14.

Schilman B., Almogi-Labin A., Bar-Matthews M., Labeyrie L., Paterne M. and Luz B. 2001. Long- and short-term carbon fluctuations in the Eastern Mediterranean during the late Holocene. *Geology* 29: 1099–1102.

Schott F. and Leaman K. D. 1991. Observations with moored acoustic Doppler current profiles in the convection regime in the golfed du Lion. *Journal Physical Oceanography* 21: 558–574.

Scrivner A. E., Vance D. and Rohling E. J., 2004. New neodymium isotope data quantify Nile involvement in Mediterranean anoxic episodes. *Geology* 32: 565–568.

Sharon D. and Kutiel H. 1986. The distribution of rainfall intensity in Israel, its regional and seasonal variations and its climatological evaluation. *Journal of Climatology* 6: 277–291.

Staubwasser M. and Weiss H. 2006. Holocene climate and cultural evolution in late prehistoric-early historic West Asia – Introduction. *Quaternary Research* 66: 372–387.

Stein M., Torfstein A., Gavrieli I. and Yechieli Y. 2010. Abrupt aridities and salt deposition in the post-glacial Dead Sea and their North Atlantic connection. *Quaternary Science Reviews* 29: 567–575.

Street F. A. and Grove A. T. 1976. Environmental and climatic implication of late Quaternary lake level fluctuation in Africa. *Nature* 261: 385–390.

Street-Perrott F. A. and Perrott R. A. 1993. Holocene vegetation, lake levels, and climate of Africa, in Wright H. E. Jr., Kutzbach J. E., Webb III, T. Ruddiman W. F., Street-Perrott F. A. and Bartlein P. J. (eds) *Global Climate Since the Last Glacial Maximum.* Minneapolis: University of Minnesota Press, pp. 318–355.

Stevens L. R., Ito E., Schwalb A. and Wright H. E. 2006. Timing of atmospheric precipitation in the Zagros mountains inferred from a multi-proxy record from Lake Mirabad, Iran. *Quaternary Research* 66: 494–500.

Thompson L. G., Mosley-Thompson E., Davis M.E., Henderson K. A., Brecher H. H., Zagorodnov V. S. et al. 2002. Kilimanjaro ice core records: evidence of Holocene climate change in tropical Africa. *Science* 298, 589–593.

Thunell R. C. and Williams D. F. 1989. Glacial-Holocene salinity changes in the Mediterranean Sea: hydrographic and depositional effects. *Nature* 338: 493–496.

Tintore J., La-Violette P. E., Blade I. and Cruzado A. 1988. A study of an intense density front in the eastern Alboran sea: The Almeria-Oran front. *Journal of Geophysical Research* 18: 1384–1397.

Torfstein A., Haase-Schramm A., Waldmann N., Kolodny Y., Stein M. 2009. U-series and oxygen isotope chronology of the mid-Pleistocene Lake Amora (Dead Sea basin). *Geochimica et Cosmochimica Acta* 73: 2603–2630.

Turner R., Roberts N. and Jones M. D. 2008. Climatic pacing of Mediterranean fire histories from lake sedimentary microcharcoal. *Global and Planetary Change* 63: 317–324.

Tzedakis P. C., McManus J. F., Hooghiemstra H., Oppo D. W. and Wijmstra T. A. 2003. Comparison of vegetation in northeast Greece with record of climate variability on orbital and suborbital frequencies over the last 450,000 years. *Earth and Planetary Science Letters* 212: 197–212.

Vaks A. 2008. Quaternary paleoclimate of northeastern boundary of the Saharan Desert: reconstruction from speleothems of Negev Desert, Israel. *Israel Geological Survey Report* GSI/14/2008, Jerusalem.

Vaks A., Bar-Matthews M., Ayalon A., Matthews A., Halicz L. and Frumkin A. 2007. Desert speleothems reveal climatic window for African exodus of early modern humans. *Geology* 35: 831–834.

Vaks A., Bar-Matthews M., Ayalon A., Schilman B., Frumkin A., Kaufman A. et al. 2003. Paleoclimate reconstruction based on the timing of speleothem growth, oxygen and carbon isotope composition from a cave located in the 'rain shado', Israel. *Quaternary Research* 59: 182–193.

Vaks A., Bar-Matthews M., Ayalon A., Matthews A., Frumkin A., Dayan U. et al. 2006. Paleoclimate and location of the border between Mediterranean climate region and the Saharo-Arabian desert as revealed by speleothems from the northern Negev Desert, Israel. *Earth Planetary Science Letters* 249: 384–399.

Vazquez-Cuervo J., Font J. and Martinez-Benjamin J. J. 1996. Observations on the circulation in the Alboran sea using ERS1 altimetry and sea surface temperature data. *Journal of Physical Oceanography* 26: 1426–1439.

Vergnaud-Grazzini C., Devaux M. and Znaid J. 1986. Stable isotope anomalies in Mediterranean Pleistocene records. *Marine Micropaleontology* 10: 35–69.

Verheyden S., Nader F. H., Cheng H. J., Edwards L. R. and Swennen R., 2008. Paleoclimate reconstruction in the Levant region from the geochemistry of a Holocene stalagmite from the Jeita cave. Lebanon. *Quaternary Research* 70: 368–381.

Viudez A., Pinot J. M. and Haney R. L. 1998. On the upper layer circulation in the Alboran sea. *Journal of Geophysical Research* 103 (C10): 21653–21666.

Volkman J. K., Eglinton G., Corner E. D. S. and Sargent J. R. 1980. Novel unsaturated straight-chain C37-C39 methyl and ethyl ketones in marine sediments and a coccolithophore Emiliania huxleyi, in A.G. Douglas and J.R. Maxwell (eds) *Advances in Organic Geochemistry.* Pergamon: New York, pp. 219–227.

Walker M. J. C., Bjørck S., Lowe J. J., Cwynar L. C., Johnsen S. J., Knudsen K. L., Wohlfarth, INTIMATE-Group, 1999. Isotopic events in the GRIP ice core: a stratotype for the late Pleistocene. *Quaternary Science Reviews* 18: 1143–1150.

Wanner H., Beer J., Bütikofer J., Crowley T.J., Cubasch U., Flückiger J., Goosse H. et al. 2008. Mid- to Late Holocene climate change: an overview. *Quaternary Science Reviews* 27: 1791–1828.

Warning B. and Brumsack H.-J. 2000. Trace metal signatures of Eastern Mediterranean sapropels. *Palaeogeography, Palaeoclimatology, Palaeoecology* 158: 293–309.

Weinstein-Evron M. 1990. Palynological history of the last pleniglacial in the levant, in Kozlowski J. K. (ed.) *Feuilles de Pierre: les Industries à Pointes Foliacées du Paléolithique Supérieur Européen.* Liège: ERAUL, pp. 9–25.

Weiss H. 2000. Beyond the Younger Dryas collapse as adaptation to abrupt climate change in ancient west Asia and the Eastern Mediterranean, in Bawden G. and Reycraft R. (eds) *Confronting Natural Disaster: Engaging the Past to Understand the Future.* Albuquerque: University of New Mexico Press, pp. 75–98.

Weiss H., Courty M. A., Wetterstrom W., Guichard F., Senior L., Meadow R. and Curnow A. 1993. The genesis and collapse of 3rd millennium north Mesopotamian civilization. *Science* 261: 995–1004.

Weldeab S., Emeis K.-C., Hemleben C. and Vennemann T. H. 2002. Sr-Nd isotope and geochemical analysis of lithogenic components in Late Pleistocene sapropels and homogeneous sediments from the Eastern Mediterranean Sea: Implications for detrital influx and climatic conditions in the source areas. *Geochimica et Cosmochimica Acta* 66: 3585–3598.

Weldeab S., Siebel W., Wehausen R., Kay-Christain E., Gerhard S., Hembleben, C. 2003. Late Pleistocene sedimentation in the western Mediterranean Sea: implications for productivity changes and climatic conditions in the catchment areas. *Palaeogeography, Palaeoclimatology, Palaeoecology* 190: 121–137.

Weninger B., Clare L., Rohling E.J., Bar-Yosef O., Bohner U., Budja M. et al. 2009. The impact of rapid climate change on prehistoric societies during the Holocene in the Eastern Mediterranean. *Documenta Praehistorica* XXXVI: 7–59.

Zanchetta G., Drysdale R. N., Hellstrom J. C., Fallick A. E., Isola I., Gagan M. K. and Pareschi M. T. 2007. Enhanced rainfall in the Western Mediterranean during deposition of sapropel S1: stalagmite evidence from Corchia cave (Central Italy) *Quaternary Science Reviews* 26: 279–286.

33

Environmental Change in the Temperate Forested Regions

Matt McGlone, Jamie Wood and Patrick J. Bartlein

1 INTRODUCTION

Recent years have seen a dramatic increase in the resolution, quality and diversity of environmental records from the Quaternary and even more spectacular improvements in databases of global or regional reach, analysis and display of results. Scientific institutes and researchers are concentrated in the temperate latitudes and, as a result, both theoretical and practical ecology – including palaeoecology – is largely focused on the temperate zone. While this should be a major advantage for this review, it presents problems in assimilating and choosing from the superabundance of material, the range of scales at which it is presented and analysed, and the issues it tackles. Concepts derived from engagement with deciduous, broad-leaved and relatively species-poor forest communities growing in previously glaciated terrain have tended to set the agenda even for evergreen, sclerophyll, species-rich communities in unglaciated terrain. Therefore, while acknowledging the huge advances made through this concerted effort over more than 100 years to understand the past dynamics of the vegetation of the land bordering the

North Atlantic, the time has come to integrate and enrich these insights with those from the species-rich temperate regions of China and Japan and the evergreen warm temperate and oceanic regions of the southern hemisphere. An excellent beginning has been made with massive global data syntheses but these invariably work by reducing the dimensions studied and restricting the scope primarily to climate drivers. While it makes good sense to use climate as the *lingua franca* of Quaternary studies, it will be to our detriment if the production of quantified climate estimates and speculations about hemispheric or global climate forcing becomes the only focus. A rich array of techniques for reconstructing past climates are now available, and increasingly climate databases can be used to test hypotheses about vegetation responses.

Trees are crucial to understanding the environments of the temperate zone. They make up most of the living biomass, capture nearly all the carbon and nutrients that subsequently pass through the various trophic layers of the ecosystem and, by virtue of their size, are strongly connected to and modify the climate at all scales from the micro to regional. We are fortunate, therefore, that trees are also by far the best represented taxa in fossil archives: wood, charcoal, leaves, seeds and specialist insect herbivores have left discontinuous but widespread clues to the composition of past forests; pollen of the major canopy trees preserved in lakes, peats and soils provides a more general, but unrivalled continuous record of their population fluctuations.

Our review will therefore concentrate to a large extent on the history of temperate forests and the climate changes that forced their continuous ebb and flow. Geomorphic change, soils, fire, extinction of the megafauna and the impact of humans will be discussed largely from the point of view of their importance for forested ecosystems.

2 TEMPERATE FORESTS: DISTRIBUTION AND ENVIRONMENT

Forests require a growing season with average soil temperatures above 5–7°C

(Körner and Paulsen, 2004) and usually more than 400 mm per year rainfall (Woodward, 1987) and moisture-stress threshold (AE/Pe alpha values) above 0.5–0.6. Low-growing shrubs and herbs of tundra, grassland, steppe and desert communities replace trees outside of this environmental range. Temperate forest is defined by the presence of two other great forest biomes: extensive tropical forests and boreal forests. All three major biomes have wet and dry facies and warmest month temperatures fall within a relatively narrow band (5–30°C). Most forests experience warmest month temperatures >20°C, and this factor only weakly discriminates biomes. Coolest month temperatures have twice the range (–30 to 26°C) and the critical factor is the coldest temperature likely to be experienced in a given year as this has a profound effect on which plants can survive (Woodward, 1987). Minimum temperature boundaries and broad physiognomic descriptors for forest biomes are given in Table 33.1. The distribution of the major biomes is depicted in Figure 33.1.

Table 33.1 Climatic descriptors for moist forest types. Climatic descriptors, annual minimum temperature and expected vegetation physiognomy (modified from Woodward, 1987)

Forest biome	Climatic factors	Minimum temp. (°C)	Vegetation physiognomy
Tropical	Warm and moist with limited dry period	> 15	Broad-leaved evergreen trees dominant; lianas, vines, palms, tree ferns, epiphytes abundant
Subtropical	Warm and moist but near zero temperatures experienced occasionally	−1 to 15	Broad-leaved evergreen trees dominant; broad-leaved tropical conifers common in some areas; lianas, vines, palms, cycads, tree ferns, epiphytes abundant
Warm temperate	Mild winters; moist warm summers	−1 to −15	Broad-leaved evergreen trees; abundant broad- and needle-leaved temperate conifers in some areas; lianas, vines, epiphytes common. Palms, cycads, tree ferns common or abundant in some areas
Oceanic temperate	Cool summers (mean < 20°C); warm winters (mean > 5°C); high rainfall	−1 to −15	Broad-leaved evergreen trees and deciduous broadleaves; abundant broad- and needle-leaved temperate conifers in some areas; tree ferns, lianas, vines, epiphytes common in some areas.
Cool temperate	Cold winters, mild summers; freezing and supercooling of plant tissue	−15 to −40	Broad-leaved deciduous dominant; boreal needle leaved conifers common.
Boreal	Very cold winters, cool summers	< −40	Evergreen and deciduous needle-leaved conifers dominant

Closed temperate forests form where there is sufficient moisture and warmth during summer to sustain a continuous forest canopy with a high leaf area index but where winter is cool enough to halt or markedly slow growth. Temperature seasonality is usually well marked because of variable day length during the year and the boundary between subtropical and warm temperate forest lies at ~25°N or S, where day length seasonality becomes insignificant and annual minima are generally above 0°C and, in the presence of sufficient moisture, growth is possible all year. At the boreal–temperate forest boundary growing season length contracts markedly.

Temperate forests are found in both hemispheres approximately between 55° and 25° (Figure 33.1). Ribbons of now fragmented temperate forests occur along the major mountain chains in all continents (e.g., afro-temperate forests along the Great Escarpment in South Africa; Mucina and Geldenhuys, 2006) but these are poorly documented from an historical perspective. Leaving these smaller areas aside, seven major regions can be clearly distinguished on the basis of their

biota and climates: western North America; eastern North America; Europe; northeastern Asia; southern South America, southeastern Australia and Tasmania; and New Zealand. The broad biogeographic outlines of these areas are set by the size of the continent they occur in, their position relative to the zonal westerly air flow and ocean currents, and orientation and height of their mountain ranges.

Where warm poleward currents flow along the coast, but high latitude continental interiors cool in winter, vast areas of closed deciduous temperate forest form, as in northern and central China and Japan, eastern North America and western Europe. The East Asian monsoon dominates central and eastern Asian climates: cold dry air masses generated in the continental interior penetrate as far south as 22°N in winter and lead to high snowfall in Japan. The Asian summer monsoon and the Pacific subtropical anticyclone advect warm humid air masses from the Pacific Ocean as far north as 54°N. The continental centre of Australia is too small and located too close to the equator to have a

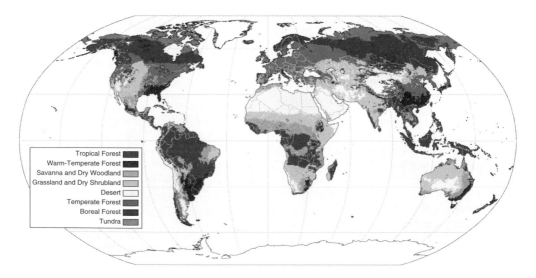

Figure 33.1   Megabiomes.

Source: Megabiome data: Kaplan et al. (2003). *Image*: Department of Geography, University of Oregon. Available at: http://geography.uoregon.edu/envchange; 0.5-degree data.

comparable effect and winter climates of the southeastern temperate coastal zone are mild and evergreen sclerophyll forests dominate. In southeastern South America, high mountains to the west starve the eastern coast of rain, leading to savannah, grassland and desert scrub. In the high pressure subtropical zone of southeastern Australia, southern China, southern Japan and the southeastern US a warm ocean with easterly windflow, augmented by monsoonal rains, creates warm, temperate, often evergreen, broad-leaved forest.

On the western flanks of the continents, where cold currents flow equatorward, a high rainfall temperate forest may form, especially where tall mountains rise above narrow coastal plains, as in southern Chile, the west coast of North America, Tasmania and New Zealand. Here, oceanic climates favour evergreen conifers, broad leaved trees and treeferns. Further equatorward, under the influence of the high-pressure subtropical cells, extremely dry conditions prevail and temperate forests give way to Mediterranean dry forest, scrub and desert as in northern Chile, southwestern Australia and the southwestern US. Summers in the western temperate regions are often dry (Chile, northwestern US) or tend to have less rainfall in summer than winter (southeastern Australia, Tasmania, New Zealand) and this favours dominance by conifers (northwestern North America, New Zealand) or by sclerophylls (central Chile, southeastern Australia, Tasmania). Rainfall is often very high in the windward zones, winter temperatures mild and summers cool, and in these regions we recognize a separate subcategory: *oceanic temperate forest*

3 TERTIARY LEGACY

Global temperatures and atmospheric carbon dioxide (CO_2) concentrations were high in the Pliocene but progressive lowering of CO_2 initiated the variable, cold environment of the Pleistocene around 2.3 million years ago

(see Chapters 8 and 17). While the climates of the current temperate zone were not tropical during the Pliocene, they tended to be warmer and less variable than now. Many plant genera identified from the Pliocene in a given region are locally extinct. Europe lost 89 tree genera in the transition from the late Tertiary to the Quaternary (Latham and Ricklefs, 1993) and, in northwestern Europe, over 30 Pliocene plant genera that survive elsewhere (~90 per cent in Asia; 65 per cent in North America) were eliminated in the course of the Pleistocene (Willis and Niklas, 2004). A similar elimination of plant genera took place in southeastern Australia (Sniderman et al., 2007), New Zealand (Lee et al., 2001) and southern South America (Markgraf and McGlone, 2005). Many of these Pliocene genera eliminated from southern Australasia thrive in highly oceanic subtropical settings or in the higher altitude tropics. Northwestern America retains conifer species common in Pliocene eastern North America but lost hardwood genera still dominant in northeastern Asia and southeastern North America (Waring and Franklin, 1979). *Metasequoia* and *Liquidambar* disappeared from the main Japanese islands and Okinawa (Fujiki and Ozawa, 2008) but persisted in China and eastern North America in areas which today have similar climates.

The ultimate cause of loss of species and genera from temperate forest is the stresses imposed by the cool, dry glacials of the Pleistocene. However, generic diversity of temperate forest trees varies markedly between separate continental areas (Huntley, 1993), with Europe being genus-poor compared with eastern North America, and eastern North America being genus-poor relative to northeastern Asia. Moreover, moist temperate forests in Asia have about three times as many species of tree as similar forests in North America (Latham and Ricklefs, 1993). Huntley (1993) argues that the ultimate cause of the greater loss of tree genera from Europe compared with elsewhere is the greater intensity of the European glaciations that left little suitable habitat for moist temperate forest, leading to stochastic elimination, only partly compensated for by speciation in remaining genera in the more diverse and regionalized landscapes of Europe. Ricklefs et al. (1999) suggest the close geographical connection between the temperate forests of Asia with eastern Asia tropical centres of diversification ensured greater taxon richness than in eastern North America and Europe, both of which have never had close connections with the moist tropics. The temperate forests of the southern hemisphere are of interest in this regard. Those of southeastern Australia and New Zealand have similar or greater numbers of tree genera and tree species than Europe or eastern North America on a constant area basis, while those of southern South America are both genus and species poor (McGlone et al., 2010). Connection with a diverse subtropical or tropical region seems to be more important in maintaining high generic diversity than any other factor. Thus Europe, northwestern North America and southern South America, all sealed off by cold oceans and land poleward, and arid subtropical deserts and steppe equatorward are less diverse than southeast Australia, northeast Asia and northeast North America which have moist coastal fringes running far equatorward into the warm temperate/subtropical zone.

4 CLIMATIC AND GEOMORPHIC CHANGE IN THE TEMPERATE ZONE

As discussed in Chapters 19 and 20, the Quaternary is defined by the cyclical alternation of glacials and interglacials. While each glacial–interglacial cycle differs somewhat, the present cycle, which began after the extreme cold, dry climates of the Last Glacial Maximum (LGM; 30–18 ka BP), followed by the transition through the Late Glacial (18–11.5 ka BP) to the current interglacial Holocene, is sufficiently typical to stand in for the rest.

4.1 Last Glacial Maximum (30–18 ka bp)

The LGM is of critical importance to our understanding of the temperate forested zone. While there is now cause to doubt that elimination of warm temperate biotic elements was as complete or as widespread as once claimed, LGM climate and geomorphic change reset the ecological clock over most of the temperate zone.

Atmospheric concentrations of CO_2, the fundamental substrate for life, were reduced globally by about one-third (see Chapter 19). Most flowering plants react to low CO_2 by increasing the number of stomata and the length of time they remain open (Ward et al 2005), thus using more water and drying soils.

Mean annual temperatures fell by 4–8°C in mid-latitudes, with centres of continents cooling more than coastal districts (Whitlock et al., 2001) and winters at high latitudes cooling very much more than summers. For instance, estimated temperature anomalies in Europe north of the Alps are –30°C for the coldest month, and –12°C on an annual basis (Peyron et al., 1998). Ice sheets grew, with two massive coalesced sheets (the Cordilleran and Laurentide) extending over North America as far south as 40°N and the Eurasian ice sheet descending to around 55–50° N in the west and 70–65° N in the east. Strong glacial anticyclonic systems developed as a consequence. Mountain glaciers expanded throughout the temperate zone, with snow lines lowering by 800–1200 m or more (see Chapter 10). Ice piedmonts formed in the southern South American and New Zealand cordilleras. Extensive zones of periglacial activity formed along the margins of the Laurentide and Eurasian ice sheets: continuous permafrost extended as far south as ~40–45°N, in the eastern US, Europe and northeast Asia (Williams et al., 1993). Glacier erosion and stripping of regolith by freeze-thaw activity provided a vast amount of sediment and rivers and streams aggraded throughout creating vast outwash plains and terraces (Bridgland and Westaway, 2008). These processes were more vigorous in the northern hemisphere as permafrost or freeze-thaw zones were limited in the southern hemisphere to areas immediately adjacent to ice margins. Loess deposits, windblown dust from exposed outwash plains and riverbeds, are extensive in the more poleward areas of the temperate forest zone and thick deposits formed in drier regions (Pye, 1995. Overall, the cool temperate zone throughout was disrupted by glacier expansion, active soil processes including stripping of regolith by freeze-thaw and rain on exposed soils, aggradation and loess deposition (Munyikwa, 2005). In contrast, outside of the major river plains and high altitude mountain areas where valley glaciers formed, the warm temperate zone retained old soils and experienced little soil rejuvenation.

Precipitation did not follow a universal trend as it depends to a large extent on the prevailing temperature as well as position within a continent, the latitude of storm tracks, monsoonal status and warmth of the adjacent ocean. Overall, the general conclusion is that temperate areas were drier than the present day, a consequence of 'ice-age boundary conditions' including the direct effects of the ice sheets themselves on albedo, in addition to lower concentrations of atmospheric CO_2, cooler oceans, equatorward displacement of the northern westerlies (but little displacement but weakening in the southern hemisphere (Rojas et al., 2009), strengthening of the Asian winter monsoon (Stebich et al. 2007) and weakening of the Asian summer monsoon (Kim et al., 2008; Yu et al., 2000). Where westerly winds intrude on the western flanks of the continents, their equatorward displacement by 5–10° latitude (Bradbury et al., 2001; Iriondo, 1999) brought increased rainfall to areas currently semiarid (southwestern US; central Chile) and maintained high rainfall along the moist temperate western coast of New Zealand while creating dry, steppe-like conditions to the east (Shulmeister et al., 2004). In western Europe, strengthened westerly winds brought very cold, dry air off the expanded sea ice in the North Atlantic (Huijzer and Vandenberghe, 1998) and southwards deflection of westerlies to the Mediterranean region led to winter rain and dry summers (Allen et al., 2000; Harrison et al., 1996). The strengthened Asian winter monsoon and weakened summer monsoon brought extreme winter cold and arid conditions out to the present coastline in northeastern China between 30 and 40°N (Yu et al. 2000).

4.2 Late Glacial: the glacial–interglacial transition (18–11.6 ka BP)

The earliest indications of warming after the LGM are seen in glacier retreat and expansion of arboreal vegetation occurring c.18 ka BP in western districts of New Zealand (Newnham et al., 2007) southern South America (Denton et al., 1999), southern Japan (Tsukada, 1988) and the southern US (Davis, 1983). In the high latitudes of the northern hemisphere, warming was registered on the terrestrial environment somewhat later at c.16 ka BP followed by a rapid warming to a warm peak (Allerød: 13.8–13 ka BP); but it was interrupted by a deep prolonged cooling (Younger Dryas: 12.9–11.7 ka BP; Steffensen et al., 2008) during which temperatures over much of eastern North America and northern Eurasia became much cooler or, in areas immediately east of the north Atlantic, returned to near glacial levels (Stocker, 2000). With the beginning of the Holocene at 11.6 ka BP, temperatures rapidly rose to current or warmer than the present day by around 10–8 ka BP (Davis et al., 2003). In the southern hemisphere there is no marked reversal in the warming trend, although a number of records show a slight cooling or plateau centred on 13.5 ka BP (the Antarctic Cold Reversal). In the southern hemisphere the period 12.6–11.6 ka BP was registered as a steady warming which culminated at the same time as in the northern hemisphere.

4.3 Holocene (11.6 ka BP to the present)

There is a long tradition of dividing this period into three phases: early, mid and late. This tripartite division has proved robust for the temperate zone, and we here follow the chronology (early, 11.6–9.0 ka BP; mid, 9.0–5.0 ka BP; and late, 5.0 ka BP to the present) of Wanner et al. (2008). North America and Europe have extensive data sets of many hundreds of sites with a variety of physical and biological proxies covering the late glacial and Holocene (Davis et al., 2003; Williams et al., 2004). Exploration of these data sets in conjunction with a range of climate and biome models has demonstrated how complex and variable the interaction between climate, land and biota has been.

The early Holocene (known as the Preboreal and Boreal in Europe) in the northern hemisphere mid latitudes had the twin influences of slowly shrinking but still influential continental ice sheets (until c.7 ka BP) and high summer and low winter insolation. Over much of the northern hemisphere this is a period of rapidly warming but lower mean annual temperatures than now (winter temperature anomalies relative to the present much larger than those of summer) and increasing rainfall. An exception is southern Europe where summer and winter temperatures fell from 11 ka BP to 8 ka BP (Davis et al., 2003; Davis and Brewer, 2009) In temperate northeast Asia, the summer monsoon became influential at c.11 ka BP and brought higher precipitation until c.5 ka BP, although the timing of its maximum influence varied from region to region (Guiot et al., 2008). In northwestern North America, the early Holocene was dry and warm (Barnosky et al., 1987).

The southern hemisphere experienced low summer insolation levels in the early Holocene, but there was a trend towards generally rising temperatures in southern South America, Australia and New Zealand. New Zealand appears to have had warmer, less seasonal climates but drier climates than the

present day(Wilmshurst et al., 2007) while Australia was continuing to warm but was still cooler than the present day (McKenzie, 2002). In southern South America where the early Holocene (11–7 ka BP) experienced a much higher level of fire (Huber et al., 2004; Whitlock et al., 2007), drier, warmer conditions occurred from 30° to 54°S (Abarzúa et al., 2004), influenced by an apparent absence of strong westerly winds. A similar westerly minimum with warmer temperatures than the present day occurred on the other side of the Pacific in the Australasian sector centred on 11–8 ka BP (Shulmeister et al., 2004; Wilmshurst et al., 2007). Mid-continental and northwestern America had drought-like conditions between 11 ka BP and 8 ka BP while eastern and southern North America experienced moist conditions (Barnosky et al., 1987; Shinker et al., 2006).

The mid Holocene (9 ka BP to about 5–6 ka BP) was the period of warmest global mean annual temperatures although not in all regions or all sites. As the northern and southern hemispheres had opposite trends in seasonal insolation, direct local effects cannot explain the global pattern. Some, but not all, southern hemisphere records suggest a warmer early Holocene than mid Holocene (Abarzúa et al., 2004; Wilmshurst et al., 2007). In Europe, northern and east central regions had a clear thermal maximum between 8 and 4 ka BP, but southern regions either had stable temperatures close to those of the present or warmed towards the present (Davis et al., 2003; Davis and Brewer, 2009), thus showing a contrary trend to the insolation forcing. In northeast Asia, peak precipitation occurred in the mid-Holocene with the maximum strength of the summer monsoon. In Europe, northern latitudes were somewhat drier (Harrison et al., 1996) but became wetter towards the present. During the mid Holocene, precipitation increased in southern South America, Australia and New Zealand, a consequence of the re-establishment of strong westerly wind flow at its current latitudes (Shulmeister et al., 2004).

The late Holocene (5–6 ka BP to preindustrial) is unambiguously cooler than the mid Holocene in many regions, with marine cores and ice cores recording strong cooling, and mountain glaciers advancing in both the northern and southern hemispheres (Wanner et al., 2008; Shulmeister et al., 2004). In North America the earlier dryness of the mid Holocene that affected the mid continent fades over this period whereas in northeast Asia, the East Asian monsoon weakened and precipitation fell and desert and steppe advanced. Southern South America saw a progressive strengthening of westerly circulation and thus wetter conditions. Australia experienced a cooler, drier, late Holocene (Shulmeister et al., 2004). In New Zealand the situation is complex, but the late Holocene was cooler overall, but with increased seasonality leading to dry warm summers and cool wet winters over much of the country (McGlone et al., 1993; Wilmshurst et al., 2007).

5 VEGETATION RESPONSES TO CHANGING INSOLATION, CLIMATES, AND CO$_2$

In Tables 33.2–33.4, vegetation response to environmental change is summarised for representative areas and sites from all seven temperate forest regions.

5.1 Last Glacial Maximum

Cold (extreme during winter close to the extended continental ice sheets), drought and lower atmospheric CO$_2$ concentrations all tended to reduce biomass productivity and plant stature and severely restricted the species that could persist in temperate regions. Steppe and tundra extended across Eurasia, with severe, near treeless conditions extending from western Europe across to the Sea of Japan. In China, steppe extended as far south as 33°N with the northern boundary of warm

temperate forest at c.25°N (Yu et al., 2000). Japan remained almost completely forested, with a range of mixed conifer, deciduous and evergreen broad-leaves in the south and boreal conifer forest in the north. In North America a tundra zone sat close to the ice margin, but *Picea* and *Pinus* parklands and woodlands dominated elsewhere.

The central Australian open sclerophyll woodlands, heaths and grassland expanded to the eastern periphery and further south, in Tasmania, montane grasslands and heaths dominated. In New Zealand *Nothofagus*-dominant mixed broad-leaved conifer forests grew north of 37°S, but various shrubland, grassland and steppe associations covered the landscape to the south. In southern South America *Nothofagus* and conifer woodlands grew in the north (36°S – 43°S), and heaths and moorland to the south.

The extent to which species currently present in a given area survived *in situ* during the LGM has been much debated, especially in Europe and eastern North America. The existence of refugia where temperate plants and animals survived the LGM has been postulated. Complex post-LGM migratory pathways have been proposed for many species in Europe on the basis of stepwise expansion interpreted from the pollen record but more recently on the basis of the distribution of haplotypes or genetic changes or for a range of taxa (Hewitt, 2004). Few refugia have been documented via macrofossils or pollen: instead, mountainous topography, the presence of key species, and the genetic diversity of those species have been used to identify likely refugial regions. In Europe, the Iberian Peninsula, Italy and the central Balkans have been often suggested as the key locations for refugia. In eastern North America, the supposition has been that populations of temperate species persisted to the south. Discovery of macrofossils of trees well north of the permafrost line and evidence for diverse conifer forest patches on the central European plain in Europe, and similar evidence for temperate forest elements far north in North America (Provan and Bennett, 2008) have

Table 33.2 North America and Europe. Vegetation history from selected sites and regions. Data from: 1. Barnosky et al. (1987); 2. Williams et al. (2001); 3. Davis (1983); 4. Berglund et al. (2008); 5. Tantau et al. (2006). 6 Willis (1994)

Standard chronology	Northwestern North America (40–55°N)[1]	Northeastern North America (40–45°N)[2]	Southeastern North America (40–25°N)[2-3]	Northern Europe (47–60°N)[4]	Southern Europe (40–47°N)[5-6]
Late Holocene 6–0 ka	As below	6 ka: Boreal elements increase; *Betula* and *Fagus* peak	6 ka: Increasing conifers (Cupressaceae, *Tsuga*)	From 3 ka: *Tilia, Corylus* and *Ulmus* decline; *Picea* and *Fagus* expand	Expansion of *Carpinus* (6 ka), *Fagus* and *Abies* (3.5 ka). Decline of *Corylus, Fraxinus* and *Ulmus*
Mid Holocene 9–6 ka	8–5 ka: *Thuja plicata* and *Tsuga heterophylla* expand	*Tsuga Betula, Fagus Carya Castanea* increase	Coastal plain Mixed deciduous broadleaf forest: *Quercus, Pinus, Liquidambar, Carya*	9–4 ka: *Pinus* slow decline: *Tilia* and *Fraxinus* spread	9–6 ka: *Corylus-Quercus* mixed deciduous forest
Early Holocene 11.6–9 ka	9.8–7.8 ka *Pseudotsuga* north of current range. Treeline 60–130 m higher. 11–10.2 ka: transition to modern forest: *Pseudotsuga-Quercus-Corylus*	*Pinus* (early peak) *Quercus* and *Betula* dominance	10–9 ka: Peak *Quercus, Ostrya/Carya* and *Fraxinus*	11.5–9 ka: Mixed deciduous forest of *Pinus, Betula, Corylus, Alnus, Quercus*	11.5–9 ka: Mixed forest *Betula, Pinus* and *Picea*; expansion of *Pistacia, Ulmus, Fraxinus, Quercus Tilia, Corylus*
Late Glacial 18–11.6 ka	13–11 ka: *Pinus* dominance. 15.7–12.4 ka: Expansion of *Pseudotsuga, Alnus* and *Pteridium*	16–11.5 ka: Progressive establishment of cool mixed deciduous forest. *Ulmus* and *Fraxinus* peak c.12 ka	16–13 ka: Decline *Pinus*; increase mesic deciduous forest on coastal plain; peak *Liquidambar*, open *Quercus* woodland Florida	12.7–11.5 ka: Decline woodland, increase steppe 13.8–12.7 ka: Decline steppe-tundra, increase *Pinus Betula-Salix-Juniperinus* woodland 14.5–13.8 ka: Highly diverse steppe-tundra – *Pinus, Betula* and *Salix* present. Pre-14.5 ice	12.4–11.5 ka. Decrease trees – re-expansion of steppe 16–12.4 ka: Open *Pinus-Betula* woodland
Full Glacial 30–18 ka	*Picea* and *Pinus* parkland; closed montane *Pseudotsuga* forest	*Picea* woodland and parkland; tundra close to ice margin	*Pinus-Quercus* open woodland and grassland	Glacial ice, permafrost	Herb steppe with patches of *Pinus* and *Betula*

Table 33.3 Asia and southern South America. Vegetation history from selected sites and areas. Data from 1. Yu et al. (2000); 2. Dearing et al. (2008); 3. Stebich et al. (2007); 4.Tsukada 1988; 5. Takahara et al. (2000); 6. Kawhata et al. (2009); 7; Haberle and Bennett (2004); 8; Abarzúa et al. (2004). 9 Rabassa et al. (2000)

Standard chronology	China (36-25°N)[1,2]	China (42°N)[1,3]	Japan (32°N)[4,5]	Japan (40-44°N)[4,5,6]	South America (41-44°S)[7,8]	South America (53°S)[9]
Late Holocene 6-0 ka	6–4.8 ka: Declining broadleaved evergreens; further decline *Tsuga*	5.2 ka: Spread of *Pinus* and *Betula*	Subalpine conifers descend; *Cryptomeria* moves northwards.	Little change from 6 ka	Mosaic Valdivian and North Patagonian forest	*Nothofagus* forest
Mid Holocene 9-6 ka	8.5–6 ka: Broadleaved evergreen forest dominant; reduced *Tsuga*	Broadleaved deciduous forest dominant; further north and higher altitudes than present	8–4.5 ka: Warm temperate broadleaves expand beneath mid temperate conifer forest. *Fagus* expands north; upper limit 400 m higher	*Castanea* and *Quercus* sbg *Cyclobalanopsis* peak	7.0 ka: Resurgence of conifers;	7 ka: Peak expansion of *Nothofagus*
Early Holocene 11.6-9 ka	*Tsuga* peaks 9.5–8.5 ka: Decline *Betula* and deciduous *Quercus*. Broad-leaved evergreens spread	11.3 ka *Ulmus*, *Quercus* and *Juglans* rise to dominance	11.5–8 ka: Broadleaved evergreens increase, but *Pinus*, *Picea*, *Tsuga* and *Betula* remain common	11.5-8 ka: Deciduous broadleaves expand; *Picea*, *Pinus* and *Betula* decline abruptly	12–7 ka: Valdivian rainforest dominant (*Eucryphia* and *Caldcluiva* and *Weinmannia*). Conifers decline	11.5 ka: Increase in *Nothofagus*
Late Glacial 18-11.6 ka	12 ka rise of *Betula* and deciduous *Quercus*	14.4–13.7 ka: Transition from steppe to *Betula Ulmus* and *Fraxinus* forest. 12.6 ka Spread of *Picea* and *Larix*	19–11.5 ka: Decline of *Abies* and *Picea*, rise of deciduous *Quercus*, *Tilia* and *Ulmus*	19–1.5 ka: Decline of *Betula*, but otherwise little change	15–12 ka: Cold tolerant podocarps expand. 17.5–15 ka: Expansion evergreen *Nothofagus* Patagonian rainforest	Grassland and heath
Full Glacial: 30–18 ka	Grassland-Artemisia steppe.	Grassland-Artemisia steppe	Mixed temperate conifer, deciduous broadleaf and warm-temperate evergreen broadleaf	Boreal conifer forest	*Nothofagus* parkland; cupressoid conifers and moorland	Grassland and heath

Table 33.4 Australasia. Vegetation history for selected sites and areas. Data from: 1. Pickett et al. (2004);. 2 Black et al. (2006); 3 McKenzie (2002); 4. Macphail (1979); 5. McGlone et al. (1993); 6 McGlone (2002)

Standard chronology	Australia: Victoria c.35–40°S [1,2,3]	Australia: Tasmania c.40–43°S [1,4]	New Zealand: North Island (38°S) [5]	New Zealand: South Island (43°S) [5]	Subantarctic Auckland Is. (51°S) [6]
Late Holocene 6–0 ka	Wet sclerophyll forest	5 ka: More open forest structure.	4 ka: Decline of moist temperate forest species	Continued expansion of *Nothofagus* in south	3 ka: Continued upslope forest movement
Mid Holocene 9–6 ka	9.5–4.5 ka: Alpine taxa gone; *Pomaderris* (understorey shrub) abundant	7 -5 ka: Shade intolerant rainforest and wet sclerophyll spread. *Pomaderris* abundant.	8 ka: Expansion of *Agathis* and *Nothofagus*.	7.5 ka: *Nothofagus* supplants podocarps	6 ka: *Metrosideros-Dracophyllum* forest
Early Holocene 11.6–9 ka	11.5–9.5 ka: Continued expansion of wet sclerophyll	10.8–7 ka: Rise to dominance of *Nothofagus*	11.5–7 ka: Increasing tree ferns, *Metrosideros* and other evergreen broadleaves	11.3–9.3 ka: Tall podocarp forest	9–6 ka: Transition to forest.
Late Glacial 18–11.6 ka	14–11.5 ka: 19.5 ka: Increase in wet sclerophyll *Eucalyptus* forest	13.4–10.8 ka: Spread of *Eucalyptus* and *Phyllocladus*. 16.5–15 ka: Local establishment of forest, *Phyllocladus*, *Nothofagus* and *Eucryphia*	18 ka: Rise of podocarp dominant conifer-broadleaf closed forest; elimination of *Nothofagus*	13.6 – 12.8 ka: Progression via conifer woodland to tall podocarp dominant forest. 16.4 ka: Expansion of forest patches	13.5 – 9.0 ka: Progression from tundra via grassland to tall shrubland
Full Glacial 30–18 ka	Tundra and steppe to open *Eucalyptus* and *Casuarina* woodland	Subalpine heathlands and herbfields	*Nothofagus*-scrub parkland	Grassland-low shrubland; scattered stands of forest	Tundra

begun to change this simple refugia-expansion model.

In the northwestern US and western Canada, the fact that the current temperate forest area is surrounded by semi-arid landscapes and was bordered by ice sheets to the north has always made LGM survival *in situ* the only tenable solution. In the case of New Zealand, southern South America and Asia, extensive tracts of conifer-broad-leaved forest either are clearly seen in the fossil record in equatorward locations or are highly likely to have existed there on the basis of other evidence. Therefore, LGM refugia, although often suggested for individual species on the basis of phylogeography, are regarded as less important than in other temperate forested areas. LGM vegetation is visualised in these areas as a poleward attenuation of temperate forest, combined with the survival of most species throughout their current range in networks of small protected patches (Markgraf and McGlone, 2005).

5.2 Late Glacial

The universal response to the late glacial climate shifts was increasing amounts and density of closed forest and progressive loss of previous tundra, steppe or parkland forests, or transformation of closed forest where it existed. The earliest forest spread into forested or nonforest glacial associations is recorded in Japan, New Zealand and southern South America shortly after 19–18 ka BP. In southern Japan, previous mixed temperate to warm temperate forests saw a decline of conifers and expansion of broad-leaves and in the north, *Betula* declined in the boreal forest. In far northern New Zealand podcarps replaced *Nothofagus*. By 13 ka BP, mid-latitude temperate forests comparable to those of the present (podocarp dominant in New Zealand; *Nothofagus*-conifer in southern South America) existed but usually consisted of light-demanding taxa favoured by immature soils. In northwestern North America, a conifer expansion beginning around 16 ka BP

led to a *Pinus* dominant phase by 11.5 ka BP, and in the east, mesic deciduous forests expanded around the same time to occupy the southern and central coastal plains and mixed deciduous forest grew further north (Shuman et al., 2009).

The situation in Europe is more complex because the Younger Dryas cooling (12.7–11.5 ka BP) led to a reversal and partial replacement by glacial tundra and steppe of the widespread *Pinus-Betula* woodlands that had developed from 16 ka BP.

5.3 Early Holocene (11.6–9 ka BP)

By the early Holocene, closed forest was dominant throughout the temperate zone and, although some cool climate taxa remained they tended to diminish rapidly. Only a few high latitude regions remained without forest (i.e., far southern South America and New Zealand subantarctic islands). In most regions, the early Holocene forest assemblages are distinctive, lacking some later elements and having others in abundance. In many areas the dominant trees of the early Holocene are those typically associated with successional communities and immature soils. For instance, in the northern hemisphere early arriving taxa such as *Alnus, Betula, Populus, Pinus* favour disturbed or stressed sites, while *Corylus, Fraxinus, Ostrya, Ulmus* favour rich soils. The distinction is not so clear in southern hemisphere sites, but in southern South America and southern New Zealand, light-demanding, long-lived pioneers such as the evergreen broad-leaved *Weinmannia trichosperma* and the tall podocarp *Prumnopitys taxifolia* that favour immature soils and disturbed sites are prominent during the early Holocene. Drier early Holocene climates in most regions, and peak summer insolation in the northern hemisphere and low summer insolation in the south, had an influence on what trees would predominate. Hence drought-resistant *Pinus* dominated in the early Holocene in the Pacific Northwest of North America and fire-tolerant *Weinmannia* dominated in southern Chile. While precipitation was still lower than the present day in New Zealand and southeastern Australia, increasing rainfall combined with low summer insolation favoured stress-intolerant understorey trees such as *Ascarina lucida* in windward and northern New Zealand, and the spread of wet sclerophyll forest in southern Australia.

Because of the rapidity with which the early-Holocene transition occurred, if there were migratory lags, they should be most evident in this period. In Europe particularly, some trees that became hyperabundant in the later Holocene (e.g., *Fagus*; Tinner and Lotter, 2006) may have been absent in the early Holocene. However, in most regions, taxa that expand later can be shown to have been present from the beginning, presumably because they persisted throughout the LGM.

5.4 Mid Holocene (9–6 ka BP)

In most northern hemisphere regions increasing rainfall and maintenance of warm summer and rising winter temperatures led to a maximal expansion of broad-leaved forest. Higher than present-day tree lines were widespread in the northern hemisphere during the mid Holocene (Wanner et al., 2008). However, as there are few reports of higher treelines in the early to mid Holocene in the southern hemisphere, treelines are probably now close to their maximum elevation. The slow expansion of shrub and forest at their limit in the subantarctic islands (McGlone, 2002) and Patagonia (e.g., Huber et al., 2004) suggests that summer temperatures and precipitation were gradually increasing over this time interval at high latitudes.

5.5 Late Holocene (6 ka BP to the present)

In the northern hemisphere there are striking indications of less favourable climates for temperate broad-leaved forest and increasing

boreal elements in northern China, northwestern and northeastern America, northern Europe and Japan. The decline of stress-intolerant broad-leaved species in Australia and New Zealand and increase of stress-tolerant sclerophylls, *Nothofagus*, and conifers, along with some evidence for rising treelines, points to increasing seasonality. In southern Chile, cooler, seasonally wet climates favoured dominance by dense *Nothofagus–Pilgerodendron* forests.

6 SOIL CHANGES AND VEGETATION

While climate alters little over short distances, soil texture, depth and nutrient status are often highly variable at the same scale. Neoecologists therefore pay a great deal of attention to soil factors and less to climate. This bias was reflected in early thinking about the glacial to late Holocene progression of forest types in mid- to high-latitude northern forests and culminated with Iversen's theory of postglacial retrogressive vegetational succession (Iversen, 1964), later developed in detail (Birks, 1986). The essence of the model is that a glacial–interglacial climatic cycle can also be understood as in part a soil development cycle. Under cool glacial climates soils are immature but base-rich (cryocratic phase), and with postglacial warming and increasing plant biomass, soils accumulate more nitrogen and phosphorus (protocratic phase) reaching maximum fertility and above ground biomass in the mesocratic phase. In the retrogressive part of the phase, decreasing soil fertility and in particular limitation of phosphorus through leaching, and cooling climate gives rise to less biomass and degrading soil structure (oligocratic) and finally open vegetation and infertile soils (telocratic).

Direct evidence for the soil-cycle hypothesis comes from palaeoliminological studies in which leaching of calcium phosphate is shown to rapidly deplete the bases in catchment soils leading to progressive lake acidification (Boyle, 2007). As the apatite mineral is the only abundant source of phosphorus, this landscape change is irreversible as long as the soils and underlying regolith remain undisturbed. Studies of plant–nutrient relationships in soil chronosequences confirm the soil trends showing a steady decrease in phosphorus, a rise and then decline of nitrogen, and decreasing biomass and plant diversity in the low phosphorus and nitrogen terminal phase (Richardson et al., 2004). Sites and regions with unstable terrain (e.g., river floodplains, earthquake-prone and mountainous areas, coastal dune fields), and those receiving regular aerial nutrient subsidies from marine, volcanic or loessic sources should show lesser soil maturation effects.

It is difficult to separate the effects of soil maturation and climate over the course of an interglacial–glacial cycle. The retrogressive phase inevitably coincides with the deteriorating climates of the end of an interglacial and boreal trees favoured by cooler climates also tend to be tolerant of low-nutrient soils. Some of the long-term vegetation changes seen in temperate areas during the Holocene are consistent with the soil-cycle hypothesis; for instance, decline of *Tilia*, *Ulmus* and *Corylus* and expansion of *Betula*, *Quercus* and *Picea* in northwest Europe and northeast Asia and decline of podocarp-broad-leaved forest and expansion of *Nothofagus* and *Agathis* in New Zealand. In some cases (e.g. *Picea* and *Nothofagus*) we cannot exclude the effects of changing seasonality (reduced seasonality in the north; increased seasonality in the south). *Fagus* is an interesting case as it expanded rapidly during the late Holocene in Europe, North America and northeast Asia (Liu et al., 2003; Huntley et al., 1989) although it has no particular affinity with poor soils, probably because it is favoured by low seasonality and high humidity.

7 FIRE

Fire is a potent factor for shaping ecosystems. Fire releases nutrients stored in above-ground biomass, creates opportunities for light-demanding plants and, depending on its frequency, favours a range of strategies such as rapid growth and early maturation, highly dispersible seeds, resprouting from below-ground tubers or rootstocks, heat- or smoke-released above-ground seedbanks, fire-resistant features such as thick bark, or fire-promoting features such as loose bark that provides fire ladders and highly flammable foliage. Fire, if frequent enough, can promote a switch from a stable, fire-resistant community to one of a different composition that is fire-prone.

Fire depends on three factors: flammable fuel, ignition and fire weather that dries fuel and promotes fire spread. Temperate forest regions have high biomass but their closed canopies reduce the drying out of fuel at ground level and thus do not favour frequent fire, but when it occurs it is more likely to be a damaging hot, crown fire (Bowman et al., 2009). During the LGM, fire frequency was low on a global level, ramped up from 18 ka BP to 10 ka BP, and has remained relatively stable ever since (Power et al., 2008). Fire appears to have peaked or stabilised at current intensity in the mid to high latitudes of the northern hemisphere between 12 and 7 ka BP, and is thus aligned with peak summer insolation. However, a similar peak is seen in southern South America over the same interval (in a summer insolation trough), while fire frequencies are low in Australia and New Zealand but peak later after 5 ka BP. Increasing fire frequencies in Australasia are linked with increasing ENSO events, decreases in South America with the position of the westerlies, and a second peak in fire in the late Holocene with weakening of the monsoon in east Asia and northeastern United States (Power et al., 2008). Recent global changes in fire frequency have been clearly mainly controlled by human activity, with a large increase in biomass burning at *c.*AD 1750 related to agricultural intensification and a decline after AD 1870 (despite warming climates and rapidly increasing populations) as a result of spread of fire management techniques and philosophy (Marlon et al., 2008).

Fire is often used to explain changes in vegetation types but recent work demonstrates both are independently and intimately associated with changes in climate and disturbance (Higuera et al., 2009). As an example, eight decades of debate have ensued as to whether the spread of *Fagus silvatica* and *Abies alba* was, among other factors, promoted by fire disturbance of pre-existing forests (Tinner and Lotter, 2006). Detailed studies show fire was a consequence of climate-driven vegetation changes, not the result of climate-generated fire regimes which transformed the vegetation. Similar conclusions have been reached over the changes in abundance of *Casuarina* and Myrtaceae in southeast Australia (Black et al., 2006).

8 DISEASE, INSECT ATTACK AND FOREST TREES

Large-scale synchronous declines and then recovery of once abundant trees have been a feature of the Holocene. Debates over whether pathogen and pests caused declines of forest canopy trees have persisted for 40 years, despite northwest Europe and the northeastern North America being the most intensively investigated regions on earth, showing how hard it is to conclusively prove single factor explanations in Quaternary biology. Whether this is because multicausal drivers are more usual, or because of limitations of the evidence that can be brought to bear, is hard to tell.

The mid-Holocene decline of *Ulmus* in Europe and *Tsuga canadensis* at about the same time in northeastern North America are the most discussed examples, and share an impressive number of hypotheses (largely variations on human interference, climate, pests and disease) as to their causation (Birks, 1986). However, the increasing prevalence of outbreaks of pest and microbial pathogens in modern forests, and the startling rapidity with which they can decimate or eliminate abundant canopy trees, has given credence to explanations involving past microbial or pest attacks. Fungal pathogens spread by a beetle is widely accepted as an explanation for the *Ulmus* decline (Clark and Edwards, 2004) but Foster et al. (2006) have demonstrated that there were synchronous changes in climate and other trees over the *Tsuga* decline interval (5.5–3.3 ka BP) and they believe that an abrupt drying of the climate is likely to be the major driver, with pathogens or pests only secondary, if that.

9 LOSS OF THE MEGAFAUNA

The late Quaternary extinctions of many large animals (often called megafauna) are well documented events. A debate as to whether extinctions arose as a result of over-hunting by humans, effects of changing climates on preferred habitats, or some combination of the two, has continued with undiminished vigour in all continents for more than four decades. Whatever the cause, the collapse of megafaunal populations in the temperate Americas, New Zealand, and probably Australia, was rapid and, in the case of New Zealand, undoubtedly human-caused. The situation in Eurasia is different as megafaunal extinctions extended over many thousands of years (Table 33.5) and possibly involved a more complex chain of causation (Barnosky 2008; Barnosky et al. 2004).

9.1 Ecological consequences of megafaunal extinctions

Large herbivorous mammals and birds are found throughout most of the world's temperate forests. The loss of many of the largest herbivores from these forests during the Pleistocene has undoubtedly had a major effect on vegetation composition, structure and ecosystem functioning (Johnson, 2009). Most important in this regard are the mega-herbivores (animals with a body mass greater

Table 33.5   Late Quaternary extinctions of megafauna from temperate regions. Dates from Barnosky (2008) and Worthy and Holdaway (2002)

	Region	Main extinction period	Examples of extinct large herbivores
Northern hemisphere	Eurasia/Beringia	Two pulses: 48–23 ka and 14–10 ka.	Giant deer, auroch, mammoth, woolly and giant rhinoceroses
	North America	Began 15.6 ka, concentrated between 13.5–11.5 ka	Mastodon, mammoth, camels, shrub-ox, giant beaver
Southern hemisphere	South-east Australia	c.50–40 ka	Giant kangaroos, *Palorchestes*, *Zygamaturus* (Diprotodon)
	Southern South America	12–8 ka	Ground sloths, *Toxodon*, *Glyptodon*, *Macrauchenia*
	New Zealand	0.75 ka	Moa (9 spp.), large flightless geese

than 1,000 kg) which escape predation regulation through sheer physical bulk, and shape their habitat through their size and browsing intensity (Owen-Smith, 1988). While all continental temperate regions had megaherbivores, some areas and all temperate islands lacked them (e.g., in New Zealand, the largest herbivores were less than 200 kg). Nevertheless, as the largest herbivore in any ecosystem will have ecological effects not replicated by much smaller herbivores, elimination of a stratum of large herbivores will have major ecosystem consequences (Hansen and Galetti, 2009).

Impacts of large herbivores on vegetation are substantial. Large browsing animals eat coarse, low-nutrient plant tissue as well as the higher nutrient and often softer components to which smaller herbivores are mostly restricted. In general, they eat more of an individual plant and more in one feeding bout. Larger animals feed from ground level to four or more metres, and thus trees and shrubs are at risk for a longer period until they grow beyond the browse zone. Being large, they have greater ranges and more general habitat use. Large herbivores repress regeneration of vulnerable plants, thus favouring better defended competitors which are often slower growing, with less nutrients and tougher foliage, and thereby alter vegetation community structure and nutrient cycling (Coomes et al., 2006). Through reducing understorey biomass and cropping grass large

herbivores will reduce fuel loads and decrease the intensity and frequency of fire (Burney et al., 2003). An increase in fire severity has been recorded following megafaunal extinction in northeastern North America c.13 ka BP (Gill et al. 2009; Robinson et al. 2005). A similar effect was proposed for Australia by Flannery (1990) but, although megafaunal extinction was followed by an increase in shrubs, there was no consistent change in fire severity or frequency (Johnson, 2006). The effects of extinct megafauna on soil structure were probably significant. Large herbivores will compact or cut-up soils in areas of concentrated grazing, and also play an important role in the cycling of nutrients via deposition of dung in that they rapidly recycle and concentrate nutrients (Harrison and Bardgett, 2008).

9.2 Evolutionary relationships

Megafauna spurred the evolution of anti-browse adaptations in plants and coevolutionary relationships such as seed dispersal. Many trees have lost their megafaunal dispersers. Janzen and Martin (1982) defined the 'megafaunal dispersal syndrome' in which trees produce large, indehiscent fruit rich in sugars, oils or protein, which lack obvious abiotic dispersal mechanisms, have thick endocarp to protect seeds during passage through a gut and the fruit fall prior to ripening to promote use by earthbound dispersers. Suggested megafaunal dispersal syndromes and defence anachronisms are most common in tropical forests or grasslands, although there are some temperate forest examples. A large fruit megafaunal dispersal syndrome has also been identified in Australia, linked to the dromornithids; giant herbivorous birds related to ducks and geese (Murray and Vickers-Rich 2004). In the forests and forest edges of eastern North America, trees with anachronistic spines include hawthorn (*Crataegus* spp.) and devil's walking stick (*Aralia spinosa*), both thought to be defences against ground sloths and mastodons (Barlow, 2000). Giant browsing birds in New Zealand may have led to the evolution of tough-stemmed, small-leaved divaricating shrubs and long-spined herbs (Lee et al., 2010).

Megafauna were also important dispersal agents for herbs with dry, small seeds held amongst their foliage where, as Janzen (1984) stated, the 'foliage is the fruit'. In natural forest situations, herbaceous plants tend to be concentrated in patchy, ephemeral habitats, and therefore seed dispersal is of particular importance. Janzen (1984) outlined how large herbivores could be critical for such plants: 'A forest bison, musk-ox, or mastodont would easily move seeds from tree fall to tree fall … while maintaining both the tree fall and riparian vegetation in a state of arrested succession by browsing, grazing, and trampling.' The same ecological relationships have been noticed in New Zealand, where coprolites from extinct ratite moa were found to be dominated by small seeds of low-statured shrubs and herbs, many of which occur in forest margin and regenerating habitats and are now rare (Wood et al., 2008).

9.3 Vegetation composition and structure

We have only indirect evidence of how extinct megafauna may have influenced temperate forest patterns as there are virtually no animals of such size left in these regions. However, as large herbivores have a significant effect on temperate zone forest composition and structure (Vázquez, 2002) it is reasonable to assume megafauna may have been even more influential. Large herbivores may have concentrated in forest clearings, thereby maintaining vegetation diversity and landscape heterogeneity and thus megafaunal extinctions would have led to continuous closed-canopy forest and reduced herbaceous diversity. Svenning (2002) showed that lowland vegetation in northwestern Europe during the climatically similar last interglacial and early Holocene periods was more heterogeneous in the former as a greater diversity of megaherbivores were present. A similar claim has been made for dryland forests in New Zealand (Lee et al., 2010). Vera et al. (2006) suggest light-demanding trees (e.g., *Quercus*, *Corylus*) in European forests relied on large forest herbivores (including aurochs, deer, bison and tarpan) for creating suitable regeneration niches.

In North America the loss of megafauna occurred during the late glacial–early Holocene interval, a time of massive vegetation restructuring, and effects due to extinction are difficult to distinguish from those due to climatic drivers. Many claim that the vegetation reorganisation was instrumental in the demise of the megaherbivores. Gill et al. (2009) show that loss of the megafauna contributed to a restructuring of forest communities and an enhanced fire regime in northeastern North America, but rule out climate-forced vegetation changes as a cause of the megafaunal collapse.

In New Zealand the extinction of megafauna in the thirteenth century left the dense evergreen forests without any herbivores larger than birds of 2–3 kg until the introduction of herbivorous mammals in the nineteenth century. During this time it has been argued that there were dramatic changes in forest composition with browse-sensitive trees and ferns, previously restricted to inaccessible sites, flourishing in forest understoreys and preventing regeneration of podocarps (Lee et al., 2010; Wardle, 1985).

There is little evidence of what impact the loss of megafauna from southern South America may have had. Late Pleistocene vegetation change in the Chilean Lake District, northern Patagonia, coincided with megafaunal extinctions, but also near contemporaneous abrupt climate change, human settlement and volcanic activity, and the relative roles of each factor in the vegetation change is not known (Moreno, 2000).

10 ANTHROPOGENIC IMPACTS

Humans have been altering landscapes through fire for at least 0.5 million years (Williams, 2008). Here we consider impacts from the late Pleistocene changes to the beginning of systematic widespread clearance for agriculture.

There has been debate in all continents about the extent to which pre-agricultural humans altered forests. In northwestern Europe, natural woodlands largely unaffected by humans are very rare and how they reached their current state is still debated. In recent decades the long-held paradigm of initial forest clearance in Europe being related to the spread of slash-and-burn Neolithic agriculturists has been challenged (e.g., by Willis and Bennett, 1994). Instead, a more complex process is postulated, beginning with forest clearance and the use of fire for hunting by people of Mesolithic culture followed by a more sedentary Neolithic society. Neolithic people depended greatly on forest resources, and as a result had a significant impact on woody communities from c.4500 BC (Williams, 2008). Such environmental impacts were initiated throughout Europe until the early Bronze Age (c.1000 BC) and included using flint and stone axes to fell trees and clear for agricultural land, spread of weedy herbs, and propagation of food trees such as walnut and olive. Initial impact of humans on the woodlands of Europe is marked by an increased abundance of early successional trees (e.g., *Betula*, *Picea*) (Williams, 2008). The dramatic decrease in *Ulmus* in Europe during the mid Holocene is often attributed to use of the foliage for forage (Parker et al., 2002).

The development of agriculture, and subsequent spread of settled communities dependent on cultivation, induced vegetation change through temperate eastern Asia. Humans settled northeast China around 8200 BC (Tarasov et al., 2006), although their exact impact is debated. Ren (2000) suggested that humans played a major role in the loss of forests from northeast China during the mid- to late-Holocene, noting relationships between the loss of forests and pattern of spreading human settlements and agriculture across the region. However, Tarasov et al. (2006) disagreed. They interpreted major vegetation changes occurring between 5.7 ka BP and 2.1 ka BP in northeast China as being induced by changes in precipitation and not humans. As in most regions, a combination of human and climate influences with the relative balance changing from site to site seems possible. From 2 ka BP, human impacts are unmistakable. In the middle Yangtze River Valley, southern China, dryland forests were cleared for expansion of rice agriculture (Jiang and Piperno, 1999). Major deforestation associated accompanied agriculture in Korea from 2 ka BP, where deciduous broadleaved forest trees (e.g., *Alnus*) were replaced by cultivated grasses, *Artemisia, Plantago* and *Typha* (Yi et al., 2008).

Humans moved into the Americas following the end of the last ice age, with the spread of Clovis culture around 13.5–12.9 ka BP (Delcourt and Delcourt, 2004). Increased charcoal abundance and reduced megafauna populations (e.g., Robinson et al., 2005) suggest human impacts, although the relative roles of climate *versus* humans is debated (e.g., Barnosky et al., 2004; Gill et al., 2009; Whitlock and Knox, 2002). The extensive temperate forests of eastern North America lack the dramatic anthropogenic alteration argued for in Europe. Nonetheless, the idea of these forests being 'untouched' prior to European contact has been abandoned. It is evident that native Americans cleared forest for building villages and agriculture (Vale, 2002), and as a result the landscape consists of a 'mosaic of fire-altered and human-selected species in difference stages of succession' (Williams, 2008). An example is in the southern Appalachian Mountains, where increased use of fire in the landscape between 3,000 and 1,000 BP resulted in increased abundance of fire-tolerant oak, chestnut and pines in upland forest communities, as well as potentially increasing ecotonal contrast and biological diversity (Delcourt et al., 1998). Williams (2008) identified three major phases of settlement associated with the eastern forests from the mid Holocene onwards. During this time, the structure of forests began to change as canopy species were cleared, and weeds began to invade the resulting open areas. However, the greatest effects on the eastern forests occurred relatively late, with the beginning of forest clearance associated with corn cultivation around AD 800–1000 (Williams, 2008).

South America was peopled at around the same time as North America, and therefore resolving the relative impacts on vegetation communities of climate change *versus* early humans is also difficult here. Markgraf and Anderson (1994) argued that there was little correlation between human activity and fire frequency in Patagonia during the late Pleistocene and Holocene. While they accepted that humans had possibly played some role in amplifying the extent of fires, they saw the main driver of increased fire frequency being climatic variability. However, on a local scale evidence for human alteration of the landscape can be seen. Humans settled in the southern Chilean Lakes District *c.*14.5 ka BP (Dillehay et al., 2008) and increases in charcoal in lake sediments and an expansion of disturbance-related taxa into the rainforest, suggest human-lit fires affected the landscape (Moreno, 2000).

Australia underwent a long-term aridification during the Neogene and El Niño southern

oscillation (ENSO)-induced climatic instability is a permanent feature of the continent (Johnson, 2006). Early Aboriginal people appear to have had little environmental influence in temperate southeast Australia. There is scant palaeoecological evidence for increased fire frequency or vegetation change associated with the first humans. On a local scale, several studies in the temperate southeast have identified increases in sedimentary charcoal potentially related to human settlement. Cook (2009) reported an early Holocene increase in charcoal from two lakes in western Victoria, following an increased presence of Aboriginal people in the local *Allocasuarina*-dominated woodland. Pollen analysis of a *c.*200 ka BP stratigraphic record from Lake Wangoom, Victoria, shows the Holocene was characterised by a greater dominance of grasses over forest trees and tree ferns compared with the penultimate and last interglacials (Harle et al., 2002). Although a drier climate may also have played a role, Harle et al. (2002) suggest anthropogenic fires may have been a major driver. However, it was not until around 6,000 years ago, a time when the human population of Australia began increasing significantly, that fire frequency on the continent appears to exceed what might be expected due to climatic influences (Johnson, 2006; Kershaw et al., 2002). By the time of initial European contact *c.*300 years ago, Aboriginal people were extensively using fire in the landscape, especially for hunting and maintaining open habitats preferred by kangaroos (Johnson, 2006). This process has been termed 'fire-stick farming', and evidence for such anthropogenic induced vegetation communities exists across southeastern Australia and Tasmania (Johnson, 2006).

Evidence of human impact on temperate forests is perhaps most evident on the last major landmass on earth to be settled by people, the islands of New Zealand. Whereas on other continents the relative roles of early humans and climate on fire frequency are difficult to determine, in New Zealand the relationship is beyond doubt. The arrival of

Polynesian settlers in New Zealand at *c*.AD 1280 (Wilmshurst et al., 2008) signalled the beginning of a period of unparalleled ecological devastation. In less than a century the moa and numerous other bird species were driven to extinction (Worthy and Holdaway, 2002) and widespread burning of eastern rainshadow forests occurred. By the time of European settlement in the early nineteenth century, approximately half the lowland forests had been destroyed (Ewers et al., 2006).

11 CONCLUSION

Environmental change over the Quaternary is coherent at a global level. Despite major differences in geography, topography, soils, climates, composition of the biota and human history, the seven temperate regions we identified have followed strikingly similar trajectories.

All areas suffered losses during the Neogene of taxa or groups now found largely in tropical or tropical–montane regions. Taxa tend to originate in the tropics and high diversity is maintained there. When resources are opened up by extinction in temperate latitudes, lineages spread poleward (Krug et al., 2009). Losses were therefore disproportionately high in northwestern North America, Europe and southern South America – all areas without easy access to moist subtropical regions.

The 20 or so major glacial–interglacial cycles of the last 2.6 million years have been particularly influential in the temperate zone because of the extremes they imposed on the biota through altering nearly all aspects of temperature, precipitation and often transforming local geomorphology. Jansson and Dynesius (2002) term the reaction of the biotas 'orbitally forced range dynamics' (ORD). Barnosky (1987) has outlined three basic reactions of biota (aside from wholesale extinction) to ORD: migration, orthoselection (extinction accompanied by speciation to produce a biota well adapted to surviving

environmental change *in situ*) and relictual survival in protected areas. Debates have continued about all three of these potential reactions, fed by the renewed enthusiasm for identification of 'refugia' following the ready availability of molecular techniques, and the associated vogue for tracing migratory pathways. However, recent reassessments raise the possibility that elements of temperate biota may have persisted at high latitudes at the glacial maximum, even in regions such as the eastern US and northern Europe where tundra, taiga or steppe dominated the landscape, and this possibility has long been orthodoxy in southern hemisphere regions. Temperate biotas are therefore largely orthoselective on a glacial–interglacial timescale, waxing and waning *in situ* in response to climate fluctuations, with migration or relictualism playing a relatively minor role, mainly in northern Europe. It seems likely that the migratory pathways and refugia now identified in increasing detail throughout the temperate zone reflect longer, multi-cycle biotic responses promoted by long-term evolutionary rather than short-term migratory processes.

Descriptive ecology relies on communities or formations as fundamental units and for many years there was a tendency to regard these as permanent features. However, recent neoecological research now focuses on mechanistic explanations for biotic function and distribution. Paleoecology has followed this trend and has been instrumental in promoting a view based on individualistic responses of taxa to changes in climate. Independent fluctuations of taxa in the course of the Quaternary and the creation and disappearance of 'no-modern-analogue' assemblages have been well documented (see, e.g., Williams et al., 2004). Indeed, it has never been entirely clear what the alternative is to individualist responses. Few taxa, other than parasitic plants, specialist herbivores and predators, are tightly bound to any particular other taxon, and independent assortment of the rest as a result of differing responses to environmental variables seems inevitable.

Nevertheless, as we have seen, climate closely controls abundances of taxa in the temperate zone and, when multiple climatic factors change in synchrony over a brief period, whole suites of species will come to dominance or fade away in a simulacrum of a tightly interconnected and interdependent assemblage.

From a short-term neoecological perspective, both ecological and abiotic relationships within these present-day assemblages seem both highly coordinated and essential. Thus soil-plant linkages and plant–mycorrhizal, plant–herbivore and predator–prey interactions seem vulnerable to even slight alterations in the prevailing climate regime. It is only with the hindsight provided by palaeoecology that we can integrate the role of climate, and thus properly understand the present and frame possible futures (Christensen, 2009). Alarming predictions of imminent loss of species with continued climate warming (e.g., (Thomas et al., 2004) need to be tempered with the actual record of little loss in the recent past under changes of equivalent or even greater magnitude.

The environmental changes associated with the late-glacial transition and the Holocene interglacial were essentially synchronous – give or take a few hundred years– throughout the temperate zone. Insolation acting on high polar latitudes and the centres of large continents in turn affects the cryosphere, oceans, monsoons and the westerlies. Rising atmospheric CO_2 concentrations fed by westerly driven upwelling in the southern ocean provides essential global warming feedback (Anderson et al., 2009). With warming temperatures, continental ice-sheets retreated from the poleward part of the zone in Europe and North America; mountain glaciers shrunk elsewhere, leaving vast areas of terrain mantled with moraine, outwash debris, extended river terraces, loess and cryogenically disturbed or erosion-stripped landscapes. Westerlies retreated poleward from their advanced position in lower latitudes. Between 18 and 10 ka BP, temperate forests reclaimed all the land they currently occupy and between 10 and 6 ka BP in most regions, under the influence of warmer or wetter climates and nutrient-rich soils, attained an interglacial peak, be that biomass, dominance by broad-leaved evergreens, absence of cool temperate elements, diversity, or northward extent. After 6 ka BP, moderate cooling (most regions of the northern hemisphere) or increasing seasonality (most apparent in the southern hemisphere), weakening monsoons, and the westerlies attaining their current position and strength, established modern environments. With few modifications, a chronostratigraphic scheme based on environmental change would be the same for all temperate regions, and this is somewhat surprising given the emphasis that has been put on local drivers such as individualistic migratory responses, disease, spread of humans, extinction of megafauna, local persistence of ice sheets, local insolation, fire and soil aging. Global climate change appears to be the major driver in all regions.

Megafaunal extinctions (or at least extinctions of the largest animals of a given biota) have now occurred in every temperate region. Possibly there is an extinction debt yet to be paid by plants that were dependent on these ecosystem engineers and food-webs; fire regimes and soils may be irrevocably altered by their absence. We are a very long way from understanding the implications of these changes but it would be indeed surprising if they were not both major and yet to be fully realised.

The debate over the timing of human arrival in much of the temperate zone has perhaps overshadowed the pervasive effects of human presence. Leaving aside the contentious issue of megafaunal extinctions, the most surprising finding has been the variation in human impact. In small Pacific landmasses, the signature of human presence in the form of widespread deforestation and sometimes geomorphic disturbance has been unmistakable. However, in most temperate areas the extent to which preagricultural people or shifting cultivators influenced their environment is vigorously debated, with alternative explanations relating to climate change, disease, soil degradation, and so on, under active consideration. As with the consequences of the extinction of the megafauna, the ecological and landscape legacy of preagricultural and early agricultural peoples remains a fascinating and incompletely explored topic of importance to our understanding of the modern world. We need to know what human legacy is and what it is not if we are to manage remaining wild landscapes effectively.

In recent years confidence in model predictions of future (greenhouse) climates has grown and ocean and cryosphere research have given us a deeper understanding of how the global system reacts to and modulates the effects of rising CO_2 concentrations. Palaeoecological research continues to be important in validating such models. However, the somewhat circular activity of deriving quantitative estimates for major climate variables from fossil assemblages, a primary focus of palaeoecology in the temperate zone in recent decades, now needs to give way to the task of establishing how global climate change affects ongoing evolutionary and ecological processes. Ecological models will be essential to provide this understanding, but they are in a relatively primitive state, reflecting the underdeveloped nature of ecological theory. Generating new ecological theory and validating models of local to global environmental change is therefore a major essential task for palaeoecologists.

Finally, it seems clear enough that greenhouse warming has effectively changed global ecological dynamics forever. If the nations of the Earth find it difficult enough to reach any consensus on preventing warming, it is almost impossible to envisage them agreeing at any time in the future to permitting cooling back to glacial or even Little Ice Age conditions. What we are doing now will influence biological dynamics far into a future that will probably never again experience ice ages. Choosing the right course will need a sound understanding of how ecological processes function at all levels and scales, and palaeoecology has a critical role in developing that understanding.

REFERENCES

Abarzúa A. M., Villagrán C. and Moreno P. I. 2004. Deglacial and postglacial climate history in east-central Isla Grande de Chiloé, southern Chile (43°S) *Quaternary Research* 62: 49–59.

Allen R. M. J., Watts W. A. and Huntley B. 2000. Weichselian palynostratigraphy palaeovegetation and palaeoenvironment; the record from Lago Grande di Monticchio southern Italy. *Quaternary International* 73/74: 91–110.

Anderson R. F., Ali S., Bradtmiller L. I., Nielsen S. H. H., Fleisher M. Q., Anderson B. E. and Burckle L. H. 2009. Wind-driven upwelling in the Southern Ocean and the deglacial rise in atmospheric CO_2. *Science* 323: 1443–1448.

Barlow C. 2000. *The Ghosts of Evolution*. New York: Basic Books.

Barnosky A. D. 2008. Megafauna biomass tradeoff as a driver of Quaternary and future extinctions. *Proceedings of the National Academy of Sciences* 105: 11543–11548.

Barnosky A. D. Koch, P.L., Feranec, R. S., Wing, S. C., Shabel, A.B.. 2004. Assessing the causes of late Pleistocene extinctions on the continents. *Science* 306: 70–75.

Barnosky C. W. 1987. Response of vegetation to climatic changes of different duration in the late Neogene. *Trends in Ecology and Evolution* 2: 247–251.

Barnosky C. W., Anderson P. M. and Bartlein P. J. 1987. The northwestern US during deglaciation; vegetational history and paleoclimatic implications, in Ruddiman W. F. and Wright H. E. J. (eds) *North America and Adjacent Oceans During the Last Deglaciation.* Boulder: Geological Society of America, pp. 289–321.

Berglund B. E., Persson T. and Björkman L. 2008. Late Quaternary landscape and vegetation diversity in a North European perspective. *Quaternary International* 184: 187–194.

Birks H. J. B. 1986. Late-Quaternary biotic changes in terrestrial and lacustrine environments with particular reference to north-west Europe, in Berglund E. (ed.) *Handbook of Holocene Palaeoecology and Palaeohydrology.* New York: John Wiley & Sons, pp. 3–65.

Black M. P., Mooney S. D. and Martin H. A. 2006. A > 43,000-year vegetation and fire history from Lake Baraba, New South Wales. *Australia Quaternary Science Reviews* 25: 3003–3016.

Bowman D. M. J. S., Balch J. K., Artaxo P., Bond W. J., Carlson J. M. Cochrane M. A. et al. 2009. Fire in the Earth system. *Science* 324: 481–484.

Boyle J. F. 2007. Loss of apatite caused irreversible early-Holocene lake acidification. *The Holocene* 17: 543–547.

Bradbury J. P., Grosjean M., Stine S. and Sylvestre F. 2001. Full and late glacial lake records along the PEP 1 transect their role in developing interhemispheric paleoclimate interactions, in Markgraf V. (ed.) *Interhemispheric Climate Linkages.* San Diego: Academic Press, pp. 265–291.

Bridgland D. and Westaway R. 2008. Climatically controlled river terraces a worldwide Quaternary phenomenon. *Geomorphology* 98: 285–315.

Burney D. A. Robinson G. S. and Burney L. P. 2003. *Sporormiella* and the Late Holocene extinctions in Madagascar. *PNAS* 100: 10800–10805.

Christensen J. 2009. Ecology lost and found. *Nature* 459: 167–168.

Clark S. H. E. and Edwards K. J. 2004. Elm bark beetle in Holocene peat deposits and the northwest European elm decline. *Journal of Quaternary Science* 19: 525–528.

Coomes D. A., Mark A. F. and Bee J. 2006. Animal control and ecosystem recovery, in Allen R. B. and Lee W. G. (eds) *Biological Invasions in New Zealand.* Berlin Heidelberg: Springer-Verlag, pp. 339–353.

Cook, EJ. 2009. A record of late Quaternary environments at lunette lakes Bolac and Turangmoroke, Western Victoria, Australia, based on pollen and a range of non-pollen palynomorphs. Review of Palaeobotany and Palynology 153:185-224.

Davis B. A. S., Brewer S., Stevenson A. C. and Guiot J. 2003. The temperature of Europe during the Holocene reconstructed from pollen data. *Quaternary Science Reviews* 22: 1701–1716.

Davis M. B. 1983. Holocene vegetational history of the eastern United States, in Wright H. E. (ed.) *Late-Quaternary Environments of the United States.* Minneapolis: University of Minnesota Press, pp. 166–181.

Davis B. A. S. and Brewer S. 2009. Orbital forcing and role of the latitudinal insolation/temperature gradient. *Climate Dynamics* 32: 143–165.

Dearing J. A., Jones R. T., Shen J., Yang X., Boyle J. F., Foster G. C. et al. 2008. Using multiple archives to understand past and present climate–human–environment interactions: the lake Erhai catchment, Yunnan Province, China. *Journal of Palaeolimnology* 40: 3–31.

Delcourt P. A. and Delcourt H. R. 2004. *Prehistoric Native Americans and Ecological Change.* Cambridge: Cambridge University Press.

Delcourt P. A., Delcourt H. R., Ison C. R., Sharp W. E. and Gremillion K. J. 1998. Prehistoric human use of fire the eastern agricultural complex and Appalachian oak-chestnut forests paleoecology of Cliff Palace Pond Kentucky. *American Antiquity* 63: 263–278.

Denton G. H., Heusser C. J., Lowell T. V., Moreno P. I., Andersen B. G. Heusser L. E. et al. 1999. Interhemispheric linkage of paleoclimate during the last glaciation. *Geografiska Annaler* 81a: 107–153.

Dillehay T. D., Ramirez C., Pino M., Collins M. B., Rossen J. and Pino-Navarro J. D. 2008. Monte Verde Seaweed Food Medicine and the Peopling of South America. *Science* 320: 784–786.

Ewers R. M., Kliskey A. D., Walker S., Rutledge D., Harding J. S. and Didham R. K. 2006. Past and future trajectories of forest loss in New Zealand. *Biological Conservation* 133: 312–325.

Flannery T. F. 1990. Pleistocene faunal loss implications of the aftershock for Australia's past and future. *Archaeology in Oceania* 25: 45–67.

Foster D. R., Oswald W. W., Faison E. K., Doughty E. D. and Hansen B. C. S. 2006. A climatic driver for abrupt mid-Holocene vegetation dynamics and the hemlock decline in New England. *Ecology* 87: 2959–2966.

Fujiki T. and Ozawa T. 2008. Vegetation change in the main island of Okinawa, southern Japan from late Pliocene to early Pleistocene. *Quaternary International* 184: 75–83.

Gill J. L., Williams J. W., Jackson S. T., Lininger K. B. and Robinson G. S. 2009. Pleistocene megafaunal collapse, novel plant communities, and enhanced fire regimes in North America. *Science* 326: 1100–1103.

Guiot J., Wu H. B., Jiang W. Y. and Yun L. L. 2008. East Asian Monsoon and paleoclimatic data analysis a vegetation point of view. *Climate of the Past* 4: 137–145.

Haberle S. G. and Bennett K. D. 2004. Postglacial formation and dynamics of North Patagonian rainforest in the Chonos Archipelago, Southern Chile. *Quaternary Science Reviews* 23: 2433–2452.

Hansen D. M. and Galetti M. 2009. The Forgotten Megafauna. *Science* 324: 42–43.

Harle K. J., Heijnis H., Chisari R., Kershaw A. P., Zoppi U. and Jacobsen G. 2002. A chronology for the long pollen record from Lake Wangoom western Victoria Australia as derived from uranium/thorium disequilibrium dating. *Journal of Quaternary Science* 17: 707–720.

Harrison K. A. and Bardgett R. D. 2008. Impacts of grazing and browsing by large herbivores on soils and soil biological properties, in Gordon I. J. and Prins H. H. T. (eds) *The Ecology of Browsing and Grazing*. Berlin: Springer, pp. 201–216.

Harrison S. P., Yu G. and Tarasov P. E. 1996. Late Quaternary lake-level record from northern Eurasia. *Quaternary Research* 45: 138–159.

Hewitt G. M. 2004. Genetic consequences of climatic oscillations in the Quaternary. *Philosophical Transactions of the Royal Society of London, Series B* 359: 183–195.

Higuera P. E., Brubaker L. B., Anderson P.M., Hu F.S. and Brown T.A. 2009. Vegetation mediated the impacts of postglacial climate change on fire regimes in the south-central Brooks Range, Alaska. *Ecological Monographs* 79: 201–219.

Liu, H., Xing Q., Ji, Z., Xu L. and Tian Y. 2003. An outline of Quaternary development of Fagus forest in China: palynological and ecological perspectives. *Flora* 198: 249–259.

Huber U. M., Markgraf V. and Schabitz F. 2004. Geographical and temporal trends in Late Quaternary fire histories of Fuego-Patagonia South America. *Quaternary Science Reviews* 23: 1079–1097.

Huijzer B. and Vandenberghe J. E. F. 1998. Climatic reconstruction of the Weichselian Pleniglacial in northwestern and central Europe. *Journal of Quaternary Science* 13: 391–417.

Huntley B. 1993. Species-richness in north-temperate zone forests. *Journal of Biogeography* 20: 163–180.

Huntley B., Bartlein P. J. and Prentice I. C. 1989. Climatic control of the distribution and abundance of beech (Fagus L.) in Europe and America. *Journal of Biogeography* 16: 551–560.

Iriondo M.H. 1999. Last Glacial Maximum and Hypsithermal in the Southern Hemisphere. *Quaternary International* 62: 11–19.

Iversen J. 1964. Retrogressive development of a forest ecosystem in the post-glacial. *Journal of Ecology* 52(suppl.): 59–70.

Jansson R. and Dynesius M. 2002. The fate of clades in a world of recurrent climatic change: Milankovitch oscillations and evolution. *Annual Review of Ecology and Systematics* 33: 741–777.

Janzen D. and Martin P. S. 1982. Neotropical anachronisms. The fruits the gomphotheres ate *Science* 215: 19–27.

Janzen D. H. 1984. Dispersal of small seeds by big herbivores: foliage is the fruit. *The American Naturalist* 123: 338–353.

Jiang Q. and Piperno D. R. 1999. Environmental and archaeological implications of a Late Quaternary palynological sequence Poyang Lake southern China. *Quaternary Research* 52: 250–258.

Johnson C. N. 2006. *Australia's Mammal Extinctions*. Cambridge: Cambridge University Press.

Johnson C. N. 2009. Ecological consequences of Late Quaternary extinctions of megafauna. *Proceedings of the Royal Society Series B*. 2509-2519.

Kaplan J. O., Bigelow N. H., Prentice I. C., Harrison S. P., Bartlein P. J., Christensen T. R. et al. 2003. Climate change and Arctic ecosystems: 2. Modeling, paleodata-model comparisons, and future projections. *Journal of Geophysical Research - Atmospheres* 108: 8171-8200.

Kawahata H., Yamamoto H., Ohkushi K., Yokoyama Y., Kimoto K., Ohshima H. et al. 2009. Changes of environments and human activity at the Sannai-Maruyama ruins in Japan during the mid-Holocene Hypsithermal climatic interval. *Quaternary Science Reviews* 28: 964–974.

Kershaw A. P., Clark J. S., Gill A. M. and D'Costa D. M. 2002. A fire history in Australia, in Bradstock R. A., Williams J. E. and Gill A. M. (eds) *Flammable Australia the Fire Regimes and Biodiversity of a Continent*. Cambridge: Cambridge University Press Cambridge, pp 3–25.

Kim S.-J., Crowley T. J., Erickson D. J., Govindasamy B., Duffy P. B. and Lee B. Y. 2008. High-resolution climate simulation of the last glacial maximum. *Climate Dynamics* 31: 1–16.

Körner C. and Paulsen J. 2004. A world-wide study of high altitude treeline temperatures. *Journal of Biogeography* 31: 713–732.

Krug A. Z., Jablonski D., Valentine J. W. and Roy K. 2009. Generation of Earth's first-order biodiversity pattern. *Astrobiology* 9: 113–124.

Latham R. E. and Ricklefs R. E. 1993. Global patterns of tree species richness in moist forests energy-diversity theory does not account for variation in species richness *Oikos* 67: 325–333.

Lee D. E., Lee W. G. and Mortimer N. 2001. Where and why have all the flowers gone? Depletion and turnover in the New Zealand Cenozoic angiosperm flora in relation to palaeogeography and climate. *Australian Journal of Botany* 9: 41–356.

Lee W. G., Wood J. R. and Rogers G. A. 2010. Legacy of avian-dominated plant-herbivore systems in New Zealand. *New Zealand Journal of Ecology* 34: 28–47.

Macphail M. K. 1979. Vegetation and climates in southern Tasmania since the Last Glaciation. *Quaternary Research* 11: 306–341.

Markgraf V. and Anderson L. 1994. Fire history of Patagonia Climate versus human cause. *Rev IG Sao Paulo* 15: 35–47.

Markgraf V. and McGlone M. S. 2005. Southern temperate ecosystem responses, in Lovejoy T. E. and Hannah L. (eds) *Climate Change and Biodiversity.* New Haven: Yale University Press, pp. 142–156.

Marlon J. R., Bartlein P. J., Carcaillet C., Gavin D. G., Harrison S. P., Higuera P. E. et al. 2008. Climate and human influences on global biomass burning over the past two millenia. *Nature Geoscience* 1: 697–702.

McGlone M. S. 2002. The Late Quaternary peat, vegetation and climate history of the Southern Oceanic Islands of New Zealand. *Quaternary Science Reviews* 21: 683–707.

McGlone M. S., Richardson S. J. and Jordan G. J. 2010. Comparative biogeography of New Zealand trees: species richness, height, leaf traits and range sizes. *New Zealand Journal of Ecology* 34: 137–151.

McGlone M. S., Salinger M. J. and Moar N. T. 1993. Palaeovegetation studies of New Zealand's climate since the Last Glacial Maximum, in Wright H. E., Kutzbach J. E., Webb T. III, Ruddiman W. F., Street-Perrrott F. A. and Bartlein P. J. (eds) *Global Climates since the Last Glacial Maximum.* Minneapolis: University of Minnnesota Press, pp. 294–317.

McKenzie G. M. 2002. The late Quaternary vegetation history of the south-central highlands of Victoria, Australia. II. Sites below 900 m. *Austral Ecology* 27: 32–54.

Moreno P. I. 2000. Climate fire and vegetation between about 13000 and 9200 ^{14}C yr BP in the Chilean lake District. *Quaternary Research* 54: 81–89.

Mucina L. and Geldenhuys C.J. 2006. Afrotemperate, subtropical and azonal forests, in Mucina L. and Rutherford M. C. (eds) *The Vegetation of South Africa, Lesotho and Swaziland.* Pretoria: South African National Biodiversity Institute. Pp 584-614.

Munyikwa K. 2005. Synchrony of Southern Hemisphere Late Pleistocene arid episodes a review of luminescence chronologies from arid Aeolian landscapes south of the Equator. *Quaternary Science Reviews* 24: 2555–2583.

Murray P. F. and Vickers-Rich P. 2004. *Magnificent Mihirungs.* Bloomington: Indiana University Press

Newnham R. M., Lowe D. J., Giles T. and Alloway B. V. 2007. Vegetation and climate of Auckland New Zealand since ca 32 000 cal yr ago support for an extended LGM. *Journal of Quaternary Science* 22: 517–534.

Owen-Smith R. N. 1988. *Megaherbivores: The Influence of Very Large Body Size on Ecology.* Cambridge: Cambridge University Press.

Parker A. G., Goudie A. S., Anderson D. E., Robinson M. A. and Bonsall C. 2002. A review of the mid-Holocene elm decline in the British Isles. *Progress in Physical Geography* 26: 1–45.

Peyron O., Guiot J., Cheddadi R., Tarasov P., Reille M., de Beaulieu J.-L. et al. 1998. Climatic reconstruction in Europe for 18,000 yr B.P. from pollen data. *Quaternary Research* 49: 183–196.

Pickett, E. J., Harrison, S. P., Hope, G., Harle, K., Dodson, J. R., Kershaw, A. P. et al. 2004. Pollen-based reconstructions of biome distributions for Australia, Southeast Asia and the Pacific (SEAPAC region) at 0, 6,000 and 18,000 14C yr BP. *Journal of Biogeography* 31: 1381–1444.

Power M. J., Marlon, J., Ortiz, N., Bartlein, P. J., Harrison, S. P., Mayle, F. E. et al. 2008. Changes in fire regimes since the Last Glacial Maximum: an assessment based on a global synthesis and analysis of charcoal data. *Climate Dynamics* 30: 887–907.

Provan J. and Bennett K. D. 2008. Phylogeographic insights into cryptic glacial refugia. *Trends in Ecology and Evolution* 23: 564–571.

Pye, K. 1995. The nature, origin and accumulation of loess. *Quaternary Science Reviews* 14: 653-667.

Rabassa J., Coronato A., Bujalesky G., Salemme M., Roig C., Meglioli A. et al. 2000. Quaternary of Tierra del Fuego, southernmost South America: an updated review. *Quaternary International* 68–71: 217–240.

Ren G. 2000. Decline of the mid- to late Holocene forests in China climatic change or human impact? *Journal of Quaternary Science* 15: 273–281.

Richardson S. J., Peltzer D. A., Allen R. B., McGlone M. S. and Parfitt R. L. 2004. Rapid development of phosphorus limitation in temperate rainforest along the Franz Josef soil chronosequence. *Oecologia* 139: 267–276.

Ricklefs R. E. Latham R. E. and Qian H. 1999. Global patterns of tree species richness in moist forests: distinguishing ecological influences and historical contingency. *Oikos* 86: 369–373.

Robinson G. S., Burney L. P. and Burney D. A. 2005. Landscape paleoecology and megafaunal extinction in southeastern New York state. *Ecological Monographs* 75: 295–315.

Rojas M., Moreno P. I., Kageyama M., Crucifix M., Hewitt C., Abe-Ouchi A. et al. 2009. The Southern Westerlies during the last glacial maximum in PMIP2 simulations. *Climate Dynamics* 32: 525–548.

Shinker J. J., Bartlein P. J. and Shuman B. 2006. Synoptic and dynamic climate controls of North

American mid-continental aridity. *Quaternary Science Reviews* 25: 1401–1417.

Shulmeister J., Goodwin I., Renwick J., Harle K., Armand L. and McGlone M. S. et al. 2004. The Southern Hemisphere westerlies in the Australasian sector over the last glacial cycle a synthesis. *Quaternary International* 118/119: 23–53.

Shuman B. N., Newby P. and Donnelly J. P. 2009. Abrupt climate change as an important agent of ecological change in the Northeast US throughout the past 15,000 years. *Quaternary Science Reviews* 28: 1693–1709.

Sniderman J. M. K., Pillans B., O'Sullivan P. B. and Kershaw A. P. 2007. Climate and vegetation in southeastern Australia respond to Southern Hemisphere insolation forcing in the late Pliocene– early Pleistocene. *Geology* 35: 41–44.

Stebich M., Arit J., Liu Q. and Mingram J. 2007. Late Quaternary vegetation history of Northeast China – recent progress in the palynological investigations of Sihailongwan maar lake. *Courier Forschungsinstitut Senckenberg* 259: 181–190.

Steffensen J. P., Andersen K. K., Bigler M., Clausen H. B., Dahl-Jensen D., Fischer H. et al. 2008. High-resolution Greenland Ice Core data show abrupt climate change happens in few years. *Science* 321: 680–684.

Stocker T. F. 2000. Past and future reorganizations in the climate system. *Quaternary Science Reviews* 19: 301–319.

Svenning J. C. 2002. A review of natural vegetation openness in north-western Europe. *Biological Conservation* 104: 133–148.

Takahara H., Sugita S., Harrison S. P. and Miyoshi N. 2000. Pollen-based reconstructions of Japanese biomes at 0, 6000 and 18,000 14C yr BP. *Journal of Biogeography* 27: 665–683.

Tantau I., Reille M., de Beaulieu J-L. and Farcas S. 2006. Late glacial and Holocene vegetation history in the southern part of Transylvania (Romania): pollen analysis of two sequences from Avrig. *Journal of Quaternary Science* 21: 49–61.

Tarasov P., Jin G. and Wagner M. 2006. Mid-Holocene environmental and human dynamics in northeastern China reconstructed from pollen and archaeological data. *Palaeogeography Palaeoclimatology Palaeoecology* 241: 284–300.

Thomas C. D., Cameron A., Green R. E., Bakkenes M., Beaumont L. J., Collingham Y. C. et al. 2004. Extinction risk from climate change. *Nature* 427: 145–148.

Tinner W. and Lotter A. F. 2006. Holocene expansions of *Fagus silvatica* and *Abies alba* in Central

Europe: where are we after eight decades of debate? *Quaternary Science Reviews* 25: 526–549.

Tsukada M. 1988. Japan, in Huntley B. and Webb T. III (eds) *Vegetation History.* Dordrecht: Kluwer Academic, pp. 459–518.

Vale T. R. (ed.) 2002. *Fire Native Peoples and the Natural Landscape.* Island Press, Washington.

Vázquez D. P. 2002. Multiple effects of introduced mammalian herbivores in a temperate forest. *Biological Invasions* 4: 175–191.

Vera F. W. M., Bakker E. S. and Olff H. 2006. Large herbivores: missing partners of western European light-demanding tree and shrub species?, in Danell K. Duncan P. Bergstrom R. Pastor J. (eds) *Large Herbivore Ecology, Ecosystem Dynamics and Conservation.* Cambridge: Cambridge University Press, pp. 203–321.

Wanner H., Beer J., Bütikofer J., Crowley T. J., Cubasch U., Flückiger J. et al. 2008. Mid- to Late Holocene climate change an overview. *Quaternary Science Reviews* 27: 1791–1828.

Ward J. K. 2005. Evolution and growth of plants in a low CO_2 world, in Ehleringer J. R., Cerling T. E. and Dearing M. D. *A history of atmospheric CO_2 and its effects on plants, animals and ecosystems Ecological Studies 177.* New York: Springer, pp. 232–257.

Wardle P. 1985. Environmental influences on the vegetation of New Zealand. *New Zealand Journal of Botany* 23: 773–788.

Waring R. H. and Franklin J. F. 1979. Evergreen coniferous forests of the Pacific Northwest. *Science* 204: 1380–1386.

Whitlock C., Bartlein P. J., Markgraf V. and Ashworth A. C. 2001. The midlatitudes of North and South America during the last glacial maximum and early Holocene similar paleoclimatic sequences despite differing large-scale controls, in Markgraf V. (ed.) *Interhemispheric Climate Linkages.* New York: Academic Press, pp. 391–416.

Whitlock C. and Knox M. A. 2002. Prehistoric burning in the Pacific Northwest human versus climatic influences, in Vale T. R. (ed.) *Fire Native Peoples and the Natural Landscape.* Island Press, Washington. Pp 195–231.

Whitlock C., Moreno P. I. and Bartlein P. 2007. Climatic controls of Holocene fire patterns in southern South America. *Quaternary Research* 68: 28–36.

Williams J. W., Shuman B. N., Webb T. III, Bartlein P. J., Leduc P. L. 2004. Late-Quaternary vegetation dynamics in North America scaling from taxa to biomes. *Ecological Monographs* 74: 309.

Williams J. W., Shumanm B. N. and Webb T. 2001. Dissimilarity analyses of Late-Quaternary vegetation and climate in eastern North America. *Ecology* 82: 3346–3362.

Williams M. 2008. A new look at global forest histories of land clearing. *Annual Review of Environment and Resources* 33: 345–367.

Williams M. A. J., Dunkerley D. L., de Deckker P., Kershaw A. P. and Stokes T. 1993. *Quaternary Environments.* London: Edward Arnold.

Willis K. J. 1994. The vegetational history of the Balkans. *Quaternary Science Reviews* 13: 769–788.

Willis K. J. and Bennett K. D. 1994. The Neolithic transition – fact or fiction? Palaeoecological evidence from the Balkans. *The Holocene* 4: 326–330.

Willis K. J. and Niklas K. J. 2004. The role of Quaternary environmental change in plant macroevolution: the exception or the rule? *Philosophical Transactions of the Royal Society of London B* 359: 159–172.

Wilmshurst J. M., Anderson A. J., Higham T. F. G. and Worthy T. H. 2008. Dating the late prehistoric dispersal of Polynesians to New Zealand using the commensal Pacific rat. *PNAS* 105: 7676–7680.

Wilmshurst J. M., McGlone M. S., Leathwick J. R. and Newnham R. M. 2007. A pre-deforestation pollen-climate calibration model for New Zealand and quantitative temperature reconstructions for the past 18 000 years BP. *Journal of Quaternary Science* 22: 535–547.

Wood J. R., Rawlence N. J., Rogers G. M., Austin J. J., Worthy T. H. and Cooper A. 2008. Coprolite deposits reveal the diet and ecology of the extinct New Zealand megaherbivore moa Aves Dinornithiformes. *Quaternary Science Reviews* 27: 2593–2602.

Woodward F. I. 1987. *Climate and Plant Distribution.* Cambridge: Cambridge University Press,

Worthy T. H. and Holdaway R. N. 2002. *The Lost World of the Moa.* Christchurch: Canterbury University Press.

Yi S., Kim J. Y., Yang D. Y., Oh K. C., Hong S. S. 2008. Mid- to Late-Holocene palynofloral and environmental change of Korean central region. *Quaternary International* 112(1): 176–177.

Yu G., Chen, X., Ni, J., Cheddai, R., Guiot, J., Han, H. et al. 2000. Palaeovegetation of China: a pollen data-based synthesis for the mid-Holocene and last glacial maximum. *Journal of Biogeography* 27: 635–664.

34

Environmental Change in the Temperate Grasslands and Steppe

Pavel E. Tarasov, John W. Williams,
Jed O. Kaplan, Hermann Österle,
Tatiana V. Kuznetsova
and Mayke Wagner

Then the entire South, even as far as the Black Sea, was a green, virgin wilderness. No plough had ever passed over the immeasurable waves of wild growth; horses alone, hidden in it as in a forest, trod it down. Nothing in nature could be finer. The whole surface resembled a golden-green ocean, upon which were sprinkled millions of different flowers... Oh, steppes, how beautiful you are! (N.V. Gogol 'Taras Bulba')

1 INTRODUCTION

1.1 Grasses and grasslands: general remarks

Presently, grasses (i.e., ~10,000 species of the family Poaceae = Gramineae) are abundant on all continents except Antarctica, and numerous organisms, not least humans, depend on them for food and habitat (Prasad et al., 2005). Besides having a high economic value throughout human history, grasslands are culturally important, inspiring many generations of writers, painters, composers and scientists in their work.·

Extensive grasslands occupy large regions of the world (Figure 34.1), where seasonally dry and semiarid to subhumid climates cause the dominant vegetation to be characterized by a more or less continuous layer of grasses and associated herbs (Coupland, 1992). At a global scale, grasslands generally occur in the zone between forest and desert. However, the patchy nature of vegetation in the forest–grassland and desert–grassland transitions makes precise drawing of the boundaries with other vegetation types a difficult task.

The earliest fossils of Poaceae are presumed grass pollen and phytolites (Prasad et al., 2005) in ~70–60 Ma old sediments from South America, India and North Africa. The first unequivocal grass macrofossils are known from ~55 Ma old sediment,

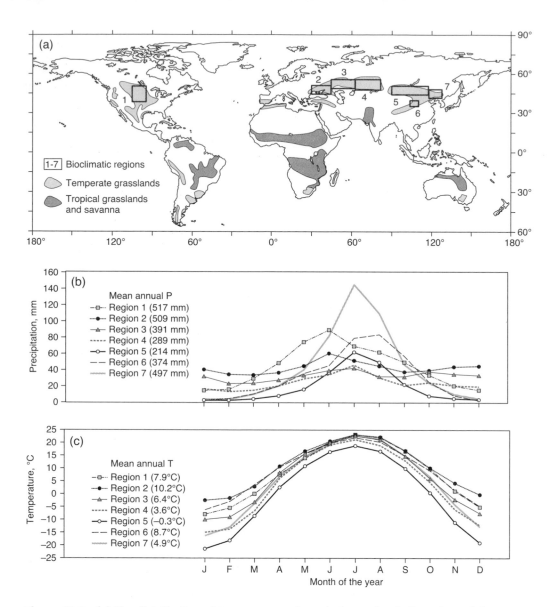

Figure 34.1 (a) The distribution of temperate and tropical grasslands (based on Ripley, 1992). Numbers 1–7 indicate selected bioclimatic regions with a predominantly temperate grassland vegetation discussed in this chapter. (b) Mean monthly values of precipitation and (c) mean monthly temperature calculated for seven selected grassland regions. Mean annual values of temperature and precipitation in each region during 1936–2007 are placed in brackets.

but grass-dominated ecosystems did not appear in the fossil record until about 15–10 Ma ago, and true temperate grasslands did not evolve until the Pleistocene; for example, ~1.8 Ma ago (Willis and McElwain, 2002, and references therein).

1.2 Definition of temperate grasslands

Various terms are used to refer to natural grasslands across the globe (Coupland, 1992). Some are specific to a geographic region; for example, campos and pampas are associated with South America, prairie with North America, veld with South Africa. Other terms, such as steppe, plains and savanna, are often used as more generalized vegetation/biome/climatic categories. The term 'steppe' often refers to grasslands of temperate and subtropical regions, while 'savanna' is reserved for the tropics, but this is not always the rule. In North America, for example 'steppe' is used to imply more arid conditions than 'prairie' (Coupland, 1992). Due to this ambiguity of terms, we advise care when summarizing publications from different authors/regions, keeping in mind that inconsistency in terminology may lead to an incorrect understanding of the primary results and conclusions.

As with other biomes, grasslands can be identified both on the basis of compositional differences and in climatic terms. The division between tropical grasslands, where C_4 grasses are common, and temperate grasslands, where C_3/C_4 mixtures are prevalent, derives from the alternate photosynthetic mechanisms of C_3 and C_4 plants: C_4 plants (less than 4 per cent of total plant species) tend to be favored in warm environments because they have a higher photosynthetic yield at higher temperatures (Willis and McElwain, 2002). C_4 plants are also more tolerant of low-CO_2 environments, which may have acted to increase their relative abundance at the Last Glacial Maximum. Of the six major tribes of Poaceae, Agrosteae, Aveneae and Festuceae are especially well represented in colder regions, while Andropogoneae, Eragrosteae and Paniceae are better spread in warmer climates (Hartley, 1950). In the BIOME1 global vegetation model (Prentice et al., 1992), the defining characteristics of cool grass/shrub (COGS) vegetation type, separating it from warm grass/shrub (WAGS) biome, are mean temperature of warmest month (MTWA) of less than 22°C and mean temperature of coldest month (MTCO) less than 5°C. Annual sum of mean-daily temperatures above 5°C (GDD5) greater than 500 degree-days and

moisture index (α, ratio of actual over potential evapotranspiration) ~0.45–0.65 are chosen to distinguish COGS from other biomes; for example, tundra, semi-desert and forest-dominated vegetation types. In BIOME1 simulations based on these criteria and a modern (1961–1990 averages) climatology (Leemans and Cramer, 1991), zonal COGS vegetation is north of 35°N in North America and Eurasia (Prentice et al., 1992). This is in line with the botanical definition of the boundary between temperate and subtropical communities in the short-grass steppe region of North America (Lauenroth and Milchunas, 1992). In the BIOME4 model spatial distribution of temperate grasslands is similar to that in BIOME1 and is further limited to the regions with relatively cold winters (Kaplan, 2001).

1.3 Scope and structure of the chapter

In this chapter we focus on zonal temperate grasslands of the northern hemisphere, which we define as grasslands in which winter temperatures fall below zero. Our definition is a reasonable compromise, which helps to link botanical, paleoecological and modeling studies. Intentionally, grasslands of the southern hemisphere (e.g., Argentinean pampas) considered as 'temperate' by some authors (e.g., Ripley, 1992) are not included in our review. The southern grasslands are much closer to subtropical and tropical savanna, in having the tussock-like form and tropical species, reflecting relatively mild winter temperatures. Mountain grasslands are not considered in this chapter, except when they are located within the steppe zone; for example, Mongolian Altai. Tundra, which consists of a mixture of dwarf shrubs, sedges, grasses and mosses, growing in high-elevation and high-latitude regions with GDD5 below 500 degree-days, is also not considered here.

We review the environmental history of temperate grasslands of the northern hemisphere for three time windows during the late

Quaternary, including the Last Glacial Maximum (LGM: ~24–18 ka BP); the Late Glacial and Holocene (~15–1 ka BP); and the last millennium with a special emphasis on the last century. Our review relies primarily on sedimentary records from lakes, mires and soils combined with higher-resolution ecological, historical and instrumental records. These three time periods are chosen because of the different proxies/records available for each time-window and because the driving mechanisms of environmental change vary by timescale. To investigate changes in the global distribution of temperate grasslands on glacial-interglacial timescales, we used the BIOME4 global vegetation model driven by a series of GCM paleoclimate scenarios. A brief summary of the present-day environments and an overview of the key archives and reconstruction methods precede the paleoenvironmental part, which starts with the LGM and then zooms in to shorter time scales.

2 REGIONAL SETTING

2.1 Temperate grasslands of North America

Grasslands of North America are commonly called prairies (from the French *prairie* – a meadow or pasture). The mid-continental Great Plains prairies are by far the most extensive, with smaller grasslands in the Central Valley of California; in eastern Washington, Oregon, and southern Idaho; and in the coastal parts of Texas and Louisiana (Coupland, 1992; Küchler, 1975; Lauenroth and Milchunas, 1992; Sims and Risser, 2000). Here we focus on the Great Plains grasslands (region 1 in Figure 34.1), where climate generally is too dry to support forests, but humid enough for the drought-resistant grasses, forbs and shrubs. The temperature and moisture gradients within the Great Plains are shaped by the presence of the Rocky Mountains and other mountain barriers in western North America, which block much of the moisture advected from the Pacific. No comparable orographic barriers are present to the north or south, which allows the incursions of the cold and dry Arctic air in winter and southeasterly flow of warm and relatively humid tropical air in summer (Ripley, 1992). The Great Plains grasslands are broadly defined as the '*Andropogon-Bouteloua-Bison-Canis* grassland biome' (Coupland, 1992). Most classifications recognize a gradient from western short-grass prairie (major grass genera: *Bouteloua*, *Buchloë*) to eastern tall-grass prairie (*Andropogon*, *Panicum*, *Sorghastrum*), with a mixed grassland (*Agropyron*, *Andropogon*, *Stipa*) occupying an intermediate position (Grimm, 2001; Küchler, 1975). The relative abundance of C_3 and C_4 grasses varies both by season and by region, with 'cool-season' C_3 grasses more abundant in the early growing season and in areas with mean annual temperatures (TANN) $<10°C$ and 'warm-season' C_4 grasses more abundant later in the growing season and in areas with TANN $>10°C$ (Sims and Risser, 2000).

The northern part of the Great Plains experiences MTCO as low as $-18°C$ and MTWA as high as $18°C$. Both the MTCO and MTWA values progressively increase southward. The western short- and mid-grass mixed prairie is drier (annual precipitation, PANN, – 300–450 mm) than the eastern tall-grass prairie, which is also characterized by less pronounced year-to-year precipitation variability and less frequent droughts.

2.2 Temperate grasslands of northern Eurasia

In Eurasia grasslands are commonly called steppes (from the Russian *step'* – a flat and dry land covered with grasses and forbs) and also occupy the interior drier part of the continent (Figure 34.1). The great size of the Eurasian landmass produces large seasonal contrasts in temperature and precipitation. In winter, a strong thermal anticyclone develops over Mongolia and Siberia, causing generally cold and dry weather in the region between the Black Sea and the Yellow Sea. With the onset of spring, the anticyclone is gradually replaced by a depression centered over northern India (Ripley, 1992), stimulating summer cyclonic activity in the steppe belt of northern Eurasia. Unlike North America, the generally low relief of mid-latitude Europe and western Asia allows penetration of westerly storm tracks as far as Lake Baikal (Tarasov et al., 2007). In eastern Asia, the summer precipitation maximum is strongly affected by the south-easterly (monsoon) circulation, which brings warm and relatively moist Pacific air into contact with Siberian air masses (Ripley, 1992). In Eurasia steppes predominate between 27 and 128°E and between 48 and 57°N (Figure 34.1). Steppes are classified into three zonal categories, replacing each other in the direction from north to south in response to the increase in aridity (Lavrenko and Karamysheva, 1993). These include (1) meadow- or forest-steppes, rich in tall bunch grasses and forbs; (2) true steppe (major grass genera: *Stipa*, *Festuca*) and (3) semi-desert steppe with relatively low species richness and vegetation cover and codominance of drought-resistant bunch grasses and dwarf half-shrubs (e.g., *Artemisia*). The structure and floristic composition of Eurasian steppes also change from west to east in response to the general decrease in winter temperatures and precipitation (Lavrenko and Karamysheva, 1993). Six bioclimatic regions are well defined (Figure 34.1), including mesic-steppe and mixed pine and broad-leaved forest (*Quercus, Acer, Carpinus, Fraxinus, Tilia, Corylus*) communities north of the Black Sea (region 2); transitional steppes and woodlands of the lower Volga basin and southern Urals (3); West Siberian and Kazakh steppes with patchy pine-birch-aspen forests growing in the northern part (4); and steppes of southern Siberia and Mongolian Plateau, which extreme-continental climate permits only the most cold/drought-resistant boreal trees (mainly *Larix*,

Betula) to grow in the locally moist habitats (5). The regions located in the central (6) and northeastern (7) parts of China (Figure 34.1) are affected by the summer monsoon. Milder winters and higher precipitation permit growth of temperate deciduous trees and shrubs there, particularly in the valleys and on east-facing mountain slopes.

3 ENVIRONMENTAL ARCHIVES AND CLIMATE PROXIES FROM TEMPERATE GRASSLANDS

Studies of the late Quaternary history in temperate grasslands exploit a wide range of natural and documentary archives and observational datasets, as well as a number of methods that allow dating and qualitative/quantitative reconstructions of past vegetation and climate. Earlier chapters in this Handbook and the four-volume *Encyclopedia of Quaternary Science* (Elias, 2007) provides a fully referenced overview of the proxies and methodological approaches used to reconstruct Quaternary environments. The aim of the following sections, therefore, is to introduce the reader to the key archives and reconstruction approaches used by the authors to highlight environmental change in their study regions.

3.1 Pollen

Many of the most informative records come from transitional forest-grassland zones, where semi-humid climate provides locally favorable environments for peat formation/preservation and continuous sediment accumulation in lakes (e.g., Grimm, 1984; Gunin et al., 1999; Kremenetski, 2003; McAndrews, 1966; Nelson et al., 2006; Umbanhowar et al., 2006). In general, North American prairies are much better sampled by pollen records than northern Eurasian steppes (e.g., Prentice et al., 2000). Most pollen records span the Holocene, and many extend into the

Late Glacial interval, but records prior to the Last Glacial Maximum are scarce. Chronologies are mainly constructed on the basis of radiocarbon dates; however, the dating quality is very variable between the sites and regions (Prentice et al., 2000, and references therein). In mid-continental North America radiocarbon dates from bulk sediments are particularly inaccurate due to hard water contamination and Accelerator Mass Spectrometry (AMS) dates of terrestrial plant material is preferable (MacDonald et al., 1991). The time resolution in most pollen records is decadal to centennial (Webb and Webb, 1988).

Pollen records from the regions occupied by temperate grasslands are among the most important archives used to reconstruct late Quaternary vegetation, climate dynamics and human–environmental interactions. Two complementary approaches have been recently used for objective reconstructions of past vegetation: (1) the biome reconstruction method (Prentice et al., 1996), which allows objective assignment of pollen taxa to plant functional types (PFTs) and to biomes, on the basis of modern ecology, bioclimatic tolerance and biogeography of pollen-producing plants; and (2) woody cover reconstruction method (Williams, 2003; Williams and Jackson, 2003; Tarasov et al., 2007), which combines satellite-based (DeFries et al., 1999) and extensive modern surface pollen datasets from North America and northern Eurasia with the best modern analogue (BMA) approach (Guiot, 1990; Overpeck et al., 1985).

The biome reconstruction approach was tested with the extensive surface pollen datasets from northern Eurasia, China and North America (Prentice et al., 2000, and references therein) where temperate grasslands occupy large areas. The number of taxa assigned to steppe in the regional studies was substantially increased (note that only *Artemisia*, Chenopodiaceae and Poaceae were used by Prentice et al., 1996), greatly improving the distinction of steppe (e.g., 97 per cent *versus* 51 per cent; Tarasov et al.,

1998a). Further modification of the method helped to improve the distinction between warm and cool steppes (Tarasov et al., 1998b). The main ecological difference between cool and warm steppes is the balance between C_4 and C_3 species (Olson et al., 1983). These are mainly grasses, which cannot be further distinguished by pollen analysts. In the modified method the lack of herbaceous taxa-indicators was compensated by using the presence/absence of arboreal taxa as an additional criterion to assign herbaceous pollen taxa to the appropriate biome.

The pollen-based quantitative approaches used to reconstruct the late Quaternary climate in the grassland regions of the northern hemisphere include: (1) the indicator-species method (Frenzel et al., 1992); (2) the biome reconstruction method (Prentice et al., 2000); (3) multiple regression transfer functions (Shi and Song, 2003; Tarasov et al., 2006); (4) the PFT-based method using artificial neural networks (Peyron et al., 1998; Tarasov et al., 1999a, 1999b); and (5) the already mentioned BMA method, which allows fossil pollen samples to be attributed to the climatic characteristics associated with their closest modern pollen analogues (Tarasov et al., 2007; Rudaya et al., 2009). As all these methods have their advantages and shortcomings, they complement each other, allowing for more robust interpretations of the fossil records.

3.2 Lake status

Lakes in grassland-dominated regions sensitively react to changes in the regional water budget (i.e., precipitation minus evaporation) by changing their depth and surface area. The changes in lake status can be derived from various geomorphic, sedimentary and biostratigraphic sources of information and have been used to reconstruct regional climate and atmospheric circulation patterns over grassland-dominated areas of northern Eurasia (Harrison et al., 1996), North America

(Harrison and Metcalfe, 1985; Shuman et al., 2002) and China (Fang, 1991). However, the earlier reconstructions and interpretations must not be taken uncritically, particularly in those regions where the data are sparse and/ or the dating control is poor; for example, in the mid-latitudinal band of eastern Europe and western Central Asia between 25 and 85°E (Tarasov and Harrison, 1998).

3.3 Faunal records

Studies of the late Quaternary mammal remains in Europe and northern Asia have been used to reconstruct changes in mammal assemblages through time and space (e.g., Frenzel et al., 1992; FAUNMAP, 1994; Markova and Puzachenko, 2007; Markova et al., 2008) and to address more specific problems; for example, late Pleistocene extinctions (Elias and Schreve, 2007) and domestication (Ludwig et al., 2009). In contrast to the Arctic regions, where permafrost guarantees excellent preservation and dating of the fossil material, the late Quaternary animal records from the grassland regions stored in the archeological, alluvial, lacustrine sediments, loess and soil sequences are less numerous and less accurately dated. In general Europe and North America have much better data coverage, while Asian mid-latitudes are poorly represented.

A clear definition of the 'steppe mammal assemblage' is necessary, but not readily available. For example, northern Eurasian steppes currently house 92 species of mammals (Formozov, 1976). Only 31 species are endemics, with the remaining also occurring in other landscapes, including deserts, forests, swamps, alpine meadows and tundra. The most characteristic representatives of the steppe assemblage are burrowing animals known for storing food or hibernating during cold time of the year (Formozov, 1976). In Eurasian steppes these include ground squirrels (*Spermophilus* sp.), steppe pika (*Ochotona pusilla*), hamsters (*Cricetus cricetus, Allocricetus eversmanni, Cricetulus*

migratorius), rodent-moles (*Myospalax* sp.), mole rats (*Spalax* sp.), mole vole (*Ellobius talpinus*), southern birch mouse (*Sicista subtilis*), steppe lemmings (*Eolagurus luteus, Lagurus lagurus*) and jerboas (*Allactaga major, Allactaga saltator*) (Markova et al., 2008). A number of predators (e.g., *Canis, Vulpes, Mustela*) and ungulates (e.g., *Procapra gutturosa, Equus, Saiga, Bison*) are commonly included in the steppe assemblage (e.g., Coupland, 1992, 1993; Markova et al., 2008). However, it should be kept in mind that the latter species have much broader ecological amplitude.

3.4 Archeological and historical data

A variety of types of human activities are represented in the archeological record. Best preserved, because they were intentionally built underground, are tombs containing physical remains of human bodies, grave goods of different material including animals and possibly tomb furniture as timber or stone. Particularly for mobile steppe inhabitants without permanent settlements, these are often the only sources of information. Interpretations concerning subsistence strategies, technical innovations, social changes and cultural developments are based on artifact classifications and typological comparisons fitted into chronological frameworks supported by stratigraphical observations and physical age determination of organic material. The application of new techniques in archeology, such as remote sensing, archeo-metallurgical, archeozoological, archeobotanical and anthropological analyses are leading to more comprehensive reconstructions of human land-use strategies and human–environmental interactions.

Written accounts by inhabitants of the Eurasian steppes themselves are not longer than a millennium. The oldest historical documents were recorded by neighbors (Greeks, Persians, Chinese and Russians) who explored and exploited the steppes.

Historical data include maps, chronicles, land surveyor's records, travelers' diaries and interviews with local inhabitants which may, when available, provide important information complementing the environmental archives. Various historical data from the temperate grasslands were extensively used to reconstruct changes in the regional climate and vegetation cover (e.g., Sheppard et al., 2004; Tarasov et al., 2006; and references therein).

To assess the impact of human activities on the global environment more accurately, several attempts have been undertaken to reconstruct historical changes in land use and land cover (Ramankutty and Foley, 1999; Ramankutty et al., 2006). Using methodological approaches of the latter studies as a starting point, Pongratz et al. (2008) reconstructed spatial changes in the global distribution of croplands and pastures over the last millennium since AD 800 and improved the existing datasets, particularly in the mid-latitudes of northern Eurasia. Their study can be used with climate and ecosystem models to access the pre-industrial human impact on the Earth system, and particularly on the grassland regions.

3.5 Meteorological observations

Low density of human population and permanent settlements in the temperate grassland regions is still one of the main reasons for the relatively low number of meteorological observatories. Weather stations with the instrumental records spanning periods of more than 150 years are available from the steppe areas of the US, former Soviet Union, Mongolia and China (e.g., Domrös and Peng, 1988; Peterson and Vose, 1997), but long records are rare. Individual station records are of limited use for inter-regional comparison and establishment of longer-term trends and shorter-term fluctuations in climatic parameters because of various complications, including gaps in observations, inconsistencies/errors in instrumental measurements,

changes in measurement techniques and so on. More objective spatial and temporal analyses are possible using datasets constructed by the interpolation of station data on the regular network of geographical coordinates. Gaps and inconsistencies in the individual station records are corrected or smoothed by averaging of the values from several stations representing one grid cell. Here we used two such archives (actualized in PIK-Potsdam) covering the global land surface at $0.5 \times 0.5°$ resolution: the CRU TS 2.1 dataset representing the period between 1901 and 2002 (Mitchell and Jones, 2005; New et al., 2000) and the Global Precipitation Climatology Centre's (GPCC) Full Data Reanalysis Product of monthly precipitation for the period 1901–2007 (Schneider et al., 2008). During the 1930s many new meteorological observatories were founded in the former Soviet Union, and the reliability of interpolated data was greatly improved, therefore we limited our analysis to the seven regions in North America and Eurasia after 1936 (Figure 34.1; Table 34.1).

4 GRASSLAND MODELING

To investigate changes in the global distribution of temperate grasslands on glacial–interglacial timescales, we used a global vegetation model driven by a series of GCM paleoclimate scenarios. The resulting maps of grassland distribution may be compared to the paleoecological record and used as boundary conditions for further climate model simulations or investigations on biogeochemical cycling, animal habitats and other studies on environmental change. Because it contains a complete process description of plant responses to climate and atmospheric CO_2 concentrations, we used the BIOME4 global vegetation model to simulate grassland area.

BIOME4 is a coupled carbon and water flux model that predicts steady-state vegetation

Table 34.1   Linear trends of the annual temperature (TANN) and precipitation (PANN) in the selected regions calculated for the observation period 1936–2007

Region	Latitude range	Longitude range	Linear trend in TANN, °C	Linear trend in PANN, mm	Relative change in PANN, %
1	38–50°N	95–105°W	0.5	46	8.9
2	44–50°N	30–45°E	1.1	40	7.9
3	50–53°N	45–55°E	1.8	53	13.4
4	47–55°N	63–82°E	1.9	10	3.5
5	44–50°N	90–120°E	1.7	−4	−2.0
6	35–39°N	104–110°E	0.9	−33	−8.9
7	42–46°N	118–128°E	1.4	−60	−12.2

distribution, structure, and biogeochemistry, taking into account interaction between these effects (Kaplan, 2001). The model is the latest generation of the BIOME series of global vegetation models, which have been applied to a wide range of problems in biogeography, biogeochemistry and climate dynamics (Haxeltine and Prentice, 1996; Jolly and Haxeltine, 1997; Kaplan et al., 2006; Prentice et al., 1992; VEMAP Members, 1995). BIOME4 has been specifically developed with the intention of improving simulation of cold-climate, high-latitude vegetation (Kaplan, 2001; Kaplan and New, 2006). While BIOME4 can be run for any area and at any spatial resolution, the model is generally designed to be used at continental to global scales.

BIOME4 is driven by climatological monthly mean time series of temperature, precipitation, cloudiness, soil texture and atmospheric CO_2 concentrations. Twelve PFTs in BIOME4 represent broad, physiologically distinct classes, ranging from cushion forbs to tropical broad-leaf trees (Kaplan, 2001). Each PFT is assigned a small number of bioclimatic limits, which determine whether it could be present in a given grid cell, and therefore whether its potential net primary productivity (NPP) and leaf area index (LAI) are calculated. The PFTs also have a set of parameter values that define its carbon and water exchange characteristics. The computational core of BIOME4 is a coupled carbon and water flux scheme that

determines the seasonal maximum LAI that maximizes NPP for any given PFT, based on a daily time step simulation of soil water balance and monthly mean calculations of canopy conductance, photosynthesis, respiration and phenological state (Haxeltine and Prentice, 1996). By simulating the coupled response of carbon uptake to temperature, light and water availability, BIOME4-modeled vegetation is sensitive to atmospheric CO_2 concentrations.

To identify the biome for a given grid cell, the model ranks the tree and nontree PFTs that were calculated for that grid cell. The ranking is defined according to a set of rules based on the computed biogeochemical variables, which include NPP, LAI, mean annual soil moisture, and an index of vulnerability to fire. The resulting ranked combinations of PFTs lead to an assignment to one of 27 global biomes of which three grassland types are a subset: tropical grasslands, temperate grasslands, and graminoid and forb tundra. The 28th cover type, ice sheets and glaciers, is prescribed. We performed a series of simulations with BIOME4 at four time slices, including the LGM (~21 ka BP), early Holocene (~10 ka BP), mid Holocene (~6 ka BP) and modern pre-industrial (~0 ka BP). The ~0 ka BP time slice uses CRU CL2.00 twentieth-century climatology; for the three paleoclimate time slices we used GCM scenarios produced by the Hadley Centre HAD-UM mixed-layer ocean–atmosphere model (see Kaplan et al., 2006, for details).

5 GRASSLAND ENVIRONMENTS DURING THE LAST GLACIAL MAXIMUM

5.1 North America

There is little direct evidence for mid-continental grassland communities at ~21 ka BP, and their extent and density must have been much reduced relative to the present (e.g., Harrison and Prentice, 2003). Much of the northern Great Plains was under ice, with the lobes of the Laurentide Ice Sheet (LIS) extending south through eastern South Dakota and into central Iowa. Even south of the LIS, a scarcity of paleoecological data limits our understanding of the composition and distribution of the Great Plains vegetation, and the distribution and composition of LGM grasslands remains a bit of a mystery.

Based on the Wolf Creek site in central Minnesota, vegetation near to the LIS was likely herbaceous or barren tundra (Birks, 1973). In the relatively well-represented east-central Great Plains of eastern Nebraska and Kansas, *Picea*-dominated woodlands and forests appear to have been widespread, although there is evidence for at least patches of grasslands. Burned *Picea* and *Pinus flexilis* macrofossils (including *Picea pungens*) dating from 21 to 17 ka BP were recovered from sites in western Kansas and southwestern Nebraska (Wells and Stewart, 1987). Rates of loess deposition across the Great Plains were very high, as indicated by the Peoria Loess (25–11 ka BP), which extends from Colorado to Ohio and may be the thickest loess unit in the world (Bettis et al., 2003). This suggests that most of the central Great Plains was characterized by high rates of erosion and sparse vegetation cover developed under cold and dry climates (Mason et al., 2007). Yet the boreal forests or woodlands in the east-central Great Plains indicate colder- and wetter-than-present conditions (Baker et al., 2009). The inference of wet conditions is supported by rich and diverse fossil assemblages of land snails from LGM sediments, containing species now mostly extirpated from the central Great Plains (Rossignol et al., 2004; Wells and Stewart, 1987). This apparent contradiction between the loess and paleoecological data could be explained by either a sharp moisture gradient in the central Great Plains, placed somewhere in eastern Nebraska, or the confinement of *Picea* forests to locally wet settings such as riparian lowlands. By contrast to the scarce evidence for the grassland biome found in the plant records, the animal records suggest wide distribution of the grassland-associated species, such as pronghorn antelope, black-tailed prairie dog and plains pocket gopher ~24–18 ka BP (FAUNMAP, 1994; Markova and Puzachenko, 2007; Markova et al., 2008). Although many grassland species doubtlessly grew either in the areas of loess deposition or in upland patches in eastern Nebraska and Kansas, an extensive grassland biome apparently did not exist at the LGM.

5.2 Northern Eurasia

The LGM vegetation maps compiled by Frenzel et al. (1992) present a great variety of open vegetation types, including loess-steppe, tundra-steppe, herb-steppe, coniferous and deciduous forests/woodlands with large steppe areas and semidesert developed under generally colder and drier-than-present full-glacial climates (Frenzel et al., 1992; Kageyama et al., 2001). In the proposed scenario, grasses were mainly associated with tundra-steppe vegetation and widely distributed south of the ice sheet in Europe and western Siberia and in areas of central and eastern Siberia currently occupied with boreal forests. Both qualitative (Prentice et al., 2000) and quantitative (Frenzel et al., 1992) reconstructions suggest that in Eurasia steppe-like communities consisting of *Artemisia*, various herbs, grasses, Chenopodiaceae and *Ephedra* species occupied a larger-than-present latitudinal belt during the LGM (e.g., Harrison and Prentice, 2003). Cool steppe was reconstructed at the

LGM sites in western Ukraine, where temperate deciduous forests grow today, and in western Siberia, where taiga and cold deciduous forests grow today and where cool steppe was graded into tundra, but scattered boreal woods grew on the northern coast of the Sea of Azov currently occupied by steppe (Tarasov et al., 1999a). The cool steppe biome is reconstructed as far south as central Greece and central Italy (Allen et al., 1999), while in western Iran the vegetation was at the boundary between cool and warm steppes (Tarasov et al., 1998b). Finds of steppe-associated animals, such as saiga antelope, steppe lemming and steppe pika, also suggest significant expansion of the steppe faunal complex ~24–18 ka BP ago (FAUNMAP, 1994; Markova and Puzachenko, 2007; Markova et al., 2008). Remains of forest animals from the East European and West Siberian Plains are relatively rare, in line with the botanical evidence for the absence of continuous forest belt. The reorganization of the animal world produced new types of 'non-analog', 'mixed', 'periglacial' mammal communities, which were distributed throughout most of northern Eurasia and inhabited the no-analogue landscapes; for example, 'mammoth steppe' (Markova and Puzachenko, 2007) or 'tundra-steppe' (Frenzel et al., 1992).

6 GRASSLAND ENVIRONMENTS DURING THE LATE GLACIAL AND HOLOCENE

6.1 North America

The distribution and composition of vegetation formations in eastern North America was transformed during the deglaciation ~18–6 ka ago (Webb, 1987; Williams et al., 2004). During this transformation, grasslands expanded tremendously, spreading from their scattered and sparse full-glacial distribution to become the dominant biome of mid-continental North America (Harrison and Prentice, 2003). The *Picea*-dominated forests and

woodlands of the LGM were succeeded first by mixed parklands containing boreal and deciduous arboreal taxa, notably *Fraxinus nigra*, *Ulmus* and *Ostrya/Carpinus* (Williams et al., 2004). After 12 ka BP the mixed parklands were replaced by a more sharply delineated vegetation zonation between grasslands to the west and forests to the east and north. During the Holocene the position of the prairie-forest ecotone shifted in response to changes in moisture availability and fire regime (Figure 34.2).

The Late Glacial to early Holocene expansion of grasslands implies a long-term warming and drying of the mid-continent; however this trend was not uniform and is overlaid by spatially and temporally complex variations in both temperature and moisture (Diffenbaugh et al., 2006; Fritz, 2008; Mason et al., 2008; Shuman et al., 2002; Williams et al., 2010). Moderately wet conditions between 14 and 11 ka BP are suggested by a

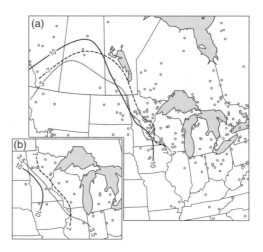

Figure 34.2   Reconstructed shifts in the prairie-forest ecotone during the Holocene, based on (a) pollen-derived position of the 20 per cent woody cover fraction at about 10, 7 and 3 ka BP (Williams et al., 2009), and (b) the 20 per cent isopoll for prairie forb taxa (the sum of *Ambrosia*, *Artemisia*, other Asteraceae, and Chenopodiaceae/Amaranthaceae) (Webb et al., 1983) at about 10, 7 and 3.5 ka BP (Williams et al., 2009).

variety of paleoclimatic proxies (Figure 34.3). Some Great Plains sites show signs of increasing aridity as early as 14 ka BP, but most sites began to dry between 10 and 8 ka BP, with western sites drying first (Williams et al., 2010). The Holocene dry interval lasted between ~9 ka BP and 4 ka BP, with signs of increases in effective moisture beginning at 6 ka BP (Figure 34.3).

This warming and drying of the North American mid-continent was driven in turn by a combination of factors, including

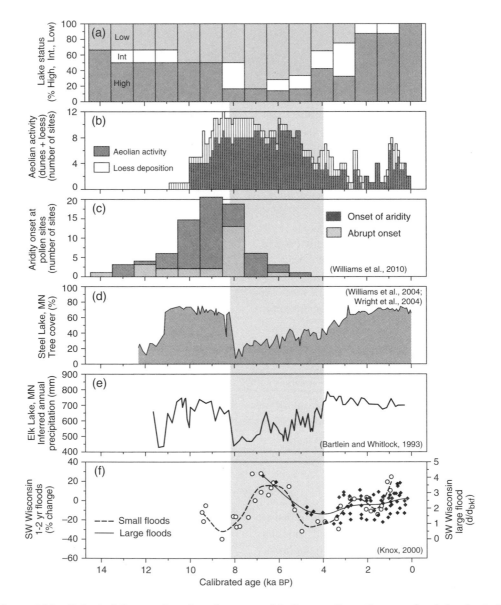

Figure 34.3 Selected time series of environmental indicators from the grassland-dominated North American mid-continent. Grey band indicates pronounced aridity phase between 8 and 4 ka BP (see Williams et al., 2010, for details and references).

Source: Cook et al. (2008) and Williams et al. (2010).

increases in summer insolation, increases in atmospheric CO_2, the retreat of the LIS, and shifts in atmospheric circulation (Harrison et al., 2003; Kutzbach et al., 1998; Shuman et al., 2006). The peak in summer insolation at 11 ka BP may have contributed to mid-continental aridity by increasing temperatures and evapotranspiration rates (Diffenbaugh et al., 2006; Kutzbach et al., 1998), and by causing increased monsoonal strength and uplift over the southwestern US resulted in increased atmospheric subsidence and drying over the North American mid-continent (Harrison et al., 2003). The coincidence between the timing and rate of LIS deglaciation and mid-continental drying also suggests a causal connection (Shuman et al., 2006; Williams et al., 2010), although the precise mechanisms remain unclear. One possibility is that sharp thermal gradient at the southern margin of the LIS may have anchored the jet stream and routed storm tracks along the southern margin of the LIS (Bromwich et al., 2005) so that as the LIS retreated north and shrank in height, the position of the jet stream shifted and became more variable, causing a reduced frequency of precipitation in the Great Plains. Land-surface feedbacks may have doubled the severity of twentieth-century droughts in the Midwest (Schubert et al., 2004), suggesting that they also may have amplified Holocene drying. However, the role of land-surface feedbacks in Holocene mid-continental North America remains understudied and speculative.

Sites from within the Great Plains show contrasting patterns of abundance among the major grass and forb pollen types (Figure 34.4; Grimm, 2001). Pollen abundances of Poaceae and *Artemisia* remained high through the Holocene with no clear temporal trend. Conversely, both *Ambrosia*-type and Chenopodiaceae/Amaranthaceae pollen abundances peak during the middle Holocene and *Ambrosia*-type abundances are highly variable (Grimm, 2001). At Kettle Lake, ND (Clark et al., 2002), these variations resolve into a strongly oscillating series of 100-year to 130-year cycles in which

periods of high *Ambrosia* abundances and low charcoal abundances alternate with periods of high Poaceae and charcoal abundances. These oscillations appear to be linked to drought variability, in which *Ambrosia* abundances are high during dry phases and grass biomass and fuel load increase during wet phases (Clark et al., 2002).

Grassland area also changed during the Holocene and has been tracked by the shifting position of the prairie-forest ecotone. Ecotonal dynamics are best described in the north (e.g., Canadian Great Plains) and in the east (e.g., Minnesota, Illinois, Iowa, and adjacent regions) (Baker et al., 2002; Nelson and Hu, 2008; Nelson et al., 2006; Webb et al., 1983; Williams et al., 2009). In the central and southern Great Plains, the history of the prairie-forest ecotone is poorly understood due to a scarcity of paleovegetation records, with many of the best data coming from soil and speleothem $\delta^{13}C$ records (Denniston et al., 2000; Nordt et al., 2007). In Minnesota and adjacent areas, prairie expanded eastward rapidly between 10 and 8 ka BP and reached a maximum eastward position between 7 and 6 ka BP (Figure 34.2). After 6 ka BP, the prairie-forest ecotone shifted westward and grassland area declined. The northern prairie-forest ecotone was apparently stable between 10 and 6 ka BP and moved southwards after 6 ka BP, indicating forest expansion and prairie retreat.

At sites located along the eastern prairie-forest ecotone, the rates of early Holocene deforestation and late Holocene reforestation are asymmetrical (Umbanhowar et al., 2006). The switch from forest to prairie typically was rapid, of the order of a few centuries or less (Williams et al., 2009). Conversely, the reforestation of ecotonal sites after 6 ka BP was gradual, of the order of several thousand years. These differing rates of deforestation and reforestation may have been directly caused by differing rates of climatic change (i.e., rapid drying and gradual wetting), or may have resulted from ecological interactions between trees and fire regime that accelerated tree mortality during drying and

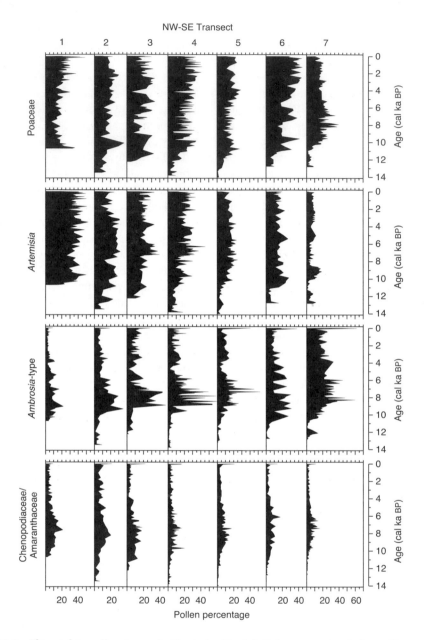

Figure 34.4   The major pollen types in the Late Glacial–Holocene records from the northern Great Plains on a NW–SE transect of sites (1 – Rice Lake, 2 – Creel Bay, 3 – Spiritwood Lake, 4 – Moon Lake, 5 – Pickerel Lake, 6 – Medicine Lake, 7 – Lake West Okoboji).

Source: Williams et al. (2010).

hindered forest expansion when effective moisture increased. Evidence exists to support both hypotheses. Abrupt shifts in aragonite/calcite ratios at West Olaf Lake and Deep Lake and $\delta^{18}O$ increases at Steel Lake (Nelson and Hu, 2008) directly indicate rapid drying in Minnesota at ~8.2 ka BP, which may have been triggered by the collapse of the Laurentide Ice Sheet (Shuman et al., 2002; Williams et al., 2010). However, the early Holocene deforestation and prairie expansion was time-transgressive (Webb et al., 1983; Williams et al., 2009) and was often abrupt even for sites where the timing of deforestation does not date to ~8.2 ka BP (Williams et al., 2009).

A close linkage among aridity, fire regime and vegetation dynamics is suggested by sites with paired charcoal and pollen records, in which early Holocene arboreal pollen abundances decreased contemporaneously with increasing abundances of charcoal (Nelson et al., 2006; Umbanhowar et al., 2006). Early Holocene drying thus may have forced the conversion from forests to grasslands both directly, by lowering the probability of juvenile tree recruitment and establishment, and indirectly, by increasing fire frequency or severity. Conversely, increasing effective moisture after 6 ka BP may have had competing effects on forest expansion: on the one hand, increased moisture may have facilitated the recruitment and establishment of juvenile trees; but on the other hand, increased moisture likely increased grassland productivity and biomass, which in turn increased fuel load and fire severity and slowed the rate of forest encroachment into grasslands (Clark et al., 2001; Nelson and Hu, 2008; Umbanhowar et al., 2006). Thus, in this hypothesis, the interactions among fire regime, aridity and fuel load accelerated the forest to grassland conversion and slowed later reforestation.

6.2 Northern Eurasia

Until now the main sources of information used to reconstruct Late Glacial and Holocene environmental dynamics in temperate grasslands of northern Eurasia are scarcely distributed pollen (e.g., Kremenetski, 2003; Rudaya et al., 2009; Tarasov et al., 2006) and lake-status records (Harrison et al., 1996). During ~14–11.6 ka BP climate amelioration and spread of forests in the present-day forest zone of Europe and Asia caused substantial reduction of the steppe-like environments

(e.g., Harrison and Prentice, 2003) and distribution areas of the steppe animals (FAUNMAP, 1994; Markova and Puzachenko, 2007; Markova et al., 2008). For example, after ~14 ka BP cool steppe was replaced by warm mixed forest in central Greece and warm steppe became fully established around Lake Zeribar in Iran (Tarasov et al., 1998b).

Well-dated postglacial pollen records representing eastern European steppe (regions 2–4) are rare (Tarasov et al., 1998a). The latest regional synthesis presented by Kremenetski (2003) suggests that in the present-day forest steppe, pine and birch dominated the patchy forests between ~12–10.5 ka BP. Warm-loving broad-leaved tree taxa, such as *Ulmus*, *Quercus*, *Tilia* and *Carpinus* were present in the southwestern part of the region (~30°E) from ~10.5 to 8.8 ka BP and reached the longitude 40°E by ~9 ka BP. Pine and birch woods likely grew in the valleys of large rivers flowing to the Black Sea and Azov Sea between 10.5 and 4.8 ka BP, and broad-leaved tree species penetrated to the Black Sea coast via the Dnepr River Valley ~8.8–8.5 ka BP (Kremenetski, 2003). The latter study describes climate of the interval between 6.8 and 4.8 ka BP as the most favorable for the maximal spread of the broad-leaved trees in the patchy and valley forests within the steppe zone of East Europe. The pollen and sedimentary data reveal several climatic oscillations during 5.2–2.6 ka BP. In the Ergeni hilly land north of the Caspian Sea pine forest declined after 5.4–5.2 ka BP and broad-leaved forest area in the lower riches of the Volga River decreased after 2.6–2.1 ka BP (Kremenetski, 2003). After 1.9 ka BP pine disappeared from the lower riches of the Dnepr River in response to the combined effect of climatic deterioration and human impact (Kremenetski, 2003). The human impact on vegetation (outside the archeological sites) can be traced in the pollen records from the forest-steppe belt in Moldova and Ukraine (region 2) since 2 ka BP and in region 3 only during the last millennium (Figure 34.5; Kremenetski, 2003).

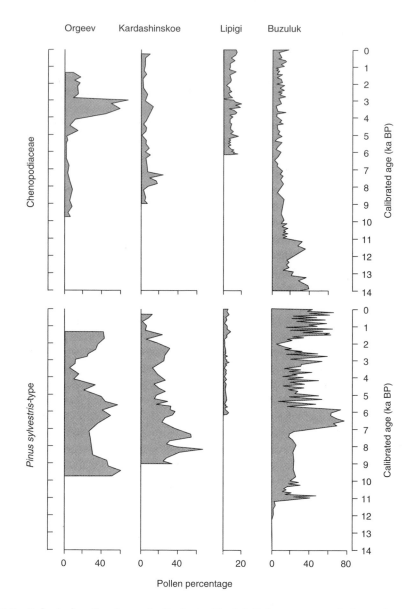

Figure 34.5   Selected pollen types in the Late Glacial–Holocene records from the grassland-dominated regions of East Europe on a W–E transect of sites (based on Kremenetski, 2003). An abrupt decrease in *Pinus sylvestris* pollen (~4–2 ka BP) and simultaneous increase in Chenopodiaceae pollen percentages in Orgeev palaeolake located north of Kishinev in Moldova likely is an indicator of human activities. Similarly, low and variable pine pollen percentages recorded in the Buzuluk peat profile might be associated with the Neolithic agriculture and early metallurgy in the Middle Volga – South Ural region after ~5.5 ka BP.

Archeological records suggest that humans began to settle the steppe north of the Black Sea, constructing villages near rivers and lakes, from ~8 ka BP (Parzinger, 2006), but full agrarian lifestyle was introduced later from the Carpathian region. The western impact intensified in 6.5–6.2 ka BP when copper objects from the Balkan region were placed as prestige goods in the elite burials in the steppes up to Volga. Rassamakin (2004) reports a cultural hiatus in the Black Sea steppe 6.1–5.9/5.8 ka BP linked to the termination of metal production on the Balkans. West of Dnepr the number of sites decreased drastically, but further east towards the Don River new cultures developed in contact with rising metal production centers in the Caucasus. Before 5.5/5.4–5.0/4.9 ka BP the area west of Dnepr again was colonized from the Balkan and the lower Don and Volga regions were under the strong influence of Caucasus (Rassamakin, 2004), which was the source of the new alloy arsenic-bronze. About 5 ka BP a radical change is visible in the archeological records from the East European and northern Kazakh steppes: a substantial increase in sites of which most were probably seasonal; fortified places in higher elevation; growing number of herding animals, such as cattle, sheep/goat, and horse; burials with wood constructions, presence of wagons with four massive wheels and metal weapons (Parzinger, 2006). The horse was domesticated in the Eurasian steppes at 5 ka BP (Ludwig et al., 2009) and soon became one of the most valuable trading goods (granting economic and military superiority) of steppe communities controlling vast grazing grounds (Sherratt, 2003).

In Kazakhstan (region 4) pollen records from Ozerki Swamp and from Pashennoe Lake (Kremenetski et al., 1997; Tarasov et al., 1997) reveal the high percentages of steppe- and desert-associated herbaceous taxa throughout the Late Glacial, early and mid Holocene. The Ozerki pollen record suggests a local presence of the riparian birch forests in the Irtysh River valley around 8.5 ka BP. However, the encroachment of the pine forests into the Kazakhstan steppe became a distinctive feature of the second half of the Holocene. Pollen records from lakes Mokhovoe and Karasye (Kremenetski et al., 1997; Tarasov et al., 1997) suggest that the pine-dominated island forests were established in the northern part of Kazakhstan about 6 ka BP and then spread southwards. Around remotely located Lake Pashennoe, the forest apparently reached its maximum

spread only during the last millennium (Figure 34.6a; Tarasov et al., 2007).

About 3.8–3.5 ka BP the desert-steppe of central Kazakhstan and mountains between Irtysh and Zeravshan became a vital source of tin supply for the first large-scale production of weapons, tools and ornaments made of tin bronze (Parzinger and Boroffka, 2003). The widespread society exploiting copper ores in the Ural mountains and the tin ores in Kazakhstan played a key role in the dissemination of metallurgy in south Siberia and China. It embraced sedentary groups in large settlements, as well as mobile herdsmen and miners closely linked to early urban centres in Bactria (Parzinger, 2008). From about 3 ka BP horse riding resumed paramount importance in warfare. Between 3 and 2 ka BP steppe societies known by different names from historical sources of their southern neighbors formed powerful political entities from the Pacific to the Black Sea coast and have decisively impacted all states on the Eurasian continent.

Pollen records from several lakes in the northwestern Mongolia (region 5) demonstrate generally parallel changes in the reconstructed environments, suggesting that the area was characterized by a relatively dry climate and cool steppe vegetation prior to 10 ka BP. Generally wetter than present conditions occurred between 10 and 5 ka BP causing a spread of boreal trees. The re-establishment of steppe as a dominant vegetation type was associated with a significant decrease in atmospheric precipitation after 5 ka BP (Grunert et al., 2000; Gunin et al., 1999; Krengel, 2000; Schlütz, 2000). Radiocarbon dating of numerous spruce, fir and larch wood fragments identified in peat deposits from Bayan-Sair in the Gobi Altai proves the disappearance of spruce and fir from the local vegetation between 5 and 3.8 ka BP, while larch was present until 3 ka BP (Gunin et al., 1999). The mid Holocene presence of taiga woods in the southern part of Mongolian Altai would require PANN to be 100–150 mm/year higher than at present (Rudaya et al., 2009).

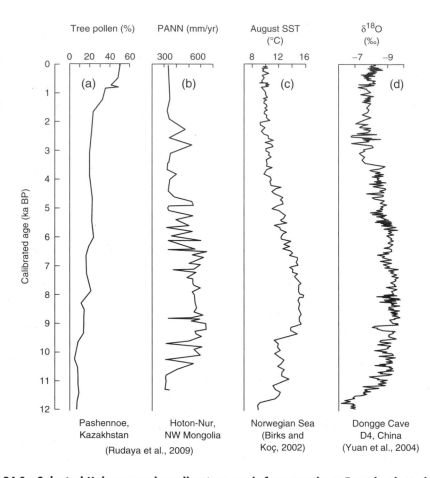

Figure 34.6   Selected Holocene palaeoclimate records from northern Eurasia plotted along the time axis. Individual graphs show: (a) Arboreal pollen percentages from Pashennoe Lake, Kazakhstan, western Central Asia indicating the late Holocene forestation of the steppe associated with the westerly-associated moisture increase. (b) Pollen-based precipitation reconstruction from Hoton-Nur, NW Mongolia, eastern Central Asia, reflecting the early Holocene strengthening and late Holocene weakening of the summer monsoon circulation. (c) Diatom-based August sea-surface temperature reconstruction from the Norwegian Sea. (d) Oxygen isotope record (e.g., indicator of the Pacific summer monsoon intensity) from the stalagmite D4, Dongge Cave, China (see Rudaya et al., 2009, for details and references).

The interpretation of changes in arboreal pollen content in terms of climate is not unambiguous, since the forest degradation and associated changes in the pollen assemblages can be also explained by human activities. However, changes in the moisture balance reconstructed from 11 Mongolian lakes (Tarasov and Harrison, 1998; Walter et al., 2003) closely resemble the pollen-based climate reconstruction from Hoton-Nur

records (Figure 34.6b) and support the assumption that the vegetation dynamic in the Mongolian Altai is primarily driven by changes in precipitation. Archeological records; for example, a great number of petroglyphic images and ritual structures in the river valleys of Tsagaan Salaa and Baga Oigor situated 75 km north of Hoton-Nur, suggest more or less continuous habitation of the Mongolian Altai from 12 ka BP

(Jacobson, 2001). Small communities subsisting on hunting of forest- and steppe-associated animals, fishing and gathering were present in the region prior to 4 ka BP. The number of images of forest animals (e.g., bear, true dear and elk) decreases after 4 ka BP and images of wild boar, an inhabitant of the riparian forests and shrubs, and wild yak, preferring open steppe landscapes, become more important in line with our environmental reconstruction (Figure 34.6b). The emergence of mounted pastoralism after 3 ka BP concluded the transition from a hunting-gathering economy to a nomadic or seminomadic pastoral economy in the Mongolian Altai (Jacobson, 2001). Wood from coniferous trees (mainly larch) was commonly used to construct tomb chambers at 2.9–2 ka BP (Parzinger, 2006); for example, only after the main transformation from woodland to steppe was completed as the result of the climate change towards aridity. It is possible that the disappearance of larch woods from the Gobi Altai after 3 ka BP (Gunin et al., 1999; Miehe et al., 2007) was initially triggered by climate change and then aggravated by human activities under conditions unsuitable for tree regrowth. However, the absence of diagnostic anthropogenic indicators in the Hoton-Nur pollen and diatom records indicates that human disturbance of soils and vegetation cover was less significant than might have been expected. Our conclusion is consistent with recent results from the Khentey Mountains (Schlütz et al., 2008), where the absence of charcoal and other anthropogenic indicators in the palynological records until modern times imply that the steppe islands are of natural origin and not a replacement of the former taiga due to burning or grazing.

In China (regions 6 and 7) pollen data have been used as a major source to reconstruct changes in the late Quaternary vegetation and climate (e.g., Winkler and Wang, 1993; Yu et al., 2000). Ren and Beug (2002) compiled 142 pollen diagrams from China in order to map Holocene changes in pollen percentages of key pollen taxa and to emphasize the spatial and temporal heterogeneity and complexity of forest distribution caused by climatic changes and by human activities.

Mid Holocene environmental and human dynamics in northeastern China were reconstructed from the Taishizhuang pollen record located in the transitional zone between regions 6 and 7, at the modern limit of summer monsoon. These results are compared with other paleoenvironmental records from northeastern China dated to ~6–2 ka BP and archeological data from 100 sites dated to prehistoric and early historic periods (Tarasov et al., 2006; Wagner, 2006).

A quantitative biome reconstruction suggests that between ~5.7 and 4.4 ka BP temperate deciduous forest dominated the vegetation cover around the Taishizhuang site. After that time the landscape became more open, and the scores of the steppe biome were always higher than those of the temperate deciduous forest, except for two oscillations dated to ~4 ka BP and ~3.5 ka BP. However, at ~3.4–2.1 ka BP the common vegetation became steppe, and the landscape was more open in comparison with the previous time interval. The results of the pollen-based precipitation reconstruction suggest that annual precipitation was ~550–750 mm (~100–300 mm higher than present) during the mid Holocene 'forest phase', and ~450–650 mm during the following 'forest–steppe phase'. From ~3.4 ka BP during the 'steppe phase', annual precipitation was similar to modern values (~300–500 mm). Archeological records prove the habitation of northeastern China from ~8.2 ka BP, but do not provide evidence of the use of wood resources intensive enough to influence the regional vegetation development and to leave traces in the pollen assemblages. Both archeological and paleoenvironmental data support the conclusion that changes in pollen composition in northeastern China between 5.7 and 2.1 ka reflect natural variations in precipitation and not major deforestation caused by humans (Tarasov et al., 2006).

Changes in humidity/precipitation derived from pollen and isotope records from different parts of northeastern China between

~112°E and 127°E show great similarity, indicating that a climatically driven shift towards drier environments was synchronous across this broad belt predominated by steppe and forest-steppe vegetation communities. Another conclusion from this comparison is that the climate aridization in northeastern China was not a gradual process. Dry episodes occurred synchronously about 5.3–5, 4.4, 3.4 and after 3 ka BP. They were interrupted by wet pulses distinguished around 5.7, 4.6 and 3.5 ka BP. To investigate the nature of these short-term climatic fluctuations and their impact on human dynamics in the area is an objective for future studies (Tarasov et al., 2006).

7 TEMPERATE GRASSLANDS DURING THE LAST MILLENNIUM

7.1 General trends

During the last millennium grasslands of northern Eurasia and North America experienced steadily increasing pressure from growing human populations; this became rapidly intensified during the last two centuries following the expansion of European settlements and agricultural practices, which converted the natural grasslands into permanent croplands and ranchland (e.g., Coupland, 1992, and references therein). The dominance of sedentary life was leading to extinctions and severe reductions of large steppe animals. The aurochs (*Bos primigenius*) became extinct in the European steppes in 1627. By the mid-eighteenth century *Bison bonansus* and by the mid-nineteenth century saiga antelope (*Saiga tatarica*) and tarpan (*Equus caballus gmelini*) disappeared from the steppes north of the Black Sea. In the nineteenth-century tarpan was exterminated all over Eurasia and Przewalski's wild horse (*Equus przewalskii*) survived only in captivity (Formozov, 1976). Intensive hunting and concurrence from livestock severely damage the populations of wild Eurasian ungulates,

including saiga, wild ass and wild horse, and pressed them into semideserts and deserts. In North America bison (*Bison bison*) too was hunted almost to extinction in the nineteenth century and its formerly large population reduced to a few hundred animals by the mid-1880s.

Estimations of total crop and pasture areas and associated decrease of natural grasslands at the global and large regional scale by Pongratz et al. (2008) show that the global extent of land covered by C_3 natural grasses decreased from 14.06×10^6 km^2 in AD 800 to 11.44×10^6 km^2 in 1992, due to conversions to cropland. The numbers are much more dramatic (e.g., 13.47×10^6 *versus* 2.82×10^6 km^2) if the pastures are included, suggesting that about 80 per cent of the area occupied by potential temperate grassland biome is modified by human activities. The reconstructed changes in three macro-regions, including the US with Canada, former Soviet Union, and China with Mongolia between AD 1000 and 1992 are summarized in Figure 34.7. All graphs demonstrate minimal changes in the grassland area in all macro-regions prior to AD 1650 and progressive strengthening of the anthropogenic land use since that time. Land use patterns have stabilized in the US and former Soviet Union regions during past decades, while natural grassland area continues to decrease in China and Mongolia.

7.2 Modern trends in temperature and precipitation dynamics

Analysis of the meteorological data (Table 34.1) shows an increase in temperature in all selected regions (Figure 34.1) through the observation period between 1936 and 2007, reflecting an anthropogenic-induced global warming. Maximal increase is recorded in regions 3 and 4. Winter temperature contributes more significantly to this warming trend than increase in summer temperature (Table 34.2). Similar patterns occur if one compares average values during the initial 25 years (1936–1960) and last 25

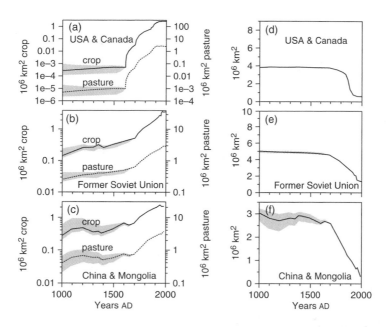

Figure 34.7   Changes in the total area of crop and pasture (a–c) and in the total area of natural grasslands (d–f) reconstructed for the three regions where potential temperate grassland vegetation occupies large area today (after Pongratz et al., 2008).

Table 34.2   Linear trends in northern hemisphere winter (December–February), spring (March–May), summer (June–August) and autumn (September–November) temperature (°C) calculated in seven bioclimatic regions for the observation period 1936–2007

Region	Season			
	Winter	Spring	Summer	Autumn
1	1.8	1.1	−0.3	−0.6
2	2.2	1.7	0.1	0.3
3	4.0	2.1	0.2	1.1
4	3.3	1.7	1.0	1.6
5	2.7	2.2	0.9	0.6
6	1.9	1.1	0.3	0.6
7	2.7	1.9	0.2	0.6

precipitation (Table 34.1) demonstrate an 8–13 percent increase in regions 1–3 (North America and East Europe) and noticeable decrease (9–12 percent) in regions 6–7 (East Asia), but do not show any clear seasonal differences. The regions occupying western (4) and eastern (5) parts of Central Asia, respectively, show only a minor (but also opposing) change in precipitation during 1936–2007 (Table 34.1).

8   MODELING RESULTS

The BIOME4 simulations of temperate and cold grasslands from the LGM to present are summarized in Table 34.4. At present the model generally captures the global patterns of known areas of natural grassland, particularly in the semi-arid 'short-grass' regions of Eurasia, North America, southern Africa and parts of South America (Figure 34.8). Graminoid and forb tundra appears on the

years (1983–2007) within the observation period (Table 34.3). If the length of the latter period is limited to an interval between 2001 and 2007, the differences become even more pronounced, demonstrating the importance of the contributions of recent years to the warming trends in the temperate grasslands of the northern hemisphere. Changes in

Table 34.3 Contribution of different seasons (numbers XII to XI indicate months from December to November) to the temperature rise between 1936 and 2007 in selected climatic regions

Region	Seasonal temperature anomalies (°C) between 1983–2007 and 1936–1960 means					Seasonal temperature anomalies (°C) between 2001–2007 and 1936–1960 means				
Number	XII–II	III–V	VI–VIII	IX–XI	Year	XII–II	III–V	VI–VIII	IX–XI	Year
1	0.3	0.3	0.0	−0.2	0.1	0.4	0.4	0.1	0.0	0.2
2	0.7	0.6	0.0	0.1	0.4	0.9	0.8	0.3	0.6	0.7
3	1.8	1.0	0.1	0.6	0.9	2.1	1.4	0.2	1.2	1.2
4	1.3	0.6	0.4	0.6	0.7	1.3	1.3	0.3	1.1	1.0
5	0.7	0.5	0.2	0.2	0.4	0.6	0.8	0.6	0.4	0.6
6	0.5	0.3	0.1	0.2	0.3	0.7	0.6	0.3	0.3	0.5
7	0.1	0.1	0.0	0.0	0.1	0.1	0.1	0.0	0.1	0.1

Table 34.4 Area of temperate grassland simulated by the BIOME4 model for four time slices in the Last Glacial and Holocene

Time	Biome area, 10⁶ km²	
	Temperate grassland	Graminoid and forb tundra
0 ka	7.52	1.71
6 ka	8.73	2.45
10 ka	10.94	5.80
21 ka	5.95	5.05

Tibetan Plateau, in Mongolia and northeastern China, and in the high Andes. BIOME4 fails to simulate extensive grassland in more humid areas, including the pampas grasslands in South America and the tall-grass prairie in central North America. During the mid and early Holocene, grassland areas expanded significantly, particularly in central and eastern Asia (Figure 34.8). At the LGM, the area of temperate grasslands was reduced by roughly 25 percent compared to the pre-industrial Holocene (PIH), but the area of graminoid and forb tundra was nearly three times larger, with significant areas of boreal Eurasia and Beringia occupied by this herbaceous tundra type. In the early Holocene, graminoid and forb tundra occupy significant areas of Eurasia, generally to the south of their location at the LGM and contiguous with temperate grasslands.

While there are some known limitations to the BIOME4 simulation of temperate grasslands, these simulations are illustrative of the changes in grassland area that have occurred over the Late Glacial and Holocene. Cold, dry climate and low atmospheric CO_2 concentrations, which affect the water use efficiency of woody plants, during the full glacial, led to the development of extensive areas of graminoid and forb tundra, a biome that has few analogues with the present-day landscape. This land-cover type is very important as it is hypothesized to have been the primary habitat of the woolly mammoth and other Pleistocene megafauna (Guthrie, 2006). In the early and mid Holocene, mid-continental aridity, particularly in Eurasia, is hypothesized to have been a result of higher summertime insolation and resulting changes in large-scale circulation. This aridity resulted in much larger areas of temperate and cold grasslands than are observed at present. Increased disturbance, from, for example, fires resulting from aridity would have also played a role in maintaining large grassland areas in the early and mid Holocene. From mid Holocene to modern pre-industrial times, grassland area does expand marginally into the northern Canadian prairies, but in most regions grassland area is reduced. Cooler and wetter conditions that prevailed in much of

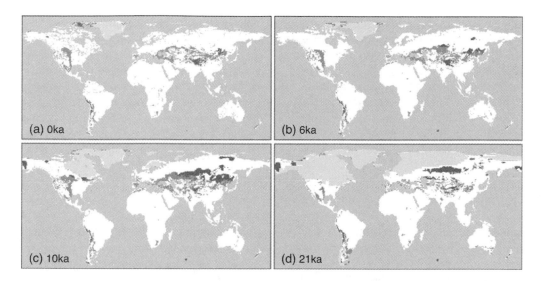

Figure 34.8 Temperate grasslands (grey) and graminoid and forb tundra (dark grey) simulated by the BIOME4 model for four time slices: (a) LGM, (b) early Holocene, (c) mid Holocene, and (d) preindustrial Holocene, driven by the HAD-UM GCM climatology. Light grey indicates areas covered with the ice sheets.

the northern hemisphere during recent times led to a ~15 percent reduction in temperate grassland area and ~30 percent reduction in graminoid and forb tundra. This reduction in grassland area was concurrent with the development of mosaic ecosystems with significant amounts of woody vegetation mixed in with grassland types; for example, on the eastern part of the Great Plains in North America.

9 CONCLUSION

There is little direct evidence for the LGM grassland communities from both North American and Eurasian mid-latitudes, and the distribution and composition of LGM grasslands remains a bit of a mystery. Pollen data from North America suggest that the extent and density of the temperate grassland biome must have been much reduced relative to the present at the expense of boreal parklands, while pollen records from Eurasia indicate greater-than-present latitudinal and

longitudinal extension of the cool steppe or COGS vegetation communities. Paleontological records from both continents point towards a greater-than-present distribution area of grassland-associated mammals. The modeling results suggest a 25 percent reduction of the temperate grassland area, but a three-fold increase in the graminoid and forb tundra area during the LGM, thus allowing a reasonable compromise in the interpretation of the environmental records from two continents.

Both plant and animal records suggest that the temperate grassland biome was shaped in its pre-industrial form during the Holocene. Synthesis of the North American and Eurasian environmental records help us to better understand the spatial structure of Holocene climate changes in a hydrological regime. In mid-continental North America the mid Holocene dry interval occurred ~9-4 ka BP and was driven by a combination of factors, including higher summer insolation, deglaciation and shifts in atmospheric circulation. The Eurasian records demonstrate spatially and temporally different climate histories, indicating that eastern Central Asian grass-lands became wetter than at present between 10 and 5 ka BP in response to the early Holocene strengthening of the summer mon-soon circulation. The re-establishment of steppe as a dominant vegetation type was associated with a decrease in atmospheric precipitation after 5 ka BP caused by a weak-ening of the summer monsoon. Grasslands of western Central Asia experienced a drier-than-present climate during the first half of the Holocene and a shift towards wetter con-ditions occurred there after 6 ka BP, when the mid-latitudinal belt, stretching from the Baltic Sea to northern Kazakhstan and south-ern Siberia, came under the control of the Atlantic air masses. In line with the data, BIOME4 model simulations show that by 10 ka BP temperate grasslands occupied their greatest area during the last 21 ka BP and were reduced to a slightly higher-than-present extent by 6 ka BP.

Records from the Great Plains show rapid climate shifts (mainly alternation in moisture balance) linked to 100-year to 130-year cycles of drought variability well-pronounced during the mid Holocene. A number of short-term dry and wet pulses were distinguished in the mid and late Holocene records from eastern and western Asia, suggesting that large-scale gradual changes in the grassland environments were complicated by the multi-decadal- and centennial-scale processes, whose mechanisms and origin are poorly known so far.

Human impacts on the Holocene grassland environments are poorly known. Regional case studies in northern China and Mongolia do not provide evidence of the use of wood resources intensive enough to influence regional vegetation development and to leave traces in the pollen assemblages prior to 1 ka BP. Both archeological and paleoenviron-mental data support the conclusion that changes in pollen composition in northeast-ern China between 5.7 and 2.1 ka BP reflect natural variations in precipitation and not major deforestation caused by humans. The absence of charcoal and other anthropogenic indicators in the palynological records from the forested areas of central Mongolia until modern times imply that the steppe islands there are of natural origin and not a replace-ment of the former taiga due to burning or grazing. The scattered woods in East European steppes and forest-steppes (espe-cially in the Black Sea and South Ural regions) were probably more affected by human practices, including farming and animal husbandry (after 7.5 ka BP) and metal production (after 5.2 ka BP). Absence of high-resolution and accurately-dated Holocene pollen records representing East European grasslands remains the main factor, which prevents more comprehensive conclusions on past climate and human-environmental inter-actions there.

The reconstruction of historical land cover demonstrates minimal changes in the grass-land area in all macro-regions prior to AD 1650 in line with the coarser-resolution envi-ronmental records. During the last few centu-ries, humans greatly contributed to the modification and impoverishment of plant and animal communities in temperate grass-lands. In 1992 global extent of land covered by C_3 natural grasses decreased by 20 per-cent due to its conversion to cropland, and by 80 percent if pastures are considered.

Meteorological data show a pronounced increase in temperature in seven selected bioclimatic regions of North America and northern Eurasia during 1936–2007, reflect-ing anthropogenic-induced global warming. Winter temperature contributes more signifi-cantly to this warming trend than summer temperatures do. Changes in precipitation demonstrate an 8–13 percent increase in the grassland regions of North America and East Europe and a 9–12 percent decrease in the regions representing East Asia. The regions occupying western and eastern Central Asia show only a minor, although opposing, change in precipitation during 1936–2007. This recorded asymmetry resembles recon-structed patterns of the Holocene climate in the mid-continental Eurasian grasslands, suggesting that hydrological parameters in western and eastern regions react differently to global-scale changes in temperature.

Records in the temperate grassland regions of North America and Eurasia indicate a significant degree of variability in the composition and extent of grasslands and in climate regime since the LGM. The broad spatial scale of these variations suggest that they are driven by hydrological variability at centennial to orbital time scales which can be complicated by human activities and herbivore population dynamics at individual sites. The relative importance of these drivers and the effects of interactions among them remains a critical area of paleoclimatic research.

The capability of the BIOME-family global vegetation models correctly to reproduce spatial patterns of natural temperate grasslands under the present-day climate, justifies their use in simulating global distribution of the cool and cold grasslands under past and future climate scenarios. The resulting maps of grassland distribution, when compared to paleoecological records, may help in the interpretation of fossil data, particularly in regions with scarce data sets. We further expect that the new BIOME5 model will help to correct some failings of BIOME4 revealed by the comparison with modern and fossil data.

ACKNOWLEDGEMENTS

We are grateful to Nikolai Gorban for his help with drawing the figures, and we thank Dr Patrick Bartlein for sending an updated version of Figure 34.3.

REFERENCES

Allen J. R. M., Brandt U., Brauer A., Hubberten H.-W., Huntley B., Keller J. et al. 1999. Rapid environmental changes in southern Europe during the last glacial period. *Nature* 400: 740–743.

Baker R. G., Bettis III E. A., Denniston R. F., Gonzalez L. A., Strickland L. E. and Krieg J. R. 2002. Holocene paleoenvironments in southeastern Minnesota – chasing the prairie-forest ecotone. *Palaeogeography Palaeoclimatology Palaeoecology* 177: 103–122.

Baker R. G., Bettis E. A. III, Mandel R. D., Dorale J. A. and Fredlund G. G. 2009. Mid-Wisconsinan environments on the eastern Great Plains. *Quaternary Science Reviews* 28: 873–889.

Bettis E. A. III, Muhs D. R., Roberts H. M. and Wintle A. G. 2003. Last glacial loess in the conterminous USA. *Quaternary Science Reviews* 22: 1907–1946.

Birks H. H. 1973. Modern macrofossil assemblages in lake sediments in Minnesota, in Birks H. J. B. and West R. G. (eds) *Quaternary Plant Ecology*. Oxford: Blackwell Scientific Publications, pp. 173–189.

Bromwich D. H., Toracinta E. R., Oglesby R. J., Fastook J. L. and Hughes T. J. 2005. LGM summer climate on the southern margin of the Laurentide Ice Sheet: Wet or dry? *Journal of Climate* 18: 3317–3338.

Clark J. S., Grimm E. C., Donovan J. J., Fritz S. C., Engstrom D. R. and Almendinger J. E. 2002. Drought cycles and landscape responses to past aridity on prairies of the northern Great Plains, USA. *Ecology* 83: 595–601.

Clark J. S., Grimm E. C., Lynch J. and Mueller P. G. 2001. Effects of Holocene climate change on the C4 grassland/woodland boundary in the northern plains, USA. *Ecology* 82: 620–636.

Cook E. R., Bartlein P. J., Diffenbaugh N. S., Seager R., Shuman B., Webb R. S. et al. 2008. Hydrological variability and change, in Clark P. U., Weaver A. J., Brook E., Cook E. R., Delworth T. L. and Steffen K. (eds) *Abrupt Climate Change*. A Report by the US Climate Change Science Program and the Subcommittee on Global Change Research. Reston: US Geological Survey, pp. 143–257.

Coupland R. T. (ed.) 1992. *Natural Grasslands Introduction and Western Hemisphere*. Ecosystems of the World 8A. Amsterdam: Elsevier.

Coupland R. T. (ed.) 1993. *Natural Grasslands Eastern Hemisphere and Resume*. Ecosystems of the World 8B. Amsterdam: Elsevier.

DeFries R. S., Townshend J. R. G. and Hansen M. C. 1999. Continuous fields of vegetation characteristics at the global scale at 1-km resolution. *Journal of Geophysical Research* 104: 16911–16923.

Denniston R. F., Gonzalez L. A., Asmerom Y., Reagan M. K. and Recelli-Snyder H. 2000. Speleothem carbon isotopic records of Holocene environments in the Ozark Highlands, USA. *Quaternary International* 67: 21–27.

Diffenbaugh N. S., Ashfaq M., Shuman B., Williams J. W. and Bartlein P. J. 2006. Summer aridity in the United States: Response to Mid-Holocene changes in insolation and sea surface temperature. *Geophysical Research Letters* 33, DOI: 10.1029/2006GL028012.

Domrös M. and Peng G. 1988. *The Climate of China.* Berlin: Springer.

Elias S. (ed.) 2007. *Encyclopedia of Quaternary Science.* Amsterdam: Elsevier.

Elias S.A. and Schreve D. 2007. Late Pleistocene megafaunal extinctions, in Elias S. (ed.) *Encyclopedia of Quaternary Science.* Amsterdam: Elsevier, pp. 3202–3216.

Fang J. Q. 1991. Lake evolution during the past 30,000 years in China, and its implications for environmental change. *Quaternary Research* 36: 37–60.

FAUNMAP. 1994. A database documenting Late Quaternary Distributions of mammal species in the United States. FAUNMAP Working Group 1994. Springfield: Illinois State Museum.

Formozov A. N. 1976. *Zveri, ptitsy i ikh vzaimosvyazi so sredoi obitaniya.* Moscow: Nauka

Frenzel B., Pecsi M. and Velichko A. A (eds) 1992. *Atlas of Palaeoclimates and Palaeoenvironments of the Northern Hemisphere.* Budapest: INQUA/Hungarian Academy of Sciences.

Fritz S. C. 2008. Deciphering climatic history from lake sediments. *Journal of Paleolimnology* 39: 5–16.

Grimm E. C. 1984. Fire and other factors controlling the Big Woods vegetation of Minnesota in the mid-nineteenth century. *Ecological Monographs* 54: 291–311.

Grimm E. C. 2001. Trends and palaeoecological problems in the vegetation and climate history of the northern Great Plains U.S.A. Biology and Environment. *Proceedings of the Royal Irish Academy* 101B: 47–64.

Grunert J., Lehmkuhl F. and Walther M. 2000. Paleoclimatic evolution of the Uvs Nuur basin and adjacent areas (Western Mongolia). *Quaternary International* 65/66: 171–191.

Guiot J. 1990. Methodology of the last climatic cycle reconstruction from pollen data. *Palaeogeography Palaeoclimatology Palaeoecology* 80: 49–69.

Gunin P. D., Vostokova E. A., Dorofeyuk N. I., Tarasov P. E. and Black C. C. (eds) 1999. *Vegetation dynamics of Mongolia.* Geobotany 26. Dordrecht: Kluwer Academic Publishers.

Guthrie R. D. 2006. New carbon dates link climatic change with human colonization and Pleistocene extinctions. *Nature* 441: 207–209.

Harrison S. P. and Metcalfe S. E. 1985. Variations in lake levels during the Holocene in North America: An indicator of changes in atmospheric circulation patterns. *Géographie Physique et Quaternaire* 39: 141–150.

Harrison S. and Prentice I. C. 2003. Climate and CO_2 controls on global vegetation distribution at the last glacial maximum: analysis based on palaeovegetation data, biome modelling and palaeoclimate simulations. *Global Change Biology* 9: 983–1004.

Harrison S. P., Kutzbach J. E., Liu Z., Bartlein P. J., Muhs D., Prentice I. C. and Thompson R. S. 2003. Mid-Holocene climates of the Americas: a dynamical response to changed seasonality. *Climate Dynamics* 20: 663–688.

Harrison S. P., Yu G. and Tarasov P. E. 1996. Late Quaternary lake-level record from northern Eurasia. *Quaternary Research* 45: 138–159.

Hartley W. 1950. The global distribution of tribes of the Gramineae in relation to historical and environmental factors. *Australian Journal of Agricultural Research* 1: 355–373.

Haxeltine A. and Prentice I. C. 1996. BIOME3: an equilibrium terrestrial biosphere model based on ecophysiological constraints, resource availability, and competition among plant functional types. *Global Biogeochemical Cycles* 10: 693–709.

Jacobson E. 2001. Tsagaan Salaa/Baga Oigor: The physical content and palaeoenvironmental considerations, in Jacobson E., Kubarev V. and Tseevendorj D. (eds) *Répertoire des Pétroglyphes D'Asie Centrale, Mémoires de la Mission Archéologique Française en Asie Centrale 6.* Paris: De Boccard, pp. 7–15.

Jolly D. and Haxeltine A. 1997. Effect of low glacial atmospheric CO_2 on tropical African montane vegetation. *Science* 276: 786–788.

Kageyama M., Peyron O., Pinot S., Tarasov P., Guiot J., Joussaume S. et al. 2001. The Last Glacial Maximum climate over Europe and western Siberia: a PMIP comparison between models and data. *Climate Dynamics* 17: 23–43.

Kaplan J. O. 2001. Geophysical applications of vegetation modeling. PhD thesis, Lund University.

Kaplan J. O. and New M. 2006. Arctic climate change with a 2°C global warming: Timing, climate patterns and vegetation change. *Climatic Change* 79: 213–241.

Kaplan J. O., Folberth G. and Hauglustaine D. A. 2006. Role of methane and biogenic volatile organic compound sources in late glacial and Holocene fluctuations of atmospheric methane concentrations. *Global Biogeochemical Cycles* 20, GB2016, 16 PP., 2006 doi:10.1029/2005GB002590.

Kremenetski K. V. 2003. Steppe and forest-steppe belt of Eurasia: Holocene environmental history, in Levine M., Renfrew C. and Boyle K. (eds)

Prehistoric Steppe Adaptation and the Horse. Cambridge: McDonald Institute for Archaeological Research University of Cambridge, pp. 11–27.

Kremenetski C. V., Tarasov P. E. and Cherkinsky A. E. 1997. Postglacial development of Kazakhstan pine forests. *Geographie Physique et Quaternaire* 51: 391–404.

Krengel M. 2000. Discourse on history of vegetation and climate in Mongolia–palynological report of sediment core Bayan Nuur I (NW-Mongolia). *Berliner geowissenschaftliche Abhandlungen A* 205: 80–84.

Küchler A. W. 1975. *Potential Natural Vegetation of the Conterminous United States.* New York: American Geographical Society.

Kutzbach J. E., Gallimore R., Harrison S. P., Behling P., Selin R. and Laarif F. 1998. Climate and biome simulations for the past 21,000 years. *Quaternary Science Reviews* 17: 473–506.

Lauenroth W. K. and Milchunas D. G. 1992. Short-grass steppe, in Coupland R. T. (ed.) *Natural Grasslands Introduction and Western Hemisphere.* Ecosystems of the World 8A. Amsterdam: Elsevier, pp. 183–226.

Lavrenko E. M. and Karamysheva Z. V. 1993. Soviet Union and Mongolia, in Coupland R. T. (ed.) *Natural Grasslands Eastern Hemisphere and Resume.* Ecosystems of the World 8B. Amsterdam: Elsevier, pp. 3–59.

Leemans R. and Cramer W. 1991. *The IIASA Climate Database for Mean Monthly Values of Temperature, Precipitation and Cloudiness on a Global Terrestrial Grid.* RR–91–18 Laxenburg: International Institute of Applied Systems Analysis.

Ludwig A., Pruvost M., Reissmann M., Benecke N., Brockmann G. A., Castaños P. et al. 2009. Coat color variation at the beginning of horse domestication. *Science* 324: 485.

MacDonald G. M., Beukens R. P. and Kieser W. E. 1991. Radiocarbon dating of limnic sediments: A comparative analysis and discussion. *Ecology* 72: 1150–1155.

Markova A. and Puzachenko A. 2007. Late Pleistocene of Northern Asia, in Elias S. (ed.) *Encyclopedia of Quaternary Science.* Amsterdam: Elsevier, pp. 3158–3175.

Markova A. K., van Kolfschoten T., Bohncke S., Kosintsev P. A., Mol J., Puzachenko A. et al. 2008. *Evolution of European Ecosystems During Pleistocene-Holocene Transition (24–8 kyr BP).* Moscow: KMK Scientific Press.

Mason J. A., Joeckel R. M. and Bettis E. A. III 2007. Middle to late Pleistocene loess record in eastern Nebraska, USA., and implications for the unique

nature of Oxygen Isotope Stage 2. *Quaternary Science Reviews* 26: 773–792.

Mason J. A., Miao X., Hanson P. R., Johnson W. C., Jacobs P. M. and Goble R. J. 2008. Loess record for the Pleistocene-Holocene transition on the northern and central Great Plains, USA. *Quaternary Science Reviews* 27: 1772–1783.

McAndrews J. H. 1966. *Postglacial History of Prairie, Savanna, and Forest in Northwestern Minnesota.* Durham, NC: Torrey Botanical Club.

Miehe G., Schlütz F., Miehe S., Opgenoorth L., Cermak J., Samiya R. et al. 2007. Mountain forest islands and Holocene environmental changes in Central Asia: A. case study from the southern Gobi Altay, Mongolia. *Palaeogeography Palaeoclimatology Palaeoecology* 250: 150–166.

Mitchell T. D. and Jones P. D. 2005. An improved method of constructing a database of monthly climate observations and associated high-resolution grids. *International Journal of Climatology* 25: 693–712.

Nelson D. M. and Hu F. S. 2008. Patterns and drivers of Holocene vegetational change near the prairie-forest ecotone in Minnesota: revisiting McAndrews' transect. *New Phytologist* 179: 449–459.

Nelson D. M., Hu F. S., Grimm E. C., Curry B. B. and Slate J. E. 2006. The influence of aridity and fire on Holocene prairie communities in the eastern Prairie Peninsula. *Ecology* 87: 2523–2536.

New M., Hulme M. and Jones P. 2000. Representing twentieth century space-time climate variability. II: Development of 1901–1996. Monthly grids of terrestrial surface climate. *Journal of Climate* 13: 2217–2238.

Nordt L., von Fischer J. and Tieszen L. 2007. Late Quaternary temperature record from buried soils of the North American Great Plains. *Geology* 35: 159–162.

Olson J. S., Watts J. A. and Allison L. J. 1983. *Carbon in Live Vegetation of Major World Ecosystems.* ORNL–5862. Oak Ridge: Oak Ridge National Laboratory.

Overpeck J. T., Webb T. III and Prentice I. C. 1985. Quantitative interpretation of fossil pollen spectra: dissimilarity coefficients and the method of modern analogs. *Quaternary Research* 23: 87–108.

Parzinger H. 2006. *Die Frühen Völker Eurasiens vom Neolithikum bis zum Mittelalter.* München: C. H. Beck.

Parzinger H. 2008. The 'Silk Roads' concept reconsidered: about transfers, transportation and transcontinental interactions in prehistory. *The Silk Road* 5: 7–15.

Parzinger H. and Boroffka N. 2003. *Das Zinn der Bronzezeit in Mittelasien I.* Mainz: Verlag Philipp von Zabern.

Peterson T. C. and Vose R. S. 1997. An overview of the Global Historical Climatology Network temperature data base. *Bulletin of the American Meteorological Society* 78: 2837–2849.

Peyron O., Guiot J., Cheddadi R., Tarasov P. E., Reille M., Beaulieu J-L. de, Bottema S. and Andrieu V. 1998. Climatic reconstruction in Europe from pollen data, 18,000 years before present. *Quaternary Research* 49 : 183–196.

Pongratz J., Reick C., Raddatz T. and Claussen M. 2008. *A Global Land Cover Reconstruction AD 800 to 1992. Technical Description.* Hamburg: Berichte zur Erdsystemforschung 51, Max Planck Institute for Meteorology.

Prasad V., Strömberg C. A. E., Alimohammadian H. and Sahni A. 2005. Dinosaur coprolites and the early evolution of grasses and grazers. *Science* 310: 1177–1180.

Prentice I. C., Cramer W., Harrison S. P., Leemans R., Monserud R. A. and Solomon A. M. 1992. A global biome model based on plant physiology and dominance, soil properties and climate. *Journal of Biogeography* 19: 117–134.

Prentice I. C., Guiot J., Huntley B., Jolly D. and Cheddadi R. 1996. Reconstructing biomes from palaecological data: a general method and its application to European pollen data at 0 and 6 ka. *Climate Dynamics* 12: 185–194.

Prentice I. C., Jolly D. and BIOME 6000 participants 2000. Mid-Holocene and glacial-maximum vegetation geography of the northern continents and Africa. *Journal of Biogeography* 27: 507–519.

Ramankutty N. and Foley J. A. 1999. Estimating historical changes in global land cover: Croplands from 1700 to 1992. *Global Biogeochemical Cycles* 13: 997–1027.

Ramankutty N., Graumlich L., Achard F., Alves D., Chhabra A., DeFries R. et al. 2006. Global land cover change: recent progress, remaining challenges, in Lambin E. and Geist H. (eds) *Land Use and Land Cover Change: Local Processes, Global Impacts.* Berlin: Springer, pp. 9–39.

Rassamakin J. J. 2004. *Die nordpontische Steppe in der Kupferzeit.* Mainz: Verlag Philipp von Zabern.

Ren G. and Beug H.-J. 2002. Mapping Holocene pollen data and vegetation of China. *Quaternary Science Reviews* 21: 1395–1422.

Ripley E. A. 1992. Grassland climate in Coupland R. T. (ed.) *Natural Grasslands Introduction and Western Hemisphere.* Ecosystems of the World 8A. Amsterdam: Elsevier, pp. 7–24.

Rossignol J., Moine O. and Rousseau D.-D. 2004. The Buzzard's Roost and Eustis mollusc sequences: Comparison between the paleoenvironments of two sites in the Wisconsinan loess of Nebraska, USA. *Boreas* 33: 145–154.

Rudaya N., Tarasov P., Dorofeyuk N., Solovieva N., Kalugin I., Andreev A. et al. 2009. Holocene environments and climate in the Mongolian Altai reconstructed from the Hoton-Nur pollen and diatom records: a step towards better understanding climate dynamics in Central Asia. *Quaternary Science Reviews* 28: 540–554.

Schlütz F. 2000. Palynological investigations in the Turgen-Kharkhiraa mountains, Mongolian Altay. *Berliner geowissenschaftliche Abhandlungen A* 205: 85–90.

Schlütz F., Dulamsuren C., Wieckowska M., Mühlenberg M. and Hauck M. 2008. Late Holocene vegetation history suggests natural origin of steppes in the northern Mongolian mountain taiga. *Palaeogeography Palaeoclimatology Palaeoecology* 261: 203–217.

Schneider U., Fuchs T., Meyer-Christoffer A. and Rudolf B. 2008. *Global Precipitation Analysis Products of the GPCC.* Offenbach am Main: Global Precipitation Climatology Centre (GPCC), Deutscher Wetterdienst.

Schubert S. D., Suarez M. J., Pegion P. J., Koster R. D. and Bacmeister J. T. 2004. On the cause of the 1930s Dust Bowl. *Science* 303: 1855–1859.

Sheppard P. R., Tarasov P. E., Graumlich L. J., Heussner K.-U., Wagner M., Österle H. and Thompson L. G. 2004. Annual precipitation since 515 BC reconstructed from living and fossil juniper growth of northeastern Qinghai Province, China. *Climate Dynamics* 23: 869–881.

Sherratt A. 2003. The horse and the wheel: the dialectics of change in the circum-pontic region and adjacent areas, 4500–1500 BC, in Levine M., Renfrew C. and Boyle K. (eds) *Prehistoric Steppe Adaptation and the Horse.* Cambridge: McDonald Institute for Archaeological Research University of Cambridge, pp. 233–252.

Shi P. and Song C. 2003. Palynological records of environmental changes in the middle part of Inner Mongolia, China. *Chinese Science Bulletin* 48: 1433–1438.

Shuman B. N., Bartlein P. J., Logar N., Newby P. and Webb T. III 2002. Parallel climate and vegetation responses to the early Holocene collapse of the Laurentide Ice Sheet. *Quaternary Science Reviews* 21: 1793–1805.

Shuman B., Huang Y., Newby P. and Wang Y. 2006. Compound-specific isotopic analyses track changes in the seasonality of precipitation in the northeastern

United States at ca. 8200 cal yr BP. *Quaternary Science Reviews* 25: 2992–3002.

Sims P. L. and Risser P. G. 2000. Grasslands, in Barbour M. G. and Billings W. D. (eds) *North American Terrestrial Vegetation*. Cambridge: Cambridge University Press, pp. 323–356.

Tarasov P. E., Cheddadi R., Guiot J., Bottema S., Peyron O., Belmonte J. et al. 1998b A method to determine warm and cool steppe biomes from pollen data; application to the Mediterranean and Kazakhstan Regions. *Journal of Quaternary Science* 13: 335–344.

Tarasov P. E., Guiot J., Cheddadi R., Andreev A. A., Bezusko L. G., Blyakharchuk T. A. et al. 1999b. Climate in northern Eurasia 6000 years ago reconstructed from pollen data. *Earth and Planetary Science Letters* 171: 635–645.

Tarasov P. E. and Harrison S. P. 1998. Lake status records from the former Soviet Union and Mongolia: a continental-scale synthesis. *Paläoklimaforschung* 25: 115–130.

Tarasov P., Jin G. and Wagner M. 2006. Mid-Holocene environmental and human dynamics in northeastern China reconstructed from pollen and archaeological data. *Palaeogeography Palaeoclimatology Palaeoecology* 241: 284–300.

Tarasov P. E., Jolly D. and Kaplan J. O. 1997. A continuous Late Glacial and Holocene record of vegetation changes in Kazakhstan. *Palaeogeography Palaeoclimatology Palaeoecology* 136: 281–292.

Tarasov P. E., Peyron O., Guiot J., Brewer S., Volkova V. S., Bezusko L. G. et al. 1999a. Last Glacial Maximum climate of the Former Soviet Union and Mongolia reconstructed from pollen and plant macrofossil data. *Climate Dynamics* 15: 227–240.

Tarasov P. E., Webb T. III, Andreev A. A., Afanas'eva N. B., Berezina N. A., Bezusko L. G. et al. 1998a. Present-day and mid-Holocene Biomes Reconstructed from Pollen and Plant Macrofossil Data from the Former Soviet Union and Mongolia. *Journal of Biogeography* 25: 1029–1053.

Tarasov P., Williams J. W., Andreev A., Nakagawa T., Bezrukova E., Herzschuh U. et al. 2007. Satellite- and pollen-based quantitative woody cover reconstructions for northern Asia: verification and application to late-Quaternary pollen data. *Earth and Planetary Science Letters* 264: 284–298.

Umbanhowar C. E. Jr, Camill P., Geiss C. E. and Teed R. 2006. Asymmetric vegetation responses to mid-Holocene aridity at the prairie-forest ecotone in south-central Minnesota. *Quaternary Research* 66: 53–66.

VEMAP Members 1995. Vegetation/ecosystem modeling and analysis project: Comparing biogeography and biogeochemistry models in a continental-scale study of terrestrial ecosystem responses to climate change and CO2 doubling. *Global Biogeochemical Cycles* 9: 407–437.

Wagner M. 2006. *Neolithikum und fruehe Bronzezeit in Nordchina vor 8000 bis 3500 Jahren. Die Nordoestliche Tiefebene (Suedteil)*. Mainz: Philipp von Zabern.

Walter M., Wünnemann B. and Tschimeksaichan A. 2003. Seen und Paläoseen in der Mongolei und Nordwestchina. *Petermanns Geographische Mitteilungen* 147: 40–47.

Webb R. S. and Webb T. III 1988. Rates of sediment accumulation in pollen cores from small lakes and mires of eastern North America. *Quaternary Research* 30: 284–297.

Webb T. III 1987. The appearance and disappearance of major vegetational assemblages: Long-term vegetational dynamics in eastern North America. *Vegetatio* 69: 177–187.

Webb T. III, Cushing E. J. and Wright Jr H. E. 1983. Holocene changes in the vegetation of the Midwest, in Wright Jr H. E. (ed.) *Late-Quaternary Environments of the United States*. Minneapolis: University of Minnesota Press, pp. 142–165.

Wells P. V. and Stewart J. D. 1987. Cordilleran-boreal taiga and fauna on the central Great Plains of North America, 14,000–18,000 years ago. *American Midland Naturalist* 118: 94–106.

Williams J. W. 2003. Variations in tree cover in North America since the last glacial maximum. *Global and Planetary Change* 35: 1–23.

Williams J. W. and Jackson S. T. 2003. Palynological and AVHRR observations of modern vegetational gradients in eastern North America. *The Holocene* 13: 485–497.

Williams J. W., Shuman B. and Bartlein P. J. 2009. Rapid responses of the prairie-forest ecotone to early Holocene aridity in mid-continental North America. *Global and Planetary Change* 66: 195–207.

Williams J. W., Shuman B., Bartlein P. J., Diffenbaugh N. S. and Webb T. III 2010. Rapid, time-transgressive, and variable responses to early Holocene midcontinental drying in North America. *Geology* 38: 135–138.

Williams J. W., Shuman B. N., Webb T. III, Bartlein P. J. and Leduc P. L. 2004. Late Quaternary vegetation dynamics in North America: Scaling from taxa to biomes. *Ecological Monographs* 74: 309–334.

Willis K. J. and McElwain J. C. 2002. *The Evolution of Plants*. Oxford: Oxford University Press.

Winkler M. G. and Wang P. K. 1993. The late-Quaternary vegetation and climate in China, in Wright H. E.,

Kutzbach J. E., Webb T. III, Ruddiman W. D., Street-Perrott F. A. and Bartlein P. J. (eds) *Global Climates Since the Last Glacial Maximum*. Minneapolis: University of Minnesota Press, pp. 221–261.

Yu G., Chen X., Ni J., Cheddadi R., Guiot J., Han H. et al. 2000. Palaeovegetation of China: a pollen data-based synthesis for the mid-Holocene and last glacial maximum. *Journal of Biogeography* 27: 635–664.

35

Environmental Change in the Arctic and Antarctic

Marianne S.V. Douglas

1 INTRODUCTION

Polar environments are unique and amongst the most extreme environments on Earth. Characterized by severe conditions, their high latitudes are typified by low seasonal temperatures – the coldest temperature (–89.2°C) has been recorded at Vostok, in Antarctica. An extreme photo period transitions from the dark winter polar night to the summer period of continuous 24 hour daylight. The summer growing season is short and nutrients are usually low. Ice is present in the form of glaciers, permafrost and sea ice. Several classic texts provide in-depth accounts of these environments, including descriptions of physical geography, climate and ecological conditions. Stonehouse (1989) and Sugden (1982) compare and contrast the two poles. Born and Böcher (2001) focus on Greenland, and French and Slaymaker (1993) describe Canada's cold environments. Hansom and Gordon (1998) focus on the Antarctic and its resources while Antarctic ecosystems are detailed by Bargagli (2005), Beyer and Bolter (2002), and Lyons et al. (1997).

Polar environments are dynamic and responsive to shifting climatic conditions.

General circulation models predict that polar regions are experiencing the highest rate of climate change (IPCC, 2007) resulting in unprecedented rates of environmental change. As outlined by the Intergovernmental Panel on Climate Change (IPCC) (2007), anthropogenic forcing, such as the increase in greenhouse gases combined with natural forcing factors (see Chapter 24 and other chapters in Section III) to effect these changes. Assessment of change during the Anthropocene requires long-term data or baseline data from which departures from the natural variability can be measured. Instrumental monitoring data sets for the polar regions are short (< 50 years), and scarce, due largely to the remoteness and associated high costs of conducting research. However, these data can be extracted from various geological records, such as cores from marine and lacustrine sediments and the cryosphere, using palaeoenvironmental methods (see Chapter 4 and several chapters in Section II).

There is an overwhelming amount of evidence demonstrating that environments in the polar regions are rapidly changing (Anisimov et al., 2007). This includes impacts to the terrestrial cryosphere, the biosphere, and marine and freshwater environments.

The degree and rapidity of changing environmental conditions compelled the Arctic Council to commission the Arctic Climate Impact Assessment report (ACIA, 2005). The impacts of climate change are reducing snow and ice content, causing reductions in permafrost, sea ice and glaciers in most regions. Animal and plant distributions are shifting polewards, introducing new species and placing others at risk. Plant productivity is increasing in response to elevated CO_2 levels and longer growing seasons. Increased ultraviolet (UV) radiation is affecting terrestrial and aquatic environments as a result of stratospheric ozone depletion. Decreasing lake levels and increased river flow is delivering pollutants into and diluting the Arctic Ocean. Eustatic sea levels are rising and coastal erosion is affecting the stability of human settlements as ground ice melts out and winter storms impact coastlines that are no longer protected by the early onset of sea ice. Contaminants continue to be introduced and deposited in Arctic environments via atmospheric and oceanic currents, affecting human wellbeing in terms of food security and health concerns. As sea ice retreats, there are heightened risks associated with increased marine shipping through the northern sea routes. Not only will marine ecosystems be affected by oil spills but greater areas of the northern reaches will also be open to resource extraction and development. These will undoubtedly compromise ecological integrity due to negative impacts of development.

A similar assessment was recently completed for the Antarctic (ACCE, 2009). Subsequent updates to the ACIA (2005) report have been completed in the NOAA State of the Arctic reports that provide tracking of recent environmental change in the Arctic (e.g., Richter-Menge et al. 2006, 2009). Implications of a changing climate on northern Canada and potential adaptation measures are provided in Furgal and Prowse (2008).

This chapter presents an account of environmental change in the Arctic and Antarctic focussing on the Anthropocene, a time period

defined by Crutzen and Stoermer (2000) as that in which human activities have had a severe impact on the environment and delimited as the last 200 years (see Chapter 1). The chapter provides an overview of the geographic Arctic and Antarctic and then a general review of polar environmental change. It deals with the cryosphere only in terms of climate amplification as ice and snow are dealt with in Chapter 10 of this book. Greater details are found in the ACIA (2005), IPCC (2007) and ACCE (2009) reports.

2 POLAR ENVIRONMENTS

2.1 The Arctic: geographic definitions

Centred over the North Pole, the Arctic region encompasses the Arctic Ocean surrounded by the northern reaches of the circumarctic continents and their associated islands. No strict definition for the southern extent of the Arctic exists although a number of delineations are in use (Figure 35.1). The Arctic Circle, situated at 66°S 33' 43" N, defines the Arctic by a fixed latitude. Other definitions, with an emphasis on environmental conditions, use the 10°C July isotherm or the tundra–forest boundary. From a marine perspective, the Arctic Marine Boundary delimits the cold Arctic Ocean surface waters from the more southern warmer and saltier waters (Stonehouse, 1989). Others use the definition of the southern extent of discontinuous permafrost that closely parallels the tundra–forest boundary (see Chapter 10). Needless to say, these last boundaries are changing, reflecting the dynamic nature of environmental conditions.

2.2 The Antarctic: geographic definitions

In contrast to the Arctic, the Antarctic is comprised of a land mass situated directly

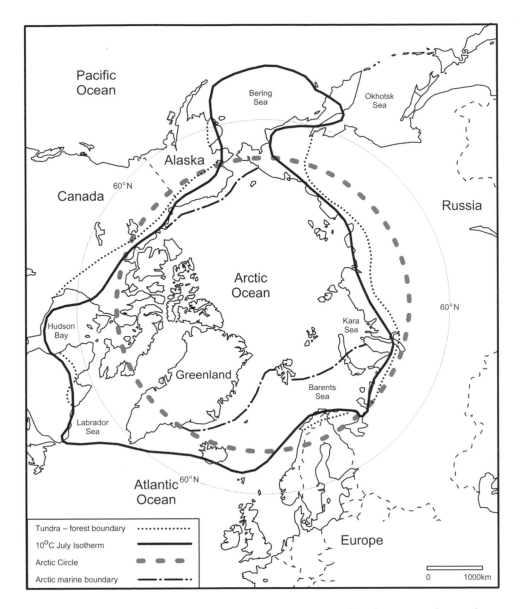

Figure 35.1   Polar projection showing various boundaries used to demarcate the Arctic.

Source: Stonehouse (1989).

over the South Pole. Divided into two parts: West Antarctica and East Antarctica, it is the fifth largest continent. Its area constitutes 13,200,000 km² of which between 98–99.5 per cent is ice-covered (Bargagli 2005; Hodgson et al., 2004). As is the case for its northern counterpart, several delineations exist for the Antarctic (Figure 35.2). Although the Antarctic Circle is situated at 66° 33' 43" S, one common definition considers the Antarctic to be the region south of 60°S. The 10°C February (southern hemisphere summer) isotherm traverses only one continent, South America (at Tierra del Fuego),

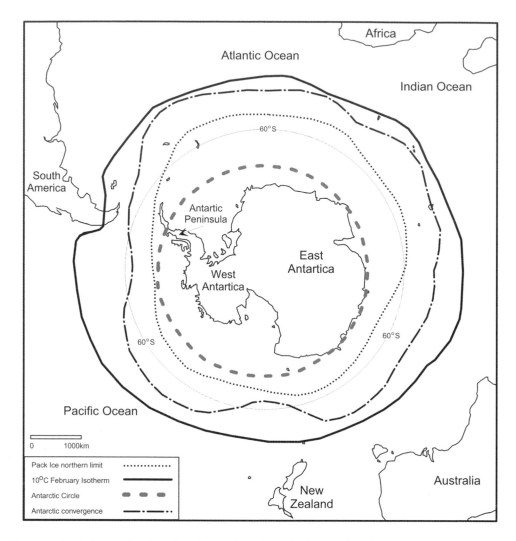

Figure 35.2   Polar projection showing various boundaries used to demarcate the Antarctic.

Source: Stonehouse (1989).

and coincides with a tree line that is affected by maritime conditions (Stonehouse, 1989). The maritime boundary delimitation for the Antarctic is called the Antarctic Convergence and is likely the most inclusive delimitation of the Antarctic region. It involves the meeting of two water currents: the cold Antarctic waters circulating eastwards around the continent and the warmer and saltier sub-Antarctic waters.

3 POLAR ENVIRONMENTAL CHANGE

Polar regions have been greatly impacted by climate change, affecting environments associated with the cryosphere, terrestrial, freshwater and marine systems (ACCE 2009; ACIA, 2005; Anisimov et al., 2007). Increases in precipitation have been concomitant with the increase in temperature that has resulted in the highest recorded temperatures in the past decades (Furgal and Prowse, 2008). Most general circulation models predict that change will be greatest at the highest latitudes (in the Arctic) and on the Antarctic Peninsula (in the Antarctic). The polar regions are drivers of global climate; they are sites for deepwater formation, which affect the thermohaline circulation of the oceans and hence heat transfer from equatorial regions towards the poles. Atmospheric circulation in polar regions is intricately tied to marine conditions and any changes affecting high latitudes will resound throughout the lower latitudes (through changing weather patterns as well as rising sea levels, in the case of global warming). The implications of shifting physicochemical environmental conditions will have profound effects on the biosphere. Polar ecosystems are being greatly affected by these changes in temperature, moisture and nutrient availability, biodiversity, competition and so forth.

3.1 Polar amplification and cryosphere

Polar temperatures are distinctly tied to the high surface albedo properties of sea ice. A positive feedback relationship amplifies the signal so that temperature increases in polar regions are almost twice as great as at lower latitudes. As the areal extent of sea ice decreases, surface albedo decreases in step with the increased surface area of open ocean waters and ocean temperatures rise. Several recent trends in Arctic observations of sea-ice extent and temperatures provide evidence for the existence of this Arctic amplification. Serreze et al. (2009) and Screen and Simmonds (2010) concluded that diminishing sea ice was a primary factor driving Arctic temperature amplification as the surface Arctic warming was consistent with heat transfers from the ocean to the lower troposphere. Similar amplification is observed as vegetation cover shifts from tundra to shrubs under increased greenhouse forcing (Chapin and Shaver, 1996; Hudson and Henry, 2009;

Swann et al., 2010) affecting surface albedo in a similar manner to decreasing sea ice.

By far the largest change in the Arctic has been changes in the cryosphere. Decreasing trends in minimum sea-ice extent in the Arctic is one of the key indicators indicating warming. Over the course of the past three decades, the September (minimum) sea-ice extent has decreased by 8.9 per cent per decade and thinned by 17 cm/year from 2004 to 2008 (Perovich et al., 2009; see also Chapters 3 and 10). Rodrigues (2009) used the length of the ice-free season (LIFS) and the inverse ice index to quantify trends in annual sea ice coverage in the Arctic. His findings confirmed the predictions that Arctic sea ice was rapidly declining. The LIFS increased from 119 to 148 days in the period from the late 1970s to 2006, an average of 1.1 days/year although for the period 2001 to 2007, this rate increased to 5.5 days/year.

Besides the temporal trends quantifying the various ice reservoirs, melting of ice can present hazards by releasing anthropogenic contaminants, but also by destabilizing gas hydrates, also known as clathrates, in submarine permafrost (Maslin et al., 2010). Depending on the amount of methane released a significant contribution to warming might occur given that methane is 20 times more effective than carbon dioxide as a greenhouse gas. Evidence of environmental change from the cryosphere, including glaciers, sea ice and permafrost is substantial and is described in detail in Chapter 10.

3.2 Terrestrial ecosystems

Changes in landscape appearance include shifts in vegetation extent and type, and geomorphological evolution. Some of the most obvious shifts can be measured, especially using remote sensing techniques such as satellites and aerial photography (Danby and Hik, 2007b; Tape et al., 2006). In the Arctic, plants appear to be showing relatively quick responses to changing conditions in temperature, precipitation, permafrost thaw and increased nutrient supply. Extended growing seasons have been tracked using remote sensing techniques (Goetz et al., 2005) and the summer warmth index (SWI) which is the sum of the monthly mean temperatures above 0°C (Walker et al., 2009). These data show that the percent change in greenness across the Arctic as measured by the normalized difference vegetation index (NDVI) increased up to 20 per cent in some parts of the Canadian Arctic and decreased by up to 5 per cent in areas associated with the Bering and West Chukchi seas. Ecozone changes such as tree-line displacement will not necessarily be a simple northwards movement in response to temperature increases alone (Callaghan et al., 2005) as moisture availability amongst other factors will be important. A threshold response is expected *versus* a gradual response to warming (Danby and Hik, 2007a) with deciduous shrubs predicted to expand the most (Forbes et al., 2010).

Measurements of tree ring widths (dendroecological techniques) have been used to reconstruct past Holocene climates in Fennoscandia (Linderholm et al., 2010), Swedish Lapland (Grudd et al., 2002), Russia (MacDonald et al., 2008), Alaska (D'Arrigo et al., 2005), Labrador (D'Arrigo et al., 2003) and North America in general (D'Arrigo et al., 2009). A relatively close fit with temperature was observed in most cases although under the current warming scenario, trees are drought-stressed and the relationship disintegrates (Barber et al., 2000, Pisaric et al., 2007, 2009). In the Antarctic Peninsula where temperature and moisture availability have increased the most, the expansion of plant ranges has also been documented (Convey and Smith, 2006).

3.3 Freshwater ecosystems

Aquatic ecosystems in the polar regions are comprised mostly of lakes, ponds, rivers and wetlands. The limnology of polar lakes and rivers is described in Vincent and Laybourn-Parry (2008) and covers many of the different habitats and systems within the polar regions including epishelf lakes as well as subglacial lakes in Antarctica. Rouse et al. (1997) discuss the limnological responses to warming conditions and Schindler and Smol (2006) summarize the cumulative effects of climate warming and other human-related stressors, including increased UV-B radiation, on Arctic and subarctic ecosystems. White et al. (2007) provide an assessment of the factors affecting the storage and cycling of the freshwater system. The 2–3°C regional warming in both polar regions (e.g., Mackenzie Valley and Antarctic Peninsula) has pushed some of these systems past ecological thresholds. For example, shallow thermokarst lakes in Russia have disappeared due to changes in permafrost-affected drainage (Smith et al., 2005) and ponds in the Canadian Arctic (Smol and Douglas, 2007) have dried up due to an increase in net evaporation (see later in this chapter). At Signy Island, in the Antarctic, the effects of a warming of 1°C in 40 years had a large impact on aquatic systems (Quayle et al., 2002). During the latter part of that 40-year period, water temperature increased by nearly as much (0.9°C) in 15 years (1980–1995), aquatic primary productivity increased fourfold, and the length of ice-free days increased by 31 days.

Palaeolimnological studies corroborate that lakes and ponds can serve as bellwethers of change. Over two decades of limnological monitoring of Arctic ponds on Ellesmere Island exhibited a similar response to that of Signy Island, namely one of warming (Smol and Douglas, 2007a, 2007b). The palaeolimnological history had signalled that an increase in temperature was the likely cause of observed algal response; that is, increased diversity (Douglas et al., 1994). The long-term limnological signal as evidenced by increased aquatic conductivity indicated a net increase in evaporation. The final ecological frontier was crossed when ponds that had existed for thousands of years dried up (Smol and Douglas, 2007b). A survey of palaeolimnological records across a latitudinal gradient in the circumarctic demonstrated that this warming signal, as captured by the diatom diversity response, was proportional to the warming pattern projected by many general circulation models, namely that warming increased with latitude (Smol et al., 2005).

Mueller et al. (2009) described threshold-dependent shifts in deep, ice-covered lakes in the high Arctic and how some lake types (e.g., epishelf lakes) associated with ice shelves on the northern coast of Ellesmere Island, are disappearing with the loss of ice shelves. Extensive associated freshwater-ice microbial ecosystems are also at risk of disappearing. Other long-term environmental changes in Arctic and Antarctic lakes and freshwaters are treated comprehensively via geographic region and proxy indicators in Pienitz et al. (2004), Hodgson and Smol (2008) and Prowse et al. (2006a).

3.4 Marine ecosystems

Polar marine ecosystems continue to be impacted by climate change and other environmental stressors. In a comprehensive review, Clarke and Harris (2003) contrast the two polar marine regions as two very different systems that face diverse challenges and which respond differently. Both regions are being impacted by heavy fishing pressures and although they are continuously accumulating pollutants from outside latitudes, critical thresholds have yet to be met. They predict that the greatest area of concern for the Arctic Ocean is that of declining sea-ice extent and its subsequent role in polar amplification as well as shifts in biological processes associated or controlled by ice presence. This is not as great a problem in the Antarctic which has not exhibited the same decline of annual sea-ice winter build-up. Instead the Antarctic's Southern Ocean faces greater threats from overexploitation of its natural fisheries. The cumulative impact of these stressors has not been assessed and the authors caution that they could act synergistically in a detrimental manner in the future as thresholds are reached.

A relatively new environmental area of concern has recently been identified in the polar oceans, namely that of acidification. Sea-water chemistry is becoming more acidic as a result of rising levels of atmospheric CO_2 that is being absorbed into the water column (Doney et al., 2009). The implications for polar waters are large and it means that ocean surface waters will likely become undersaturated in aragonite, an important implication for marine biota which have calcified shells (Fabry et al., 2009).

The Southern Ocean has an important biogeophysical role in nutrient and carbon cycling (see Chapter 23). Paleoceanographical research can help to reconstruct the natural baseline variability for past conditions. Paleoceanographical studies have also provided the necessary background on past sea-ice conditions (Ledu et al., 2008) and ocean acidification (Pelejero et al., 2010).

Animals associated with marine environments have been affected by change. Past animal populations have been tracked using palaeolimnological techniques. In the Antarctic, seal populations were tracked (Hodgson et al., 1998) based on the number of seal hairs in the sediments. Futhermore, a 3,000-year record of penguin populations was inferred based upon the influx of nutrients (phosphorus) and some elements (sulphur, calcium, copper, zinc, fluorine, barium, strontium). Penguin populations appeared to peak during periods of high precipitation and were at a low during cold periods (Sun et al., 2000). Penguins are sensitive to changing climatic conditions and this may be linked to food sources such as krill as well as shifts in climate. Present day observations and the paleoecological record suggest that penguins will adapt to shifting ice conditions by relocating to appropriate environmental conditions (Forcada and Trathan, 2009). Keystone species such as Antarctic krill are showing a long-term decline in numbers and a replacement by salps (Atkinson et al., 2004). In the Southern Ocean, new species invasions are occurring including the benthos (Aronson et al., 2007). Some organisms, thought to be extinct since the Miocene, such as lithodid crabs, are reappearing (Thatje et al., 2005). Smale and Barnes (2008) suggest that the biodiversity of the marine ecosystems of the Antarctic Peninsula will increase as a result of augmented tourism and due to the increase in disturbed habitats in the shallow coastal regions resulting from increased scouring by icebergs. In the Arctic, those animals whose life cycles include sea ice as platforms are showing negative impacts (e.g., the polar bear; Regehr et al. 2010).

3.5 Contaminants

Given the sparse human population across the Arctic and relatively low level of industrial development, the presence and concoction of contaminants identified at high latitudes was initially surprising, until long-range transport mechanisms were understood. Although some pollutants associated with industrial activities originated from within the polar regions, a significant amount of contaminants originated from afar. Heightened concern for the impact of pollutants on the Arctic ecosystems led to the creation of the Arctic Monitoring and Assessment Programme (AMAP) in 1991. This program incorporated pre-existing national and international monitoring activities and provided an in-depth assessment of the Arctic environment in light of pollution issues (AMAP, 1998). Additional programs such as the Conservation of Arctic Flora and Fauna (CAFF, 2001) were also initiated. In the Antarctic, where there are no permanent settlements other than scientific research stations, measurements of pollutants are ongoing and described later in this chapter.

Pollutants originating from outside the Arctic include the family of anthropogenically synthesized persistent organic pollutants (POPs). This encompasses the organochlorine pesticides such as hexachlorocyclohexanes (HCHs), polychlorinated biphenyls (PCBs) and combustion products such as chlorinated dioxins and polycyclic aromatic hydrocarbons (PAHs) (AMAP, 1998). Because many of these chemicals have a human fingerprint and were only used at lower latitudes, their transport was deduced to be via atmospheric and marine circulation flowing northwards (Lockhart, 1995). Wania and Mackay (1993) described how many such pollutants are transported via the atmosphere towards higher latitudes through distillation and fractionation, also known as cold condensation or the grasshopper effect. Other pollutants of concern include radioactivity, heavy metals, petroleum hydrocarbons, and sulphur and nitrogen compounds associated with acidification (AMAP, 1998, Gordeev, 2002). These are transported to the Arctic through various routes although atmospheric and ocean currents are the primary carriers. Northwards-flowing rivers are also important contributors while glaciers, permafrost, sea ice and the ocean are important sources that can release stored contaminants previously deposited within their reservoirs.

As more information regarding the long-range transport and fate of pollutants in polar regions is gathered, it is possible to refine the pathways effecting the distribution and biological consequences of the certain pollutants. For example, Li and Macdonald (2005) described the complex pathways that exist for HCHs, dischlorodiphenyltrichloroethane (DDT), toxaphene and endosulphan. Introduced in the 1940s, global applications of these pollutants, with the exception of endosulphan, decreased by 2000 as a result of legislation such as the Stockholm Convention. Close examination of the available data revealed that isomers of α-HCH and β-HCH reached the Arctic via different pathways. According to their individual physical and chemical properties and the process of solvent switching, the pollutant pathway transfers from air to water or vice versa. For the most part, α-HCH's pathway northwards was initially atmospheric but over distance switched to cold marine waters. β-HCH followed a similar pattern except that its partition into cold water was faster and was completed before the air masses reached the Arctic, hence the final access is solely via ocean currents. Despite a better understanding of these entrance pathways into the Arctic, the biological concentration of the two isomers across the circumarctic is complex and heterogenous.

As additional studies are completed, and the environment continues to change, new contamination pathways are discovered, adding to the complexity. For example, as climate warms and sea ice is at a minimum, HCH is being re-emitted from the warming ocean back into the atmosphere (Li and Macdonald, 2005). Biological vectors are also important. Nesting seabirds from colonies located along the Arctic coasts are capable of reintroducing contaminants such as HCH, DDT and PAH back to the terrestrial system. Through feeding activities, these birds bioconcentrated and biomagnified contaminants from their marine-based diet. Nutrients and contaminants derived from guano at the nesting sites were released into nearby freshwater ponds (Blais et al., 2005). Individual bird species, such as terns and eiders, bring back different contaminants (mercury and cadmium *versus* lead, aluminium and manganese, respectively) based upon their feeding strategies (in this case, fish *versus* benthic communities) (Michelutti et al., 2010).

The impacts and effects of contaminants on ecosystems are still uncertain in many instances and may be underestimated as potential thresholds have yet to be crossed. Even after POPs and other pollutants, including lead and cadmium had been detected, it wasn't clear initially if the concentrations were sufficiently high to impact biological communities (Fisk et al. 2005; Lockhart, 1995). Fisk et al. (2005) concluded that, with the exception of polar bears, there was not enough evidence that contaminants had reached critical levels for most other Arctic organisms, including fish and birds. One exception involved Yukon burbot and Greenland sharks in which excessive levels of PCBs and DDT had been measured. They also cautioned that as new chemicals were

constantly being introduced to the Arctic, potentially harmful conditions could still occur. The cumulative impact of biological magnification was shown to affect the health of higher-level organisms in the food web, including humans. The spatial and temporal trends of POPs and mercury in marine biota was reviewed by Braune et al. (2005). In this synthesis, regional differences in distribution of contaminants were apparent. Organochlorine concentrations in marine biota in general were higher in the European and Greenland Arctic, lower in the Canadian Arctic and the lowest in Alaska. Mercury concentrations were highest in the Canadian Arctic although the spatial and temporal trends are highly variable. Although the values for legacy organochlorines, such as DDT, PCBs and HCHs, continue to be monitored, 'new-use' chemicals such as perfluorooctane sulphonate, used in stain-repellents and fire-fighting foams, are now present in Arctic environments and their toxicity and impacts are being scrutinized.

The difficulty in predicting the future distribution of pollutants in the Arctic is compounded when considering the impacts of climate change. Macdonald et al. (2005) noted that climate change and contaminant science were not well integrated as each field's priorities focused on the need for fundamental baseline knowledge. Their interdisciplinary study discusses future trends that will result from the changing climatic conditions in the Arctic. They identify the importance of the Northern Annular Mode, also known as the Arctic Oscillation (see Chapter 22) in Arctic climate, affecting wind patterns and intensity, temperature, precipitation and runoff and sea ice extent and drift trajectory, amongst other changes. The resulting increase in precipitation will increase the concentration of heavy metals such as lead, cadmium and zinc. In the case of mercury, a more complex pattern emerges as natural factors, such as methylation, cold temperature related scavenging (Jitaru et al., 2009) and human activity are at play. In addition, glaciers, sea ice and permafrost act as reservoirs for contaminants.

As these melt, pollutants stored within will be released (Blais et al., 1998; Pfirman et al., 1995).

In the Antarctic, the concentration of contaminants remains low. Bargagli (2005) provides a very detailed synthesis of environmental contamination, climate change and human impact on Antarctic ecosystems. These studies are broken down into examination of the atmosphere, Southern Ocean, and terrestrial regions including freshwaters. Given its remoteness, the Antarctic is the least contaminated environment on Earth. Examination of the snow and ice core records show that a variety of contaminants likely originating from South America were deposited on the continent via atmospheric transport. These include radionuclides, persistent organic pollutants (DDT, HCHs, PCBs) and metals (lead, copper, chormium, zinc). Although data are scarce, trends indicate that DDT and its derivatives increased from the early 1960s to the late 1980s before exhibiting a decline as usage of the chemical decreased. Pollutants from local scientific research bases are mainly the products of combustion, such as PAHs, trace metals and hydrocarbon spills, and are not widely dispersed. The Protocol on Environmental Protection to the Antarctic Treaty details approved methods to minimize any further impacts. The seasonal pack ice behaves as a sink for persistent contaminants which are released during spring and summer melt. These are transferred to the marine ecosystem where they move up through the food chain. Increased precipitation resulting from climate warming may enhance the mobility of these contaminants within soils and release any cryogenically preserved contaminants.

The levels of contaminants in the Arctic, as reported by AMAP (1998), and the Antarctic (Bargagli 2005) are low in comparison to the rest of the polluted world, and the Arctic and Antarctic can be described as clean, with the latter being the least impacted. However, regions or areas of concern exist, especially with regard to certain ecosystems, some of which are linked with human populations.

Muir et al. (2005) refer to the Northern Contaminants Program (NCP), a Canadian Government program that links in with international partners, as a means to communicate the risks associated with contaminant exposure. As greater understanding of the transport pathways and behavior of pollutants in high latitudes is gained, so will our understanding of the temporal and spatial heterogeneity of these pollutants in marine, terrestrial and freshwater biological systems.

3.6 Stratospheric ozone depletion and Arctic haze

Polar atmospheres are impacted by the long-distance transport of pollutants through atmospheric currents. In both the Arctic and Antarctic, the atmosphere acts as a collector for pollutants during the course of the cold, dark winter season. These pollutants are then either deposited through precipitation the terrestrial, aquatic and cryosphere systems; or, through a series of photochemical reactions, these pollutants can modify the atmosphere's chemical composition and decrease its effectiveness as a protective shield against incoming harmful UV-B radiation. In addition, these changes to the atmosphere are affecting the global climate.

Pollution of the Arctic atmosphere was likely first noticed in the mid 1950s by airborne researchers observing polar weather patterns (Shaw, 2003). At that time, it was suggested that the haze observed at high latitudes was due to the scattering of light by small particles (i.e., aerosols) which originated from far away. However, it wasn't until 20 years later that the extent of this phenomenon was recognized. It is now known that Arctic haze is the manifestation of particulates such as sulphate, particulate organic matter, ammonium, nitrate, dust, black carbon and heavy metals and gases originating from industrial sources primarily from Eurasia and secondarily from North America (Quinn et al., 2007). Other contributing natural sources include desert dust as well as carbon from forest fires (Quinn et al., 2007). The pollutants build up in the polar atmosphere column due to atmospheric long distance transport. Because the cold air masses are relatively stable, the residence time of the pollutants is higher due to the low rates of precipitation and scavenging that would normally clean out the air, as is the case at lower latitudes. These pollutants appear to have a moderating impact on climate: elevated levels of black carbon and other particles in the haze scatter and absorb incoming sunlight, with a warming effect of the atmosphere. Conversely, the surface temperatures cool due to the scattering of light. Although these climate modifications are thought to be slight, Arctic haze aerosols can act as cloud condensation nuclei and introduce new cloud parameters whose effects are difficult to predict at this time. From the 1990s onwards, it appears that Arctic haze concentrations were decreasing across the Canadian, Norwegian and Finnish Arctic, while increasing in Alaska; however, given the long distance source of Arctic haze components and their complex interactions in cloud dynamics and consequent climate effects, the consistency of these detected trends is uncertain (Quinn et al., 2007).

In the case of the Antarctic, stratospheric ozone depletion is the result of photochemical reactions by derivatives of anthropogenic chlorofluorocarbons (CFCs), which exhibit long atmospheric residency times of the order of decades to centuries. Although often referred to as the 'ozone hole', in reality it is a thinning of the stratospheric ozone layer to values below 220 DU. Destruction of stratospheric ozone was predicted by the 1995 Nobel chemistry laureates Paul Crutzen, Mario Molina and Sherwood Rowland; however, 10 years earlier, in 1985, Farman and colleagues reported its existence based on observations beginning in 1957 at the Antarctic Peninsula using ground-based spectrophotometers (Farman et al., 1985). The combination of unique conditions in the Antarctic provided the perfect conditions for the photochemical destruction of ozone

molecules by free chlorine and bromine radicals. These conditions include the polar vortex that forms as a result of the high winds and currents that circumnavigate the continent, collecting ozone-depleting substances such as CFCs during the polar night, as well as the extreme cold temperatures and the formation of polar stratospheric clouds. Photochemical destruction of stratospheric ozone commences with the austral spring. These photochemical processes and ozone chemistry are detailed in Seinfeld and Pandis (2006).

Once the ozone hole had been identified, the global community moved remarkably quickly to mitigate the situation. The Montreal Protocol was signed in 1989 and banned the use of ozone-depleting substances (ODS). Some cynics note that international cooperation was quick because substitute chemicals were commercially available resulting in little lobbying against the ban; other cynics will point out that some of the substitute chemicals turned out to be environmentally damaging. Regardless, it is promising example of national cooperation.

The ozone hole grew at an alarming rate, reaching a maximum size of 29.6 km^2 in September 2006 (Newman, 2010) in the Antarctic as well as also manifesting in the Arctic. Volcanic eruptions such as Mount Pinatubo in 1991, contributed to the increase in size of the ozone hole (Solomon et al., 2005). Because of the long residence time of CFCs, repair of the stratospheric ozone hole will not be immediate despite the ban on ODS. As of 2009, no definite recovery could be reported (Hofmann et al., 2009), although the size of the hole appears to have stabilized. Newman et al. (2006) predict that recovery of the ozone hole will not occur prior to about 2024. Thinning of the Arctic stratospheric ozone is variable and was at its highest in the mid 1990s; however, due to a weaker Arctic vortex, the extent of depletion is not as large as in the Antarctic (Balis et al., 2009; Solomon et al., 2007). During spring, temperatures warm sufficiently to disperse polar stratospheric clouds and slow photochemical destruction of stratospheric ozone.

The depletion of stratospheric ozone affects (global) climates in a complex relationship that is yet not fully understood. Several studies have shown that ozone loss is causing a cooling of the lower stratosphere (Ramaswamy et al., 2001) and this, combined with increased greenhouse gases, has caused an increasing frequency of westerlies in the Antarctic (Perlwitz et al., 2008). This has resulted in differential warming in the Antarctic with the Antarctic Peninsula warming and the interior of the continent cooling (ACCE, 2009; Thompson and Solomon, 2002).

4 SCIENCE IN THE CIRCUMPOLAR REGIONS

Science and research within the Arctic and Antarctic are administered differently. The Antarctic is governed by the international Antarctic Treaty. The Scientific Committee on Antarctic Research (SCAR) is an interdisciplinary body concerned with ongoing science in the Antarctic. It provides advice at the annual Antarctic Treaty Consultative Meetings of the Antarctic Treaty System that governs the Antarctic and suggests key research programmes are needed to better understand the polar environment. No such body exists for the Arctic although the International Arctic Science Committee (IASC), an international nongovernmental entity, serves a similar purpose by encouraging and directing international scientific endeavours. It takes direction from working groups within the Arctic Council and as demonstrated by the extensive summary assessments (e.g., ACIA, 2005 and ACCE, 2009), the importance of international cooperation to obtain the necessary spatial and temporal data needed to monitor these remote yet dynamic polar environments cannot be overstated. The Arctic Council, made up of the eight circumarctic nations, permanent participants and observers, has directed several important initiatives so as to assess the state

of the Arctic, including ACIA (2005) and AMAP (1998). Overall, stewardship of the polar regions is based upon sound science and understanding of global linkages down to the ecosystem level.

5 CONCLUSION

A prime reason for seeking to understand environmental change is to enable better projections and scenarios relating to the future of the Earth's ecosystems. There are large implications for the wellbeing of northern communities as well as globally. Adaptation has already started but, unless the nature and rate of environmental changes can be predicted, it will be very difficult to engineer for future scenarios. By putting a human face on climate change (and environmental change more generally), global policy makers will be better positioned to effect change.

> We have to give climate change a human face – it is not all about 'sinks', 'emission trading schemes' and technology. Climate change is about people, children, families and of our relationship with the world around us. To Inuit it is a question of our very survival as a hunting people and a hunting culture. Our human rights – to live our traditional way – are being violated by human-induced climate change (Sheila Watt-Cloutier, chair, Inuit Circumpolar Conference, November 2004).

While the ACIA (2005) provided an assessment of changing Arctic environments as a whole, the Arctic Human Dimension Report (2004) provided an overview of the state of human development across the Arctic. While it recognized the great impact of climate change on northern environments, it suggested that adaptation to these conditions by northern societies would also require immediate adaptation to other areas such as globalization, socioeconomic and political transitions (Larsen and Fondahl 2010). A follow-up study highlighted a small set of social indicators that could be used to monitor human adaptation to environmental change across northern communities (ASI, 2010).

These included monitoring indicators of human health, food security, socioeconomic conditions, and education and governance details.

The need for long-term monitoring of polar environments is highlighted in many of the summary assessments of the polar regions mentioned throughout this chapter. These monitoring programs are faced with the challenge of covering vast geographic areas that are sparsely (if at all) populated. The costs for such programs are high and require a coordinated international effort, such as that espoused by Sustaining *Arctic* Observing Networks (SAON), an international consortium formed in 2007 at the directive of the Arctic Council (SAON 2010). Given the complex interactions between the components that collectively form the polar environments (marine, terrestrial, freshwater and cryosphere), an exhaustive list of indicators needs to be monitored. This includes continuing to monitor the rate of climate change in the polar regions, the distribution of snow and ice, sea-level rise, impacts on ocean circulation, biodiversity changes, methane release, impacts of global pollution transport to polar regions and the health and well-being of Arctic societies (ICSU/WMO 2009). As more remote sensing capabilities are developed and put into place, such as CryoSat 2 which was launched by the European Space Agency in 2010, or maintained, such as RADARSAT-2, monitoring capabilities are becoming more efficient. The International Geosphere-Biosphere Programme (IGBP) proposed an index that would provide an easy-to-understand metric for policy makers (Gaffney, 2009). The IGBP Climate-Change Index integrates the annual changes of four metrics – sea level, global average temperature, atmospheric CO_2 and Arctic sea ice cover – into one cumulative annual change number. Normalized to between −100 and +100, these annual values should provide a simple comprehensive assessment. Time will tell if this index effectively communicates the essential information on global trends to the public at large or if it is too simple.

However, not all environments can be effectively monitored remotely and there remains the need for on-the-ground measurements to provide information at a regional level or to ground-truth date obtained remotely.

Of course, environmental impacts are immediate for those living within the polar regions. Northern societies have always adapted to their surrounding environmental conditions, even as these changed. But, until recently, the degree and rate of change have not been as great as they are today. There are global implications for these changes. Just as many of the pollutants in the polar regions originated from outside the region, changes within the polar regions will have long-lasting effects felt throughout the globe, including rising global sea levels and shifting climate patterns, amongst others.

REFERENCES

ACCE. 2009. *Antarctic Climate Change and the Environment*. Scientific Committee on Antarctic Research. Cambridge: Victoire Press.

ACIA. 2005. *Arctic Climate Impact Assessment: Impacts of a Warming Arctic*. Cambridge: Cambridge University Press.

AHDR. 2004. *Arctic Human Development Report*. Akureyri Stefansson: Arctic Institute.

AMAP. 1998. *AMAP Assessment Report: Arctic Pollution Issues*. Oslo: Arctic Monitoring and Assessment Programme.

Anisimov O. A. Vaughan D. G. Callaghan T. V. Furgal C. Marchant H. Prowse T. D. et al. 2007. Polar regions (Arctic and Antarctic), in Parry M. L., Canziani O. F., Palutikof J. P., van der Linden P. J. and Hanson C. E. (eds) *Climate Change 2007: Impacts, Adaptation and Vulnerability. Contribution of Working Group II to the Fourth Assessment Report of the Intergovernmental Panel on Climate Change*. Cambridge: Cambridge University Press.

Aronson R. B., Thatje S., Clarke A., Peck L. S., Blake D. B., Wilga C. D. and Seibel B. A. 2007. Climate change and invisibility of the Antarctic benthos. *Annual Review of Ecology, Evolution and Systematics* 38: 129–154.

ASI. 2010. *Arctic Social Indicators: a follow up to the AHDR*. Akureyri Stefansson: Arctic Institute.

Atkinson A., Siegel V., Pakhomov E. and Rothery P. 2004. Long-term decline in krill stock and increase in salps within the Southern Ocean. *Nature* 432: 100–103.

Barber V. A., Juday G. P. and Finney B. P. 2000. Reduced growth of Alaskan white spruce in the twentieth century fro temperature-induced drought stress. *Nature* 405: 668–673.

Balis D., Bojkov R., Kleareti T. and Zerefos C. 2009. Characteristics of the ozone decline over both hemispheres. *International Journal of Remote Sensing* 30: 3887–3895.

Bargagli R. 2005. *Antarctic Ecosystems Environmental Contamination, Climate Change and Human Impact.* Ecological Studies 175. New York: Springer.

Belatos S. and Prowse T. 2009. River-ice hydrology in a shrinking cryosphere. *Hydrological Processes* 23: 122–144.

Beyer L. and Bolter M. (eds) 2002. *Geoecology of Antarctic ice-free Coastal Landscapes.* Ecological Studies 154. Berlin: Springer.

Blais J. M., Kimpe L. E., McMahon D., Keatley B. E., Mallory B. E., Douglas M. S. V. and Smol J. P. 2005. Arctic seabirds transport marine-derived contaminants. *Science* 309: 445.

Blais J. M., Schindler D. W., Muir D. C. G., Kimpe L. E., Donald D. B. and Rosenberg B. 1998. Accumulation of persistent organochlorine compounds in mountains of western Canada. *Nature* 395: 585–8.

Born E. W. and Böcher J. 2001. The ecology of Greenland. Nuuk, Greenland: Ministry of Environment and Natural Resources.

Braune B. M. Outridge P. M. Fisk A. T. Muir D. C. G. Helm P. A. Hobbs K. et al. 2005. Persistent organic pollutants and mercury in marine biota of the Canadian Arctic: An overview of spatial and temporal trends. *Science of the Total Environment* 351–352: 4–56.

CAFF (Conservation of Arctic Flora and Fauna) 2001. Arctic flora and fauna: status and conservation. Helsink: Edita.

Callaghan T. V. Björn L. O. Chapin F. S. III Chernov Y. Christensen T. R. Huntley B. et al. 2005. *Arctic tundra and polar desert ecosystems in Arctic Climate Impact Assessment* London: Cambridge University Press, pp. 243–252.

Chapin F. S. and Shaver G. R. 1996. Physiological and growth responses of arctic plants to a field experiment simulating climatic change. *Ecology* 77: 822–840.

Clarke A. and Harris C. M. 2003. Polar marine ecosystems: major threats and future change. *Environmental Conservation* 30: 1–25.

Convey P. and Smith R. I. L. 2006. Responses of terrestrial Antarctic ecosytems to climate change. *Plant Ecology* 182: 1–10.

Crutzen P. J. and Stoermer E. F. 2000. The 'Anthropocene'. *Global Change Newsletter* 41: 17–18.

Danby R. K. and Hik D. S. 2007a. Variability, contingency and rapid change in recent subarctic alpine tree line dynamics. *Journal of Ecology* 95: 352–363.

Danby R. K. and Hik D. S. 2007b. Evidence of recent tree line dynamics in southwest Yukon from aerial photographs. *Arctic* 60: 411–420.

D'Arrigo R., Buckley B., Kaplan S. and Woollett J. 2003. Interannual to multidecadal modes of Labrador climate variability inferred from tree rings. *Climate Dynamics* 20: 219–228.

D'Arrigo R., Jacoby G., Buckley B., Sakulich J., Frank D., Wilson R., Curtis A. and Anchukaitis K. 2009. Tree growth and inferred temperature variability at the North American Arctic treeline. *Global and Planetary Change* 65: 71–82.

D'Arrigo R., Mashig E., Frank D., Wilson R. and Jacoby G. 2005. Temperature variability over the past millennium inferred from Northwestern Alaska tree rings. *Climate Dynamics* 24: 227–236.

Doney S. C., Balch W. M., Fabry V. A. and Feely R. A. 2009. Ocean acidification: a critical emerging problem for the ocean sciences. *Oceanography* 22: 16–25.

Douglas M. S. V., Smol J. P. and Blake W. Jr 1994. Marked post-18th century environmental change in high Arctic ecosystems. *Science* 266: 416–419.

Fabry V. J., McClintock J. B., Mathis J. T. and Grebier J. M. 2009. Ocean acidification at high latitudes: the bellwether. *Oceanography* 22: 160–171.

Farman J. C., Gardiner B. G. and Shanklin J. D. 1985 Large losses of total ozone in Antarctica reveal seasonal ClO_x /NO_x interaction. *Nature* 315: 207–210.

Fisk A. T., de Wit C. A., Wayland M., Kuzyk Z. Z., Burgess N., Letcher Braune B. et al. 2005. An assessment of the toxicological significance of anthropogenic contaminants in Canadian arctic wildlife. *The Science of the Total Environment* 351–352: 57–93.

Forcada J. and Trathan P. N. 2009. Penguin responses to climate change in the Southern Ocean. *Global Change Biology* 15: 1618–1630.

French H. M. and Slaymaker O. (eds) 1993. *Canada's Cold Environments.* Montreal and Kingston: McGill–Queen's University Press.

Furgal C. and Prowse T. D. Northern Canada. From impacts to adaptation, in Lemmen D. S. Warren F. J.

Lacroix J. and Bush E. (eds) Canada: a changing climate. Ottawa: Government of Canada, pp. 57–118.

Gaffney O. 2009. Climate change in a nutshell *Global Change* 74: 14–15.

Gajewski K. and MacDonald G. M. 2004. Palynology of North American arctic lakes, in Pienitz R. Douglas M. S. and Smol J. P. (eds) *Long-term Environmental Change in Arctic and Antarctic Lakes*. Developments in Paleoenvironmental Research 8. Dordrecht: Springer, pp. 89–116.

Goetz S. J., Bunn A. G., Fiske G. J. and Houghton R. A. 2005. Satellite-observed photosynthetic trends across boreal North America associated with climate and fire disturbance. *Proceedings of the National Academy of Science USA* 102: 13521–13525.

Grudd H., Briffa K. R., Karlen W., Bartholin T. S., Jones P. D. and Kromer B. 2002. A 7400-year tree-ring chronology in northern Swedish Lapland: natural climatic variability expressed on annual to millennial timescales. *Holocene* 12: 657–665.

Hansom J. D. and Gordon J. E. 1998. *Antarctic Environments and Resources: A Geographical Perspective*. Harlow: Addison Wesley Longman.

Hodgson, D. A., Doran, P. T., Roberts, D. and McMinn, A. 2004. Paleolimnological studies from the Antarctic and sub Antarctic islands. In Pienitz, R., Douglas, M. S. V. & Smol, J. P. (eds) Developments in Paleoenvironmental Research. Volume 8. Dordrecht: Springer, pp. 419–474.

Hodgson D. A., Johnston N. M., Caulkett A. P. and Jones V. J. 1998. Paleolimnology of Antarctic fur seal *Arctocephalus gazella* populations and implications for Antarctic management. *Biological Conservation* 83: 145–154.

Hodgson D. A. and Smol J. P. 2008. High-latitude paleolimnology, in Vincent W. F. and Laybourn-Parry J. (eds) *Polar Lakes and Rivers*. New York: Oxford University Press, pp. 43–64.

Hofmann D. J., Johnson B. J. and Oltmans S. J. 2009. Twenty-two years of ozonesonde measurements at the South Pole. *International Journal of Remote Sensing* 30: 3995–4008.

Hudson J. M. G. and Henry G. H. R. 2009. Increased plant biomass in a High Arctic heath community from 1981 to 2008. *Ecology* 90: 2657–2663.

ICSU/WMO. 2009. The state of polar research: a statement from the International Council for Science/World Meteorological Organization Joint Committee for the International Polar Year 2007–2008. Available at: www.wmo.int

IPCC. 2007. Summary for Policymakers, in Solomon S., Qin D., Manning M., Chen Z., Marquis M., Averyt K. B.,

Tignor M. and Miller H. L. (eds) *Climate Change 2007: The Physical Science Basis Contribution of Working Group I. to the Fourth Assessment Report of the Intergovernmental Panel on Climate Change*. Cambridge: Cambridge University Press.

Jitaru P., Gabrielli P., Marteel A., Plane J. M. C., Planchon F. A. M., Gauchard P.-A. et al. 2009. Atmospheric depletion of mercury over Antarctica during glacial periods. *Nature Geoscience* 2: 505–508.

Larson J. N. and Fondahl G. 2010. *Introduction: human development in the Arctic and Arctic social indicators in AHDR 2004*. Arctic Human Development Report. Akureyri Stefansson: Arctic Institute, pp. 11–28.

Ledu D., Rochon A., de Vernal A. and St-Onge G. 2008. Palynological evidence of Holocene climate oscillations in the Eastern Arctic: a possible shift in the Arctic Oscillation at a millennial time scale. *Canadian Journal of Earth Sciences* 45: 1363–1375.

Li Y. F. and Macdonald R. W. 2005. Sources and pathways of selected organochlorine pesticides to the Arctic and the effect of pathway divergence on HCH trends in biota: a review. *Science of the Total Environment* 342: 87–106.

Linderholm H. W., Björklund J. A., Seftigen K., Gunnarson B. E., Grudd H., Jeong J.-H., Drobyshev I. and Liu Y. 2010. Dendroclimatology in Fennoscandia – from past accomplishments to future potential. *Climate of the Past* 6: 93–114.

Lockhart W. L. 1995. Implications of chemical contaminants for aquatic animals in the Canadian Arctic: some review comments. *Science of the Total Environment* 160/161: 631–641.

Lyons W. B., Howard-Williams C. and Hawes I. (eds) 1997. *Ecosystem Processes in Antarctic Ice-free Landscapes*. Rotterdam: A. A Balkema.

MacDonald G., Kremenetski K. V. and Beilman D. W. 2008. Climate change and the northern Russian treeline zone. *Philosophical Transactions of the Royal Society B – Biological Sciences* 363: 2285–2299.

Michelutti N. Blais J. M. Mallory M. L. Brash J. Thienpont J. Kimpe L. E. et al. 2010. Trophic position influences the efficacy of seabirds as metal biovectors. *Proceedings of the National Academy of Science USA* 107: 10543–10548.

Muir D. C. G. and Rose N. L. 2004. Lake sediments as records of arctic and antarctic pollution, in Pienitz R., Douglas M. S. and Smol J. P. (eds) *Long-term Environmental Change in Arctic and Antarctic Lakes*. Developments in Paleoenvironmental Research 8. Dordrecht: Springer, pp. 209–239.

Macdonald R. W., Harner T. and Fyfe J. 2005. Recent climate change in the Arctic and its impact on contaminants pathways and interpretation of temporal trend data. *Science of the Total Environment* 342: 5–86.

Maslin M., Owen M., Betts R., Day S., Jones T. D. and Ridgwell A. 2010. Gas hydrates: past and future geohazard? *Philosophical Transactions of the Royal Society A* 368: 2369–2393.

Mueller D. R., Van Hove P., Antoniades D., Jeffries M. O. and Vincent W. F. 2009. High Arctic lakes as sentinel ecosystems: cascading regime shifts in climate, ice-cover and mixing. *Limnology and Oceanography* 54: 2371–2385.

Muir D. C. G., Shearer R. G., Oostdam J. V., Donaldson S. G. and Furgal C. 2005. Contaminants in Canadian arctic biota and implications for human health: conclusions and knowledge gaps. *Science of the Total Environment* 351–352: 539–546.

Newman P. A. 2010. Ozone hole watch. Available at: http: //ozonewatch.gsfc.nasa.gov/

Newman P. A., Nash E. R., Kawa S. R., Montzka S. A. and Schauffler S. M. 2006. When will the Antarctic ozone hole recover? *Geophysical Research Letters* 33: L12814.

Pelejero C., Calvo E. and Hoegh-Gudberg O. 2010. Paleo-perspectives on ocean acidification *Trends in Ecology and Evolution* 25: 332–344.

Perlwitz J., Pawson S., Fogt R. L., Nielsen J. E. and Neff W. D. 2008. Impact of stratospheric ozone hole recovery on Antarctic climate. *Geophysical Research Letters* 35: L08714.

Perovich D., Kwok R., Meier W., Nghiem S. and Richter-Menge J. 2009. Sea ice cover, in Richter-Menge J. and Overland J. E. (eds) *Arctic Report Card 2009.* Available at: http: //www.arctic.noaa.gov/reportcard

Peterson B. J., Holmes R. M., McClelland J. W., Vorosmarty C. J., Lammers R. B., Shiklomanov A. I. et al. 2002. Increasing river discharge to the Arctic Ocean. *Science* 298: 2171–2173.

Pfirman S. L., Eicken H., Bauch D. and Weeks W. F. 1995. The potential transport of pollutants by Arctic sea ice. *The Science of the Total Environment* 159: 129–146.

Pienitz R., Douglas M. S. V. and Smol J. P. 2004. *Long-term Environmental Change in Arctic and Antarctic Lakes.* Developments in Paleoenvironmental Research 8. Dordrecht: Springer.

Pisaric M. F. J., Carey S. K., Kokelj S. V. and Youngblut D. 2007. Anomalous 20th century growth, Mackenzie deta, Northwest Territories, Canada. *Geophysical Research Letters* 34: L05714.

Pisaric M. F. J., St-Onge S. M. and Kokelj S. V. 2009. Tree-ring reconstruction of early-growing season precipitation from Yellowknife, Northwest Territories,

Canada. *Arctic Antarctic and Alpine Research* 41: 486–496.

Prowse T. D., Wrona F. J., Reist J. D., Gibson J. J., Hobbie J. E., Levesque L. M. J. and Vincent W. F. 2006a. Historical changes in arctic freshwater ecosystems. *Ambio* 35: 339–346.

Prowse T. D., Wrona F. J., Reist J. D., Hobbie J. E., Lévesque L. M. J. and Vincent W. F. 2006b. General features of the Arctic relevant to climate change in freshwater ecosystems. *Ambio* 35: 330–338.

Quayle W. C., Peck L. S., Peat H., Ellis-Evans J. C. and Harrigan P. R. 2002. Extreme Responses to climate change in Antarctic lakes. *Science* 295: 645.

Quinn P. K., Shaw G., Andrews E., Dutton E. G., Ruoho-Airola T. and Gong S. 2007. Arctic haze: current trends and knowledge gaps. *Tellus* 59B: 99–114.

Ramaswamy V., Chanin M. L., Angell J., Barnett J., Gaffen D., Gelman M. et al. 2001. Stratospheric temperature trends: observations and model simulations. *Review of Geophysics* 39: 71–122.

Regehr E. V., Hunter C. M., Caswell H., Amstrup S. C. and Stirling I. 2010. Survival and breeding of polar bears in the southern Beaufort Sea in relation to sea ice. *Journal of Animal Ecology* 79: 117–127.

Richter-Menge J. and Overland J. E. (eds) 2009. Arctic Report Card 2009. Available at: http: //www.arctic. noaa.gov/reportcard

Richter-Menge J. Overland J. Proshutinsky A. Romanovsky V. Bengtsson L. Brigham L. et al. 2006. *State of the Arctic Report.* NOAA OAR Special Report. Seattle WA: NOAA/OAR/PMEL

Rodrigues J. 2009. The increase in the length of the ice-free season in the Arctic. *Cold Regions Science and Technology* 59: 78–101.

Rouse W. R., Douglas M. S. V., Hecky R. E., Kling G. W., Lesack L., Marsh P. et al. 1997. Effects of climate change on freshwaters of Region 2: arctic and subarctic North America. *Hydrological Processes* 11: 873–902.

SAON. 2010. Sustaining Arctic Observing Networks. Available at: http://www.arcticobserving.org/

Schindler D. W. and Smol J. P. 2006. Cumulative effects of climate warming and other human activities on freshwaters of Arctic and subarctic North America. *Ambio* 35: 160–168.

Screen J. A. and Simmonds I. 2010. The central role of diminishing sea ice in recent Arctic temperature amplification. *Nature* 464: 1334–1337.

Serreze M. C., Barrett A. P., Stroeve J. C., Kindig D. N. and Holland M. M. 2009. The emergence of surface-based Arctic amplification. *Cryosphere* 3: 11–19.

Seinfeld J. H. and Pandis S. N. 2006. *Atmospheric Chemistry and Physics – From Air Pollution to Climate Change*, 2nd edition. New York: John Wiley & Sons. Available at: http: //knovel.com/web/portal/browse/display?_EXT_KNOVEL_DISPLAY_bookid= 2126&VerticalID=0

Shaw G. 2003. Arctic haze, in Holton J. R., Curry J. A. and Pyle J. A. (eds) *Encyclopedia of Atmospheric Sciences*. Amsterdam: Elsevier, pp. 155–159.

SheilaWatt-Cloutier, 2004. Inuit Circumpolar Conference, Human Face of Climate Change: Weather out of Its Mind, CBC NEWS, March 24, 2005. Available at http: //www.cbc.ca/news/background/climatechange/weather.html

Smale D. A. and Barnes D. K. A. 2008. Likely responses of the Antarctic benthos to climate-related changes in physical disturbance during the 21st century, based primarily on evidence from the West Antarctic Peninsula region. *Ecography* 31: 289–305.

Smith L. C., Sheng Y., MacDonald G. M. and Hinzman L. D. 2005. Disappearing arctic lakes. *Science* 308: 1429.

Smol J. P. and Douglas M. S. V. 2007a. From controversy to consensus: Making the case for recent climatic change in the Arctic using lake sediments. *Frontiers in Ecology and the Environment* 5: 466–474.

Smol J. P. and Douglas M. S. V. 2007b. Crossing the final ecological threshold in high Arctic ponds. *Proceedings of the National Academy of Sciences USA* 104: 12395–12397.

Smol J. P. and Schindler D. W. 2006. Cumulative effects of climate warming and other human activites on freshwaters of Arctic and Subarctic North America. *Ambio* 35: 160–168.

Smol J. P., Wolfe A. P., Birks H. J. B., Douglas M. S. V., Jones V. J., Korhola A. et al. 2005. Climate-driven regime shifts in the biological communities of arctic lakes. *Proceedings of the National Academy of Sciences US* 102: 4397–4402.

Solomon S., Portmann R. W., Sasaki T., Hofmann D. J. and Thompson D. W. J. 2005. Four decades of ozonesonde measurements over Antarctica. *Journal of Geophysical Research-Atmospheres* 110: D21311.

Solomon S., Portmann R. W. and Thompson D. W. J. 2007. Contrasts between Antarctic and Arctic ozone depletion. *Proceedings of the National Academy of Sciences USA* 104: 445–449.

Stonehouse B. 1989. *Polar Ecology*. New York: Chapman and Hall.

Sugden D. 1982. *Arctic and Antarctic. A Modern Geographic Synthesis*. Totowa Barnes and Noble Books.

Sun L., Xie Z. and Zhao J. 2000. A. 3,000 year record of penguin populations. *Nature* 407: 858.

Swann A. L., Fung I. Y., Levis S., Bonan G. B. and Doney S. C. 2010. Changes in Arctic vegetation amplify high-latitude warming through the greenhouse effect. *Proceedings of the National Academy of Sciences US* 107: 1295–1300.

Tape K., Sturm M. and Racine C. 2006. The evidence for shrub expansion in Northern Alaska and the Pan-Arctic. *Global Change Biology* 12: 686–702.

Thatje S., Anger K., Calcagno J. A., Lovrich G. A., Pötner H.-O. and Arntz W. E. 2005. Challenging the cold: crabs reconquer the Antarctic. *Ecology* 86: 619–625.

Thompson D. W. J. and Solomon S. 2002. Interpretation of recent Southern Hemisphere climate change. *Science* 296: 895–899.

Vincent W. and Laybourn-Parry J. 2008. *Polar Lakes and Rivers: Limnology of Arctic and Antarctic Aquatic Ecosystems*. New York: Oxford University Press.

Walker D. A., Bhatt U. S., Raynolds M. K., Comiso J. E., Epstein H. E., Jia G. J. et al. (eds) Arctic Report Card 2009. Available at: http: //www.arctic.noaa.gov/reportcard

Wania F. and Mackay D. 1993. Global fractionation and cold condensation of low volatility organochlorine compounds in polar regions. *Ambio* 22: 10–18.

White D., Inzman L., Alessa L., Cassano J., Chambers M., Falkner K. et al. 2007. The arctic freshwater system: changes and impacts. *Journal of Geophysical Research* 112: G04S54.

36

Environmental Change in Mountain Regions

Martin Beniston

1 INTRODUCTION

Mountains and upland features (Figure 36.1) occupy close to 25 percent of continental surfaces (Kapos et al., 2000) and, although only about 26 percent of the world's population resides within mountains or in the foothills of the mountains (Meybeck et al., 2001), mountain-based resources indirectly provide sustenance for over half. Moreover, 40 percent of the global population lives in the watersheds of rivers originating in the planet's different mountain ranges. Because mountains are regions of primary importance in terms of the resources that they provide, any change in their physical, biological and socioeconomic characteristics is likely to have long-term repercussions, both within the mountains themselves and in the more populated lowland areas beyond.

This chapter will inventory the reasons why mountains are important to humankind, before reviewing how mountain environments have evolved in the past, prior to the more recent decades where direct anthropogenic activities have been accelerating change. The chapter then reviews some of the natural and human drivers of environmental change before summarizing, in the

fourth part, the possible or plausible changes that many mountain domains are likely to face in the future, because of the multiple stress factors imposed by human activities. Finally, the chapter closes with a brief review of cross-cutting issues and possible adaptation and mitigation strategies.

1.1 Importance of mountain regions

Mountains are often recognized as being among the most spectacular features of our planet. Possibly the most important resource that mountains provide is water; indeed, the source of over half the world's major rivers are located in high mountain areas; they provide the principal water catchment areas, or watersheds, for much larger geographical regions. Water resources for populated lowland regions are highly influenced by mountain climates and vegetation; snow feeds into the hydrological basins and acts as a control on the timing of water runoff in the spring and summer months (Beniston, 2004). Forests delay the period of snow-melt and extend the period in which water is available for river flow and also enhance the infiltration capacity of soils. Regions such as southern and eastern

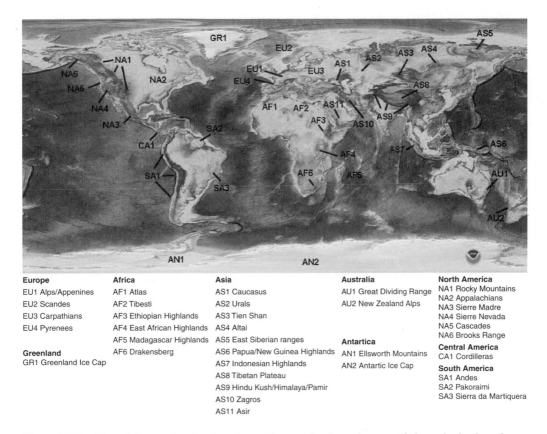

Europe	Africa	Asia	Australia	North America
EU1 Alps/Appenines	AF1 Atlas	AS1 Caucasus	AU1 Great Dividing Range	NA1 Rocky Mountains
EU2 Scandes	AF2 Tibesti	AS2 Urals	AU2 New Zealand Alps	NA2 Appalachians
EU3 Carpathians	AF3 Ethiopian Highlands	AS3 Tien Shan		NA3 Sierre Madre
EU4 Pyrenees	AF4 East African Highlands	AS4 Altai		NA4 Sierre Nevada
	AF5 Madagascar Highlands	AS5 East Siberian ranges		NA5 Cascades
Greenland	AF6 Drakensberg	AS6 Papua/New Guinea Highlands	**Antartica**	NA6 Brooks Range
GR1 Greenland Ice Cap		AS7 Indonesian Highlands	AN1 Ellsworth Mountains	**Central America**
		AS8 Tibetan Plateau	AN2 Antartic Ice Cap	CA1 Cordilleras
		AS9 Hindu Kush/Himalaya/Pamir		**South America**
		AS10 Zagros		SA1 Andes
		AS11 Asir		SA2 Pakoraimi
				SA3 Sierra da Martiquera

Figure 36.1 Mountains and upland regions. The numbering of some of the principal regions is given clockwise for each continent

Asia, home to almost half the world's population, depends largely upon water originating in the Himalayas for economic activities such as agriculture, industry and energy production. Changes to mountain environments, such as shifts in precipitation regimes, reduced snow and ice cover, or deforestation, may significantly alter the flow patterns in major rivers, such as the Ganges or the Yangtze, and thereby disrupt patterns of water use and water management across large areas.

Mountains also represent unique areas for the detection of climatic change and the assessment of climate-related impacts. One reason for this is that, as climate changes rapidly with height over relatively short horizontal distances, so does vegetation and hydrology (Whiteman, 2000). Certain mountain chains

have been referred to as 'islands' rising above the surrounding plains (Hedberg, 1964), such as those in East Africa.

Mountains are not only passive elements of terrestrial environments; they also play a key role in global systems, such as climate. They contribute to the triggering of cyclogenesis in mid-latitudes, through the disruption of large-scale atmospheric flows by the physical presence of mountain barriers. The seasonal blocking episodes experienced in many regions of the world, with large associated anomalies in temperature and precipitation, are closely linked to the presence of mountains. Major mountain building episodes in the recent geologic past, such as uplift of the Tibetan Plateau, fundamentally altered global climate (Ruddiman, 2001).

Mountain regions are particularly sensitive to a wide range of environmental stresses on various spatial and temporal scales. Erosion is a constant feature of mountain environments, from the long-term effects of glaciological, hydrological and chemical weathering acting on time scales from centuries to millennia, to abrupt manifestations (i.e., minutes to hours) of major natural catastrophes such as rock falls, mud slides, avalanches or outburst floods. Whatever the source of the disturbance, mountains are composed of a number of inherently fragile systems.

1.2 Socioeconomic aspects of mountains and uplands

Human populations have lived in most mountain regions for centuries and even millennia. In certain regions of the world, however, notably those settled by Europeans who displaced indigenous populations, the history of today's predominant population dates back only to recent centuries. Grötzbach (1988) has described these two broad categories as, respectively, 'old' and 'young' mountains. The latter, in North America, Australia, and New Zealand, are further characterized by sparse settlement, extensive market-oriented pastoral agriculture and forestry and, over the past few decades, the rise of tourism as a major economic force (Beniston, 2003; Price, 1994).

'Old' regions, notably the European mountains, are relatively densely settled and can be further divided into three broad categories. The first of these is characterized by the decline of traditional agriculture and forestry, which subsist essentially through government subsidies (e.g., in Switzerland; Beniston, 2004). During the postwar period, tourism grew rapidly, and is in many regions today the principal income-earner for numerous mountain communities and even at the national level (e.g., Austria and Switzerland; OECD, 2006). The most recent stage of development is characterized by a globalized

economy, where under the mounting pressure exerted on the agricultural sector to deregulate, even highly protected and subsidized mountain agriculture is being forced to adapt. Many problems specific to mountains, such as the transportation of goods across the Alps, demand solutions that can nowadays only be found within a supranational context. Current policies relevant to the European Alpine region are no longer determined exclusively within the Alps themselves, but increasingly in the EC administration in Brussels. While in some regions, such as Western Europe and Japan, mountains are experiencing depopulation (Price, 1994; Yoshino et al., 1980), in general land-use pressures are increasing because of competition between refuge use, mineral extraction and processing, recreation development, and market-oriented agriculture, forestry, and livestock grazing (Ives and Messerli, 1989; Messerli, 1989).

The second category of 'old' mountain regions includes most of the mountains of developing nations. Traditional subsistence agriculture is one of the main economic activities in the upland areas of countries such as Nepal, Bhutan, New Guinea and Peru. The inhabitants of upland zones of many developing countries are essentially mountain peasants, who practice some form of pastoral migration (or transhumance), in particular in the Andes and the Himalaya. In other parts of the developing world, nomadic agriculture is practiced even today, although there is a clear tendency towards sedentary lifestyles; nomadic practices are observed in the High Atlas of Morocco, parts of the Middle East, and in the uplands of Pakistan, for example. In all of these regions, tourism is becoming economically important in small, well-defined areas, often with considerable effects up to the national scale (Price, 1993). In some countries such as Nepal, Kenya, or some of the Andean states, tourism is the primary source of foreign exchange, because of the attractiveness of the high and remote mountains for hikers and climbers, among other factors.

2 PAST ENVIRONMENTAL CHANGE IN MOUNTAINS

Mountain areas are the locale for a variety of excellent natural archives that document former environmental changes with differing temporal and parameter resolution. These records include physical systems such as glaciers and biological indicators such as trees that respond to environmental stress factors, often in close juxtaposition, thereby providing complementary insights into former environmental changes. Using various archives stored in tree rings, lake sediments and glaciers, and supplementing these with historical sources and new data allows a documentation of the magnitude and timing of past environmental changes and provide a benchmark natural variability. Understanding and modeling this variability for decades to centuries is a critical element in our ability to detect any anthropogenic signal in the relatively short instrumental climate record. It defines the background variability and trends upon which any future changes will be superimposed. The many methods for reconstructing past climates and environments are discussed elsewhere in this volume (see especially, Chapter 4 and Section II), and thus this chapter will not focus on methods but more on results pertaining to mountain regions.

Glaciers exist on all continents except Australia and at virtually all latitudes from the tropics to the poles, and mountain glaciers are valuable indicators of climate change. The volume of ice in a glacier, and correspondingly its surface area, thickness, and length, is determined by the balance between inputs (accumulation of snow and ice) and outputs (melting and calving); these are largely controlled by temperature, humidity, wind speed, and other factors like slope angle and ice albedo (Lemke et al., 2007). As climate changes, this balance may be altered, resulting in a change in thickness and the advance or retreat of the glacier

Numerous studies are available in the various mountain areas of the world on glacier variations and past glacier extent (Haeberli et al., 1989; Karlén, 1988). The lower altitudinal limits of ice cover can be mapped and dated by, for example, radiocarbon techniques for the last glaciation and lichenometric methods for the last few centuries. For the more recent glacial advances or still stands, ice thickness, and therefore volume, may be estimated from trimlines on the valley sides (e.g., Thompson et al., 2006a). High-altitude ice cores have exhibited significant increases in temperature over the last few decades, and in some places glaciers and ice caps have disappeared altogether (Paul et al., 2007). These records, with additional evidence from other small ice cores from high altitudes, point to a dramatic climatic change in recent decades (Haeberli and Beniston, 1998), prompting concern over the possible loss of these unique archives of paleo-environmental history.

For the last glacial cycle, global climates have been reconstructed from mountainous regions using a variety of paleoclimatic data extracted from cores of lake and bog sediments, ice layers, tree rings, and so on. Glacial-geomorphologic evidence of former glacier extent and downward displacement of vegetation zones in tropical mountain areas of the order of almost 1 km imply continental cooling of 5–6°C; for example, in New Guinea (Bowler et al., 1976) and Hawaii (Porter, 1979). In the highest mountains, ice caps and glaciers are dominant features and in a few areas, ice cores have been recovered from mountain ice caps, providing a set of unique, high-resolution records extending back for approximately 100 to over 10,000 years. High elevation ice cores provide paleoenvironmental information that complements and expands upon that obtained from polar regions. For example, in ice cores from Huascarán, Peru, located at 6,048m above sea-level, the lowest few meters contain ice from the last glacial maximum, with the isotopic ratio $\delta^{18}O$ approximately 8 percent lower than Holocene levels, and a much higher dust content (Thompson et al., 1995). The lower $\delta^{18}O$ suggests that tropical temperatures were 8–12°C lower during the Last Glacial Maximum (LGM).

Ice marginal positions are recorded by moraines, and trim-lines on valley walls. In ideal situations, it may be possible to identify a series of overlapping or nested moraines representing former glacier positions. However, more commonly the most recent advance of ice (the exact timing of which may have varied but is generally considered to lie within the AD 1300–1850 timeframe) was the most extensive for several millennia (and in some glaciers, the most extensive since the last ice age) and has obliterated evidence of earlier advances. These Little Ice Age moraine systems are the latest in a series of glacier advances which began in the late Holocene. Dating such advances is problematic, relying principally on radiocarbon dating of organic material buried by the advancing moraine, exposure dating of boulder surfaces, or by lichens growing on the moraine itself, once it has stabilized. Clearly, such evidence can only be episodic and does not provide the kind of high-resolution, continuous data that is favored in paleoclimatic analysis. However, closely related records may be obtained from glacier-fed lakes, which may register the growth of ice and the deterioration of mountain climates by a reduction in organic matter, and an increase in silt input to the lake-bottom sediments (Nesje et al., 1991). In ideal circumstances, such records may be annually laminated, providing very high-resolution insight into past climatic conditions (e.g., Leonard 1986). Pollen and other microfossils in lake sediments, or in high-altitude bogs, can be interpreted in terms of former tree-line movements and hence provide a framework for other proxy records in the mountains. Based on a composite view of such data, Karlèn (1993) argues that glaciers (in the more continental parts of Scandinavia) advanced to positions comparable to those of the Little Ice Age around 3,000, 2,400, 2,000 and 1,200 years BP. Many of these glaciers had completely disappeared in the early to mid Holocene, only reforming within the last 3,000 years (Nesje et al. 1994).

Since 1850 the glaciers of the European Alps have lost about 30 to 40 percent of their surface area and about half of their volume (Haeberli and Beniston, 1998). Similarly, glaciers in the Southern Alps of New Zealand have lost 25 percent of their area over the last 100 years (Chinn, 1996), and glaciers in several regions of central Asia have been retreating since the 1950s (Thompson et al., 2006b). The seven-year average rate of ice loss for several glaciers monitored in the US Pacific Northwest was higher for the period since 1989 than for any other period studied (Hodge et al., 1998). Glacial retreat is also prevalent in the higher elevations of the tropics, and Mount Kenya and Kilimanjaro have lost over 60 percent of their ice cover in the last century (Thompson et al., 2002); accelerated retreat has also been reported for the tropical Andes (Thompson et al., 2006b). Figure 36.2 illustrates the rate of retreat, on average, of mountain glaciers on all continents in terms of changes in total mass balance (following

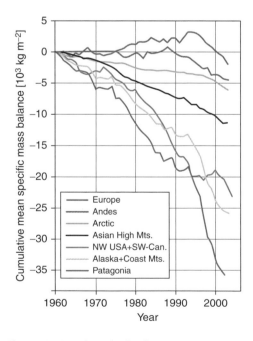

Figure 36.2   Changes in the average mass balance of mountain glaciers of different continents.

Dyurgerov and Meier, 2005), and serves to highlight the fact that glacier dieback is not a local, but a global phenomenon.

Beyond the realm of snow and ice and alpine tundra, the tree-line defines an important climate-related ecotone. Although the tree-line itself varies in structure and composition from one mountain region to another, and is subject to many potentially limiting ecological constraints (Körner, 2007b), climate is the dominant control, at least away from the oceanic margin. Consequently evidence of past changes in tree-line position is generally interpreted in terms of variations in summer temperature. Radiocarbon-dated macrofossils (tree stumps, or wood fragments) from above the modern tree-line can thus provide testimony of warmer conditions in the past. This is well illustrated in the northern Urals where now-dead trees beyond the modern tree-line have been dated using dendrochronology techniques to obtain information on the timing of past tree growth at high elevations. This reveals that most of the trees were growing in the tenth to twelfth century AD; no trees were found to date from the late eighteenth and nineteenth centuries, indicating that the tree-line had retreated by that time. This evidence is strongly supported by tree-ring studies in nearby forests, where maximum ring widths were found at the time the forest advanced and minimum ring widths were characteristic of the eighteenth and nineteenth centuries.

Subfossil wood from above the present tree limit has been found over wide areas of Scandinavia, (Kvamme, 1993), Eurasia (MacDonald et al., 2000), the mountains of the western US (Rochefort et al., 1994), and Canada (MacDonald et al., 1998). In both areas, there is strong evidence that the upper tree limit was well above modern levels, especially before about 5000 BP. In parts of the western US and Western Canada, trees were growing as much as 150 m above modern limits in the period from 8,000 to 6,000 BP (Rochefort et al., 1994) and similarly, in Scandinavia, trees were up to 300 m above modern limits in the early Holocene

suggesting that summer temperatures were 1.5–2 °C above modern levels (Karlèn, 1993) In both areas, it appears that climate deteriorated after 5,000 BP, leading to a decline in tree limits. This corresponds to both pollen records and the evidence from glacier moraines that temperatures became lower, especially after about 3,500 BP, marking the onset of late Holocene neoglaciation. Minor oscillations of the tree-line have taken place since then, culminating in the coldest episodes in the sixteenth to mid nineteenth century. At that time, temperature in the mountains of southern Sweden were around 1°C colder than in the mid twentieth century (Kullman, 1989).

3 ANTHROPOGENIC ENVIRONMENTAL CHANGE IN MOUNTAINS

3.1 Human impacts in prehistory and historical times

Probably the impacts of early human communities in mountain regions were fairly low and limited to very local areas where sedentary populations lived. It is more likely that humans responded to environmental and climatic change by out-migration under adverse conditions and some measure of local adaptation.

Archaeological evidence indicates that land routes across Central Asia, and coastal routes into Australia and along the shores of the Americas circumvented the extensive ice-caps during the last glacial maximum. These population movements led to the settlement of populations with enhanced technological capabilities (more efficient tools; clothing to resist the cold; more efficient hunting weapons; boats capable of navigating in coastal waters) in different parts of the world (Monastersky, 2000).

While the general boundary conditions of climate did not change dramatically over the last 6,000 years (Wanner et al., 2008) compared to glacial/interglacial periods, for example, there are nevertheless a number of climatic shifts that have been identified in the paleoclimatic record. Some of these are related to changes in orbital forcing and the redistribution of solar energy at the Earth's surface. During the Holocene optimum, which lasted from about 6,000 to 5,000 years BP, temperatures were between 2 and 4°C warmer than currently; precipitation patterns were also markedly different from those of today. In Iraq, India and China, for example, river valleys flooded every spring due to the enhancement of monsoon circulations, and replaced the minerals of the soil, thereby allowing soils to remain fertile for many years and enabling villages to grow into cities. The settled way of life led to an increase in food production and population. Technologies developed rapidly, from farming tools, clay pottery and bricks, wheels, to reading and writing.

The relative stability of environmental conditions, allowing rapid progress in human technology and culture, was not the only influencing factor for reducing or limiting out-migration. As food supplies became sufficient for larger numbers of people due to farming practices and the storage of food to avert famine during more difficult times, and because technology and shelter were available to a much larger extent, the size of populations grew far more rapidly than in the past, both in lowlands and in uplands (Cohen, 1977). As a result the carrying capacity of the environment often became insufficient beyond a certain critical threshold of population. If demographic growth began to outpace food supply, then a time inevitably came when there was insufficient food supply to sustain an entire, growing population, and out-migration would become once again an option. Strife or epidemics, which could sometimes curtail increasing populations and allow the survivors to remain in their usual environment, could in some instances avert the necessity for out-migration (Armelagos, 1991).

3.2 Current anthropogenic impacts and their drivers: climate change

Climate characterizes the location and intensity of biological, physical and chemical processes and thus warrants a particular focus in this chapter. Mountain climates consist of four major factors (Barry, 2008), namely continentality, latitude, altitude and topography. Continentality refers to the proximity of a particular region to an ocean. The diurnal and annual amplitude of temperatures in a maritime climate are lower than inland regions due to the large thermal capacity of the ocean, which warms and cools far less rapidly than land. Onshore winds and particular conditions of atmospheric stability also result in higher precipitation levels in many maritime mountains, because of the readily available source of moisture. Latitude determines to a large extent the amplitude of the annual cycle of temperature and, to a lesser extent the amount of precipitation which a region experiences. because of local site characteristics and the degree of continentality.

Altitude, however, is the most distinguishing and fundamental characteristic of mountain climates. Atmospheric density, pressure and temperature decrease with height in the troposphere. The altitudinal controls on mountain climates also exert a significant influence on the distribution of ecosystems. Indeed, there is such a close link between mountain vegetation and climate that vegetation belt typology has been extensively used to define climatic zones and their altitudinal and latitudinal transitions (e.g., Klötzli, 1991; Ozenda, 1985).

Topographic features also play key role in determining local climates, in particular due to the slope, aspect, and exposure to weather elements. These factors are important determinants for radiative flux exchange as well as precipitation, which is highly sensitive to local site characteristics. Indeed, precipitation amounts are generally amplified in the presence of the mountains due to the topographic uplift of moist air that, according to the atmospheric flow situation, can be forced to rise above the mountains.

Major mountain ranges of the world can trigger large-amplitude and gravity waves that extend around the globe. These waves represent a major mechanism for the propagation of energy within the atmospheric flows from one part of the world to another.

3.3 Acid deposition and atmospheric pollution

Many mountain regions, particularly those close to centers with high levels of industrial activity, are vulnerable to atmospheric pollution. In particular, the presence of sulfur and nitrogen compounds in the air can lead to acid deposition through the scavenging of these compounds by precipitation (e.g., Vitousek et al., 1997). While in the 1980s and 1990s, air-pollution abatement measures addressed essentially sulfur compounds in North America and Europe, nitrogen emissions resulting from oxidation of fossil fuels (NO_x) and ammonium (NH_x) related to agricultural activities remain today a primary source of acidification (as much as 50 percent of the acidity of water and snow, according to the US Environmental Protection Agency, 2000).

Acid deposition has widely varying consequences for environmental systems, notably plant species, stimulating growth through enhanced fertilization effects in some, and destroying others due to toxic levels and critical pH threshold. Certain species are thus more robust than other in the face of acid deposition (Stevens et al., 2004), thus creating an imbalance in natural ecosystems that lead to ecosystem disruption and shifts in species distribution. Soils also experience long-term effects of acidification; elements in the soils that are capable of neutralizing acidity will progressively lose this capability through leaching when exposed to long-term deposition (Burns, 2004). Such soils become nutrient-depleted and are less able to sustain

plants and trees. Aquatic ecosystems are also negatively impacted upon by acidification of surface waters, many species being extremely sensitive to small changes in pH levels (e.g., Clow et al., 2003). In mountain regions that are subject to wintertime snow cover, peaks of acidity can be detected in surface waters during the period of seasonal snow melt.

The degree of recovery of mountain ecosystems to sulfur emission reductions in Europe and North America shows mixed results, varying from region according to climate and soil types (Driscoll et al., 2001), but on the whole recovery is either inexistent or far slower than anticipated. This is possibly due to the fact that nitrogen deposition, unlike sulfur deposition, has not decreased and continues to accumulate acid compounds at the surface (Fenn et al., 2003).

Other forms of pollution that affect mountain ecosystems include ozone and aerosols of industrial origin. A notable continental-scale source of pollution, known as the Asian Brown Cloud (ABC) affects both the Indian and the Chinese sides of the Himalayas. ABC is the consequence of the intensive burning of fossil fuels and biomass that in some parts of India and China are compounding greenhouse-gas warming because of the presence of black carbon, soot particles and ozone that tend to absorb solar radiation (Ramanathan et al., 2005). In other regions, the aerosol content of the brown cloud partially offsets greenhouse warming through the direct reflection of solar energy.

The presence of the Himalayas acts as a barrier to pollutant dispersal from both the Chinese and Indian sides. Both the natural and human environments are inevitably subject to this pollution because the mean altitude of the ABC is generally well below that of the mountains themselves. Cryospheric and hydrologic systems are particularly affected, since the ABC effect is thought to be as important as greenhouse warming at high elevations, negatively affecting glaciers through warmer temperatures and energy absorption by black carbon deposition on snow and ice (UNEP, 2008). The acceleration of glacier retreat since the 1970s is also influencing runoff in the critical headwater region of the Ganges, which supplies water to 400 million persons in northern India and whose basin encompasses 40 percent of Indian irrigated cropland. Mountain agriculture at the altitudes affected by the ABC is already responding negatively to high levels of ozone, and many plants retain the many acidic and toxic particles that are deposited on leaves and on the topsoils. While it is difficult to estimate absolute numbers specifically in the Himalayan region, UNEP (2008) states that there are numerous health effects resulting from the aggressive mix of gaseous and particulate pollutants contained within the brown cloud.

3.4 Deforestation

The overexploitation of forest resources over centuries, often in the face of population pressures and the consequent need to provide land on which to grow food, has led to a situation where many forested areas of the world in general, and mountain regions in particular are now in rapid retreat. For centuries, forest destruction has been at the expense of dwindling species habitat and, in some cases, species extinction. For example, the Himalayan watershed covering Northern India, Nepal, and Bangladesh had lost 40 percent of its forest by 1980. Costa Rica has lost one-third of its forests and within decades they may only exist in highly fragmented biosphere reserves.

Deforestation and associated biomass burning have both direct and indirect effects on mountain environments. The direct effects of reduced mountain-tree cover include enhanced erosion of denuded slopes and loss of topsoil that is detrimental to mountain agriculture, as experienced in places like Nepal or Madagascar. These are likely to increase in tropical mountain regions where higher rates of precipitation may occur and thus accelerate slope denudation processes, with higher rates of silting in rivers. Intensified seasonal

flooding with resultant loss of lives and property, or water shortages during the dry season, may be on the increase because of the combined stresses of climatic change and the loss of vegetation cover. Changing temperature and precipitation regimes will also impact upon specific forest ecosystems such as the unique cloud forests of Central America, Papua New Guinea, and East Africa.

Indirect effects of deforestation include enhanced air pollution when trees are removed by burning (e.g., slash-and-burn practices in the tropical world). Biomass burning releases significant quantities of gaseous and particulate combustion products into the atmosphere. Combustion gases include carbon dioxide, carbon monoxide, hydrocarbons, and oxides of nitrogen, methyl chloride, and methyl bromide (Granier et al., 2004). Biomass burning is a major global source of elemental carbon particulates, which absorb and scatter incoming solar radiation, and hence impact the Earth's radiation budget and global climate. In addition, biomass burning impacts the biogenic production and emission of nitrogen and carbon gases from the soil, and therefore intervene in the hydrological cycle.

3.5 Land-use and socioeconomic change

Mankind is adding a new dimension to the global environment in general, and mountain regions are no exception. 'Human interference' is generally perceived as being detrimental to mountain environments; while this may often be the case, there are examples where man has augmented the safety or the esthetics of mountains. Terraced agriculture as practiced in many parts of the world (Asia, Africa or South America) is considered by many to enhance the beauty of a region; in addition, this form of land-use acts to stabilize mountain slopes.

In many mountain environments, human activity has modified natural systems such as tree-lines in the Alps (e.g., Gehrig-Fasel et al., 2007), making it difficult to discriminate between natural and anthropogenic controls of recent changes. In certain mountain regions such as the Canadian Rockies, for example, large areas were set aside within National Parks prior to significant human impact, and observed changes can be attributed solely to natural forcing (e.g., Klasner, 2002). In addition, the absence of significant human disturbance has allowed the preservation of physical evidence of change that has been destroyed elsewhere. However, this absence of human activity also limits access and the period of instrumental or documentary records: contrary to the situation in Europe, there are scant documentary observations for the Canadian Rockies prior to 1900 and few areas have been studied in detail. Land-use change in many mountain regions is today linked to economic pressures; the return of natural vegetation in regions that were used for agriculture a few years or a few decades ago is a sign of changing economic priorities and resources, forcing mountain farmers in many regions to search for income elsewhere (e.g., Garcia-Ruiz et al., 1996).

Anthropogenic stresses on mountain regions have increased significantly in the second half of the twentieth century. Rapid technological developments and sinking costs of transportation have increased human mobility into what used to be remote areas with difficult access. In more developed countries, signs of encroachment into the mountain environments include industrialization of mountain valleys, communications infrastructure, buildings for recreation and tourism. In the past, human interference with mountain environments was essentially linked to agricultural practices; in regions with increasing populations and demands on food and resources, overgrazing and deforestation have transformed mountain environments. Some of the most overgrazed regions over the past 2,000 years have been the eastern Mediterranean mountains, where pastoral practices have transformed once-forested mountains into arid and semi-arid regions.

4 FUTURE ENVIRONMENTAL CHANGE AND ITS IMPACTS

4.1 Climate projections

Regional-scale climate information must by essence be available at the time and space scales useful for climate impacts research and shaping policy response. Processes acting on local or regional scales often feed into the climate system, so that investigations of scale-interactions are crucial to furthering our understanding of the climate system and the role of the smaller scales in its spatial and temporal evolution. One technique that was pioneered by Giorgi and Mearns (1991) and has since become commonplace in the climate modeling arena (e.g., Giorgi and Mearns, 1999), to address the scale problems related to insufficient grid resolution in global climate models (GCMs) is that of 'nested modeling'. Results from a GCM are used as initial and boundary conditions for a regional climate model (RCM) that operates at much higher resolution and with generally more detailed physical parameterizations. A regional model is in this case an 'intelligent interpolator', since it is based on the physical mechanisms governing climatic processes rather than a mathematical interpolation technique, and serves to enhance the regional detail lacking in global climate models. Such a procedure becomes particularly attractive for remote regions or those with topographic elements, whose complexity is unresolved by the coarse structure of a GCM grid, and where observational data is often sparse or nonexistent.

Today, RCMs have even the capability of 'cascade self-nesting' (e.g., Goyette et al. 2003; Laprise et al., 1998) by which the resolution of the same basic model is refined in order to capture the salient features of both atmospheric phenomena and the geographical characteristics (coastlines, relief) that are fundamental in shaping, amplifying or damping regional climate mechanisms. As the spatial definition of a grid increases, so does the detail of topography, but at the expense of much larger computational resources. Because of the problems of model resolution in regions of complex terrain, specific details of climatic change in mountain areas are less certain than in lowland areas, even though considerable progress has been made over the past few years (e.g., Beniston, 2007; Déqué et al., 2005) as to the use and interpretation of RCM results for a changing climate. Model biases with respect to observations often, but not always, tend to decrease with increasing resolution or decreasing RCM domain size. RCMs may in some instances provide more realistic information on climatological processes than GCMs, however, because they tend to capture more realistically the regional detail of key elements such as topography.

Changes in temperature are often more marked in high latitudes than over the equatorial zone, and many mountain regions that already today experience stronger warming trends than the lowlands may continue to be subject to stronger warming (e.g., Beniston, 2006). One reason for this may be found in the positive feedback effects of changing snow and ice surfaces in mid- and high-latitude mountain chains.

There are few mountain-specific regional climate projections, because the improvements of some of the aforementioned problems have intervened only in recent years. However, based on the IPCC models (IPCC, 2007) and major research projects such as the EU 'PRUDENCE' project (Christensen et al., 2002), some major trends can be ascertained:

1 During the course of the twenty-first century, mid- and high-latitude mountains will experience warming both in winter and in summer, the latter being strongest in regions where prolonged summer drought is expected to occur (e.g., western and central Europe). At low latitudes, warming will also occur but will probably be of a somewhat lower amplitude. In all cases, the actual amplitude and rate of change will depend on the greenhouse-gas emissions pathway;

2 Mid- and high-latitude mountains will see increased overall precipitation; most of the increase will occur in wintertime, and some regions may see significant shortfalls over current precipitation levels during summer. At tropical latitudes, drier conditions than today are expected to prevail throughout the year, while close to the equator and in regions subject to monsoons, increased precipitation is likely as a result of the acceleration of the hydrological cycle resulting from a warmer climate,

4.2 Environmental impacts on natural systems: snow and ice

Snow and ice are, for many mountain ranges, a key component of the hydrological cycle, and the seasonal character and amount of runoff is closely linked to cryospheric processes. In addition, because of the sensitivity of mountain glaciers to temperature and precipitation, the behavior of glaciers provides some of the clearest evidence of atmospheric warming and changes in the precipitation regime, both modulated by atmospheric circulation and flow patterns over the past decades (Haeberli and Beniston, 1998; WGMS, 2000).

Changes in climate have been shown to result in shifts in seasonal snow pack (Stewart et al., 2004); glacier melt influences discharge rates and timing in the rivers that originate in mountains. In temperate mountain regions, the snow-pack is often close to its melting point, so that it may respond rapidly to apparently minor changes in temperature. As warming progresses in the future, regions where snowfall is the current norm will increasingly experience precipitation in the form of rain. For every degree Celsius increase in temperature, the snowline rises by about 150 m. In many instances, increased winter precipitation in mid-latitude mountains will not compensate for the temperature increase, as seen in Figure 36.3 (Beniston et al., 2003), that shows for two Swiss locations the duration of snow on the ground as a function of average wintertime temperature and precipitation. Reduced snow cover will have a number of implications, in particular

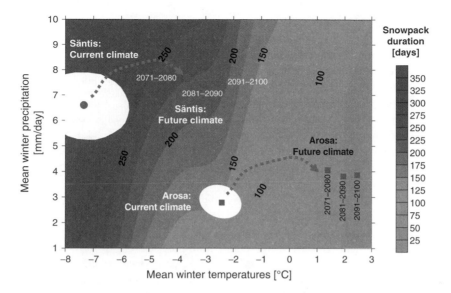

Figure 36.3   Snow-cover duration as a function of winter (DJF) minimum temperature and precipitation for two Swiss locations (Säntis, 2500 m ASL and Arosa, 1600 m ASL), under current climate and for the last three decades of the twenty-first century (IPCC A2 emissions scenario). The ellipses show the 2σ range of DJF minimum temperature and precipitation.

for early seasonal runoff (Dettinger and Cayan, 2005), and the triggering of the annual cycle of mountain vegetation (Cayan et al., 2001).

Empirical and energy-balance models indicate that 30–50 percent of existing mountain glacier mass could disappear by 2100 according to the amplitude of global warming scenarios (Lemke et al., 2007). The smaller the glacier, the faster it will respond to changes in climate. With an upward shift of 200–300 m in the equilibrium line altitude (ELA, which represents the level below which ablation rates exceed accumulation), the reduction in ice thickness of temperate glaciers could reach 1–2 m per year. As a result, many glaciers in temperate mountain regions would lose most of their mass within decades (Lemke et al., 2007). Shrinking glaciers will lead to changes in the hydrological response of certain regions compared to today; as glaciers melt rapidly, they will provide enhanced runoff, but as the ice mass diminishes, the runoff will wane.

4.3 Impacts on hydrology

Because mountains are the source region for over 50 per cent of the globe's rivers, the impacts of climatic change on hydrology are likely to have significant repercussions not only in the mountains themselves but also in populated lowland regions that depend on mountain water resources for domestic, agricultural, energy and industrial purposes. Water resources for populated lowland regions are influenced by mountain climates and vegetation; snow feeds into the hydrological basins and acts as a control on the timing of water runoff in the spring and summer months (Graham et al., 2007). Significant shifts in climatic conditions will also have an effect on social and economic systems in many regions through changes in demand, supply and water quality. In regions that are currently sensitive to water stress (arid and semi-arid mountain regions), any shortfalls in water supply will enhance competition for water use for a wide range of economic, social, and environmental applications. In the future, such competition will be sharpened as a result of larger populations, leading to heightened demand for irrigation and perhaps also industrialization, at the expense of drinking water (Noble and Gitay, 1998). Armed disputes over water resources may well be a significant social consequence in an environment degraded by pollution and stressed by climatic change (Beniston, 2002).

4.4 Impacts on mountain vegetation

Plant life at high elevations is primarily constrained by direct and indirect effects of low temperatures, radiation, wind and storminess or insufficient water availability (Körner, 2007a). Plants respond to these climatological influences through a number of morphological and physiological adjustments such as stunted growth forms and small leaves, low thermal requirements for basic life functions, and reproductive strategies that avoid the risk associated with early life phases. In some instances, diverse stress factors can enhance positive interactions between many alpine plant species, as shown by Callaway et al. (2002).

Because temperature decreases with altitude by 5–10°C/km, a first-order approximation regarding the response of vegetation to climate change is that species in regions that will experienced sustained warming will migrate upwards to find climatic conditions in tomorrow's climate which are similar to today's (Peters and Darling, 1985). According to this paradigm, the expected impacts of climate change in mountainous nature reserves would include the loss of the coolest climatic zones at the peaks of the mountains and the linear shift of all remaining vegetation belts upslope. Because mountaintops are smaller than bases, the present belts at high elevations would occupy smaller and smaller areas, and the corresponding species population would drop, making them potentially more vulnerable to genetic and environmental pressures. However, the migration hypothesis may not always be applicable because of the different climatic tolerance of species involved, including genetic variability between species, different longevities and survival rates, and the competition by invading species (Engler et al., 2009).

It is expected that, on a general level, the response of ecosystems in mountain regions will be most important at ecoclines (the ecosystem boundaries if these are gradual), or ecotones (where step-like changes in vegetation types occur). Brönnimann et al. (2007) note that ecological changes at ecoclines or ecotones will be amplified because changes within adjacent ecosystems are juxtaposed. In steep and rugged topography, ecotones and ecoclines increase in quantity but decrease in area and tend to become more fragmented as local site conditions determine the nature of individual ecosystems. Even though the timberline is not a perfect ecocline in many regions, it is an example of a visible ecological boundary that may be subject to change in coming decades (Bugmann and Pfister, 2000). There are instances, however, where ecotones are the result of disturbance rather than climate. For example, the current level of many forests in the European Alps is lower than its potential limit because of pastoral practices; in such cases, ecotones may be the drivers of local climatic gradients rather than the other way round (Becker and Bugmann, 2001).

In regions where climatic change may lead to warmer and drier conditions, mountain vegetation could suffer as a result of increased evapotranspiration. This is most likely to occur in mountain climates under the influence of continental and Mediterranean regimes. Even in tropical regions, however, there are indications that plants are already sensitive to water stress on mountains, such as Mt Kinabalu in the Malaysian part of Borneo (Kitayama, 1996).

The length and depth of snow cover, often correlated with mean temperature and precipitation, is one of the key climatic factors in alpine ecosystems (Körner, 1999). Snow cover provides frost protection for plants in winter, and water supply in spring. Alpine plant communities are characterized by a very short growing season (i.e., the snow-free period) and require water to begin their growth cycle. Keller et al. (2005) have shown that vegetation communities which live in snow beds and in hollows will be the most vulnerable to change, because their usual habitat will be strongly disturbed.

Forest model simulations applied to many mid-latitude mountains have shown that certain upward-moving forest ecosystems could actually disappear from their potential habitats because of the lack of winter cooling, vital for regeneration and the robustness of trees, and a greater sensitivity to droughts and frosts. Tropical mountain forests, particularly cloud forests, could be highly sensitive because of the likely change in the levels of persistent condensed moisture, leading to a possible collapse of the genetically abundant ecosystems that gravitate around cloud forests (Benzing, 1998). In all forest impact studies, both in latitudinal and altitudinal terms, climatic change as projected by the IPCC (2007) will be more rapid than the migration capacity of forests.

4.5 Impacts on fire hazard

Fire is an element which is of particular importance in many ecological systems; it plays a vital role in the recycling of organic material and the regrowth of vegetation, but can be devastatingly destructive. Changing climatic conditions are likely to modify the frequency of fire outbreaks and intensity, but other factors may also play a major role; for example, changes in fire-management practices and forest dieback leads to a weakening of the trees in response to external stress factors (King and Neilson, 1992). Fire suppression policies, as carried out until recently in North America, resulted in a substantial increase in biomass compared to natural levels. When this is the case, forests tend to

transpire most of the available soil moisture, so that catastrophic fires can occur as a result of the greater sensitivity of trees to seemingly minor changes in environmental conditions (Stocks et al., 2001). A combination of deadwood accumulation and prolonged drought resulted in the spectacular fire outbreak in Yellowstone National Park in the United States during the summer of 1988.

If a warmer climate is also accompanied by prolonged periods of summer drought, which the IPCC suggests may affect many mid-latitude mountains, this would transform areas already sensitive to fire into regions of sustained fire hazard. The coastal ranges of California, the Blue Mountains of New South Wales (Australia), Mount Kenya, and mountains on the fringes of the Mediterranean Sea, already subject to frequent fire episodes, would be severely affected. Fires are also expected to occur in regions which are currently relatively unaffected, as critical climatic, environmental and biological thresholds for fire outbreaks are exceeded.

Because many fire-sensitive regions are located close to major population centers, damage to infrastructure and disruption of economic activities may be expected at the boundaries of cities such as Los Angeles in California, Sydney in Australia, and coastal resorts close to the mountains in Spain, Italy and southern France. This has already occurred in the past and is likely to become more frequent in the future because of enhanced fire hazards and greater urban sprawl.

4.6 Impacts on managed systems

Shifting precipitation patterns by season and sharply curtailed glacier mass in the mountains will lead to modifications in hydrological regimes and will also mean glaciers will no longer feed water into river catchments at a time of the year when precipitation amounts are low and the snow-pack has completely melted. These changes will have significant impacts on several critical socioeconomic sectors in mountain regions, particularly since these are also subjected to various other forces that influence their viability. There will in addition be cascading effects on downstream areas. Climatic change will affect overall land use patterns, which in turn feed back into effects on water and carbon fluxes.

Climatic change will increase competition over water that will be available at different times and in different quantities. Water is difficult to allocate because of its public good features, which are aggravated by upstream potential to capture the resource and by the fact that flowing water may cross internal and international borders. Adding further to the potential for conflict are socio-economic changes that modify existing distribution schemes. Changing social patterns and economic incentives have resulted in major land-use changes in many mountains of the world (e.g., Begueria et al., 2006), which will generate significant shifts in the amount and seasonality of water resources. For example, deforestation causes an increase in the average annual discharge, and an acceleration of runoff during rainstorm events, also enhancing erosion and downstream sediment supply. Conflicting water use (e.g., between agriculture and hydropower, or between hydropower and tourism) as the resource diminishes through reduced precipitation in some areas and glacier retreat in others, will add further social and economic stress in most mountain regions vulnerable to climatic change. New water resource management practices will not be just a matter of adjusting to shifts in the physical environment but will also be strongly associated with social changes.

Tourism is an industry that has exhibited the most sustained growth of any global industry since the 1970s. It accounts for 10 percent of the world's net financial output, with many countries in the developing world dependent on tourism as their main source of income. In the developing countries, tourism of all types contributes roughly US$50 billion annually (Perry, 2000). Patterns of tourism have become more diversified, with new activities added to more traditional recreational activities and destinations. As a consequence, even remote natural areas, in particular mountain regions in the Himalayas, the Andes and East Africa, are attracting increasing numbers of tourists, with a parallel boom in the development of tourism infrastructure and construction. This infrastructure is often located in attractive cultural and natural landscapes, often with negative impacts for those landscapes and the sensitive ecosystems that they support (Godde et al., 2000). Tourism is thus both a significant economic driver for many mountain communities, but also an industry capable of adversely affecting the environmental quality of mountains and uplands. Climate change is likely to have both direct and indirect impacts on tourism in mountain areas. Direct impacts refer to changes in the climatic conditions necessary for specific activities. Indirect changes may result from both changes in mountain landscapes, and wider-scale socio-economic changes such as patterns of demand for specific activities or destinations (WTO, 2000). Several studies focusing on the ski industry, particularly in the European Alps (e.g., OECD, 2006) suggest that many resorts may be subject to economic downturns as rain would take over from snow at low to medium elevations in a warmer climate. Such impacts might be partially offset by new opportunities in the summer season and also by investments in technology, such as snow-making equipment, as long as climatic conditions remain within appropriate bounds. Mountaineering and hiking may provide compensation for reduced skiing, and thus certain mountain regions would remain attractive destinations.

Mountain runoff (electricity supply) and electricity consumption (demand) are both sensitive to changes in precipitation and temperature. Long-term changes in future climate will have a significant impact on the seasonal distribution of snow storage, runoff from hydro-electric catchments and aggregated electricity consumption. According to the future climate scenario used the seasonal variation of electricity consumption may be less pronounced than at present, with largest changes in winter which is the time of peak energy use for heating. The sensitivity of mountain hydrology to climate change is a key factor that needs to be considered when planning hydro-power infrastructure.

5 ADAPTATION, MITIGATION AND POLICY

Environmental change involves involve a complex set of issues and a subtle mix of social, economic, political, and technological driving forces. Novel answers to social and economic issues such as changes in social arrangements among mountain populations and their downstream neighbors, energy production in regions where water is underused as a means of helping abate greenhouse-gas emissions, and environmental issues such as the future evolution of water supply for use in domestic, tourist, or agricultural sectors will need to be addressed. The future challenges are significant, especially because of the growing interconnection among regions and among economic and social sectors. The relationships with downstream areas are especially critical, notably in cases where upstream and downstream entities concern separate sovereign countries.

The possibilities for adaptation and, even more so, mitigation, are limited in mountain regions, because of difficulties linked to the complexity of the terrain, the much higher costs for developing protection infrastructure (e.g., against landslides or avalanches) and the compounded effect of conflicting time scales related to economic policies, that are on the short term (days to months), and the management of the planetary environment, climate, and resources, that by essence needs to be planned on the very long term (decades or more).

6 CONCLUSION

This chapter has provided a succinct overview of some of the issues related to the drivers of environmental change and the plausible impacts of future change in a world where human activities are accelerating or exacerbating many of these changes. In order to keep up with the pace of environmental change, including climate change, more work is needed to better understand the intricacies of the causes and consequences of change that could be detrimental to natural and managed systems in mountains (Beniston, 2004). Among these factors, the following warrant particular attention:

- Long-term monitoring and analysis of indicators of environmental change in mountain regions, with a particular focus on cryospheric indicators, watershed hydrology, and terrestrial and aquatic ecosystems;
- Integrated model-based studies of environmental change in different mountain regions, in particular coupled ecological, hydrological and land-use models; models allowing the study of feedbacks between land surfaces and the atmosphere and thus enabling an assessment of the sensitivity of vegetation, snow, ice, and water resources to a range of forcings; and integrated (physical, biological and economic) analyses of environmental change for policy purposes.
- Sustainable land-use and natural resource management, with priorities for changes in forest resources, shifts in mountain agriculture and food security, and perturbations to water resources as a result of economic and demographic factors.

NOTE

This work draws on text that was published in "Climatic Change" in 1997 (Beniston, Diaz and Bradley: Climatic Change at High Elevation Sites: A Review; Vol. 36, pp. 233–251) and 2003 (Beniston: Climatic Change in Mountain Regions: A Review of Possible Impacts; Vol. 59, pp. 5–31), as well as from Beniston, 2004: Climatic Change and Impacts: An Overview Focusing on Switzerland (Kluwer Academic Publishers, Dordrecht; selected text from pages 6, 9, 134, 145, 146 and 158), reproduced with kind permission from Springer Science+Business Media B.V.

REFERENCES

Armelagos G. J., 1991. The origins of agriculture: population growth during a period of declining health. *Population and Environment* 13: 9–22.

Barry R. G., 2008. *Mountain Weather and Climate*, 3rd edition. Cambridge: Cambridge University Press.

Becker A., and Bugmann H. (eds) 2001. *Global Change and Mountain Regions. The Mountain Research Initiative.* IGBP Report 49. Stockholm: IGBP.

Beguería S., López Moreno J. I., Gómez Villar A., Rubio V., Lana-Renault N. and García Ruiz J. M. 2006. Fluvial adjustment to soil erosion and plant cover changes in the Central Spanish Pyrenees. *Geografisker Annaler* 88A: 177–186.

Beniston M. (ed.) 2002. *Climatic Change. Implications for the Hydrological Cycle and for Water Management.* Advances in Global Change Research. Dordrecht: Kluwer Academic Publishers.

Beniston M. 2003. Climatic change in mountain regions: a review of possible impacts. *Climatic Change* 59: 5–31.

Beniston, M., Keller, F., Koffi, B., and Goyette, S., 2004: Estimates of snow accumulation and volume in the Swiss Alps under changing climatic conditions. *Theoretical and Applied Climatology*, 76, 125–140

Beniston M. 2004. *Climatic Change and Impacts; A Review Focusing on Switzerland.* Dordrecht: Kluwer Academic.

Beniston M. 2006. The August 2005 intense rainfall event in Switzerland: not necessarily an analog for strong convective events in a greenhouse climate. *Geophysical Research Letters* 33: L5701.

Beniston M. 2007. Entering into the 'greenhouse century': recent record temperatures in Switzerland are comparable to the upper temperature quantiles in a greenhouse climate. *Geophysical Research Letters* 34: L16710.

Benzing, D. H. 1998. Vulnerabilities of tropical forests to climate change: the significance of resident epiphytes. *Climatic Change* 39: 519–540.

Bowler, J.M., Hope, G.S., Jennings, J.N., Singh, G., and Walker, D., 1976: Late Quaternary climates of Australia and New Zealand. *Quaternary Research*, 6, 359–399.

Brönnimann O., Treier U. A., Müller Schärer H., Thuiller W., Peterson A. T. and Guisan A., 2007. Evidence of climatic niche shift during biological invasions. *Ecology Letters* 10: 701–709.

Bugmann H. and Pfister C. 2000. Impacts of interannual climate variability on past and future forest composition. *Regional Environmental Change* ¾: 112–125.

Burns D. A. 2004. The effects of atmospheric nitrogen deposition in the Rocky Mountains of Colorado and southern Wyoming, USA – a critical review. *Environmental Pollution* 127: 257–269.

Callaway R. M. R. W., Brooker P., Choler Z., Kikvidze C. J., Lortie R., Michalet L. et al. 2002. Positive interactions among alpine plants increase with stress. *Nature* 417: 844–848.

Cayan D. R., Kammerdierner S. A., Dettinger M.D., Caprio J.M., and Peterson D.H. 2001. Changes in the onset of spring in the western United States. *Bulletin of the American Meteorological Society* 82: 399–415.

Chinn T. 1996. New Zealand glacier responses to climate change of the past century. New Zealand. *Journal of Geology and Geophysics* 39: 415–428.

Christensen J. H., Carter T. R. and Giorgi F. 2002. PRUDENCE employs new methods to assess European climate change. *American Geophysical Union Newsletter* 83: 13.

Clow D.W., Sickman J.O., Striegl R.G., Krabbenhoft D.P., Elliott J.G., Dornblaser M.M. et al. 2003. Changes in the chemistry of lakes and precipitation in high-elevation National Parks in the western United States, 1985–99. *Water Resources Research* 39: 6.

Cohen M. 1977. *The Food Crisis in Prehistory.* New Haven: Yale University Press.

Déqué M. R. G., Jones M., Wild F., Giorgi J. H., Christensen D. C., Hassell P. L. et al. 2005. Global high resolution versus Limited Area Model climate change projections over Europe: quantifying confidence level from PRUDENCE results. *Climate Dynamics* 25: 653–670.

Dettinger, M. D. and Cayan, D. R.: 1995: Large-scale atmospheric forcing of recent trends toward early snowmelt runoff in California', *Journal of Climate*, 8, 606–623

Driscoll C.T., Lawrence G.B., Bulger A.J., Butler T.J., Cronan C.S., Eager C. et al. 2001. Acid Rain Revisited: advances in scientific understanding since the passage of the 1970 and 1990. Clean Air Act Amendments. Hubbard Brook Research Foundation Science Links Publication. Vol. 1, No. 1.

Dyurgerov M. and Meier M.F. 2005. *Glaciers and the Changing Earth System: A 2004 Snapshot.* Occasional

Paper 58. Boulder: Institute of Arctic and Alpine Research, University of Colorado.

Engler R., Randin C.F., Vittoz P., Czáka T., Beniston M., Zimmermann N.E. and Guisan A., 2009. Predicting future distributions of mountain plants under climate change: does dispersal capacity matter? *Ecography.* 32, 34–45.

Fenn M. E., Haeuber R., Tonnesen G. S., Baron J. S., Gorssman-Clarke S., Hope D. et al. 2003. Nitrogen emissions, deposition, and monitoring in the western United States. *Bioscience* 53(4): 391–403.

Garcia-Ruiz J. M., Lasanta T., Ruiz-Flano P., Ortigosa L., White S., Gonzalez C., and Marti C., 1996. Land-use changes and sustainable development in mountain areas : a case study in the Spanish Pyrenees. *Landscape Ecology* 11: 267–277.

Gehrig-Fasel J., Guisan A. and Zimmermann N. E. 2007. Tree line shifts in the Swiss Alps. Climate change or land abandonment? *Journal of Vegetation Science* 18: 571–582.

Giorgi F. and Mearns L.O. 1991. Approaches to the simulation of regional climate change: a review. *Reviews of Geophysics* 29: 191–216.

Giorgi F., and Mearns L. O. 1999. Regional climate modeling revisited. *Journal of Geophysical Research* 104: 6335–6352.

Godde P., Price M.F. and Zimmermann F.M. (eds) 2000. *Tourism and Development in Mountain Regions.* Wallingford: CABI Publishing.

Goyette S., Brasseur O. and Beniston M. 2003. Application of a new wind gust parameterisation ; multi-scale case studies performed with the Canadian RCM. *Journal of Geophysical Research* 108: 4374–4389.

Graham L. P., Hagemann S., Jaun S. and Beniston M. 2007. On interpreting hydrological change from regional climate models. *Climatic Change* 81: 97–122.

Granier C., Artaxo P. and Reeves C. E. (eds) 2004. Emissions of atmospheric trace compounds. Dordrecht: Kluwer Academic.

Grötzbach E. F. 1988. High mountains as human habitat, in Allan N. J. R., Knapp G.W. and Stadel C. (eds.) *Human Impact on Mountains.* Totowa: Rowman and Littlefield, pp. 24–35.

Haeberli W. and Beniston M. 1998. Climate change and its impacts on glaciers and permafrost in the Alps. *Ambio* 27: 258–265.

Haeberli W., Muller P., Alean J. and Bösch H. 1989. Glacier changes following the Little Ice Age. A survey of the international data base and its perspectives, in Oerlemans J. (ed.) *Glacier Fluctuations and Climate.* Dordrecht: Reidel, pp. 77–101.

Hedberg O. 1964. The phytogeographical position of the afroalpine flora. *Recent Advances in Botany*, 49, 914–919.

Hodge, S.M., D.C. Trabant, R.M. Krimmel, T.A. Heinrichs, R.S. March, and E.G. Josberger, 1998: Climate variations and changes in mass of three glaciers in western North America. *Journal of Climate*, 11, 2161–2179

IPCC. 2007. *Climate Change. The Intergovernmental Panel on Climate Change (IPCC) Fourth Assessment Report.* Cambridge: Cambridge University Press.

Ives, J.D., and B. Messerli, 1989: The Himalayan Dilemma, Routledge Publishing Company, London, 336 pp.

Kapos V., Rhind J., Edwards M., Ravilious C. and Price M. 2000. Developing a map of the world's mountain forests, in Price M.F. and Butt N. (eds.) Forests in a Sustainable Mountain Environment. Wallingford: CAB International, pp. 4–9

Karlén W. 1988. Scandinavian glacial and climatic fluctuations during the Holocene. *Quaternay Science Review* 7: 199–209.

Karlén W. 1993. Glaciological, sedimentological and palaeobotanical data indicating Holocene climatic change in Northern Fennoscandia, in Frenzel B., Eronen M., Vorren K-D. and Gläser B. (eds) *Oscillations of the Alpine and Polar Tree Limits in the Holocene.* Stuttgart: Gustav Fischer Verlag, pp. 69–83.

Keller F., Goyette S. and Beniston M. 2005. Sensitivity analysis of snow cover to climate change scenarios and their impact on plant habitats in alpine terrain. *Climatic Change* 72: 299–319.

King G. A. and Neilson R. P. 1992. The transient response of vegetation to climate change : a potential source of CO_2 to the atmosphere. *Water, Air and Soil Pollution* 64: 365–383.

Kitayama, K., 1996 : Climate of the summit region of Mount Kinabalu (Borneo) in 1992, an El Niño year. *Mountain Research and Development*, 16(1), 65–75.

Klasner F. L. 2002. A half century of change in Alpine patterns at Glacier National Park, Montana U.S.A. *Arctic, Antarctic and Alpine Research* 34: 49–56.

Klötzli F. 1991. Longevity and stress, in Esser G. and Overdiek D. (eds) *Modern Ecology: Basic and Applied Aspects.* Amsterdam: Elsevier, pp. 97–110.

Körner C. 1999. *Alpine Plant Life.* Heidelberg: Springer-Verlag.

Körner C. 2007a. Climatic treelines: Conventions, global patterns, causes. *Erdkunde* 61: 316–324.

Körner C. 2007b. The use of 'altitude' in ecological research. *Trends in Ecology Evolution* 22: 569–574.

Kullman L. 1989. Tree-limit history during the Holocene in the Scandes mountains, Sweden, inferred from subfossil wood. *Review of Palaeobotany and Palynology* 58: 163–171.

Kvamme M. 1993. Holocene forest limit fluctuations and glacier development in the mountains of southern Norway, and their relevance to climate history, in Frenzel B., Eronen M., Vorren K-D. and Gläser B. (eds) *Oscillations of the Alpine and Polar Tree Limits in the Holocene*. Stuttgart: Gustav Fischer Verlag, pp. 99–113.

Laprise R., Caya D., Giguère M., Bergeron G., Côte H., Blanchet J.-P. et al. 1998. Climate and climate change in western Canada as simulated by the Canadian regional climate model. *Atmosphere–Ocean* 36: 119–167.

Lemke P. J., Ren R. B., Alley I., Allison J., Carrasco G., Flato Y. et al. 2007. Changes in Snow, Ice and Frozen Ground, in Solomon S. D., Qin M., Manning Z. et al. (eds.) *Climate Change 2007: The Physical Science Basis. Contribution of Working Group I. to the Fourth Assessment Report of the Intergovernmental Panel on Climate Change*. Cambridge: Cambridge University Press.

Leonard E. 1986. Use of lacustrine sedimentary sequences as indicators of Holocene glacier history, Banff National Park, Alberta, Canada. *Quaternary Research* 26: 218–231.

MacDonald G. M., Szeicz J. M., Claricoates J. and Dale K. A. 1998. Response of the central Canadian treeline to recent climatic changes. *Annals of the Association of American Geographers* 88: 183–208.

MacDonald G. M., Velichko A. A., Kremenetski C. V., Borisova O. K., Goleva A. A., Andreev A. A. et al. 2000. Holocene treeline history and climate change across northern Eurasia. *Quaternary Research* 53: 302–311.

Messerli, P., 1989: Mensch und Natur im alpinen Lebensraum: Risiken, Chancen, Perspektiven. Haupt-Verlag, Bern and Stuttgart

Meybeck M., Green P. and Vörösmarty C. 2001. A new typology for mountains and other relief classès: an application to global continental water resources and population distribution. *Mountain Research and Development* 21: 34–45.

Monastersky S. 2000. Drowned lands hold clues to first Americans. *Science News* 157: 85–90.

Nesje A., Kvamme M., Rye N. and Løvlie, R. 1991. Holocene glacier and climate history of the Jostedalsbreen region, western Norway: evidence from lake sediments and terrestrial deposits. *Quaternary Science Reviews* 10: 87–114.

Nesje A. S., O. Dahl, Løvlie R. and Sulebak J. R. 1994. Holocene glacier activity at the southwestern part of Hardangerjøkulen, central-south Norway: evidence from lacustrine sediments. *The Holocene* 4: 377–382.

Noble I. and Gitay H. 1988. Climate change in desert regions, in Watson R. T., Zinyowera M., and Moss R. (eds.) *IPCC: The Regional Impacts of Climate Change*. Cambridge: Cambridge University Press, pp. 191–217.

OECD. 2006. *Adapting to Climatic Change in the European Alps*. OECD Report Series. Paris, OECD.

Ozenda P. 1985. *La Végétation de la Chaine Alpine dans l'Espace Montagnard Européen*. Paris: Masson.

Paul F., Kääb A. and Haeberli W. 2007. Recent glacier changes in the Alps. Global and Planetary Change, 56, 111–122.

Perry A. H. 2000. Impacts of climate change on tourism, in Parry M. L. (ed.) *Assessment of Potential Effects and Adaptations for Climate Change in Europe; the ACACIA Report*. Brussels/Norwich: EU Publications/Jackson Environment Institute, pp. 217–226.

Peters R. L. and Darling J. D. S. 1985. The greenhouse effect and nature reserves: global warming would diminish biological diversity by causing extinctions among reserve species. *Bioscience* 35: 707–717.

Porter S. C. 1979. Hawaiian glacial ages. *Quaternary Research* 12: 161–187.

Price, M.F., 1993: Patterns of the development of tourism in mountain communities, Mountains at Risk: Current Issues in Environmental Studies, N.J.R. Allan, (ed.), Kluwer, Dordrecht

Price M. F. 1994. Should mountain communities be concerned about climate change?, in M. Beniston (ed.) *Mountain Environments in Changing Climates*. London: Routledge, pp. 431–451.

Ramanathan V., Chung C., Kim D., Bettge T., Buja L. et al. 2005. Atmospheric brown clouds: Impacts on South Asian climate and hydrological cycle. *PNAS* 102: 5326–5333.

Rochefort R. R., Little R. L., Woodward A. and Peterson D. L. 1994. Changes in sub-alpine tree distribution in western North America: a review of climatic and other causal factors. *The Holocene* 4: 89–100.

Ruddiman W. F. 2001. *Earth's Climate, Past and Future*. New York: Freeman Publishers.

Stevens, C.J., Dise, N.B., Mountford, J.O., and Gowing, D.J., 2004: Impact of nitrogen deposition on the species richness of grasslands, *Science* 303, 1876–1879

Stewart I., Cayan D. R. and Dettinger M. D. 2004. Changes in snowmelt runoff timing in Western North America under a 'Business as Usual' climate change scenario. *Climatic Change* 62: 217–232.

Stocks B. J., Wotton B. M., Flannigan M. D., Fosberg M. A., Cahoon D. R., and Goldammer J. G. 2001. *Boreal Forest Fire Regimes and Climate Change*. Advances in Global Change Research, 7. Dordrecht: Kluwer Academic Publishers, pp. 233–246.

Thompson L. G., Mosley-Thompson E., Davis M. E., Lin P-N., Henderson K. A., Dai J. et al. 1995. Late Glacial Stage and Holocene tropical ice core records from Huascarán, Peru. *Science* 269: 46–50.

Thompson L. G., Mosley-Thompson E., Davis M. E., Henderson K. A., Brecher H. H., Zagorodnov V. S. et al. 2002. Kilimanjaro ice core records: Evidence of Holocene. *Science* 298: 589–593.

Thompson L. G. E., Mosley-Thompson H., Brecher M., Davis B., Leon D., Les P.-N. et al. 2006a. Abrupt tropical climate change: past and present. *Proceedings of the National Academy of Sciences* 103: 10536–10543.

Thompson L. G., Mosley-Thompson E., Davis M. E., Mashiotta T. A., Henderson K. A., Lin P. N. and Yao T. D. 2006b. Ice core evidence for asynchronous glaciation on the Tibetan Plateau. *Quaternary International* 154: 3–10.

UNEP. 2008. *Dirty Brown Clouds Impact Glaciers, Agriculture and the Monsoon*. UNEP Publications, Nairobi

US Environmental Protection Agency. 2000. *National Air Pollutant Emission Trends, 1900–1998*. Report EPA – 454/R–00–002. Washington, DC: US Environmental Protection Agency.

Vitousek P.M., Aber J.D., Howarth R.W., Likens G.E., Matson P.A., Schindler D.W. et al. 1997. Human alteration of the global nitrogen cycle: Sources and consequences. *Ecological Applications* 7: 737–750.

Wanner H., Beer J., Bütikofer J., Crowley T.J., Cubasch U., Flückiger J. et al. 2008. Mid- to late Holocene climate change – an overview. *Quaternary Science Reviews* 27: 1791–1828.

WGMS. 2000. Glacier Mass Balance, Bulletin, in Haeberli W. and Hoelzle M. (eds.) *World Glacier Monitoring Service*. Switzerland: ETH-Zurich

Whiteman D. 2000. *Mountain Meteorology*. Oxford: Oxford University Press.

WTO. 2000. *Compendium of Tourism Statistics*, 2000 edition. Madrid: World Tourism Organization Publications.

Yoshino M., Horie T., Seino H., Tsujii H., Uchijima T. and Uchijima Z. 1988. The effects of climatic variations on agriculture in Japan, in Parry M. L., Carter T. R. and Konijn N. T. (eds), *The Impact of Climatic Variations on Agriculture. Vol 1: Assessments in Cool Temperature and Cold Regions*. Dordrecht: Kluwer, pp. 723–868.

37

Environmental Change in Coastal Areas and Islands

Patrick Nunn

1 INTRODUCTION

The environments of coastal areas are more changeable than many because of their location at the interface of the lithosphere, hydrosphere and atmosphere. Coastal areas are affected by changes in any of these, ranging from regular minor change linked to short-term perturbations to rapid change associated with extreme events. Syntheses of coastal environmental changes have emphasized the uniqueness of the processes involved in these locations (Bird, 1985. Harff et al., 2007; Woodroffe, 2002).

Islands exhibit considerable environmental diversity, depending largely on their size. In evaluating the causes of environmental changes, most smaller oceanic islands can be regarded as entirely coastal while most parts of larger oceanic islands will typically experience environmental changes arising from maritime influences (Menard, 1986; Nunn, 1994).

Sea-level change is perhaps the most important cause of long-term environmental change in coastal areas and islands (Tooley and Shennan, 1987). Late Quaternary sea-level oscillations transformed such environments, often to the point of shifting coastlines considerable distances laterally or causing islands to alternately emerge (as sea level fell) or become submerged (as sea level rose). Future sea-level rise will impact coastal areas and islands profoundly.

In acknowledgement of their often abundant food resources, modern humans (*Homo sapiens*) have been routinely interacting with coastal areas for more than 10,000 years, something that increased in most such places during the Holocene as population densities rose and sea-level rise reduced habitable coastal area (Nunn, 2007a). Within the past few hundred years, more people have come to occupy many coastal areas than can be sustained by their environments. High coastal population densities have led to a range of deleterious environmental impacts that will be exacerbated in the future by further population concentration in such areas and by climate change (Nicholls et al., 2007).

Islands, particularly those in the world's oceans far from continental coasts, were generally settled much later than these and by people whose subsistence strategies were focused on coastal resources. Today many island populations are clustered along island coasts with interior areas usually sparsely populated. The vulnerability of many island populations will therefore increase in the future for reasons that are similar to those applying to non-island coastal areas (Mimura et al., 2007).

2 NON-HUMAN CAUSES OF ENVIRONMENTAL CHANGE IN COASTAL AREAS AND ISLANDS

The environments of coastal areas and islands change at a range of scales for many reasons but it makes sense, particularly in terms of contemporary management solutions, to separate natural (non-human) causes of morphodynamic change from direct human impacts, which are considered in a separate section. This section deals with natural causes of change which are separable into low-magnitude changes, which occur routinely and usually slowly enough so as not to abruptly disrupt the form of coastal areas and islands, and extreme events which are by definition more disruptive.

2.1 Low-magnitude changes

Coastal areas are subject to a range of low-magnitude changes that may over long periods of time alter the location and the nature of a particular coastline. While any judgement of the magnitude of change is ultimately a question of scale, it is fair to say that sea-level changes are the foremost cause of coastal change at scales that are of relevance to coastal populations. This section also discusses tectonic (land-level) changes and atmospheric changes.

The level of the ocean surface – mean sea level – changes at a variety of scales for a variety of reasons, ranging from daily tidal (lunar) cycles through subannual and annual (orbital) cycles to interannual changes. The latter range from those lasting 1–3 years associated with phenomena like the El Niño Southern Oscillation (ENSO) through decadal changes to those with centennial–millennial

periods that appear to be largely associated with solar cycles to those that recur every 70,000–100,000 years as a result of orbital (Milankovitch) changes. All such changes result in changes to coastal environments that alter both their form and the nature of their resident ecosystems.

For example, coral-reef death since the 1970s attributable to a sustained low sea level during El Niño (ENSO-negative) events has profoundly altered process regimes in many coastal areas (Reyes-Bonilla et al., 2002), particularly on tropical islands (Goreau et al., 2008). In addition, recent work shows that 18.6-year nodal tidal cycles are the dominant control over coastal-wetland sedimentation (French, 2006); a field study of mudbanks along the French Guiana coast showed that this cycle was the principal control on their centennial-scale development (Gratiot et al., 2008).

It is also clear that solar cycles with 1300/1500-year periods have brought about environmental changes which have altered coastal ecosystems in ways that have reduced food availability for their human inhabitants, something that is implicated in models of cultural transformation (Hsü, 2000). For instance, the environments of some Kurile Islands (northwest Pacific) were altered in phase with late Holocene climatic and sea-level perturbations (Razjigaeva et al., 2004) while for much of the past 5000 years an unambiguous relationship between sea-level fall and cultural decline in the Changjiang (Yangtze) Delta has been established (Zhang et al., 2005).

Yet it is longer-term (glacio-eustatic) sea-level changes that produced the most widespread and best understood changes to coastal areas and islands. Figure 37.1a shows the nature of late Quaternary sea-level changes in the world's oceans. The general picture of low sea level (–120/130 m) during the Last Glacial Maximum and high sea level today having caused almost all the world's coasts to be submerged. Some of the most profound changes were to the geography of oceanic islands region (Figure 37.1b).

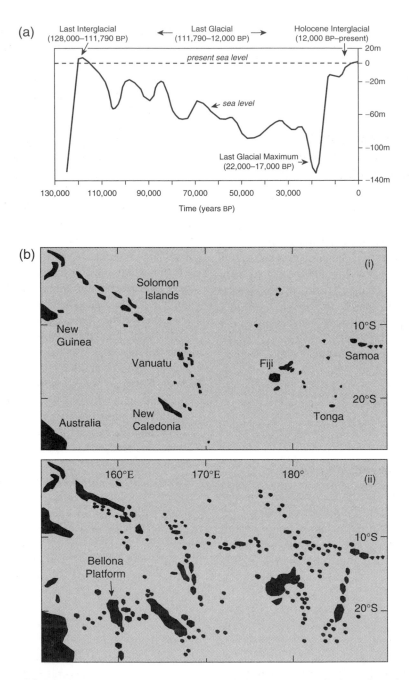

Figure 37.1   (a) Late Quaternary sea-level changes (after Nunn, 2007b; based on data provided by John Chappell). (b) Changes in the geography of southwest Pacific islands comparing (i) today with (ii) the Last Glacial Maximum 18,000 BP (based on Nunn, 2009). The upper map (i) shows the southwest Pacific region as it is today, the lower map (ii) the way it appeared 18,000 years ago during the Last Glacial Maximum when sea level was 120 m lower. The difference is dramatic. Not only were there more islands, but those that still exist were larger and closer together. Some large islands like the Bellona Platform have disappeared from view altogether.

While such sea-level changes were accompanied by climate and vegetation changes that had their own impacts on coastal areas and islands, the effects of sea-level rise were generally of a larger magnitude and are therefore easiest to isolate. Most of the world's coastal areas experienced net submergence as a result of late Quaternary sea-level rise and many also experienced an increase in the length and sinuosity of their coastlines. This is because during the low sea level of the Last Glaciation, subaerial landscapes continued to evolve then, as sea level rose subsequently, the irregularities that had been created as a result were submerged.

A good example is that of river valleys that formerly met the sea along a straight coast across which the ocean then rose and flooded the coastal plain but also ran upvalley creating a coastal embayment where none previously existed (Hijma et al., 2009). Such changes also had implications for coastal ecosystems, particularly the spread and recession of mangrove forests in tropical areas, as well as for coastal populations whose subsistence strategies may have altered as a result of such environmental changes (Tossou et al., 2008; Yulianto et al., 2004). A good example of the latter is the increasing importance of nearshore marine subsistence for the people of Japan as their islands were gradually reduced in size and distanced from mainland Asia by Holocene sea-level rise (Nunn et al., 2006).

There is a difference between the sea-level histories of the Atlantic and Pacific Oceans that is responsible for contrasting scenarios of Holocene coastal evolution. In general terms, sea level has been rising continuously in the Atlantic since the Last Glacial Maximum (around 18,000 BP) whilst in the Pacific it exceeded its present level 4000–5000 BP and has since fallen (Nunn, 2004). This difference is responsible for variations in late Holocene coastal histories that culminate in the interpretation of modern process regimes.

In the Atlantic, there was also a deceleration of sea-level rise about 7000 BP that led in many coastal areas to a reduced influence of those environmental changes driven by sea-level rise and a concomitant increase in those driven by terrestrial sedimentation (Milne et al., 2005). The most widespread response to this along continental coasts was the development of sediment barriers in shallow water and consequent infilling of the areas behind. Good examples come from the west coast of Portugal (De Carvalho et al., 2006; Freitas et al., 2002).

By contrast, in the Pacific, there was a continuous sea-level rise until about 6000 BP, then a 2000-year period of relative stability, followed by a sea-level fall of around 2 m. Most Pacific coasts therefore show signs of long-term submergence followed by short-term emergence, typically manifested by incision of river channels into coastal lowlands. Closely-argued examples come from the Cook Islands (Moriwaki et al., 2006) and Fiji (Nunn and Peltier, 2001). Another comes from the Hwajinpo Lagoon on the east coast of the Korean Peninsula where sediment coring revealed a history of sea-level change and lake-level change that has been interpreted in terms of regional climate change and its environmental response (Yum et al., 2004). More recent work in the same area confirms the mid-Holocene sea-level maximum for the Pacific (Munyikwa et al., 2008).

Tectonic changes affect many coastal areas but typically at rates that are an order of magnitude less than those of glacio-eustatic changes. Yet there are instances where tectonic change has had comparable effects on coastal areas and islands. Those coastal areas that have been subject to isostatic rebound as a result of postglacial land-ice melt have risen considerably (Stocchi et al., 2005), sometimes more than sea level since the Last Glacial Maximum, exhibiting landforms of net emergence (Fjeldskaar et al., 2000). Similarly many islands close to convergent plate boundaries have risen comparatively rapidly and show diagnostic landforms like emerged coastal notches, platforms and reef terraces (Merritts et al., 1998; Ota and Chappell, 1999).

Finally it is important to acknowledge that atmospheric changes, whether long-term (climatic) or shorter-term (meteorological), can also affect the environments of coastal areas and islands. The principal variable responsible for such changes is precipitation, an increase of which may increase coastal sedimentation (lowland aggradation and shoreline progradation) while a decrease of which may reduce coastal sedimentation, facilitating increased shoreline erosion. Good examples come from many of the world's largest deltas that, while experiencing net submergence for much of the early Holocene, began to grow seawards during the wetter middle Holocene even though sea level continued to rise (Preoteasa et al., 2009; Zong et al., 2009).

2.2 Abrupt and extreme events

A renewal of interest over the past few decades in the role of abrupt and extreme events in changes to the environments of coastal areas and islands has come about largely from observations of the effects of such events. This section reviews these in the categories of changes in sea level, climate, tectonics, and volcanic activity.

In terms of sea-level impacts, the 2004 Indian Ocean tsunami was perhaps the most influential such event in that it made scientists and coastal managers worldwide begin to treat such extreme phenomena as possibilities. The impacts of this tsunami on coastal areas were variable but interest has focused on reducing their future vulnerability (Bishop et al., 2005) just as research has focused on understanding the variable effects of this tsunami on islands (Kench et al., 2006). Many coastal areas and islands close to convergent plate boundaries, where tsunamigenic earthquakes may occur regularly, are strewn with large boulders dumped onland by these powerful waves (Figure 37.2). Some coasts also bear the erosional scars of phenomenally high waves, such as those along part of eastern Australia (Young and Bryant, 1993) and the Canary Islands (Perez-Torrado et al., 2006).

The 2004 tsunami also impelled coastal scientists to examine sedimentary sequences for traces of similar events with a view to contributing data to models of tsunami recurrence for particular coasts. Key studies are that of Hori et al. (2007) who reported on the nature of the sedimentary signature of this event and Jankaew et al. (2008) who found traces of earlier analogues along Indian Ocean coasts.

Viewed from a distance in time, many comparatively rapid climate and sea-level changes appear abrupt. During the early Holocene, rapid cooling during the 8,200-BP event altered human activities in many low-lying coastal areas (Berger and Guilane, 2009). Around AD 1300, a sea-level fall of some 80 cm impacted nearshore marine food resources to such a degree that many island peoples suffered a food crisis at the time that led to widespread societal collapse (Nunn, 2007b; Nunn et al., 2007). More recently, along the coast of Canada, an ENSO-mediated sea-level rise of 40 cm caused as much as 12 m of shoreline retreat (Barrie and Conway, 2002).

Abrupt climate change is the major cause of rapid sea-level change but has also brought about changes to coastal areas and islands independently. Over long time periods, changes in irregular and interannual climatic phenomena are the main causes of these. In particular, changes in tropical-storm frequency and intensity have been linked to coastal changes (Nott, 2004) while ENSO frequencies have varied and also been linked to environmental changes in such places (Sandweiss, 1996).

An uncommonly exposed coastal area lies at the apex of the Bay of Bengal where the massive rivers Brahmaputra, Ganges and Meghna all come together to form a delta covering an above-sea area of more than 100,000 km². Around 63 per cent of the nation of Bangladesh occupies this delta, with approximately 150 million people living on it. This delta is long established but its

Figure 37.2   Tossed as though it were a pebble, this giant piece of reefrock (named Kasakanja) was thrown 30 m up onto the clifftop of southern Okinawa Island, Japan. After it landed it shielded the underlying rock surface from erosion, which is the origin of its pedestal, the land surface all around having since been reduced in level. The wave that deposited it was a tsunami, generated along the nearby Ryukyu Trench.

location makes it among the most vulnerable of its kind. Storms that move north through the Bay of Bengal funnel water towards its apical delta, making the storm surges penetrate far further inland than might otherwise be expected. This situation is expected to worsen significantly in the next century (Karim and Mimura, 2008). This region is also tectonically active; the earthquake that triggered the devastating 2004 Indian Ocean tsunami also raised the floor of the Bay of Bengal causing a global sea-level rise of 0.5 mm (Bilham et al., 2005).

Tectonic (coseismic) change is often abrupt in those parts of the world close to convergent plate boundaries. Such changes may abruptly raise or drop a coastline several metres, changing its appearance and process dynamics almost instantly. An example comes from Samos Island in the Aegean Sea (Greece) where a small part of the island's coast has been regularly uplifted during the

past 5000 years while the remainder exhibits signs of submergence attributable to postglacial sea-level rise (Stiros et al., 2000).

Subsidence may also be abrupt or rapid. Many oceanic islands are now recognized as having been shaped largely by flank collapses (Keating and McGuire 2000); for instance, Hawaii and Tahiti (Figure 37.3). River deltas are also now understood as places that are subject to both rapid subsidence, largely as a result of sediment compaction but also by occasional catastrophic failure of their outer margins. Delta subsidence, often exacerbated by groundwater extraction, averages 1 cm/year in the Mississippi Delta (Tornqvist et al., 2004). Parts of the Chinese city of Tianjin, which was built on the old Huanghe (Yellow River) delta, sank 3 m between 1959 and 1994, partly as a result of historical groundwater extraction but also as result of delta compaction (Shi et al., 2007). Catastrophic failure is

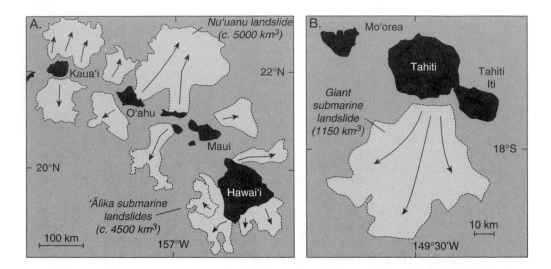

Figure 37.3   Examples of submarine landslides resulting from island-flank collapses (based on Nunn, 2009). Note that the two maps have different scales. (a) The main submarine landslides around the southeast Hawaiian Islands. One of the most recent of these landslides, the two-phase Ālika Slide, is believed to have occurred about 100,000 years ago and produced a giant wave that washed back over nearby islands. (b) The giant submarine landslide that occurred 650,000–850,000 years ago off the south coast of Tahiti Island, French Polynesia. The form of the landslide is not readily seen on Tahiti itself because the concavity at its head became filled with lavas from eruptions of the Tahiti volcanoes after the slide took place.

less frequent but studies of some of the world's largest and oldest deltas leave no doubt that this occurs. Some of the giant sediment fans along the seaward edge of the Amazon Delta have collapsed, travelling more than 200 km out to sea, dropping in depth by 1.5 km, and forming bodies of sediment as large as 15,000 km² (Maslin et al., 2005).

While some coastal areas were created by large volcanoes and remain within their shadow, environmental changes associated with volcanic activity are more common on islands. Some islands may be formed by a single volcano and destroyed (or radically altered) by its eruption. Many such islands appear to experience cycles of rebirth and destruction. One example comes from the volcanic island Santorini in the Aegean Sea which was previously a much larger island (Stronghyle) (Vespa et al., 2006). Its well-documented explosive eruption 3577–3550 BP created a caldera, largely underwater, in

which islands have subsequently began to build and are likely to eventually produce another island like Stronghyle. Another example comes from the Marquesas Island group in the Pacific where the volume of ocean-floor volcanic debris greatly exceeds that within the modern islands, leading to the conclusion that these are only the latest in a series of islands to have alternately been built up and then collapsed (Wolfe et al., 1994).

3 UNDERSTANDING ENVIRONMENTAL HISTORIES OF COASTAL AREAS AND ISLANDS

There are numerous reasons for deciphering long-term environmental histories of coastal areas and islands. Many reasons relate to providing a context for current and future management of these areas, others to employing such histories to gain insights into associated questions such as those relating to ecosystem development or earth rheology. This section looks at some of the methods that have been used to reconstruct late Quaternary environmental histories of coastal areas and islands and gives three thematic examples; barrier coasts, coral-reef coasts and atoll islands.

The choice of methods to be employed in the reconstruction of coastal history depend on the nature of the coast in question. Low coasts are generally formed from soft sediments which, despite often containing a precise record of the processes that led to their formation, are often difficult to decipher. For their study of the coastal plain at Marathon in Greece, Pavlopoulos et al. (2006) dug four trenches and sunk two long boreholes from which they recognized 24 categories of sediment, each with its own unique palaeoenvironmental significance. From their study they were able to identify the nature and chronology of the climate changes and level changes (sea-level and tectonic changes) that have affected the area within the past 5800 years. Their conclusion that during recent millennia, the effects of sea-level change have been almost wholly balanced by tectonic change makes this coast effectively stable (neither submergent nor emergent), a rare situation.

Advances in luminescence dating over the past two decades have revolutionized the reconstruction of soft-coast histories (Jacobs, 2008). For example, a precise chronology for the past 1700 years was obtained by Brooke et al. (2008) from optically stimulated luminescence (OSL) dating of a coastal sand ridge complex in eastern Australia. In many soft-coastal areas, compounding the difficulties of sediment interpretation from cores is the problem of sediment reworking: the ways in which unconsolidated sediments can be disturbed and re-deposited, often numerous times, by large waves or large floods. This is perhaps the most profound aspect of soft-coast 'sensitivity', which involves a range of processes that frustrate the straightforward

interpretation of coastal sediment sequences in many places (Hansom, 2001).

Hard-rock coastal histories are equally challenging to unravel but for different reasons. Within this category might be included uplifted coral-reef coasts, which abound along coasts around convergent plate margins in tropical regions, and are discussed in a separate section later in this chapter. Yet where a coastline is bare rock, erosion dominates and much of the detailed information needed to reconstruct its history is apparently missing. Techniques known as exposure dating have revolutionized the study of hard-rock coasts, as they enable ages of wave-cut (and other) surfaces to be determined (Figure 37.4).

The overwhelming control of the nature and process regime of many modern coasts appears to be antecedence, the inherited geological character of these. An important study was that of Cooper (2006) who argued that variations in the gross form of river estuaries in Ireland were largely attributable to their geological history. The effects of antecedence on island environments are even more marked because many oceanic islands are young (compared to continents) and are located in comparatively dynamic geological settings. The coasts of islands that have been rising (or subsiding) for long time periods are relatively easy to interpret; good examples are the studies by Stiros et al. (2000) and Pedoja et al. (2006). Other island coasts exhibit signs of alternate emergence and submergence and their histories are consequently more difficult to unravel (Nunn, 1995, Ota and Chappell, 1999).

3.1 Barrier coasts

The environments of barrier coasts allow some of the most comprehensive studies of Holocene coastal changes because they contain some of the most complete sedimentary records of these, from both marine and terrestrial sources. For example, recent multiproxy analyses of sediments in back-barrier

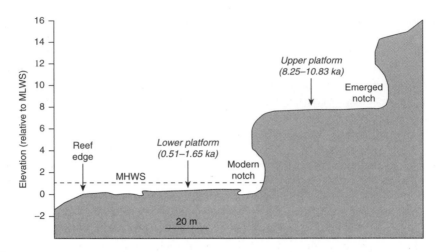

Figure 37.4　The coast of Niue Island (central Pacific Ocean) exhibits evidence for recent uplift associated with movement up the flank of a lithospheric flexure. At Namukulu on the northwest coast, there are two erosional platforms cut in bedrock limestone that were dated using exposure dating. The ages shown were calculated from measurements of ^{36}Cl for the lower platform (257.6–273.2 ppm) and upper platform (54.9–142.9 ppm) made by John Stone and Keith Fifield in cooperation with the author who obtained the samples.

locations in New Zealand allowed Ogden et al. (2006) and Nichol et al. (2007) to derive millennial-scale coastal histories that included information about sea-level change, volcanic activity, and a variety of human impacts.

Well-preserved sediment sequences along the northern part of the east coast of the US have allowed fine-grained chronologies of last-millennium sea-level change to be developed (van der Plassche et al., 1998) and its role in shaping environments to be cogently assessed (Willard et al., 2003). Some innovative studies in this region have employed ground-penetrating radar to understand the last-millennium development of estuarine and back-barrier sediments obscured by vegetated dunes (Buynevich, 2006; Culver et al., 2006).

3.2 Coral-reef coasts

The presence of a coral reef off a particular shoreline has implications for its evolution. For instance, an offshore reef prevents large waves from impacting the shoreline: the reason why there was comparatively little erosion of the Maldives islands (atolls) when the 2004 Indian Ocean tsunami passed by (Kench et al., 2006). An offshore reef is often also an important supplier of sediment (biogenic) to coastal areas, and this is typically the sole constituent of atoll islands (see the following section) (Collen and Garton, 2004).

The Quaternary histories of many coasts and islands have been calibrated using coral reefs, whether uplifted (Ota and Chappell, 1999) or sampled at or below modern sea level. Among the latter are the submerged coral reefs around the Hawaiian Islands, most of which are thought to have been submerged by net sea-level rise (Webster et al., 2007).

3.3 Atoll islands

The only above-sea part of an atoll is a ring of coral reef, often so broad that reef rubble, driven up on to the reef platform by large waves, has accumulated to form islands.

These islands are narrow, often sinuous, made largely from unconsolidated materials, and do not rise more than a few metres above sea level. Every part of these islands is coastal which underlines their close relationship to Holocene sea-level changes. For such islands mostly occur only in those parts of the world where in the last few millennia sea level has fallen slightly, exposing a core of emerged reef around which islands have accumulated (Figure 37.5). Ephemeral and transient, atoll islands are now considered to be at the front line of twenty-first-century climate change because some appear liable to disappear as a result of future sea-level rise (Woodroffe, 2008).

Deep drilling of atolls has proved a robust method of understanding the histories of these island types (Hongo and Kayanne, 2009) as well as helping in the understanding of sea-level and earth-rheology changes in the ocean basins (Lincoln and Schlanger, 1987, Titschack et al., 2008).

4 RECENT HUMAN IMPACTS ON COASTAL AREAS AND ISLANDS

Within the past 200 years or so, most coastal areas and islands have been affected by human activities to the extent that natural (non-human) environmental changes have become increasingly difficult to identify. Studies of locations marginal for human settlement have been carried out to try and detect this, some of the most successful being in the Canadian Arctic where environmental evidence has been found for warming post-1800 as well as a marked acceleration of this in recent decades (Lim et al., 2008; Thomas et al., 2008).

Island sedimentary environments also provide some of the clearest environmental signatures for a colonizing human presence (Ljung et al., 2006; Stevenson and Hope, 2005) although there are concerns about the uncritical interpretation of these (Nunn, 2001; Nunn et al., 2007).

Figure 37.5 Eroding shoreline of Luamotu Island, Funafuti Atoll, Tuvalu. Like most atoll islands in this region, Luamotu has a foundation of resistant fossil reef (produced during the mid-Holocene sea-level maximum), shown here rising 40 cm above low-tide level. This reef-rock foundation was formerly buried beneath a covering of surficial deposits, mostly sand along this lagoon-facing coast. As sea level has been rising recently, so the high-tide level has started to rise above the reef-rock foundation, leading to the rapid erosion of its sand cover, as shown by its bare surface and the leaning trunks of coconut palms at this location. (Details from Dickinson (1999), base photograph courtesy of William R. Dickinson).

Human occupation of coastal areas (including island fringes) has usually had two main environmental effects. First has been the conversion of natural vegetation from high-diversity ecosystems to low-diversity ones – effectively monospecific in places. Loss of natural biodiversity (marine and terrestrial) has frequently been exacerbated by pollution of various kinds, which may increase eutrophication (Scharin, 2002) or otherwise contaminate water sources in ways that inhibit desirable organic growth (Jayaraju et al., 2009). Second have been physical environmental changes intended to facilitate high-density human occupation or increase environmental resilience. Examples include the artificialization of natural shorelines, the construction of levees (dykes) and tsunami barriers, and novel solutions such as the re-evaluation of the efficacy of shoreline protection methods focused on stability (Nordstrom et al., 2007).

Low densities of humans along coasts and on islands have often led to environmental change, particularly when human activity is geared towards production of food surpluses, as with aquaculture (Cole et al., 2009) or lowland rice agriculture (Huan et al., 1999). Yet the most profound environmental changes in such places have occurred where they are occupied by large numbers of people, typically in large urban areas, the growth of which has required management. Particularly during the nineteenth and twentieth centuries, many coastal cities expanded rapidly with scant regard for the environmental impacts of this process. This has led to a situation where modern management must not only to adapt to natural environmental changes, particularly extreme events, but also needs to invest in neutralizing the environmental problems that have arisen from rapid expansion in the past (Smith et al., 2006; Wenzel et al., 2007).

The ocean frontages of coastal (and island) cities are among their most vulnerable parts. In the Indonesian megacity Jakarta (Jabotabek), the seventh largest such agglomeration in the world, the coast is experiencing

a range of problems, including littering, water pollution, saltwater intrusion, land subsidence, seafood contamination and wildlife habitat loss, which programmes of community management are considered best able to reduce (Nur et al., 2001).

5 PROSPECTS FOR THE ENVIRONMENTS OF COASTAL AREAS AND ISLANDS

Arguably the two most profound future threats to coastal areas and islands are population increase and climate change. There are of course many others but these are generally more localized and/or predicted to be less extreme in their environmental effects. Population increase and climate change are not mutually exclusive in terms of their future environmental effects but they are separable. Most would agree that integrated coastal management (ICM) provides the best theoretical context for future coastal development but there are concerns that this will continue to be subordinate to short-term accommodation of stresses in some key population centres (Smith et al., 2006).

Future population growth is expected to accelerate, particularly in poorer countries. Associated with this, urban populations are expected to increase more than rural populations in developing countries and, since most urban centres in such places are coastal, it is clear that these trends will increase environmental problems for coastal areas and islands. Studies have highlighted the imperative of integrated approaches but have also acknowledged that there are specific ways of reducing coastal population pressure in both developing and developed countries (Conway, 2005; Scott et al., 2006; Sekhar, 2005).

Owing largely to future sea-level rises, it will become increasingly difficult for large numbers of people to continue to occupy areas they currently occupy. Most future management strategies still focus on accommodation; those for New Orleans and

Shanghai (Liu et al., 2006) and the Pearl River delta in China (Huang et al., 2004) are representative. Yet strategies which recognize that future sea-level rise cannot be accommodated in such vulnerable locations by simply building sea defences higher are now being discussed (Costanza et al., 2006). Recognition of the imperative of resettlement from vulnerable locations is also becoming more common; examples come from the east coast of England (Milligan and O'Riordan, 2007) and Tangiers Bay in Morocco (Snoussi et al., 2009). It is likely that this represents a change in emphasis that will be sustained for at least several decades.

REFERENCES

Barrie J. V. and Conway K. W. 2002. Rapid sea-level change and coastal evolution on the Pacific margin of Canada. *Sedimentary Geology* 150: 171–183.

Berger J. F. and Guilane J. 2009. The 8200 cal BP abrupt environmental change and the Neolithic transition: a Mediterranean perspective. *Quaternary International* 200: 31–49.

Bilham R., Engdahl R., Feldl N. and Satyabala S. P. 2005. Partial and complete rupture of the Indo-Andaman plate boundary 1847–2004. *Seismological Research Letters* 76: 299–311.

Bird E. C. F. 1985. *Coastline Change: A Global Review.* New York: Wiley.

Bishop P., Sanderson D., Hansom J. and Chaimanee N. 2005. Age-dating of tsunami deposits: lessons from the 26 December 2004 tsunami in Thailand. *The Geographical Journal* 171: 379–384.

Brooke B., Lee R., Cox M., Oiley J. and Pietsch T. 2008. Rates of shoreline progradation during the last 1700 years at Beachmere, southeastern Queensland, Australia, based on optically stimulated luminescence dating of beach ridges. *Journal of Coastal Research* 24: 640–648.

Buynevich I. V. 2006. Coastal environmental changes revealed in geophysical images of Nantucket Island, Massachusetts, USA. *Environmental and Engineering Geoscience* 12: 227–234.

Cole D. W., Cole R., Gaydos S. J., Gray J., Hyland G., Jacques M. L. et al. 2009. Aquaculture: environmental, toxicological, and health issues. *International Journal of Hygiene and Environmental Health* 212: 369–377.

Collen J. D. and Garton D. W. 2004. Larger foraminifera and sedimentation around Fongafale Island, Funafuti Atoll, Tuvalu. *Coral Reefs* 23: 445–454.

Conway T. M. 2005. Current and future patterns of land-use change in the coastal zone of New Jersey. *Environment and Planning B – Planning and Design* 32: 877–893.

Cooper J. A. G. 2006. Geomorphology of Irish estuaries: inherited and dynamic controls. *Journal of Coastal Research Special Issue* 39: 176–180.

Costanza R., Mitsch W. J. and Day J. W. 2006. A new vision for New Orleans and the Mississippi delta: applying ecological economics and ecological engineering. *Frontiers in Ecology and the Environment* 9: 465–472.

Culver S. J., Ames D. V., Corbett D. R., Mallinson D. J., Riggs S. R., Smith D. G. and Vance D. J. 2006. Foraminiferal and sedimentary record of Late Holocene barrier island evolution, Pea Island, North Carolina: the role of storm overwash, inlet processes, and anthropogenic modification. *Journal of Coastal Research* 22: 836–846.

De Carvalho G. S., Granja H. M., Loureiro E. and Henriques R. 2006. Late Pleistocene and Holocene environmental changes in the coastal zone of northwestern Portugal. *Journal of Quaternary Science* 21: 859–877.

Dickinson W. R. 1999. Holocene sea-level record on Funafuti and potential impact of global warming on central Pacific atolls. *Quaternary Research* 51: 124–132.

Fjeldskaar W., Lindholm C., Dehls J. F. and Fjeldskaar I. 2000. Postglacial uplift, neotectonics and seismicity in Fennoscandia. *Quaternary Science Reviews* 19: 1413–1422.

Freitas M. C., Andrade C. and Cruces A. 2002. The geological record of environmental changes in southwestern Portuguese coastal lagoons since the Lateglacial. *Quaternary International* 93–4: 161–170.

French J. R. 2006. Tidal marsh sediment trapping efficiency and resilience to environmental change: exploratory modeling of tidal, sea-level and sediment supply forcing in predominantly allochthonous systems. *Marine Geology* 235: 119–136.

Goreau T. J., Fisher T., Perez F., Lockhart K., Hibbert M. and Lewin A. 2008. Turks and Caicos Islands 2006. coral reef assessment: large-scale environmental and ecological interactions and their management implications. *Revista de Biologia Tropical* 56: 25–49.

Gratiot N., Anthony E. J., Gardel A., Gaucherel C., Proisy C. and Wells J. T. 2008. Significant contribution of the 18.6 year tidal cycle to regional coastal changes. *Nature Geoscience* 1: 169–172.

Hansom J. D. 2001. Coastal sensitivity to environmental change: a view from the beach. *Catena* 42: 291–305.

Harff J., Hay W. H. and Tetzlaff D. M. 2007. *Coastline Changes: Interrelation of Climate and Geological Processes.* Boulder: Geological Society of America.

Hijma M. P., Cohen K. M., Hoffmann G., Van der Spek A. J. F. and Stouthamer E. 2009. From river valley to estuary: the evolution of the Rhine mouth in the early to middle Holocene (western Netherlands, Rhine-Meuse delta). *Netherlands Journal of Geosciences – Geologie en Mijnbouw* 88: 13–53.

Hongo C. and Kayanne H. 2009. Holocene coral reef development under windward and leeward locations at Ishigaki Island, Ryukyu Islands, Japan. *Sedimentary Geology* 214: 62–73.

Hori K., Kuzumotot R., Hirouchi D., Umitsu M., Janjirawuttikul N. and Patanakanog B. 2007. Horizontal and vertical variation of 2004. Indian tsunami deposits. *Marine Geology* 239: 163–174.

Hsü K. J. 2000. *Climate and Peoples: A Theory of History.* Zurich: Orell Fussli.

Huan N. H., Mai V., Escalada M. M. and Heong K. L. 1999. Changes in rice farmers' pest management in the Mekong Delta, Vietnam. *Crop Protection* 18: 557–563.

Huang Z. G., Zong Y. Q. and Zhang W. Q. 2004. Coastal inundation due to sea level rise in the Pearl River Delta, China. *Natural Hazards* 33: 247–264.

Jacobs Z. 2008. Luminescence chronologies for coastal and marine sediments. *Boreas* 37: 508–535.

Jankaew K., Atwater B. F., Sawai Y., Choowong M., Charoentitirat T., Martin M. E. and Prendergast A. 2008. Medieval forewarning of the 2004. Indian Ocean tsunami in Thailand. *Nature* 455: 1228–1231.

Jayaraju N., Reddy B. C. S. R. and Reddy K. R. 2009. Metal pollution in coarse sediments of Tuticorin coast, southeast coast of India. *Environmental Geology* 56: 1205–1209.

Karim M. F. and Mimura N. 2008. Impacts of climate change and sea-level rise on cyclonic storm surge floods in Bangladesh. *Global Environmental Change-Human and Policy Dimensions* 18: 490–500.

Keating B. H. and McGuire W. J. 2000. Island edifice failures and associated tsunami hazards *Pure and Applied Geophysics* 157: 899–955.

Kench P. S., McLean R. F., Brander R. W., Nichol S. L., Smithers S. G., Ford M. R. et al. 2006. Geological effects of tsunami on mid-ocean atoll islands: the Maldives before and after the Sumatran tsunami. *Geology* 34: 177–180.

Lim D. S. S., Smol J. P. and Douglas M. S. V. 2008. Recent environmental changes on Bank Island (NWT., Canadian Arctic) quantified using fossil diatom assemblages. *Journal of Paleolimnology* 40: 385–390.

Lincoln J. M. and Schlanger S. O. 1987 Miocene sea-level falls related to the geologic history of Midway atoll. *Geology* 15: 454–457.

Liu D. F., Shi H. D. and Pang L. 2006. Disaster prevention design criteria for the estuarine cities: New Orleans and Shanghai – the lesson from Hurricane Katrina. *Acta Oceanologia Sinica* 25: 131–142.

Ljung K., Bjorck D., Hammalund D. and Barnelow L. 2006. Late Holocene multi-proxy records of environmental change on the South Atlantic Island Tristan da Cunha. *Palaeogeography Palaeoclimatology Palaeoecology* 241: 539–560.

Maslin M., Vilela C., Mikkelsen N. and Grootes P. 2005. Causes of catastrophic sediment failures of the Amazon Fan. *Quaternary Science Reviews* 24: 2180–2193.

Menard H. W. 1986. *Islands.* New York: Scientific American.

Merritts D., Eby R., Harris R., Edwards R. L. and Chang H. 1998. Variable rates of Late Quaternary surface uplift along the Banda Arc-Australian plate collision zone, eastern Indonesia. *Coastal Tectonics* 146: 213–224.

Milligan J. and O'Riordan T. 2007. Governance for sustainable coastal futures. *Coastal Management* 35: 499–509.

Milne G. A., Long A. J. and Bassett S. E. 2005. Modelling Holocene relative sea-level observations from the Caribbean and South America. *Quaternary Science Reviews* 24: 1183–1202.

Mimura N., Nurse L., McLean R. F., Agard J., Briguglio L., Lefale P. et al. 2007. Small islands, in Parry M. L., Canziani O. F., Palutikof J. P., van der Linden P. J. and Hanson C. E. (eds) *Climate Change 2007: Impacts, Adaptation and Vulnerability. Contribution of Working Group II to the Fourth Assessment Report of the Intergovernmental Panel on Climate Change.* Cambridge: Cambridge University Press, pp. 687–716.

Moriwaki H., Chikamori M., Okuno M. and Nakamura T. 2006. Holocene changes in sea level and coastal environments on Rarotonga, Cook Islands, South Pacific Ocean. *The Holocene* 16: 839–848.

Munyikwa K., Choi J. H., Choi K. H., Byun J. M., Kim J. K. and Park K. 2008. Coastal dune luminescence chronologies indicating a mid-Holocene highstand along the east coast of the Yellow Sea. *Journal of Coastal Research* 24 (2B suppl.): 92–103.

Nicholls R. J., Wong P. P., Burkett V. R., Codignotto J. O., Hay J. E., McLean R. F. et al. 2007. Coastal systems and low-lying areas, in Parry M. L., Canziani O. F., Palutikof J. P., van der Linden P. J. and Hanson C. E. (eds) *Climate Change 2007: Impacts, Adaptation and Vulnerability. Contribution of Working Group II to the Fourth Assessment Report of the Intergovernmental Panel on Climate Change.* Cambridge: Cambridge University Press, pp. 315–356.

Nichol S. L., Lian S. B., Horrocks M. and Goff J. R. 2007. Holocene record of gradual, catastrophic, and human-induced sedimentation from a backbarrier wetland, northern New Zealand. *Journal of Coastal Research* 23: 605–617.

Nordstrom K. F., Lampez R. and Jackson N. L. 2007. Increasing the dynamism of coastal landforms by modifying shore protection methods: examples from the eastern German Baltic Sea Coast. *Environmental Conservation* 34: 205–214.

Nott J. 2004. The tsunami hypothesis – comparisons of the field evidence against the effects, on the Western Australian coast, of some of the most powerful storms on Earth. *Marine Geology* 208: 1–12.

Nunn P. D. 1994. *Oceanic Islands.* Oxford: Blackwell.

Nunn P. D. 1995. Holocene tectonic histories for five islands in the south-central Lau group, South Pacific. *The Holocene* 5: 160–171.

Nunn P. D. 2001. Ecological crises or marginal disruptions: the effects of the first humans on Pacific Islands. *New Zealand Geographer* 57: 11–20.

Nunn P. D. 2004. Understanding and adapting to sea-level change, in Harris F. (ed.) *Global Environmental Issues.* Chichester: Wiley, pp. 45–64.

Nunn P. D. 2007a. Holocene sea-level change and human response in Pacific Islands. *Transactions of the Royal Society of Edinburgh: Earth and Environmental Sciences* 98: 117–125.

Nunn P. D. 2007b. *Climate, Environment and Society in the Pacific During the Last Millennium.* Amsterdam: Elsevier.

Nunn P. D. 2009. *Vanished Islands and Hidden Continents of the Pacific.* Honolulu: University of Hawai'i Press.

Nunn P. D. and Peltier W. R. 2001. Far-field test of the ICE–4G (VM2) model of global isostatic response to deglaciation: empirical and theoretical Holocene sea-level reconstructions for the Fiji Islands, Southwest Pacific. *Quaternary Research* 55: 203–214.

Nunn P. D., Keally C. T., King C., Wijaya J. and Cruz R. 2006. Human responses to coastal change in the Asia-Pacific region, in Harvey N. (ed.) *Global Change and Integrated Coastal Management: The Asia-Pacific Region.* Berlin: Springer, pp. 117–161.

Nunn P. D., Hunter-Anderson R., Carson M. T., Thomas F., Ulm S. and Rowland M. 2007. Times of plenty, times of less: chronologies of last-millennium societal disruption in the Pacific Basin. *Human Ecology: An Interdisciplinary Journal* 35: 385–401.

Nur Y., Fazi S., Wirjoatmodjo N. and Han Q. 2001. Towards wise coastal management practice in a tropical megacity – Jakarta. *Ocean and Coastal Management* 44: 335–353.

Ogden J., Deng Y., Horrocks M., Nichol S. and Anderson S. 2006. Sequential impacts of polynesian and European settlement on vegetation and environmental processes recorded in sediments at Whangapoua Estuary, Great Barrier Island, New Zealand. *Regional Environmental Change* 6: 25–40.

Ota Y. and Chappell J. 1999. Holocene sea-level rise and coral reef growth on a tectonically rising coast, Huon Peninsula, Papua New Guinea. *Quaternary International* 55: 51–59.

Pavlopoulos K., Karkanas P., Triantaphyllou M., Karymbalis E., Tsourou T. and Palyvos N. 2006. Paleoenvironmental evolution of the coastal plain of Marathon, Greece, during the Late Holocene: Depositional environment, climate, and sea level changes. *Journal of Coastal Research* 22: 424–438.

Pedoja K., Dumont J. F., Lamothe M., Ortlieb L., Collot J. Y., Ghaleb B. et al. 2006. Plio-Quaternary uplift of the Manta Peninsula and La Plata Island and the subduction of the Carnegie Ridge, central coast of Ecuador. *Journal of South American Earth Sciences* 22: 1–21.

Perez-Torrado F. J., Paris R., Cabrera M. C., Schneider J. L., Wassmer P., Carracedo J. C. et al. 2006. Tsunami deposits related to flank collapse in oceanic volcanoes: The Agaete Valley evidence, Gran Canaria, Canary Islands. *Marine Geology* 227: 135–149.

Preoteasa L., Roberts H. M., Duller G. A. T. and Vespremeanu-Stroe A. 2009. Late-Holocene coastal dune system evolution in the Danube Delta, NW Black Sea Basin. *Journal of Coastal Research* 56: 347–351.

Razjigaeva N. G., Grebennikova T. A., Ganzey L. A., Mokhova L. M. and Bazarova V. B. 2004. The role of global and local factors in determining the middle to late Holocene environmental history of the South

Kurile and Komandar Islands, northwestern Pacific. *Palaeogeography Palaeoclimatology Palaeoecology* 209: 313–333.

Reyes-Bonilla H., Carriquiry J. D., Leyte-Morales G. E. and Cupul-Magana A. L. 2002. Effects of the El Nino-Southern Oscillation and the anti-El Nino event (1997–1999) on coral reefs of the western coast of Mexico. *Coral Reefs* 21: 368–372.

Sandweiss D. H. 1996. Environmental change and its consequences for human society on the central Andean coast, in Reitz E. J., Newsom L. A. and Scudder S. J. (eds) *Case Studies in Environmental Archaeology*. New York: Plenum, pp. 127–146.

Scharin H. 2002. Nutrient management for coastal zones: a case study of the nitrogen load to the Stockholm Archipelago. *Water Science and Technology* 45: 309–315.

Scott G. I., Holland A. F. and Sandifer P. A. 2006. Managing coastal urbanization and development in the twenty-first century: the need for a new paradigm, in Kleppel G. S. DeVoe M. R. and Rawson M. V. (eds) *Changing Land Use Patterns in the Coastal Zone: Managing Environmental Quality in Rapidly Developing Regions*. New York: Springer, pp. 285–299.

Sekhar N. U. 2005. Integrated coastal zone management in Vietnam: present potentials and future challenges. *Ocean & Coastal Management* 48: 813–827.

Shi C. X., Zhang D., You L. Y., Li B. Y., Zhang Z. L. and Zhang O. Y. 2007. Land subsidence as a result of sediment consolidation in the Yellow River delta. *Journal of Coastal Research* 23: 173–181.

Smith T. F., Alcock D., Thomsen D. C. and Chuebpagdee R. 2006. Improving the quality of life in coastal areas and future directions for the Asia-Pacific region. *Coastal Management* 34: 235–250.

Snoussi M., Ouchani T., Khouakhi A. and Niang-Diop I. 2009. Impacts of sea-level rise on the Moroccan coastal zone: quantifying coastal erosion and flooding in the Tangier Bay. *Geomorphology* 107: 32–40.

Stevenson J. and Hope G. 2005. A comparison of late Quaternary forest changes in New Caledonia and northeastern Australia. *Quaternary Research* 64: 372–383.

Stiros S. C., Laborel J., Laborel-Deguen F., Papageorgiou S., Evin J. and Pirazzoli P. A. 2000. Seismic coastal uplift in a region of subsidence: Holocene raised shorelines of Samos Island, Aegean Sea, Greece. *Marine Geology* 170: 41–58.

Stocchi P., Spada G. and Cianetti S. 2005. Isostatic rebound following the Alpine deglaciation: impact on the sea level variations and vertical movements in the Mediterranean region. *Geophysical Journal International* 162: 137–147.

Thomas E. K., Axford Y. and Briner J. P. 2008. Rapid 20th century environmental change on northeastern Baffin Island, Arctic Canada inferred from a multi-proxy lacustrine record. *Journal of Paleolimnology* 40: 507–517.

Titschack J., Nelson C. S., Beck T., Freiwald A. and Radtke U. 2008. Sedimentary evolution of a Late Pleistocene temperate red algal reef (Coralligene) on Rhodes, Greece: correlation with global sea-level fluctuations. *Sedimentology* 55: 1747–1776.

Tooley M. J. and Shennan I. 1987. *Sea-Level Changes*. Oxford: Blackwell.

Tornqvist T. E., Gonzalez J. L., Newsom L. A., van der Borg K., de Jong A. F. M. and Kurnik C. W. 2004. Deciphering Holocene sea-level history on the US Gulf Coast: A. high-resolution record from the Mississippi Delta. *Geological Society of America Bulletin* 116: 1026–1039.

Tossou M. G., Akoegninou A., Ballouche A., Sowunmi M. A. and Akpagana K. 2008. The history of the mangrove vegetation in Benin during the Holocene: a palynological study. *Journal of African Earth Sciences* 52: 167–174.

van de Plassche O., van der Borg K. and de Jong A. F. M. 1998. Sea level – climate correlation during the past 1400 yr. *Geology* 26: 319–322.

Vespa M., Keller J. and Gertisser R. 2006. Interplinian explosive activity of Santorini volcano (Greece) during the past 150,000 years. *Journal of Volcanology and Geothermal Research* 153: 262–286.

Webster J. M., Clague D. A. and Braga J. C. 2007. Support for the giant wave hypothesis: evidence from submerged terraces off Lanai, Hawaii. *International Journal of Earth Sciences* 96: 517–524.

Wenzel F., Bendimerad F. and Sinha R. 2007. Megacities – megarisks. *Natural Hazards* 42: 481–491.

Willard D. A., Cronin T. M. and Verardo S. 2003. Late-Holocene climate and ecosystem history from Chesapeake Bay sediment cores, USA. *The Holocene* 13: 201–214.

Wolfe C., McNutt M. and Detrick R. S. 1994. The Marquesas archipelagic apron: seismic stratigraphy and implications for volcano growth, mass wasting and crustal underplating. *Journal of Geophysical Research* 99: 13591–13608.

Woodroffe C. 2002. *Coasts: Form, Process and Evolution*. Cambridge: Cambridge University Press.

Woodroffe C. D. 2008. Reef-island topography and the vulnerability of atolls to sea-level rise. *Global and Planetary Change* 62: 77–96.

Young R. W. and Bryant E. A. 1993. Coastal rock platforms and ramps of Pleistocene and Tertiary age in
southern New South Wales, Australia. *Zeitschrift für
Geomorphologie* 37: 257–272.

Yulianto E., Sukapti W. S., Rahardjo A., Noeradi D.,
Siregar D. A., Suparan P. and Hirakawa K. 2004.
Mangrove shoreline responses to Holocene
environmental change, Makassar Strait, Indonesia.
Review of Palaeobotany and Palynology 131:
251–268.

Yum J. G., Yu K. M., Takemura K., Naruse T., Kitamura A.,
Kitagawa H. and Kim J. C. 2004. Holocene
evolution, of the outer lake of Hwajinpo Lagoon on
the eastern coast of Korea; environmental changes
with Holocene sea-level fluctuation of the East Sea
(Sea of Japan). *Radiocarbon* 46: 797–808.

Zhang Q., Zhu C., Liu C. L. and Jiang T. 2005.
Environmental change and its impacts on human
settlement in the Yangtze Delta P.R. China. *Catena*
60: 267–277.

Zong Y., Huang G., Switzer A. D., Yu F. and Yim W.
W-S. 2009. An evolutionary model for the Holocene
formation of the Pearl River delta, China. *The
Holocene* 19: 129–142.

Past, Present and Future Responses of People to Environmental Change

Testing the Role of Climate Change in Human Evolution

Simon P. E. Blockley, Ian Candy
and Stella M. Blockley

1 BACKGROUND

The potential importance of both long and short-term climatic change on the evolution of our species has long been acknowledged. Indeed environmental change has under-pinned many of the proposed mechanisms behind models of human evolution, including population movement, isolation, specific adaptation and speciation. One significant problem with testing these ideas, however, is identifying the appropriate environmental evidence for a region and directly relating this evidence to the archaeological record. This chapter looks at some of the ongoing scientific debates into the role of climate on the evolutionary history of our species. In scope it cannot encompass the entire evolutionary history of the hominid line, nor outline the whole history of scientific research into evolution and climate forcing, as the literature on these topics is vast. This chapter will instead focus on key areas where a link between evolution and climate has been suggested, tested and debated, and will try to point to more general reviews along the way.

This chapter will be divided into three parts, each one focusing on a major period of hominid evolution, dispersal or extinction.

The significance of each of these periods will be described, the archaeological/palaeonto-logical evidence reviewed and the climatic background discussed. Section 2 will focus on the early period of hominid evolution and the impact of climate history on bipedalism, migration out of Africa and the colonisation of Europe. Significantly, although it has been suggested that bipedal apes may have existed as far back as the Miocene, *Orronin tuganensis*, this discussion will examine the last 4.4 million years, beginning with the earliest well-dated Pliocene hominid *Ardipithecus ramidus*. Later sections (3 and 4) will consider the middle and upper Palaeolithic, particularly the evolution and dispersal of anatomically modern humans (AMH) and the extinction of archaic hominid species during this period.

2 EARLY HOMINID EVOLUTION AND MIGRATION

2.1 Climatic setting for early hominid evolution

As discussed elsewhere in this handbook, the long-term history of environmental change is recorded in the $\delta^{18}O$ composition of marine sediment records and ice-core sequences (EPICA, 2004; Lisiecki and Raymo, 2005). These are dated often with reference to models of orbital forcing of global climate as predicted by Milankovitch (Imbrie et al., 1993). For East Africa, where the first ancestral ape and hominid species evolved, there is the additional advantage that long marine sediment archives, taken as part of the international Ocean Drilling Programme (ODP) can, in this case, be directly correlated to the hominid record by the presence of volcanic ash in both the marine sediment sequence and the archaeology-bearing deposits. Given the inherent dating uncertainties in both types of records over long time ranges, these isochronous markers improve the security of exploring the relationship between hominid activity and environmental change. These volcanic ash deposits are also suitable for dating using potassium–argon (K–Ar) and argon–argon ($^{40}Ar/^{39}Ar$) techniques and thus provide the chronological underpinning for the evolutionary record.

In general terms, the record of global climate over the last 4 million years is one of progressive cooling and a gradual increase in global ice volume, overprinted by glacial–interglacial cycles. A transition to cooler climates occurs at around 3.2 Ma BP which becomes intensified at the onset of the Quaternary Period, from 2.6 Ma BP onwards. Stratigraphic procedure divides the Quaternary into the Pleistocene (2.6 Ma BP to 11,500 BP) and the Holocene (11,500 BP to the present). The Pleistocene is further subdivided into the early Pleistocene (2.6 million years to 780,000 BP), the middle Pleistocene (780,000 to 130,000 BP) and the late Pleistocene (130,000 to 11,500 years BP). During the Pleistocene, glacial–interglacial cycles are a key component of the climate system. However, a major change occurs with respect to the duration of these cycles at $c.$ 1 Ma BP. In the early Pleistocene the magnitude of glacial–interglacial cycles is relatively subdued and driven by the 41,000-year obliquity cycle. At $c.$ 1 Ma BP the intensity of

glacial–interglacial cycles increases and they are driven by the 100,000-year eccentricity cycle. This transition is known as the Mid-Pleistocene Revolution (MPR) of the Mid-Pleistocene Transition (MPT) (Imbrie et al., 1993; Ruddiman et al., 1989).

In many of the regions of the world where evidence for early human evolution is found, namely Africa, the Mediterranean and the near East, the long-term and cyclic climate changes are not simply expressed through temperature changes but also through changes in the degree of environmental aridity. For example, marine cores from around the coast of Africa, such as ODP 1084 (Marlowe et al., 2001), record the long term patterns of global cooling for the last 4 Ma, as outlined above, along with evidence for aridifcation, as preserved in records of atmospheric dust production. Sea-surface temperature (SST) records from ODP 1084 record an onset of cooling at $c.$ 3.2 Ma BP, an intensification of cooling at $c.$ 2.2–2.0 Ma BP and the occurrence of high intensity glacial-interglacial cycles from c.0.6 Ma BP onwards (Marlowe et al., 2001). The impact of increasing cooling is recorded in numerous marine cores and frequently occurs in conjunction with increasing aeolian dust concentrations, suggesting increasing aridity of the continent across the Quaternary Period (see deMenocal, 1995, 2004; Marlowe et al., 2000).

One note of caution to be made, however, when comparing fossil hominid records and various marine archives of long term climate change, is that there are broad uncertainties in both the age of the deposits which contain key hominid fossils and the chronologies associated with marine sediment records (Henderson and Slowey, 2000). Whereas chronologies associated with marine sediment cores are typically constructed through the tuning of the climatic record that they contain to the record of orbitally driven insolation fluctuations, the chronology of the terrestrial record in East Africa is based upon the radiometric dating of volcanic ash layers and the palaeomagnetic dating of the deposits themselves. Both orbital tuning and radiometric dating will have associated uncertainties of varying magnitude (see Chapter 5). An additional complication is also generated because of the highly fragmented and discontinuous nature of the terrestrial record in comparison to the relatively continuous nature of the marine sequence (deMenocal, 1995, 2004).

2.2 Climatic and other environmental influences on the evolution of early hominids

The general story of hominid evolutionary history and the general debates therein are discussed elsewhere (Elias, 2007) and an excellent review of the current debates over the role of climate in hominid evolution is available in deMenocal (2004). This section will specifically focus on key periods in the evolution of early African hominids where a clear link between changing climates and evolutionary trends has been proposed. Within the general story of the evolution of the hominid line there has always been a role for testing the influences of climate change on species evolution. One of the key mechanisms behind evolution is the suitability of an organism for its environment and the potential role of climatic change has long been apparent. In the case of hominid evolution one of the key traits of hominids and ancestral apes that has been proposed to be environmentally driven was the first appearance of bipedalism. This was seen by many researchers (e.g., Dart, 1925) as being an evolutionary response to the development of savannah landscapes in Africa. In recent years, however, discoveries of older and older bipedal hominids has shown that bipedalism has a much longer history than once thought, and begins before the appearance of broad savannah landscapes (deMenocal, 2004). In this case, modern evidence suggests that a proposed link between long term environmental change and a specific evolutionary trait does not stand up to scrutiny.

There is, however, an ongoing debate over the role of the environmental transitions discussed above on the geographic range of hominids and other fauna. Within the general evolutionary trend of hominids, there are clear periods of significant change and it is for these that an environmental forcing mechanism has been proposed. During the mid to late Pliocene and the early Pleistocene, a number of bipedal ape and hominid species appear in the fossil record: these include *Ardepithicus Ramidus* (the earliest Pliocene bipedal ape), *Australopithecus anamensis, A. afarensis* and *Kenyanthropus platyops*. Table 38.1 provides a list of the key species and the current age estimates of earliest known appearances and extinctions (after deMenocal, 2004). These Pliocene hominid species all become extinct by ~2.9 Ma BP and palaeoenvironmental evidence suggests that the preferred habitat of all of these species is characterised by relatively wet climates and wooded landscapes.

The period between 2.9 and ~1 million years ago sees an increase in the number of hominid species with a later gracile *A. africanus*, as well as the appearance of cranially more robust bipedal apes, *Paranthropus boisei, P. robustus* and *P. aeopithicus*. In addition, the first appearance of species defined as the beginning of the *Homo* lineage begins ~2.4 Ma BP with *H. Rudolfensis* followed by *H. hablis* at around 2 Ma BP, and *H. erectus* ~1.8 Ma BP. Moreover the first stone tool use begins around 2.6–2.3 Ma BP (deMenocal, 2004), whilst the first movement of hominids out of Africa occurs very soon after the appearance of *H. Erectus* at ~1.8 Ma BP in both Asia and Eastern Europe (Swisher et al., 1994; Gabunia et al., 2000, 2001; Zhu et al., 2008). Some scholars regard *H. erectus* as only a non-African species and classify all African specimens as *H. ergastor*. There is then some evidence for hominid evolution, extinction, adaptation and dispersal in the late Pliocene and early Pleistocene that could relate to environmental changes in Africa associated with the onset of major global cooling and glacial–interglacial cycles.

Table 38.1 General timeline of the main species in hominid evolution. The exact details and some of the dates and even the definition of species are a matter of ongoing debate – after deMenocal 2004

Species	Approximate ages	Technology	Known range
Ardipithecus ramidus	4.4 Ma	na	East Africa
Australopithecus anamensis	4.2–3.9 Ma	na	East Africa
Australopithecus afarensis	3.9–2.9 Ma	na	East Africa
Kenyanthropus platyops	~3.5 Ma	na	East Africa
Australopithecus africanus	2.9–2.2 Ma	na	Southern Africa
Paranthropus aethiopicus	~2.7–2.3	na	East Africa
Paranthropus robustus	2.0–1.5 (1.2?) Ma	na	Southern Africa
Paranthropus boisei	2.3–1.2	Oldawan?	East Africa
Homo rudolfensis	2.4?/1.9–1.8 Ma	Oldawan?	East Africa
Homo habilis	2.4?–1.4	Oldawan?	East and Southern Africa
Homo erectus/ergastor	1.9–~0.50 Ma (0.2 in Africa?)	Achulean	Old World
Homo Heidelbergensis/ archaic hominids	700–200 ka?	Achulean/Mousterian	Africa, Europe, Near East
Homo neanderthalensis	350 ka? 130–30ka	Mousterian/Chatelperronian	Europe, Near East
Homo sapiens	200–0 ka	Aurignacian and others	Global

DeMenocal (1995), Vrba (1980 and Vrba et al. 1989) developed hypotheses relating to hominid response to changing habitats in Africa during the periods of environmental transition outlined above. These focus on the development of the African savannah and open drier environments after ~3.2 Ma BP as a driver of hominid and other faunal speciation. Vrba in particular developed a model of increased rates of turnover (appearance and disappearance) of faunal species associated with shifts in African climates at ~2.8, 1.8 and ~1.0 Ma BP. These proposed changes in the turnover of faunal species do broadly coincide with important transitions in the hominid record, namely the first appearance of the genus *Homo*, the expansion of the number of *Homo* taxa, the appearance of *H. erectus* and the migration out of Africa at ~1.8 Ma BP. There are problems with this hypothesis, however, as terrestrial evidence of African environments is less abundant and thus the hypothesis is harder to test, while the marine records of environmental change do not show unequivocal evidence of these phases of climatic instability alone. While at ODP 1084 there is evidence for changes in SST at these times, far more variability is present in ODP records of African dust input, suggesting instead a pattern of high frequency climatic variability with a general trend to increasing aridity (deMenocal, 1995). Other evidence for species response to climate change from a case study in the Omo basin (Bobe et al., 2002) suggests more gradual processes. Analysis of the faunal assemblages from the period 3.6–2.4 Ma BP suggests a gradual decrease in obligate woodland species during this period with a more rapid increase in open grassland species after ~2.6 Ma BP.

An additional problem relates to the nature of Pliocene and Pleistocene climate variability, which after ~3.2 Ma BP is dominated by cycles of warming and cooling on relatively short timescales. It has been suggested in the variability selection hypothesis (Potts, 1999) that the rapidity of changing environments, linked to glacial–interglacial cycles, was the driving force of species evolution. In an extended study of faunal remains from excavations from the Omo basin and a comparison to species turnover events at a number of sites, Bobe et al. (2002) suggest that there was a peak in species turnover at ~2.8 Ma BP, and significant instability in turnover rates between ~2.5 and 2.2 Ma BP. If we are to understand how these climate cycles, both their occurrence and the progressive changes in their magnitude and frequency (i.e., before and after the MPR), may have effected hominid species, we require secure and high resolution correlations between archaeological and palaeoenvironmental data. Moreover, hominid sites tend to be palimpsests containing evidence for multiple episodes of relatively short periods of occupation at individual sites. Thus, debates over the exact relationship between early African hominid evolution and climatic forcing are likely to remain unsettled until: (1) the numbers of sites and the numbers of specimens of each *Homo* species are at a stage where the picture of hominid evolution is robustly understood; (2) occupation events can be reliably constrained in time; and (3) the age constraints are at a resolution comparable to the periodicity of the glacial–interglacial cycles that occur during these time periods.

A recent study of terrestrial palaeoclimate from a speleothem in Buffalo Cave, South Africa (Hopley et al., 2007), provides oxygen and carbon ($\delta^{18}O$ and $\delta^{13}C$) isotopic evidence, thought to relate to monsoon strength and vegetation change, particularly the balance between C3 (wooded) and C4 (grassland) vegetation. The speleothem record covers the time period from 2.0 to ~1.5 Ma BP and is overlain by sedimentary deposits, which are thought to date through to ~250,000 BP. The record is dated partly by palaeomagnetic stratigraphy and by matching the isotopic record to insolation changes at a latitude of 24° south. While the reservations regarding orbitally tuned chronologies remain, this site does provide direct terrestrial proxy evidence for climate change during a crucial time interval. These data suggest that between 1.8 and 1.7 Ma BP there was an increase in dryland grasses, as reflected in an increased C4 signal in the Buffalo $\delta^{13}C$ isotopic record. Within dating uncertainties it is possible that the appearance of *H. erectus*, suggested to be the first dry open habitat hominid, and the subsequent onset of migration out of Africa

relate to increased aridity in Africa (Ruff, 1991). Interestingly, the work of Bobe and Behrensmeyer (2004) suggest that while grassland indicator species were increasing in relative abundance from around 2.4 Ma BP, there was a sharp increase between ~1.8 and 1.6 Ma BP in the Omo sequence. These records are relatively well dated by intrastratification with radiometrically dated volcanic deposits and do match well with the Buffalo cave data. Thus, there is independently dated evidence for an increased open grassland from 1.8 Ma BP, possibly connected to the intensification of northern hemisphere glaciations.

2.3 Early hominid migration out of Africa

Climatic and environmental change has been cited as one of the key factors behind the migration of the first hominids out of Africa from ~1.8 Ma BP onwards. There is now good evidence for migration into Eastern Europe, Indonesia and China, often based on radiometric dating of volcanic deposits, although as always in this time period dating uncertainties remain. The evidence is, however, sparse and a model of episodes of sporadic migration of *Homo erectus* from ~1.8 to 0.7 Ma BP has been developed (Bar-Yosef and Belfer Cohen, 2001). The increased drying and extension of savannah landscapes has been suggested as one key driver behind the increased geographical range of *H. erectus*, both as a push factor, with periodic expansion and reduction in areas available for exploitation by hominids (Potts, 1998), and as a factor in allowing exploitation of the opening landscapes. For example, palaeoenviroenmental evidence from one of the earliest sites with *H. erectus* outside Africa, Dminisi in the Caucuses (Gabunia et al., 2001) suggests the existence of sparse vegetation and a semiarid landscape (Gabunia et al., 2000). This may have suited a species that has been considered to be an aridadapted hominid. Similarly, lowered sea levels during glacial episodes after 1.8 Ma BP are thought likely to have facilitated *H. erectus* migration into Indonesia, and glacial periods may have also provided more open woodland and grassland mosaic environments for Javan *H. erectus* to exploit (Bettis et al., 2009).

As the data on the earliest out of Africa migrations is rather sparse, other explanations are, however, possible. The first appearance of *H. erectus* outside Africa is essentially identical in age to the age of the first *H. erectus/ergastor* specimens within Africa. This has led some to suggest that the driving factor behind dispersal was evolutionary rather than ecological with the adoption of meat protein into hominid diets, differences in both cranial and postcranial morphology, and technological innovations being proposed as key components in the increased geographical range. This has neatly been summarised by Foley (2002) as a shift in the evolutionary grade between *H. erectus/ergastor* and the earlier *Homo* and *Australopithicenes*. In fact, in the light of the difficulties in assigning species and the sparse geographical coverage of *H. erectus* fossils, Dennell and Roebroeks (2005) have postulated a potential non-African origin of *H. erectus* and pointed out the poor preservation or exposure potential for fossils away from fine-grained volcanic regions such as Africa's rift valley, and the fact that most early lithic material ascribed to *H. erectus* outside Africa are done so on the assumption that *H. erectus* was the only species there. While admitting that there is little supporting evidence they point out that fossil evidence in this time is so sparse that an earlier hominid migration out of Africa is at least possible and that *H. erectus/ergastor* evolved in Asia and migrated into Africa. Indeed they recognise that the division of Africa and Asia is a modern invention and that actually if savannah landscapes are key to evolutionary developments in the genus *Homo* then it may be significant that open grassland existed in the Plio-Pleistocene from West Africa across into East Asia.

Thus, as the above review makes clear there is a role for the study of climate change in understanding the early evolution of our species, from *Australopithicines* to the early *Homo*. There is reasonable evidence for shifts in the climate of the late Pliocene and the early Pleistocene, possibly connected to the onset and then intensification of northern hemisphere glaciations, which correlate with changes in both nonhominid fauna and in the record of evolution and dispersal of early hominids. This is, however, a period where fossil specimens are rare and there are significant preservation biases between different regions, with the East African rift valley being a geographical hotspot for preservation discovery and dating of hominid finds. This relatively sparse evidence and potential geographical bias, coupled with the role of other important factors in the evolution of our species, such as increases in brain size and technological innovations, mean understanding the interplay between environment and evolution require significant caution.

2.4 Earliest hominid migration into southern and northern Europe

The role of climate on hominid migration and the complexities associated with the related debate can be seen in current ideas about the earliest human occupation in southern and northern Europe. Traditionally, archaeologists have referred to a 'long chronology' of occupation in southern Europe and a 'short chronology' in northern Europe (Roebroeks, 2001, 2006; Dennell, 2003). This is based around the occurrence of evidence for human occupation in southern Europe, at sites such as Atapuerca (Spain; Carbonell et al., 1995), Orce (Spain; Scott and Gibert, 2009) and Ceprano (Italy; Manzi, 2004), prior to 780,000 BP but the earliest known human occupation in Europe north of the Alps being restricted to 500,000 BP or younger (at sites such as Mauer, Germany, and Boxgrove, Britain; Roberts et al., 1994).

One of the suggestions for the discrepancy between the occupation history for these regions has been the role of climate with the suggestion that the colonisation of northern Europe offered a different set of challenges to the colonisation of southern Europe, which, in terms of thermal regime, flora and fauna, may have offered a similar environment to parts of Africa and the near east. Expansion of hominids into the Mediterranean potentially, therefore, required no major adaptation or evolution. Northern Europe, characterised by cool summers and, potentially, extremely cold winters along with a different assemblage of flora and fauna, offered a different set of environmental problems to overcome before the region could be colonised (see discussion in Dennell, 2003; Dennell and Roebroeks, 1996; Roebroeks, 2001, 2006).

The idea of a long and short chronology has been challenged by the discovery, at the site of Pakefield on the east coast of Britain, of flint artefacts that, through combined litho-, bio- and magneto-stratigraphy, have been dated to >650,000 BP possibly to the interglacial MIS 19, c.780,000 BP (Parfitt et al., 2005). These ages suggest that there is no major discrepancy between the colonisation of northern and southern Europe, questioning the role of climate in hominid migration. Clearly, given the particularly early age of human activity recorded at Pakefield, the climates and environments of this episode of occupation are important. Environmental reconstruction based on biological (Coope, 2006; Parfitt et al., 2005), sedimentological and geochemical data (Candy et al., 2006) has suggested that climates of the time were characterised by summer temperatures that were up to 5°C warmer than the present (Coleopteran MCR reconstruction and the presence of thermophile plant taxa such as *Slavinia natans* and *Trapa natans*) and winter temperatures that were at least as warm as, if not several degrees warmer than, the present (through the presence of Mediterranean species such as *Hippopotamus*, frost insensitive insect

species and plant taxa which require winter temperatures in excess of 0°C such as *Ilex* and *Hedera*). The precipitation regime that existed during the accumulation of the Pakefield deposits was also significantly different from the present day with palaeosol profiles indicating strongly seasonal rainfall with the summer months being characterised by extended drought (Candy et al., 2006).

The reconstructed climate for human occupation at Pakefield, with hot summers, mild winters and a seasonally dry precipitation regime, is more typical of modern-day southern Europe (i.e., a Mediterranean climate) than it is of temperate northwest Europe (Candy et al., 2006; Parfitt et al., 2005). Consequently, although the early occurrence of flint artefacts at Pakefield suggests that the idea of the long and short chronology is flawed the environmental reconstruction indicates that the earliest known human occupation in Europe north of the Alps occurred under periods of extreme warmth, when northwest Europe was characterised by a 'Mediterranean'-style climate (Parfitt et al., 2005). It is possible, therefore, that early human occupation in northern Europe was aided by the existence, albeit for a relatively short period of time (i.e., the duration of an interglacial), of climates to which humans were already adapted. Climatic oscillations may still have been an important factor in the colonisation of northern Europe with short-lived periods of extreme warmth offering 'windows of opportunity' during which the first migration into this region could have occurred.

Although the earliest known colonisation of northern Europe occurred during an interglacial of unusual warmth it is important to note that subsequent episodes of human occupation during the early middle Pleistocene (780,000 to 500,000 BP) occurred under a range of climatic conditions. At British sites such as Happisburgh I (Ashton et al., 2008), High Lodge (Ashton et al., 1992) and Boxgrove (Roberts et al., 1994) evidence for human occupation is found in association with climatic conditions that were significantly cooler than the present day. Deposits at these sites possibly record sedimentation during the late part of an interglacial (during the climatic deterioration into the subsequent glacial) or during interstadial events. Therefore, although extreme warmth may have aided the first colonisation of northern Europe either progressive adaptation or the migration of different hominid species meant that subsequent occupation events do not seem to be strongly controlled by the degree of warmth in warm intervals. The ability of hominids, during this time period, to survive episodes of extreme climate cooling is not currently known because of issues regarding our understanding of whether these colonisers were a seasonal or a persistent presence.

3 HOMINID EVOLUTION DURING THE MIDDLE AND LATE PLEISTOCENE

3.1 Middle and late Pleistocene climate change

The final stages of the evolution of the genus *Homo* occur in the last three glacial cycles (from marine isotope stage MIS 10 onwards). The chronology of northern hemisphere glacial cycles, based on the chronology derived from Astronomical Theory (see Chapter 19), features warm interglacial episodes during MIS 9, 7 and 5, cold glacial episodes during MIS 10, 8 and 6, with subdivisions of the last glacial into glacial cold maxima in MIS 4 and 2, and a warmer substage MIS 3. These glacial cycles form the backdrop to the evolution of hominids in Europe, the Levant and Africa, although the local climatic effects differ between cold and warm intervals in northern Europe and drier or wetter phases in lower latitudes. For the last glacial cycle there is good evidence from a number of regional palaeoclimate proxies, including ice core, marine and terrestrial archives (Allen and Huntley, 2009; Shackleton et al., 2004; Svensson et al., 2006) for abrupt shifts in

climate superimposed upon the record of glacial to interglacial climates. Figure 38.1 shows a composite of two Greenland ice core records (NGRIP and GISP2) covering the last glacial cycle and gives an indication of the rapid transitions as recorded in Greenland.

While the absolute and relative timing of these abrupt changes across different regions are debated, it is apparent that during the last glacial period, at least, a relationship exists between rapid transitions from short-lived warm episodes (interstadials) to much colder periods (stadials) recorded in the North Atlantic, North America, Europe and much of the Mediterranean (e.g., Allen and Huntley, 2009; Shackleton et al., 2004; Svensson et al., 2006). These transitions, Dansgaard–Oesheger (D/O) cycles (see Chapter 20), are thought generally to be driven by cyclical growth and collapse of the northern hemisphere ice sheets, with meltwater pulses into the Atlantic triggering a slowing or shutdown of Atlantic thermohaline circulation (e.g., Bond et al., 1993). During these cycles, marine records as far south as the Iberian margin show evidence of ice-rafted debris being incorporated into marine sediments. These sediments, termed Heinrich layers, are thought to correlate to the coldest part of repeated cycles of warming and cooling as reported from marine and ice core records and there have been recent attempts to correlate these Heinrich Events with cold episodes on land. One key question that has been raised recently by a number of researchers is to what extent do these cold Heinrich Events impact upon late Pleistocene hominid populations?

3.2 Climate change and the middle and late Pleistocene evolution, dispersal and replacement of hominids

In the middle Pleistocene, *H. heidelbergensis* was present as the dominant form of hominid in Africa and Eurasia. However, there is

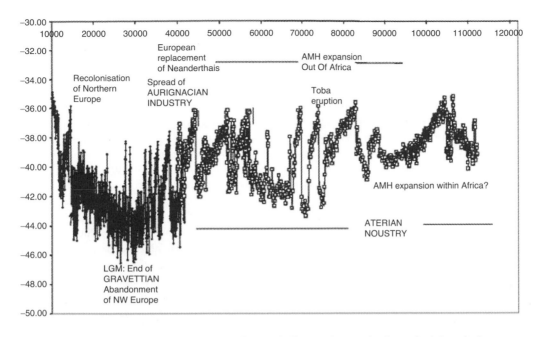

Figure 38.1 Key events in human evolution and dispersal over the last glacial cycle for Africa and Eurasia plotted against the NGRIP (Svensson et al., 2006) ice core record and the GRIP ice core record (before 40,000 ka BP; Johnsen et al., 1997).

evidence of increased species diversity in the late middle Pleistocene, with evidence for non *H. heidelbergensis* forms in Africa, and late surviving *H. erectus* in East Asia (Tattersall and Shwartz, 2009). There is an increasing development in technology in the late middle Pleistocene, with evidence for sustained use of fire and nonlithic technology. During the late middle to late Pleistocene, from ~500,000 to 11,000 years ago, there is an increased diversification in hominid forms, with regional evolution and adaptation, such as the European and Near Eastern *H. neanderthalensis* (Neanderthals) and further developments in technology, and what we would consider modern cultural traits. This culminates in the evolution of *H. sapiens* in sub-Saharan Africa, their migration out of Africa and their subsequent dispersal across the world.

While still contentious, the dominant 'out of Africa' hypothesis sees this process of dispersal by AMH resulting in the replacement in the fossil record of all other hominid forms. Strong genetic evidence now

suggests that modern *H. sapiens* are descended from a small sub-Saharan AMH population and that the migration out of Africa, sometime after ~120,000 years ago was made by a small founder population. Although it is possible that the dispersing AMH may not have actually encountered the hominids they were destined to replace in the fossil record, there is evidence in Europe at least for crossover of technological innovations between AMH and Neanderthals.

3.3 Neanderthal anatomy and evolution

The timing of the first appearance of Neanderthals is not clear. During the late middle Pleistocene from sometime after 400,000 years ago (possibly earlier), some Neanderthal traits appear within the archaic hominid (e.g., *H. heidelburgensis*) record. However, by ~130,000 BP, fully Neanderthal forms are present in the fossil record. Neanderthals have traditionally been seen as a cold-adapted species (see Caldwell, 2008; Holliday, 1997, 1999). The debate draws on Bergmann's (1847) and Allen's (1877) rules, which describe human adaptation to tropical and arctic climes and state that humans (and other species) adapted to cold climates have a short, stocky trunk and proportionally short lower limbs, in order to conserve body temperature. Populations adapted to hotter climates, in contrast, have a relatively slender trunk and long lower limbs, to facilitate heat loss. The argument for Neanderthal cold-adaptation is, in part, based on the calculation of crural and brachial indices, a measure of proportion of the upper and lower limbs. Neanderthals have a low crural index compared with most AMH populations, and have a robust, stocky anatomy, consistent with observations of present-day cold-adapted human populations. The broad nasal aperture displayed by Neanderthals has also been cited as a mark of their being cold-adapted (it allows air to be warmed on inhalation). The anatomical basis for Neanderthal cold adaptation, however, has been challenged for both crania and postcrania (Hubbe et al., 2009; Rae et al., 2009).

Further complications arise from the difficulty of ascribing definite species to some specimens, giving rise to the use of the blanket term 'archaic humans'; thus comparisons between species can be difficult. The Middle Pleistocene Boxgrove tibia, for example, can be ascribed to 'non-modern human' only, and shows robusticity comparable to Neanderthal anatomy, either as a result of activity or cold adaptation; however the taxonomic difficulties surrounding the specimens makes direct comparison impossible (Stringer et al., 1998). An interesting point to note in the idea of Neanderthal's being cold adapted, is that an early individual from Pontnewydd Cave in Wales, dating to *c*.230,000 years ago, has been identified as possibly being an early Neanderthal, as it displays a trait unique to Neanderthal dental morphology, taurodontism (Green, 1984). This individual was present during a warm phase in Wales, giving rise to the argument that Neanderthals were

evolving during a warm phase, which would mean they were unlikely to be adapted to the cold. The lack of postcrania in this specimen makes it difficult to assess such claims.

3.4 Neanderthal distribution and extinction

Neanderthal distribution in Europe and the Near East has also been cited as being influenced by climatic and environmental factors (Finlayson and Carrion, 2007; Gilligan, 2007; Jimenez-Espejo et al., 2007; see van Andel, 2002). This has been complicated by debates about speciation, and the classification of some populations by some researchers as 'archaic humans' and ascribing some individuals to other species (such as *H. heidelbergensis*). Nevertheless, the distribution of Neanderthals has been suggested to reflect climate change, with populations occupying southern refugia during cold phases (Carrion et al., 2003; Jimenez-Espejo et al., 2007). This suggests a climatic driver to their distribution, while perhaps arguing against their being a cold-adapted species, successfully exploiting fully arctic climate zones. A further argument against Neanderthals being cold adapted, is their presence as a stable species for thousands of years in the near east; records show a stable presence of Neanderthals in that region during a warm phase (see Bar Yosef et al., 1992; Schwartz et al., 1989; Valladas, 1991).

What is clear, however, is that Neanderthal populations had adapted to the broad geographical range covering Europe to Israel and that they persisted in this range during a series of glacial and interglacial episodes. Importantly, during this time, Neanderthals were not competing with other hominids. The extinction of Neanderthals, during the middle to late part of the last glacial stage, with the last known Neanderthal populations dating to *c*.28,000 BP (Gibraltar; Finlayson et al., 2006), is a key research question in Palaeolithic archaeology. Many factors have been cited as contributing to the extinction of the Neanderthal as a distinct species, such as competition with AMH populations after AMH dispersal from Africa (Banks et al., 2008), breeding rates and climatic change (Finlayson and Carrion, 2007; Gilligan, 2007; Jimenez-Espejo, 2007). These need not be mutually exclusive and the most robust explanations are likely to be multi-factorial. The final extinction of the Neanderthals and their interaction with both AMHs and late Pleistocene climate change are discussed in more detail later in this chapter.

4 ANATOMICALLY MODERN HUMANS AND CLIMATE CHANGE

4.1 Climate change and the evolution and dispersal of anatomically modern humans

The relationship between the evolution and subsequent dispersal of AMH and its relationship to climate change has seen significant debate. A key issue has been understanding the environmental influences on AMH migration out of sub-Saharan Africa and subsequently around the globe. In addition, the replacement of Neanderthal forms in the European archaeological record and the potential roles of both AMH and abrupt climate change are also now a key focus of debate within archaeology and Quaternary science.

Until relatively recently, there was a significant debate over the origin of *H. sapiens*, with conflicting views over an African origin or a gradual coregional evolution across the old world, with gene flow between different populations allowing gradual appearance of AMH forms. A particular problem was the difficulty in dating the fossil remains to examine the relative timing of the appearance of AMH forms. While dating remains a problem, a combination of improved chronological control, new fossil discoveries, and particularly the impact of a range of DNA studies on modern and ancient populations, has led to a level of consensus that places the

first appearance of AMH in sub-Saharan Africa prior to the last interglacial. The earliest known sites so far are Omo I and II in the Kibish formation in Ethiopia (McDougall et al., 2008). These sites are now radiometrically dated by ^{40}Ar/^{39}Ar dating of intastratified volcanic Tuffs to a most likely age of 195,000 ± 5,000 BP, with an absolute minimum age of 104,000 BP. On current evidence sub-Saharan Africa is seen as the evolutionary start point of *H. sapiens* with subsequent dispersal from this region. Understanding the later spread of this species within Africa and the development of modern behaviour is a key question. Crucial to this argument are: (1) the timing, nature and route of this dispersal, with several migration corridors proposed; and (2) late Quaternary climate change, which is seen as a potential driver. Of the two possible migration routes proposed, a southern route envisages AMH crossing the Red Sea, while the northern route sees AMH colonising eastern North Africa from routes along the Nile corridor. The latter model is supported by early dates from the Levant that place modern humans at Qafzeh at 92,000 ± 5,000 BP (Valadas et al., 1998), based on thermoluminescence ages of burnt flint, and at Skuhl Cave at 88,000–117,000 BP, based on a combination of several dating techniques (Grun et al., 2005). Both of these sites are found in modern Israel. The recent discovery of a site with technology associated with AMH at Jebel Faya, United Arab Emirates, dated to as 127000–95,000 BP (Armitage et al., 2011), however, lend support to an early southern dispersal route. Either way there is now good evidence for human expansion into northern Africa and the near East during the last interglacial.

The distribution of dates between sites north and south of the Sahara may indicate a post-MIS 5e age for the dispersal of AMH northwards across this region. Mellars (2006) has suggested, however, that while there is evidence of an AMH presence in North Africa and the Levant at this time other evidence suggests this was not a successful colonisation. Recent DNA evidence examining the gene variability within modern populations suggests that only one ancestral population of AMH migrated out of Africa. This is based on the analyses, in part, of the variability in gene types of mitochondrial DNA. Mitochondria are small subunits of cells that supply the majority of the energy to the cell. In human reproduction, only the female mitochondria pass on to the offspring and thus mitochrondrial DNA (mtDNA) represent the genetic signature of a particular female (haplotype) derived from the female lineage of her genetic history. This allows an assessment of natural mutation over time in generations descending along the female line (Forster, 2004). Thus, the mtDNA of individuals in modern populations can be structured as a genetic tree and the numbers of mutations from earlier haplotypes can be estimated. Moreover, if the rate of genetic variations between haplotypes is assumed to be constant and can be reliably estimated, then the timing of genetic mutations can be reconstructed (the genetic clock). The estimation of the rate of mutation is difficult and is based on either studies of past speciation events, which are radiometrically dated, studies of several generations of modern populations, or by looking at diversity in populations where the initial date of the founding of that population is known. These estimates all have inherent uncertainties and impart both quantifiable errors and some element of additional uncertainty over the reliability of the genetic clock. Nevertheless, when such studies are applied to the timing of the first successful AMH dispersals out of Africa (e.g., Forster 2004) the genetic clock, based on mtDNA evidence, suggests that an initial AMH L1 haplotype was present in parts of Africa between 200,000 and 100,000 BP (in line with archaeological evidence). There are only a limited number of African populations that retain the initial lineage of the earliest AMH haplotype, and these are restricted to a small number of East, West and southern African populations (Forster, 2004). Genetic dating suggests that sometime between 80,000 and 60,000 BP that a second wave of

African colonisation took place with the spread of L1, L2 and L3 populations within Africa (haplotypes still present in African population gene trees). Non of these haplotypes is found outside African genetic trees, but two new haplotypes, M and N which derive from the L3 haplotype, appear in populations in the near East at this time, and it is thought that all non-African populations are descended from the L3 haplotype via M and N (Forster, 2004).

Building on the genetic evidence for a major dispersal sometime between 80,000 and 60,000 BP, Mellars (2006) has examined the published archaeological evidence from South Africa at this time and suggests that between 75,000 and 55,000 BP there was a significant increase in technological diversity during the African Middle Stone Age and evidence for cultural complexity, such as the presence of pierced shell beads. Mellars suggests that both the population dispersal and technological evolution are related to major environmental transitions. This hypothesis suggests that the transition between the warm, interglacial episode MIS 5 and the cold, glacial episode MIS 4, dated to ~73,000 BP, possibly coupled with the effects of the Toba super eruption (~74,000 BP) may have induced adaptive stress on African populations and stimulated both technological and cultural developments along with population movements, with the final phase of this being the successful migration out of Africa between 70,000 and 60,000 BP.

This is an interesting and important hypothesis that requires due consideration. The genetic dating puts the dispersal of AMH between 80,000 and 60,000 BP and in Mellars' model the climatic transitions between 80,000 and 70,000 BP coincide with technological and economic changes in East and southern Africa, followed by expansion between 70,000 and 60,000 BP. This would suggest that a potentially important driver of the technological developments was the transition between interglacial (MIS 5) and glacial (MIS 4) conditions. For African environments, this transition is characterised by increased aridity and reduced rainfall, based on orbitally tuned palaeorainfall data from southern Africa (Partridge et al., 1997). Some support for a coincidence between this transition and technological innovations is the dating by the optically stimulated luminescence (OSL) technique for new technological innovations at Blombos cave to 76,800 ± 3,100 BP and by thermoluminescence on burnt material to 74,000 ± 5,000 BP (Villa et al., 2009). Both of these age estimates overlap with the orbital chronology for the MIS 5/4 transition. This would then mean however that population expansion in Africa then followed during MIS 4.

There are however, some issues with this hypothesis. The first is that MIS 4, with the onset of drier conditions, is not necessarily the most likely setting for population expansion. Widespread aridification in Africa would, in all likelihood, lead to the expansion of the Sahara and make population re-expansion into northern Africa difficult. Moreover, due to the inherent dating uncertainties it is not possible to know if the population expansion and technological innovations took place in MIS 5 or MIS 4. An equally possible scenario is technological innovation and population expansion before the MIS5/4 transition within Africa, with a subsequent movement of populations northwards out of Africa, perhaps driven by increasing aridity in MIS 4. In order to consider this issue, and the nature of AMH occupation of Africa and subsequent movements out of Africa, it is necessary to first consider the North African archaeological evidence for this time period.

4.2 North African archaeology and environment: human dispersal and adaptation in MIS 5

The archaeological evidence from North Africa suggests that despite genetic evidence linking the first successful moves out of Africa with late MIS 5 and early MIS 4, there were well-established AMH populations in North Africa during MIS 5. Recent excavations

in sites in Morocco and Libya, building on work dating back to the 1950s, have questioned both the timing of human dispersal and the timing and geographical range of the cultural developments in Africa. These studies focus on (admittedly sparse) skeletal evidence for human presence across North Africa during MIS 5, but also the appearance in the archaeological record of an important technological industry known as the Aterian (Barton et al., 2009), which is widespread across North Africa. Importantly, this technology is linked with the presence of the shell bead ornaments (Bouzouggar et al., 2007), echoing the evidence for cultural innovations in southern Africa. Recently, new dating programmes in western North Africa, particularly Morocco, have focused on dating the first appearance of the Aterian. At one key site, Dar es-Soltan I (Rabat, Morocco) a new suite of OSL dates through the sequence place the earliest identifiably Aterian tools at between 106,000 ± 7,400 and 119,100 ± 6,400 BP, with a direct age on Aterian bearing sediments of 114,700 ± 6,400 BP. These dates place the appearance of the Aterian squarely in MIS 5 and are supported by geomorphological evidence, with the first appearance of lithic material in the site being found immediately on top of a marine highstand deposit, indicating that occupation occurred shortly after MIS 5e or 5c.

Other sites in Morocco are also providing important evidence of early and geographically widespread cultural innovations in Africa. The site of Grotte des Pigeons (Taforalt, Morocco) has yielded stratigraphically secure evidence of shell bead manufacture within Middle Palaeolithic strata. Recent dating of this sequence, again by OSL, places the shell bead deposits throughout a stratigraphical horizon dating from 84,500 to 60,100 BP. Importantly, the shell beads first appear at the base of this sequence and are bracketed by OSL ages of 84,500 ± 4,400 to 74,500 ± 5,700 BP (Bouzouggar et al., 2007). Taking these sites together, along with the spread across North Africa of the Aterian industry, Barton et al. (2009) have argued that modern behaviour and indeed the (admittedly limited) AMH skeletal material associated with the Aterian, suggest an earlier and much more geographically widespread development of technological and cultural innovations than in the model proposed by Mellars (2006). Importantly from the point of view of this study, this activity is all within the warmer and wetter conditions prevailing during MIS 5, and in the case of the Aterian much earlier than the 60,000–80,000 BP (Forster, 2004) genetic timescale for modern human expansion within and immediately outside Africa.

As well as the archaeological evidence for AMH presence and cultural development in North Africa during MIS 5, recent palaeoenviroenmental studies have examined MIS 5 environments to test potential migration routes into North Africa. Many studies have hypothesised either a route from South and East Africa along the Nile corridor, or alternatively a migration via a southern route into Arabia. Given, however, the dating and the geographical range of the Aterian across North Africa and in the Libyan Sahara (Osborne et al., 2008) new studies are examining the possibility that environmental conditions in North Africa may have facilitated AMH migration across the Sahara itself. It is well known that humid conditions have prevailed in the Sahara in the past, with evidence for major palaeo-lake basins occurring across this region (Armitage et al., 2007). OSL dating of remnant shoreline features within the catchment of these major lakes reveals evidence for extensive Saharan drainage network formation during MIS 5, between c.100,000 and 110,000 BP (Armitage et al., 2007). Although this is clear evidence for a wetter Sahara there is as yet no clear evidence of the major lake formation that appears in the early Holocene. Nevertheless, it is evidence of a wetter Sahara and is supported by recent isotopic analyses on marine core ODP Hole 971A, in the eastern Mediterranean (Osborne et al., 2008). This study suggests that during peak oxygen isotopic excursions correlated to the warmest

part of the last interglacial (MIS 5e: 120,000–125,000 BP), there is a significant shift in neodymium isotopes. By comparing these values to neodymium isotopes from shells in lake sediments from Libya, the authors suggest that during MIS 5 there is evidence for increased runoff into the Mediterranean from an extensive catchment draining a significant part of the Sahara (Osborne et al., 2008). Thus, there is growing palaeoenvironmental data to support the idea of wet corridors across the Sahara during the last interglacial, coupled with evidence for major human expansion into North Africa within MIS 5. Furthermore, when the recent dating of the Aterian is taken into consideration, this expansion seems to be quite early in the last interglacial, much earlier than suggested by the genetic dating and certainly much earlier than MIS 4, as suggested by Mellars (2006). Consequently, it is possible that numerous dispersal routes into North Africa and beyond may have existed and that these may have facilitated a migration out of Africa earlier than previously proposed.

Clearly there is a key question here over the environmental forcing on AMH dispersal and subsequent migrations out of Africa. The simplest solution would be an MIS 5 expansion stimulated by more humid interglacial conditions and a subsequent push into coastal zones during the return of dry conditions at the onset of MIS 4. While such an argument may seem valid it is necessary to consider the evidence of AMH presence in the immediate periphery of Africa, see below, but first we have to consider the discrepancy between the genetic dating of human expansion with the North African archaeological evidence. The archaeological evidence suggests that population expansion, as evidenced by the Aterian, and cultural innovations such as pierced shell beads occurred during MIS 5 and in all probability early in this warm episode. The genetic evidence does not point to a major expansion until late in MIS 5 or indeed into MIS 4, with a definite age of successful expansion out of Africa at 54,000 ± 8000 BP (Forster et al., 2001). Part of the problem may simply be a case of comparing 'apples and oranges', genetic evidence for population expansion does not necessarily imply recognisable archaeological change. Indeed the skeletal hominid record linking the Aterian with modern humans is relatively slim and not entirely accepted (Barton et al., 2009). It does, however, seem surprising that a major genetic expansion leaving footprints on modern populations would not leave an imprint on the archaeological record (Barton et al., 2009). Although there are later cultural changes in North Africa, the appearance of the Aterian and cultural developments such as the use of shell beads are the closest in time to the proposed genetic age of expansion. Moreover, there is clear evidence of humans outside Africa in the Levant in very good agreement with the Aterian appearance. Perhaps then the issue is one of dating and archaeological visibility.

The dating of genetic events has inherent assumptions over rates of genetic mutation and the current estimate represents a best available calibration. It is possible that the calibration of the genetic clock is simply incorrect and that revision in the future will bring it into line with the archaeological evidence in North Africa and the Levant. As Forster (2004) points out, however, the genetic mutation rates are calculated using archaeological evidence as a calibration. Therefore, changes in archaeological ages will result in changes in the genetic date for other colonisation events, such as the peopling of the Americas, which is relatively secure. Nevertheless, this remains a possibility. There is also of course the problem that the archaeological chronologies themselves may change if improvements in direct dating of cave sediments, such as the single grain method in OSL dating, force a revision of the archaeological chronologies. It is still possible to have a MIS 5 onset for the Aterian in North Africa, as the stratigraphy of sites such as Dar es-Soltan I clearly imply, and a fit with the older range of the genetic chronology for the expansion within Africa, even if the dating of the first non-African sites in the

Levant cannot be reconciled. It is clear, however, that this is a crucial area of research for archaeology and palaeoenvironmental studies and it is likely that future studies tracing the exact timing of the spread of the Aterian within northern Africa, as well as extensive dating studies on Aterian and other sites across North Africa are likely to shed more light on this question.

4.3 Anatomically modern human' migration out of Africa

As with the above debate over the role of environmental change within Africa and the earliest evidence for AMH immediately outside Africa, there are also significant questions over the timing of human migration around the globe. The same is true of the role of environmental change on the apparent replacement of archaic hominid lineages, such as the Neanderthals, by AMH. Genetic evidence, based again on genetic differences in modern population lineages, suggests that during the dispersals both within Africa and out of Africa that human populations may have gone through genetic bottlenecks, where the spread of AMH is based on substantial population reduction followed by an expansion from a small group (see Ambrose, 1998). Archaeological models for such a process can follow a model of a weak population dispersal that is prone to contraction due to environmental change (Ambrose, 1998) or multiple population expansion events from isolated groups at different times.

Reviewing genetic and archaeological evidence alongside palaeogeographical and environmental evidence, Lahr and Foley (1998) have suggested that more than one genetic bottleneck may have taken place in the later evolution of our species. They argue for an initial population bottleneck in Africa during MIS 6, sometime between 200,000 and 130,000 BP. This, they suggest, is followed by population expansion within Africa and on to the Levant during the first part of MIS 5, after 130,000 BP. This proposed expansion explains the MIS 5 occupation of North Africa by modern humans, and the spread of the Aterian, as well as AMH presence immediately outside Africa. This is followed by a second contraction in population and another bottleneck, leaving isolated groups within Africa. They note, for example, that in the Levant AMH populations are replaced by Neanderthal populations during MIS 4, with Neanderthal remains and associated lithic material from this region dating to 70,000–53,000 BP. Following this population decline, Lahr and Foley (1998) have suggested two isolated population centres with AMH groups occurring to the north and south of the Sahara. This, it is suggested, led to two migration routes from Africa, with an expansion along a southern route from sub-Saharan Africa taking a route through Asia and eventually to Australia and a later expansion after 50,000 BP. While there is little direct evidence in the form of skeletal remains, they argue that such a model, driven largely by environmental variability, best explains both the genetic diversity of modern populations and the affinities between Asian tool technology and the African Middle Stone Age industries. Future testing of this hypothesis may be difficult but information relevant to this hypothesis may be gleaned by ongoing work to reliably date the timing of dispersal across Asia and ongoing investigations into cultural continuity and the patterning of site occupation in the rich North African sites in Morocco, Tunisia, Libya and Egypt.

4.4 Anatomically modern human genetic bottlenecks, volcanic super-eruptions, and abrupt global cooling

A significant addition to the debate over the possible role of genetic bottlenecks in the pattern of human dispersal and the associated role of environmental forcing comes in the form of a hypothesis first advanced by Ambrose (1998). In this hypothesis, Ambrose acknowledges the genetic evidence for a

bottleneck and the idea of a southern dispersal route across the Red Sea, which he suggests would most likely occur during MIS 4 (74,000–60,000 BP), when lowered global sea levels would have made this more likely. He then proposes a bottleneck resulting in isolated populations within and outside Africa, which subsequently expanded, thus explaining the genetic diversity of modern populations. Ambrose (1998, 2003) has suggested that it is likely that the driver of this population bottleneck may have been partly the onset of cooling during MIS 4 but that an additional significant factor was possibly the impact of the largest known volcanic eruption in the last 2 million years, the Toba Ash, or Youngest Toba Tuff, an Indonesian eruption dated to 73,000 ± 5,000 BP by argon–argon dating. The eruptions left widespread ash deposits from the South China to Arabian seas (Petraglia et al., 2007). The scale of this eruption has led some to suggest that it may have induced a volcanic winter early in MIS 4.

Ambrose suggests that several palaeoenvironmental archives support this. In particular, the occurrence of a large sulphate peak (indicative of a major eruption event) in the Greenland ice core GISP2, has been dated by ice core counting to ~71,000 BP. The sulphate peak occurs at the end of a short lived warm interval and is immediately followed by rapid cooling as indicated by oxygen isotope records in the ice core. This evidence is suggestive of a significant cooling event brought about by the Toba eruption. This argument was developed further by Williams et al. (2009), who addressed some of the chronological issues associated with the initial argument. The GISP2 chronology has significant errors by this time and has been 'calibrated' by matching the GISP2 climate record to Chinese speleothem climate data which is independently dated by uranium series dating. On this basis, Williams et al. (2009) argue for a close synchronisation between the sulphate peak, the onset of ~2,000-year cooling in the GISP2 record and the argon–argon age for the Toba Ash. This, along with palaeoenvironmental data from a marine core from the

Indian Ocean, led Williams et al. (2009) to reinforce the suggestion of a volcanically induced climatic deterioration caused by the Toba eruption.

This important hypothesis has generated significant controversy. There are several key issues on which the hypothesis rests that are open to question. The first is the exact role of the Toba eruption in climate change. Rapid cooling events are not uncommon during the last glacial stage with oscillations between short-lived warm and cold stages, lasting usually 500–2000 years, occurring throughout this period (Dansgaard/Oeschgar cycles; see Chapter 20). Exactly how these temperature changes, documented in the North Atlantic, affect low latitudes is a matter of much debate, but the key point here is that the cooling recorded immediately after the Toba Ash is not significantly greater than that associated with other stadial cooling events that are known to have occurred during the late Quaternary. Williams et al. (2009) argue that, in fact, Greenland stadial 20, which is postulated as being influenced by Toba, is significantly colder than later stadial episodes. However, this is the first stadial within the GISP2 that is recognised within MIS 4 and thus occurs during an orbitally driven glacial stage. Later stadial episodes, such as stadial 13 cited by Williams et al. (2009) sit within the warmer MIS 3, where long-term orbitally driven climate will influence the magnitude of abrupt climatic cycles. In fact, the cooling that occurs during stadial 20 does not appear to be significantly colder than other cold events within MIS 4 or during the most recent glacial stage, MIS 2. Moreover, it is always difficult to be sure how reliably climatic teleconnections between regions can be established. The tying of the Greenland ice core chronologies to records from caves in China, assumes a long-range climatic synchronisation that has yet to be established. Comparison for example of the most recent Greenland ice core, NGRIP, on an independent layer counted chronology, with the Chinese speleothem data (Svensson et al., 2007) suggests that there are both similarities

and differences between the records with respect to both the timing of climatic events and the pattern of the climatic signal.

The most significant response to the Toba hypothesis has, however, come directly from the archaeological record. Working in the Jwalapuram area of southern India, Petraglia et al. (2007) have examined the archaeological record of the region to test the impact of the Toba eruptions directly. At several sites they present lithic assemblages with strong affinities to the African Middle Stone Age tools. These industries are found immediately below and above deposits of the Toba ash and are dated by OSL to 77,000 ± 6,000 BP below the Toba ash and 74,000 ± 7,000 BP above the Toba ash. Moreover, there is no evidence for cultural discontinuity in the lithic assemblages, indeed the layers immediately above the ash layer were still rich in volcanic material, suggesting no depositional hiatus. All of this suggests that AMH were in India at the time of the MIS 5/4 transition, placing the first southern dispersal evidence to 77,000–74,000 BP, significantly after the first evidence of AMH in the Levant, but significantly earlier than dispersals further north into Eurasia. Most importantly, however, there is clear evidence in southern India for cultural continuity either side of the Toba eruption, significantly at odds with ideas of a volcanically induced cooling leading to a genetic bottleneck.

Where does this leave ideas of an environmentally induced population bottleneck to explain the genetic evidence? One answer lies with the phased dispersal model of Lahr and Foley (1998). Clearly, though only in a small number of sites at present, this model provides evidence for human presence as far away as India as early as c.77,000–74,000 BP. If confirmed by a significant number of additional sites it is possible that their model of isolated groups of AMH dispersed in different directions will be supported. Indeed around this time the onset of MIS 4 is likely to have limited the potential for trans-Sahara migration, perhaps leading to the isolation of northern and southern African populations which may, in turn explain the wide genetic diversity in modern humans. The possibility that this occurs at the end of MIS 5, after the African expansion, along with the North African Aterian industry and the evidence for cultural innovations such as pierced shell beads, the onset of which is dated in North Africa to somewhere between ~115,000 BP for the Aterian and 84,500 ± 4,400 to 74,500 ± 5,700 BP for the North African shell beads (Barton et al., 2009; Bouzouggar et al., 2007) may suggest a North African colonisation of Eurasia. The Aterian ages do not exactly agree with the genetic clock age for the spread of L1–3 haplotypes or the idea of a major, second within-African migration, but at least it seems likely that this migration was an MIS 5 phenomena, with the first southern evidence for migration at the end of MIS 5. The evidence, however, for a northern dispersal, possibly from a North African origin, is much later and has led to one of the most long-running debates about the relationship between climate change and the interaction, or otherwise of different hominid species.

4.5 Out of Africa and into Europe: anatomically modern humans and Neanderthals in the last glacial

The exact timing and nature of AMH dispersal into Eurasia and the role our species played in the disappearance from the archaeological record of Neanderthal skeletal forms has been the subject of debate for nearly a century. For much of this time the debate has partly focussed on the question of whether AMH are the sole genetic ancestors of modern humans or whether there is evidence for some genetic admixture between populations of different *Homo* species. While some studies still perpetuate this debate, such as the apparent hybrid AMH/Neanderthal child from Lagar Velho, in Portugal (Duarte et al., 1999), the majority of modern genetic and ancient DNA studies, along with a now significant body of skeletal evidence, suggest that

Neanderthals are a separate species of human to AMH (Stringer, 2002). How the appearance of AMH has influenced the apparent regional replacement of Neanderthals remains a topic of significant debate and the role of climate change has been just as contentious.

The exact timescale for the replacement of Neanderthals by AMH is debated, not least because the appearance of AMH in Europe dates to somewhere before 40,000 BP, which is towards the maximum limit of the radiocarbon dating technique, and in European contexts this is still the most appropriate dating method. Radiocarbon or ^{14}C dating is highly suitable for examining the presence or absence of human groups as it is a method suited to directly dating organic carbonates, based on the radioactive decay of the ^{14}C isotope (carbon has three isotopes ^{12}C, ^{13}C and ^{14}C). Of these only ^{14}C is radioactive and is produced by cosmic ray interaction in the atmosphere. The radioactive ^{14}C quickly combines to produce $^{14}CO_2$ and through the carbon cycle is rapidly incorporated into the atmosphere, biosphere and oceans. The half-life of ^{14}C is known and modern accelerator techniques can measure the ratios of the three isotopes of carbon with high precision, even on small samples. In theory this is, therefore, the ideal technique for examining the relative timing of AMH appearance and Neanderthal disappearance. Moreover, as an absolute technique it should be possible to compare these dates with high precision climatic data. There are, however, two problems with ^{14}C dating that are particularly relevant here. First, the production of ^{14}C in the atmosphere is not constant over time. Due to the variability of solar activity and the amount of cosmic rays reaching the atmosphere (due in part to changes in the strength of the Earth's magnetic field) the production of ^{14}C varies significantly and rapidly.

This severely limits the ^{14}C technique as an independent absolute dating technique. Fortunately, by comparing radiocarbon dates on materials of known age, such as tree rings, a calibration curve can be constructed that can be used to correct for this change in production. Unfortunately for Palaeolithic archaeology, the calibration curve beyond the global tree ring chronology (~12,500 BP) is not fully resolved. Until recently, an international consensus curve (INTCAL04; Reimer et al., 2004) based on paired uranium series and ^{14}C dates on corals, and radiocarbon dates on laminated marine sediments gave a calibration back to ~24,000 calendar years BP (calibrated radiocarbon dates are usually referred to as cal. BP). This curve was not necessarily correct but it was internationally accepted because there was consensus between different calibration data sets. Beyond that limit researchers have been using a mix of different curves that have not been internationally ratified, and can have differences of thousands of years. A new curve, ratified to ~50,000, is now available (INTCAL09; Reimer et al., 2009) but this has not had time yet to effect the discussion, although much of the underlying data used in this curve has been used in some of the studies on the AMH/Neanderthal transition, which in Europe is often termed the Middle/Upper Palaeolithic Transition. There is a problem even here, however, as the curve may still not be correct and is likely to be revised in future. Thus, the discussion of chronology is conditional on the revision of the calibration curve.

A second and equally important problem for radiocarbon dating in this period is that radiocarbon has a relatively short half-life (5,750 years) and is a very low abundance isotope at modern levels. This means that the levels of contamination of a sample by younger, modern carbon, required to return a false date, are very small. Radiocarbon specialists have been attempting to improve their ability to remove carbon contamination from samples by a range of methods, particularly on bone, for decades and the dates available for both hominid bone, either of Neanderthal or AMH skeletal material, or more usually cut-marked animal bone, as well as dates on charcoal from hearths, have a range of different pre-treatment strategies of varying quality. The importance of this variability is only recently being realised by many researchers in the field but has significant consequences for the Middle/Upper Palaeolithic.

4.6 Final anatomically modern human dispersal and the demise of Neanderthals in Europe

Currently, the best available chronology for the appearance of AMH in Europe places their arrival in southeast Europe somewhere shortly before 40,000 cal. BP. While the chronology should be treated with caution, due to the issues outlined above, there appears to be a fairly rapid subsequent colonisation of Europe, with similar ages for the first appearance of upper Palaeolithic tool industries, known as Aurignacian, which are often taken as the first definitive appearance of AMH populations (e.g., Mellars, 2006). There is some evidence for an overlap in time between AMH and Neanderthal populations as the final ages for Neanderthal (late middle Palaeolithic) tool industries and sites overlap with the first appearance ages of the upper Palaeolithic, associated with AMH. Due to the difficulties of radiocarbon dating materials that are close in age to the effective radiocarbon limit the situation is currently unclear (Joris and Street, 2008). In some regions, there is accepted evidence for some technological similarities between elements of AMH Aurignacian technology and the final toolkits that are thought to be middle Palaeolithic and generally associated with the Neanderthals.

In France, the Chatelperronnian industry was initially seen as upper Palaeolithic but has recognisable middle Palaeolithic features and is now widely regarded as a product of the final European Neanderthals (e.g., Mellars et al., 2007). Whether the Chatelperronian and other transitional industries represent some form of interaction between late Neanderthals and AMH is a much-debated issue. Some argue that dating uncertainties mean that many sites that suggest an overlap between these industries, and by extension interaction between the species, are unreliable. Joris and Street (2008) argue that taking these dating uncertainties into account the current best model is a rapid replacement of Neanderthals by AMH. They point out that there are few directly dated early AMH or late Neanderthal fossils and examining the evidence in the light of recent advances in radiocarbon dating the pattern suggests no overlap between the fossil remains of the two species. They also suggest that all of the transitional industries predate or only very slightly overlap the first appearance of the upper Palaeolithic.

A key site in this debate is Grotte des Fées de Châtelperron, the type site for the French Chatelperronian. Gravina et al., (2005) and later Mellars et al. (2007) published evidence for intrastratification of Chatelperronian and Aurignacian levels, suggesting sequential periods of use by Neanderthal and AMH groups, dated to ~41,000 to 35,000 radiocarbon BP, based on new radiocarbon dates on material excavated from the site during the 1950s. Zilhao et al. (2006) have conversely argued that there are stratigraphic uncertainties at the site and they suggest that some or all of the dates at the site are from samples recovered from the back-dirt of previous investigations. This has been refuted by Mellars et al. (2007) who discuss a range of factors including the reporting of the original excavation and the consistency of their chronology with expected dates from Chatelperronian sites. Similar questions have also arisen in understanding the regional distribution of late Neanderthal sites, with some dates for northern Iberia suggesting an earlier AMH occupation of this region than neighbouring France, while southern Iberia has been suggested by some as a late refuge for the Neanderthals, with very late ages from Gorhams cave, Gibraltar (Finalyson et al., 2006). While the early ages for northern Iberia have been criticised by D'Errico and Sanchez-Goni (2003) they accept the late dates for Neanderthals in the south, and moreover suggest that this distribution was partly influenced by the abrupt climatic downturn

associate with a major Heinrich Event at around 40,000 BP.

4.7 Neanderthal extinction and Heinrich Events

While the previous section indicates that the fine detail of the Neanderthal AMH transition is fraught with chronological uncertainties, the process across most of Europe took place somewhere around 35,000 radiocarbon BP. The recent publication of a new calibration curve, IntCal09 (Reimer et al., 2009), now allows these early AMH and late Neanderthal ages to be calibrated and gives a broad calendar time estimate for the transition of ~43,000–42,000 BP for the appearance of the last Neanderthal industries such as the Chatelperronian and ~41,000–39,000 BP for the appearance of the Aurignacain across most of Europe, with a late Neanderthal population surviving in southern Iberia possibly as late as 28,000–26,000 radiocarbon BP (~33,000–30,000 cal BP). During this time period, well-known episodes of significant climate cooling, Heinrich (H) Events 2, 3 and 4, are reported in marine archives, and are thought to correspond to cool episodes recorded in the latest Greenland ice core record (NGRIP; Svensson et al., 2006). The latest ice core age for the cooling associated with H4 is ~40,000–38,000 BP. Several authors have examined the potential role of the H4 event in the Neanderthal AMH transition. Joris and Street (2008) suggest that the H4 event is coincident with the onset of the Aurignacian and that all of the transitional or early upper Palaeolithic technologies are recorded prior to this prolonged period of cooling. They and others (Fedele et al., 2008) suggest that the H4 event would have led to considerable environmental impact on continental Europe and restricted hominid ranges. For Iberia, however, D'Errico and Sanchez-Goni (2003) suggest that the impact of H4 on AMH populations delayed their arrival into southern Iberia, as the reduced biomass of arid southern Iberia may have limited their ability to occupy this region, and thus may have led to the prolonged Neanderthal presence in the region. In their view the final move by AMH into southern Iberia after the H4 event led to the final disappearance of the Neanderthals. Conversely it has been argued by Jimenez-Espejo et al. (2007) that the final disappearance of Neanderthals from Iberia may coincide with a climatic downturn correlated with the later H2 cold event. This has been heavily criticised by Tzedakis et al. (2007) who compare the range of late dates for Neanderthals from Gorham's Cave, Gibraltar, with the radiocarbon ages from the high-resolution marine palaeoclimate archive from the Carriaco basin. Their comparison suggests no correlation, even on the youngest (and contentious) southern Iberian ages and the H2 event, and no real correlation between any of the late Neanderthal ages and Heinrich events. This lends support to the argument of D'Errico and Sanchez-Goni (2003) that the incursion into southern Iberia by AMH after the H4 event led to the final demise of Neanderthals.

A major problem in all of these hypotheses is the aforementioned dating uncertainties, both in the archaeological and environmental records. While it may remain difficult, however, to fully address the role of climate or otherwise on the final extinction of the Neanderthals there is more hope of resolving the question of their disappearance from Europe outside of Iberia and the role of the H4 event in this disappearance. This is due to the fortunate coincidence of the H4 event with the widespread dispersal of volcanic ash (tephra) from a large volcanic eruption the Campanian Ignimbrite (CI). This eruption emanates from the Campe Flegrei volcanic complex in Italy and is very precisely dated to $39,282 \pm 110$ BP by $^{40}Ar/^{39}Ar$ of multiple proximal outcrops of the eruption (de Vivo et al., 2001). Tephra correlated to this eruption has been found in several important archaeological sites containing middle Palaeolithic, transitional and Aurignacian industries. Moreover, tephra from the eruption is located in marine records across the Mediterranean which contain important climatic records of this time. Within the marine sequence the CI occurs in association with the early part of a prolonged cooling episode. Comparison of the $^{40}Ar/^{39}Ar$ age of the CI with the latest Greenland ice core records indicates that this is the cooling usually associated with the H4 event (Figure 38.2). In sites where this ash is reported the pre-Aurignacian industries are stratigraphically below the CI layer. This has been taken by Joris and Street (2008) as support for the idea that the H4 event is coincident with the transition to the Aurignacian and the disappearance of Neanderthal populations. While the occurrences of the CI as a visible ash layer are currently limited by site formation factors, this relationship holds in sites as far afield as Italy and Kostenki in Russia.

Another important lesson from this tephra, however, is that dating uncertainties are significant even for relatively modern and normally acceptable radiocarbon samples. In a comparison of radiocarbon ages from archaeological sites associated with the CI and the radiocarbon ages and $^{40}Ar/^{39}Ar$ ages from proximal deposits of the ash, there were significant discrepancies between the archaeological ages and the expected ages based on the tephra (Blockley et al., 2008). This relationship was apparent using all currently available radiocarbon calibration data. This study suggested that either the calibration data for this time period, which is now widely used, are incorrect, or that ^{14}C ages, which in any site not containing the CI would be acceptable, are systematically too young. Confirmation of the latter problem has been found by Higham et al. (2009) who re-dated sites in northern Italy using refined methods for cleaning old charcoal samples. The revised dates demonstrated a systematic difference between samples pre-treated using standard methods and the revised protocol, with the standard samples being systematically younger. This, taken with the study of Blockey et al. (2008), suggests that there are serious concerns over the radiocarbon chronology for the middle to upper Palaeolithic transition

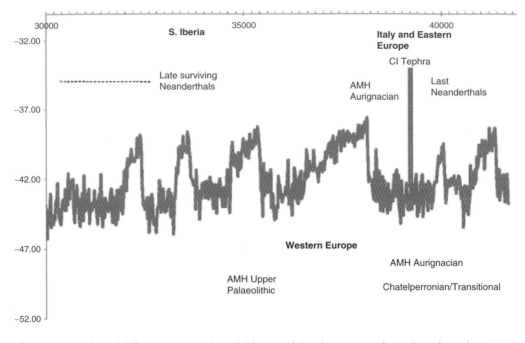

Figure 38.2   The middle to upper Palaeolithic transition in Europe plotted against the NGRIP ice core record.

across Europe, where many sites have chronologies based on either charcoal prepared using standard techniques, or dates on samples of faunal and occasionally hominid bone that are at present based on a mix of different pretreatment strategies. While the work on the CI and northern Italy points to the need for systematic improvement in the dating of charcoal, equally important developments in the dating of archaeological bone are also having a significant impact on the chronology of the middle and upper Palaeolithic (e.g., Jacobi and Higham, 2008).

While it will be some time until re-dating of sites across Europe allows a more confident assessment of the timing of this transition, as well as ongoing improvements in radiocarbon calibration data, the CI and its stratigraphical relationship to both climatic transitions and cultural transitions from sites where it is present, suggest a key hypothesis to be tested. Figure 38.2 follows the arguments of Joris and Street (2008) extrapolating from the sites where the CI is present, and suggests that for Italy and Eastern Europe the middle to upper Palaeolithic transition

and the rapid appearance of the Aurignacian was closely associated with the CI eruption and, by extension, the H4 event. Following the normal logic that these two industries are useful proxies for AMH and Neanderthals this suggests that within much of Europe the disappearance of Neanderthals was coincident with two events, the arrival of AMH and the onset of a significant ~2,000-year-long cold episode. This is still only a hypothesis at this stage, which requires testing with improved dates and the identification of more examples of the CI tephra in archaeological sites and terrestrial palaeoenvironmental sites within Europe (Lowe et al., 2007). Nevertheless, it does raise the possibility that one of the most iconic transitions between hominid species was in part driven by late Pleistocene environmental change.

5 CONCLUSION

After the period discussed previously, AMH were the only hominids on the planet, with the exception of the geographically isolated *H. Florensis*, which appeared to continue in isolation until the late Quaternary. For the rest of the globe, however, our species was the only extant member of our genus. The evolution of our genus and species over the preceding 4 Ma, at least, has taken place against a backdrop of major environmental change, which intensified during the Quaternary after the onset of northern hemisphere glaciations. While this chapter has highlighted the significant difficulties and debates often associated with delineating the precise role of environmental change in the process of evolution, speciation, dispersal and extinction, it is clear that this is one of the most interesting and important questions in archaeological and environmental research. Moreover, while it is beyond the scope of this chapter, it is worth noting that abrupt environmental change has continued to have an impact on human societies. In the late Pleistocene, key events such as the abandonment of northern Europe during the Last Glacial Maximum and the current and later recolonisation towards the end of the Pleistocene and the onset of the Holocene is seen as clear evidence of a major environmental influence on established AMH population (Blockley et al., 2006). Moreover in the late Pleistocene and the Holocene epoch, other important events, such as the colonisation of the America's by AMH in the late Pleistocene, the onset of agriculture around the Pleistocene/Holocene transition, and even the later collapse of fairly complex societies have all been linked to climatic and environmental change.

REFERENCES

Allen J. A. 1877. The influence of physical conditions in the genesis of species. *Radical Review* 1: 108–140.

Allen J. R. M. and Huntley B. 2009. Last Interglacial palaeovegetation, palaeoenvironments and chronology: a new record from Lago Grande di Monticchio, southern Italy. *Quaternary Science Reviews* 28: 1521–1538.

Ambrose S. H. 1998. Late Pleistocene human population bottlenecks, volcanic winter, and differentiation of modern humans. *Journal of Human Evolution* 34: 623–651.

Ambrose S. H. 2003. Did the super-eruption of Toba cause a human population bottleneck? Reply to Gathorne-Hardy and Harcourt-Smith. *Journal of Human Evolution* 45: 231–237.

Armitage, S. J., Jasim, S. A., Marks, A. E., Parker, A. G., Usik, V. I., and Uerpmann, H. P. 2011. The Southern Route "Out of Africa": Evidence for an Early Expansion of Modern Humans into Arabia. *Science*, 331: 453–456.

Armitage S. J., Drake N. A., Stokes S., El-Hawat A., Salem M. J., White K., Turner P. and McLaren S. J. 2007. Multiple phases of North African humidity recorded in lacustrine sediments from the Fazzan Basin, Libyan Sahara. *Quaternary Geochronology* 2: 181–186.

Ashton N. M., Cook J., Lewis S. G. and Rose J. 1992. *High Lodge: Excavations by G. de G. Sieveking, 1962–68 and J. Cook, 1988*. London: British Museum Press.

Ashton N., Parfitt S. A., Lewis S. G., Coope G. R. and Larkin N. 2008. Happisburgh Site 1 (TG388307), in Candy I., Lee J. R. and Harrison A. M. *The Quaternary of Northern East Anglia: Field Guide*. Quaternary Research Association.

Banks W. E., d'Errico F., Peterson A. T., Kageyama M., Sima A. and Sanchez-Goni M. F. 2008. Neanderthal extinction by Competitive Exclusion. *PLOS ONE* 3 (12): e3972.

Barton R. N. E., Bouzouggar A., Collcutt S. N., Scwenninger J. L. and Clark-Balzan L. 2009. OSL dating of the Aterian levels at Dar es-Soltan I. (Rabat, Morocco) and implications for the dispersal of modern *Homo sapiens*. *Quaternary Science Reviews* 28: 1914–1931.

Bar-Yosef O. and Belfer-Cohen A. 2001. From Africa to Eurasia early dispersals. *Quaternary International* 75: 19–28.

Bar-Yosef O., Van Der Meersch B., Arensburg B., Belfercohen A., Goldberg P., Laville H. et al. 1992. The excavations in Kebara Cave, mt-Carmel. *Current Anthropology* 33: 497–550.

Bergmann C. 1847. Ueber die Verhaltnisse derWarmeokonomie der thiere zu ihrer grosse. *Gottinger Studien* 3: 595–708.

Bettis A., Milius E., Carpenter A.K., Larick R., Zaim Y., Rizal R. et al. 2009. Way out of Africa: Early Pleistocene paleoenvironments inhabited by Homo erectus in Sangira. Java. *Journal of Human Evolution* 56: 11–24.

Blockley S. P. E., Ramsey C. B. and Higham T. F. G. 2008. The Middle to Upper Palaeolithic Transition: dating, stratigraphy and isochronous markers. *Journal of Human Evolution* 55: 764–771.

Bobe R. and Behrensmeyer K. 2004. The expansion of grassland ecosystems in Africa in relation to mammalian evolution and the origin of the genus Homo. *Palaeogeography Palaeoclimatology Palaeoecology* 207: 399–420.

Bobe R., Behrensmeyer K. and Chapman R. E. 2002. Faunal change, environmental variability and late Pliocene hominin evolution. *Journal of Human Evolution* 42: 475–497.

Bond G., Broecker W., Johnsen S., McManus J., Labeyrie L., Jouzel. J. and Bonani G. 1993. Correlations between climate records from North-Atlantic sediments and Greenland Ice. *Nature* 365: 143–147.

Bouzouggar A., Barton N., Vanhaeren M., d'Errico F., Collcutt S., Higham T. et al. 2007. 82,000-year-old shell beads from North Africa and implications for the origins of modern human behaviour. *Proceedings of the National Academy of Sciences of the USA* 104: 9964–9969.

Caldwell D. 2008. Are Neanderthal portraits wrong? Neanderthal adaptations to cold and their impact on palaeolithic populations. *Rock Art Research* 25(1): 101–116.

Candy I., Rose J. and Lee J. R. 2006. A seasonally 'dry' interglacial climate in Eastern England during the early Middle Pleistocene: Palaeopedological and stable isotopic evidence from Pakefield, UK. *Boreas* 35: 255–265.

Carbonell E., Bermudez de Castro J. M., Arsuaga J. L., Diez J. C., Rosas A., Cuenca-Bescos G. et al. 1995. Lower Pleistocene hominids and artifacts from Atapuerca-TD6 (Spain). *Science* 269: 826–830.

Carrion J. S., Yll E. I., Walker M. J., Legaz A. J., Chain C. and Lopez A. 2003. Glacial refugia of temperate, Mediterranean and Ibero-North African flora in south-eastern Spain: new evidence from cave pollen at two Neanderthal man sites. *Global Ecology and Biogeography* 12: 119–129.

Coope G. R. 2006. Insect faunas associated with Palaeolithic industries from five sites of pre-Anglian age in central England. *Quaternary Science Reviews* 25: 1738–1754.

D'Errico F. and Sanchez-Goni M. F. 2003. Neandertal extinction and the millennial scale climatic variability of OIS 3. *Quaternary Science Reviews* 22: 769–788.

Dart R. A. 1925. *Australopithecus africanus*: The man ape of South Africa. *Nature* 115: 195–199.

Dennell R. 2003. Dispersal and colonisation, long and short chronologies: how continuous is the Early Pleistocene record for hominids outside Africa. *Journal of Human Evolution* 45: 421–440.

Dennell R. and Roebroeks W. 1996. The earliest colonisation of Europe the short chronology revisited. *Antiquity* 70: 535–542.

Dennell R. and Roebroeks W. 2005. An Asian perspective on early human dispersal from Africa. *Nature* 438: 1099–1104.

deMenocal P.B. 1995. Plio-Pleistocene African climate. *Science* 270: 53–59.

deMenocal P. B. 2004. African climate change and faunal evolution during the Pliocene-Pleistocene. *Earth and Planetary Science Letters* 220: 3–24.

de Vivo B., Rolandi G., Gans P. B., Calvert A., Bohrson W. A., Spera F. J. and Belkin H. E. 2001. New constraints on the pyroclastic eruptive history of the Campanian volcanic plain (Italy). *Mineralogy and Petrology* 73: 47–65.

Duarte C., Mauricio J., Pettitt P.B., Souto P., Trinkaus E., van der Plicht H. and Zilhao J. 1999. The early Upper Paleolithic human skeleton from the Abrigo do Lagar Velho (Portugal) and modern human emergence in Iberia. *Proceedings of the National Academy of Sciences of the USA* 96: 7604–7609.

EPICA Community. 2004. Eight glacial cycles from an Antarctic ice core. *Nature* 429: 623–628.

Fedele F. G., Giaccio B. and Hajdas I. 2008. Timescales and cultural process at 40,000 BP in the light of the Campanian Ignimbrite eruption, Western Eurasia. *Journal of Human Evolution* 55: 834–857.

Finlayson C. and Carrion J. S. 2007. Rapid ecological turnover and its impact on Neanderthal and other human populations. *Trends in Ecology & Evolution* 22: 213–222.

Finlayson C., Giles Pacheco F., Rodriguez-Vidal J., Fa D. A., Guiterrez Lopez J. M. et al. 2006. Late survival of Neanderthals at the southernmost extreme of Europe. *Nature* 443: 850–853.

Foley R. 2002. Adaptive radiations and dispersals in hominin evolutionary ecology. *Evolutionary Anthropology* 1: 32–37.

Forster P. 2004. Ice Ages and the mitochondrial DNA chronology of human dispersals: a review. *Philosophical Transactions of the Royal Society of London B* 359: 255–264.

Gabunia L., Anton S. C., Lordkipanidze D., Vekua A., Justus A. and Swisher C. C. 2001. Dmanisi and dispersal. *Evolutionary Anthropology* 10: 158–170.

Gabunia L., Vekua A. and Lordkipanidze D. 2000.The environmental contexts of early human occupation of Georgia (Transcaucasia). *Journal of Human Evolution* 38: 785–802.

Gilligan I. 2008. Neanderthal extinction and modern human behaviour: the role of climate change and clothing. *World Archaeology* 39: 499–514.

Gravina B., Mellars P. and Ramsey C. B. 2005. Radiocarbon dating of interstratified Neanderthal and early modern human occupations at the Chatelperronian type-site. *Nature* 438: 51–56.

Green H. S. 1984. *Pontnewydd Cave: A Lower Palaeolithic Hominid Site in Wales: The First Report.* Cardiff: National Museum of Wales.

Grun R., Stringer C., McDermott F., Nathan R., Porat N., Robertson S. et al. 2005. U-series and ESR analyses of bones and teeth relating to the human burials from Skhul. *Journal of Human Evolution* 49: 316–334.

Henderson G. M. and Slowey N. C. 2000. Evidence from U-Th dating against Northern Hemisphere forcing of the penultimate deglaciation. *Nature* 404: 61–66.

Higham T., Brock F., Peresani M., Broglio A., Wood R. and Douka K. 2009. Problems with radiocarbon dating the Middle to Upper Palaeolithic transition in Italy. *Quaternary Science Reviews* 28: 1257–1267.

Holliday T. W. 1997. Postcranial evidence of cold adaptation in European Neandertals American. *Journal of Physical Anthropology* 104: 245–258.

Holliday T. W. 1999. Brachial and crural indices of European Late Upper Paleolithic and Mesolithic humans. *Journal of Human Evolution* 36: 549–566.

Hopley P. J., Weedon G. P., Marshall J. D., Herries A. I. R., Latham A. G. and Kuykendall K. L. 2007. High- and low-latitude orbital forcing of early hominin habitats in South Africa. *Earth and Planetary Science Letters* 256: 419–432.

Hubbe M., Hanihara T. and Harvati K. 2009. Climate signatures in the morphological differentiation of worldwide modern human populations. *Anatomical Record-Advances in Integrative Anatomy And Evolutionary Biology* 292: 1720–1733.

Imbrie J., Berger A., Boyle E. A., Clemens S. C., Duffy A., Howard W. R. et al. 1993. On the structure and origin of major glaciation cycles. Part 2: The 100,000-year cycle. *Paleoceanography* 8: 699–735.

Jacobi R. M. and Higham T. F. G. 2008. The 'Red Lady' ages gracefully: new ultrafiltration AMS determinations from Paviland. *Journal of Human Evolution* 55: 898–907.

Jimenez-Espejo F. J., Martínez-Ruiz F., Finlayson C., Paytan A., Sakamoto T., Ortega-Huerta M. et al. 2007. Climate forcing and Neanderthal extinction in Southern Iberia: insights from a multiproxy marine record. *Quaternary Science Reviews* 26: 836–852.

Johnsen S. J. H. B., Clausen W., Dansgaard N. S., Gundestrup C. U., Hammer U., Andersen K. K. et al. 1997. The 18O record along the Greenland Ice Core Project deep ice core and the problem of possible Eemian climatic instability. *Journal of Geophysical Research - Oceans* 102: 26397–26410.

Joris O. and Street M. 2008. At the end of the C–14 time scale-the Middle to Upper Paleolithic record of western Eurasia. *Journal of Human Evolution* 55: 782–802.

Lahr M. M and Foley R. A. 1998. Towards a theory of modern human origins: geography, demography, and diversity in recent human evolution. *American Journal of Physical Anthropology* (suppl.) 27: 137–176.

Lisiecki L. E. and Raymo M. E. 2005. A Pliocene-Pleistocene stack of 57 globally-distributed benthic $\delta^{18}O$ records. *Paleoceanography* 20: PA1003.

Lowe J. J. et al. 2007. RESET – response of Humans to Abrupt Environmental Transitions. Available at: http://c14.arch.ox.ac.uk/reset/embed.php?File=

Marlow J. R., Lange C. B., Wefer G. and Rosell-Mele A. 2000. Upwelling intensification as a part of the Pliocene-Pleistocene climate transition. *Science* 290: 2288–2291.

Manzi G. 2004. Human evolution at the Matuyama-Brunhes boundary. *Evolutionary Anthropology* 13: 11–24.

McDougall I., Brown F. H. and Fleagle J. G. 2008. Sapropels and the age of hominins Omo I. and II., Kibish, Ethiopia. *Journal of Human Evolution* 55: 409–420.

Mellars P. 2006a. Archeology and the dispersal of modern humans in Europe: Deconstructing the 'Aurignacian'. *Evolutionary Anthropology* 15: 167–182.

Mellars P. 2006b. Why did modern human populations disperse from Africa ca. 60,000 years ago? A new model. *Proceedings of the National Academy of Sciences of the USA* 103: 13560–13560.

Mellars P., Gravina B. and Ramsey C. B. 2007. Confirmation of Neanderthal/modern human interstratification at the Chatelperronian type-site. *Proceedings of the National Academy of Sciences of the United States of America* 104: 3657–3662.

Osborne A. H., Vance D., Rohling E. J., Barton N., Rogerson M. and Fello N. 2008. A humid corridor across the Sahara for the migration of early modern humans out of Africa 120,000 years ago. *Proceedings of the National Academy of Sciences of USA* 105(43): 16444–16447.

Parfitt S. A., Barendregt R. W., Breda M., Candy I., Collins M. J., Coope G.R., et al. 2005. The earliest humans in Northern Europe: artefacts from the Cromer Forest-bed Formation at Pakefield, Suffolk, UK. *Nature* 438: 1008–1012.

Partridge T. C., Demenocal P. B., Lorentz S. A., Paiker M. J. and Vogel J. C. 1997. Orbital forcing of climate over South Africa: A 200,000-year rainfall record from the Pretoria Saltpan. *Quaternary Science Reviews* 16: 1125–1133.

Petraglia M., Korisettar R.,. Boivin N., Clarkson C., Ditchfield P., Jones S. et al. 2007. Middle paleolithic assemblages from the Indian subcontinent before and after the Toba super-eruption. *Science* 317: 114–116.

Potts R. 1998. Environmental hypotheses of hominin evolution. *Yearbook of Physical Anthropology* 41: 93–136.

Potts R. 1999. Variability selection in hominid evolution. *Evolutionary Anthropology: Issues, News and Reviews* 7: 81–96.

Rae T. C., Koppe T. and Stringer C. B. 2009. The Neanderthal face is not cold adapted. *American Journal of Physical Anthropology* 48(suppl.): 216–216.

Reimer P. J., Baillie M. G. L., Bard E., Bayliss A., Beck J. W., Blackwell P. G. et al. 2009; IntCal09 and Marine09 radiocarbon age calibration curves, 0–50,000 years cal BP. Radiocarbon 51(4): 1111–1150.

Reimer P. J., Baille M. G. L., Bard E., Bayliss A.,Warren B. J., Chanda B. J. H. et al. 2004. IntCal04 terrestrial radiocarbon age calibration, 0–26 cal kyr BP. *Radiocarbon* 46: 1029–1058.

Renne P., Walter R., Verosub K., Sweitzer M. and Aronson J. 1993. New data from Hadar (Ethiopia) support orbitally tuned time scale to 3.3 Ma. *Geophysical Research Letters* 20: 1067–1070.

Roberts M. B., Stringer C. B. and Parfitt S. A. 1994. A hominid tibia from Middle Pleistocene sediments at Boxgrove, UK. *Nature* 369: 311–313.

Roebroeks W. 2001. Hominin behaviour and the earliest occupation of Europe: an exploration. *Journal of Human Evolution* 41: 437–461.

Roebroeks W. 2006. The human colonisation of Europe: where are we? *Journal of Quaternary Science* 21: 425–235.

Ruddiman W. F., Raymo M. E., Martinson D. G., Clement B. M. and Backman J. 1989. Pleistocene evolution: Northern Hemisphere ice sheets and North Atlantic Ocean. *Paleoceanography* 4: 353–412.

Ruff C. B. 1991. Climate and body shape in hominid evolution. *Journal of Human Evolution.* 21: 81–105.

Schwarcz H. P., Buhay W. M., Grun R., Valladas H., Tchernov E., Bar-Yosef O. and Vandermeersch B. 1989. ESR dating of the Neanderthal site, Kebara cave, Israel. *Journal of Archaeological Science* 16: 653–659.

Scott G. R. and Gibert L. 2009. The oldest hand-axes in Europe. *Nature* 461: 82–85.

Shackleton N. J., Fairbanks R. G., Chiu T. C. and Parrenin F. 2004. Absolute calibration of the Greenland time scale: implications for Antarctic time scales and for Delta C–14. *Quaternary Science Reviews* 23: 1513–1522.

Steegmann A. T., Cerny F. J. and Holliday T. W. 2002. Neanderthal cold adaptation: Physiological and energetic factors. *American Journal of Human Biology* 14: 566–583.

Stringer C. 2002. New perspectives on the Neanderthals. Conference Information: Centenary Congress of the Zurich Anthropological Institute and Museum. *Evolutionary Anthropology* 11: 58–59.

Stringer C. B., Trinkaus E., Roberts M. B., Parfitt S. A. and Macphail R. I. 1998. The Middle Pleistocene human tibia from Boxgrove. *Journal of Human Evolution* 34: 509–547.

Svensson A., Andersen K.K., Bigler M., Clausena H.B., Dahl-Jensena D., Davies S.M. et al. 2007. The Greenland ice core chronology 2005, 15–42 ka. part 2: comparison to other records. *Quaternary Science Reviews* 25: 3258–3267.

Swisher C. C., Curtis G. H., Jacob T., Getty A. G. and Suprijo Widiasmoro A. 1994. Age of earliest known hominids in Java, Indonesia. *Science* 263: 1118–1121.

Tattersall I. and Schwartz J. H. 2009. Evolution of the genus Homo. *Annual Review of Earth and Planetary Science* 37: 67–92.

Tzedakis P. C., Hughen K. A., Cacho I., Harvati K. 2007. Placing late Neanderthals in a climatic context. *Nature* 449: 206–208.

Valladas H., Valladas G., Bar-Yosef O. and Vandermeersch B. 1991. The excavations in Kebara Cave, Mt Carmel. *Endevour* 15: 115–119.

van Andel T. H. 2002. The climate and landscape of the middle part of the Weichselian glaciation in Europe: The Stage 3 Project. *Quaternary Research* 57: 2–8.

Villa P., Soressi M., Henshilwood C. S. and Mourre V. 2009. The Still Bay points of Blombos Cave (South Africa). *Journal Of Archaeological Science* 36: 441–460.

Vrba E. S. 1980. Evolution, species, and fossil: how does life evolve? *South African Journal of Science* 76: 61–84.

Vrba E. S., Denton G. H. and Prentice M. L. 1989. Climatic influences on early hominid behavior. *Ossa* 14: 127–156.

Williams M. A. J., Ambrose S. H., van der Kaars S., Ruehlemann C., Chattopadhyaya U., Pal J. and

Chauhan P. R. 2009. Environmental impact of the 73 ka Toba super-eruption in South Asia. *Palaeogeography Palaeoclimatology Palaeoecology* 284: 295–314.

Zhu R.X., Potts R., Pan Y.X., Yao H.T., Lu L. Q., Zhao X. et al. 2008. Early evidence of the genus Homo in East Asia. *Journal of Human Evolution* 55: 1075–1085.

Zilhao J., d'Errico F., Bordes J.-G., Lenoble A., Texier J.-P. and Rigaud J.-P. 2006. Analysis of Aurignacian interstratification at the Châtelperronian type-site and implications for the behavioral modernity of Neanderthals. *Proceedings of the National Academy of Sciences USA* 103: 12643–12648.

The Origins and Spread of Early Agriculture and Domestication: Environmental and Cultural Considerations

Deborah M. Pearsall and
Peter W. Stahl

1 INTRODUCTION

More than six billion humans enter the new millennium overwhelmingly reliant on a narrow range of domesticated foods for their survival. Although thousands of plants and animals are known to have been domesticated since the onset of Holocene conditions over ten thousand years ago, contemporary agriculture focuses on a highly restricted inventory of plant and animal species whose production occupies the labor of almost one-third of the world's workforce and covers much of its land surface.

Modern agriculture is responsible for massive alterations of the global landscape and shares complicity in climatic modification. Agricultural activities such as burning, deforestation, irrigation, use of fertilizers and increased mechanization lead to loss of vegetative biomass, decreased biodiversity, soil depletion, erosion, conversion to grassland, desertification, pollution, eutrophication, salinization and augmented fossil fuel consumption. In turn, these effects can lead both directly and indirectly to: changing concentrations of atmospheric gases, especially carbon dioxide, methane and nitrous oxide; increased amounts of atmospheric aerosols, ozone depletion, thermal pollution, and surface albedo; and alterations in precipitation and hydrology (Clay, 2004; Goudie, 1994). The global record of human landscape modification and its possible impact on climate systems is extensive and deep (Oldfield, 2008). It may be possible to link early agricultural activities with climatic change far back into the Holocene (Ruddiman, 2005).

What is agriculture, and what characteristics did humans seek in the plants and animals they domesticated? Where and when did humans become agricultural? Why did humans become agricultural, and what role might environmental change have played in the adoption of an agricultural lifestyle, particularly as its earliest manifestations appear around a boundary between geological epochs associated with major climatic shifts? This chapter discusses the process of domestication and approaches for identifying it, outlines the geography and timing of early agriculture around the world, examines explanations offered for its adoption, and assesses the role of environmental change in the origin and spread of early agriculture.

2 DOMESTICATION

2.1 Domestication and agriculture

Minimally from the initial stages of the Holocene, humans selectively bred plants and animals to provide food and medicine, clothing and companionship, draft and transportation, tools and weapons, and fertilizer. Domestication can increase the efficiency and reliability of food procurement, which facilitates an increase in human population size and density. Although some human groups that rely on hunting and gathering are sedentary and others that are agricultural may be mobile, food production often provides the necessary foundation for larger populations to live in fixed settlements for longer periods of time.

Domestication is best considered within a continuum of activities ranging from selective exploitation and manipulation of wild populations to their integration into cultural settings in which reproduction is controlled for human benefit. Meadow (1989) characterizes animal domestication as a selective process of change in relationships which shifts the focus of humans from dead to living animals, especially their progeny. Humans incorporate plants and animals into food production systems through a process that is both cultural and biological. Domestication can be characterized as biological mutualism, which increases the survivorship of its participants (Rindos, 1980). The resultant dependency is bidirectional as humans and domesticates become reliant upon each other for survival. Although humans modify plants and animals through selectively encouraging those characteristics that they want, the process of domestication may be unintentional as well as deliberate (Harlan, 1995). It can result in a population that is genetically altered such that it would experience difficulty flourishing in the wild (Figure 39.1).

2.2 Identifying plant and animal domestication

As humans select for specific qualities, certain plants and animals might be considered preadapted to domestication. Both seed and vegetatively propagated plants have been incorporated into food production systems since at least the early Holocene. Harlan (1995) summarizes the effects of human selective pressure on seed plants: reduced dormancy, increased synchronous ripening, suppression of natural abscission, recovery of fertility in reduced or rudimentary flowers, larger seeds, and increased fitness for survival in anthropogenic landscapes. He also notes that vegetatively propagated plants that are sexually infertile or nearly so have an advantage as candidates for domestication over their seed counterparts because the selected clone is instantly domesticated.

Archaeobotanists utilize direct and indirect evidence to identify domestication in the archaeological record. Direct evidence derives from the identification and morphometric analyses of macroremains (e.g., seeds, fruits, nuts, roots, tubers), and microremains (e.g., pollen, phytoliths, starch), and study of ancient DNA. Indirect evidence might include plant impressions and materials and landforms (terraces, raised-beds, irrigation canals) associated with plant cultivation. Reduction of tree species, increases in disturbance plants, and evidence of fire are among the earliest indications of agriculture in paleoenvironmental records. Cultivation, as distinct from gathering, can be identified in the absence of morphologically altered remains

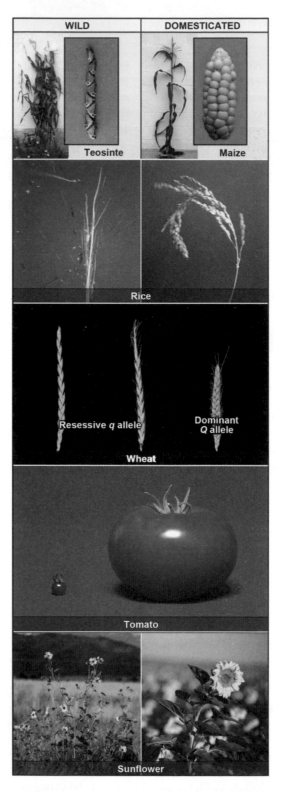

Figure 39.1 Phenotypes of some crops (right) and their progenitors (left) illustrating changes in fruit form, seed dispersal, size, and plant architecture. Reproduced with permission of Elsevier Inc. from Doebley et al. (2006).

by two criteria: conspicuous quantities of seeds (in amounts greater than natural stands would supply) and admixture of predomesticated taxa and plants that grow as weeds in contemporary fields (Weiss et al., 2006).

For a variety of locally contingent reasons, only a small fraction of animals were domesticated since the end of the Pleistocene. Diamond (1999) has isolated traits which characterize the few principally herbivorous large mammals that are today economically important domesticates. Each wild ancestor must occupy a low trophic level with a greater efficiency of converting energy to food biomass. Each must be able to breed in captivity, grow quickly, and be relatively docile or not easily alarmed. Ancestral candidates often live in herds with marked dominance hierarchies that occupy overlapping home ranges.

By controlling animal reproduction, humans select desired traits and thereby potentially increase or decrease variation within the ancestral population. Early stage domestication often produces directional changes in both size and morphology. These could be related to selected characteristics within the founding populations or become manifest through the conditions of early stage confinement. Body size of larger animals is often reduced from their ancestral stock, whereas the opposite may be true for smaller animals. Sexual differences in markedly dimorphic taxa may become exaggerated, as can changes in body proportions through neoteny or retention of juvenile traits. Changes in size and shape can alter the morphology of skeletal elements to produce potentially distinguishing features in archaeological specimens.

Archaeologists also recognize domesticates from materials associated with animal husbandry and by recovering known domesticates in geographical contexts where wild ancestors are lacking. Evidence for early domestication is also based on changing age profiles or male/female ratios whose proportions differ from wild populations, although neither technique can rule out selective

culling of wild populations. The reliability of each kind of evidence for plant or animal domestication is strengthened when corroborated by multiple lines of evidence (Bökönyi, 1969).

3 GEOGRAPHY OF EARLY DOMESTICATION AND THE PALEOENVIRONMENTAL CONTEXT

Plant domestication was earliest and virtually simultaneous in West Asia, East Asia and the Neotropics. We focus on Late Glacial to mid Holocene records in these regions, and to facilitate comparisons use calibrated BC dates. Dates designated cal BC* were estimated graphically using Bartlein et al. (1995). Figure 39.2 summarizes the geography of plant domestication.

3.1 West Asia

The earliest evidence of domestication in West Asia (Figure 39.3) comes from the Levant. Plants domesticated by sedentary or near sedentary foragers include the cereals barley (*Hordeum vulgare*), einkorn (*Triticum monococcum*) and emmer (*T. turgidum*), the pulses pea (*Pisum sativum*), chickpea (*Cicer arietinum*), lentil (*Lens culinaris*) and bitter vetch (*Vicia ervilia*) and fiber flax (*Linum usitatissimum*) (Bellwood, 2005; Harris, 1996; Zohary and Hopf, 2000). Genetic evidence suggests that the 'founder crop' cereals were domesticated once or very few times.

Rosen (2007) synthesizes climatic and environmental proxy data, including oxygen and carbon isotopes, pollen, stream activity, paleosols and lake levels. Isotopes from speleothems in Soreq Cave (Bar-Matthews et al., 1997) are key to the argument that cooling was associated with drying and warming with increased precipitation during the Pleistocene–Holocene transition. This is the interpretation of Rosen and others (but see Issar and Zohar, 2007) and appears consistent

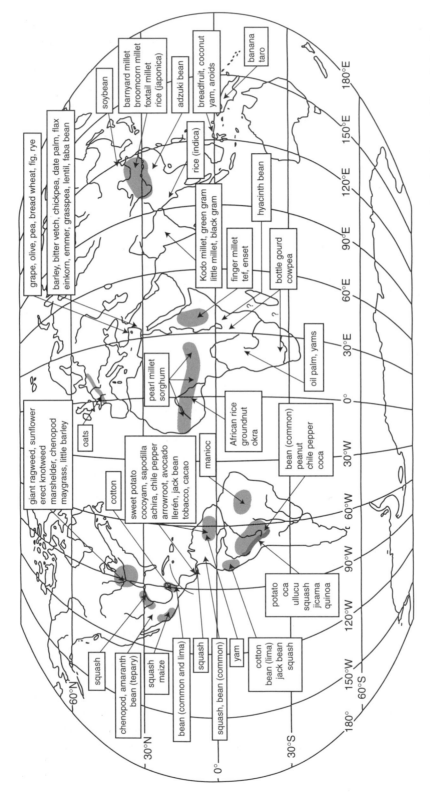

Figure 39.2 Overview of the geography of plant domestication. Figure composed by Kristin Smart. Reproduced with permission of Elsevier Inc. from Pearsall (2008).

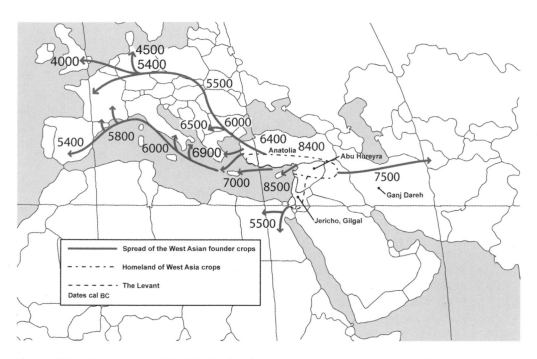

Figure 39.3 The spread of the West Asian founder crops and sites discussed in the text. (Figure composed by Joel Fecke and Howard Wilson).

with the data. Climate in the southern Levant and neighboring regions was very dry and cool following the Late Glacial Maximum, with dominance of steppes and restriction of trees to around springs and marshes. From 15,000 to 12,000 cal BC, rainfall and temperature increased steadily with forest and grassland expansions. Climatic amelioration was followed by a reversal to cold, dry Younger Dryas (YD) conditions from 10,700 to 9,500 cal BC. Climatic effects in the southern Levant were less severe than further north, but throughout the region forests declined, grasslands expanded, and streams downcut their beds. The early Holocene (9,500–5,500 cal BC) began with steady change towards warmer, wetter conditions. Forests recovered more slowly in the southern Levant than the north, due perhaps to human management of open habitats favored by game animals. The early Holocene was also marked by rainfall fluctuations, with two very wet periods (6,400 cal BC, 5,600 cal BC), and a cool dry period (6,200 cal BC). Rosen (2007) argues that in

arid regions like West Asia, increased, even rainfall, perennial river flows, and high water tables created more predictability and lower risks for farmers.

Subsistence among Natufian hunter-gatherers (12,500–9,600 cal BC) was characterized by intensive use of wild plants, including cereals, lentil, bitter vetch, flax and tree fruits (Zohary and Hopf, 2000). Domesticated rye grains (*Secale cereale*) were directly dated at Abu Hureyra to 10,700 cal BC (Hillman et al., 2001). A wide array of wild animals was utilized, including spur-thighed tortoise (*Testudo graeca*), cape hare (*Lepus capensis*), chukar partridge (*Alectoris chukar*) and other birds, and small to large ungulates, including gazelle (*Gazella gazella*), fallow deer (*Dama mesopotamica*) and auroch (*Bos primigenius*) (Stutz et al., 2009). Unusual demographic profiles of sheep (*Ovis aries*) bones from sites in Iraq and Anatolia suggest a degree of management of wild sheep populations began as early as 10,000 cal BC (Redding, 2005; Zeder, 2008).

The Natufian spanned the latter part of the late Pleistocene climatic amelioration and most of the YD. The Early Natufian (12,500–11,000 cal BC) was characterized by increasing sedentism, with an increase in tools for harvesting and processing grasses and other small seeds, as well as higher ranked resources like nuts (Rosen, 2007). Expansion of both woodlands and grasslands (into driest areas) in the late Pleistocene provided a diverse subsistence base. This focus continued into the Late Natufian (11,000–9,600 cal BC), during the latter part of which the cooling and drying of the YD reduced woodlands and expanded grasslands, resulting in increased availability of cereals. Mobile late Natufian populations expanded into new areas. By Final Natufian, near the end of the YD, many sites were abandoned. Rosen (2007) argues that Natufian societies successfully shifted to lower-ranked resources of the expanding grasslands of the YD, with which they were already familiar and for which they had the appropriate processing technology.

The subsequent pre-pottery Neolithic A (PPNA; 9,700–8,500 cal BC) and early Pre-pottery Neolithic B (PPNB) (8,500–8,100 cal BC), rather than the Natufian, were pivotal in the emergence of crop-based subsistence (Bellwood, 2005). The PPNA began late in the YD, and continued into the warmer, wetter conditions of the early Holocene, when the first true villages appeared in well-watered locations, an apparent resumption of the early Natufian pattern of aggregation in favorable locations (Rosen, 2007). Evidence for early domestication during the PPNA is sparse and controversial; domesticated barley and emmer/einkorn occurred at two sites dated 10,000–8,800 cal BC (PPNA) and at four dated 8,800–8,400 cal BC (early PPNB) (Colledge et al., 2004); at Abu Hureyra domesticated einkorn, emmer, barley, lentil and chickpea occurred in 8,750–7,500 cal BC* contexts (Zohary and Hopf, 2000).

Evidence is more widespread for the early PPNB; for example, domesticated einkorn, emmer, barley, lentil, peas, chickpea and flax at Jericho (8,300–7,500 cal BC*) and at sites in Anatolia between 8,400 and 7,500 cal BC (Harris, 1996; Zohary and Hopf, 2000). This was well into the improved conditions of the early Holocene. Rosen (2007) argues for a social *push* towards agriculture from growing hunter-gatherer populations coupled with a climatic *pull* from improving, increasingly predictable and low risk conditions.

PPNA and early PPNB populations potentially cultivated morphologically wild cereals and pulses. A granary dating 9,400–9,200 cal BC at the Gilgal site, Jordan valley, contained over 260,000 wild barley (*Hordeum* sp.) and 120,000 wild oat grains (*Avena* sp.), more than wild stands today provide (Weiss et al., 2006). This PPNA site falls within the early Holocene warming interval, during which the range of wild barley may have expanded (Willcox et al., 2009). Weiss et al. (2006) suggest that much smaller quantities of wild lentils (*Lens* sp.) represent cultivation and loss of seed dormancy since low germination rates and seed production otherwise make wild lentil cultivation unproductive (but see Miller, 1996, for dung fuel burning as a potential source of some seeds).

Willcox et al. (2009) argue that long-term trends in the occurrence of foods at sites in the middle Euphrates region dating 11,250–7,300 cal BC correlate with climatic change, taking into account the response of individual species to environmental conditions. Wild rye, a cool climate cereal, was common prior to and during YD conditions that would have favored an expansion of its range, while wild barley and emmer only become common as warmer Holocene conditions favorable to their growth became established. They feel small-scale rye cultivation during the late Pleistocene was unsustainable because of unstable climatic conditions and argue with Weiss et al. (2006) for a long period of pre-domestication cultivation of wild grasses and pulses during the more stable early Holocene, with a transition to cultivation of domesticated cereals around 8,000 cal BC. This 'long domestication' scenario contrasts with Hillman and Davies's (1990) experimental findings that the nonshattering trait can be selected in wild wheat and barley within 20–30 years.

Clear evidence for domesticated plants and farming appear in the later PPNB (8,100–6,250 cal BC), during which the founder crop suite was adopted in its entirety or in subsets throughout the region, spreading into the uplands of Anatolia (eastern Turkey), the Zagros mountains (Iraq, Iran), by sea to Cyprus and Crete, and into the Balkans and Greece. This nutritionally balanced crop suite was suitable for early expansion under favorable climatic conditions by movements of farmers who replaced or assimilated local foragers (Bellwood, 2005; Harris, 1996), or interacted and exchanged with them.

Sex-specific age curves of goat (*Capra hircus*) remains dated to 7,900 cal BC from the PPNB Ganj Dareh site, western Iran, indicate selective harvest of young males and delayed slaughter of females, a pattern indicative of herd management. Reduction in body size and changes in size and shape of horns occurred 500–1,000 years later, when goats were moved out of their natural habitat (Zeder, 2008; Zeder and Hesse, 2000). mtDNA analysis of wild and domesticated goats suggests early domestication in the Central Iranian Plateau and Southern Zagros region, and a second domestication in Eastern Anatolia that gave rise to modern domesticated goats (Naderi et al., 2008).

From 9,250 to 7,500 cal BC*, increasingly intensive sheep (*Ovis aries*) hunting developed into management and domestication. Management of morphologically unchanged sheep is documented at 8,000 cal BC* in central Anatolia and on Cyprus at 8,400–8,000 cal BC*. Domesticated sheep arrived in the southern Levant between 7,500 and 6,750 cal BC* (Zeder, 2005). Indications of pig (*Sus scrofa*) management occur by 8,500–8,000 cal BC; taurine cattle (*Bos taurus*) by 9,000–8,000 cal BC (Zeder, 2008).

Most evidence indicates that plant and animal domestication arose in West Asia from the sustained interactions between humans and wild populations in the steadily improving conditions of the early Holocene. Earlier indications (domesticated rye at Abu Hureyra; sheep and pig management) were apparently unsustainable subsistence behaviors. Plant domestication and management and domestication of animals in West Asia appear to be near contemporaneous.

The history of Near Eastern domesticated plants and animals continued with the spread of farming into Western, Central, and Northern Europe and Northern Africa between 6,500–4,000 cal BC. Neolithic sites appeared suddenly in northern Africa and the Nile valley between 5,500 and 5,000 cal BC. To the south the monsoonal summer rainfall climate of Sudan inhibited the spread of the founder crops for millennia. Plant domestication in sub-Saharan Africa was apparently late, some 3,000 years after Near Eastern crops appeared in Egypt (Pearsall, 2008).

3.2 Eastern Asia

Agriculture arose independently in different Eastern Asian biomes, from the arid steppes of northwest China to forested New Guinea (Figure 39.4). Understanding the potential role of climate change in shaping these processes is complicated by this diversity, the paucity of paleoenvironmental records for some regions, dating issues, and debate about wild or domesticated status of some plant remains. Progress has been made toward understanding the environmental and cultural contexts in northwest China of broomcorn millet (*Panicum miliaceum*) domestication; in eastern central China of foxtail millet (*Setaria italica*) and *japonica* rice (*Oryza sativa*) domestication in the middle Yellow River and middle Yangtze River valleys, respectively; and in New Guinea of *Colocasia* taro, yam (*Dioscorea*), and *Musa* (*Eumusa* section) banana domestication.

Northwest China

A key controlling feature of Holocene climate in eastern Asia was the establishment of summer monsoon rains around 11,000 cal BC.

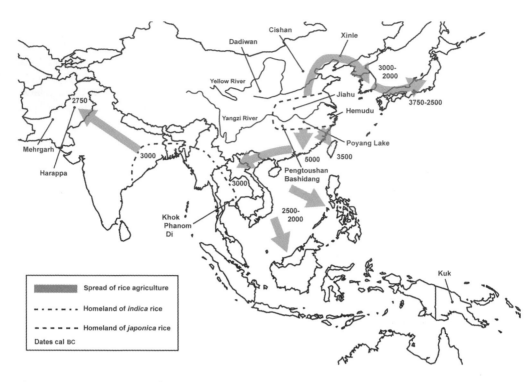

Figure 39.4   The spread of rice agriculture in East Asia and sites discussed in the text. (Figure composed by Joel Fecke and Howard Wilson).

Monsoon intensity differed regionally, as did the location of the monsoon margin through time (Madsen et al., 2007). Early Holocene (12,600–5,100 cal BC) and periods of deglaciation in the western China steppes were characterized by humid monsoonal climate, with a dry mid Holocene (5,100–1,800 cal BC) resulting from a weaker monsoon. Both humid and dry periods featured climatic oscillations of 1,500 years, including a dry YD (10,800–9,400 cal BC) (Fa-Hu et al., 2007). There were also differences in Holocene climate within arid and semiarid western China. The early Holocene was warm but dry in some areas, with a moister mid Holocene. Monsoons, westerlies and topography all played a role in this variability. The Holocene Climatic Optimum (highest moisture) occurred during the early to mid Holocene in arid and semiarid China (Yan et al., 2007).

The transition to agriculture in northwest China was rapid, focusing on domestication of fast ripening, cold and drought tolerant broomcorn millet during the Middle Neolithic Dadiwan complex (6,000–5,000 cal BC). Bettinger et al. (2007) argue that Dadiwan populations did not represent a development from local hunter-gatherers, as antecedent sites are lacking, but rather from hunter-gatherers who migrated from the Upper Yellow River, attracted to wetland habitats that developed during the mid Holocene megahumid period (6,500–4,000 cal BC*). Based on evidence of hunter-gatherer small seed use (milling stones) during an early Holocene warm, moist period, and the selection of suitable habitats for plant cultivation by the newcomers, prior low level food production on the desert margins north of the Yellow River is proposed (Bettinger et al., 2007). Madsen et al. (2007) suggest a complementary model, that broomcorn and foxtail millet (*Setaria italica*) were domesticated along the northern and western margins of the summer monsoon, where short-term monsoonal fluctuations would have a great effect on wild stands, with domestication emerging to maintain stands during drier periods.

Eastern Central China

The modern biomes of the middle and lower Yellow and Yangzi River valleys are deciduous forest and mixed deciduous-evergreen forest, with the Yellow River valley drier (Lu, 2007; Yan et al., 2007). In an overview of records from the northern Yellow River, Zhou et al. (2001) find synchrony with major global climatic events: late glacial warm period (12,900–10,900 cal BC), YD (10,900–9600 cal BC), early to mid Holocene warming (9,600–3,700 cal BC, with abrupt reversal at 6,200 and Holocene optimum 6,200–3,700 cal BC), and a later Holocene cold, drier period (after 3,700 cal BC). Climate change was driven by the quantity of monsoon rain. In most cases, warming was associated with increased moisture, but during the YD, there was fluctuation between cold-dry and cold-wet conditions.

Records from Poyang Lake, south of the middle Yangtze River, document late Pleistocene reduced forest cover (13,000–10,500 cal BC*), reflecting cooler and drier conditions than present, with increasing frequencies of forest elements during much of the interval, indicating increasing temperature and moisture (Jiang and Piperno, 1999; Zhao and Piperno, 2000). *Oryza* (wild rice) phytoliths were present, but rare in late Pleistocene sediments. *Oryza* is a genus of tropical grasses favoring semiaquatic habitats; wild rice expansion into the middle Yangtze would be expected during early Holocene warmer and wetter conditions. Following a depositional hiatus (10,500–2500 cal BC*), the Poyang record documented conditions warmer and moister than today. Numerous domesticated rice phytoliths indicated that rice agriculture was well developed (Zhao and Piperno, 2000).

The Eastern Central China mid Holocene (5,000–2,500 cal BC) was very warm with high precipitation and high sea levels between 4,000 and 3,000 cal BC (Lu, 2007). Mixed deciduous-broadleaved evergreen and subtropical broadleaved evergreen forests expanded northwards into the region. There was considerable variability in the onset of the Holocene Climatic Optimum, with some records placing it just after 6,200 cal BC (Lu, 2007; Zhou et al., 2001). Sedimentary records from the lower Yangtze delta (Atahan et al., 2008; Itzstein-Davey et al., 2007) document highly dynamic environmental conditions. Indicators of human activity began at 5,000 cal BC, contemporary with the earliest cultural record for the region. Atahan et al. (2008) argue that early food production in the delta was constrained by frequent inundation, and populations combined agriculture with hunting, gathering, and cultivation of wild rice.

Neolithic farming communities were present in the middle Yellow River valley by 6,500 cal BC (Jiahu and Cishan complexes), and in the middle Yangtze River valley by 7,000–6,000 cal BC (Bashidang and Pengtoushan complexes). There is evidence for continuity in lithic toolkits in both regions from the terminal Pleistocene to the early Holocene. Wild cereal exploitation – foxtail millet in the Yellow River region and wild rice in the Yangtze – characterized subsistence prior to agriculture. Pottery, dating 10,000–9,800 cal BC, predated agriculture, and facilitated wild cereal cooking (Lu, 1999).

The earliest Yellow River Neolithic sites were large: hundreds of pits were excavated at Jiahu and Cishan (Lu, 1999). The cultural complexity of Jiahu (6,561–5,529 cal BC) suggests earlier farming communities will eventually be found. Foxtail millet was domesticated locally from green foxtail (*Setaria viridis*); domesticated millet was found at Jiahu, with wild and domestic rice, and at Cishan. Domesticated dog (*Canis familiaris*) and pig (*Sus scrofa*) were present. The occurrence of rice at Jiahu, directly dated to 7,000–6,000 cal BC (Crawford, 2006), and at two other sites dating 5,200–4,500 cal BC, suggests cultural influences from the Yangtze region, one homeland of rice (Lu, 2007). Early farming villages in the Yellow River were founded during the warmer, moister conditions of the early to mid Holocene, and show continuity through an abrupt cooling/drying event at 6,200 cal BC.

Pengtoushan and Bashidang, early farming communities in the middle Yangtze, were also substantial villages. Wild or domesticated rice was present at Pengtoushan by 6,995–6,000 cal BC, and domesticated rice, in substantial quantities, at Bashidang by 7,000–5,500 cal BC, with domesticated ox and pig (Lu, 1999; Liu et al., 2007; but see Fuller et al., 2007, who argue for a longer period of wild rice use and domestication closer to 4,000 BC). Rice was domesticated by 9,250–8,000 cal BC*, as indicated by the Diatonghuan phytolith study (Zhao, 1998), which with the Poyang record provide evidence that wild rice grew in the middle Yangtze region. Genetic research indicates that rice was domesticated twice from the *Oryza rufipogon-O. nivera* complex: *japonica* rice in China, *indica* rice independently in Eastern India, Myanmar or Thailand (Londo et al., 2006). If, as Zhao and Piperno (2000) argue, wild rice expanded in range in the Yangtze region during the early Holocene and become more productive because of increased CO_2 levels, it, with other plant resources, would have provided a valuable resource for early Holocene forager/cultivators. The large farming communities of the middle Yangtze were founded during the warmer, moister conditions of the early to mid Holocene, and show continuity through the abrupt cooling/drying event at 6,200 cal BC. Farmers moved into the lower Yangtze delta region later. Domesticated rice is documented at the Hemudu site, south of the Yangtze mouth, at 5,000–4,500 cal BC (Crawford, 2006).

Millet spread into northeast China from the middle Yellow River region, beginning around 5,750 cal BC*, and became more important in 5,000–3,750 cal BC* (Jia, 2007). Domesticated broomcorn millet was documented at the 5,500 cal BC Xinle site (Jia,

2007), which is contemporary with its earliest occurrence in the Dadiwan complex (6,000–5,000 cal BC) of northwest China, raising the possibility of independent domestications (Bettinger et al., 2007). Rice agriculture spread north from east central China into the Korean Peninsula by 3,000–1,000 cal BC, where it was predated by foxtail and broomcorn millet (3,360 cal BC). Rice, barley, and several millets were introduced into Japan around 3,750–2,500 cal BC (Crawford, 2006).

New Guinea

With extensive areas of continental shelf, the Sahul land mass did not separate into New Guinea, Australia and Tasmania until 9,000–6,000 cal BC (Anderson et al., 2007). The extent to which lowland Late Glacial Maximum landscapes were grass- *versus* forest-covered is debated; interpretations range from relatively little reduction to restriction of tropical forests to refugia. Grasslands dominated the highest portions of the New Guinea highlands and some intermontane valleys from 16,000 to 7,000 cal BC, in association with high fire incidence, but mixed montane forests expanded beginning at 12,500 cal BC, with the onset of warmer, moister conditions, and some intermontane valleys were probably forested throughout the LGM. Reforestation was a complex process, in which different forest taxa attained dominance at different times, and grassland indicators dropped and rebounded. High charcoal levels in Late Glacial and early Holocene records resulted from a combination of climatic and human factors; conditions were not uniformly favorable to forests until after 7,000 cal BC, when it was warmer and wetter than today (Denham and Haberle, 2008; Haberle, 2007). Rainforest reached its maximum extent from 6,500 to 3,000 cal BC (Anderson et al., 2007). Beginning at 5,000 cal BC, records from highland New Guinea show sustained and gradually increasing forest clearing and burning; the valley in which the Kuk site is located may never have completely reforested (Haberle, 2007).

There is no clear YD signal in New Guinea (Denham and Haberle, 2008). Anderson et al. (2007) argue that there was significant suppression of ENSO (El Niño southern oscillation) and a semi-permanent La Niña state (warm, moist) in the early Holocene western Pacific. Starting *c*.3,000 cal BC, there were forest reductions, interpreted variously as cooling, drought associated with modern ENSO onset, and clearance by humans (Kershaw et al., 2007). Denham and Haberle (2008) identify a dry period beginning at 4,000 cal BC in New Guinea; evidence from a variety of sources documents the onset of modern ENSO at 5,000–2,000 cal BC (Anderson et al., 2007). Evidence for swiddening is documented in Sumatra between 7,000 and 8,300 cal BC, and in New Guinea between 7,000 and 3,000 cal BC (Kershaw et al., 2007).

Evidence from the Kuk site, located on a wetland margin in the Wahgi Valley, suggests that plant domestication began independently in Highland New Guinea during the early to mid Holocene, prior to agricultural influences from Southeast Asia (Denham et al., 2003). Taro (*Colocasia esculenta*) and yam (*Dioscorea*) starch was recovered from stone tools, and *Musa* (*Eumusa* section) banana phytoliths from sediments affiliated with a single use cultivated plot dating between 8,200 and 7,900 cal BC (Denham et al., 2003; Fullagar et al., 2006). While taro, yam and banana microfossils cannot be identified definitively as domesticated or wild, the evidence as a whole is consistent with shifting cultivation. Populations were mobile, broad-spectrum foragers who maintained and created open patches to increase the availability of useful plants, and used the wetland margin during a drier period (Denham and Haberle, 2008; Denham et al., 2003).

Distinctive mounded cultivation beds make up with other features a later paleosurface (5,000–4,500 cal BC). A spike in Musaceae phytoliths indicated local cultivation of banana, and taro starch was present on tools. Swidden cultivation, as evidenced by forest clearance, fire, and disturbance, was widespread after this time, creating an anthropogenic landscape (Denham and Haberle, 2008; Denham et al., 2003; Fullagar et al., 2006). Anderson et al. (2007) suggest that increased variability in precipitation associated with modern ENSO and greater use of fire both contributed to these changes. Whether increased instability associated with ENSO led to the innovation of wetland gardening is difficult to determine: the timing of modern ENSO onset is inexact, and rainforests are at their maximum extent, denoting warm, wet conditions. Creation and maintenance of forest openings and edges was a long-standing cultural practice. It is difficult to correlate local- (archaeological site) and landscape- (burning and deforestation) scale human activities with regional paleoclimate models and proxy data.

South Asia

Agriculture has ancient roots in South Asia (India and Pakistan). A distinctive crop complex originated in four regions: the Near East (einkorn, emmer, bread wheat (*Triticum aestivum*), barley), Africa (sorghum (*Sorghum bicolor*), pearl millet (*Pennisetum glaucum*), cowpea (*Vigna unguiculata*), hyacinth bean (*Lablab niger*), finger millet (*Eleusine coracana*)), East Asia (rice, foxtail millet, broomcorn millet), and locally (the pulses black and green gram (*Vigna mungo, V. radiata*), kodo and little millet (*Paspalum scrobiculatum, Panicum sumatrense*)) (Harlan, 1995; Weber, 1998). Near Eastern founder crops and caprines dispersed eastwards across the Iranian Plateau to Central and Southern Asia between 8,000 and 6,750 cal BC* (Bellwood, 2005). By 7,000 cal BC Near Eastern crops are documented at the preceramic Mehrgarh site in western Pakistan, with native Zebu cattle, sheep, and goat appearing near the end of the preceramic. African finger millet (*Eleusine coracana*) appeared in South Asia at 3,250 cal BC*, joining indigenous little millet and Asian foxtail millet (Weber, 1998).

Domesticated rice was likely present in eastern India by 3,000 cal BC, but wild rice grows there, complicating its archaeological identification. Domesticated rice was present after 2,750 cal BC* at Harappa (Bellwood, 2005). Environmental records from Thailand indicate that food production predated the Neolithic, and that agricultural intensification varied temporally among regions. Burning and disturbance were documented at 6,750–5,750 cal BC* in northeast and central Thailand, and at 5,000–3,750 cal BC* in southern Thailand. In the Khok Phanom Di area, disturbance indicators are dated to 9,250–6,750 cal BC*, with rice and rice weeds identified at 8,000–6,750 cal BC* (Pearsall, 2008).

3.3 The Americas

The earliest plant domestication in the Americas (Figure 39.5) occurred in the tropics (Piperno and Pearsall, 1998). Understanding Late Glacial and early to mid Holocene low-latitude paleoenvironments in the northern and southern hemisphere remains challenging. Proxy records are sparse for some critical regions; climatic linkages between the northern and southern hemispheres are incompletely understood; potential environments range from deserts to evergreen forests to high elevation grasslands.

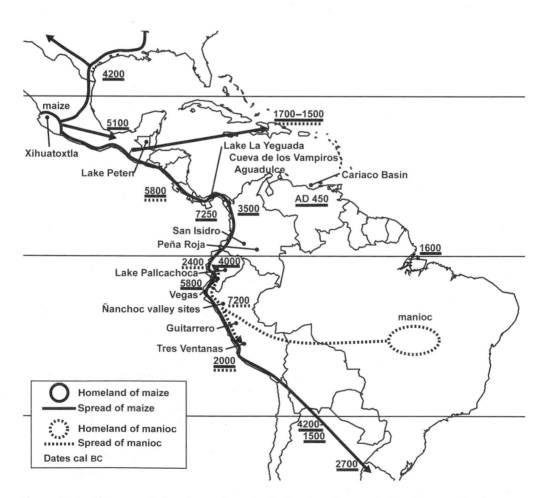

Figure 39.5   The spread of maize and manioc in the Americas and sites discussed in the text. (Figure composed by Joel Fecke and Howard Wilson).

Rainfall is a critical aspect of seasonality in the warm low latitudes. Seasonal rainfall is caused by migration of the intertropical convergence zone (ITCZ), which brings alternating wet and dry conditions to the southern and northern hemisphere tropics depending upon its position. Large-scale precipitation patterns are also influenced by elevation, distance from moisture sources, and rain-shadow effects, and vary year to year. The ENSO is a coupled ocean–atmosphere phenomenon that accounts for much climatic variation on interannual time scales; however, it features decadal and longer variation in strength and frequency and no two cycles are alike (Maasch, 2008). Today, a 'typical' ENSO, occurring at 3–7 year frequencies, brings drought to the northern tropics, and heavy rains, flooding and incursions of warm offshore water as far as 10°S because the ITCZ extends further to the south for a longer period. The tropical Pacific is characterized by two broad oceanic modes, a cold and oscillating ENSO, and a warm, steady mode (Rodbell et al., 1999). Pronounced changes in Neotropical climates would result from decreased ENSO frequency, reduced seasonal contrasts, or changes in ITCZ mean latitudinal position, which may have occurred during the early and mid Holocene.

Based on the recovery of warm water mollusks in coastal Peruvian archaeological sites and other proxy indicators, Richardson and Sandweiss (2008) propose a history of ENSO frequency. The cold and oscillating ENSO mode and a northern extension of coastal aridity is re-established after 3,800 cal BC, with modern frequency appearing after 1,200 cal BC. Sediment records from Lake Pallcachoca, near the southern edge of current ITCZ migrations, provide a record of moderate/severe ENSO events (Rodbell et al., 1999; Moy et al., 2002). A weak early Holocene ENSO signal suggests a warmer eastern Pacific, with modern (and above modern), highly variable periodicity established after 3,000 cal BC. Although Rodó and Rodriguez-Arias (2004) agree with the earlier relaxation of ENSO, they argue that this

earlier weakening was strongly overestimated. Pollen records from Pallcacocha and nearby Lake Chorreras (Hansen et al., 2003) suggest that the onset of the Holocene at 9,000–8,000 cal BC, was warmer, but abundant charcoal indicates pronounced dry seasons. The data indicate warm, moist, relatively stable conditions in the early to mid Holocene (8,000–5,500 cal BC) when ENSO frequencies were a third of modern or less. These results, in combination with data summarized by Richardson and Sandweiss (2008), support a somewhat weakened ENSO signal and reduced seasonality for much of the early Holocene, with a southward extension of the warm water province to around 10°S latitude.

Moist, warm, early Holocene conditions are also documented for the northern tropics from Venezuela's Cariaco Basin. Hodell et al. (2008) found similar patterning in the Late Glacial and early Holocene portion of a long core from Lake Peten, Guatemala, under the same seasonal controls as coastal Venezuela (northern ITCZ edge). Both the Cariaco and Peten records are strongly correlated to the Greenland ice core, suggesting strong coupling of high-latitude temperature and northern Neotropical moisture regimes, controlled by mean latitudinal ITCZ position (Haug et al., 2001; Hodell et al., 2008). A moist early Holocene implies a more northerly mean annual position for the ITCZ relative to the later Holocene. The opposite condition should prevail in the southern tropics (dry Holocene thermal maximum). ENSO events are characterized by a mean southward shift of the ITCZ; such a shift after 3,400 cal BC would correlate with strengthened ENSO.

While data from the northern tropics indicate that mean ITCZ position shifted north during the early to mid Holocene, moister, not drier, conditions existed to 10°S. The drier reversal identified in the Cariaco sequence at 6,300–5,800 cal BC is not long enough to account for this entire interval, and it is unclear whether relaxed ENSO would create a broad latitudinal zone of higher rainfall or a localized coastal effect. In general, Andean pollen records north of 11°S document increased precipitation beginning 8,000–6,000 cal BC, while to the south a dry period began that continued until c.2,000 cal BC (Bush et al., 2007b). The latter is the opposite southern hemisphere effect predicted by a northern ITCZ migration during the thermal maximum (Haug et al., 2001). In the south central Andes, a humid beginning Holocene was followed by a dramatic decrease in effective moisture, resulting in conditions significantly drier than today (Grosjean et al., 2007).

Reconstructing Amazonian Holocene rainfall records is complex. The refugial hypothesis, that Amazonian forests greatly declined during the last glacial, has largely been replaced by the presence of forests with admixtures of montane elements across a late Pleistocene Amazonia that was drier than today (Bush et al., 2007a). During the Holocene a dry event interrupts many records, but its timing and severity are not uniform. In broad terms, dryness peaked between 5,000 and 2,000 cal BC (Bush et al., 2007a), which is corroborated by Mayle and Power's (2008) synthesis of pollen and charcoal records across the Amazonian lowlands. The eastern Andes showed widespread evidence for the interval with latitudinal differences in timing, the driest conditions appearing later in the south. Seasonal drying was, and is today most pronounced on ecotonal margins of the Amazon, and forests were replaced by drier vegetational formations on its northern, eastern and southern margins (Mayle and Power, 2008). Following the dry interval, increased precipitation and forest expansion resumed earliest in the north, and sites in eastern and western Amazonia showed signs of human forest disturbance. In general, forest biomes were very resilient to significantly drier conditions and widespread evidence of burning (Mayle and Power, 2008). In broad terms, precipitation patterning for the Amazon parallels that of the south central Andes in showing drying/increased probability of drought during most of the thermal maximum.

The onset of drying is later in the lowlands, and effects depend greatly on latitude and ecotonal position.

Human occupation of the Neotropics began in the late Pleistocene (13,400–9,400 cal BC). For some regions this was a considerably drier period. Landscapes currently with potential vegetation of deciduous or drier semi-evergreen forests (e.g., Petén, Guatemala, coastal Pacific Central America, northern South America) would have been more open woodlands, scrublands, or savannas, while areas with evergreen and semi-evergreen forests today (e.g., coastal Caribbean Central America; various locations in South America) would have remained forested. By 10,900–9,400 cal BC, when sites are numerous enough to show habitat choices, humans occupied diverse environments and in some cases modified them by fire (Piperno, 2007; Piperno and Pearsall, 1998).

Research at Xihuatoxtla shelter, Central Balsas River valley, SW Mexico, where the wild ancestor of maize, *Zea mays* ssp. *parviglumis*, is native, documented maize (*Zea mays*) in a dry tropical forest setting by 6,700 cal BC (Piperno et al., 2009). Xihuatoxtla confirms early Holocene maize domestication, as indicated by genetic studies and numerous examples of early maize spread within Mesoamerica and into Central and South America.

Domesticated plants were early in Central America, notably Pacific coastal Panama. Central America and Northern South America form a macro-region that is the likely origin area for domesticated arrowroot (*Maranta arundinacea*), llerén (*Calathea allouia*), cocoyam (*Xanthosoma sagittifolium*), sweet potato (*Ipomoea batatas*), achira (*Canna edulis*), avocado (*Persea americana*), one species of chile (*Capsicum frutescens*), jack bean (*Canavalia ensiformis*), and lima bean (*Phaseolus lunatus*). Many wild plants were used throughout prehistory in Central America, and some were likely cultivated: burning of forests and small-scale land clearance date to 11,050 cal BC* at Lake La Yeguada. Arrowroot was the first documented domesticate, dating 7,800 cal BC* at Cueva de los Vampiros and 5,800 cal BC* at Aguadulce. By 5,800 cal BC* maize and gourd (*Lagenaria siceraria*; an Old World plant brought into the Americas) were introduced, and llerén and squash (*Cucurbita* sp.) were present. Manioc (*Manihot esculenta*) was introduced 5,800–5,050 cal BC* (Piperno and Pearsall, 1998).

Plant domestication began before 8,500 cal BC in southwest coastal Ecuador. Squash phytoliths were recovered from terminal Pleistocene and early Holocene strata at Vegas sites (11,800–5,300 cal BC). Phytoliths recovered from the earliest levels were smaller, in the range of wild squash, than those from later levels. Domesticated-size squash phytoliths were directly dated to 9,840–8,555 cal BC, in association with *Calathea*, and characterize the later part of the sequences (Piperno and Stothert, 2003). Other Vegas domesticates included gourd and llerén (by 8,300 cal BC*), and maize, introduced before 5,800 cal BC*. Maize was grown at Valdivia tradition farming villages (4,500–2,250 cal BC), as was cotton (*Gossypium* sp.), jack bean (*Canavalia* sp.), achira, manioc, chili pepper (*Capsicum* sp.), llerén, and arrowroot (Chandler-Ezell et al., 2006). Coastal agriculture remained broad-based for many millennia, incorporating wild/managed tree fruits, domesticated roots, tubers, pulses and maize. Cotton and jack bean were likely domesticated in southwest coastal Ecuador or northern coastal Peru.

Domesticated arrowroot is documented at San Isidro (9,250–8,500 cal BC*) in the Colombian Andes. Palms and avocado were also present, but whether domesticated or wild/managed is unknown. Pollen records documented maize in association with forest clearance and disturbance beginning at 7,250 cal BC* in one core, and in several sequences from 5,500 cal BC* and after. Palm and domesticated squash, llerén, and gourd were direct-dated to 8,250–6,500 cal BC* at the Peña Roja site, eastern Colombia (Piperno and Pearsall, 1998). At sites in the northern Peruvian Andean Ñanchoc valley, occupied

9,500–3,200 cal BC, initial direct dates on domesticates with primitive morphologies, including manioc and peanut (*Arachis hypogaea*), were modern, but new direct dates documented squash at 8,283 cal BC, peanut at 6,538 cal BC, and cotton at 4,113 cal BC, and confirmed early occurrence of manioc (Dillehay et al., 2007). Starch from *Phaseolus* and *Inga feuillei* seeds, squash flesh, and peanut was recovered from dental calculus of numerous teeth dated 7,163–5,744 cal BC (Piperno and Dillehay, 2008). Plant domestication may have been equally ancient in the central Peruvian sierra, but dating ambiguities exist for desiccated macroremains. Oca (*Oxalis tuberosa*), chili pepper (*Capsicum* sp.), lucuma (*Pouteria lucuma*), and common and lima beans (*Phaseolus vulgaris, P. lunatus*) were recovered from Guitarrero Cave in strata dated 9,250–8,500 cal BC*, but beans were direct-dated much younger. Several root crops were recovered from Tres Ventanas Cave in equally ancient strata, but one was direct-dated 5,800 cal BC* (Pearsall, 2008; Piperno and Pearsall, 1998). Domestication of a diverse array of local Andean tubers, pulses and quinoa (*Chenopodium quinoa*) was likely underway before 5,800 cal BC*.

All earliest records of domesticated plants in the northern low latitudes fall within the thermal maximum (8,500–3,400 cal BC, Cariaco Basin), a period wetter than present, some prior to the reversal to colder, drier conditions (6,300–5,800 cal BC) (maize, arrowroot), others at the end or shortly after that reversal (squash, llerén, manioc). Records of domesticated plants began earlier in the southern low latitudes, with arrowroot, llerén, squash and gourd present before the thermal maximum (8,000–5,500 cal BC, southern Ecuador), during the first millennium of the Holocene or perhaps earlier (date ranges are broad). The list of domesticates in the southern tropics expands greatly just before and during the thermal maximum and the millennia during which ENSO was weak (6,800–3,800 cal BC) (maize, peanut, cotton, *Phaseolus*, jackbean, achira, manioc, chili, potato (*Solanum tuberosum*)). Domesticates were also moved during this warm interval; for example, maize from west Mexico into Central and South America, and manioc from the southern edge of the Amazon, its likely area of origin, to Peru (with peanut, also likely from the southern Amazon edge), Ecuador and Panama. Based on current dates for their 'arrivals' outside Amazonia, both manioc and peanut were domesticated prior to the Amazonian early to mid Holocene dry interval (6,000/5,000–2,000 cal BC). Too little is known of the areas of origin of many crops to trace early movements; many early finds are starch or phytolith residues from artifacts, and artifacts of comparable ages have not been studied from potential areas of origin. Drier conditions than present existed in the South Central Andes until *c*.1,500 cal BC (Grosjean et al., 2007). Llamas (*Lama glama*) and alpacas (*Vicugna pacos*) were domesticated around 3,500 cal BC, during this arid interval, but millennia after the beginning of harsh conditions.

4 THEORIES OF AGRICULTURAL ORIGINS AND ADOPTION

Throughout their existence agricultural societies have unquestioningly believed that their particular mode of subsistence was universally preferable. In teleological fashion, agricultural origins were attributed to acts of divine intervention or the logical consequences of inevitable discovery by inquisitive humans seeking to better themselves. In either case, dependence on domesticated plants and animals was considered an inherently superior lifestyle over foraging, minimally due to the productivity and predictability that agriculture provided.

Current ideas on why humans became agricultural reject overtly ethnocentric assumptions about unique events awaiting inevitable discovery. Indeed modern foragers possess quite sophisticated knowledge of their surroundings, including an intimate awareness of plants and animals that is often more informed than their agricultural counterparts. Contemporary foragers know about agriculture; most have probably practiced it in the past, or will use it in the future, depending upon their unique historical circumstances. Sauer (1952) reasonably suggested that humans had long been aware of agriculture and only adopted it under certain circumstances. This idea was strongly reinforced by studies that debunked the notion of agriculture as a necessarily superior subsistence option. Although agriculture could increase production and mitigate uncertainty, it was not necessarily more efficient in terms of labor expenditure and return, suggesting to some that hunter-gatherers may have been the original affluent society (Lee and DeVore, 1968). Moreover, biological evidence preserved in human skeletal samples suggests a clear decline in health after the adoption of agriculture and increasingly sedentary lifestyles (Larsen, 2006; Steckel et al., 2002).

Most modern theories consider the early origin of agriculture as a response to some set of initiating or driving factors. The latter can be considered as forces that pushed, or motivations that pulled, humans into agricultural dependence either separately or in tandem (Stark, 1986). Most tend to emphasize adjustments in resource availability that were prompted by the interactions of environmental change, demographic pressure, and/or cultural mechanisms which eventually set human groups on a path toward food production and a dependence on domesticated plants and animals.

4.1 Environmental change

Although archaeologists have documented the presence of technologically sophisticated human groups at the close of the Pleistocene, global evidence for domestication and the appearance of agriculture is restricted to the onset of Holocene environmental conditions. Indeed, the option of agricultural life may only have become possible at a time when modern humans first encountered the earliest environments conducive to domestication. Climate proxy data from ice cores indicate that Pleistocene environments could have experienced greater temperature variations at much shorter time scales and with greater aridity than during the Holocene. In some areas, the risks involved in agricultural production may have outstripped existing coping strategies, whereas in others agriculture may have become entirely impracticable. Plant productivity would also have been limited by lower concentrations of atmospheric carbon dioxide, which along with increased aridity would have contributed to the reduction of suitable agricultural habitat. Only with the onset of relatively warm, wet, stable and carbon dioxide-rich conditions could humans pursue agriculture as a viable subsistence option. As a more efficient way to use land, agriculture would thereafter become compulsory and expanding populations would extend its frontier at the expense of foraging populations (Richerson et al., 2001).

Early interest in the origins of agriculture had sought a connection between plant and animal domestication and the changing Holocene environment. Childe (1929) had emphasized Holocene desiccation as the prevailing factor behind the increased concentration of resources into waterside oases surrounded by growing tracts of desert. Here, humans would become masters of their own food supply through an increased familiarity with, and possession of, cereals and animals appropriate for domestication.

Subsequent treatments used this idea of resource propinquity during the Holocene as an important condition for adopting agriculture either with or without the necessary inducement of environmental change. Collaborative interdisciplinary field research in the Near East suggested that agriculture arose in, and eventually diffused from, particularly propitious nuclear areas endowed with a natural wealth of cereals and animals that were eventually domesticated by an increasingly differentiating and specializing Holocene population in the absence of desiccation (Braidwood, 1960). Sauer (1952) had also suggested that agriculture likely arose amongst sedentary populations that inhabited resource-rich locales under secure circumstances which afforded humans the time and luxury to experiment with local resources. In particular, he suspected that these early developments were most likely to have appeared in wooded areas by specialized groups living in mild climates alongside fresh water sources that supplied abundant and stable aquatic resources. Like Childe before him, Henry (1989) invoked climatic stress through desiccation, by suggesting that early human settlements had become established in highly productive areas of wild cereals and nut trees during earlier pluvial periods. With a subsequent return to drier conditions, populations that were closer to reliable water sources began domesticating cereal grains in order to continue exploiting their resource base while others reverted to foraging.

Nonetheless, the global timing of agriculture is not synchronous with the onset of the Holocene as it emerges independently at different times in different places. Nor should it be assumed that the appearance of Holocene conditions was uniform as it is believed to have been quite variable. Shorter periods of environmental change in the form of augmented annual seasonality are also considered to be crucial for the appearance of agriculture in localized geographical settings. The idea that a transition to agriculture required reliable regimes of rain and sunshine which encouraged fast-ripening grain, particularly in restricted regions where seasonal outmigration of early agriculturalists was discouraged, is certainly as old as Childe's oasis hypothesis (Peake and Fleure, 1927).

It is argued that the end of ice-age circulation and increased continental summer insolation had markedly increased the seasonality of precipitation. A cyclical regime of warmer, dry summer Mediterranean climates favored annual plants, especially wild cereals, and enabled their expansion into newly available habitats. Here they were exploited by already sedentary populations that had modified the local surroundings by creating disturbed mosaic habitats which favored the growth of annuals. Reminiscent of other scenarios, it has also been suggested that storage and sedentism arose in response to seasonal shortages created by increased summer aridity and desiccation, which when combined with the ensuing strain on local resources, prompted harvesting intensification and plant domestication (McCorriston and Hole, 1991, Wright, 1993).

The origins of tropical vegeculture have been sought in climatic zones associated with intermediate-length dry seasons. The substantial storage organs of root crops are adapted for plant survival during protracted periods of desiccation or cold, only to mature quickly with the onset of optimal growing conditions. Their wild ancestors were therefore sought in intermediate dry zones characterized by seasons with more than 2.5 and less than 7.5 dry months (Harris, 1969).

Risk aversion under unpredictable environmental circumstances was the goal behind a series of coping strategies proposed by Flannery (1986) to understand the adoption of early agriculture in Mexico. With the close of the Pleistocene, human populations could no longer rely on emigration as an effective strategy for coping with uncertainty in an environment that was being encroached upon by Holocene floras in a context of modern annual and seasonal climatic variation. Humans were forced to cope with environmental unpredictability, specifically annual variations in rainfall, which was achieved by diversifying their subsistence to include a broader spectrum of resources at lower trophic levels. Technological and behavioral changes were also implemented, such as: the increased use of storage to extend seasonal resource availability; the development of ground stone and fishing technology; and an increased emphasis on logistically based collection in which resources were moved to the consumer. Food production, the 'ultimate collecting strategy' of cultivating weedy perennials which were regular pioneers of disturbances created by humans, was included as another adjustment to increase resource predictability in the face of environmental extremes.

4.2 Population pressure

Some models have considered relentless demographic pressure from burgeoning human populations, with or without the proximate assistance of environmental change, as the prime motivating factor that tipped most of us into agricultural subsistence. Demographic pressure has been considered both as the consequence of an increased productivity that pulled its expansion, and as the causal variable that relentlessly pushed humans into agricultural production. Likewise, demographic upset as the result of agricultural productivity could subsequently become the causal variable necessitating increased intensification; however, it has been more common to consider a natural propensity for human population growth as the important independent variable.

The concept of demographic increase is founded upon early assumptions posited by Thomas Malthus (1798) that unchecked human population growth increases in a geometrical ratio compared to an arithmetical increase of subsistence resources. It is important to emphasize that Malthus developed his propositions of increased population, subsequent misery and vice in the context of political debates and philosophical reaction to existing utopian beliefs about the perfectibility of humankind and attempts at reshaping the Poor Laws of Georgian England; this is clear in the full title of his essay. Warning of the tendency for actual population growth to outstrip food supply, Malthus argued that any attempt to improve the condition of the poor was pointless (Boorstin, 1986: 474).

Although they still remain assumptions, his propositions have had a tremendous impact on subsequent thought. Perhaps the most influential outgrowth of these ideas was expressed by Danish agricultural economist Esther Boserup (1965) who argued against the Malthusian assumption that population growth was basically dependent upon changes in an inherently inelastic agricultural productivity. She treated population growth as the independent variable, countering that the inevitable Malthusian trap could be surmounted by demographically stressed human populations that either placed more land into agriculture, and/or intensified their productive efforts through fertilizing, irrigation or crop improvement.

The relatively recent shift to agricultural dependence has been viewed as the necessary consequence of prehistoric population explosion. The inevitable strain on resources produced by steadily increasing numbers of humans that successfully relied on big game hunting could be mitigated through colonization of new and uninhabited areas. All available areas were eventually saturated with hunter-gatherers, and human populations necessarily shifted to a broader spectrum of less palatable resource options. Unrelenting population increase further pushed these populations into agricultural dependence. Its relatively late appearance was not due to ignorance; rather, farming was seen as an inferior alternative to hunting, fishing and foraging, as it required increased labor effort. The only advantage of agriculture was its increased efficiency in terms of unit/area production, and its appearance was delayed until it remained the only viable option left for burgeoning human populations (Cohen, 1977).

Nonetheless, others have argued that incessant population increase was not an obligatory characteristic of hunter-gatherers as these groups regularly restricted their demographic growth through cultural practices. Such practices could be considered as homeostatic mechanisms which kept the human demand on resources below local supplies. If static populations were in no need of expanding local resources, then some form of disequilibrium was required to initiate the process of domestication. Changes in staple resources or procurement strategies could lead to reduced mobility and increased sedentism, eventual population growth and demographic upset. High productivity in particularly rich areas is considered to be the most likely causal reason; however, it was not in these locations where the shift to agricultural production was undertaken as these areas were already endowed with sufficient resources. Constant population density was being maintained through emigration of groups or individuals who settled adjacent and potentially marginal territories. Newly settled areas that may not have been readily compatible to the immigrant's mode of subsistence, yet which possessed resources amenable to domestication, were prime locations for the early adoption of agricultural production. Furthermore, plant cultivation and animal herding may have developed together as a strategy to overcome environmental instability by 'banking' surplus production (Binford, 1968; Flannery, 1969; Harris, 1977).

4.3 Social factors

Demographic and technoenvironmental forces that ultimately pressured humans into committing to agriculture and producing more food have been typically considered as external variables; however, neither is independent of human systems – both are products of social structures (Bender, 1978). Competitive feasting amongst complex, economically specialized and socially differentiated hunting-gathering groups living in stable and resource-rich areas may have been a driving force behind increased food production (Hayden, 1990). Such groups are believed to appear no earlier than the Upper Paleolithic, thereafter becoming increasingly widespread, particularly in abundant areas where subsistence was specialized on r-selected resources. The latter employ a reproductive strategy that predominates in unstable conditions and which can feature high fecundity, early maturation, and wide dispersal potential. Here, food production may have begun as competition between accumulating individuals who could maximize their own power base by persuading others under their control to increase the production of specific foods used in waging competitive feasts. The process of domestication was dependent upon the relative influence of the accumulator and the local availability of potentially domesticable feasting plants and animals which could fulfill some desired curiosity. In this respect, the production of inebriants may have been the motivation behind early cereal domestication (e.g., Katz and Maytag, 1991).

4.4 Coevolution, historical ecology and human behavioral ecology

Over a half century ago, Anderson (1952: 150) wrote that 'the history of weeds is the history of man'. While considering Sauer's proposal that sedentary riverine settings were likely sites for early agriculture, Anderson emphasized the key role that refuse heaps played in the appearance of cultivated plants. Open habitats, especially dump heaps, were common human signatures on the landscape and locations where early intermingling and subsequent domestication of introduced weedy flora took place.

The importance of weedy heliophytes colonizing open and disturbed habitats was echoed in later coevolutionary models of early domestication (Rindos, 1980). Through exploiting fire, soil disturbance, forest clearance and dump heaps, humans created niches to which early domesticated plants were preadapted. Domestication appeared before the development of agricultural systems as the result of a symbiotic relationship between humans and plants. A predator that feeds on a plant's propagules is in turn utilized as its dispersal agent. Agriculture is seen as an outgrowth of mutualism wherein the predator's actions indirectly increase its own survival chances as they also directly increase

those of its prey. Subsequent weedy invaders that are subject to the same selective forces as domesticates can become established as 'secondary' domesticates, but over time the dependence on an increasingly smaller subset of domesticated resources intensifies at the expense of others. Agriculture can be seen as a new type of climax formation composed of specialized colonizers adapted to early stage succession which remains stable only as long as humans interfere with subsequent successional processes. The actual moment when a transition to agricultural dependence takes place remains contingent upon unique circumstances. Burgeoning agricultural populations assume a competitive advantage over other modes of subsistence as population growth is enhanced by an agricultural production that must also keep pace with the demands of its escalating predatory population through increased sophistication, innovation and environmental control.

The coevolutionary model is a powerful and informative perspective from which to view the possibilities of early domestication. Although undervaluing the degree to which intentional behavior is important or even necessary in human/domesticate mutualism, it remains relevant to viewpoints which stress human intentions. Historical ecology is an anthropocentric research program that considers humans as keystone species who adapt environments to their particular requirements through landscape-scale manipulations. In particular, sustained levels of environmental disturbance ensure resilience in anthropogenic landscapes (Balée and Erickson, 2006). Humans select and manage a range of incidentally, incipiently, semi and fully domesticated phenotypes that coevolve in anthropogenically disturbed habitats of promoted, managed and cultivated landscapes (Clement, 1999a). In biologically rich tropical settings, intentionally experimenting and inquisitive humans likely took the initial steps of domestication before the onset of Holocene conditions (Clement, 1999b). No connotation of orthogenesis is implied, as subsistence decisions are historically contingent;

humans opt into or out of food production as immediate environmental or social demands dictate.

Optimal foraging models, although developed for nonagricultural subsistence, have been increasingly applied to the origin and adoption of food production. Human Behavioral Ecology (HBE) operates under an assumption of constrained optimization that is the outcome of predator-based decisions which assess the relative cost and benefit of pursuing alternative actions given a range of conditions. These decisions are founded on the changing values of resources (marginal valuation), return gained from pursuing an option against the costs of ignoring it (opportunity cost), evaluating future reward against present cost (discounting), and long-term averaging of stochastic resources (risk-sensitive behavior). In its emphasis on human behavioral responses to changing socioecological conditions, a virtue of the HBE approach is seen as its ability to integrate any and all of the principal initiating or driving factors that are considered important to the onset of agricultural production (Winterhalder and Kennett, 2006).

For example, an HBE approach views the transition to food production in the neotropics as the replacement of foraging by an increasingly more advantageous agricultural strategy during a time of environmental change, demographic growth and technological innovation. The onset of Holocene conditions signaled a decreasing density of high ranking Pleistocene resources and the expansion of forests. Foraging return rates decreased as human populations increased their diet breadth by emphasizing resources at lower trophic levels. These included abundant and rapidly producing plant protodomesticates of seasonal forests which gained coevolutionary advantage through enhanced yield via human manipulation. A subsistence strategy reliant on plant cultivation commenced between 9,250 and 8,000 cal* BC when the net returns from plant cultivation began to exceed those from forest foraging (Piperno, 2006).

5 ASSESSING THE ROLE OF ENVIRONMENTAL CHANGE IN THE ORIGIN AND SPREAD OF EARLY AGRICULTURE

Long considered as a fortuitous event awaiting discovery by human populations, early plant and animal domestication came to be considered as a process by which humans behaviorally changed themselves, their domesticates and the environments they inhabited. It is entirely likely that at least since the appearance of anatomically modern humans and long before the onset of modern Holocene environments, our ancestors had sufficient knowledge and experience to understand the rudiments of plant and animal domestication. Nevertheless, on a broader level, the current evidence suggests that domestication was relatively synchronous in at least three portions of the globe at the earliest onset of the Holocene. We might consider the longer-term environmental changes that accompanied the Pleistocene–Holocene transition as the ultimate causal factors behind the development of food production. Indeed, the appearance of Holocene environmental conditions may have been the only time that food production could have become a viable option, certainly in terms of enhanced predictability and lowered risk for the producer.

Nonetheless, we can also include a variety of initiating or driving factors that pushed and/or pulled humans into domesticating plants and animals which could be considered proximate to the ultimate causes associated with a change to modern Holocene climatic conditions. Our earliest evidence for domestication suggests that it took place in a variety of environments, ranging from dry to wet and temperate to tropical, and in the context of changing climatic conditions, namely increasing warmth and moisture. In arid and semiarid regions like West Asia and Northwest China, increased precipitation and more reliable river flows created favorable conditions for plant cultivation while changing distributions and densities of wild resources, and perhaps their dietary ranking or desirability. Expansion of forests in the Neotropics and New Guinea similarly changed plant distributions, closing formerly open woodlands and altering edge habitats favored by many starch-rich root/tuber species. Creating and maintaining open habitats for favored plants, altering mobility patterns as resources expand or contract, changing diet in response to changing availabilities of foods, increasing densities of desirable plants and animals by cultivation/management: all these are among the human responses indicated by the record of early agriculture. Evidence that increasing dryness, drought frequency or climatic unpredictability led people to domesticate plants and animals is largely lacking.

The identification of proximate causation in specific cases is much more elusive and conjectural as environmental and cultural factors are difficult to disentangle, at least from the perspective of the archaeologist's toolkit. Many of these factors are old and long-standing departures for debate, such as the timing of seasons, the onset of appropriate temperature and moisture regimes, and the expansion of favorable biomes. In different places and at different times during the Holocene, these factors likely interacted with human demographic expansion, the development of technological sophistication for increasing predictability and averting risk, and local pressures to share and/or compete with other humans. With clear evidence that the roots of plant domestication lie in the early Holocene, the challenge now facing us is to expand the paleoenvironmental and archaeological records of this process, and to understand better the nature of change by improving our knowledge of human–plant–animal interrelations during the Late Glacial period.

REFERENCES

Anderson A., Gagan M. and Sulmeister J. 2007. Mid-Holocene cultural dynamics and climatic change in the western Pacific, in Anderson D. G., Maasch K. A. and Sandweiss D. H. (eds) *Climate Change and Cultural Dynamics: A Global Perspective on Mid-Holocene Transitions*. Amsterdam: Elsevier, pp. 265–296.

Anderson E. 1952. *Plants, Man and Life*. Boston:Little, Brown and Company.

Atahan P., Itzstein-Davey F., Taylor D., Dodson J., Qin J., Zheng H. and Brooks A. 2008. Holocene-aged sedimentary records of environmental changes and early agriculture in the lower Yangtze, China. *Quaternary Science Reviews* 27: 556–570.

Balée W. and Erickson C. L. 2006. Time, complexity, and historical ecology, in Balée W. and Erickson C. L. (eds) *Time and Complexity in Historical Ecology*. New York: Columbia University Press, pp. 1–17.

Bar-Matthews M., Ayalon A. and Kaufman A. 1997. Late Quaternary paleoclimate in the Eastern Mediterranean region from stable isotope analysis of speleothems at Soreq Cave, Israel. *Quaternary Research* 47: 155–168.

Bartlein P. J., Edwards M. E., Shafer S. L. and Barker E. D. Jr 1995. Calibration of radiocarbon ages and the interpretation of paleoenvironmental records. *Quaternary Research* 44: 417–424.

Bellwood P. 2005. *First Farmers. The Origins of Agricultural Societies*. Malden, MA: Blackwell.

Bender B. 1978. Gatherer-hunter to farmer: a social perspective. *World Archaeology* 10: 204–222.

Bettinger R. L., Barton L., Richerson P. J., Boyd R., Hui W. and Won C. 2007. The transition to agriculture in northwestern China, in Madsen D. B., Fa-Hu C. and Xing G. (eds) *Late Quaternary Climate Change and Human Adaptation in Arid China*. Amsterdam: Elsevier, pp. 83–101.

Binford L. R. 1968. Post-Pleistocene adaptations, in Binford S. R. and Binford L. R. (eds) *New Perspectives in Archaeology*. Chicago: Aldine, pp. 313–341.

Bőkőnyi S. 1969. Archaeological problems and methods of recognizing animal domestication, in Ucko P. J. and Dimbleby G. W. (eds) *The Domestication and Exploitation of Plants and Animals*. Chicago: Aldine Atherton, pp. 219–229.

Boorstin D. J. 1986. *The Discoverers. A History of Man's Search to the World and Himself*. New York: Random House.

Boserup E. 1965. *The Conditions of Agricultural Growth. The Economics of Agrarian Change Under Population Pressure*. New York: Aldine.

Braidwood R. J. 1960. The agricultural revolution. *Scientific American* 203(3): 130–148.

Bush M. B., Gosling W. D. and Colinvaux P. A. 2007a. Climate change in the lowlands of the Amazon

Basin, in Bush M. B. and Flenley J. R. (eds) *Tropical Rainforest Responses to Climate Change.* Berlin: Springer, pp. 55–76.

Bush M. B., Hanselman J. A. and Hooghiemstra H. 2007b. Andean montane forests and climate change in, Bush M. B. and Flenley J. R. (eds) *Tropical Rainforest Responses to Climate Change.* Berlin: Springer, pp. 33–54.

Chandler-Ezell K., Pearsall D. M. and Zeidler J. A. 2006. Root and tuber phytoliths and starch grains document manioc *(Manihot esculenta)*, arrowroot *(Maranta aundinacea)*, and llerén *(Calathea* sp.) at the Real Alto site, Ecuador. *Economic Botany* 60: 103–120.

Childe V. G. 1929. *The Most Ancient East: The Oriental Prelude to European History.* New York: Alfred A. Knopf.

Clay J. 2004. *World Agriculture and the Environment.* Washington DC: Island Press.

Clement C. R. 1999a. 1492 and the loss of crop genetic resources. I. The relation between domestication and human population decline. *Economic Botany* 53: 188–202.

Clement C. R. 1999b. 1492 and the loss of Amazonian crop genetic resources. II. Crop biogeography at contact. *Economic Botany* 53: 203–216.

Cohen M. N. 1977. *The Food Crisis in Prehistory.* New Haven: Yale University Press.

Colledge S., Conolly J. and Shennan S. 2004. Archaeobotanical evidence for the spread of farming in the Eastern Mediterranean. *Current Anthropology* 45(Suppl.): 35–58.

Crawford G. W. 2006. East Asian plant domestication, in Stark M. T. (ed.) *Archaeology of Asia.* Malden: Blackwell, pp. 77–95.

Denham T. P., Haberle S. G., Lentfer C., Fullagar R., Field J., Therin M. et al. 2003. Origins of agriculture at Kuk swamp in the highlands of New Guinea. *Science* 301: 189–193.

Denham T. and Haberle S. 2008. Agricultural emergence and transformation in the Upper Wahgi valley, Papua New Guinea, during the Holocene: theory, method, and practice. *Holocene* 18: 481–496.

Diamond J. 1999. *Guns, Germs, and Steel. The Fates of Human Societies.* New York: W. W. Norton.

Dillehay T. D., Rossen J., Andres T. C. and Williams D. E. 2007. Preceramic adoption of peanut, squash, and cotton in northern Peru. *Science* 316: 1890–1893.

Doebley J. F., Gaut B. S. and Smith B. D. 2006. The molecular genetics of crop domestication. *Cell* 127: 1309–1321.

Fa-Hu C., Bo C., Hui Z., Yu-Xin F., Madsen D. B. and Ming J. 2007. Post-glacial climate variability and drought events in the monsoon transition zone of western China, in Madsen D. B., Fa-Hu C. and Xing G. (eds) *Late Quaternary Climate Change and Human Adaptation in Arid China.* Amsterdam: Elsevier, pp. 25–39.

Flannery K. V. 1969. Origins and ecological effects of early domestication in Iran and the Near East, in P. J. Ucko and G. W. Dimbleby (eds) *The Domestication and Exploitation of Plants and Animals.* Chicago: Aldine Atherton, pp. 73–100.

Flannery K. V. (ed.) 1986. *Guilá Naquitz. Archaic Foraging and Early Agriculture in Oaxaca, Mexico.* Orlando: Academic Press.

Fullagar R., Field J., Denham T. and Lentfer C. 2006. Early and Mid Holocene tool-use and processing of taro *(Colocasia esculenta)*, yam *(Dioscorea* sp.) and other plants at Kuk Swamp in the highlands of Papua New Guinea. *Journal of Archaeological Science* 33: 595–614.

Fuller D. Q., Harvey E. and Qin L. 2007. Presumed domestication? Evidence for wild rice cultivation and domestication in the fifth millennium BC of the lower Yangtze region. *Antiquity* 81: 316–331.

Goudie A. 1994. *The Human Impact on the Natural Environment,* 4th edition. Cambridge, MA: MIT Press.

Grosjean M., Santoro C. M., Thompson L. G., Núñez L. and Standen V. G. 2007. Mid-Holocene climate and culture change in the South Central Andes, in Anderson D. G., Maasch K. A. and Sandweiss D. H. (eds) *Climate Change and Cultural Dynamics: A Global Perspective on Mid-Holocene Transitions.* Amsterdam: Elsevier, pp. 51–115.

Haberle S. G. 2007. Prehistoric human impact on rainforest biodiversity in highland New Guinea. *Philosophical Transactions of the Royal Society* 362: 219–228.

Hansen B. C. S., Rodbell D. T., Seltzer G. O., León B., Young K. R. and Abbott M. 2003. Late-glacial and Holocene vegetational history from two sites in the western Cordillera of southwestern Ecuador. *Palaeogeography, Palaeoclimatology, Palaeoecology* 194: 79–108.

Harlan J. R. 1995. *The Living Fields: Our Agricultural Heritage.* Cambridge: Cambridge University.

Harris D. R. 1969. Agricultural systems, ecosystems and the origins of agriculture, in P. J. Ucko and G. W. Dimbleby (eds) *The Domestication and Exploitation of Plants and Animals.* Chicago: Aldine Atherton, pp. 3–15.

Harris D. R. 1977. Alternative pathways toward agriculture, in C. A. Reed (ed.) *Origins of Agriculture*. The Hague: Mouton, pp. 179–243.

Harris D. R. 1996. The origins and spread of agriculture and pastoralism in Eurasia: an overview, in Harris D. R. (ed.) *The Origins and Spread of Agriculture and Pastoralism in Eurasia*. Washington DC: Smithsonian Institution Press, pp. 552–573.

Haug G. H., Hughen K. A., Sigman D. M., Peterson L. C. and Rohl U. 2001. Southward migration of the intertropical convergence zone through the Holocene. *Science* 293: 1304–1308.

Hayden B. 1990. Nimrods, piscators, pluckers, and planters: the emergence of food production. *Journal of Anthropological Archaeology* 9: 31–66.

Henry D. O. 1989. *From Foraging to Agriculture: the Levant at the End of the Ice Age*. Philadelphia: University of Pennsylvania Press.

Hillman G. and Davies M. 1990. Measured domestication rates in wild wheat and barley. *Journal of World Prehistory* 4: 157–222.

Hillman G., Hedges R., Moore A., Colledge S. and Pettitt P. 2001. New evidence of Late Glacial cereal cultivation at Abu Hureyra on the Euphrates. *Holocene* 11: 383–393.

Hodell D. A., Anselmetti F. S., Ariztegui D., Brenner M., Curtis J. H., Gilli A. et al. 2008. An 85-ka record of climate change in lowland Central America. *Quaternary Science Reviews* 27: 1152–1165.

Issar A. S. and Zohar M. 2007. *Climate Change— Environment and History of the Near East*, 2nd edition. Berlin: Springer.

Itzstein-Davey F., Atahan P., Dodson J., Taylor D. and Zheng H. 2007. A sediment-based record of Late Glacial and Holocene environmental changes from Guangfulin, Yangtze delta, eastern China. *Holocene* 17: 1221–1231.

Jia W. M. 2007. *Transition from Foraging to Farming in Northeast China*. BAR International Series 1629. Oxford: Archaeopress.

Jiang Q. and Piperno D. R. 1999. Environmental and archaeological implications of a Late Quaternary palynological sequence, Poyang Lake, southern China. *Quaternary Research* 52: 250–258.

Katz S. H. and Maytag F. 1991. Brewing an ancient beer. *Archaeology* 44: 24–27.

Kershaw A. P., van der Kaars S. and Flenley J. R. 2007. The Quaternary history of far eastern rainforests, in Bush M. B. and Flenley J. R. (eds) *Tropical Rainforest Responses to Climate Change* 2007. Berlin: Springer, pp. 77–115.

Larsen C. S. 2006. The agricultural revolution as environmental catastrophe: implications for health and lifestyle in the Holocene. *Quaternary International* 150: 12–20.

Lee R. B. and DeVore I. (eds) 1968. *Man the Hunter*. Chicago: Aldine.

Liu L., Lee G-A., Jiang L. and Zhang J. 2007. Evidence for the early beginning (c. 9000 cal. BP) of rice domestication in China: a response. *The Holocene* 17: 1059–1068.

Londo J. P., Chiang Y-C., Hung K-H., Chiang T-Y. and Schaal B. A. 2006. Phylogeography of Asian wild rice, *Oryza rufipogon*, reveals multiple independent domestications of cultivated rice, *Oryza sativa*. *Proceedings of the National Academy of Sciences* 103: 9578–9583.

Lu T. L-D. 1999. *The Transition from Foraging to Farming and the Origin of Agriculture in China*. BAR International Series 774. Oxford: John and Erica Hedges.

Lu T. L-D. 2007. Mid-Holocene climate and cultural dynamics in eastern Central China, in Anderson D. G., Maasch K. A. and Sandweiss D. H. (eds) *Climate Change and Cultural Dynamics: A Global Perspective on Mid-Holocene Transitions*. Amsterdam: Elsevier, pp. 297–329.

Maasch K. A. 2008. El Niño and interannual variability of climate in the western Hemisphere, in Sandweiss D. H. and Quilter J. (eds) *El Niño, Catastrophism, and Culture Change in Ancient America*. Washington DC: Dumbarton Oaks Research Library and Collection, pp. 33–55.

Madsen D. B., Fa-Hu C. and Xing G. 2007. Changing views of Late Quaternary human adaptation in arid China, in Madsen D. B., Fa-Hu C. and Xing G. (eds) *Late Quaternary Climate Change and Human Adaptation in Arid China*. Amsterdam: Elsevier, pp. 227–232.

Malthus T. 1798. *An Essay on the Principle of Population, as it Affects the Future Improvement of Society with Remarks on the Speculations of Mr. Goodwin M. Condorcet, and Other Writers*. London: J. Johnson.

Mayle F. E. and Power M. J. 2008. Impact of a drier Early-Mid-Holocene climate upon Amazonian forests. *Philosophical Transactions of the Royal Society of London Biological Sciences* 363: 1829–1838.

McCorriston J. and Hole F. 1991. The ecology of seasonal stress and the origins of agriculture in the Near East. *American Anthropologist* 93: 46–69.

Meadow R. H. 1989. Osteological evidence for the process of animal domestication, in J. Clutton-Brock (ed.) *The Walking Larder. Domestication, Pastoralism and Predation*. London: Unwin Hyman, pp. 80–90.

Miller N. F. 1996. Seed eaters of the ancient Near East: human or herbivore? *Current Anthropology* 37: 521–528.

Moy C. M., Seltzer G. O., Rodbell D. T. and Anderson D. M. 2002. Variability of El Niño/Southern Oscillation activity at millennial timescales during the Holocene epoch. *Nature* 420: 162–165.

Naderi S., Rezaei H-R., Pompanon F., Blum M. G. B., Negrini R., Naghash H-R. et al. 2008. The goat domestication process inferred from large-scale mitochondrial DNA analysis of wild and domestic individuals. *Proceedings of the National Academy of Sciences* 105: 17659–17664.

Oldfield F. 2008. The role of people in the Holocene, in R. W. Battarbee and Binney H. A. (eds) *Natural Climate Variability and Global Warming: a Holocene Perspective.* Oxford: Blackwell Publishing, pp. 58–97.

Peake H. and Fleure H. J. 1927. *Peasants and Potters.* New Haven: Yale University Press.

Pearsall D. M. 2008. Plant domestication, in Pearsall D. M. (ed.) *Encylopedia of Archaeology.* Oxford: Academic Press, pp. 1822–1842.

Piperno D. R. 2006. The origins of plant cultivation and domestication in the Neotropics. A behavioral ecological perspective, in Kennett D. J. and Winterhalder B. (eds) *Behavioral Ecology and the Transition to Agriculture.* Berkeley: University of California Press, pp. 137–166.

Piperno D. R. 2007. Prehistoric human occupation and impacts on Neotropical forest landscapes during the Late Pleistocene and Early/Middle Holocene, in Bush M. B. and Flenley J. R. (eds) *Tropical Rainforest Responses to Climate Change.* Berlin: Springer, pp. 193–218.

Piperno D. R. and Dillehay T. D. 2008. Starch grains on human teeth reveal early broad crop diet in northern Peru. *Proceedings of the National Academy of Sciences.* 105: 19622-19627

Piperno D. R. and Pearsall D. M. 1998. *The Origins of Agriculture in the Lowland Neotropics.* San Diego: Academic Press.

Piperno D. R., Ranere A. L., Holst I., Iriarte J. and Dickau R. 2009. Starch grain and phytolith evidence for early ninth millennium B.P. maize from the Central Balsas River Valley, Mexico. *Proceedings of the National Academy of Sciences.* 106: 5019–5024

Piperno D. R. and Stothert K. E. 2003. Phytolith evidence for early Holocene *Cucurbita* domestication in southwest Ecuador. *Science* 299: 1054–1057.

Redding R. W. 2005. Breaking the mold: a consideration of variation in the evolution of animal domestication, in Vigne J. P., Peters J. and Helmer D. (eds)

First Steps of Animal Domestication. New Archaeozoological Approaches. Oxford: Oxbow, pp. 41–48.

Richardson III J. B. and Sandweiss D. H. 2008. Climate change, El Niño, and the rise of complex society on the Peruvian coast during the Middle Holocene, in Sandweiss D. H. and Quilter J. (eds) *El Niño, Catastrophism, and Culture Change in Ancient America.* Washington DC: Dumbarton Oaks Research Library and Collection, pp. 59–75.

Richerson P. J., Boyd R. and Bettinger R. L. 2001. Was agriculture impossible during the Pleistocene but mandatory during the Holocene? A climate change hypothesis. *American Antiquity* 66: 387–411.

Rindos D. 1980. Symbiosis, instability, and the origins of agriculture. *Current Anthropology* 21: 751–772.

Rodbell D. T., Seltzer G. O., Anderson D. M., Abbott M. B., Enfield D. B. and Newman J. H. 1999. An ~15,000-year record of El Niño-driven alleviation in southwestern Ecuador. *Science* 283: 516–520.

Rodó X. and Rodriguez-Arias M-A. 2004. El Niño-Southern Oscillation: absent in the early Holocene? *Journal of Climate* 17: 423–426.

Rosen A. M. 2007. *Civilizing Climate. Social Responses to Climate Change in the Ancient Near East.* Lanham: AltaMira Press.

Ruddiman W. F. 2005. *Plows, Plagues, and Petroleum. How Humans Took Control of the Climate.* Princeton: Princeton University Press.

Sauer C. O. 1952. *Agricultural Origins and Dispersals.* New York: American Geographical Society.

Stark B. L. 1986. Origins of food production in the New World, in Meltzer D. J., Fowler D. D and Sabloff J. A. (eds) *American Archaeology Past and Future.* Washington DC: Smithsonian Institution Press, pp. 277–321.

Steckel R. H., Rose J. C., Larsen C. S. and Walker P. L. 2002. Skeletal health in the western hemisphere from 4000 B.C. to the present. *Evolutionary Anthropology* 11: 142–155.

Stutz A. J., Munro N. D. and Bar-Oz G. 2009. Increasing the resolution of the Broad Spectrum Revolution in the Southern Levantine Epipaleolithic (19–12 ka). *Journal of Human Evolution* 56: 294–306.

Weber S. A. 1998. Out of Africa: the initial impact of millets in South Asia. *Current Anthropology* 39: 267–274.

Weiss E., Kislev M. E. and Hartmann A. 2006. Autonomous cultivation before domestication. *Science* 312: 1608–1610.

Willcox G., Buxo R. and Herveux L. 2009. Late Pleistocene and early Holocene climate and the beginnings of cultivation in northern Syria. *Holocene* 19: 151–158.

Winterhalder B. and Kennett D. J. 2006. Behavioral ecology and the transition from hunting and gathering to agriculture, in Kennett D. J. and Winterhalder B. (eds) *Behavioral Ecology and the Transition to Agriculture*. Berkeley: University of California Press, pp. 1–21.

Wright H. E. Jr 1993. Environmental determinism in near eastern prehistory. *Current Anthropology* 34: 458–469.

Yan Z., Zicheng Y., Fa-Hu C. and Chengbang A. 2007. Holocene vegetation and climate changes from fossil pollen records in arid and semi-arid China, in Madsen D. B. Fa-Hu C. and Xing G. (eds) *Late Quaternary Climate Change and Human Adaptation in Arid China*. Amsterdam: Elsevier, pp. 51–65.

Zeder M. A. 2005. A view from the Zagros: new perspectives on livestock domestication in the Fertile Crescent, in Vigne J. P., Peters J. and Helmer D. (eds) *First Steps of Animal Domestication. New Archaeozoological Approaches*. Oxford: Oxbow, pp. 125–146.

Zeder M. A. 2008. Domestication and early agriculture in the Mediterranean Basin: Origins, diffusion, and impact. *Proceedings of the National Academy of Sciences* 105: 11597–11604.

Zeder M. A. and Hesse B. 2000. The initial domestication of goats *(Capra hircus)* in the Zagros Mountains 10,000 years ago. *Science* 287: 2254–2257.

Zhao Z. 1998. The middle Yangtze region in China is one place where rice was domesticated: Phytolith evidence from the Diaotonghuan cave, northern Jiangxi. *Antiquity* 72: 885–897.

Zhao Z. and Piperno D. R. 2000. Late Pleistocene/Holocene environments in the Middle Yangtze River valley, China and rice *(Oryza sativa* L.) domestication: The phytolith evidence. *Geoarchaeology* 15: 203–222.

Zhou W., Head M. J. and Deng L. 2001. Climate changes in northern China since the late Pleistocene and its response to global change. *Quaternary International* 83–85: 285–292.

Zohary D. and Hopf M. 2000. *Domestication of Plants in the Old World*, 3rd edition. Oxford: Oxford University Press.

40

Complexity, Causality and Collapse: Social Discontinuity in History and Prehistory

Georgina Endfield

1 INTRODUCTION

Collapse is a recurrent feature of complex human societies. The term itself is somewhat contested, meaning different things to different people, but has been generally defined as a process of marked sociopolitical simplification from an established level of complexity (Tainter, 1990: 4). Societal collapse can be manifest in regional abandonment, population decline, the long-term fragmentation of sociopolitical and economic structures, the replacement of one subsistence base by another or conversion to a lower energy sociopolitical organisation (Weiss and Bradley, 2001). While some have argued that the process must be rapid, taking place over time scales of a few decades (Tainter, 1990), longer-term trends of socioeconomic deterioration and decline over periods of centuries have also been recognised (Greer, 2005). Either way, societies have been repeatedly proven to be 'fragile and impermanent entities' (Tainter, 1990: 4) and, as Weiss and Bradley (2001: 609) have suggested, 'the archaeological and historical record is replete with evidence for prehistoric, ancient and pre-modern collapse' at a variety of temporal and spatial scales.

Collapse scenarios have long been of interest (Tainter, 1990: 3). The fact that 'civilisation can die because it has already died once' (Mazzarino, 1966, cited in Tainter, 1990: 2), has been a source of anxiety (Randers, 2008) and a focus of scholarly and public fascination (Greer, 2005). Moreover, concerns over the disintegration of societies and social order have gained momentum in recent decades as a result of two key developments: first, a number of high-profile publications which have adopted what are being termed 'neo-environmental deterministic perspectives'; and second, a contemporary environmental zeitgeist typified by apocalyptic climate change scenarios and predictions of resultant global crisis (Randers, 2008).

Diamond's (1997) *Guns, Germs and Steel*, which posits an environmental explanation for Europe's apparent superiority in history, and more specifically his 'sequel' (2005) text, *Collapse: How Societies Choose to Fail or Succeed*, consider the role of environment in shaping and framing global human histories and the ecological self-destruction of past and present societies (Hornborg, 2005: 94). In the case of the former text, Diamond argues that 'the fates of human societies' and whether they have become in effect rich or poor societies, have been determined by differences in the environment (Sluyter, 2003). In *Collapse*, Diamond introduces a suite of examples to illustrate how the disintegration of ancient civilisations such as the Maya, Anasazi and Akkadian can be linked to climatic variability and to sociopolitical changes in the way in which societies use, regulate and protect resources. As the subtitle of *Collapse* suggests, Diamond places emphasis on how societies effectively 'choose' to fail or succeed. Although he draws on 12 examples of collapse from around the world, covering both industrial and preindustrial case studies, Diamond's 'template' for collapse is the history of Easter Island prior to European contact and the fact that 'in just a few centuries, the people of Easter island wiped out their forest, drove their plants and animals to extinction and saw their complex society spiral into chaos and cannibalism'. He highlights the apparently 'chillingly obvious' similarities between Easter Island and modern society (Diamond, 2005: 19). Such parables of environmental morality, it has been argued, make for compelling narratives and have highlighted what for some are 'disquieting parallels' between past case studies and societies under threat from unprecedented global change today (deMenocal and Cook, 2005).

Diamond is not alone in effectively reviving environmentally deterministic arguments. Citing scholarship produced by the palaeoenvironmental community, for example, geographers, Coombes and Barber (2005) discuss various examples of scholarship where abrupt climate change has been invoked to explain catastrophe and civilisation collapse. The fall of the Akkadian Empire and the end of the Egyptian Old Kingdom around 4200 BP, have both been attributed to climate change and regional desiccation (Cullen et al., 2000;

Hassan, 1997; Weiss, 1997). The demise of the Tiwanaku State in the Titicaca Basin around 1,000 BP has also been linked to drought (Kolata et al., 2000; Ortloff and Kolata, 1993), as has the collapse of the Mayan civilisation around AD 900 (Haug et al., 2003). The role of climatic forcing in such scenarios has led some to argue that, in a context of contemporary global concern over anthropogenic climate change, 'collapse is even more relevant for the future than we would like to believe' (deMenocal and Cook, 2005: 92).

Yet this work has also faced considerable criticism. Blaut (1999), for example, has catalogued a sequence of logical and factual errors in *Guns Germs and Steel*, while Sluyter (2003) has slated the work as 'junk science'. The central thesis has also been criticised for 'overselling' geography as an explanation of history (McNeill, 2005a: 161), while it has been argued that such texts mark a worrying 'new trend' for simplistic and essentially Eurocentric, environmentally deterministic perspectives on human history (Blaut, 1999). There is a fear that palaeoenvironmental work drawing links between climate change and collapse, is also representative of this 'new paradigm' in environmental determinism, from which the geographical community at least is still, to some extent, reeling (Coombes and Barber, 2005; Judkins et al., 2008). Such approaches, it has been argued, fail 'to take into account many of the advances in human–environmental thought since the early twentieth century' (Judkins et al., 2008: 17) and particularly those shifts away from reductionist, deterministic thinking toward much more complex socioeconomic, cultural and political explanations of social vulnerability and discontinuity. Increasingly, direct one-to-one cause and effect relationships between environment and human action are thought to be providing a 'false lead' (McCann, 1999) and, while the majority of human history could be cast in terms of the emergence, fluorescence and decline of great civilisations and the impacts of extreme events and natural disasters, this tends to obscure the social, political, demographic, ecological and climatic contexts which might have helped shape or frame these events (Costanza et al., 2007: 522). Moreover, as Butzer (2005) has recently illustrated, human perception and, in particular, ecological behaviour and coping strategies are absolutely pivotal to understanding cause and effect relationships.

The purpose of this chapter is to explore various examples of societal collapse and social discontinuity in global history and prehistory, but in so doing to highlight the importance of complexity and context in understanding collapse scenarios. It will be argued that in order to understand any scenario of collapse, it is vitally important to acknowledge a society's relative vulnerability, that is to say its temporal and spatially specific social, cultural, demographic and political context and how this changes over time, as well as the environmental and climatic changes that might have contributed to and combined to result in cultural discontinuity. As will be illustrated, the absolute cause of discontinuity or collapse cannot always be linked to a major event in history of prehistory, so much as a combination of events or cumulative developments acting against particular circumstances. Moreover, it will be suggested that environmental changes, and particularly climate events, do not always result in negative impacts but can promote social complexity as well as reduce it (Brooks, 2006; Redman, 1999). Indeed, in line with what Danish economist Boserup argued, times of stress can lead to innovative developments (Boserup, 1965). It follows that environmental changes can act as a kind of 'trigger' to adaptation and can play an important role in the possible emergence and development, as well as the collapse, of civilisations. Considering a society's adaptive capacity at any point in time is thus critical to understanding its propensity for collapse. By means of a theoretical introduction to understanding the idea of collapse, it is important to first consider the meaning of, and approaches to, both vulnerability and adaptation.

2 THE CONTEXT OF COLLAPSE: TRAJECTORIES OF VULNERABILITY IN HISTORY AND PREHISTORY

2.1 Vulnerability, adaptive capacity and the propensity for collapse

Vulnerability can be broadly defined as the potential for loss (Cutter, 1996: 529), the 'state of susceptibility to harm from exposure to stresses associated with environmental and social change and from the absence of capacity to adapt' (Adger, 2006: 268), or 'the degree to which human and environmental systems are likely to experience harm due to a perturbation or stress' (Luers et al., 2003: 255). Though frequently referred to and widely used in the risk, hazards and disaster literature (Blaikie et al., 1994; Burton et al., 1993; Kasperson and Kasperson, 2001), vulnerability is becoming an increasingly important concept in the fields of environmental and climate change (Cutter, 2006) and is fundamental to understanding how and why societies may have been susceptible to collapse at different points in time. However, there have been relatively few historical treatments of vulnerability (Cutter, 1996: 533) or investigations of how the vulnerability of particular societies has changed over time. To some extent, this relative neglect may be a function of the multiplicity of definitions of and approaches to vulnerability, a lack of consensus on the meaning of the term (Luers et al., 2003: 255), and the fact that vulnerability is a concept, not an observable or tangible phenomena, rendering it difficult to identify or monitor in the historical record.

Various genres of literature have focused on vulnerability from the perspective of purely biophysical threats and hazards (Blaikie et al., 1994; Burton et al., 1993; Hewitt and Burton, 1971). Many 'naturalist' explanations of disasters, for example, place total blame for societal disruption on 'the violent forces of nature' (Blaikie et al., 1994: 11; Foster, 1980; Frazier, 1979). These approaches position people or societies as being implicitly vulnerable, owing to their presence in particular fragile or precarious environments. In recent years, however, there has been a move away from attention on the stressors in vulnerability studies to the stressed system, its component parts and its ability to respond (Clarke et al., 2000; Ribot, 1995). Social vulnerability, for instance, focuses on the 'susceptibility of social groups or society at large to potential losses' from hazardous events and disasters (Cutter, 2006: 72). In these approaches, as MacNaghten and Urry (1998) suggest, 'cultural criteria are implicated in the definition and trajectories of even the most apparently physical environmental issues' (cited in Jones, 2002: 248). The nature of a hazardous event or condition is taken as a given or at the very minimum viewed as a social construct, not a biophysical condition (Cutter, 1996). This perspective highlights human coping strategies and responses, including societal resilience to environmental change or hazard, and positions biophysical threats as socially constructed phenomena.

Interdisciplinary research teams have now begun to explore vulnerability to environmental changes as a function of exposure, sensitivity and adaptive capacity, manifested within interactions of both the social and ecological system (Turner et al., 2003). Vulnerability in this body of literature is conceived both as a biophysical risk as well as a social response (Cutter, 1996: 533) and the complex interaction between the environment and human society represents a constraint that not only influences how livelihoods in a region may be vulnerable to disruption, and the way in which environmental change is experienced, but also how different social systems and groups can successfully respond or adapt to such change (Oliver-Smith and Hoffman, 1999: 73). As Kasperson et al. (1988) and Kasperson and Kasperson (1996), have suggested, risks and hazards interact with cultural, social and institutional processes to either temper or aggravate public response and adaptive capacity (Cutter, 2006). On this theme, Vogel (2001) has suggested that greater emphasis needs to be

placed on the 'double-sided structure' of vulnerability, that is to say, studies that consider adaptive capacity, or society's ability to successfully adjust to changed environmental circumstances, 'that take place naturally within biological systems and with some deliberation or intent in social systems' (Adger et al., 2009: 337). Adaptation to climate change, for example, can represent 'an adjustment in ecological, social or economic systems in response to observed or expected changes in climatic stimuli and their effects or impacts in order to alleviate adverse impacts of change or take advantage of new opportunities' (Adger et al., 2005: 78). Environmental changes and stresses can in this way act as 'significant stimuli' for social and technological innovation, and have done throughout history (Adger et al., 2009: 336).

Vulnerability and adaptability may change according to different social, community and institutional scenarios and over time (Kasperson et al., 1995). Messerli et al. (2000), for example, have suggested that the vulnerability of a particular society changes in conjunction with complexity, that is to say a system that develops adaptive mechanisms and adjustments, some of which may in fact render society more vulnerable in the long run. They argue that there is an historical 'trajectory of vulnerability' through which all societies pass as they develop economically, socially and technologically. The position of a society on this trajectory influences its relative vulnerability. 'Nature dominated' hunting and gathering societies, for example, are considered to be the most vulnerable to environmental changes and stresses. Societies who have modified their environments in order to buffer themselves from these changes, for example, who have developed productive agrarian–urban systems based on these adaptations and so become more socially complex, are thought to be more resilient. With population expansion, overexploitation, and heavy reliance upon these buffer systems, the vulnerability of these societies to stresses and challenges, and also to collapse to a lower level of complexity, increases such that, as Van der Leeuw (2008: 481) has argued, society has invested in such a way of life as to be unable to 'innovate itself out of difficulty.' Similarly, Brunk (2002) has argued that increased complexity brings heightened vulnerability, but considers that collapses can best be understood in terms of 'cascades' in self-organising systems. In brief, as civilisations grow in complexity, this brings increases in the interconnectivity of economic and sociocultural systems, or greater self-organising complexity. Societal development is accompanied by a growing interdependency between these different systems. Increased sophistication may lead to a society that as a whole becomes less flexible in its ability to respond to or adapt to rapid environmental changes (Coombes and Barber, 2005). Moreover, with greater interconnectivity, even slight external changes can cause major disruption by 'cascading' through the entire system.

The purpose of the following section is to consider a series of regionally focused case studies drawn from the Americas, Asia, the Mediterranean and Scandinavia and from different periods in history and prehistory in order to explore the case for environmentally driven collapse. It will be argued that while there is considerable archaeological and palaeoclimatic evidence which might be used to draw links between environmental deterioration and the demise or collapse of a complex society, scholarship investigating trajectories of vulnerability reveals that a combination of social, economic and demographic circumstances may have rendered societies effectively more vulnerable to collapse, and less able to adapt, at particular points in time. The purpose is to illustrate that in many cases, environmental change and particularly climate events often act as 'triggers' or 'tipping points' of collapse in history and prehistory rather than as major causes. Consideration is given to the degree to which society is able to learn from and adapt to environmental stresses, as well as the way in which climatic deterioration was a factor in the increasing

complexity and possible emergence, rather than the collapse, of societies.

2.2 Triggers, tipping points or background noise: the role of environment and climate in collapse scenarios

As Cronon (1992: 1347) has highlighted, a key dilemma for scholars of environmental change is that the past is filled with 'many stories, of many places in many voices, pointing towards many ends'. Historical and prehistorical events are in effect assembled into causal sequences and narratives, that 'order and simplify those events to give new meanings' to the past. Scholars try to find meaning in 'an overcrowded and disordered chronological reality' such that inevitably 'we divide the causal relationships of an ecosystem with a rhetorical razor that defines included and excluded, relevant and irrelevant, empowered and disempowered' (Cronon, 1992: 1349). Particular aspects of an historical event can thus be emphasised, elaborated, obscured, omitted or ignored in order to tell a particular story. It follows that sociopolitical discontinuities in the historical and prehistorical record can be interpreted very differently depending on the availability, choice and interpretation of evidence. The difficulties of dealing with complex processes of biosphere and the human realm, or anthroposphere, mean that 'data are often open to multiple interpretations and ... it is difficult to decide between the different scenarios' (Van der Leeuw, 2008: 477).

At various stages during the Holocene, whole empires collapsed and their populations were reduced or became more dispersed, migrated and adapted to new modes of subsistence (deMenocal, 2001: 292). The most striking of studies involve the apparently 'sudden' collapse of empires (Coombes and Barber, 2005: 304), though there are very many other examples from the historical and prehistorical record. Indeed, as Rothman (2005: 575) has recently pointed out,

'[M]any more societies … have graced the face of the earth and have then fallen, never to be seen again.' These collapse scenarios can be framed by very different narratives, privileging, for example, predominantly cultural or environmental explanations of change and discontinuity. In some cases, however, and as shall be demonstrated, much more complex explanations may need to be considered. As Van der Leeuw (2008: 481) has suggested, for example, periods of crisis in the Holocene can be 'triggered by both social and environmental changes or (most often) by a combination of both'.

Climate change has often been invoked as a key cause or driving force of collapse events in a considerable body of archaeological literature. Weiss and Bradley (2001: 609), for example, highlight how the accumulation of high-resolution palaeoclimatic data is providing 'an independent measure of the timing, amplitude and duration of past climate events', some of which have been highly disruptive, and contributed to collapse. They have argued on the basis of numerous examples that societies at a range of different spatial scales were highly vulnerable to climate disturbances, and they forward evidence to identify climate as a key agent in their apparent collapse. The extent to which past climatic fluctuations and perturbations have influenced past human societies, however, is an area of considerable debate. It has been argued, for example, that a causal link cannot and should not necessarily be inferred 'merely by observing correlations between environmental and cultural variabilities. Such associations may be purely coincidental' (Coombes and Barber, 2005: 305). Catto and Catto (2004: 123) highlight a number of examples where climate may act 'as a supporting player' or merely as 'background noise' rather than a direct cause in collapse scenarios. Alternative explanations may be found in cultural, economic or political causes, and environmental explanations are not always sufficient to account for societal change. Attention must be given to the nature of the event but also the societal context within which this acts. As Figure 40.1 suggests, low-amplitude low-frequency events may have little effect on resilient societies

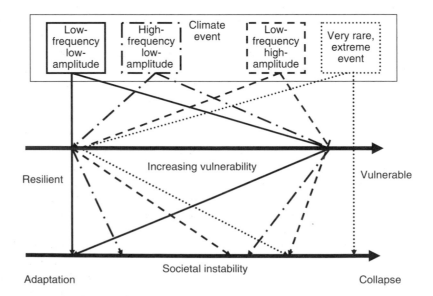

Figure 40.1 Hypothetical relationships between climatic events, vulnerability and societal adaptation or collapse.

but may affect the most vulnerable societies depending on the demographic and socioeconomic context in which they act. High-frequency low-amplitude events, or sequential climate events may cause some disruption for resilient societies but acting against a vulnerable context may lead to more societal disruption. Low-frequency high-amplitude events, however, can cause disruption even for more resilient societies, though the latter may be better placed to respond through adaptation than the most vulnerable. The rare, extreme event can cause disruption for even the most resilient of societies. Collapse is, however, the exception rather than the norm in the historical and prehistorical record and where links are made between climate and collapse, the context in which the climate event acts needs necessarily to be taken into account. Moreover, notwithstanding developments in high-resolution dating techniques, causal relationships can still be particularly difficult to unravel in palaeoenvironmental reconstructions (Matthews, 2005: 1103). Meyer et al. (1998: 230) suggest, conclusions over environmental explanations of collapse are often be reached simply because external triggering events, such as climate, are more 'conspicuous' in the surviving data relative to more complex underlying socioeconomic or political problems which can be difficult to identify and decipher.

Recent archaeological and palaeoenvironmental investigations of past collapse scenarios are serving to highlight some of the problems with predominantly environmental explanations. Archaeological evidence has shown, for example, that the Akkadian empire, with its highly developed civilisation, complex systems of water management and grain storage designed to buffer against periodic drought, collapsed abruptly around 4,170 BP. This phenomenon has been related to a shift to more arid conditions around this time (Cullen et al., 2000). Based on excavation of a mound at Tel Leilan, in Northeast Syria, Weiss et al. (1993) have argued that a significant and protracted drought, which may have lasted over several centuries, might

account for the fate of this, and also possibly other civilisations in Greece, Egypt and the Indus Valley around the same time. Archaeological evidence certainly points to the widespread abandonment of urban centres around this period, coincident with 'dramatic influxes' of people into Southern Mesopotamia (deMenocal, 2001: 669). Cullen et al. (2000), meanwhile, have identified an increase in windblown or aeolian dust of Mesopotamian provenance in a deep-sea sediment core from the Gulf of Oman. The presence of wind-borne volcanic glass at Leilan and adjacent sites has also been attributed to a major volcanic eruption thought to be in Anatolia. This, Weiss et al. (2003) argue, may have triggered a 'volcanic winter' accounting for climatic disruption and prolonged drought. In short, the combined lines of evidence would point towards tectonism as a key forcing factor of socioeconomic discontinuity.

As Meyer et al. (1998: 231) argue, while there is evidence for widespread disruption and discontinuity in this period, the post-Leilan collapse strata 'document a catastrophic biological event rather than an abrupt climate change in the strict sense of the term'; the volcanic glass is thought to have resulted from a regional rather than a global event judging by a lack of evidence for any globally effective eruption in ice cores from Greenland. Pollen cores from the Zagros Mountains indicate that there were significant vegetation changes around 4,150 years BP corresponding to a decline in oak forest and an increase in grassland. There is also evidence, in the form of reduced discharge on the Euphrates River, that would suggest drier climatic conditions. This, Meyer et al. (1998) argue, points towards a regional drying event in the Upper Euphrates Basin and Zagros mountains. That said, the widespread political discontinuity that was experienced is thought to have been real, but related more to difficulties cascading through the long-distance trade dependencies with the Akkadian state. Indeed, as Meyer et al. (1998: 233) argue 'the collapse of one state within a chain to commercially integrated regions … can have repercussions for all the others' and thus may provide one alternative explanation for the widespread collapse despite the apparently regional climate events that have been invoked.

Other long-studied examples have invoked climate as a cause of collapse. The complex civilisations of Ancient Egypt, for example, were 'inextricably linked' to the River Nile (Meyer et al., 1998). The agricultural cycle depended on the summer floods and the year could be divided into three distinctive periods: the season of flooding, peaking in August; the season of drought, with the river reaching its lowest levels in May; and the period in between May and September that witnessed an increase in the level of the flood water (Hassan, 2005). Following the adoption of an agricultural way of life, societies began to coalesce across the region until around 5,250 BP when Egypt had been unified into a single state. This was a period of urban expansion, pyramid building and cultural fluorescence. Around 4,150 BP, however, the centralised government of this Old Kingdom began to disintegrate and, as Tainter (1990: 8) has argued, there followed a period of considerable strife, famine and hunger, theft, killing, revolution and social anarchy, all of which contributed to this being 'one of the darkest episodes in Egyptian history'. This was linked at the time to the failure of the Nile (Bell, 1971), a theory that has been supported by evidence of lake-level changes in the Faiyum Depression (Hassan, 2005). Butzer (1976), however, has argued that Nile fluctuations were a contributory factor rather than cause of collapse of the Old Kingdom in Egypt, in combination with other problems, specifically, political weaknesses, poor leadership and over taxation, that effectively increased social vulnerability. All these factors, it has been argued, combined to contribute to social disintegration, with climate being a trigger, but not a direct cause of the ultimate collapse (Tainter, 1990: 48).

This event, which is thought to have been relatively short-lived, did nevertheless stimulate various adaptations. Within a quarter of a century the country began to take steps towards recovery and the development of a hydraulic society. Dams were built, canals were dug and reservoirs were excavated to regulate the flow of water into the Faiyum Depression such that during years of drought, such as those that contributed to the collapse of the Old Empire, water could still be provided for the population. It is somewhat ironic, therefore, that the Middle Kingdom is thought to have been weakened by a series of devastating Nile floods around 3,620 BP. The New Kingdom (3,100–3,020 BP) came to a similarly dramatic end, but at the same time as other states in the Near East collapsed, leading some to argue once again for a predominantly climatic cause. Certainly there is evidence that would indicate reduced Nile flood levels, high food prices and considerable social unrest at this time (Butzer, 1976, 1980 and 1984). As Meyer et al. (1998) demonstrate, however, there may again have been other contributory factors which reduced social resilience and flexibility to adapt at the time, not least a series of financially draining battles with barbarian peoples, the consequent loss of external trading relationships, and shifts in the balance of power from pharaoh to priesthood. The Egyptian state had also been 'characterized by a top-heavy socio-political structure' since the late fifteenth century BC which was placing pressure on an already stretched agricultural sector (Meyer et al., 1998: 235). Against these underlying structural factors, environmental stress in the form of reduced Nile levels, could have represented the final straw or 'trigger' to collapse.

Among the most spectacular of collapse scenarios is that of the Mayan civilisation in the Southern Lowlands of Mesoamerica. The Maya represented a particularly complex and sophisticated society which constructed elaborate palaces, temples and entire cities. They occupied an area that equates with approximately 25,000 km² of the karstic lowlands of central America and enjoyed 'an uninterrupted legacy' from around 400 BC to around

AD 900 (Scarborough, 2008: 51). Notwithstanding the often difficult and frequently inhospitable environment, the civilisation reached its zenith around AD 700 with an estimated population of 10 million people. According to Scarborough (2008: 53), this society represented the 'first experiment in state-like complexity' and its 'associated material trappings'. This was a society that built pyramids and ball courts and which significantly modified their landscape to centralise their water needs in an area that was (and is) subject to drought. Following a theory first forwarded by the environmental determinist, Ellsworth Huntington in 1917, climate has been invoked to explain the 'sudden' demise of this sophisticated society around AD 800–900 when urban centres faced depopulation and trading decreased across a broad region (Curtis et al., 1996). It is known that droughts regularly destabilised agricultural production and affected food security and social and economic well being in various parts of the country throughout history and prehistory (Florescano, 1980; Hodell et al., 1995; Liverman, 1999), and still do today. Palaeolimnological analysis of lake sediments from Punta Laguna in the Yucatan Peninsula (and also from Lake Peten Itza in the lowlands of Guatemala) (Curtis et al., 1996, 1998; Hodell et al., 2005a, 2005b) revealed that the period between 1,785 and 930 BP was particularly dry, with exceptionally arid events centred around AD 862, 986 and 1051. Based on measurements of the bulk titanium content of undisturbed sediment in the Cariaco Basin of the southern Caribbean, which reflect variations in riverine input and the hydrological cycle over northern tropical South America, Haug et al. (2003) have also identified an extended dry period, punctuated by intense multiple year droughts centred at approximately AD 810, 860, and 910. Moreover, tree ring investigations support evidence of a prolonged period of drought, or megadrought, around the same time (Acuña-Soto et al., 2005). There is, therefore, little doubt that climate fluctuated in this area in the past and there is very strong

support for sustained drought around the time of the Maya collapse.

The context against which these drought events took place, however, needs to be considered, particularly 'from the vantage of a fragile biophysical setting' that was heavily engineered by the Maya (Scarborough, 2008: 51). This was a 'high density, stressed population, practicing intensive agriculture, living largely in political centres, supporting both an elite class and major public works programs and competing for scarce resources' (Tainter, 1990: 169). To sustain what was a substantial population for such a long period of time necessitated significant landscape modification (Meyer et al., 1998: 236). Anthropological and archaeological evidence points to problems of nutritional stress and disease, particularly among the poorer classes, and demographic instability, agricultural intensification, mono-cropping and degradation of the agricultural landscape (Stanley et al., 1986). All of these factors may have rendered the population potentially vulnerable to climatic fluctuations, be they shifts to wetter or drier conditions or prolonged and occasionally intense drought.

Much of Mexican history and prehistory is characterised by cycles of growth and collapse of cultural centres. Although some instances of collapse and cultural demise coincide with climatic anomalies, other examples are more ambiguous. Moreover, there are plenty of examples of collapse from elsewhere in Mexico where, on the basis of existing evidence at least, climate may not have been a contributory factor so much as population pressure and environmental degradation. The collapse of the civilisation in Monte Alban, Oaxaca around AD 750 represents a case in point. The rise of Monte Alban in Oaxaca, Southern Mexico, around 2,450 BP coincided with a growth in public and ceremonial buildings at many other sites and there are numerous secondary centres in the valleys of Oaxaca, Tlacolula, Etla and Zimatlán (Kowalewski et al., 1989; Marcus and Flannery, 1996). While Monte Alban retained is unparalleled political and administrative importance in the region, there is evidence to suggest there were other important population centres in the valley by the Early Classic period (AD 200–450) which became much better integrated both politically and economically (Feinman et al., 1985). Settlements on defendable hilltop locations had begun to emerge in the southern end of the valley by AD 450, though the population of the Etla valley had begun to decline by this stage. However, sometime around AD 600–700, there was an increase in the amount of land that was exploited, including areas classed as agriculturally marginal or high risk (Sanders and Nichols, 1988), associated perhaps with a threefold increase in population during Monte Alban's period of supremacy (Feinman et al., 1985). When the city reached its maximum population of 15,000–30,000 sometime after about AD 500, it is thought that population pressure may have begun to exceed the supporting capacity of the surrounding area. Greater exploitation of less favourable areas of land accelerated soil erosion and this in turn may have contributed to a period of cultural collapse known as the Epiclassic demographic decline in the Valley of Oaxaca about AD 750 (Balkansky et al., 2000: 368). Population began to decline and settlements were dramatically abandoned, including those at distances of up to 20 km from Monte Alban (Feinman et al., 1985). Ultimately, the administration in the valley is thought to have disintegrated into fragmented political entities, governed by indigenous elites (Balkansky et al., 2000).

Among the more controversial collapse scenarios is that of Tiwanaku, a complex state that developed on the southern shores of Lake Titicaca, Bolivia, in the central-southern Andes (Ortloff and Kolata, 1993). This has been acknowledged as one of the most studied and most contested archaeological sites in South America (Calaway, 2005). It has been argued that periods of cultural change in the area, including the emergence, intensification and abandonment of agriculture, coincided with changes in the water balance of the Lake Titicaca drainage basin (Binford et al., 1997: 242). Indeed, on the basis of a combination of palaeolimnological and archaeological evidence, the emergence of complex societies in the Lake Titicaca area is thought to have been associated with an increased availability of moisture around 3,450 BP, leading to a rise in lake levels. Between AD 100 and AD 1,100, however, after an estimated seven centuries of growth and expansion and 'remarkable cultural adaptability within a seemingly harsh environment' (Binford et al., 1997: 235), involving adaptations such as raised field cultivation, the Tiwanaku State is thought to have disintegrated as a regional political centre. Collapse is thought to have been 'sequential' across the agricultural system, with the irrigation based agriculture of the lower altitudes being disrupted first, followed by the ground water based raised fields of the altiplano (Ortloff and Kolata, 1993: 195). One supposed cause is a prolonged (c.300 year) period of drought which had severe impacts on these intensive agricultural systems upon which the society had become dependent (Binford et al., 1997; Kolata, 1993; Kolata et al., 2000; Ortloff and Kolata, 1993). Certainly, this period of aridity is thought to be evident in the Quelcaya Ice Cap which lies in the north of Lake Titicaca's watershed (Thompson et al., 1988).

This thesis has been criticised. Erickson (1999), for example, has suggested that the argument was a product of 'neo-environmental determinism' and underplays the role of human agency in the collapse, leading to rebuttals from Kolata et al. (2000) that such criticism is based on 'a pastiche of ethnographic analogy and currently fashionable theory' rather than sound empirical palaeoenvironmental data (Kolata et al., 2000: 425–6). Calaway's (2005) analysis of the scientific evidence presented in the palaeoenvironmental studies on Tiwanaku, however, has highlighted the complexities of using ice core data, which may provide insight into the precipitation history of a particular summit rather than a region per se. Furthermore, he argues that palaeoclimatologists and archaeologists have 'failed to distinguish between the timescales appropriate to environmental and archaeological evidence' (Calaway, 2005: 779). Thus, whereas palaeoclimatologists discuss climatic change on a scale of millennia, archaeologists interested in the record of human activity work towards more narrow date margins; the interpretation of data from each discipline can be compromised when attempts are made to compare them. Calaway argues this to be the case in the Tiwanaku case study. Climatic events and social history do not necessarily tally and 'fluctuating lake levels or ice accumulation do not necessarily mean that a significant environmental threshold has been compromised or crossed' (Calaway, 2005: 787). In fact, he concludes that there is little unambiguous evidence of a drought having caused the collapse of the Tiwanaku State. Other scholars, moreover, have suggested that the abandonment of urban centres and raised fields in the Tiwanaku State represent separate events and are not necessarily related (Owen, 2005), leading to questions over the degree to which agricultural decline as a result of climatic change led to the collapse of the central state.

There are more recent examples of collapse from elsewhere which may again be a function of more complex interaction between environmental and socioeconomic parameters. The period between around AD 800 and 1,000, for example, witnessed the expansion of Nordic settlement and colonisation of the islands of the North Atlantic (Vesteinsson et al., 2002), producing complex interactions of culture and nature (McGovern et al., 2007: 27). Nordic pirates, traders and settlers first colonised the islands in the east (Shetland, Orkney, Hebrides, Man, Ireland and the Faroes) between AD 800 and 850, They expanded into Iceland around AD 874, Greenland around AD 985 and across to the Newfoundland–Gulf of the St Laurence region, where for a short period the Vinland

Colony was established around AD 1,000. By the mid eleventh century, however, the Vinland Colony had been abandoned and sometime around the mid fifteenth century, the Greenland colony had become 'completely extinct' (McGovern et al., 2007: 25). Iceland, which had become a central cultural and literary foci of Scandinavia, deteriorated into 'an impoverished marginal backwater by 1500' (McGovern, 1990: 333). The reasons for the demise of these parts of the Nordic empire have long been investigated by philologists, medieval archaeologists and historians (Frioriksson, 1994); the first systematic investigations in this respect began in the nineteenth century (McGovern, 1990). Different disciplinary approaches have resulted in a variety of explanations for the apparent collapse, 'producing overly simplistic explanations of complex phenomena by privileging environmental or social explanations, or taking grand evolutionary or local historical perspectives according to the scholarly fashion prevalent among the investigators' (Vesteinsson et al., 2002: 99). Jones (1986) has argued, for example, that changes in politics and market forces in Europe played a central role in bringing an end to the 'North Atlantic Saga', while the impacts of humans on island ecosystems has also been forwarded as an explanation. Certainly, broad similarities have been drawn with Polynesian collapse trajectories and the Nordic island enterprises (McGovern, 1990). This demise has also been taken as a prime example of the impact of changing climate on human populations (Arneborg et al., 1999; Barlow et al., 1997; Catto and Catto, 2004). As Mann (2002) explains, the cooling that accompanied the Little Ice Age may have led to more extensive coverage of sea ice in the Atlantic, which will inevitably have created problems for the fishermen in Iceland and across Scandinavia, as well as for the Norse settlements in Greenland and Iceland. Previously accessible trade routes with Europe, upon which some of the Norse settlements relied, will also have been closed off. Recent interdisciplinary initiatives combining archival documentary, palynological, archaeobotanical and zooarchaeological approaches with settlement surveys, and soil and geomorphological investigations, have produced a richer understanding of the Norse migrations and settlement and have afforded clearer insight into their effective environmental and economic contexts (Arneborg and Gronnow, 2005; Ogilvie and McGovern, 2000). Dugmore et al. (2007: 12), for example, have illustrated that changing patterns of trade, rather than climatic deterioration per se might have 'marginalised the Norse Greenland settlements and effectively sealed their fate'. In this case, long-term increases in vulnerability caused by economic change, albeit in combination with ongoing climatic changes associated with the onset of colder conditions, may have led to difficulties which then cascaded through what had become an interdependent settlement system, ultimately resulting in collapse. Others have posited that a combination of factors, including climate variability, declining trading relationships with Norway, Norse impact on the environment, disease, hostile contact with the Inuit and also possibly inbreeding, might explain why the settlements failed (see Hunt, 2009).

Although there are many examples of past societies that have collapsed under external stresses, there are many more examples which have been positively transformed under similar threats (Steffen, 2008: 512). Indeed, it should be remembered that social demise and disintegration represent only one response to crisis. Human societies respond to change via multiple pathways, including adaptation and innovation (Redman, 1999). As Costanza et al. (2007: 523) have argued, extreme drought, for instance, has 'triggered both collapse and ingenious management of water through irrigation'. Adaptation might also be manifest in the establishment and implementation of trading relationships and market exchanges, and might be achieved through individual or institutional action and decision making, through legal and legislative intervention, regulation or technological

change. It might also involve the marshalling and extending of social networks and relationships (Adger et al., 2005; Smit et al., 2000), and the emergence, rather than the decline, of complex societies. Knowledge of drought, flood and disease events in the past, for instance, can condition how society comprehends and responds to the problems of uncertainty and how that society not only conceptualises the likely risk of events, but also anticipates the impacts of those of the future (Koselleck, 1985). Experience or knowledge of such events and their impacts in the recallable past can effectively become part of the cultural and infrastructural fabric of thought, discourse and practice of a society or community and can thus prompt a variety of remedial or mitigating actions, coping strategies and adaptations (Hassan, 2000; Redman, 1999). In short, stresses can improve, as well as challenge, societal resilience. Indeed, although considerable literature has focused on the ways in which declining environmental conditions, drought or desiccation in the past may have contributed to social disintegration and collapse, abrupt changes in climate or environmental stresses should not always be associated with interruptions or reversals in the progressive development of human societies. Citing examples of the emergence of complex societies, that is to say highly organised state level societies in Egypt, Mesopotamia, South Asia, northern China and coastal Peru, for example, Brookes has argued that 'pronounced increases in social complexity in the middle Holocene coincided with climatic and environmental deterioration, and in particular with increased aridity' throughout the global monsoon belt around 6,000 BP (Brooks, 2006: 30). What this demonstrates is that even in a context of significant environmental stress, collapse is by no means an inevitability. In contrast, human societies have demonstrated an inherent adaptability throughout history and prehistory and have developed institutions and cultural coping strategies to deal with the impacts of environmental changes at a range of temporal and spatial scales (Fraser et al., 2003; Hassan, 2000).

3 DISCUSSION: LESSONS OR 'STORIES' FROM THE PAST?

There has been a growing interest in the impacts of environmental change on past civilisations (Catto and Catto, 2004). Benefiting from high-resolution environmental data over the past two decades (Weiss and Bradley, 2001), recent scholarship has striven to demonstrate strong links between environmental, particularly climatic, stresses and cultural collapse. It has been argued that this is a key reason to 'resurrect' environmentally deterministic explanations of social history and prehistory (Leroy, 2006: 7). Purely environmental explanations of collapse, however, are difficult to infer and remain controversial. As Leroy (2006: 5) has argued, 'a large cause (environmental hazard) may produce only a small effect (disaster) and vice versa. Moreover, societal responses to external forces are non-linear in nature; hence, in the geoarchaeological record, assumption of causality between cultural transition and natural hazard is often questionable and must be used with care.' Moreover, of major environmental catastrophes, 'only a fraction has caused societal collapse' (Leroy, 2006: 7). Alternative explanations do exist. In fact many of the cases discussed in this chapter, and in the collapse literature more generally, have been at one time interpreted 'solely in terms of human factors ... such as warfare, overpopulation, deforestation and resource depletion' rather than being ascribed an environmental or climatic cause (deMenocal, 201: 672). It should also be remembered that major events, and particularly those which leave discontinuities in the historical record, do not always require major causes (Coombes and Barber, 2005: 303); rather a suite of social, economic, political, demographic and environmental factors that have the potential to coalesce at a particular point in time to cause dislocation, and invariably only then for a particular culture group or sector of society. As Butzer (2005: 1798) has argued, 'cause-and-effect ... is ultimately about real people, living communities', suggesting that the mediating context in which an event acts must be taken into account in any study of societal discontinuity or collapse. The physical environment more often provides a setting or context in which social change occurs, providing not only constraints but also opportunities for social, cultural and technological innovation (Brooks, 2006: 30).

However, it has been argued that for global society 'to survive we must learn from the past, adapt to environmental changes, and modify our lifestyles in front of a darkening nature' (Leroy, 2006: 11). In short we are encouraged to tap into what historians have referred to as 'a usable past' (McNeill, 2005b), that is to say a past that 'must be used to understand the present' (deMenocal, 2001: 667). Whether proscriptive lessons can be learned from the failures of the past, however, is open to question. Deriving insights about the contemporary and future implications from the past is problematic (McNeill, 2005a: 178). Limited chronological resolution, insufficient data and oversimplification can hamper the value of using such historical examples as analogues to glean insights about the potential implications of contemporary or predicted environmental or climatic scenarios (McNeill, 2005a: 178; Meyer et al., 1998). The fact that past societies differed significantly from those in the modern world makes simple parallels unrealistic and unworkable (Ingram et al., 1981: 5; Meyer et al., 1998). The world population, for example, is larger than ever and growing exponentially. More people are living in fragile environments than ever before, rendering them potentially more vulnerable. Societies today are inherently more complex, and display less flexibility, than those of the past (Torrence, 2002, cited in Leroy, 2006: 7). Furthermore, today's societies rely more on a constructed or modified environment, are generally less resilient and less able to adapt in the same way as less complex communities. As McNeill (2005a: 178) has suggested, therefore, 'taking the extra step to get from historical examples to insights about the contemporary world is always tricky'. There will, moreover, always remain gaps in the record and uncertainty with regard to narrative description and explanations (Dearing, 2007).

Yet while there is a good deal of uncertainty, for example, with respect to climatic futures, the value of looking to the historical record to learn about societal vulnerability to past environmental changes should not be underestimated (Bradley, 1999). We do have the benefits of hindsight (Matthews, 2005), and what makes such events relevant to modern times is the fact that they document both resilience and vulnerability of large societies to environmental stresses (deMenocal, 2001: 672). Knowledge of successes and failures in adaptation to past environmental change and climatic variability might possibly increase the ability to respond to the threats of long-term climate changes (Tompkins and Adger, 2004). In particular, exploring the experiences of individuals, groups and places in the past could help us understand the relative flexibilities of societies in dealing with environmental changes and stresses, and demonstrate how societies can 'shift to lower subsistence levels by reducing social complexity, abandoning urban centres, and reorganizing systems of supply and production' (deMenocal, 2002: 672). Moreover, they might also help us to identify the most vulnerable societies and places (Meyer et al., 1998: 218; Swetnam et al., 1999). Coordinated and interdisciplinary efforts in this respect are being conducted under initiatives such as the 'Integrated History and Future of the People on Earth,' or IHOPE, which seeks to map the integrated records of biophysical and human change on earth over the past few thousand years with the ultimate goal of creating 'a better, more sustainable and desirable future' (Costanza et al., 2007: 526). Computational modelling has also recently been highlighted as a useful

tool with which to derive contemporary lessons from past human–environment interactions (Axtell et al., 2002; Dean et al., 1999; Oldfield, 2008). Axtell et al. (2002) for example, have developed a 'multiagent computational model' of the Kayenta Anasazi in Long House Valley in northeastern Arizona. This reproduces the main features of the society's cultural and environmental history between AD 800 to 1300, including changing demographic conditions, settlement patterns and also the society's ultimate rapid decline. Based on the model, the authors argue that environmental conditions alone may not explain the cultural collapse and that additional push or pull factors may have contributed to the abandonment of the valley sometime after AD 1300. The severely depressed state of the population, for example, may have been 'insufficient to maintain cultural institutions', while alternatively a combination of environmental, demographic and epidemiological factors may have been responsible for cultural demise.

Finally, it should be remembered that the majority of human history and prehistory, is characterised by human adaptability. As Butzer argues, 'people can and do adapt to uncertainty and change', and human societies have, in general, proved resourceful. Collapse is not the norm but rather, and as Tainter (1995: 398) has suggested 'the rich historical records of societies that have collapsed represents to us, not the normal destiny of complex societies but a set of anomalies needing to be explained'. A focus on collapse events and a general contemporary environmental pessimism have obscured the fact that vulnerability to environmental and climatic change does have an orienting function and that human societies are adaptable and have developed institutions and cultural coping strategies to deal with the impacts of such changes (Fraser et al., 2003; Hassan, 2000). Even where societies did face disintegration, 'new civilisations rise in the very places that the old ones failed' (Rothman, 2005: 576) and as Brooks' research demonstrates 'societies seem to ebb and flow into one another' and this 'melding of cultures and the accretion that accompanies it are crucial dimensions in the evolution of human societies' (Rothman, 2005: 576). Indeed, although it could be argued that society today confronts global challenges of an unprecedented scale (Tainter, 1995: 404), and that our highly interconnected globalised world means that social or environmental failure in one region could threaten society at a much greater scale, as Matthews (2005:1103) has succinctly stated, 'successful long term maintenance of past society suggests there are grounds for optimism in the future.'

REFERENCES

Acuña-Soto R., Stahle D. W., Cleaveland M. K. and Therrell M. D., Chavez S. G. and Cleaveland M. K. 2005. Drought, epidemic disease and the fall of the classic period cultures in Mesoamerica (A 750–950). Hemorrhagic fevers as a cause of massive population loss. *Medical Hypotheses* 65(2): 405–409.

Adger W. N. 2006. Vulnerability. *Global Environmental Change* 16: 268–281.

Adger W. N., Arness N. W. and Tompkins E. L. 2005. Successful adaptation to climate change across scales. *Global Environmental Change* 15(2): 77–86.

Adger W. N., Dessai S., Goulden M., Hulme M., Lorenzoni I., Nelson D. R. et al. 2009. Are there social limits to adaptation to climate change? *Climatic Change* 93(3–4): 335–354.

Arneborg J. and Gronnow B. 2005. Dynamics of northern societies. *Proceedings of the SILA/NABO conference on Arctic and North Atlantic archaeology, Copenhagen May 10–14, 2004, Vol. 10.* National Museum Studies in Archaeology and History. pp. 363–372

Arneborg J., Heinemeier J., Lynnerup N., Nielsen H. L., Rud N., Sveinbjornsdottir E. A. 1999. Change of diet of the Greenland Vikings determined from stable isotope analysis and 14C dating of their bones. *Radiocarbon* 41: 157–168.

Axtell R. L., Epstein J. M. and Dean J. S. et al. 2002. Population growth and collapse in a multi-agent model of the Kayenta Anasazi in Long House valley. *Proceedings of the National Academy of Sciences* 99: 7275–7279.

Balkansky A. K., Kowalewski S. A., Perez-Rodriguez V., Pluckhahn T. J., Smith C. A., Stiver L. R. et al. 2000. Archaeological survey in the Mixteca Alta of Oaxaca, Mexico. *Journal of Field Archaeology* 27(4): 365–389.

Barlow L. K., Sadler J. P., Ogilvie A. E. J., Buckland P. C., Amorosi T., Ingimundarson J.H. et al. 1997. Interdisciplinary investigations of the end of Norse western settlement in Greenland. *The Holocene* 7: 489–499.

Bell B. 1971. The Dark Ages in ancient history, 1: the first Dark Age in Egypt. *American Journal of Archaeology* 75: 1–26.

Binford M. W., Kolata A. L., Brenner M. Janusek J. W., Seddon M. T. et al. 1997. Climate variation and the rise and fall of an Andean civilization. *Quaternary Research* 47: 235–248.

Blaikie P., Cannon T., Davis I. and Wisner B. 1994. *At Risk. Natural Hazards, People's Vulnerability And Disaster.* London: Routledge.

Blaut J. 1999. Environmentalism and Eurocentricism. *The Geographical Review* 89(3): 391–408.

Boserup E. 1965. *The Conditions of Agricultural Growth: The Economics of Agrarian Change Under Population Pressure.* London: G. Allen and Unwin.

Bradley R. S. 1999. *Paleoclimatology: Reconstructing Climates of the Quaternary.* San Diego: Harcourt Academic Press.

Brooks N. 2006. Cultural responses to aridity in the Middle Holocene and increased social complexity. *Quaternary International* 151: 29–49.

Brunk G. G. 2002. Why do societies collapse? A theory based on self-organised criticality. *Journal of Theoretical Politics* 14: 195–230.

Burton I., Kates R. W. and White G. F. 1993. *The Environment as Hazard.* New York: Guildford Press.

Butzer K. W. 1976. *Early Hydraulic Civilization in Egypt.* Chicago University Press, Chicago.

Butzer K. W. 1980. Civilisations organisms or systems? *American Scientist* 68: 517–523.

Butzer K. W. 1984. Long-term Nile flood variation and political discontinuities in Pharaonic Egypt, in Clark J. D. and Brandt S. A. (eds) *From Hunters To Farmers.* Berkeley: University of California Press.

Butzer K. 2005. Environmental history in the Mediterranean world: cross-disciplinary investigation of cause-and-effect for degradation and soil erosion. *Journal of Archaeological Science* 32: 1773–1800.

Calaway M. J. 2005. Ice cores, sediments and civilization collapse: a cautionary tale from lake Titicaca. *Antiquity* 79 (306): 778–790.

Catto N. and Catto G. 2004. Editorial. *Quaternary International* 123–125: 7–10.

Clark, W. C., Jaeger J., Corell R., Kasperson R, McCarthy J. J., Cash D, Cohen S. J. et al. (2000) Assessing Vulnerability to Global Environmental Risks. Discussion Paper 2000–12, Environment and Natural Resources Program, Belfer Center for Science and International Affairs, Harvard Kennedy School.

Coombes P. and Barber K. 2005. Environmental determinism in Holocene research: casualty or coincidence? *Area* 37(3): 303–311.

Costanza R., Graumlich L., Steffen W., Crumley C., Dearing J., Hibbard K. et al. 2007. Sustainability or collapse: what we can learn from integrating the history of humans and the rest of nature. *Ambio* 36(7): 522–527.

Cronon W. 1992. A place for stories: nature, history and narrative. *The Journal of American History* 78(4): 1347–1376.

Cullen H. M., deMenocal P. B., Heming S., Hemming G., Brown F. H., Guilderson T. and Sirocko F. 2000. Climate change and the collapse of the Akkadian Empire: evidence from the deep sea. *Geology* 28: 379–382.

Curtis J. H., Hodell D. A. and Brenner M. 1996. Climate variability on the Yucatan peninsula (Mexico) during the past 3500 years and implications for Maya cultural evolution. *Quaternary Research* 56(1): 37–47.

Cutter S. L. 1996. Vulnerability to environmental hazards. *Progress in Human Geography* 20: 4: 529–539.

Cutter S. 2006. (ed.) *Hazards, Vulnerability and Environmental Justice*. London: Earthscan.

Dean J. S., Gumerman G. J., Epstein J. M. et al. 1999. Understanding Anasazi culture change through agent-based modeling, in Kohler T. and Gumerman G. (eds) *Dynamics of Human and Primate Societies*. Oxford: Oxford University Press, pp. 179–204.

Dearing J. 2007. Human-environmental interactions. Learning from the past, in Costanza R., Graumlich L. J. and Steffen W. (eds) *Sustainability or Collapse. An Integrated History and Future of People on Earth*. Dahlem Workshop Reports. Cambridge, MA: MIT Press, pp. 19–37.

deMenocal P. B. 2001. Cultural responses to climate change during the Late Holocene. *Science* 292: 667–673.

deMenocal P. B. and Cook E. P. 2005. Agents of Collapse: Megadroughts in the American West (Book review: *Collapse*, by Jared Diamond). *Current Anthropology* 46(5): 91–100.

Diamond J. 1997. *Guns, Germs and Steel. The Fates of Human Societies*. New York: W.W. Norton.

Diamond J. 2005. *Collapse: How Societies Choose to Fail or Succeed*. New YorK: Viking Press.

Dugmore A. J., Keller C. and McGovern T. H. 2007. Norse Greenland settlement: reflections on climate change, trade, and the contrasting fates of human settlements in the North Atlantic islands. *Arctic Anthropology* 44(1): 12–36.

Erickson C. 1999. Neo-environmental determinism and agrarian 'collapse' in Andean prehistory. *Antiquity* 73: 634–642.

Feinman G. M., Kowalewski S. A., Finsten L., Blanton R. E and Nicholas L. 1985. Long-term demographic change: a perspective from the Valley of Oaxaca. *Journal of Field Archaeology* 12(3): 333–362.

Florescano E. 1980. Una historia olvidada: la sequia en Mexico. *Nexos* 32: 9–13.

Foster H. 1980. *Disaster Planning: the Preservation of Life and Property*. New York: Springer Verlag.

Fraser E. D. G., Mabee W. and Slaymaker O. 2003. Mutual vulnerability, mutual dependence. The reflexive relation between human society and the environment. *Global Environmental Change* 13: 137–144.

Frazier K. 1979. The Violent Face of Nature. New York: William Morrow.

Frioriksson A. 1994. *Sagas and Popular Antiquarianism in Icelandic Archaeology*. Aldershot, Avesbury: Samson Publishers.

Greer J.M. 2005. *How Civilisations Fall: A Theory of Catabolic Collapse*. www.xs4all.nl/~wtv/power-down/greer/htm, accessed 21 June 2006.

Haug G. H., Gunthe D., Peterson C., Sigman D. M., Hugeb K. A and Aeschlimann B. 2003. Climate and the collapse of the Maya civilisation. *Science* 299: 1731–1735.

Hassan F. 1997. The dynamics of a riverine civilization: a geoarchaeological perspective on the Nile Valley, Egypt. *World Archaeology* 29(1): 51–74.

Hassan F. 2000. Environmental perception and human responses in history and prehistory, in McIntosh R. J., Tainter J. A. and McIntosh S. K (eds) *The Way the Wind Blows: Climate, History and Human Action*. New York: Columbia University Press, pp. 121–140.

Hassan F. 2005. A river runs through Egypt: Nile floods and civilization. *Geotimes*, 5th April.

Haug G. H., Gunther D., Peterson L. C., Sigman D. M., Hughen K. A. and Aeschlimann B. 2003. Climate and the collapse of the Maya civilization. *Science* 299: 1731–1735.

Hewitt K. and Burton I. 1971. *The Hazardousness of a Place: A Regional Ecology of Damaging Events*. Toronto: University of Toronto.

Hodell D. A., Curtis J. H. and Brenner M. 1995. Possible role of climate in the collapse of the Classic Maya Civilisation. *Nature* 375: 391–394.

Hodell D. A., Brenner M. and Curtis J.H. 2005a, Terminal Classic drought in the northern Maya lowlands inferred from multiple sediment cores in Lake Chichancanab. *Quaternary Science Reviews* 24: 1413–1427.

Hodell D. A., Brenner M., Curtis J. H., Medina-González R., Idelfonso Chan Can E., Abornaz-Pat A. and Guilderson T. P. 2005b. Climate change on the Yucatan Peninsula during the Little Ice Age. *Quaternary Research* 63: 109–121.

Hornborg A. 2005. Review of J. Diamond, 'Collapse: How Societies Choose to Fail or Succeed'. *Current Anthropology* 46: 94–95.

Hunt B. G. 2009. Natural climatic variability and the Norse settlements in Greenland. *Climatic Change.* 97(3-4): 389-407

Ingram M. J., Farmer G., and Wigley T. M. L 1981. Past climates and their impact on man: a review, in Wigley T. M. L., Ingram M. J and Farmer G. (eds) *Climate and History. Studies in Past Climates and Their Impact on Man.* Cambridge: Cambridge University Press, pp. 3–49.

Jones, G. (1986) The North Atlantic Saga: Being the Norse Voyages of Discovery and Settlement to Iceland, Greenland, and North America. Oxford University Press.

Jones S. 2002. Social constructionism and the environment: through the quagmire. *Global Environmental Change* 12(4): 247–251.

Judkins G., Smith M. and Keys E. 2008. Determinism within human-environment research and the rediscovery of environmental causation. *The Geographical Journal* 174(1): 17–29.

Kasperson R. E. and Kasperson J. X. 1996. The social amplification and attenuation of risk. *Annals of the Academy of Social and Political Science* 545: 95–105.

Kasperson J. X. and Kasperson R. E. 2001. *Global Environmental Risk.* London: Earthscan.

Kasperson R. E., Renn O., Solvic P., Brown H. S., Emel J., Goble R., Kasperson J. X. and Ratick S. 1988. The social amplification of risk: a conceptual framework. *Risk Analysis* 8 (2): 177–187.

Kasperson J. X., Kasperson R. E. and Turner III. B. L. 1995. *Regions at Risk: Comparisons of Threatened Environments.* Tokyo: United Nations University Press.

Kolata A. L. 1993. *The Tiwanaku.* Oxford: Blackwell.

Kolata A. L., Binford M. W., Brenner M., Janusek J. W. and Ortloff C. 2000. Environmental thresholds and the empirical reality of state collapse: a response to Erickson 1999). *Antiquity* 74: 424–426.

Koselleck R. 1985. *Futures Past: On the Semantics of Historical Time* (K. Tribe, trans.). Cambridge MA: MIT Press.

Kowalewski S. A., Feinman G. M., Finsten L., Blanton R. E. and Nichoals L. M. 1989. *Monte Alban's hinterland, Part II. Prehispanic settlement patterns in Tlacolula, Etla and Ocotlan, the Valley of Oaxaca, Mexico.* Memoirs of the Museum of Anthropology. Ann Arbor, Michigan: University of Michigan.

Leroy S. A. G. 2006. From natural hazard to environmental catastrophe: past and present. *Quaternary International* 158: 4–12.

Liverman D. 1999. Vulnerability and adaptation to drought in Mexico. *Natural Resources Journal* 39: 99–115.

Luers A. L., Lobell D. B., Sklar L., Addams C. L. and Matson P. A. 2003. A method for quantifying vulnerability, applied to the agricultural system of the Yaqui Valley, Mexico. *Global Environmental Change* 13: 255–267.

MacNaghton P. and Urry J. 1998. *Contested Natures.* London: Sage.

Mann M. E. 2002. Little Ice Age, in MacCracken M. C. and Perry J. S. (eds) The Earth system: physical and chemical dimensions of global environmental change, in Munn T. (ed.) *Encyclopedia of Global Environmental Change.* Chichester: John Wiley and Sons.

Marcus J. and Flannery K. V. 1996. *Zapotec Civilization.* London: Thames and Hudson.

Matthews J. 2005. Review of Diamond J. 2005. Collapse: how societies choose to fail or succeed. Viking Press. *The Holocene* 115(7): 1103.

Mazzarino S. 1966. *The End of the Ancient World* (Holmes G., trans.). London: Faber and Faber.

McCann J. C. 1999. Climate and causation in African history. *The International Journal of African Historical Studies* 32(2–3): 261–279.

McGovern T. 1990. The archaeology of the Norse North Atlantic. *Annual Review of Anthropology* 19: 331–351.

McGovern T. H., Vesteinsson O., Fridriksson A., Church M., Lawson I., Simpson I. et al. 2007. Landscape of settlement in northern iceland: historical ecology of human impact and climate fluctuation on the millennial scale. *American Anthropologist* 109 (1): 27–51.

McNeill J. R. 2005a. Diamond in the rough: is there a genuine environmental threat to security. A review essay. *International Security* 30 (1): 178–195.

McNeill J. R. 2005b. A usable past. (Review of Diamond J. 2005. Collapse: how societies choose to fail or succeed. Viking Press. *American Scientist* 93(2): 1.

Messerli B., Grosjean M., Hofer T., Nuñez L. and Pfister C. 2000. From nature dominated to human dominated environmental changes. *Quaternary Science Reviews* 19: 459–479.

Meyer W. B., Butzer K. W., Downing T. E., Turner B. L (II)., Wenzel G. and Wescoat J. 1998. Reasoning by Analagy, in Raynor S. and Malone E. L. (eds) *Human Choice and Climate Change, no. 3, Tools for Policy Analysis*. Columbus, Batelle Press, pp. 218–289.

Ogilvie A. E. J. and McGovern T. H. 2000. Sagas and science: climate and human impacts in the North Atlantic, in Fitzhugh W. W. and Ward E. I. (eds) *Vikings: the North Atlantic Saga*. Washington: Smithsonian Institution Press. pp. 385–393

Oldfield F. 2008. The role of people in the Holocene, in Battarbee R. W. and Binney H. (eds) *Natural Climate Variability and Global Warming: a Holocene Perspective*. Oxford: Blackwell Scientific Publishing, pp. 58–97.

Oliver-Smith A. and Hoffman S. M. (eds) 1999. *The Angry Earth. Disasters in Anthropological Perspective*. London: Routledge.

Ortloff C. R. and Kolata A. L. 1993. Climate and collapse: agro-ecological perspectives on the decline of the Tiwanaku State. *Journal of Archaeological Science* 20: 195–221.

Owen B. 2005. Distant colonies and explosive collapse: the two stages of the Tiwanaku disapora in the Osmore drainage. *Latin American Antiquity* 16(1): 45–80.

Randers J. 2008. Global collapse – fact or fiction? *Futures* 40: 853–864.

Redman C. L. 1999. *Human Impact on Ancient Environments*. Tuscon: University of Arizona Press.

Ribot J. C. 1995. The casual structure of vulnerability: its application to climate impacts analysis. *Geo-Journal* 35: 119–122.

Rothman H. 2005. So many civilizations. So few pages: a review essay (review of Diamond J. 2005. Collapse: How societies choose to fail or succeed). *Population and Development Review* 31(3): 573–580.

Sanders W. T. and Nichols D. L. 1988. Ecological theory and cultural evolution in the Valley of Oaxaca. *Current Anthropology* 29(1): 33–80.

Scarborough V. L. 2007. The rise and fall of the ancient Maya. A case study in political ecology, in Costanza R., Graumlich L. J. and Steffen W. 2007. *Sustainability or collapse. An integrated history and future if people on earth*. Dahlem Workshop reports. Cambridge, MA: MIT Press, pp. 51–59.

Scarborough V. L. 2008. Rate and process of societal change in semitropical settings: the ancient Maya and the living Balinese. *Quaternary International* 184: 24–40.

Sluyter A. 2003. Neoenvironmental determinism, intellectual damage control and Nature/ Society Science. *Antipode* 35(4): 813–817.

Smit B., Burton I., Kelin R. J. T and Wandel J. 2000. An anatomy of adaptation to climate change and variability. *Climatic Change* 45: 223–251.

Stanley R. S., Killion T. W. and Lycett M. T. 1986. On the Maya collapse. *Journal of Anthropological Research* 42(2): 123–159.

Steffen W. 2008. Looking back to the future. Ambio Special Report 14: 507–513.

Swetnam T. W., Allen C. D. and Betancourt J. L. 1999. Applied historical ecology: using the past to manage for the future. *Ecological Applications* 9(4): 1189–1206.

Tainter J. 1990. *The Collapse of Complex Societies*. Cambridge: Cambridge University Press.

Tainter J. A. 1995. Sustainability of complex societies. *Futures* 27: 397–407.

Tompkins E. L. and Adger W. N. 2004. Does adaptive management of natural resources enhance resilience to climate change? *Ecology and Society* 9(2): 10.

Thompson L. G., Davis M. E., Mosley-Thompson E. and Liu K. B. 1988. Pre-Incan agricultural activity recorded in dust layers in two tropical ice cores. *Nature* 226 (22/29): 763–765.

Torrence R. 2002. What makes a disaster? A long term view of volcanic eruptions and human responses in Papua New Guinea, in Torrence R. and Grattan J. (eds) *Natural Disasters and Culture Change*. London: Routledge, pp. 292–310.

Turner B. L. P. A., Matson J. J., McCarthy R. W., Corell L., Christensen N., Eckley G. K. et al. 2003. Illustrating the coupled human-environment system for vulnerability analysis: three case studies. *Proceedings of the National Academy of Sciences* 100: 8080–8085.

Van der Leeuw S. 2008. Climate and society: lessons from the past 10,000 years. *Ambio Special Report* 14: 476–481.

Vesteinsson O., McGovern T. H. and Keller C. 2002. Enduring impacts: social and environmental aspects of Viking age settlement in Iceland and Greenland. *Archaeologia Islandica* 2: 98–136.

Vogel C. 2001. Vulnerability and global environmental change. Draft Paper for the Human Dimensions of Global Change Meeting, Rio de Janeiro, October 2001.

Weiss H. 1997. Late third millennium abrupt climate change and social collapse in West Asia and Egypt, in Dalfes H. N., Kukla G. and Weiss H. (eds) *Third Millennium BC Climate Change and Old World Collapse*, NATO ASI Series Vol. I. 49. Berlin: Springer-Verlag, pp. 711–723.

Weiss H. and Bradley R. S. 2001. What drives societal collapse? *Science* 291: 609–610.

Weiss H., Courty, M-A., Wetterstron W., Guichard F., Senior L., Meadow R. and Curnow A. 1993. The genesis and collapse of third millennium North Mesopotamian civilization. *Science* 261: 995–1004.

41

Vulnerabilities and the Resilience of Contemporary Societies to Environmental Change

Donald R. Nelson

1 INTRODUCTION

The archaeological record is replete with evidence of societies that were unable to adequately cope with environmental change. Populations dispersed, died off, or otherwise vanished from the material record as environmental conditions changed. Humans still populate much of the Earth, providing evidence for successful adaptation at some level; however, in light of the magnitude of current change there remains concern over the long-term viability of particular contemporary human populations. Changes in the climate, the nitrogen cycle and biodiversity loss, among many others changes, call attention to the way that humans influence and are influenced by the environment. Thus, there is a need to understand the current impacts of environmental change and to anticipate the development trajectories of societies within a context of rapid change. Two veins of research that address this need are vulnerability and resilience studies. Together they advance our

understanding of some of the real challenges in environmental change research, including exploration of cross-scale and cross-level challenges and defining the relationships of humans and the natural environment.

Research on environmental change can focus at an individual level (e.g., a person, an organization, an institution, etc.) or at a system level. In public discourse, environmental change is often equated with impacts at a system level. The future state of society may be called into question and alarmists may question the continuation and persistence of humankind. Loss of ecosystems and the consequent impacts on neighboring environments and people that depend on the systems are highlighted. At the same time, particularly in human populations, there is ongoing concern with the precariousness of subpopulations and individuals. Lives and livelihoods of many individuals are at risk due to ongoing change. The responses of these individuals to stressors within their local context scale-up to influence larger-scale processes and systems.

To respond better to change it is necessary to have the capability to work across these levels – acknowledging the importance of system persistence, but recognizing that systems are made up of individuals with values and needs.

The search to understand processes of change has a long history of discipline-based research, as evident from the chapters in this handbook. Frequently there is a significant divide in approaches that disciplines apply to this research. Natural scientists tend to place people into a box within a model of the environment and social scientists tend to place a 'natural environment' box into their model of social systems (Turner et al., 2003; Westley et al., 2002). Nevertheless, while acknowledging the importance of disciplinary research, trying to understand ecosystems without understanding the social dimensions, or vice versa, leads to only incomplete conclusions and poorly elaborated recommendations (Folke and Rockstrom, 2009). In isolation, neither view emphasizes humans as embedded in nature and the need to account for both ecosystems and social systems in seeking answers in how to deal with change.

There is a growing demand for a science that brings about a more robust, integrated perspective, which addresses social structure, agency and the environment (McLaughlin and Dietz, 2008). The concepts of vulnerability and social–ecological resilience contribute in this direction (Eakin and Luers, 2006; Janssen and Ostrom, 2006). Individually, these two concepts broadly reflect the current tension in environmental change research between promoting the importance of actors on one hand, and a focus on system-level processes and states, on the other. Within the framework of resilience science, researchers are incorporating much of the state-of-the-art knowledge from vulnerability studies. The merging of vulnerability and resilience studies provides an analytical lens that prioritizes both the current state of particular populations and the future trends and trajectories of societies in light of internal and external

change processes. A resilience framework is based on the concept that humans and their environments are linked through what is termed a social–ecological system (SES). This systems focus reflects the notion that human action and organization are an integral part of nature, and that divisions between the two are arbitrary. While specific formulations of an SES may privilege the social over the environmental or environmental over the social, the resilience framework is predicated on linkages and feedbacks between the two. Thus, a resilience framework contributes to our ability to understand how humans influence and respond to environmental change.

The intellectual and disciplinary histories of vulnerability and resilience studies are quite distinct from each other. The following section provides a discussion of the histories and the intellectual evolution within each area of study. This is followed by a section which presents ways in which the two areas are being brought to bear on each other within studies of adaptation to environmental change. Although vulnerability and resilience studies help address some key questions in environmental change research, there remain significant challenges to be explored, which are discussed in the third section. Finally, the last section briefly highlights some of the future areas of work in vulnerability and resilience studies, as researchers continue to identify pathways for the long-term viability of human populations.

2 EVOLUTION OF THE CONCEPTS

Vulnerability and resilience studies developed over multiple decades within various disciplines and draw upon a range of theories, worldviews and methodologies. Today, there is no one particular conceptual model of vulnerability or resilience that is universally accepted. Rather, the different applications and objectives of research and practice have contributed to an expanded understanding of these concepts, indicating intellectual vigor and the value of the concepts (Adger, 2006). At the same time, competing uses and approaches create challenges to talking across disciplines and researchers (Vogel et al., 2007). This section describes the intellectual development of the two concepts and provides descriptions of the types of problems they have been used to address.

2.1 Vulnerability

Vulnerability focuses on particular groups to assess risk in relation to multiple and interacting stresses (McLaughlin and Dietz, 2008). Though not exclusively, much of the vulnerability research emerged in response to perceived inequities in the impacts of natural hazards and environmental change (Eakin et al., 2009) and sought to explain socially differentiated impacts within and between populations. Vulnerability, which is often defined as susceptibility, assesses the attributes of sensitivity, exposure and adaptive capacity (Adger, 2006; Eakin and Luers, 2006). Adaptive capacity refers to the ability to respond to stresses. It encompasses the preconditions necessary to adapt to change, including social and physical elements and the ability to mobilize these elements (Nelson et al., 2007). Within environmental change literature, vulnerability is an approach often predicated on a moral and ethical responsibility to groups and populations that are more susceptible to, and unable to cope with, change. There have been a number of analyses of histories, trends and approaches to analyzing the vulnerability of human populations (Adger, 2006; Cutter, 2003; Eakin and Luers, 2006; Füssel, 2007; Füssel and Klein, 2006; McLaughlin and Dietz, 2008). Below, the lineages have been grouped into three categories: risk-hazard, political economy/political ecology and integrated approaches.

The risk-hazard approach is based on the identification of biophysical threats to a particular 'exposure unit', which are populations or valued elements. It is used to assess levels of risk, in light of the potential for loss.

In this model, risks are the outcomes of two distinct factors. The hazard is the potential biophysical event, characterized by location, intensity, frequency and probability. Vulnerability is concerned with the amount of potential damage due to the severity of the hazard. Thus, this approach, based on the theoretical contributions of (Burton et al., 1978; Kates, 1985; White, 1973) and others, seeks to explain to what a population or valued element is exposed (potential hazards), what the impacts could be (vulnerability), and timing and location of the impacts. This approach is focused primarily on economic valuations of loss.

The analysis of vulnerability under a political economy/political ecology approach focuses on people. It looks to answer questions of who is vulnerable, how are they vulnerable, and why? (Adger and Kelly, 1999; Ribot et al., 1996). In contrast with the risk-hazard approach, which is designed to assess vulnerability in terms of outcomes, the political economy/ecology approaches are context based and interpret vulnerability as a process, not just an outcome (O'Brien et al., 2007). The approach recognizes that individuals and populations are enmeshed in wider networks and contexts that influence levels of susceptibility. Thus the perspective underscores the importance of the social, cultural, and political factors which differentiate vulnerability within and between populations. In this sense, vulnerability in fact defines the relationship of people with the wider environment and to broader political and socioeconomic forces (Oliver-Smith, 2004). The entitlement concepts of Sen (1986, 1981) have linked food security with vulnerability and environmental change research in vulnerability draws heavily on his ideas. While this sometimes underplays the role of physical processes and ecological risks, political ecologists stress the explanatory power of natural processes as well.

Integrated approaches bridge and extend the risk-hazard model and political economy approaches. Turner et al. (2003) provide a framework that addresses a criticism of earlier models that did not do enough to address the vulnerability of biophysical subsystems and which provide little detail on the structure of the hazard's causal sequence. Their coupled vulnerability framework combines external hazards with internal factors of a vulnerable system. The hazard-of-place model (Cutter, 1993, 2003) also combines the risk-hazard and political economy approaches. Her model defines vulnerability through the interactions of hazards within a particular place with the social profiles of communities. Another example which highlights the combination of bio-physical and social factors is the work by O'Brien and others (2004) exploring 'double exposure' in India. Their work explores vulnerability to global climate change as mediated by other multiple stressors.

As vulnerability has increased in application and scope, comparisons are increasingly called for. The Intergovernmental Panel on Climate Change (IPCC) and individuals working on issues of adaptation to climate change have sought to identify the most vulnerable regions and populations (Brooks and Adger, 2005). In order to do so, it is necessary to derive indicators or metrics that allow for comparison across time and space. Among the challenges is the need to capture the dynamic nature of vulnerability. The vulnerability of a particular group or individual changes constantly and is rooted in the actions of multiple human actors. It is constructed at more than one scale and is defined by multiple stressors. Thus trying to identify indicators that work across contexts is a significant challenge (Downing et al., 2005). Some argue that that these challenges are unlikely to be overcome (Eakin and Luers, 2006). Nevertheless, currently there in much research dedicated to resolving these questions (Birkmann, 2006; Downing et al., 2005).

Vulnerability is a relative concept. All populations are vulnerable to change to some degree. But the differences within populations are inherently related to cultural, social, political, and geographic context. Humans can adapt to environmental change, and have been doing so since our species first emerged. The ability and success to do so depend to large degree on the level of vulnerability. However, a key lesson from vulnerability studies is that vulnerabilities do not exist independently of society. Humans create their own vulnerabilities (Orlove, 2005). They are created and maintained, intentionally or not, through human action and inaction. Because they are created, in principle they can also be reduced through purposeful action.

2.2 Resilience

At its core, resilience science is about understanding change processes at a system level. Scientists have been using the concept of system resilience for many years (e.g., Holling, 1973; Vayda and McCay, 1975). In both ecology and social sciences, the use of the term came about in response to contemporary worldviews that promoted equilibrium models of ecological and social systems. In ecology, there was recognition that a natural system could reside in two or more fundamental states. A system could, for example, shift between grasslands and shrub dominated landscapes, or reef systems could switch between kelp- and urchin-dominated systems. In the social sciences, resilience and flexibility were used to interrogate the types of adaptations that societies developed in response to dynamic physical environments. Although both social and ecological sciences used the term, resilience science as we know it today was elaborated primarily within ecology and complex system sciences. Recently there has been a concerted effort to bring the social back into a resilience framework.

Resilience is the extent to which a system can undergo change and still retain the function, structure, identity, feedbacks and ability to develop (Carpenter et al., 2001; Walker et al., 2002). Initially there was strong focus on the ability of a system to maintain its structure and function. Research was directed to understanding the amount of change that a system was able to withstand prior to moving into an alternative stable state. The focus was on the dynamics of the system when it is disturbed far from its modal state. The 'state space' of a system is characterized by the state variables and the relationships that constitute the system. These are regulated by controlling parameters (e.g., soil structures, water salinity, climate, etc.). A system can change states in one of two ways (Beisner et al., 2003), either by significant changes in the state variables (e.g., overgrazing of rangeland leads to degradation and shrub encroachment), or through changes in parameters (e.g., changes in the rainfall regime lead to a moister climate and changes in vegetation patterns). The delineation of different states is marked by thresholds. Crossing a threshold indicates movement into an alternative state.

Beyond this, resilience is also concerned with the ability to cope with, adapt to and manage change in a manner that does not reduce future response options and therefore contributes to the process of sustainability (Berkes et al., 2003; Folke, 2006). Within global change studies, a resilience framework is attractive precisely because it is concerned with the inevitability of change. This does not imply that change occurs continuously, but rather that there can be relatively stable periods followed by periods of significant disturbance. Resilience highlights that the long-term stability of systems is marked by disruptions and cycles. The adaptive cycle provides a framework for characterizing disturbance in SESs (Holling, 1986). The cycles include disruption, reorganization, renewal and growth. The disruption can come from either ecological or social disturbances such fire or political elections and the end result may either be a similar or a new, alternative state (Walker et al., 2004).

The concept of resilience creates the need for the additional concepts of adaptability and transformability (Walker et al., 2004). Adaptability refers to the ability of actors to influence the resilience of a system. It is defined by social, human, natural, manufactured and financial capitals, and the types of governance and institutions that are in place (Walker et al., 2006a). The concept is very similar to adaptive capacity discussed previously. It is specifically concerned with reflexive action and planning for the future. Chapin et al. (2009) identify four components of adaptability in a resilience framework. These include biological, economic and cultural diversity; the capacity of individuals and groups to learn about the system; the ability to experiment to support the learning process; and the capacity to govern effectively.

No system is resilient to all types of shocks and the level of resilience is not equal across a system. Thus there are always trade-offs in terms of system performance and vulnerabilities (Anderies et al., 2006). Systems can be identified as having either specific or general resilience (Leslie and Kinszig, 2009; Walker et al., 2009). Specific resilience is the capacity of a system to respond to a particular perturbation. It may have high resilience in relation to that perturbation but remain vulnerable to other types of shocks. General resilience refers to the ability to absorb a broad range of disturbances. However, although resilient to a number of disturbances, the capacity to deal with any specific perturbation will be lower than if there had been investment towards that perturbation. Investing in the capacity of a system to be resilient to a variety of stressors creates vulnerabilities within portions of the system.

A resilient system is one that avoids crossing thresholds. However, there are times when the magnitude of change is outside the ability of actors to control, and systems do collapse. A changing climate may mean that cropping conditions have changed so that traditional livelihoods are no longer viable. Or changes in rainfall regimes may change stream runoff and affect the sustainability of fisheries. Additionally, there are times in which a particular system may not be desirable and transformation required. For example, irrigated agriculture may be phased out if it begins to interfere with socially desirable riparian areas, or mining interests may give way to tourism industry due to pollutant loads in the water system. Transformability refers to the ability of actors to reconceptualize and create a new system state. Similar to collapse, the system is fundamentally changed. However, in contrast to a situation where thresholds are unwittingly and unwillingly crossed, transformation refers to purposeful management of that change, or at least a negotiation of the process (Olsson et al., 2006). Understanding the capacities necessary for transformation and the ability to manage this type of change are a current topic of focus in the resilience literature.

3 CONVERGENCE AND OVERLAP

Vulnerability and resilience research have several points of interest in common – the types of shocks that an SES is exposed to; the types of responses to the stresses; and the capacity for adaptive actions (Adger, 2006). Thus, both are concerned with impacts of stresses and individual and system responses. Resilience focuses primarily on how to manage the change, while vulnerability emphasizes how to reduce the impacts while doing so (Liu et al., 2007). Because of their relevance for environmental change studies, both concepts are now in the lexicon of adaptation studies. In fact, the IPCC Chapter 17 explicitly states that 'adaptation to climate change takes place through adjustments to reduce vulnerability or enhance resilience in response to observed or expected changes in climate and associated extreme weather events' (Adger et al., 2007). In this definition, adaptation is a general concept. The concepts of resilience and vulnerability help identify underlying system dynamics and the structural characteristics of the system, thus providing explanatory value and analytical rigor (Young et al., 2006). Currently, adaptation studies provide context to much of the vulnerability and resilience research. This section discusses some of the ways in which

the merger of vulnerability and resilience research contributes to better understanding of how to respond to and manage change. Specifically it highlights the importance of systems-level research, that all systems demonstrate some vulnerability, and that resilience itself is an insufficient outcome. Goals and values must go in to defining a desirable system.

3.1 Adapting to environmental change

The goal of adaptation is to be able to live comfortably with current and future change. However, what adaptation entails is very much a decision based on individual and cultural choices. One significant issue is that we don't have a clear understanding of what future change may bring. Climate model projections for example, provide indication of trends and tendencies and potential magnitude of change, but provide much less information on frequencies and levels of variability, which is useful knowledge for long-term planning. The future environment also promises to provide numerous surprises. Surprises are simply outcomes that are either novelties, increasing frequencies in known events or unexpected outcomes (Gunderson, 2003). The adaptation process requires the capacity to learn from previous experiences to cope with current climate and to apply these lessons to cope with future climate, including surprises (Brooks and Adger, 2005). These factors pose significant challenges for preparing for the future and taking anticipatory action.

However, much contemporary adaptation conceptualizes the future from a narrow definition of vulnerability and risk. Responses continue to use a model similar to the risk-hazard approach described above, evaluating a specific threat and analyzing the best way to respond (Serewitz et al., 2003). For example, the possibility of rising sea levels leads to the development of a large coastal defense. Similarly, reduced rainfall leads to increased investment in groundwater pumping for irrigation. Both of these responses may be adequate for the perceived problem, but they fail to consider the long-term or spatially distant implications of the actions – which may include changed coastal morphology down the coast or depleted aquifers and saline soils. This approach entails a preference for dealing with problems that are well defined and that are close in time. Applying a risk-hazard model to uncertain and distant problems does not fit well within the economic calculus used in cost–benefit analyses. The implications of this narrow approach are two-fold. The first is that there is little preparation for unexpected events. The second is that these types of actions have potential to undermine resilience of the system and decrease sustainability.

3.2 Implications of adaptation for system resilience

A study by Adger et al. (2011) assesses the implications of adaptation to climate change for system resilience in nine well-documented cases around the world. The cases include examples of fisheries, forest resource management, agriculture, livestock herders and coastal management. Across the cases the analysis draws attention to three factors that affect system resilience; problem framing, sensitivity to feedbacks and governance structures. For the cases in which resilience declined, the key system stress was identified and framed within a narrow and technological perspective. Problems included variability on water levels, reduced forage, and destruction of forest resources by disease, and drought events. Actions included restricting fishing access, fencing off pasture commons, harvesting timber, and providing farm subsidies. These narrow responses undermined and in many cases degraded local response capacities, reduced ecological diversity, and provided perverse incentives for management innovation and effective governance – all factors identified by Chapin

et al. (2009) as critical to maintenance of system resilience. In contrast, responses that incorporated the role of multiple stresses and worked across scales tended to enhance overall resilience.

Resilience is not a normative concept, but rather a framework that is used to understand processes of change in SESs. But people live within these systems and it is imperative that people become part of the analysis. The fact that a system is resilient or persistent does not mean that it is necessarily desirable or equitable. So while a resilience framework is not normative, our evaluation of the system itself is. The concept of vulnerability contributes in two ways. First, it can assist the evaluation of what the priorities are for a system. What should the system look like; what are the vulnerabilities that will exist, and who will they affect? Although not the only factor in assessing the value of alternative system states, vulnerability is one way to add a normative component to the assessment of desirability. In addition, vulnerability analysis can also help to identify where the threats to a system exist and identify the location of thresholds.

The Brazilian northeast provides a context that illustrates these issues. This region is characterized by frequent droughts with significant socioeconomic impacts on the population dependent on dryland agriculture. In response to the environmental vagaries, society developed a system of social relations based on a patron–client model, which continues today. In the highly precarious and uncertain environment of traditional rural society, patron–client relationships were an essential aspect of survival and provided the platform for unequal exchange. The clients depended on their patrons for protection during drought episodes, and the patrons in turn counted on labor and political support. As part of the historical process the patronage institution demonstrates high resilience in the face of periodic drought and a changing socioeconomic and political context. But the system is founded on poverty, inequities and the persistent vulnerability of much of the rural population (Nelson and Finan, 2009). This case highlights a resilient system with built-in vulnerabilities. A resilience perspective also calls attention to the limits of the current system to respond to a changing climate. The current system has a reduced response diversity, control is top-down and inefficient and stifles innovation in response. These factors which limit adaptive capacity would suggest the need for a system transformation. In fact, such a transformation is slowly taking place, primarily in response to internal drivers.

4 KEY CHALLENGES FOR RESILIENCE AND VULNERABILITY STUDIES

Resilience and vulnerability studies clearly have much to offer for research in environmental change. While these concepts have advanced our understanding of how change occurs and how impacts are distributed, there remain significant challenges. For example, vulnerability studies recognize that risk can be biophysical in nature, but are less effective at understanding how social systems influence natural systems, thereby impacting levels of vulnerability. A society may show great ability to cope with change and adapt if analyzed only through the social dimension lens. But this success may be at the expense of changes in the capacity of ecosystem to sustain the adaptation (Folke, 2006). In addition, differing perspectives of vulnerability and resilience directly influence the livelihoods and wellbeing of real-time households and communities and thus bear critical social and moral implications (Nelson and Finan, 2009). This section highlights challenges in these two areas: functional links between ecological and social systems across scales and the implications of resilience and vulnerability research for effective policy.

Anderies et al. (2006) suggest that a theory to explain SESs must account for information transfers between elements of the system, the actions of individual agents, and the way that these actions influence the environment. They argue that such a theory doesn't exist now but there are extensive efforts in this area (cf. Ostrom, 2009). Most SES frameworks that are used today to explore relationships of humans and their environments are primarily organizational models that provide the researcher with ideas of which elements in systems might be related but without providing any type of explanation of how the relationship functions or what are outcome expectations.

The first steps in developing a theory of SES are to bound the coupled system and identify the scale of interest. Studies of the human and natural components are challenging individually, and when they are combined these difficulties increase. In ecology and geography, scale is usually defined in terms of spatial and temporal dimensions. Social scale also incorporates space and time, but adds ideas about representation and organization. Humans, for example, have the power to influence scales beyond what might be expected. This is in part due to the human ability to construct and manipulate symbols. We think in the abstract, are self-reflexive and forward looking (Westley et al., 2002). One of the problems humans have in managing their environments and successfully adapting to change reflects a mismatch between the scale that humans are working and the scale of the ecosystems that are relevant (Cumming et al., 2006). Recognizing where these mismatches occur is only one part of the challenge. The second is to develop institutions that function at appropriate scales, which are determined by the empirical reality of the phenomenon and our ability to observe and collect data.

Agency is the capacity of actors to project alternative future possibilities and then to actualize those possibilities (McLaughlin and Dietz, 2008). Social scientists have long been aware of the gap and the trade-offs in the explanatory power of agency and systems (Brumfiel, 1992; Giddens, 1979; Moran, 1982). Bridging vulnerability and resilience studies offers a possible pathway to overcome some of these constraints (Nelson et al., 2007). Vulnerability provides analysis of the processes of access, negotiation, decision-making and action; and resilience provides an understanding of the implications of these actions on the system, both now, and in the future. While this doesn't inherently offer precise predictive powers, merging of the two concepts does help to understand the linkages between individuals and systems and offers a starting point for further exploration.

Nevertheless, currently there is limited understanding of how individual autonomous responses scale up to influence the overall sustainability of an SES (Eakin and Wehbe, 2009; Lambin, 2005). Researchers and policy makers recognize the influence of controlling parameters on the range of decisions that individuals choose from. Labor and commodity markets, credit opportunities and available resources, all influence the choices that individuals make in their daily lives. However, the ability to influence change in the system is not resident only at higher levels. Individuals make decisions in order to manage perceived risk and reduce vulnerability. The aggregate results of individual actions have the potential to affect the larger vulnerabilities and trajectories of sustainable systems. One proposed strategy for addressing these potential conflicts recommends analyzing the role of information exchange, the motivations of individuals, and their capacities (Eakin and Wehbe, 2009). Further exploration of the way in which autonomous action is determined will inform policy that provides the necessary flexibility for individuals to make decisions that are in line with maintaining a trajectory of sustainable development.

How is a policy or decision maker to make sense of the research on managing and responding to environmental change? Although presented here as complementary and synergistic concepts, vulnerability and resilience aren't always presented in this light. Currently, there are three primary typologies of responses to environmental change that are used within academic and lay literatures. These are vulnerability, adaptation and resilience (Table 41.1). In this context, vulnerability refers to a political ecology or integrated framework as described above and adaptation entails a risk-hazard approach. The adoption of a particular framework by policy makers will entail trade-offs with consequences for populations affected by policy. These trade-offs exist in policy processes and outcomes. For example, governments do not always equally protect their populations from environmental change. Usually the most vulnerable populations have the least voice. This argues for the importance of fair processes which enable participation of affected communities (Paavola and Adger, 2006).

More importantly for the discussion here is that in the context of policy makers (within nongovernmental organizations (NGOs) and the public sector) adaptation, vulnerability and resilience perspectives each have different takes on problem parameters including spatial scale, temporal emphasis, the policy goal and targeted outcome (Table 40.1; Eakin et al., 2009). For example, the targeted outcome of responses to environmental change include maximized loss reduction at lowest

Table 41.1   Priorities for environmental change policy: Comparing three response typologies (based on Eakin et al., 2009)

Problem parameters	Adaptation (risk-hazard)	Vulnerability (political ecology)	Resilience
Spatial scale	Scale of risk of focus	Scale of population of focus	Social-ecological system
Temporal scale	Near-term	Past and present	Mid to long-term
Policy goal	Reduce known risks	Protect vulnerable populations	Maintain sustainability
Targeted outcome	Maximum risk reduction at lowest cost	Minimized social inequity	Minimized probability of undesirable and irreversible change

cost under the adaptation model; minimized social inequity in impacts under the vulnerability model and minimized probability of undesirable and irreversible change under a resilience model. The temporal emphasis is also significantly different. The adaptation approach stresses short-term to medium-term risks. The vulnerability perspective stresses past and present vulnerabilities and a resilience perspective prioritizes the long-term future. There is no clear *a priori* best approach. Reducing a particular vulnerability can result in a reduction of resilience later in time, or in another geographical space, but a long-term focus on the resilience of a system can leave in place dramatic social and economic inequities in resources and exposure to change. Obviously, science alone cannot make these decisions. However, resilience-based approaches can improve management of systems, particularly if merged with a vulnerability perspective. A resilience perspective does not provide a mechanism to predict the impact of human actions, but rather calls attention to particular system attributes that play important roles in the dynamics of SESs (Anderies et al., 2006).

5 SUSTAINABILITY SCIENCE

The concepts of vulnerability and resilience contribute to an emerging perspective in environmental change research, which focuses on developing system resilience and sustainable development trajectories rather than adaptation per se (Nelson et al., 2007). The justification for this approach is based on the understanding that humans and ecosystems are not only affected by a changing environment, but that adaptive responses themselves have repercussions throughout society and the environment (Walker et al., 2006b). Thus, any evaluation of a response to change must consider the response not only in relation to the particular change, but also in relation to the consequences on the long-term resilience of the system. Furthermore, resilience and

vulnerability can only be fully understood by considering the cross-level linkages between and within humans and their environments. This requires understanding not only how larger-level processes circumscribe individual action, but also how individual actions can affect larger-level processes.

Analyses of human history provide numerous examples of human responses and strategies that were successful over the short term, but that in the long-term undermined the environment and human response capacity (Redman and Kinzig, 2003). These findings highlight a shortcoming in past adaptation research, which privileges a focus on the proximate, rather than the ultimate causes of change. Adaptation, narrowly defined, is designed to reduce risk to specific hazards rather than address the ultimate causes that create vulnerabilities to change. In many cases, this oversight results in actions that exacerbate future vulnerabilities or pass on risks to other populations (Adger et al., 2011). Increased awareness and understanding of how current actions constrain or enable future action will permit more adequate policy designed to reduce vulnerability to a changing climate.

Resilience and vulnerability studies fit into a larger domain of research referred to as sustainability science. Within this domain are calls for a fundamentally different type of science: one that is truly integrative of disciplinary approaches and that works across scales and nations. Moran (2010) highlights several key areas in sustainability science, two of which are of particular relevance to reliance and vulnerability studies. These are improved understanding of how human institutions influence resource use and increased understanding of how socioeconomic and political context and change are reflected in vulnerabilities to environmental change. Environmental change research and sustainability research are converging in a number of areas. Two of these include recognition of the importance of understanding flows of, and access to, information as well as the human and ecological capacities to respond to change (Lambin, 2005). In order to promote a sustainable development trajectory there is need for systemic perspective that recognizes the linkages and relationships between individuals, society and the environment, which stresses flexibility and provides a framework for evaluating responses to change within a forward looking frame.

REFERENCES

Adger W. N. 2006. Vulnerability. *Global Environmental Change* 16: 268–261.

Adger W. N., Agrawala S., Mirza M. M. Q., Conde C., O'Brien K., Pulhin J. et al. 2007. Assessment of adaptation practices, options, constraints and capacity. in Parry M. L., Canziani O. F., Palutikof J. P., van der Linden P. J. and Hanson C. E. (eds) *Climate Change 2007: Impacts, Adaptation and Vulnerability. Contribution of Working Group II to the Fourth Assessment Report of the Intergovernmental Panel on Climate Change*. Cambridge: Cambridge University Press, pp. 717–743.

Adger W. N. and Kelly P. M. 1999. Social vulnerability to climate change and the architecture of entitlements. *Mitigation and Adaptation Strategies for Global Change* 4: 253 - 266.

Adger W. N., Brown K., Nelson D. R., Berkes F., Eakin H., Folke C., Galvin K. et al. (2011). Resilience implications of responses to climate change. *Wiley Interdisciplinary Reviews Climate Change*.

Anderies J. M., Walker B. H. and Kinzig A. P. 2006. Fifteen weddings and a funeral: case studies and resilience-based management. *Ecology and Society* 11(1): 21.

Beisner B. E. D. T. H. and Cuddington K. 2003 Alternative stable states in ecology. *Frontiers in Ecology* 1(7): 376–382.

Berkes F., Colding J. and Folke C. (eds) 2003. *Navigating Social-Ecological Systems: Building Resilience for Complexity and Change*. Cambridge: Cambridge University Press.

Birkmann J. (ed.) 2006. *Measuring Vulnerability to Natural Hazards – Towards Disaster Resilient Societies*. Tokyo: United Nations University Press.

Brooks N. and Adger W. N. 2005 Assessing and enhancing adaptive capacity, in Lim B., Spanger-Siegfried E., Burton I., Malone E. and Huq S. (eds) *Adaptation Policy Frameworks for Climate Change: Developing Strategies, Policies and Measures.*

Cambridge: Cambridge University Press, pp. 165–181.

Brumfiel E. M. 1992. Distinguished lecture in archeology: breaking and entering the ecosystem – gender, class, and faction steal the show. *American Anthropologist* 94(3): 551–567.

Burton I., Kates R. W. and White G. F. 1978. *The Environment as Hazard*. Oxford: Oxford University Press.

Carpenter S. R., Walker B., Anderies M. J. and Abel N. 2001 From metaphor to measurement: resilience of what to what? *Ecosystems* 4: 756–781.

Chapin III. F. S., Kofinas G. and Folke C. (eds) 2009. *Principles of Ecosystem Stewardship: Resilience-Based Natural Resource Management in a Changing World*. New York: Springer.

Cumming G. S., Cumming D. H. M., and Redman C. L. 2006. Scale mismatches in socio-ecological systems: causes, consequences and solutions. *Ecology and Society* 11: 14.

Cutter S. L. 1993. *Living With Risk : The Geography of Technological Hazards*. London/New York: E. Arnold/ Routledge.

Cutter S. L. 2003. The vulnerability of science and the science of vulnerability. *Annals of the Association of American Geographers* 93: 1–12.

Downing T. J., Aerts J., Barthelemy S. O., Bharwani S., Ionescu C., J. Hinkel J. et al. 2005. *Integrating Social Vulnerability into Water Management*. Stockholm: Stockholm Environment Institute.

Eakin H. and Luers A. L. 2006. Assessing the vulnerability of social-environmental systems. *Annual Review of Environment and Resources* 31: 365–394.

Eakin H., Tompkins E., Nelson D. R. and Anderies J. M. 2009. Hidden costs and disparate uncertainties: trade-offs in approaches to climate policy, in Adger W. N., O'Brien K. L., and Lorenzoni I. (eds) *Adapting to Climate Change: Limits, Thresholds and Governance*. Cambridge: Cambridge University Press, pp. 212–226..

Eakin H. and Wehbe M. B. 2009. Linking local vulnerability to system sustainability in a resilience framework: two cases from Latin America. *Climatic Change* 93: 355–377.

Folke C. 2006. Resilience: The emergence of a perspective for social-ecological systems analyses. *Global Environmental Change* 16: 253–267.

Folke C. and Rockstrom J. 2009. Turbulent times. *Global Environmental Change* 19: 1–3.

Füssel H.-M. 2007. Vulnerability: A generally applicable conceptual framework for climate change research. *Global Environmental Change* 17: 155–167.

Füssel H.-M. and Klein R. J. T. 2006. Climate change vulnerability assessments: an evolution of conceptual thinking. *Climatic Change* 75(3): 301–329.

Giddens A. 1979. *Central Problems in Social Theory: Action, structure and contradiction in social analysis.* London: Macmillan Press.

Gunderson L. H. 2003. Adaptive dancing: interactions between social resilience and ecological crises, in Berkes F., Colding J. and Folke C. (eds) *Navigating Social-Ecological Systems: Building Resilience for Complexity and Change.* Cambridge: Cambridge University Press, pp. 33–52.

Holling C. S. 1986. The resilience of terrestrial ecosystems: local surprise and global change, in Clark W. C. and Munn R. E. (eds) *Sustainable Development of the Biosphere.* Cambridge: Cambridge University Press, pp. 292–317.

Holling C. S. 1973. Resilience and stability of ecological systems. *Annual Review of Ecology and Systematics* 4: 1–21.

Janssen M. A. and Ostrom E. (eds) 2006. Resilience, Vulnerability, and Adaptation: A Cross-Cutting Theme of the International Human Dimensions Programme on Global Environmental Change *Global Environmental Change* 16: 3.

Kates R. W. 1985. The interaction of climate and society, in Kates R. W., Ausubel H. and Berberian M. (eds) *Climate Impact Assessment.* Chichester: Wiley, pp 3 – 36.

Lambin E. F. 2005. Conditions for sustainability of human-environment systems: Information, motivation, and capacity. *Global Environmental Change* 15: 177–180.

Leslie H. M. and Kinszig A. P. 2009. Resilience science, in McLeod K. and Leslie H. (eds) *Ecosystem-Based Management for the Oceans.* Washington DC: Island Press, pp. 55–73.

Liu J., Dietz T., Carpenter S. R., Folke C. Alberti M. , Redman C. L. et al. 2007. Coupled human and natural systems. *Ambio* 36(8): 639–649.

McLaughlin P. and Dietz T. 2008. Structure, agency and environment: Toward an integrated perspective on vulnerability. *Global Environmental Change* 18: 99–111.

Moran E. F. 1982. *Human Adaptability: An Introduction to Ecological Anthropology.* Boulder, CO: Westview Press.

Moran E. F. 2010. *Environmental Social Science: Human-Environment Interactions and Sustainability.* West Sussex: Wiley-Blackwell.

Nelson D. R. and Finan T. J. 2009. Praying for drought: Persistent vulnerability and the politics of patronage in Ceará, Northeast Brazil. *American Anthropologist* 111(3): 302–316.

Nelson D. R., Adger W. N. and Brown K. 2007. Adaptation to environmental change: Contributions of a resilience framework. *Annual Review of Environment and Resources* 32(11): 395–420.

O'Brien K. L., Eriksen S., Nygaard L. P. and Schjolden A. 2007. Why different interpretations of vulnerability matter in climate change discourses. *Climate Policy* 7: 73–88.

O'Brien K. L., Leichenko R. M., Kelkar U., Venema H., Aandahl G., Tompkins H. et al. 2004. Mapping vulnerability to multiple stressors: Climate change and globalization in India. *Global Environmental Change* 14: 303–313.

Oliver-Smith A. 2004. Theorizing vulnerability in a globalized world: a political ecological perspective, in Bankoff G., Frerks G., and Hilhorst D. (eds) *Mapping Vulnerability: Disasters, Development and People.* London and Sterling, VA: Earthscan, pp. 10–24.

Olsson P., Gunderson L. H., Carpenter S. R., Ryan P., Lebel L., Folke C. and Holling C. S. 2006. Shooting the rapids: navigating transitions to adaptive governance of social-ecological systems. *Ecology and Society* 11(1): 18.

Orlove B. 2005. Human adaptation to climate change: a review of three historical cases and some general perspectives. *Environmental Science and Policy* 8: 589–600.

Ostrom, E. 2009. General framework for analyzing sustainability of social-ecological systems. *Science* 325: 419.

Paavola J. and Adger W. N. 2006. Fair adaptation to climate change. *Ecological Economics* 56: 594–609.

Redman C. L. and Kinzig A. 2003. Resilience of past landscapes: resilience theory, society, and the *longue durée. Conservation Ecology* 7(1): 14.

Ribot J. C., Najam A. and Watson G. 1996. Climate variation, vulnerability and sustainable development in the semiarid tropics, in Ribot J. and Magalhães A. R. (eds) *Climate Variability, Climate Change and Social Vulnerability in the Semi-arid Tropics.* Cambridge: Cambridge University Press, pp. 13–51.

Sen, A. 1986. *Food, Economics And Entitlements*: Helsinki: World Institute For Development Economics Research of the United Nations University.

Sen A. K. 1981. *Poverty and Famines: An Essay on Entitlement and Deprivation.* Oxford/New York: Clarendon Press/Oxford University Press.

Serewitz D. R., Pielke R. and Keykhah M. 2003. Vulnerability and risk: some thoughts from a political and policy perspective. *Risk Analysis* 23: 805–810.

Turner B. L., Kasperson R. E., Matson P. A., McCarthy J. J., Corell R. W., Christensen L. et al. 2003a. Framework for Vulnerability Analysis in Sustainability Science. *Proceedings of the National Academy of Sciences of the United States of America* 100(14): 8074–8079.

Vayda A. P. and McCay B. J. 1975. New directions in ecology and ecological anthropology. *Annual Review of Anthropology* 4: 293–306.

Vogel C., Moser S. C., Kasperson R. E. and Dabelko G. D. 2007. Linking vulnerability, adaptation, and resilience science to practice: Pathways, players, and partnerships. *Global Environmental Change* 17: 349–364.

Walker, B., Abel N., Anderies J. M. and Ryan P. 2009. Resilience, adaptability and transformatility in the Goulburn-Broken Catchment, Australia. *Ecology and Society* 14(1): 12.

Walker B., Carpenter S. R., Anderies J. M., Abel N., Cumming G. S., Janssen M. A. et al. 2002. Resilience management in social-ecological systems: a working hypothesis for a participatory approach. *Conservation Ecology* 6(1): 14.

Walker B., Gunderson L. H., Kinzig A., Folke C. and Schultz L. 2006a. A handful of heuristics and some propositions for understanding resilience in social-ecological systems. *Ecology and Society* 11(1): 13.

Walker B. H., Anderies J. M., Kinzig A. P. and Ryan P. 2006b. Exploring resilience in social-ecological systems through comparative studies and theory development: introduction of the special issue. *Ecology and Society* 11(1): 12.

Walker B., Holling C. S., Carpenter S. R. and Kinzig A. 2004. Resilience, adaptability and transformability in social-ecological systems. *Ecology and Society* 9(2): 5.

Westley F., Carpenter S. R., Brock W. A., Holling C. S. and Gunderson L. H. 2002. Why systems of people and nature are not just social and ecological systems, in Gunderson L. H. and Holling C. S. (eds) *Panarchy: Understanding Transformations in Human and Natural Systems*. Washington DC: Island Press, pp. 103–120.

White G. 1973. Natural hazards research, in Chorely R. (ed.) *Directions in Geography*. London: Methuen, pp. 193–216.

Young, O. R., Berkhout F., Gallopín G. C., Janssen M. A., Ostrom E. and van der Leeuw S. 2006. The globalization of socio-ecological systems: an agenda for scientific research. *Global Environmental Change* 16(3): 304–316.

42

Disease, Human and Animal Health and Environmental Change

Matthew Baylis and Andrew P. Morse

1 INTRODUCTION

The impact of infectious diseases of humans and animals seems as great now as it was a century ago. While many threats have disappeared or dwindled, at least in the developed world, others have arisen to take their place. Important infectious diseases of humans that have emerged in the last 30 years, for a range of reasons, include *acquired immune deficiency syndrome* (AIDS), variant Creutzfeldt–Jakob disease (vCJD), multidrug resistant tuberculosis, severe acute respiratory syndrome (SARS), *E. coli* O157, avian influenza, swine flu, West Nile fever and Chikungunya (Morens et al., 2004; Jones et al., 2008). The same applies to diseases of animals; indeed, all but one of the aforementioned human diseases have animal origins – they are zoonoses – and hence the two subjects, usually studied separately by medical or veterinary scientists, are intimately entwined. Considering the animals themselves, in the developed world infectious diseases still cause animal suffering, harm the environment, cause financial losses and threaten food security. In the developing world, infectious human diseases remain major causes of mortality and morbidity, while diseases of animals, livestock in particular, limit productivity, constrain development and exacerbate poverty (Perry et al., 2002).

What will be the global impact of infectious diseases at the end of the twenty-first century? Any single disease is likely to be affected by many factors that cannot be predicted with confidence, including changes to human demography and behaviour, new scientific or technological advances, pathogen evolution, livestock management practices and developments in animal genetics, and changes to the physical environment. A further, arguably more predictable, influence is climate change.

Owing to anthropogenic activities, there is widespread scientific agreement that the world's climate is warming at a faster rate than ever before (IPCC, 2001), with concomitant changes in precipitation, flooding, winds and the frequency of extreme events such as El Niño. Innumerable studies have demonstrated links between infectious diseases and climate, and it is unthinkable that a significant change in climate during this century will not impact on at least some of them.

How should we react to predicted changes in diseases ascribed to climate change? The answer depends on the animal populations and human communities affected, whether the disease changes in incidence or spatio-temporal distribution and, of course, on the direction of change: some diseases may spread but others may retreat in distribution. It also depends on the relative importance of the disease. If climate change is predicted to affect mostly diseases of relatively minor impact on human society, while the more important diseases are refractory to climate change's influence, then our concerns should be tempered.

2 CLIMATE AND DISEASE

Many diseases are affected directly or indirectly by weather and climate. These links may be *spatial*, with climate affecting distribution, *temporal* with weather affecting the timing of an outbreak, or relate to the *intensity* of an outbreak. Here we present a selection of these associations, which is by no means exhaustive but is, rather, intended to demonstrate the diversity of effects.

- Anthrax is an acute infectious disease of most warm-blooded animals, including humans, with worldwide distribution. The causative bacterium, *Bacillus anthracis*, forms spores able to remain infective for 10–20 years in pasture. Temperature, relative humidity and soil moisture all affect the successful germination of anthrax spores, while heavy rainfall may stir up dormant spores. Outbreaks are often associated with alternating heavy rainfall and drought, and high temperatures (Parker et al., 2002).
- Cholera, a diarrhoeal disease which has killed tens of millions of people worldwide, is caused by the bacterium *Vibrio cholerae*, which lives amongst sea plankton (Eiler et al., 2006). High temperatures causing an increase in algal populations often precede cholera outbreaks.

Disruption to normal rainfall helps cholera to spread further, either by flooding, leading to the contamination of water sources, such as wells, or drought which can make the use of such water sources unavoidable. Contaminated water sources then become an important source of infection in people.

- Salmonellosis is a serious food-borne disease caused by *Salmonella* bacteria, most often obtained from eggs, poultry and pork. Salmonellosis notification rates in several European countries have been shown to increase by about 5–10 per cent for each 1°C increase in ambient temperature (Kovats et al., 2004). Salmonellosis notification is particularly associated with high temperatures during the week prior to consumption of infected produce, implicating a mechanistic effect via poor food handling.
- Foot-and-mouth disease (FMD) is a highly contagious, viral infection of cloven-footed animals, including cattle, sheep and pigs. Most transmission is by contact between infected and susceptible animals, or by contact with contaminated animal products. However, FMD can also spread on the wind. The survival of the virus is low at relative humidity (RH) below 60 per cent (Donaldson, 1972), and wind-borne spread is favoured by the humid, cold weather common to temperate regions. In warmer, drier regions, such as Africa, wind-borne spread of FMD is considered unimportant (Sutmoller et al., 2003).
- Peste des petits ruminants (PPR) is an acute, contagious, viral disease of small ruminants, especially goats, which is of great economic importance in parts of Africa and the Near East. It is transmitted mostly by aerosol droplets between animals in close contact. However, the appearance of clinical PPR is often associated with the onset of the rainy season or dry cold periods (Wosu et al., 1992), a pattern that may be related to viral survival. The closely related rinderpest virus survives best at low or high relative humidity, and least at 50–60 per cent (Anderson et al., 1996).
- Several directly transmitted human respiratory infections, including those caused by rhinoviruses (colds) and seasonal influenza viruses (flu) have, in temperate countries, seasonal patterns linked to the annual temperature cycle. There may be direct influences of climate, such as the effect of humidity on survival of the virus in aerosol (Soebiyanto et al., 2010), or indirect

influences via, for example, seasonal changes in the strength of the human immune system or more indoor crowding during cold weather (Lowen et al., 2007).

- Haemonchosis – infection with the nematode *Haemonchus contortus* – occurs worldwide in the guts of sheep and cattle. It can cause significant economic loss in terms of reduced productivity or, with heavy infestations, mortality. Eggs are excreted in droppings. Survival of the eggs and larvae, until they are ingested by another animal, depends on temperature and moisture: under appropriate conditions of warmth and moderate humidity, the larvae can survive for weeks or months.

- Fasciolosis, caused by the *Fasciola* trematode fluke, is of economic importance to livestock producers in many parts of the world and also causes disease in humans. The disease is a particular problem where environmental conditions favour the intermediate host, lymnaeid snails. These conditions include low-lying wet pasture, areas subject to periodic flooding, and temporary or permanent bodies of water (Hall, 1988).

- Plague is a flea-borne disease caused by the bacterium *Yersinia pestis*; the fleas' rodent hosts bring them into proximity with humans. In Central Asia climate forcing synchronizes the rodent population dynamics over large areas (Kausrud et al., 2007), allowing population density to rise over the critical threshold required for plague outbreaks to commence (Davis et al., 2004).

- African horse sickness (AHS), a lethal infectious disease of horses, is caused by a virus transmitted by *Culicoides* biting midges. Large outbreaks of AHS in the Republic of South Africa over the last 200 years are associated with the combination of drought and heavy rainfall brought by the warm-phase of the El Niño southern oscillation (ENSO) (Baylis et al., 1999).

- Rift Valley Fever (RVF), an important zoonotic viral disease of sheep and cattle, is transmitted by *Aedes* and *Culex* mosquitoes. Epizootics of RVF are associated with periods of heavy rainfall and flooding (Davies et al., 1985; Linthicum et al., 1987, 1999) or, in East Africa, with the combination of heavy rainfall following drought associated with ENSO (Anyamba et al., 2002; Linthicum et al., 1999). ENSO-related floods in 1998, following drought in 1997, led to an epidemic of RVF (and some other diseases) in the Kenya/Somalia border area and the deaths of more than 2,000 people and two-thirds of all small ruminant livestock (Little et al., 2001). Outbreaks of several other human infections, including malaria and dengue fever have, in some parts of the world, been linked to ENSO events.

- Diseases transmitted by tsetse flies (sleeping sickness, animal trypanosomosis) and ticks (such as anaplasmosis, babesiosis, East Coast fever, heartwater) impose a tremendous burden on African people and their livestock. Many aspects of the vectors' life cycles are sensitive to climate, to the extent that their spatial distributions can be predicted using satellite-derived proxies for climate variables (Rogers and Packer, 1993).

- Mosquitoes (principally *Culex* and *Aedes*) transmit several viruses of birds that can also cause mortality in humans and horses. Examples are West Nile fever (WNF) and the viral encephalitides such as Venezuelan, western and eastern equine encephalitis (VEE, WEE and EEE respectively) (Weaver and Barrett, 2004). The spatial and temporal distributions of the mosquito vectors are highly sensitive to climate variables.

3 CLIMATE CHANGE AND DISEASE

There is a substantial scientific literature on the effects of climate change on health and disease, but with strong focus on human health and vector-borne disease (Githeko et al., 2000; Hay et al., 2002; Kovats et al., 1999, 2001; Lines, 1995; McMichael and Githeko, 2001; Patz and Kovats, 2002; Randolph, 2004; Reeves et al., 1994; Reiter et al., 2004; Rogers and Packer, 1993; Rogers and Randolph, 2000; Rogers et al., 2001; Semenza and Menne, 2009; Sutherst, 1998; WHO, 1996; Wittmann and Baylis, 2000; Zell, 2004). By contrast, the effects of climate change on diseases spread by other means, or animal diseases in general, have received comparatively little attention (but with notable exceptions: (Cook, 1992; Gale et al., 2008, 2009; Harvell et al., 1999, 2002). Given the global burden of diseases that are not vector-borne, and the contribution made by animal diseases to poverty in the developing world (Perry et al., 2002), attention to these areas is overdue.

The previous section demonstrates the range of climate influences upon infectious disease. Such influences are not the sole preserve of vector-borne diseases: food-borne, water-borne and aerosol-transmitted diseases are also affected. A common feature of non-vector-borne diseases affected by climate is that the pathogen or parasite spends a period of time outside of the host, subject to environmental influence. Examples include the infective spores of anthrax; the wind-borne aerosol droplets that spread FMD viruses in temperate regions; the *Salmonella* bacteria that contaminate food products; and the moisture-dependent survival of the parasites causing haemonchosis.

By contrast, most diseases transmitted directly between humans and between animals in close contact (e.g., human childhood viruses, sexually transmitted diseases, tuberculosis; and, considering animal infections, avian influenza, bovine tuberculosis, brucellosis, Newcastle's disease of poultry, rabies) have few or no reported associations with climate. Clear exceptions are the viruses that cause colds and seasonal flu in humans, and PPR in small ruminants; these viruses are spread by aerosol between individuals in close contact but are nevertheless sensitive to the effects of ambient humidity and possibly temperature.

The influence of climate on diseases that are not vector-borne appears to be most frequently associated with their timing or seasonal occurrence rather than their spatial distribution. There are examples of such diseases that occur only in certain parts of the world (for example, PPR) but most occur worldwide. By contrast, the associations of vector-borne diseases with climate are equally apparent in time and space, with very few vector-borne diseases being considered a risk worldwide. This is a reflection of the strong influence of climate on both the spatial and temporal distributions of intermediate vectors.

In the scientific literature many processes have been proposed by which climate change might affect infectious diseases. These processes range from the clear and quantifiable to the imprecise and hypothetical. They may affect pathogens directly or indirectly, the hosts, the vectors (if there is an intermediate host), epidemiological dynamics or the natural environment. A framework for how climate change can affect the transmission of pathogens between hosts is shown in Figure 42.1. Only some of the processes can be expected to apply to any single infectious disease.

3.1 Effects on pathogens

Higher temperatures resulting from climate change may increase the rate of development of certain pathogens or parasites that have one or more lifecycle stages outside their human or animal host. This may shorten generation times and, possibly, increase the total number of generations per year, leading to higher pathogen population sizes (Harvell et al., 2002). Conversely, some pathogens are sensitive to high temperature and their survival may decrease with climate warming.

Phenological evidence indicates that spring is arriving earlier in temperate regions (Walther et al., 2002). Lengthening of the warm season may increase or decrease the number of cycles of infection possible within one year for warm or cold-associated diseases respectively. Arthropod vectors tend to require warm weather so the infection season of arthropod-borne diseases may extend. Some pathogens and many vectors experience significant mortality during cold winter conditions; warmer winters may increase the likelihood of successful overwintering (Harvell et al., 2002; Wittmann and Baylis, 2000).

Pathogens that are sensitive to moist or dry conditions may be affected by changes to precipitation, soil moisture and the frequency of floods. Changes to winds could affect the spread of certain pathogens and vectors.

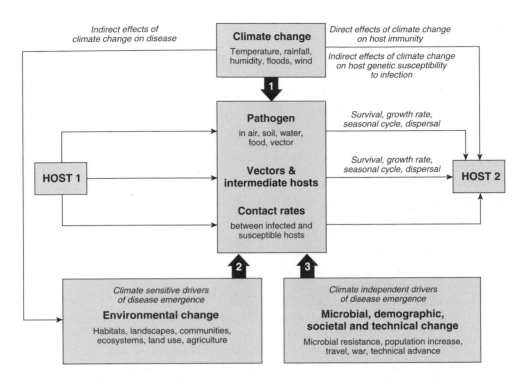

Figure 42.1   A schematic framework of the effects of climate change on the transmission of diseases of humans and animals. Climate change can act directly on pathogens in a range of external substrates, or their vectors and intermediate hosts, thereby affecting the processes of survival, growth, seasonality and dispersal. It can also directly affect hosts themselves or the contact rates between infected and susceptible individuals. Climate change can have indirect effects on disease transmission via its effects on the natural or anthropogenic environment; and via the genetics of exposed populations. Environmental, demographic, social and technical change will also happen independently of climate change and have as great, or much greater, influence on disease transmission as climate change itself. The significance of climate change as a driver of disease will depend on the scale of arrow 1, and on the relative scales of arrows 2 and 3.

3.2 Effects on hosts

A proposed explanation for the tendency for human influenza to occur in winter is that the human immune system is less competent during that time; attributable to the effects of reduced exposure to light on melatonin (Dowell, 2001) or vitamin D production (Cannell et al., 2006). The seasonal light/dark cycle will not change with climate change, but one might hypothesize that changing levels of cloud cover could affect exposure in future. A second explanation, the tendency for people to congregate indoors

during wintertime, leads to a more credible role for a future influence of climate change.

Mammalian cellular immunity can be suppressed following heightened exposure to ultraviolet B (UV-B) radiation – an expected outcome of stratospheric ozone depletion (Aucamp, 2003; de Gruijl et al., 2003). In particular, there is depression of the number of Type 1 helper lymphocytes, cells which stimulate macrophages to attack pathogen-infected cells and, therefore, the immune response to intracellular pathogens may be particularly affected. Examples include many viruses, rickettsia (such as *Cowdria* and *Anaplasma*, the causative agents of heart water and anaplasmosis), *Brucella, Listeria monocytogenes* and *Mycobacterium tuberculosis*, the bacterial agents of brucellosis, listeriosis and tuberculosis respectively, and the protozoan parasites *Toxoplasma gondii*, and *Leishmania* which cause toxoplasmosis and visceral leishmanosis (kala-azar) respectively in humans (Jankovic et al., 2001).

A third host-related effect worthy of consideration is genetic resistance to disease. Some human populations, and many animal species have evolved a level of genetic resistance to some of the diseases to which they are commonly exposed. Malaria presents a classic example for humans, with a degree of resistance to infection in African populations obtained from heterozygosity for the sickle-cell genetic trait. Considering animals, wild mammals in Africa may be infected with trypanosomes, but rarely show signs of disease; local Zebu cattle breeds, which have been in the continent for millennia, show some degree of trypanotolerance (resistance); by contrast, recently introduced European cattle breeds are highly susceptible to trypanosomosis. In stark contrast, African mammals proved highly susceptible to rinderpest which swept through the continent in the late nineteenth century, and which they had not previously encountered. It seems unlikely that climate change will directly affect genetic or immunologic resistance to disease in humans or animals. However, significant shifts in disease distributions driven by climate-change pose a greater threat than simply the exposure of new populations. Naïve populations may, in some cases, be particularly susceptible to the new diseases facing them.

Certain diseases show a phenomenon called *endemic stability*. This occurs when the severity of disease is less in younger than older individuals, when the infection is common or endemic and when there is life-long immunity after infection. Under these conditions, most infected individuals are young, and experience relatively mild disease. Counter-intuitively, as endemically stable infections become rarer, a higher proportion of cases are in older individuals (it takes longer, on average, to acquire infection) and the number of cases of severe disease rises. Certain tick-borne diseases of livestock in Africa, such as anaplasmosis, babesiosis and cowdriosis, show a degree of endemic stability (Eisler et al., 2003). If climate change drives such diseases to new areas, nonimmune individuals of all ages in these regions will be newly exposed, and outbreaks of severe disease could follow.

3.3 Effects on vectors

Much has already been written about the effects of climate change on invertebrate disease vectors. Indeed, this issue, especially the effects on mosquito vectors, has dominated the debate so far. It is interesting to bear in mind, however, that mosquitoes are less significant as vectors of animal disease than they are of human disease (Table 42.1). Mosquitoes primarily, and secondarily lice, fleas and ticks, transmit between them a significant proportion of important human infections. By contrast, biting midges, brachyceran flies (e.g., tabanids, muscids, myiasis flies, hippoboscids), ticks and mosquitoes (and, in Africa, tsetse) all dominate as vectors of livestock disease. Therefore, a balanced debate on the effects of climate change on human and animal disease must consider a broad range of vectors.

There are several processes by which climate change might affect disease vectors. First, temperature and moisture frequently impose limits on their distribution. Often, low temperatures are limiting because of high winter mortality, or high temperatures because they involve excessive moisture loss. Therefore, cooler regions which were previously too cold for certain vectors may begin to allow them to flourish with climate change. Warmer regions could become even warmer and yet remain permissive for vectors if there

Table 42.1 The major diseases transmitted by arthropod vectors to humans and livestock. Adapted from Mullen and Durden (2002)

Vector	Diseases of humans	Diseases of livestock
Phthiraptera (lice)	Epidemic typhus Trench fever Louse-borne relapsing fever	
Reduvidae (assassin bugs)	Chagas' disease	
Siphonaptera (fleas)	Plague Murine typhus Q fever Tularaemia	Myxomatosis
Psychodidae (sand flies)	Leishmanosis Sand fly fever	Canine leishmanosis Vesicular stomatitis
Culicidae (mosquitos)	Malaria Dengue Yellow fever West Nile Filiariasis Encephalitides ((WEE, EEE, VEE, Japanese encephalitis, Saint Louis encephalitis) Rift Valley Fever	West Nile fever Encephalitides Rift Valley fever Equine infectious anemia
Simulidae (Black flies)	Onchocercosis	Leucocytozoon (birds) Vesicular stomatitis
Ceratopogonidae (Biting midges)		Bluetongue African horse sickness Akabane Bovine ephemeral fever
Glossinidae (tsetse flies)	Trypanosomosis	Trypanosomosis
Tabanidae (horse flies)	Loiasis	Sura Equine infectious anaemia *Trypanosoma vivax*
Muscidae (muscid flies)	Shigella *E. coli*	Mastitis Summer mastitis Pink-eye (IBK)
Muscoidae, Oestroidae (myiasis-causing flies)	Bot flies	Screwworm Blow fly strike Fleece rot
Hippoboscoidae (louse flies, keds)		Numerous protozoa
Acari (mites)	Chiggers Scrub typhus (tsutsugamushi) Scabies	Mange Scab Scrapie?
Ixodidae (hard ticks) Argasidae (soft ticks)	Human babesiosis Tick-borne encephalitis Tick fevers Ehrlichiosis Q fever Lyme disease Tickborne relapsing fever Tularaemia	Babesiosis East coast fever (theileriosis) Louping ill African Swine Fever Ehrlichiosis Q fever Heartwater Anaplasmosis Borreliosis Tularaemia

is also increased precipitation or humidity. Conversely, these regions may become less conducive to vectors if moisture levels remain unchanged or decrease, with concomitant increase in moisture-stress.

For any specific vector, however, the true outcome of climate change will be significantly more complex than that just outlined . Even with a decrease in future moisture levels, some vectors, such as certain species of mosquito, could become more abundant, at least in the vicinity of people and livestock, if the response to warming is more water-storage and, thereby, the creation of new breeding sites. Equally, some vectors may be relatively insensitive to direct effects of climate change, such as muscids which breed in organic matter or debris, and myiasis flies which breed in hosts' skin.

Changes to temperature and moisture will also lead to increases or decreases in the abundance of many disease vectors. This may also result from a change in the frequency of extreme weather events such as ENSO. Outbreaks of several biting midge and mosquito-borne diseases, for example, have been linked to the occurrence of ENSO (Anyamba et al., 2002; Baylis et al., 1999; Gagnon et al., 2001, 2002; Hales et al., 1999; Kovats, 2000) and mediated, at least in part, by increase in the vector population size in response to heavy rainfall, or rainfall succeeding drought, that ENSO sometimes brings (Anyamba et al., 2002; Baylis et al., 1999). Greater intra- or interannual variation in rainfall, linked or unlinked to ENSO, may lead to an increase in the frequency or scale of outbreaks of such diseases.

The ability of some insect vectors to become or remain infected with pathogens (their vector competence) varies with temperature (Kramer et al., 1983; Purse et al., 2005). In addition to this effect on vector competence, an increase in temperature may alter the balance between the lifespan of an infected vector, its frequency of feeding, and the time necessary for the maturation of the pathogen within it. This balance is critical, as a key component of the risk of transmission

of a vector-borne disease is the number of blood meals taken by a vector between the time it becomes infectious and its death (Macdonald, 1955). Accordingly, rising ambient temperature can increase the risk of pathogen transmission by shortening the time until infectiousness in the vector and increasing its feeding frequency at a faster rate than it shortens the vector's lifespan, such that the number of feeds taken by an infectious vector increases. However, at even higher temperatures this can reverse (De Koeijer and Elbers, 2007) such that the number of infectious feeds then decreases relative to that possible at lower temperatures. This point is extremely important, as it means that the risk of transmission of vector-borne pathogens does not uniformly increase with rising temperature, but that it can become too hot and transmission rates decrease. This effect will be most important for short-lived vectors such as biting midges and mosquitoes (Lines, 1995).

Lastly, there may be important effects of climate change on vector dispersal, particularly if there is a change in wind patterns. Wind movements have been associated with the spread of epidemics of many *Culicoides*- and mosquito-borne diseases (Sellers, 1992; Sellers and Maarouf, 1991; Sellers and Pedgley, 1985; Sellers et al., 1977, 1978, 1982).

3.4 Effects on epidemiological dynamics

Climate change may alter transmission rates between hosts by affecting the survival of the pathogen or the intermediate vector, but also by other indirect forces that may be hard to predict with accuracy. Climate change may be one of the forces that leads to changes in future patterns of international trade, local animal transportation and farm size, all of which may affect the chances of an infected animal coming into contact with a susceptible one. For example, a series of droughts in East Africa between 1993 and 1997 resulted in pastoral communities moving their cattle to graze in areas normally reserved for wildlife. This resulted in cattle infected with a mild lineage of rinderpest transmitting disease both to other cattle and to susceptible wildlife, causing severe disease, for example, in buffalo, lesser kudu and impala, and devastating certain populations (Kock et al., 1999).

3.5 Indirect effects

No disease or vector distribution can be fully understood in terms of climate only. The supply of suitable hosts, the effects of coinfection or immunological cross-protection, the presence of other insects competing for the same food sources or breeding sites as vectors (Davis et al., 1998), and parasites and predators of vectors themselves, could have important effects (Harvell et al., 2002). Climate change may affect the abundance or distribution of hosts or the competitors/predators/parasites of vectors and influence patterns of disease in ways that cannot be predicted from the direct effects of climate change alone.

Equally, it has been argued that climate change-related disturbances of ecological relationships, driven perhaps by agricultural changes, deforestation, the construction of dams and losses of biodiversity, could give rise to new mixtures of different species, thereby exposing hosts to novel pathogens and vectors and causing the emergence of new diseases (WHO, 1996). A possible 'example in progress' is the re-emergence in the UK of bovine tuberculosis, for which the badger (*Meles meles*) is believed to be a carrier of the causative agent, *Mycobacterium bovis*. Farm landscape, such as the density of linear features like hedgerows and so on, is a risk factor for the disease, affecting the rate of contact between cattle and badger (White et al., 1993). Climate change will be a force for modifying future landscapes and habitats, with indirect and largely unpredictable effects on diseases.

4 OTHER DRIVERS OF DISEASE

The future disease burden of humans and animals will depend not only on climate change and its direct and indirect effects on disease, but also on how other drivers of disease change over time. Even for diseases with established climate links, it may be the case that in many, or perhaps most, examples these other drivers will prove to be more important than climate. A survey of 335 events of human disease emergence between 1940 and 2004 classified the underlying causes into twelve categories (Jones et al., 2008). One of these, 'climate and weather' was only listed as the cause of ten emergence events. Six of these were noncholera *Vibrio* bacteria which cause poisoning via shellfish or exposure to contaminated seawater; the remaining four were a fungal infection and three mosquito-borne viruses. The other eleven categories included, however, 'land use changes' and 'agricultural industry changes', with 36 and 31 disease emergence events respectively, and both may be affected by climate change. The causes of the remaining 77 per cent of disease emergence events – 'antimicrobial agent use', 'international travel and commerce', 'human demography and behaviour', 'human susceptibility to infection', 'medical industry change', 'war and famine', 'food industry changes', 'breakdown of public health' and 'bushmeat' – would be expected to be either less or not indirectly influenced by climate change. Hence, climate change's indirect effects on human disease may exceed its direct effects, while drivers unsusceptible to climate change may be more important still at determining our disease future.

5 EVIDENCE OF CLIMATE CHANGE'S IMPACT ON DISEASE

Climate warming has already occurred in recent decades. If diseases are sensitive to such warming, then one might expect a number of diseases to have responded by changing their distribution, frequency or intensity. A major difficulty, however, is the attribution of any observed changes in disease occurrence to climate change because, as shown above, other disease drivers also change over time. It has been argued that the minimum standard for attribution to climate change is that there must be known biological sensitivity of a disease or vector to climate, and that the change in disease or vector (change in seasonal cycle, latitudinal or altitudinal shifts) should be statistically associated with observed change in climate (Kovats et al., 2001). This has been rephrased as the need for there to be change in both disease/vector and climate at the same time, in the same place, and in the 'right' direction (Rogers and Randolph, 2003). Given these criteria, few diseases make the standard; indeed, only a decade ago one group concluded that the literature lacks strong evidence for an impact of climate change on even a single vector-borne disease, let alone other diseases.

This situation has changed. One disease in particular, bluetongue, has emerged dramatically in Europe over the last decade and this emergence can be attributed to recent climate change in the region. It satisfies the right time, right place, right direction criterion (Purse et al., 2005), but in fact reaches a far higher standard – a model for the disease, with variability in time driven only by variation in climate, produces quantitative estimates of risk which fit closely with the disease's recent emergence in both space and time.

5.1 Bluetongue

Bluetongue is a viral disease of sheep and cattle. It originated in Africa, where wild ruminants act as natural hosts for the virus, and where a species of biting midge, *Culicoides imicola*, is the major vector (Mellor et al., 2000). During the twentieth century bluetongue spread out of Africa into other, warm parts of the world, becoming endemic in the Americas, southern Asia and Australasia; in most of these places, indigenous *Culicoides* became the vectors. Bluetongue also occurred very infrequently in the far extremes of southern Europe: once in the southwest (southern Spain and Portugal, 1955–1960), and every 20–30 years in the southeast (Cyprus, 1924; Cyprus, 1943–1946; Cyprus and Greek islands close to Turkey (1977–1978); the presence of *C. imicola* was confirmed in both areas and this species was believed to be the responsible vector. Twenty years after this last outbreak, in 1998, bluetongue once again reappeared in southeastern Europe (Mellor and Wittmann, 2002). Subsequent events, however, are unprecedented.

Between 1998 and 2008 bluetongue accounted for the deaths of more than one million sheep in Europe: by far the longest and largest outbreak on record. Bluetongue has occurred in many countries or regions that have never previously reported this disease or its close relatives. There have been at least two key developments. First, *C. imicola* has spread dramatically, now occurring over much of southern Europe and even mainland France. Second, indigenous European *Culicoides* species have transmitted the virus. This was first detected in the Balkans where bluetongue occurred but no *C. imicola* could be found. In 2006, however, bluetongue was detected in northern Europe (the Netherlands) from where it spread to neighbouring countries, the UK and even Scandinavia. The scale of this outbreak has been huge, yet the affected countries are far to the north of any known *C. imicola* territory.

Recently, the outputs of new, observation-based, high spatial resolution (25 km) European climate data, from 1960 to 2006 have been integrated within a model for the risk of bluetongue transmission, defined by the basic reproduction ratio R_0 (Guis et al., 2011). In this model, temporal variation in transmission risk is derived from the influence of climate (mainly temperature and rainfall) on the abundance of the vector species, and from the influence of temperature alone on the ability of the vectors to transmit the causative virus. As described in section 3.3, this arises from the balance between vector longevity, vector feeding frequency and the time required for the vector to become infectious. Spatial variation in transmission risk is derived from these same climate-driven influences and, additionally, differing densities of sheep and cattle which determines the probability of feeding on one or the other. This integrated model successfully reproduces many aspects of bluetongue's distribution and occurrence, both past and present, in Western Europe, including its emergence in northwest Europe in 2006. The model gives this specific year the highest positive anomaly (relative to the long-term average) for the risk of bluetongue transmission since at least 1960, but suggests that other years were also at much higher than average risk. The model suggests that the risk of bluetongue transmission increased rapidly in southern Europe in the 1980s and in northern Europe in the 1990s and 2000s.

These results indicate that climate variability in space and time are sufficient to explain many aspects of bluetongue's recent past in Europe and provide the strongest evidence to date that this disease's emergence is, indeed, attributable to changes in climate. What then of the future? The same model was driven forwards to 2050 using simulated climate data from two ensembles of 11 regional climate models. The risk of bluetongue transmission in northwestern Europe is projected to continue increasing up to at least 2050 (Figure 42.2). Given the continued presence of susceptible ruminant host populations, the models suggest that by 2050, transmission risk will have increased by 30 per cent in northwest Europe relative to the 1961–1999 mean. The risk is also projected to increase in southwest Europe, but in this case only by 10 per cent relative to the 1961–1999 mean.

The matching of observed change in bluetongue with quantitative predictions of a

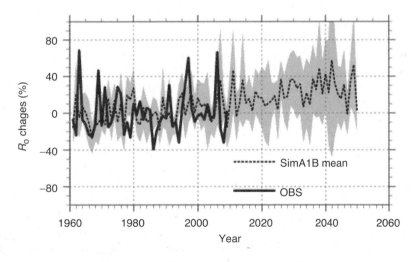

Figure 42.2 Projection of the effect of climate change on the future risk of transmission of bluetongue in Northern Europe. The *y*-axis shows relative anomalies (%) with respect to the 1961–1999 time period for the risk of bluetongue transmission, during August–October in northwest Europe, as defined by the basic reproductive ratio, R_0. R_0 was estimated from climate observations (OBS – thick black line), and an ensemble of 11 future climate projections (SimA1B), for which the dashed line presents the mean and the grey envelope the spread (adapted from Guis et al., 2011). Guis, H, Caminade, C., Calvete, C., Morse, A., Tran, A., Baylis, M. (2011) Modelling the effects of past and future climate on the risk of bluetongue emergence in Europe, J.R. Soc. Interface, doi: 10.1098/rsif.2011.0255

http://rsif.royalsocietypublishing.org/content/early/2011/06/22/rsif.2011.0255.full

climate-driven disease model provides evidence for the influence of climate change far stronger than the 'same place, same time, right direction' criterion described earlier. Indeed, it probably makes bluetongue the most convincing example of a disease that is responding to climate change. In this respect, bluetongue differs remarkably from another vector-borne disease, malaria.

5.2 Malaria

Some 3.2 billion people live with the risk of malaria transmission, between 350 and 500 million clinical episodes of malaria occur each year and the disease kills at least 1 million people annually (WHO, 2005). Of these, each year about 12 million cases and 155,000 to 310,000 deaths are in epidemic areas (Worrall et al., 2004). Inter-annual climate variability primarily drives the timing of these epidemics.

Malaria is caused by *Plasmodium spp.* parasites. Part of the parasite's life cycle takes place within anopheline mosquitoes, at rates driven by ambient temperature, while the remainder of the life cycle occurs within the human host. The parasite and mosquito life cycles can be driven by weather and climate information (namely, rain, temperature and humidity), allowing the disease to be modelled by a range of approaches including the production of dynamic daily time step models (Hoshen and Morse, 2004; Jones and Morse, 2010) or statistical models using observed malaria data and seasonal climatic anomalies (Thomson et al., 2006; Jones et al., 2007). These differing modelling approaches have been driven by seasonal forecasts from ensemble prediction systems (Palmer et al., 2004) to make forecasts of potential malaria risk with lead times of up to 4–6 months (Morse et al., 2005; Jones and Morse, 2010).

The global importance of malaria and the current concerns relating to climate change has led to a degree of polarization of views about the overall importance of climate or indeed climate change on the transmission of malaria. The two ends of this spectrum are,

one, that climate variability or change is the primary actor in any changing transmission pattern of malaria and, two, that any changing patterns today or in the foreseeable future are not related to climate but are due to other factors (Epstein, 2010; Lafferty, 2009). A recent study shows that there has been a net reduction in malaria in the tropics over the last 100 years and, albeit with a simplified climate analysis, that temperature or rainfall change observed so far cannot explain this reduction (Gething et al., 2010). Malaria is demonstrated to have moved from being highly climate-sensitive (an increasing relationship between ambient temperature and the extent of malaria transmission) in the days before disease interventions were widely available; to a situation today where regions with malaria transmission are warmer than those without, but within the malaria-affected region, warmer temperatures no longer mean more disease transmission. Instead, other variables affecting malaria, such as good housing, the running of malaria control schemes, or ready access to affordable prophylaxis, now play a greater role than temperature in determining whether there are higher or lower amounts of transmission.

This would suggest that the importance of climate change in discussion of future patterns of malaria transmission is likely to have been significantly overplayed. However, we should remain wary of drawing conclusions from oversimplified climate analyses, especially ones based on mean temperatures. For example, temperature variability must be taken into account when understanding shifts in malaria transmission (Pascual et al., 2009) and this is highlighted in the sensitivity of transmission to diurnal variability especially around temperature means of 21°C (Paaijmans et al., 2009).

What is clearly recognized, by all sides in the malaria and climate debate, is that mosquitoes need water to lay their eggs in, and for larval development; and that adult mosquitoes need to live long enough in an environment with high humidity, and with sufficiently high temperature for transmission to be possible to the human host. Hence, while the spatial distribution of higher *versus* lower degrees of malaria transmission appears to have become, in a sense, divorced from ambient temperature, it seems likely that the weather plays as important a role as ever in determining when seasonal transmission will start and end. Climate change may therefore still have a role to play in malaria, not so much affecting where it occurs but, via changing rainfall patterns and mosquito numbers, when or for how long people are most at risk.

Malaria has only recently become confined to the developing world and tropics. It is less than 40 years since malaria was eradicated in Europe and the US; and the 15°C July isotherm was the northern limit until the mid nineteenth century (Reiter, 2008). Changes in land use and increased living standards, in particular, acted to reduce exposure to the mosquito vector in these temperate zones, leading ultimately to the final removal of the disease. In the UK, a proportion of the reduction has been attributed to increasing cattle numbers and the removal of marshland (Kuhn et al., 2003). In Finland, changes in family size, improvements in housing, changes in farming practices and the movement of farmsteads out of villages led to the disappearance of malaria (Hulden, 2009); where it had formerly been transmitted indoors in winter. While future increases in temperature may, theoretically, lead to an increased risk of malaria transmission in colder climes than at present (Kuhn et al., 2003; Lindsay et al., 2010), our much-altered physical and natural environment may preclude this risk increasing to a level that merits concern. Once again, a more important future driver of malaria risk, in the UK at least, may be the pressure to return some of our landscape to its former state, such as the reflooding of previously drained marshland.

6 PROSPECTS FOR THE FUTURE

Climate change is widely considered to be a major threat to human and animal health via its effects on infectious diseases. How realistic is this threat? Will most diseases respond to climate change, or just a few? Will there be a net increase in disease burden or might as many diseases decline in impact as increase?

The answers to these questions are important, as they could provide opportunities to mitigate against new disease threats, or may provide the knowledge-base required for policy makers to take necessary action to combat climate change itself. However, we currently lack both the methodology to accurately predict climate change's effects on diseases and, in most cases, the data to apply the methodology to a sufficiently wide range of pathogens.

The majority of pathogens, particularly those not reliant on intermediate hosts (such as arthropod vectors) for transmission, either do occur, or have the potential to occur, in most parts of the world already. Climate change has the capacity to affect the frequency or scale of outbreaks of these diseases: good examples would be the frequency of food poisoning events from the consumption of meat (such as salmonellosis) or shellfish (caused by *Vibrio* bacteria).

Vector-borne diseases are usually constrained in space by the climatic needs of their vectors, and such diseases are therefore the prime examples of where climate change might be expected to cause distributional shifts. Warmer temperatures usually favour the spread of vectors to previously colder environments, thereby exposing possibly naïve populations to new diseases.

However, altered rainfall distributions have an important role to play. Many pathogens or parasites, such as those of anthrax, haemonchosis and numerous vector-borne diseases, may in some regions be subject to opposing forces of higher temperatures promoting pathogen or vector development, and increased summer dryness leading to more pathogen or vector mortality. Theoretically, increased dryness could lead to a declining risk of certain diseases. A good example is fasciolosis, where the lymnaeid snail hosts of the *Fasciola* trematode are particularly dependent on moisture. Less summer rainfall and reduced soil moisture may reduce the permissiveness of some parts of the UK for this disease. The snail and the free-living fluke stages are, nevertheless, also favoured by warmer temperatures and, in practice, current evidence is that fasciolosis is spreading in the UK (Pritchard et al., 2005).

One way to predict the future for disease in a specific country is to learn from countries that, today, are projected to have that country's future climate (Rogers et al., 2001; Sutherst, 1998). At least some of the complexity behind the multivariate nature of disease distributions should have precipitated out into the panel of diseases that these countries currently face.

For example, in broad terms, the UK's climate is predicted to get warmer, with drier summers and wetter winters, becoming therefore increasingly 'Mediterranean'. It would seem reasonable, therefore, to predict that the UK of the future might experience diseases currently present in, or that occur periodically in, southern Europe. For humans, the best example would be leishmanosis (cutaneous and visceral) (Dujardin et al., 2008), while for animals examples include West Nile fever (Gould et al., 2006), *Culicoides imicola*-transmitted bluetongue and African horse sickness (Wittmann and Baylis, 2000), and canine leishmanosis (Shaw et al., 2008). The phlebotomid sandfly vectors of the latter do not currently occur in the UK but they are found widely in southern continental Europe, including France, with recent reports of their detection in Belgium (Depaquit et al., 2005).

However, the contrasting examples of bluetongue and malaria – one spreading because of climate change and one retreating despite it – show that considerations which focus entirely on climate may well turn out to be false. Why are these two diseases, both vector-borne and subject to the similar epidemiological processes and temperature dependencies, so different with respect to climate change? The answer lies in the relative importance of other disease drivers. For bluetongue, it is difficult to envisage epidemiologically relevant drivers of disease transmission, other than climate, that have changed significantly over the time period of the disease's emergence (Purse et al., 2005). Life on farm for the midges that spread bluetongue is probably not dramatically different today from the life they enjoyed 30 years ago. Admittedly, changes in the trade of animals or other goods may have been important drivers of the increased risk of introduction of the causative viruses into Europe, but after introduction, climate change may be the most important driver of increased risk of spread.

For malaria, change in drivers other than climate, such as land use and housing, the availability of prophylaxis, insecticides and, nowadays, insecticide-treated bed nets, have played far more dominant roles in reducing malaria occurrence than climate change may have played in increasing it. Two key reasons, then, for the difference between the two diseases are, first, that life for the human hosts of malaria has changed more rapidly than that of the ruminant hosts of bluetongue; and second, the human cost of malaria was so great that interventions were developed; while the (previously small) economic burden of bluetongue did not warrant such effort and our ability to combat the disease five years ago was not very different from that of 50 years before. The very recent advent of novel inactivated vaccines for bluetongue may now be changing this situation.

We began this section by asking whether climate change will affect most diseases or just a few. The examples of malaria and bluetongue demonstrate that a better question may be as follows: of those diseases that are sensitive to climate change, how many are relatively free from the effects of other disease drivers such that the pressures brought by a changing climate can be turned into effects?

7 CONCLUSIONS

Oh, happy he who still can hope in our day
to breathe the truth while plunged in seas of error!

What we don't know is really what we need, and what we know is of no use whatever! (Lamentation on the difficulty of halting the plague: from Goethe's *Faust*).

What we do know is that the spatiotemporal distributions and incidences of many diseases are associated with climate, and that at least some of these infections will respond to climate change. The most responsive diseases will be those where the causative agent spends a period of time outside an animal host, exposed to environmental influence, the prime examples being vector-borne, food-borne and water-borne diseases.

What we do not know, at least with confidence, is what the responses will be. Climate change will affect diseases through forces operating on the pathogen, the host, the vector, epidemiological processes and other, indirect routes. Other forces will also be at play. The number and variety of diseases, our poor understanding of many of them, the number of climate change (and other) processes affecting them, the uncertainties in climate change predictions, and the spatial heterogeneity of these predictions, all combine to make attempts to predict our disease future bewilderingly complex.

A key issue will be the relative importance of other drivers of the same diseases, and to what extent those drivers will change over the long time scales by which climate change is measured. This is far from clear for all but a handful of diseases, but knowledge of drivers is an important and necessary step towards building a more accurate picture of our future human and animal disease burdens.

REFERENCES

Anderson J., Barrett T. and Scott G. R. 1996. *Manual of the Diagnosis of Rinderpest*. Food and Agriculture Organization of the United Nations. Rome.

Anyamba A., Linthicum K. J., Mahoney R., Tucker C. J. and Kelley P. W. 2002. Mapping potential risk of Rift Valley fever outbreaks in African savannas using vegetation index time series data. *Photogrammetric Engineering and Remote Sensing* 68: 137–145.

Aucamp P. J. 2003. Eighteen questions and answers about the effects of the depletion of the ozone layer on humans and the environment. *Photochemical & Photobiological Sciences* 2: 9–24.

Baylis M., Mellor P. S. and Meiswinkel R. 1999. Horse sickness and ENSO in South Africa. *Nature* 397: 574.

Cannell J. J., Vieth R., Umhau J. C., Holick M. F., Grant W. B., Madronich S., Garland C. F. and Giovannucci E. 2006. Epidemic influenza and vitamin D. *Epidemiology and Infection* 134: 1129–1140.

Cook G. 1992. Effect of global warming on the distribution of parasitic and other infectious diseases: a review. *Journal of the Royal Society of Medicine* 85: 688–691.

Davies F., Linthicum K. and James A. 1985. Rainfall and epizootic Rift Valley fever. *Bulletin of the World Health Organization* 63: 941–943.

Davis A. J., Jenkinson L. S., Lawton J. H., Shorrocks B. and Wood S. 1998. Making mistakes when predicting shifts in species range in response to global warming. *Nature* 391: 783–786.

Davis S., Begon M., De Bruyn L., Ageyev V. S., Klassovskiy N. L., Pole S. B. et al. 2004. Predictive thresholds for plague in Kazakhstan. *Science* 304: 736–738.

de Gruijl F. R., Longstreth J., Norval M., Cullen A. P., Slaper H., Kripke M. L. et al. 2003. Health effects from stratospheric ozone depletion and interactions with climate change. *Photochemical & Photobiological Sciences* 2: 16–28.

De Koeijer A. A. and Elbers A. R. W. 2007. Modelling of vector-borne disease and transmission of bluetongue virus in North-West Europe, in Takken W. and Knols B. G. J. (eds) *Emerging Pests and Vector-borne Diseases in Europe*. Wageningen: Wageningen Academic Publishers, pp. 99–112.

Depaquit J., Naucke T. J., Schmitt C., Ferté H. and Léger N. 2005. A molecular analysis of the subgenus *Transphlebotomus* Artemiev, 1984 (Phlebotomus, Diptera, Psychodidae) inferred from ND4 mtDNA with new northern records of *Phlebotomus mascittii* Grassi, 1908. *Parasitology Research* 95: 113–116.

Donaldson A. I. 1972. The influence of relative humidity on the aerosol stability of different strains of foot-and-mouth disease virus suspended in saliva. *Journal of General Virology* 15: 25–33.

Dowell S. F. 2001. Seasonal variation in host susceptibility and cycles of certain infectious diseases. *Emerging Infectious Diseases* 7: 369–374.

Dujardin J. C., Campino L., Canavate C., Dedet J. P., Gradoni L., Soteriadou K. et al. 2008. Spread of vector-borne diseases and neglect of leishmaniasis, Europe. *Emerging Infectious Diseases* 14: 1013–1018.

Eiler A., Johansson M. and Bertilsson S. 2006. Environmental influences on *Vibrio* populations in northern temperate and boreal coastal waters (Baltic and Skagerrak Seas). *Applied And Environmental Microbiology* 72: 6004–6011.

Eisler M. C., Torr S. J., Coleman P. G., Machila N. and Morton J. F. 2003. Integrated control of vector-borne diseases of livestock-pyrethroids: panacea or poison? *Trends In Parasitology* 19: 341–345.

Epstein P. 2010. The ecology of climate change and infectious diseases: comment. *Ecology* 91: 925–928.

Gagnon A. S., Bush A. B. G. and Smoyer-Tomic K. E. 2001. Dengue epidemics and the El Nino Southern Oscillation. *Climate Research* 19: 35–43.

Gagnon A. S., Smoyer-Tomic K. E. and Bush A. B. G. 2002. The El Nino Southern Oscillation and malaria epidemics in South America. *International Journal of Biometeorology* 46: 81–89.

Gale P., Adkin A., Drew T. and Wooldridge M. 2008. Predicting the impact of climate change on livestock disease in Great Britain. *Veterinary Record* 162: 214–215.

Gale P., Drew T., Phipps L. P., David G. and Wooldridge M. 2009. The effect of climate change on the occurrence and prevalence of livestock diseases in Great Britain: a review. *Journal Of Applied Microbiology* 106: 1409–1423.

Gething P. W., Smith D. L., Patil A. P., Tatem A. J., Snow R. W. and Hay S. I. 2010. Climate change and the global malaria recession. *Nature* 465: 342–344.

Githeko A. K., Lindsay S. W., Confalonieri U. E. and Patz J. A. 2000. Climate change and vector-borne diseases: a regional analysis. *Bulletin of the World Health Organization* 78: 1136–1147.

Gould E. A., Higgs S., Buckley A. and Gritsun T. S. 2006. Potential arbovirus emergence and implications for the United Kingdom. *Emerging Infectious Diseases* 12: 549–555.

Guis H., Caminade C., Calvete C., Morse A. P., Tran A. and Baylis M. Modelling the effects of past and future climate on the risk of bluetongue emergence in Europe, *Journal of the Royal Society Interface*, doi: 10.1098/rsif.2011.0255.

Hales S., Weinstein P., Souares Y. and Woodward A. 1999. El Niño and the dynamics of vectorborne disease transmission. *Environmental Health Perspectives* 107: 99–102.

Hall H. T. B. 1988. *Diseases and Parasites of Livestock in the Tropics.* Harlow: Longman Scientific and Technical.

Harvell C. D., Kim K., Burkholder J. M., Colwell R. R., Epstein P. R., Grimes D. J. et al. 1999. Review: Marine ecology – Emerging marine diseases – Climate links and anthropogenic factors. *Science* 285: 1505–1510.

Harvell C. D., Mitchell C. E., Ward J. R., Altizer S., Dobson A. P., Ostfeld R. S. and Samuel M. D. 2002. Ecology – Climate warming and disease risks for terrestrial and marine biota. *Science* 296: 2158–2162.

Hay S. I., Cox J., Rogers D. J., Randolph S. E., Stern D. I., Shanks G. D. et al. 2002. Climate change and the resurgence of malaria in the East African highlands. *Nature* 415: 905–909.

Hoshen M. B. and Morse A. P. 2004. A weather-driven model of malaria transmission. *Malaria Journal* 3: 32.

Hulden L. 2009. The decline of malaria in Finland – the impact of the vector and social variables. *Malaria Journal* 8: 94.

IPCC. 2001. *Climate Change 2001: The Scientific Basis. Intergovernmental Panel on Climate Change.* Cambridge: Cambridge University Press,

Jankovic D., Liu Z. G. and Gause W. C. 2001. Th1-and Th2-cell commitment during infectious disease: asymmetry in divergent pathways. *Trends in Immunology* 22: 450–457.

Jones A. E. and Morse A. P. 2010. Application and validation of a seasonal ensemble prediction system using a dynamic malaria model. *Journal of Climate* 23: 4202–4215.

Jones A. E., Wort U. U., Morse A. P., Hastings I. M. and Gagnon A. S. 2007. Climate prediction of El Nino malaria epidemics in north-west Tanzania. *Malaria Journal* 6: 162.

Jones K. E., Patel N. G., Levy M. A., Storeygard A., Balk D., Gittleman J. L. and Daszak P. 2008. Global trends in emerging infectious diseases. *Nature* 451: 990–U994.

Kausrud K. L., Viljugrein H., Frigessi A., Begon M., Davis S., Leirs H., Dubyanskiy V. and Stenseth N. C. 2007. Climatically driven synchrony of gerbil populations allows large-scale plague outbreaks. *Proceedings of the Royal Society B – Biological Sciences* 274: 1963–1969.

Kock R. A., Wambua J. M., Mwanzia J., Wamwayi H., Ndungu E. K., Barrett T., Kock N. D. and Rossiter P. B. 1999. Rinderpest epidemic in wild ruminants in Kenya 1993–97. *Veterinary Record* 145: 275–283.

Kovats R. S. 2000. El Nino and human health. *Bulletin of the World Health Organization* 78: 1127–1135.

Kovats R. S., Campbell-Lendrum D. H., McMichael A. J., Woodward A. and Cox J. S. 2001. Early effects of climate change: do they include changes in vector-borne disease? *Philosophical Transactions of the Royal Society of London Series B – Biological Sciences* 356: 1057–1068.

Kovats R. S., Edwards S. J., Hajat S., Armstrong B. G., Ebi K. L. and Menne B. 2004. The effect of temperature on food poisoning: a time-series analysis of salmonellosis in ten European countries. *Epidemiology and Infection* 132: 443–453.

Kovats R. S., Haines A., Stanwell-Smith R., Martens P., Menne B. and Bertollini R. 1999. Climate change and human health in Europe. *British Medical Journal* 318: 1682–1685.

Kramer L., Hardy J. and Presser S. 1983. Effect of temperatures of extrinsic incubation on the vector competence of *Culex tarsalis* for Western Equine Encephalomyelitis virus. *American Journal of Tropical Medicine and Hygiene* 32: 1130–1139.

Kuhn K. G., Campbell-Lendrum D. H., Armstrong B. and Davies C. R. 2003. Malaria in Britain: Past, present, and future. *Proceedings of the National Academy of Science, USA* 100: 9997–10001.

Lafferty K. D. 2009. The ecology of climate change and infectious diseases. *Ecology* 90: 888–900.

Lindsay S. W., Hole D. G., Hutchinson R. A., Richards S. A. and Willis S. G. 2010. Assessing the future threat from vivax malaria in the United Kingdom using two markedly different modelling approaches. *Malaria Journal* 9: 70.

Lines J. 1995. The effects of climatic and land-use changes on insect vectors of human disease. *Insects in a Changing Environment.* 157–175.

Linthicum K. J., Anyamba A., Tucker C. J., Kelley P. W., Myers M. F. and Peters C. J. 1999. Climate and satellite indicators to forecast Rift Valley Fever epidemics in Kenya. *Science* 285: 397–400.

Linthicum K. J., Bailey C. L., Davies G. and Tucker C. J. 1987. Detection of Rift Valley fever viral activity in Kenya by satellite remote sensing imagery. *Science* 235: 1656–1659.

Little P. D., Mahmoud H. and Coppock D. L. 2001. When deserts flood: risk management and climatic processes among East African pastoralists. *Climate Research* 19: 149–159.

Lowen A. C., Mubareka S., Steel J. and Palese P. 2007. Influenza virus transmission is dependent on relative humidity and temperature. *PLoS Pathogens* 3: e151.

Macdonald G. 1955. The measurement of malaria transmission. *Proceedings of the Royal Society of Medicine – London* 48: 295–302.

McMichael A. J. and Githeko A. K. 2001. Human health, in Canziani James O. F., McCarthy J., Leary N. A., Dokken D. J. and White K. S. (eds) *Climate Change 2001: Impacts, Adaptation, and Vulnerability. Contribution of Working Group II to the Third Assessment Report of the Intergovernmental Panel on Climate Change.* Cambridge: Cambridge University Press, pp. 453–485.

Mellor P. S., Boorman J. and Baylis M. 2000. Culicoides biting midges: their role as arbovirus vectors. *Annual Review of Entomology* 45: 307–340.

Mellor P. S. and Wittmann E. J. 2002. Bluetongue virus in the Mediterranean Basin 1998–2001. *The Veterinary Journal* 164: 20–37.

Morens D. M., Folkers G. K. and Fauci A. S. 2004. The challenge of emerging and re-emerging infectious diseases. *Nature* 430: 242–249.

Morse A. P., Doblas-Reyes F. J., Hoshen M. B., Hagedorn R. and Palmer T. N. 2005. A forecast quality assessment of an end-to-end probabilistic multi-model seasonal forecast system using a malaria model. *Tellus Series a-Dynamic Meteorology and Oceanography* 57: 464–475.

Mullen G. and Durden L. 2002. *Medical and Veterinary Entomology.* Orlando: Academic Press.

Paaijmans K. P., Read A. F. and Thomas M. B. 2009. Understanding the link between malaria risk and climate. *Proceedings of the National Academy of Sciences of the United States of America* 106: 13844–13849.

Palmer T. N., Alessandri A., Andersen U., Cantelaube P., Davey M., Delecluse P. et al. 2004. Development of a European multimodel ensemble system for seasonal-to-interannual prediction (DEMETER). *Bulletin of the American Meteorological Society* 85: 853–872.

Parker R., Mathis C., Looper M. and Sawyer J. 2002. *Guide B–120: Anthrax and livestock: Cooperative Extension Service, College of Agriculture and Home Economics.* Las Cruces: University of New Mexico.

Pascual M., Dobson A. P. and Bouma M. J. 2009. Underestimating malaria risk under variable temperatures. *Proceedings of the National Academy of Sciences of the United States of America* 106: 13645–13646.

Patz J. A. and Kovats R. S. 2002. Hotspots in climate change and human health. *British Medical Journal* 325: 1094–1098.

Perry B. D., Randolph T. F., McDermott J. J., Sones K. R. and Thornton P. K. 2002. *Investing in Animal*

Health Research to Alleviate Poverty. Nairobi: International Livestock Research Institute.

Pritchard G. C., Forbes A. B., Williams D. J. L., Salimi-Bejestani M. R. and Daniel R. G. 2005. Emergence of fasciolosis in cattle in East Anglia. *Veterinary Record* 157: 578–582.

Purse B. V., Mellor P. S., Rogers D. J., Samuel A. R., Mertens P. P. C. and Baylis M. 2005. Climate change and the recent emergence of bluetongue in Europe. *Nature Reviews Microbiology* 3: 171–181.

Randolph S. E. 2004. Evidence that climate change has caused .'emergence' of tick-borne diseases in Europe? *International Journal of Medical Microbiology* 293: 5–15.

Reeves W. C., Hardy J. L., Reisen W. K. and Milby M. M. 1994. Potential effect of global warming on mosquito-to-borne arboviruses. *Journal of Medical Entomology* 31: 323–332.

Reiter P. 2008. Global warming and malaria: knowing the horse before hitching the cart. *Malaria Journal* 7: (Supp1):S3.

Reiter P., Thomas C. J., Atkinson P. M., Hay S. I., Randolph S. E., Rogers D. J. et al. 2004. Global warming and malaria: a call for accuracy. *Lancet Infectious Diseases* 4: 323–324.

Rogers D. J. and Packer M. J. 1993. Vector-borne diseases, models, and global change. *Lancet* 342: 1282–1284.

Rogers D. J. and Randolph S. E. 2000. The global spread of malaria in a future, warmer world. *Science* 289: 1763–1766.

Rogers D. J. and Randolph S. E. 2003. Studying the global distribution of infectious diseases using GIS and RS. *Nature Reviews Microbiology* 1: 231–237.

Rogers D. J., Randolph S. E., Lindsay S. W. and Thomas C. J. 2001. *Vector-borne Diseases and Climate Change*. London: Department of Health.

Sellers R. F. 1992. Weather, *Culicoides*, and the distribution and spread of bluetongue and African horse sickness viruses. in Walton T. E. and Osburn B. I. (eds) *Bluetongue, African horse sickness and related Orbiviruses*. Boca Raton: CRC Press, pp. 284–290.

Sellers R. F. and Maarouf A. R. 1991. Possible introduction of epizootic hemorrhagic disease of deer virus (serotype 20) and bluetongue virus (serotype 11) into British Columbia in 1987 and 1988 by infected *Culicoides* carried on the wind. *Canadian Journal of Veterinary Research* 55: 367–370.

Sellers R. F. and Pedgley D. E. 1985. Possible windborne spread to Western Turkey of bluetongue virus in 1977 and of Akabane virus in 1979. *Journal of Hygiene* 95: 149–158.

Sellers R. F., Pedgley D. E. and Tucker M. R. 1977. Possible spread of African horse sickness on the wind. *Journal of Hygiene* 79: 279–298.

Sellers R. F., Pedgley D. E. and Tucker M. R. 1978. Possible windborne spread of bluetongue to Portugal, June-July 1956. *Journal of Hygiene* 81: 189–196.

Sellers R. F., Pedgley D. E. and Tucker M. R. 1982. Rift Valley fever, Egypt 1977: Disease spread by windborne insect vectors? *Veterinary Record* 110: 73–77.

Semenza J. C. and Menne B. 2009. Climate change and infectious diseases in Europe. *Lancet Infectious Diseases* 9: 365–375.

Shaw S. E., Langton D. A. and Hillman T. J. 2008. Canine leishmaniosis in the UK. *Veterinary Record* 163: V-VI.

Soebiyanto R. P., Adimi F. and Kiang R. K. 2010. Modeling and Predicting Seasonal Influenza Transmission in Warm Regions Using Climatological Parameters. *PLoS ONE* 5: e9450.

Sutherst R. W. 1998. Implications of global change and climate variability for vector-borne diseases: generic approaches to impact assessments. *International Journal for Parasitology* 28: 935–945.

Sutmoller P., Barteling S. S., Olascoaga R. C. and Sumption K. J. 2003. Control and eradication of foot-and-mouth disease. *Virus Research* 91: 101–144.

Thomson M. C., Doblas-Reyes F. J., Mason S. J., Hagedorn R., Connor S. J., Phindela T. et al. 2006. Malaria early warnings based on seasonal climate forecasts from multi-model ensembles. *Nature* 439: 576–579.

WHO 1996. *Climate Change and Human Health*. Geneva: World Health Organisation.

WHO 2005. *World Malaria Report, Rollback Malaria Programme*. Geneva: World Health Organisation.

Walther G. R., Post E., Convey P., Menzel A., Parmesan C., Beebee T. J. C. et al. 2002. Ecological responses to recent climate change. *Nature* 416: 389–395.

Weaver S. C. and Barrett A. D. T. 2004. Transmission cycles, host range, evolution and emergence of arboviral disease. *Nature Reviews Microbiology* 2: 789–801.

White P. C. L., Brown J. A. and Harris S. 1993. Badgers (*Meles meles*), cattle and bovine tuberculosis (*Mycobacterium bovis*) a hypothesis to explain the influence of habitat on the risk of disease transmission in southwest England. *Proceedings of the Royal Society of London Series B – Biological Sciences* 253: 277–284.

Wittmann E. J. and Baylis M. 2000. Climate change: effects on *Culicoides*-transmitted viruses and implications for the UK. *The Veterinary Journal* 160: 107–117.

Worrall E., Rietveld A. and Delacollette C. 2004. The burden of malaria epidemics and cost-effectiveness of interventions in epidemic situations in Africa. *American Journal of Tropical Medicine and Hygiene* 71: 136–140.

Wosu L. O., Okiri J. E. and Enwezor P. A. 1992. Optimal time for vaccination against peste des petits ruminants (PPR) disease in goats in the humid tropical zone in southern Nigeria, in Rey B., Lebbie S. H. B. and Reynolds L. (eds) *Proceedings of the Small Ruminant Research and Development in Africa. Proceedings of the First Biennial Conference of the African Small Ruminant Research Network* International Laboratory for Research in Animal Diseases (ILRAD).

Zell R. 2004. Global climate change and the emergence/re-emergence of infectious diseases. *International Journal of Medical Microbiology* 293: 16–26.

43

Policy and Management Options for the Mitigation of Environmental Change

Katie Moon and Chris Cocklin

1 INTRODUCTION

Mitigation involves the development and implementation of new or improved technologies to reduce the extent of human-induced environmental change. Awareness of the detrimental environmental effects of human activity and the consequent need for mitigation has increased progressively since the 1960s. Rachel Carson's seminal publication *Silent Spring* highlighted the need to minimise the adverse environmental effects of human activity noting that 'the rapidity of change and the speed with which new situations are created follow the impetuous and heedless pace of man rather than the deliberate pace of nature' (Carson, 1962: 6). Since that time, there has been increasing scientific interest in the environment and the links between human activity and environmental change. Over recent decades, the attention of scientists and governments shifted from the national to the international arena as global effects of human activity began to manifest. Acid rain, ozone depletion and climate change respected no boundary and accordingly, demanded global action through

cooperation among nations. There have been successes; acid deposition in North America and Europe has been reduced and the rate of ozone depletion has been arrested due to international cooperation, for example.

The avoidance of climate change has not met with the same success and presents an interesting context in which to explore the complexities of mitigation policy, strategy and action for environmental change. The key causes of anthropocentric climate change are well understood: carbon dioxide emissions result from fossil fuel use and changes in land use; elevated methane levels are generated largely by ruminant animals, rice cultivation, biomass burning and changes in the temperature and hydrology of wetland environments; and nitrous oxide, largely generated through application of manufactured fertilisers and land use change (Solomon et al., 2007). Since 1750, it has been estimated that fossil fuels and land use change have contributed 66 and 33 per cent of anthropocentric carbon dioxide emissions, respectively (Solomon et al., 2007). Agricultural activity has been largely responsible for increases in methane emissions during this period. Mitigation policy, in the context of climate change, aims to reduce global greenhouse gas emissions primarily through new or advanced emissions reduction technology or through increases in the extent and effectiveness of carbon sinks and storage.

Responses to climate change are being defined through the development and implementation of public policy. For the purpose of organising this chapter, public policy is defined in terms of a hierarchy of actions (O'Riordan, 1995):

- *Policy*: the level of commitment to mitigation (and adaptation) action, which sets the direction and broad intent of government. Policy makers need to consider the relationships between mitigation and adaptation, the need to make trade-offs and identify synergies, and aim to achieve equitable outcomes, accommodate uncertainty and manage risks.
- *Program*: the strategic framework to implement policy objectives, which includes mitigation activities such as abatement or reduction in emissions, carbon sink establishment, geoengineering and adaptation. The appropriate use of policy tools will support effective mitigation outcomes.
- *Plan*: a specific element or aspect of a program that defines the actions that are required to meet the policy objectives. Elements can include institutional, technological, behavioural, and financial modifications and improvements. Plans to deliver mitigation outcomes must be developed with an understanding of technology development and change, the availability of current and near-future technological options, and the barriers to technology deployment and behaviour change.

2 MITIGATION POLICY: THE COMMITMENT

Mitigation of the drivers and effects of environmental change relies on the formulation and implementation of policy, a process which can be as complex as the environmental systems that policy seeks to protect. This section will present elements that may be considered during mitigation policy formulation,

including the potential to substitute mitigation and adaptation responses, the existence of tradeoffs and synergies, the need to manage inequalities within and between nations, and the uncertainty of science to predict the extent and magnitude of environmental change. The challenges of international cooperation are discussed, as well as the policy tools that may assist in the delivery of efficient and effective mitigation policy outcomes. We begin with a review of the challenges to policy formulation, thereby setting the broad context within which effective mitigation policy must emerge.

2.1 Challenges to policy formulation

Challenges to policy formulation arise due to the existence of four main barriers (Connelly and Smith, 2003). First, policy development relies on *collective action*; that is, agreement that there is a problem, what can be done to fix the problem, and then how to deliver the agreed solution through policy. Collective action can be compromised when some individuals or groups are unwilling to act unless all others act, the result being complete inaction (Connelly and Smith, 2003). Collective action can also be constrained when different actors are reluctant to agree on policy objectives.

Second, public interest in environmental problems can influence the formulation of policy, as characterised by the *issue-attention cycle* (Downs, 1972). According to this representation, the pre-problem stage begins when interest groups or experts become concerned about an environmental problem such as climate change. 'Alarmed discovery' and 'euphoric enthusiasm' follow, whereby the public looks for a 'quick fix' to solve the problem. However, when the public realise the financial cost of solving the problem and the sacrifices that individuals or organisations may need to make to find a solution, there is a gradual decline in public interest. Moreover, people become discouraged, bored and/or threatened by thinking about solutions to the problem. Finally, the post-problem stage involves replacement of the original problem with a new problem. In many instances, though, the creation of new policies, programs and institutions that were implemented to redress the original problem can persist.

Third, policy making can be constrained by *complexity*, *uncertainty* and *bounded rationality*. Environmental systems are complex and so the responses of the environment to human activity are not always completely understood. Effective policy can account for some of this uncertainty, for example via the application of the precautionary principle. However, uncertainty can force policy formulation to operate within 'bounded rationality' whereby the decisions are bounded by the knowledge and capacity of policy makers and of the organisation. Bounded rationality typically arises during complex situations because 'our logical apparatus ceases to cope – our rationality is "bounded"' (Arthur, 1994: 406). The result is decision making at the departmental level, based on rational administration and technical options that overlook irrational, coordinated, creative and imaginative alternatives that may generate better outcomes.

Fourth, the successful formulation of policy depends on the *power and influence* of individuals, groups or organisations. Lukes 1974) proposed three dimensions of power that are useful to consider in the context of policy making. These dimensions involve overt conflicts where the tangible effects of the power of one over another can be observed (e.g., party politics); covert conflicts, where the effects of power are immediate but not observed (e.g., non-decision making by one party against another); and latent conflicts whereby the effects of the power of one over another is observed at a later date (Connelly and Smith, 2003).

These challenges to policy formulation are not always immediately obvious to policy makers, or those groups or individuals who are affected by policies to mitigate the effects of environmental change. Yet, these challenges can confound action on environmental problems and so must be considered as part of mitigation and adaptation policy formulation.

2.2 Relationships between mitigation and adaptation

Mitigation and adaptation serve fundamentally different purposes. Mitigation may increase species survival and the functioning of ecosystems through a commitment to reduce atmospheric greenhouse gas concentrations that will reduce the extent and severity of climatic and environmental changes. In contrast, adaptation focuses principally on human survival and quality of life and aims to preserve them through adjustment in human or natural systems. According to one view, adaptation to the effects of a changed environment is tantamount to 'passive acceptance'. In this sense, adaptation strategies have the potential to 'lend an impression, rightly or wrongly, that one was against mitigation activities and in a broader sense anti-environmental' (Pielke 1998: 162).

Nonetheless, mitigation and adaptation responses to environmental change are intimately linked. If a business-as-usual scenario was taken; that is, no mitigation of carbon emissions, then an adaptation strategy would need to accommodate more severe environmental change. If adaptive strategies fail to meet their objectives, then human and non-human species and the functioning of ecosystems may be threatened. Mitigation therefore presents both a precautionary approach to minimise the extent of environmental change and a reduced need to rely solely on the potential of adaptive responses. Moreover, mitigation is particularly important for environmental systems or communities that have low 'adaptive capacity'; that is, their ability to adjust to environmental change to (a) moderate harm, (b) cope with the consequences of environmental change, or (c) take advantage of any opportunities of a changed environment (Garnaut, 2008).

Mitigation actions can have both positive and negative consequences for adaptation (Figure 43.1). For example, energy-efficient housing minimises energy consumption (mitigation) and also provides a dwelling that will be less affected by temperature extremes (adaptation). However, large investments in mitigation action can divert funds from adaptation. In nonindustrialised nations, mitigation resources can sometimes be better spent on adaptation, such as reductions in infectious disease-related incidents that result from climate change (Tol, 2005). Adaptation can have positive or negative consequences for the mitigation of environmental change. For example, afforestation activities may be required in some areas to stabilise soil where increased rainfall, flooding and landslides are predicted to occur. Carbon sequestered by the new plantations will contribute to climate change mitigation. However, adaptation actions such as increased use of air conditions, contributes to greenhouse gas emissions where fossil fuels are used. Generally speaking, the more mitigation action that is taken, the less adaptation will be required. External processes, synergies and tradeoffs

also affect mitigation and adaptation policy and action.

The extent of mitigation will hinge upon the credibility of climate science and the perceived economic, social and environmental cost of a warmer climate. The ability of regions to adapt to environmental change may also influence a nation's commitment to mitigate change and their pressure on other nations to mitigate.

In addition to the influence of mitigation and adaptation actions on one another, there are several main differences between these two approaches that can influence policy formulation (Table 43.1). First, mitigation minimises the cause and ultimately the effect of environmental change, while adaptation is commonly a reaction to observed environmental change. Second, spatially, adaptation activities are typically implemented at a local and regional level because this is often the most appropriate scale at which to manage the effects of environmental change. For example, adaptation may include the installation of flood prevention infrastructure or the use of drought-resistant plant species. In contrast, mitigation policy will typically be

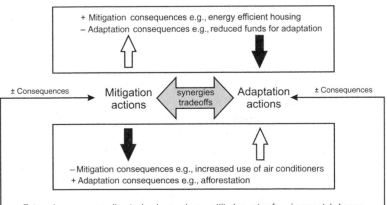

Figure 43.1 The four main interrelationships between mitigation and adaptation action: consequences of mitigation action on adaptation; consequences of adaptation action on mitigation (+ = positive, − = negative); synergies and tradeoffs that exist between mitigation and adaptation; and consequences of external processes on mitigation and adaptation.

Table 43.1   Similarities and differences between mitigation and adaptation action

	Mitigation action	Adaptation action
Action	Proactive action, long term reduction of climate effects	Reactive action, iterations based on realised effects of climate change
Temporal effect	Benefits to later generations	Benefits usually to generation that bears the costs
Spatial effect	Global benefits, but varying across regions	Primarily local and regional benefits
Sectoral effect	Homogenous, focus on fossil fuel industry	Heterogeneous, some stress on agriculture
Stakeholders	Ministries of Energy, Finance, Transport, Agriculture, Forestry, Environment	Local health, water, and coastal zone managers, farmers, tourism operators, architects, energy suppliers
Cooperation	Global	Local, regional and national
Uncertainty	Regular policy and strategy review to account for new climate change projections	Delay of action in areas of high uncertainty because of irreversibility of some adaptation actions and the need to justify spend
Equity	Action of some nations can create benefits for those countries that do not act, potentially those less vulnerable to climate change (free-riding)	Countries most vulnerable to climate change will need to rely on (other) large carbon contributors to implement mitigation actions

Source: Adapted from Dang et al. (2003).

formulated at the national and international level to improve energy efficiency, develop and deploy non-fossil fuel-based technology, and commit to increase global carbon sinks. Third, temporally, the environmental costs or benefits of mitigation action will be experienced by later generations, while environmental benefits of adaptation are characteristically realised primarily by those who bear the financial costs. Fourth, the relevance of environmental change to different sectors influences policy formulation. Mitigation activities focus largely on abatement of fossil fuel use and therefore improvements within the energy sector are primarily sought. Adaptation activities, on the other hand, must deal with the effects of a changed climate to human and natural systems (Dang et al., 2003) and may include improvements in agricultural efficiencies, and natural disaster planning. Fifth, different stakeholders will be engaged, based on whether mitigation or adaptation policies are being formulated. Mitigation policy will involve heads of state and global agreement, such as through the Kyoto Protocol, while adaptation will more commonly involve local and regional stakeholders and the development of regional management plans.

The identification of tradeoffs and synergies between mitigation and adaptation can assist in policy formulation. Tradeoffs balance 'adaptation and mitigation when it is not possible to carry out both activities fully at the same time' (Klein et al., 2007: 749). Tradeoffs aim to balance risk and uncertainty that result from imperfect scientific information, the inability of analytical models to accommodate all variables, and the intrinsic complexity of natural systems that can be difficult, if not impossible, to assess (Halsnæs et al., 2007). Tradeoffs between mitigation and adaptation activities can result in either a neutral or a negative effect on the concentration of greenhouse gases. For example, agricultural diversification, crop insurance, and coordinated plant breeding are adaptation responses that do not necessarily have mitigation outcomes (neutral effect). Alternatively, increased energy consumption of fossil fuels required to adapt to a changed environment would increase greenhouse gas emissions (negative effect).

In contrast to tradeoffs, a synergy is 'the interaction of adaptation and mitigation so that their combined effect is greater than the sum of their effects if implemented separately' (Klein et al., 2007: 749). For example, reduced water availability can stimulate an increase in irrigation efficiency (adaptation), which uses less energy (mitigation). Watershed management policies can involve reforesting areas to reduce the effects of flooding from a changed climatic regime (adaptation) and sequester carbon (mitigation). The extent of synergies should be calculated before management plans are implemented so that synergies can be maximised (Dang et al., 2003).

2.3 Uncertainty, precaution and equity

The precautionary principle aims to minimise the extent and severity of environmental change when the effects of human activities are uncertain. Uncertainty can manifest in three main ways: inadequate data, deficient models or unmeasurable system components (O'Riordan, 1995). Indeed 'it is better to be roughly right in due time, bearing in mind the consequences of being very wrong, than to be precisely right too late' (NAVF, 1990, in O'Riordan and Jordan, 1995: 194). Policy makers can accommodate uncertainty via consideration of the timing and sequencing of decisions that may provide for increased knowledge of environmental effects of climatic change; the anticipated effect of climate change, be that nonlinear and sudden, or gradual and proportionate to the level of anthropocentric greenhouse gas emissions; the costs of mitigation and whether costs change with time and extent of environmental change, or if they are subjected to thresholds that make them nonlinear; and the distribution of effects, whether clustered in some regions or global in extent (Halsnæs et al., 2007). Precautionary approaches therefore exist along a continuum, from weak formulations that tend to protect the existing systems, norms and behaviours, through to strong formulations that require significant social and institutional change (O'Riordan and Jordan, 1995).

The precautionary principle consists of several core intellectual and policy elements (O'Riordan and Jordan, 1995: 195).

1 *Pro-action:* willingness to take action in advance of scientific proof, or in the face of fundamental ignorance of possible consequences, on the grounds that further delay or thoughtless action could ultimately prove far more costly than the 'sacrifice' of not carrying on right now.

2 *Cost-effectiveness of action:* the application of proportionality of response to show that there should be a regular examination of identifiable social and environmental gains arising from a course of action that justifies the costs.

3 *Safeguarding ecological space:* a fundamental notion underlying all interpretations of the precautionary principle is how far natural systems and social organisations are resilient or vulnerable to further change or alteration.

4 *Legitimising the status of intrinsic value:* the stronger formulations of the precautionary principle are consistent with a bioethic; that vulnerable, or critical natural systems, namely those close to thresholds, or whose existence is vital for natural regeneration, should be protected as a matter of moral right.

5 *Shifting the onus of proof:* the precautionary principle suggests that the burden of proof could shift onto the protodeveloper to show 'no reasonable environmental harm' to such sites or processes, before development of any kind could proceed.

6 *Meso-scale planning:* the meso-scale is the period, roughly 25 to 100 years from now, over which any major decision will have an influence, yet where the normal tools for foresight and decision analysis are simply not workable. Cost–benefit analyses rarely take into account the likely costs and benefits of various courses of action during this period. Similarly, legal rules for compensation or obligation to take care are still ill-developed.

7 *Paying for ecological debt:* precaution is essentially forward looking, but there is a case for considering a burden-sharing responsibility for those not being cautious or caring in the past.

Equitable mitigation policy accounts for both contemporary society (intragenerational equity) and for future generations (intergenerational equity). Intragenerational equity is particularly important in the consideration of the activities of both the industrialised and the nonindustrialised world. Nations differ in their historical and projected greenhouse gas emission contribution, their vulnerability to climate change and their ability to finance mitigation actions (Cazorla and Toman, 2000). Article 3 of the United Nations Framework Convention on Climate Change 1992 (UNFCCC, 1992), considers such inequities in the development of mitigation action: 'The Parties should protect the climate system for the benefit of present and future generations of humankind, on the basis of equity and in accordance with their common but differentiated responsibilities and respective capabilities. Accordingly, the developed country Parties should take the lead in combating climate change and the adverse effects thereof.' Commitment to mitigation action based on responsibility for the differential contributions to climate change will, in an important sense, provide for intragenerational equity.

Intergenerational equity is commonly anchored in the concept of sustainable development, where it is defined as 'development that meets the needs of the present without compromising the needs of the future' (Bruntland, 1987). Current and near-future mitigation action of current generations can influence the distribution of environmental, economic and social costs and benefits to future generations (Halsnæs et al., 2007). If the effects of a changed climate deplete the diversity of ecosystems and the life within, the range of opportunities available to future generations to manage their problems and satisfy their needs may be limited. Such considerations should be made in the development of mitigation policy, more so than for adaptation policy.

2.4 Policy development in an international context

The international action that is required to mitigate climate change is confounded by the differentiated effects of predicted changes and a strong sense of differentiated responsibility (Connelly and Smith, 2003). The environmental effects of climate change are predicted to vary geographically because some areas will experience greater change than others. This situation is graphically illustrated by the island nations of the Pacific that are predicted to be inundated as sea levels rise. Nations that are less affected by climate change may be less willing than nations more affected by change to minimise their greenhouse gas emissions, to the detriment of those nations that will experience moderate to severe environmental change. The responsibility to mitigate climate change is differentiated because industrialised nations have relied on the combustion of fossil fuels to underpin their economic and social advancement. These nations have contributed more greenhouse gas emissions than less industrialised nations and so are more responsible for the anthropocentric greenhouse effect. Mitigation strategies that expect all nations to take the same level of responsibility for climate change will prospectively limit the economic prospects of less industrialised nations as a result of restricted access to cheap fossil fuels that have been essential to the development of the industrialised nations. According to the notion of differentiated responsibility, the legacy of fossil fuel use must be accounted for by the industrialised nations, calling for a relatively higher level of investment in mitigation strategies by those nations. The distribution of the costs and benefits of mitigation is an essential input to climate change policy.

Although the threat of predicted climate change effects requires cooperation of government and society at a global scale, it may not always be in the best interests of a nation to commit to a strong mitigation plan. Nations that implement ambitious mitigation plans will possibly do so at their exclusive cost. Meanwhile, nations that do not adopt strong greenhouse reduction strategies can 'free-ride' on those nations that have committed to mitigation action. In some cases, nations can gain a trade benefit, referred to as 'carbon leakage'. Carbon leakage occurs when one nation reduces their greenhouse gas emissions or their provision of goods and services that rely on the use of fossil fuels. A nation that acts to meet the demand for those goods and services through fossil fuel-based production impedes achievement of a net mitigation benefit (Halsnæs et al., 2007).

O'Riordan (1995) offers a concise structure to explain situations in which international agreement is achievable and situations when agreement is difficult and complex. To begin, success in international policy negotiations relies on nations surrendering their national sovereignty: '[T]he freedom to act as a sovereign state must be circumscribed by an obligation to respect the legitimate interests of other sovereign states' (O'Riordan, 1995: 349). Next, there are three conditions that determine the willingness of a nation to cooperate: *mutual advantage*, that requires all nation states to recognise that cooperation and compliance with a collective agreement is preferable to noncompliance of one or more states; *credible threat*, an acknowledgement that there exists accepted science of the effects of environmental change and that either individual or collective noncompliance is not in the interest of any nation; and *credible enforcement*; each nation recognises that compliance will be enforced and noncompliance penalised.

This framework can be used to explain why international agreement was achieved in the case of ozone depletion, but why delays over mitigation action for anthropocentric climate change continue. In the case of the ozone depletion, no nation was better off if the hole in the ozone layer increased in size (mutual advantage), the science was largely irrefutable (credible threat), and enforcement and penalisation were realistic management options (credible enforcement). In contrast, anthropogenic climate change does not meet these conditions. For example, the condition of mutual advantage does not hold; it is expected that some nations will be less affected by climate change, while other nations, such as low-lying island nations in the tropical zone, may suffer greatly. Moreover, scientific uncertainty (lack of an agreed credible threat) has provided an excuse for inaction. The differential effects on nations and differentiated responsibility of nations to act also complicates global cooperation.

An additional matter of interest in regard to international cooperation is that of policy culture and style. A policy style is the 'national way of doing things', the typical processes that a government employs, irrespective of the characteristics of the problem (Sathaye et al., 2007). National policy style is derived from two interacting components: (a) the approach that government takes to problem solving and (b) the nature and function of relationships between government and other actors during the policy making process (Richardson et al., 1982). These components may be explored through a framework that presents three decision-making cultures that adopt different decision-making approaches and interact with other actors in different ways (Rayner, 1991) (Table 43.2). Although governments and other institutions do not consistently hold to these precise positions, they may, in a general sense, frame a government's internal discourse. The predominant policy style can also influence governments' response strategies and selection of policy instruments (see later in this chapter). Similarly, national decision-making can be anticipatory or reactive and the political context can be either consensus-based or impositional (Sathaye et al., 2007). Europe, for example, is considered to take a more cooperative approach to environmental protection, while the US tends to be more confrontational (Sathaye et al., 2007). Similarly, European nations rely on the EU to determine an appropriate policy approach to environmental, economic and social matters, while 'the US places much of its faith in free-market methods' (Dessai, 2001: 143).

Deliberative democracy, a shared responsibility for the design of policy, provides one option for cooperative engagement and effective action to respond to environmental change at the international level. According to this concept, policy spaces are not characterised by hierarchical orders; instead, opportunities

Table 43.2　Summary of three decision-making cultures

	Decision-making culture		
	Market	*Hierarchical*	*Egalitarian Collectivist (collective equality)*
Sovereignty	Consumer	Institution	Collective
Procedures applied	Skills	Rules	Ethical standards
Criteria	Experience	Evidence	Argument
Consent	Revealed	Hypothetical	Consensus
Liability	Loss spreading	Redistributive taxation	Strict-fault
Trust	Successful individuals	Procedures	Participation
Time depth	Short	Long	Compressed
Future generations	Self-sufficient	Resilient	Fragile
Policy preference	Market-based incentives, fiscal incentives, RD&D support	Regulation, command and control, incentive regulation	Voluntary and informational, command and control information
Nature myths and institutional cultures	Adaptation – nature robust	Sustainable development – nature robust within limits	Prevention – nature fragile

Source: Rayner (1991).

are present for a variety of forms of public-private cooperation, policy networks, formal and informal consultation, and working across scales from local to multinational levels (O'Riordan and Stoll-Kleeman, 2002). In the context of international climate negotiations, Rothkin (2005: 31) advocates the use of democratic deliberation, due to the inherent difficulties created by the unevenly distributed costs and benefits of mitigation, and where there is an absence of shared ideas about 'fairness and equal partnership'. She proposes, specifically, 'a deliberative and administrative body to which all countries commit; [that] enacts a stream of decisions, accommodating all countries' needs fairly and reasonably' (Rothkin, 2005: 32). An advantage of this deliberative arrangement is that 'fairness is not a prior principle, but rather a property of specific allocations of costs and benefits aggregated over time, across many decisions' (Rothkin, 2005: 32).

2.5　Policy instruments

There is a range of instruments that policy-makers can select from to support mitigation policy, programs and plans. Typically, policy instruments fall into four main categories: (1) regulation (command and control); (2) market-based (economic) instruments; (3) voluntary approaches; and (4) information and education. These instruments are not mutually exclusive; for example, they all depend to some extent on the provision of education and information. Moreover, an economic incentive program may be supported by government regulation. The multifaceted nature of environmental change often warrants the use of a mix of policy instruments (Table 43.3).

Regulation is a long-standing policy instrument that has been applied to environmental problems. Regulations can directly control activities and process outputs, and can also be effective in the creation of minimum performance standards. However, innovation can be stifled because an organisation may only operate according to the minimum performance standards and may not seek to improve their performance beyond set standards because there is no incentive to do so. Moreover, regulations are often inadequately enforced and penalties are often insufficient to deter environmentally damaging behaviour. Nonetheless, many environmental gains have been made through the application of

Table 43.3   Summary of key greenhouse gas mitigation policy instruments (based on Gupta et al., 2007)

Policy instrument	Use	Constraints
Regulations and standards	• Can provide clarity about emission levels • Can be useful in the absence of price signals • Can be tailored to sectors	• Can inhibit innovation • Can depend on stringent emission levels to achieve environmental effectiveness
Market-based: Taxes and charges	• Sets a price for carbon but does not limit emissions • Applies to commodities that reflect emissions, such as energy	• Can be politically unpopular • Can become the burden of low-income earners • Can be difficult for some nations to institutionalise
Market-based: Tradable permits	• Establishs price for carbon • Can be tailored to sectors	• Needs accurate emissions monitoring • Can have high transaction costs • Can depend on stringent emission levels to achieve environmental effectiveness • Can be difficult for some nations to institutionalise
Market-based: Financial incentives	• Can stimulate development and diffusion of new technologies through direct and indirect financial incentives, such as subsidies and tax credits	• Can increase emissions levels because of subsidies and incentives provided to the fossil fuel industry • Can cause spread of agricultural activities into marginal lands
Voluntary agreements	• Raises awareness • Moves industry towards best practice	• Can have low environmental performance targets • Can fail to achieve significant emissions reductions
Information instruments and R&D	• Can lead to better-informed decision-making • Can improve technology and reduce deployment costs through research and development (R&D)	• Can be difficult to measure the greenhouse gas emissions reductions that result from information provision • Behaviour change

Source: Adapted from Gupta et al., (2007)

direct regulation, which continues to be an important policy instrument.

Market-based instruments have become a popular policy instrument because they reward best practice, are flexible, and can attend to market externalities. 'Polluter-pays' instruments, for instance, provide a financial incentive to minimise pollution (reward) and do not necessarily prescribe the activity or technology an industry must adopt to reduce emissions (flexibility). Such instruments can correct market failure because the external environmental costs are incorporated into the cost of production and consumption which places a value on the environment and thereby increases efforts to minimise environmental harm. That is, market-based instruments can internalise the external environmental costs

of production, consumption and disposal, not traditionally reflected in the price of many goods and services.

Market-based instruments typically fall into two main categories. Price-based instruments (e.g., taxes and charges) use cost mechanisms as an incentive for businesses to reduce their emission levels. Rights-based instruments assign entitlements to pollute or to use resources, through the allocation of tradeable permits. These schemes encourage innovation and can effectively operate within current market frameworks. Tradeable schemes are less useful when there are too many or too few permit holders in the market (Connelly and Smith, 2003) and are ineffective when emission levels are not set at a level that prevents serious or irreversible environmental damage. Spatially and temporally unrestricted permit schemes may be environmentally detrimental when excess permit use occurs at the same time in the same location.

Voluntary policy instruments aim to shape more environmentally sensitive behaviour without the use of regulation or financial incentives. There are two main voluntary instruments. The first group of instruments includes voluntary standards, information and education, and includes eco-labelling, education campaigns and environment management systems. The second group includes negotiated agreements which are voluntary commitments to reduce environmental harm or emissions levels and are commonly held between government and industry. Negotiated agreements aim to improve efficiencies by reducing regulatory burdens.

2.6 Policy instrument choice

The selection of policy instruments must be sensitive to a number of considerations. First, the effectiveness of policy instruments will depend on the implementation timeframe. Policy objectives for short-term strategies will differ from policy objectives for long-term strategies. For example, short-term policy may involve reductions in emissions targets through the use of emissions trading schemes, or increases in industrial energy efficiency. In contrast, long-term policy will require mechanisms that extend beyond taxes and traditional government policy (Hasselmann et al., 2003). For instance, market incentives may be required to support the adoption of new technologies and infrastructure and may also involve investment from both private industry and government (Hasselmann et al., 2003). Therefore, long-term policy should have the ability to support strict emissions reductions targets that are not necessarily achievable through short-term policy. Efficiencies may be gained when short and long-term strategies are designed to work together. For example,

short-term emission reductions can make immediate gains and inform the development of long-term policy that supports stricter emission targets.

Second, the properties of policy instruments require careful consideration before they are used in policy design and implementation. For example, regulatory standards comprise two dominant categories: performance standards and technology; when implemented, these two instruments will result in different outcomes. Performance standards prescribe 'specific environmental outcomes per unit of product' such as parts per million of a pollutant, while technology prescribes the use of pollution abatement technology or the method of production such as the use of wet scrubbers in incinerators to control particulate emissions (Gupta et al., 2007: 754). Therefore, performance standards limit the amount of greenhouse gas emissions, while technology regulation would specify the capture and storage methods to be used (Gupta et al., 2007). The applicability of these instruments may vary from one context to another.

Third, the attitudes of stakeholders are fundamental to the effective implementation of environmental policy. Policy outcomes rely on the willingness and ability of individuals, land managers, government and industry to fulfil policy objectives. Therefore, policy that is developed in consultation with stakeholders can reveal differing views on the rate and extent of predicted environmental change, appropriate responses to environmental problems and preferences for different policy instruments. This information can be used to inform policy development so that environmental outcomes may be maximised.

Finally, governments must have the capacity to operate the policy instruments they select. Less industrialised nations can be constrained by a lack of finances and administrative staff and so must exercise caution in the selection of policy instruments. Even industrialised nations must consider if they can successfully deliver policy outcomes through their policy instrument choice.

Ideal conditions for the successful application of a suite of policy instruments include (O'Riordan et al., 1998: 414):

- a well-developed institutional infrastructure to implement regulation;
- an economy that is likely to respond well to fiscal policy instruments because it possesses certain characteristics of the economic models of the free market;
- a highly developed information industry and mass communications infrastructure for educating, advertising, and public opinion formulation; and
- a vast combined public and private annual research, development and deployment budget for reducing uncertainties and establishing pilot programmes.

Other criteria that can be used to evaluate policy instrument choice include (Connelly and Smith, 2003: 177):

- *Effectiveness*: each method can be effective if used appropriately. However, it is worth noting that if the goal is to reduce a damaging activity quickly, it is often better to use regulations as incentives; education programmes and negotiated agreements take time to formulate, introduce and to become effective.
- *Motivation*: taxes and permits provide a motive to constantly improve environmental performance; command-and-control does not. Voluntary approaches not only aim to alter behaviour but also the attitudes and character of polluters.
- *Administrative cost*: command-and-control tends to have high administrative costs; taxes and permits generally reduce these costs.
- *Efficiency*: the efficiency of each type of measure is a function of effectiveness, motivation and cost: the method which maximises effectiveness while minimising costs and providing a motive to avoid environmentally harmful acts in each particular sphere is the one which should be chosen.
- *Political acceptability*: some solutions might be theoretically sound but hard to implement because they are, for example, seen as 'giving a license to pollute'. Irrespective of the truth of this claim, or the claimed effectiveness of the policy, political sensitivities might lead to a reluctance to employ certain approaches.

- *Distributional impact*: methods affect different groups differentially. For example, taxes tend to be regressive in that they have a greater impact on the poor. Thus they might be politically or morally sensitive and best avoided in some cases, or, where they are deemed appropriate, additional measures may need to be taken to offset the impact on those most affected by them.

3 MITIGATION PROGRAMS: THE STRATEGY

Policy sets broad objectives and commitments while programs (strategies) provide a framework to achieve those policy objectives. The program can be achieved through institutional, informational, technological, behavioural and financial interventions (Jepma et al., 1995). The strategy adopted to mitigate climate change will predominantly rely on the deployment of current technology, advancement of technology, and behavioural change. Technology may be defined as a 'broad set of competences and tools covering know-how, experience and equipment, used by humans to produce services and transform resources' (Barker et al., 2007: 36). Modern society relies on nonrenewable fossil fuels as its main energy source, which constitutes 81 per cent of global energy supply (Barker et al., 2007), so a shift toward renewable low- and zero-carbon energy supplies and behavioural change is required as part of any climate change mitigation response. This section presents technological and behavioural aspects that may be considered as part of a strategic mitigation response.

3.1 Mitigation technology options

Population and global energy consumption will continue to increase over the coming decades. Energy demand is a function of factors such as population, the nature and level of human activities and demographic change. Future population projections for 2100 are between 8 and 12 billion people (Fisher et al., 2006; Lutz et al., 2001). Future energy consumption projections are complex and depend on energy intensity (gigajoules/gross domestic product) and carbon intensity of the energy system (carbon dioxide/gigajoules). Primary energy demand for 2050 is estimated to increase by 40–150 per cent and electricity demand in particular, is anticipated to increase by 110–260 per cent for the same period (Sims et al., 2007). Between 1900 and 2000, the world's primary energy use increased more than ten-fold; during the same period, the world's population increased only four-fold from 1.6 billion to 6.1 billion (Sims et al., 2007).

Low- and zero-carbon technologies will be required to mitigate climate change. Currently, many mitigation technologies are available and technologies have been advanced in all major sectors, including energy supply, transport, agriculture and waste management (Table 43.4). These technologies will need to be deployed and diffused to realise environmental benefits.

The mitigation potential of improved technologies has been summarised by the Intergovernmental Panel on Climate Change (IPCC). The IPCC presents forecasted mitigation potential for various sectors, based on the implementation of a carbon tax which will stimulate the use of improved technologies

Table 43.4   Key mitigation technologies and practices by sector.

Sector	Examples of key mitigation technologies and practices currently commercially available
Energy supply	• Fuel switching from coal to gas • Nuclear power • Renewable heat and power (hydropower, solar, wind, geothermal and bioenergy) • Early applications of carbon capture and storage (CCS)
Transport	• More fuel efficient vehicles including hybrids and biofuels • Modal shifts from road transport to rail and public transport systems and non-motorised transport • Land-use and transport planning
Buildings	• Efficient lighting, electrical appliances and heating and cooling devices • Improved insulation • Passive and active solar design for heating and cooling • Recovery and recycle of fluorinated gases
Industry	• More efficient end-use electrical equipment • Heat and power recovery • Material recycling and substitution • Wide array of process-specific technologies
Agriculture	• Improved crop and grazing land management to increase soil carbon storage • Improved rice cultivation techniques and livestock and manure management to reduce CH_4 emissions • Improved nitrogen fertilizer application techniques to reduce N_2O emissions • Restoration of cultivated peaty soils and degraded lands • Dedicated energy crops to replace fossil fuel use
Forests/forestry	• Afforestation, reforestation and reduced deforestation • Harvested wood product management • Forestry products for bioenergy to replace fossil fuel use
Waste management	• Waste minimisation and recycling • Composting of organic wastes • Waste incineration with energy recovery • Landfill methane recovery • Controlled waste water treatment

Source: IPCC (2007).

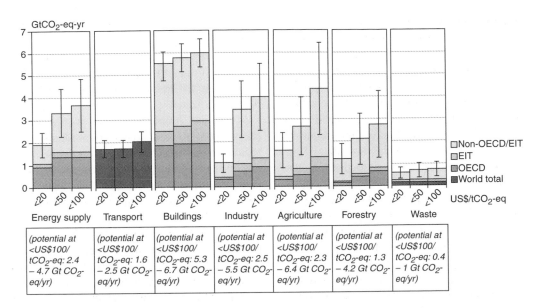

Figure 43.2   Estimated sectoral economic potential for global mitigation *for* different regions as a function of carbon price in 2030 from bottom-up studies, compared to the respective baselines assumed in the sector assessments. Note that only the world total is available for transport.

Source: IPCC (2007).

(Figure 43.2). The greatest reductions in greenhouse gas emissions are available within the residential and commercial building sector. These reductions may be achieved through increased energy efficiency, reduced embodied energy in building materials and increased use of renewable energy. The energy supply, industrial and agricultural sectors can also make significant reductions in greenhouse gas emissions. Although waste management mitigation actions present the lowest mitigation potential, technologies such as the conversion of methane (produced during anaerobic decomposition of biodegradable waste) to carbon dioxide through gas flaring, are already available and cost-effective. For each sector, an increased carbon price increases the mitigation potential, particularly for industry and agriculture. The economic and social ramifications of such actions will need to be considered during policy formation.

For new technologies to emerge, three broad drivers must interact: research and development, learning-by-doing and spill overs (Halsnæs et al., 2007). Research and development occurs when government, industry and other entities specifically invest in new technological developments or knowledge. Learning-by-doing refers to the increase in efficiency of technology use that comes from using that technology. This driver feeds back into research and development because technology use can identify areas for improvement and impediments to these technological improvements. Spillovers refer to the transfer of knowledge from one entity to another.

3.2 Barriers to development, deployment and diffusion of new technology

There are both internal and external barriers to the development, deployment and diffusion of new technology. Internal barriers include an absence of commitment and leadership at senior levels, corporate culture, organizational structure, the specific charac-teristics of plant and operations and the costs to employ new technology (Howes, 2005). External barriers include a persistent empha-sis in the markets on short-term profit, accountability systems that ignore social and environmental performance, the inability to adequately influence supply chain relation-ships, and the risks to the individual firm of enacting fundamental change while the mainstream remains locked into the neo-classical growth model (Howes, 2005). Additionally, regulatory barriers can include 'regimes that do not encourage firms to go beyond compliance, that do not provide a constant long-term direction for change or are not backed up with a consistent political will of enforcement' (Howes 2005: 163). Additionally, social and environmental costs of production, distribution and disposal are largely externalized by firms (Sharma and Starik, 2004). These externalities could be internalised by full-cost accounting of the environmental and social effects of an organ-isation's operations and the life cycle of their products.

The uncertainty of the extent of climate change can stall investment in abatement technologies. Given the uncertainty in some climate models, organisations may be reluc-tant to make large, irreversible investments in technology. Instead, businesses may elect to delay or stage their investment, or purchase smaller or modular infrastructure that amor-tises the cost of technological change (Sullivan and Blyth, 2006). Delay in govern-ment commitment to climate change mitiga-tion will further stifle private investment in low- or zero-carbon technology which will limit the diffusion of abatement technology.

Behavioural change is therefore required to both mitigate and adapt to climate change. Drivers of behaviour change include experi-ence with mitigation activity, education, access to resources, receiving support, access to professional networks and community leadership (Mazur et al., 2009). Governments and industries perhaps have greater potential to engage in mitigation activities because

they can access resources, conduct research and development outcomes and invest in pilot projects. Government and industry also have the ability to deliver mitigation infra-structure to the community, such as renewa-ble energy schemes.

Mitigation action is already occurring in many parts of the world. Individuals are reducing household energy use, reducing car use and recycling. However, barriers to fur-ther mitigation activity include not knowing how to reduce greenhouse gas emissions; not believing that individual actions make a dif-ference; and not having enough time or money to engage in carbon-reducing activi-ties (Semenza et al., 2008). People have been found to erect psychological barriers to explain why they won't act individually or collectively to mitigate climate change (Stoll-Kleemann et al., 2001). Social–psychological denial mechanisms were associated with individuals' cost of moving away from their comfortable lifestyles, blame on the inaction of others and the uncertain and distant effects of climate change (Stoll-Kleemann et al., 2001).

Adaptation behaviours include preventing and tolerating personal loss, changing uses or activities, changing locations and invest-ment in environmental and infrastructural restoration (Burton et al., 1993). The level of adaptive behaviour will depend to a large extent on the adaptive capacity of a system or community. Determinants of adaptive capac-ity include awareness, technology, resources, institutions, human and social capital, risk management and information management (Reid et al., 2007; Smit and Pilifosova, 2001). For example, the structure of institutions will determine the allocation of decision-making authority and the processes through which information is obtained, assessed and disseminated. Moreover, the education, skills, and experience of individuals (human capital), combined with social networks (social capital) will influence the willingness and ability of community members to be involved in adaptation activities (Yohe and Tol, 2002). Mitigation and adaptation strategies will need to remove the barriers of behavioural change, as well as provide necessary infrastructure and social institutions.

4 MITIGATION PLANS: THE ACTION

Mitigation plans achieve policy objectives through action and in accordance with the strategic framework. The plan that a government develops will be based on the mix of available technologies that are selected to achieve mitigation outcomes. This final section outlines three approaches to mitigate environmental change and includes a discussion on the merits and limitations of each.

In the context of climate change, mitigation action can be defined as 'an anthropocentric intervention to reduce the sources or enhance the sinks of greenhouse gases' (Klein et al., 2007: 750). Mitigation measures involve a reduction or modification of the human activities that cause environmental change. Mitigation measures broadly comprise: *abatement*, which reduces the emission levels of greenhouse gas; *carbon sequestration* which increases the volume of greenhouse gases sequestered by natural means or otherwise; and *geoengineering,* technologies that could reduce the level of global climate change primarily through storage methods and reflection devices.

4.1 Abatement

Abatement activity aims to reduce the amount of greenhouse gas emissions. Reductions can be made via three abatement alternatives. First, renewable energy relies on the conversion of natural energy sources to electricity and can be derived from a multitude of sources, such as hydro, tidal, wave, solar, wind and geothermal resources. Biomass and biofuels are also considered to form part of the renewable energy portfolio. Continued advancement of renewable technology will ensure they remain an important part of the global energy supply and that constraints to their deployment and use, efficiency, reliability and availability, are reduced or removed. Changes in global markets and the establishment of a carbon tax may increase the price competitiveness of these energy alternatives. Second, low- and zero-carbon energy sources, such as natural gas and nuclear power, result in less greenhouse gas emissions. Increases in business opportunities and policy intervention will be required to foster the development and deployment of zero- and low-carbon technologies. Such interventions must be sensitive to the security of energy supply and must remove the market advantages currently available for fossil fuels (Sims et al., 2007). Third, clean coal technology is gaining favour as an abatement activity and comprises five main methods: carbon capture and storage (see below), coal preparation, pollution removal, particulate emissions reduction and gasification. Coal preparation involves the removal of impurities, such as the mineral content of mined coal, which improves burning efficiency. Pollutants generated during combustion are removed through wet scrubbers that remove pollutants such as sulphur dioxide and nitrous oxides from flue gas. Electrostatic precipitators can remove particulate matter from flue gas through the creation of an electrical field which negatively charges the particles that are then attracted to positively-charged collection plates. Gasification combines coal with oxygen and steam to form 'syngas', comprised of hydrogen, carbon monoxide and other gases (Garnaut, 2008), which is burned to create a gas that powers a steam turbine. Impurities are removed, (hydrogen sulphide, ammonia and particulate matter), which are used to power a combustion turbine. Used in combination, this infrastructure is known as an integrated gas combined cycle system and results in high power generation efficiencies of 60 per cent, compared to traditional coal-based boiler plants that operate at 33–40 per cent efficiency (DOE, 2008).

4.2 Carbon sequestration and biosequestration

Carbon sequestration can be via geosequestration or biosequestration. Geosequestration involves the injection of carbon dioxide gas into geological formations underground (on land) or below the seabed. Geological formations include depleted oil and gas reservoirs, deep saline aquifers and deep coal seams (Garnaut, 2008). This process relies on carbon capture and storage technology of carbon dioxide gas from low concentration sources, typically from existing coal-fired plants, or from high concentration sources, which are usually produced in new plants that have been designed to capture carbon dioxide during the combustion or gasification process. Carbon capture and storage (sequestration) has been criticised because it does not necessarily foster the development of low- or zero-carbon energy technologies and may present risks to future generations if containment is compromised.

Biosequestration is the removal of greenhouse gas emissions from the atmosphere via biological processes or by storage in marine reserves. Biological processes include sink establishment such as avoided deforestation, reforestation and afforestation. Sink establishment provides a technically feasible mitigation option that can result in net sequestration of carbon dioxide, even when accounting for future deforestation projections (Fisher et al., 2007). However, large-scale plantations can result in negative social effects (Barker et al., 2007), such as community displacement or replacement of agricultural production needed to sustain regional communities. Biochar, the waste product of the gasification of biomass, is another potential biosequestration option (Lehmann et al., 2006). Biochar is a charcoal product that is high in carbon and degrades slowly. It can be used to build soil carbon and can be used as a fertiliser when infused with nutrients, such as ammonium biocarbonate. However, to produce biochar with a high carbon sequestration potential, some energy production may be sacrificed during gasification (Johnson et al., 2007). Algal biofuels can also be used for biosequestration whereby they absorb emissions from coal-fired power stations and metal smelters (Garnaut, 2008). Many biosequestration opportunities exist within the agricultural sector, such as reductions in enteric (methane) emissions from livestock through the use of biological or chemical control of rumen bacteria, dietary changes, or a shift to nonruminant animal production. Changed land use practices can also increase biosequestration. Examples include practices that minimise soil loss and provide for increased soil carbon storage; less fertiliser use and therefore less nitrous oxide emissions; and less direct carbon dioxide emissions from improved fire management regimes (Garnaut, 2008).

Another biosequestration technology is ocean fertilisation that allows biological storage in marine reserves. Iron, an important element for phytoplankton growth, does not exist in sufficient concentrations in 30 per cent of Earth's oceans. If iron concentrations are elevated, increased phytoplankton populations will purportedly sequester higher concentrations of carbon dioxide as particulate organic carbon (Barker et al., 2007). However, the maximum sequestration potential of phytoplankton is less than 10 per cent, accounting for respiration. Moreover, the effectiveness of iron fertilisation can be limited by insufficient concentrations of silicic acid and inefficient vertical transfer of carbon in the water column (Boyd et al., 2004). Negative environmental effects of iron fertilisation may include increased production of nitrous oxide and methane, as well as deoxygenation of local waters, and the increased likelihood of phytoplankton blooms as a result of changes in community composition (Barker et al., 2007).

4.3 Geoengineering

Geoengineering offers another approach to reduce the level of global climate change, without reducing the amount of greenhouse gas emissions (Victor et al., 2009). This reduction may be achieved through an increase in airborne particulates and the use of orbiting mirrors, both alternatives which reflect solar radiation and reduce warming potentials (Hammitt and Harvey, 2002). Amongst the arguments in favour of the use of geoengineering technologies is a current lack of availability of carbon-free technologies that could be rapidly deployed to produce humanity's energy requirements (Hoffert et al., 1998). Deployment of geoengineering technology would reduce humanity's need to find energy alternatives. However, geo-engineering options may be inadequate to accommodate the dynamic diurnal, seasonal and regional nature of Earth's systems. For example, radiative forcing from greenhouse gases continues across a 24-hour cycle, yet, reflection devices would be effective only in daylight (Govindasamy and Caldeira, 2000). Undesirable environmental consequences include the effects of reduced sunlight on ecosystem function, while ethical and political concerns centre on the need to implement geoengineering solutions at a global level that may negatively affect regions that experience less extreme climatic changes (Govindasamy and Caldeira, 2000). Currently, geoengineering also presents significant logistical problems; to counteract estimated increases in anthropocentric greenhouse gas emissions, a disc of approximately 1,600 kilometres in diameter would need to be placed in space (Govindasamy and Caldeira, 2000).

5 CONCLUSION

Policy sets the direction and the level of commitment to mitigate environmental change, programs provide a framework, and plans set the actions to be undertaken. Policy, program and plan formulation must account for the complexity and uncertainty of natural and social systems and should seek to maximise synergies. However, this complex process relies on multiple inputs to support decisions that will affect multiple actors, and varied sectoral, temporal and spatial implications of policy implementation confound policy formulation. Uncertain science and vested interests also pose threats to mitigation policy. Greater understanding and commitment to the precautionary principle and incorporation of intragenerational and intergenerational equity may combat these threats. Education, information provision and ongoing research will support efforts to deliver mitigation policy that can bring about positive technological, social, behavioural and institutional change to mitigate the environmental effects of human action.

REFERENCES

Arthur W. B. 1994. Inductive reasoning and bounded rationality (The El Farol Problem). *American Economic Review (Papers and Proceedings)* 84: 406.

Barker T., Bashmakov A., Alharthi M., Amann L., Cifuentes J., Drexhage M. et al. 2007a. Mitigation from a cross-sectoral perspective, in Metz B., Davidson O. R., Bosch P. R., Dave R. and Meyer L. A. (eds) *Climate Change 2007: Mitigation. Contribution of Working Group III to the Fourth Assessment Report of the Intergovernmental Panel on Climate Change.* Cambridge: Cambridge University Press.

Barker T., Bashmakov L., Bernstein J. E., Bogner P. R., Bosch R., Dave O. R. et al. 2007b. Technical summary, in Metz B., Davidson O. R., Bosch P. R., Dave R. and Meyer L. A. (eds) *Climate Change 2007: Mitigation. Contribution of Working Group III to the Fourth Assessment Report of the Intergovernmental Panel on Climate Change.* Cambridge: Cambridge University Press.

Boyd P. W., Law C. S., Wong Y., Nojiri A., Tsuda M., Levasseur S. et al. 2004. The decline and fate of an iron-induced subarctic phytoplankton bloom. *Nature* 428(6982): 549–553.

Bruntland G. 1987. *Our Common Future: The World Commission on Environment and Development.* Oxford: Oxford University Press.

Burton I. R., Kates W. and White G. F. 1993. *The Environment as Hazard.* New York: Guildford Press.

Carson R. 1962. *Silent Spring.* London: Ebenezer Baylis and Son.

Cazorla M. and Toman M. 2000. *International Equity and Climate Change Policy.* Climate Issue Brief. Washington DC: Resources for the Future.

Connelly J. and Smith G. 2003. *Politics and the Environment: From Theory to Practice.* London: Routledge.

Dang H. H., Michaelowa A. and D. D. Tuan 2003. Synergy of adaptation and mitigation strategies in the context of sustainable development: the case of Vietnam. *Climate Policy* 3(Suppl. 1): S81–S96.

Dessai S. 2001. Why did The Hague Climate Conference fail? *Environmental Politics* 10(3): 139–144.

DOE. 2008. How Coal Power Gasification Plants Work. Available at: http://fossil.energy.gov/programs/powersystems/gasification/howgasificationworks.html.

Fisher B. S., Jakeman G., Pant H. M., Schwoon M. and Tol R. S. J. 2006. CHIMP: A simple population mode for use in integrated assessment of global environmental change. *The Integrated Assessment Journal* 6(3): 1–33.

Fisher B. S., Nakicenovic N., Alfsen K., Corfee J., Morlot F., de la Chesnaye J.-C. et al. 2007. Issues related to mitigation in the long term context, in Metz B., Davidson O. R., Bosch P. R., Dave R. and Meyer L. A. (eds) *Climate Change 2007: Mitigation. Contribution of Working Group III to the Fourth Assessment Report of the Intergovernmental Panel on Climate Change.* Cambridge: Cambridge University Press.

Garnaut R. 2008. *The Garnaut Climate Change Review.* Port Melbourne: Cambridge University Press.

Govindasamy B. and Caldeira K. 2000. Geoengineering Earth's radiation balance to mitigate COe-induced climate change. *Geophysical Research Letters* 27(14): 2141–2144.

Gupta S., Tirpak D. A., Burger N., Gupta J., Höhne N., Boncheva A. I. et al. 2007. Policies, instruments and co-operative arrangements, in *Climate Change 2007: Mitigation. Contribution of Working Group III to the Fourth Assessment Report of the Intergovernmental Panel on Climate Change.* Cambridge: Cambridge University Press.

Halsnæs K., Shukla P., Ahuja D., Akumu G., Beale R., Edmonds J. et al. 2007. Framing issues, in Metz B., Davidson O. R., Bosch P. R., Dave R. and Meyer L. A. (eds) *Climate Change 2007: Mitigation. Contribution of Working Group III to the Fourth Assessment Report of the Intergovernmental Panel on Climate Change.* Cambridge: Cambridge University Press.

Hammitt J. K. and Harvey C. M. 2002. Equity, efficiency, uncertainty, and the mitigation of global climate change. *Risk Analysis* 20(6): 851–860.

Hasselmann K., Latif M., Hooss G., Azar C., Edenhofer O., Jaeger C. C. et al. 2003. The challenge of long-term climate change. *Science* 302(5652): 1923–1925.

Hoffert M. I., Caldeira K., Jain Haites A. K., Haites E. F., Harvey L. D. D., Potter S. D. et al. 1998. Energy implications of future stabilization of atmospheric CO_2 content. *Nature* 395(6705): 881–884.

Howes M. 2005. *Politics and the Environment: Risk and the Role of Government and Industry.* Sydney: Allen and Unwin.

IPCC 2007. Summary for policymakers, in Metz B., Davidson O. R., Bosch P. R., Dave R. and Meyer L. A. (eds) *Climate Change 2007: Mitigation. Contribution of Working Group III to the Fourth Assessment Report of the Intergovernmental Panel on Climate Change.* Cambridge: Cambridge University Press.

Jepma C. J., Asaduzzaman M., Mintzer I., Maya R. S. and Al-Monef M. 1995. *A Generic Assessment of Response Options.* Cambridge, Cambridge University Press.

Johnson J. M.-F., Franzluebbers A. J., Lachnicht Weyers S. and Reicosky D. C. 2007. Agricultural opportunities to mitigate greenhouse gas emissions. *Environmental Pollution* 150: 107–124.

Klein R. J. T., Huq S., Denton F., Downing T. E., Richels R. G., Robinson J. B. and Toth F. L. 2007. Inter-relationships between adaptation and mitigation, in Parry M. L., Canziani O. F., Palutikof J. P., van der Linden P. J. and Hanson C. E. (eds) *Climate Change 2007: Impacts, Adaptation and Vulnerability. Contribution of Working Group II to the Fourth Assessment Report of the Intergovernmental Panel on Climate Change.* Cambridge: Cambridge University Press, pp. 745–777.

Lehmann J., Guant J. and M. Rondon 2006. Bio-char sequestration in terrestrial ecosystems – a review. *Mitigation and Adaptation Strategies for Global Change* 11: 403–427.

Lukes S. 1974. *Power: A Radical View.* London: Macmillan Press.

Lutz W., Sanderson W. and Scherbov S. 2001. The end of world population growth. *Nature* 412(6846): 543–545.

Mazur N., Curtis A., Thwaites R. and Race D. 2009. *Rural landholders adapting to climate change: Social research perspectives.* Canberra: Department of the Environment, Water, Heritage and the Arts.

NAVF 1990. Sustainable Development, Science and Policy: The Conference Report Bergen, 8–12 May 1990. Oslo, Norwegian Research Council for Science and the Humanities (NAVF).

O'Riordan T. 1995a. *Environmental Science for Environmental Management.* Harlow: Longman.

O'Riordan T. 1995b. Introduction. *Environmental Science for Environmental Management.* Harlow: Longman Scientific and Technical, pp. 1–13.

O'Riordan T., Cooper C. L., Jordan A., Rayner S., Richards K. R., Runci P.and Yoffe S. 1998. Institutional frameworks for political action, in Rayner S. and Malone E. L. (eds) *Human Choice and Climate Change: Volume I: The Societal Framework.* Columbus, OH: Battelle Press, pp. 345–439.

O'Riordan T. and A. Jordan 1995. The precautionary principle in contemporary environmental politics. *Environmental Values* 4: 191–212.

O'Riordan T. and Stoll-Kleeman S. 2002. Deliberative democracy and participatory biodiversity, in O'Riordan T. and Stoll-Kleeman S. (eds) *Biodiversity, Sustainability and Human Communities: Protecting Beyond the Protected.* Cambridge: Cambridge University Press, pp. 87–112.

Pielke R. A. 1998. Rethinking the role of adaptation in climate policy. *Global Environmental Change* 8(2): 159–170.

Rayner S. 1991. A cultural perspective on the structure and implementation of global environmental agreements. *Evaluation Review* 15(1): 75–102.

Reid S., Smit B., Caldwell W. and Belliveau S. 2007. Vulnerability and adaptation to climate risks in Ontario agriculture. *Mitigation and Adaptation Strategies for Global Change* 12(4): 609–637.

Richardson J., Gustafsson G. and Jordan G. 1982. The concept of policy style, in Richardson J. (ed.) *Policy Style in Western Europe.* London: George Allen & Unwin: 11–16.

Rothkin K. 2005. Fairness among countries. International Workshop on Human Security and Climate Change. Oslo, Norway.

Sathaye J., Najam A., Cocklin C., Heller T., Lecocq F., Llanes-Regueiro J. et al. 2007. Sustainable development and mitigation, in Metz B., Davidson O. R., Bosch P. R., Dave R. and Meyer L. A. (eds) *Climate Change 2007: Mitigation. Contribution of Working Group III to the Fourth Assessment Report of the Intergovernmental Panel on Climate Change.* Cambridge: Cambridge University Press.

Semenza J. C., Hall D. E., Wilson D. J., Bontempo B. D., Sailor D. J. and George L. A. 2008. Public perception of climate change: voluntary mitigation and barriers to behavior change. *American Journal of Preventive Medicine* 35(5): 479–487.

Sharma S. and Starik M. (eds) 2004. *Stakeholders, The Environment, and Society.* Cheltenham: Edward Elgar.

Sims R. E. H., Schock R. N., Adegbululgbe A., Fenhann J., Konstantinaviciute I., Moomaw W. et al. 2007. Energy supply, in Metz B., Davidson O. R., Bosch P. R., Dave R. and Meyer L. A. (eds) *Climate Change 2007: Mitigation. Contribution of Working Group III to the Fourth Assessment Report of the Intergovernmental Panel on Climate Change.* Cambridge: Cambridge University Press.

Smit B. and Pilifosova O. 2001. Adaptation to climate change in the context of sustainable development and equity, in McCarthy J. J., Canziani O. F., Leary N. A., Dokken D. J. and White K. S (eds) *Climate Change 2001: impacts, adaptation and vulnerability—contribution of Working Group II to the Third Assessment Report of the Intergovernmental Panel on Climate Change.* Cambridge: Cambridge University Press, pp. 876–912.

Solomon S., Qin D., Manning M., Alley R. B., Berntsen T., Bindoff N. L. et al. 2007. Technical summary, in *Climate Change 2007: The Physical Science Basis. Contribution of Working Group I to the Fourth Assessment Report of the Intergovernmental Panel on Climate Change.* Cambridge: Cambridge University Press.

Stoll-Kleemann S., O'Riordan T. and Jaeger C. C. 2001. The psychology of denial concerning climate mitigation measures: evidence from Swiss focus groups. *Global Environmental Change* 11(2): 107–117.

Sullivan R. and Blyth W. 2006. *Climate Change Policy Uncertainty and the Electricity Industry: Implications and Unintended Consequences.* Chatham House.

Tol R. S. J. 2005. Adaptation and mitigation: trade-offs in substance and methods. *Environmental Science & Policy* 8: 572–578.

UNFCCC. 1992. *United Nations Framework Convention on Climate Change.* New York: United Nations.

Victor D., Morgan M. G., Apt J., Steinbruner J. and Ricke K. 2009. The Geoengineering option: A last resort against global warming? *Foreign Affairs* 88(2): 64–76.

Yohe G. and Tol R. S. J. 2002. Indicators for social and economic coping capacity – moving toward a working definition of adaptive capacity. *Global Environmental Change* 12: 25–40.

44

Socioeconomic Adaptation to Environmental Change: Towards Sustainable Development

Chris J. Barrow

1 INTRODUCTION

Environmental change presents an ongoing and multifaceted complex of challenges and opportunities. Some disciplines focus on past changes and others on current or future shifts; some are interested in social and others economic or physical variations. Viewing one or a few of these aspects of change in isolation is insufficient for adequate environmental management, yet a broad, holistic approach is likely to be too demanding.

It is something of a cliché that an organism faced with environmental change must move, adapt, or die. For humans on a crowded Earth, moving is becoming less of a practical proposition, which means adaptation or mitigation of change are important. These activities depend on people perceiving that change is happening and then deciding fast enough that it is worth worrying about and having sufficient natural socioeconomic and technical resources available to effectively respond. When citizens, politicians or researchers talk of environmental change they often focus on one element currently attracting attention

and other important threats or opportunities may be missed. Today, concern for global warming is especially strong; however, nitrification of oceans, global pollution with toxic compounds, stratospheric ozone depletion, social breakdown, and many other facets of environmental and socioeconomic change also deserve attention and care is needed to ensure concern for them is not sidelined.

Environmental change encompasses both gradual trends and relatively sudden shocks, the latter including for example, volcanic eruptions, tsunamis, warfare, economic depressions and epidemics (MacCracken et al., 2007). Environmental change can also involve fluctuations that are relatively regular (e.g., quasi-periodic or seasonal shifts), but even regular, gradual and perceived shifts may cause an ecosystem, group of organisms, society or economy to reach a threshold at which it possibly becomes a sudden threat. Identifying and watching for critical thresholds is thus important when monitoring or modelling environmental change. There tends to be an emphasis on the adverse impacts rather than opportunities offered by environmental change. Sometimes environmental change has no real impact, yet people involved learn things and in future may react differently to challenges (this is an outcome rather than an impact). Environmental change is a complex and broad field and people or bodies interested in it can have differing goals and concerns; for example, the United Nations Framework Convention on Climate Change (UNFCCC) focuses on international policy, while the Intergovernmental Panel on Climate Change (IPCC) is more concerned with scientific assessment.

1.1 Sensitivity, adaptation, mitigation, resilience and vulnerability

Researchers in a number of disciplines have been exploring indicators and benchmarks of environmental change (see Janssen et al., 2006; Parry et al., 2007). *Sensitivity* is the degree to which a system such as the environment is affected, either adversely or beneficially. There have been many attempts to define *adaptation*; one in wide usage is that it is a process of deliberate change in anticipation of, or in reaction to, stimuli or stress (Nelson et al., 2007). Another is 'any adjustment that reduces the risks associated with ... change, or vulnerability to ... change impacts, to a predetermined level, without compromising economic, social and environmental sustainability' (De França Doria et al., in press: 16). Adaptation is response (conscious or unwitting) to actual or expected impacts of environmental change; it may be precautionary or reactive but it is not really preventative. Some adaptation is temporary because it is based on technology or organisation, which could fail or be overtaken by further environmental change. Adaptive capacity is the potential or ability to respond to change (see Figure 44.1). *Mitigation* of environmental change can be defined as efforts to control or soften the impacts caused by the change. However, in many cases mitigation is not possible because the change is not adequately perceived, is already underway or cannot be controlled. Mitigation has received greater attention than adaptation from scientists and policy makers (Füssel and Klein, 2006: 304).

Adaptation and mitigation (see later) can take place before, during, but more often after environmental change through mechanisms

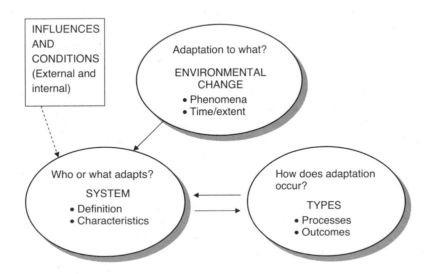

Figure 44.1   Adaptation to environmental change and variability (partly based on McCarthy et al., 2001).

like learning (individual learning, social learning, policy learning). There is a tendency to see adaptation as involving costs and resulting in negative impacts, but some people and organisms may find opportunities to their benefit. There is a rapidly growing volume of research on adaptation to environmental changes, such as climate fluctuations and adverse weather events (Kane and Yohe, 2000; Pandey, 2005; Schipper and Burton, 2008). Aid bodies have become aware of the need to support improvement of adaptation and reduction of vulnerability. For example, the Asian Development Bank, like many funding agencies, has integrated climate change adaptation into its grant and loan procedures (UNEP, 2006: 12).

Adaptation and mitigation are frequently distorted by politics, procrastination, corruption and bureaucracy, so can be piecemeal, delayed, unsustainable and unfair. Climate change, stratospheric ozone depletion, acid deposition, drought, land degradation, pollution and many other challenges have generated interest in adaptation and mitigation measures. Adaptation and mitigation costs are often more manageable if action is undertaken gradually and as soon as a challenge is perceived (Bouwer and Aerts, 2006). That was a core message of the *Stern Review* presented to the UK Parliament: that if effective adaptation and mitigation of climate change are delayed there will be very serious economic, environmental and social impacts later (Stern, 2007). Environmental change adaptation and mitigation efforts have direct costs but may also impose opportunity costs in that the expenditure is not available for something else (Stern and Patel, 2009; UNEP, 2006). The scale of some possible environmental change adaptation and mitigation measures are huge. Calculations suggest global carbon dioxide emissions attributable to the present day ($\sim$7 billion) human population is around 28 billion tonnes per year. The IPCC has suggested that to stabilise atmospheric carbon dioxide at a 'reasonably safe' level (mitigation) emissions must be reduced to 5 billion tonnes per year by 2050,

by which time global population will probably have risen to over 9 billion. To reach that target, all the world's people will have to have a carbon footprint lower than that of the average Indian today (ELDIS Climate Change Adaptation[1]).

Resilience is a concept originally developed by ecologists (Holling, 1973), which seeks to explain what determines the persistence of relationships within an ecosystem (or society) and to measure its ability to absorb change, reorganise, regain key functions and recover. Caution is needed because a system might reorganize and recover key functions in ways that differ from previous patterns; thus the past is not a wholly reliable indicator of future developments. A resilience perspective is often adopted as an approach for understanding and modelling environmental and socioecological systems. When people have little resilience, change may shift them into a more vulnerable condition (this is widely termed 'marginalization'); for example, delayed rains may mean farmers plant less and so are more likely to suffer hunger the following year, become weakened and again plant too little even if environmental conditions improve.

Vulnerability is a very broad term, which refers to the ability of a group or ecosystem to cope with disturbances from within or outside. The disturbance or innovation may be social, cultural, economic or environmental (or a mix) (Adger, 2006; Füssel, 2005; Füssel and Kiein, 2006: 305, 324; Kelly and Adger, 2004;). Vulnerability is a function of the character, magnitude and rate of environmental variation to which a system is exposed, its sensitivity and its adaptive capacity. Care is needed to ascertain whether vulnerability is an inherent property of a system or contingent upon a particular scenario of external and internal challenges and responses. Environmental change often impacts on already problematic conditions generated by interacting elements of the environment: culture, economy and so forth. There may be unexpected, perhaps indirect, cumulative effects, which are difficult to predict.

The concept of vulnerability was first developed by ecologists and hazard researchers and is now also used by social scientists and aid agencies (see Chapter 41). Vulnerability assessment may be undertaken to improve scientific understanding, to inform specific targets or to develop adaptation strategies in general (Birkmann, 2006; Patt et al., 2008). Vulnerability assessment can focus upon: (1) the exposure of a particular group, place, or system to threat(s) associated with environmental change; (2) the sensitivity of a group, place, or system to the threat(s) and (3) the adaptive capacity of the group, place, or system to resist impacts, cope with challenges and regain functions. Unfortunately, efforts to reduce vulnerability often fail to address (or even identify) root causes so do little to prevent recurrence and just treat symptoms.

The linkages between vulnerability, resilience and adaptation are complex (Gallopin, 2006; Handmer et al., 2004). Because there may be unperceived threats to address it may be wiser to seek to reduce vulnerability to a range of possible dangers than rely on adaptation or mitigation. Nowadays, many citizens lack the skills, strengths and social capital to survive serious environmental change or social unrest in the way their ancestors could. They assume technology and governance protects them, but things like globalization, population growth and reliance on petroleum and growing pollution probably mean they are more vulnerable to environmental change.

1.2 Sustainable development

Development can sometimes prove to be mal-development, or if there are benefits they may be temporary or restricted to a few groups. There is wide agreement that improvement should be secure, sustained and not achieved at the cost of future generations. This is known as sustainable development and has three inter-related goals: (1) economic progress, especially poverty reduction; (2) social improvement and

(3) environmental protection. However, there is no single widely accepted firm definition of sustainable development. Some see it as a broad concept and guiding principle, like liberty or justice, open to some range of interpretation, often elusive, but very much worth trying to achieve (International Institute for Sustainable Development[2]). Others view it as achievable in practice and are working to develop laws, policies, tools, and so on. Sustainable development has at its core a theoretical conflict: a goal of development, which if not carefully addressed means ongoing environmental demands, *versus* the limitations of a finite and vulnerable world. The solution might be to alter development to something less environmentally demanding, draw upon better management and technology and alter what people seek. Most of those involved in sustainable development stress the need to ensure future generations suffer no loss of options or resources because of present developments. This is unlikely to be a costless strategy – there have to be present-to-future trade-offs and stakeholders may find it difficult to forego benefits now in order to protect people and the environment in years to come. One problem with pursuing sustainable development is that the future needs to have a voice today.

The concept of sustainable development is evolving rapidly and currently there are two common interpretations: 'strong sustainable development' and 'weak sustainable development' (Mansfield, 2009). The latter is now mainstream and accepts human welfare may demand use of finite resources and it may be permissible not to pass on undiminished options, provided some sustainable benefit is established. For example, exploiting a finite resource like petroleum is admissible if pollution is controlled and the profits pay for useful sustainable infrastructure. The strong interpretation holds that existing stocks of natural capital must be maintained or improved, even if there is human need and development must rely only on natural capital that regenerates. This may demand significant ethical and lifestyle changes.

Sustainable development is a challenge under stable and predictable environmental and socioeconomic conditions. However, growing population and environmental and socioeconomic change can disrupt access to resources and destabilise institutions. To stand a chance of success, sustainable development needs adaptable approaches, early warning of problems like environmental change, duplication and dispersal of key resources, and resilient institutions.

2 THINKING ABOUT ENVIRONMENTAL CHANGE AND DEVELOPMENT

2.1 Perception of environmental change

For roughly the last two decades growing concern about global warming has come to dominate perception of environmental change (Davoudi et al., 2009; Philander, 2008). One of the key bodies involved is the Intergovernmental Panel on Climate Change (IPCC, 2001). IPCC Group II focuses on the impacts of climate change and on vulnerability and adaptation, while Group III deals with mitigation. Group II published three important assessment reports in 2007, which examine impacts, adaptation and vulnerability (Parry et al., 2007), mitigation of climate change (Metz et al., 2007) and the physical science basis of climate change (Solomon et al., 2007). Several well-regarded 'business as usual' scenarios suggest a 2° or 3°C global mean temperature increase and roughly a doubling of current atmospheric carbon dioxide levels by AD 2050, unless there is sufficient avoidance, mitigation, or negative feedback (UNFCC, 2007: 5). Yohe (2007) provides tables of possible degree of change and the implications. He argues that if, in the next 50 years or so, warming exceeds 2°C above present levels, it will pose serious threats and make sustainable development very difficult to achieve. A rise of 4°C over

that time span, he warns, could threaten human survival (*The Stern Review* suggests a 50 per cent chance of 5°C); a 2°C rise by 2100, Yohe suggests, could place 20–30 per cent of plant and animal species at risk of extinction.

Global warming dominates environmental change concern at present but there is a threat from volcanic eruptions, strikes by asteroids and similar bodies, tsunami, solar radiation fluctuations, land degradation, nitrification, acidification, stratospheric ozone depletion, human and crop disease, nuclear, chemical or biological warfare, and many other things. For example, America and Europe went hungry around AD 1815–1817 when there were cold and wet summers, possibly due to an eruption in Indonesia. Citizens and governments are not sufficiently aware of the full range of likely threats or able to assess the need for precautions, so environmental managers and researchers need to ensure they are adequately alerted. Some years ago researchers warned that oceanic nitrogen accumulation and acidification pose a serious threat but these warnings seem to have had limited effect (Galloway and Cowling, 2002). The 2004 tsunami disaster in Southeast Asia had been foreseen by a few researchers who lobbied for preparedness and early-warning systems, which would have cost relatively little. Virtually nothing happened until after the disaster struck. A rust fungus (Ug99) currently shows signs of spreading from eastern Africa to strike the world's key crop, wheat. However, few people who depend on that cereal are aware of this. The world's electricity grids and electronics could be suddenly destroyed by a solar storm, termed a Carrington Event (Brookes, 2009). The likelihood is high over a 100-year timescale with greatest likelihood at times of high flare activity during the 11-year solar cycle. The next time of peak risk is 2012. The eponymous Carrington Event in 1859 disabled most of the world's telegraph systems. (It should be noted that, compared with modern electrical systems, Victorian equipment was very robust.) Climate change is just one of many environmental change threats that humans should prepare for (Beg et al., 2002).

Global warming-related sea-level rise has caught the attention of low-lying states (Barnett, 2005; Barnett and Campbell, 2009) and has prompted the formation of the Association for Small Island States (AOSIS) and the publication of the Pacific Islands Framework for Action on Climate Change 2006–2015. Currently around 25 per cent of the world's population lives in areas vulnerable to sea-level rise.

Adaptation can demand citizens are motivated to respond promptly to what are imperfectly predicted problems. People and governments tend especially to view rare, possibly catastrophic events as 'unlikely' and resist spending on preparations. There is also a tendency to respond to warnings by waiting to see if more evidence appears, or in the hope others will resolve things (at their cost), or to allow technology to develop. Those hopes may not materialise. Many threats are difficult to predict and change may not be gradual and gentle, so it is wise to seek adaptability and resilience and reduce vulnerability. But, in reality, development often results in inflexible and vulnerable policies and infrastructure.

Over the last 20 or so years, proactive approaches have been spreading. The central concept is the precautionary principle (PP) which plays an important role in environmental management and has been working its way into legislation and planning. The PP should help stimulate interest in preparing for environmental change (Applegate, 2000) and has four main elements:

1 seek preventative action in the face of uncertainty;
2 shift the burden of proof to those seeking/planning development/change;
3 explore a wide range of alternatives to pursue the best option; and
4 increase public participation in decision-making.

The PP demands that if a significant threat is established, regulatory action can precede

full evidence. European and international law have already incorporated the PP to a considerable extent (Barrow, 2006: 32).

2.2 Facing environmental change in developed and developing countries

Low- and high-latitude countries may be impacted differently by environmental change and the ability of each state to respond will be shaped by their access to funds, technology, supportive social institutions and so on (Markandya and Halsnaes, 2002; Watson et al., 2001; Yohe et al., 2006). The majority of developing countries are at low latitude, their soils are often deeply weathered and poor in nutrients, with evapotranspiration and temperatures higher than those of developed countries. Table 44.1 lists some Asian rivers which are meltwater-fed; similar rivers are also important in Latin America, Western Asia, Morocco and the southern parts of the former Soviet Union. As meltwater and precipitation patterns change there could be serious impacts on irrigation in a number of developing countries and on their water supplies for many cities. There is a need to ensure laws relating to resources like rivers shared by more than one country

Table 44.1 Seven Asian rivers, which are highly dependent on glaciers/snowmelt to maintain summer flows, with estimated basin populations (in all more than one-sixth of total global population)

River	Basin population (millions)	Meltwater dependence
Syrdarya	8	Very high
Amudarya	20	Very high
Indus	178	Very high
Ganges	407	High
Brahmaputra	118	High
Yangtze	368	High
Yellow	147	High

Source: based on data extracted from UNEP 2007: Global outlook for ice and snow, United Nations Environment Programme, Nairobi.

are sensitive to environmental change. Some progress has been made, but it is nowhere near sufficient (Churchill and Freestone, 1991).

Tropical and extra-tropical storm patterns may alter as climate changes and could have enhanced impact because of raised sea-levels, removal of mangroves and the loss of coral reefs. A significant proportion of the world's population, infrastructure and food production is increasingly to storm surges and must start adapting (IPCC, 2007a; Parry et al., 2004, 2007).

Peoples of richer nations and affluent groups in poor countries might be able to buy adaptation. The poor, although vulnerable through poverty, may prove hardy. The EU and US already receive migrants and the numbers could rise through environmental change (eco-refugees forced off the land by drought, sea-level rise, or soil degradation). Developing countries may have less sophisticated infrastructure than developed countries but be more resilient; it is also possible that the powers of some developed nations will decline. The impact of environmental change is not easy to predict, but will certainly not affect all countries to the same extent and different peoples will respond with varying effectiveness. Foreign aid is increasingly targeted at assisting poorer countries to adapt and to support environmental change mitigation (Agarwal et al., 2008; USAID, 2007).

Adaptation and reaction to environmental challenges is not always improved by national wealth. Cuba responded quite effectively to hurricane damage in 2009, while in contrast, New Orleans, with all the resources of the US, suffered serious disorder for a long time after Hurricane Katrina struck. Repeated challenges may weaken response or provide the experience to improve, predicting which is not easy. Responses may vary over time; in 1962–1963, the UK was hit by a severe winter and managed relatively well. Nowadays, similar conditions would probably have a greater impact, because in the 1960s coal was locally stockpiled and services were less centralized and more robust (Smith, 1968). Some countries have adopted vulnerable and poorly resilient practices, such as high-tech data handling; dependency on energy supplies from overseas; purchase of seeds from foreign companies; and construction of large schemes, which cannot easily adapt to environmental changes and depend on imported spare parts.

A crucial and neglected adaptation to environmental change is to improve world food security and sustainability. For decades food production improvements have mainly been in terms of yield whilst security and sustainability have been neglected. Ideally, all countries should have access to reserves sufficient to cushion one or more failed harvests on a continental scale, but that is far from being reality. *The World Development Report 2008* examined how agriculture might adapt to and even help mitigate climate change and how it could be made more environmentally sustainable (World Bank, 2007: 181–221). The report also noted that, if global mean temperature increases 3°C by 2100, there would be severe agricultural losses, especially in the tropics.

Not only are food reserves inadequate, there is also a global trend toward reliance on fewer food-supply regions and less-diverse crops. The Green Revolution has helped feed the world, but it has focused too little on food security, agricultural resilience and sustainability (see, for example, the special issue of *Environmental Science and Policy* 12(4)). A growing number of agriculturalists are therefore calling for a Doubly Green Revolution, which is food production that seeks greater yields and aims for sustainability and reduction of environmental damage. Green Revolution reliance on agrochemicals has already caused the abandonment of some successful food production regions because of pollution; and poor land husbandry is also taking land out of production.

The photosynthetic strategies of some important high- and low-latitude crops, and their weeds differ. Thus, as atmospheric carbon dioxide increases, there is a possibility that temperate and tropical farming will face different challenges.

Some areas may become wetter and some drier; however, beneficial warming or increased rainfall will not improve crops if there is not also suitable soil, favourable weather and conditions which allow crops to adapt better than plant diseases and pests. The response of agriculture to environmental changes is complex and has not been adequately modelled. There is much uncertainty. If change is gradual, developed countries could be advantaged because they have the resources to support adaptation. However, if there is sudden change or economic troubles, warfare or terrorism, it might be the poorer, hardier developing countries that prove more adaptable.

2.3 Learning from the past

Valuable experience is often lost as was the case with the Dust Bowl disaster of the1930s, which hugely disrupted the American Midwest. Many of the lessons learnt are now largely forgotten. Learning from the past is patchy and it is easy to be misled and it is vital to seek more than one line of evidence to reduce misinterpretation. It can be difficult to reliably reconstruct environmental cause-and-effect because of complex webs of physical and human cumulative and indirect causation (Doyel and Dean, 2004). Further complications arise because explanation of causation can be shaped by the outlook of those making assessments (Davis, 2007). It is all too easy to engage in environmental determinism and make simplistic or false assumptions about past and ongoing adaptation.

There are predictions that diseases like malaria will spread with global warming (Griffiths et al., 2009; McMichaels et al., 1996; Rogers et al., 2006). Yet malaria was common in England and Europe during the Little Ice Age when conditions were colder

than now. A range of factors, not just temperature, influence transmission and infection: housing has improved and employment has changed reducing exposure, pollution and drainage probably reduced mosquito breeding and quinine became available (see Chapter 42). Diamond (2005) and Fagan (2008) have explored why a number of past societies collapsed. Although such studies lack reproducibility, they not only suggest that past environmental changes may have been sufficient to end cultures but also encourage thinking about adaptation. It may be possible to learn from ways people responded in the past; however, faced with a repeat of a challenge, individuals and groups may react differently (see Chapter 40). Furthermore, future changes may be faster and are likely to differ in other ways from those of the past. Military commanders have argued that researching history is vital preparation even if it proves irrelevant because it encourages a review of present situations and thinking about possible options.

2.4 Learning by studying a region or ecosystem

Insights can often be provided by studying national, regional, city or local ecosystem scenarios that have a manageable scale and level of complexity. An example is provided by the study of adaptation in Vietnam by Adger et al. (2001). Ecosystems and regional units can be modelled to see how they change as controlling factors are varied. The impact of things like an increase in atmospheric carbon dioxide can also be explored using tools such as controlled-chamber experiments.

2.5 Key thresholds and early warning

Avoidance, mitigation and adaptation are difficult if there is no clear and reliable warning sufficiently in advance of significant change.

Early warning should not only focus on environmental change but also on social and economic issues that could affect adaptation (see Chapter 7). Various disciplines and agencies have been exploring ways to assess vulnerability and resilience to environmental change; one study tool is the sustainable livelihood approach. There is no precise definition, but a livelihood can be seen as the capabilities, assets and activities required for a means of living; it may be sustainable if it copes with and can recover from stresses and shocks. The sustainable livelihoods approach or framework recognises 'capitals' (natural, physical, human, social and financial) and argues that an understanding of livelihoods is vital if they are to be adapted and sustained in the face of environmental change. The poor are widely seen to be more vulnerable and improvement of livelihoods can be viewed as providing a route to reduction of vulnerability and better adaptation.

Adaptation to environmental change needs financing and will have profound impacts on taxation, banking, trade and aid (Bouwer and Aerts, 2006). There is an expanding literature on environmental change, economics and sustainability.[3] Adaptation also requires suitable supportive social institutions, so early warning of social change is important, but thus far is little developed. If a group undergoes a decline in its social capital (see section 3.2) or loses a coping strategy, then people may suffer. It could be useful to monitor such changes and, if needs be, establish suitable institutions to strengthen coping. To set up early-warning demands awareness of critical thresholds, availability of practical indicators and workable means of measuring change (Adger, 2003b; Brooks et al., 2005; Niemeyer et al., 2005).

3 RESPONSE TO THREATS AND OPPORTUNITIES

Response is influenced by the outlook of advisors, decision makers and citizens, and is shaped by a complex of other factors.

3.1 Reacting to challenges

Füssel and Klein (2006) examined climate impact assessment, vulnerability assessment and adaptation policy assessment. One trend they discussed is the inclusion of nonclimatic determinants in vulnerability assessments to try to better model complex chains of causation. While knowledge about climate change trends is improving there are still difficulties. Table 44.2 suggests some reasons why there is often a failure to adequately assess and advise on adaptation.

Climate change stresses may coincide with other challenges such as economic problems, disease, warfare, land degradation and population increase, so forecasting is a challenge (Parry et al., 2007: 811–841). False alarms can discredit those issuing warnings and those who act. Understandably many decision makers seek 'win–win' solutions, which they hope will pay-off whether predictions are right or wrong.

Reactions to threats depend on the perceptions and power of stakeholders, policy makers, voters, resources users, and so on. Stakeholders are often easily distracted from a problem and may be reluctant to spend on what they deem to be based on shaky evidence, especially if any outlay does not give immediate and more-or-less personal returns.

Some groups or systems are more sensitive than others and some are more exposed to threats than others (O'Brien and Leichenko, 2007). The worst scenario is being exposed and vulnerable. Groups can draw on a range of 'capitals' to reduce vulnerability: human capital (organizational and character strengths, experience, skills, knowledge); social capital (institutions, governance, obligations, sense of community, social structures, traditions); economic capital (funds to prepare or adapt) and technical capital (knowledge, research capability, engineering skills). Some environments, societies and economies are unpredictable and unstable, which makes it more difficult to model adaptation; for example, 'boom and bust'

Table 44.2 **Some reasons for failure to adapt to environmental and other challenges: a shorthand label (italicized) and definition for each reason is followed by possible remedies (in bold)**

(1) *Ignorance* – the threat is not perceived or well-enough known **[Improve research]**

(2) *False analogy* – a problem is judged by observer(s) against their (perhaps inadequate) experience so crucial issues are missed. **[There should be a careful assessment of hindsight]**

(3) *Insufficient detail* – researchers, planners, managers have over generalized; and/or the problem is more widespread or developed faster than expected; and/or resilience was not as good as expected; and/or the response to environmental change took an unexpected route; and/or study was too general, or the coverage patchy, or there was difficulty in making measurements. **[Carefully check data]**

(4) *Observation over too short a time-span* – with slow processes there may be poor records, and faulty memories; and/or funding and the administrative situation may not permit adequate study (a common problem is the need to decide before there is proof or even reasonable information). **[Seek better modelling and/or information from historians, palaeoecology, etc.]**

(5) *Observer/manager detachment* – elite make decisions and are socially and spatially separate. As decision-making is increasingly urban, there may be insufficient awareness of non-urban situations. Failure to get local information or to inspire local activists. **[Improve accountability and try to empower a wider range of people]**

(6) *Evidence not questioned* – published ideas may not be adequately checked and become established although perhaps erroneous (see, e.g., Fairhead and Leach, 1996; Lomborg, 2001). **[Question established assumptions]**

(7) *Reactions are ill-timed* – decision makers wait for backing from voters, or for money to become available, or for the problem to develop so that it evident it is surely happening. **[Adopt the precautionary principle]**

(8) *The problem is somebody else's* – decision makers judge they can delay while someone else acts (indeed, it might even weaken or distract a rival). **[Improve accountability and public oversight]**

(9) *Inappropriate beliefs or ethical/political outlook* – for example, a land degradation problem may be wrongly attributed to climate by inept local peoples, but in reality causation may be globalization or national taxation **[Seek dispassionate, objective assessment]**

(10) *Symptom mistaken for cause* – is needed to identify causation **[Objective and thorough research, and ideally multiple independent lines of evidence]**

conditions (CSIRO, 2008). Social capital has been described as 'the glue' holding a group together. It may help people withstand physical, economic and social challenges and can support improvements. Understanding and monitoring social capital is important for any group or region wishing to reduce vulnerability or sustain development (Pelling and Heigh, 2005). However, there is debate as to whether social capital is something individuals maintain or whether it is more a group function. A salient question is, 'can social capital be steered to enable humans to better manage environmental change?' To date, there have been only a few studies of social capital and environmental change.

Vulnerability research has been conducted by academics interested in hazards and the political economy of development. Insurance companies and civil defence bodies are also interested in vulnerability research. In the

1980s, El Niño Southern Oscillation (ENSO) events caused sufficient disruption to stimulate interest in environmental threats and continued awareness has been stimulated by the Mount St Helens eruption and fears of a disastrous earthquake in California. Floods in 2007 in the UK and Europe reinforced public interest in environmental threats. Hurricane Katrina and the flooding of New Orleans showed the risks of insensitive and inadequate solutions to environmental challenges and the need to carefully assess threats and make appropriate adaptations. The New Orleans floods also showed that accurate early warnings combined with insensitive preparations can be of limited value if people are too poor, lawless, elderly or fearful of looting to leave their homes for safe refuge.

Although a generalization, it is reasonable to expect the poorest to be more vulnerable. However, any link between poverty and vulnerability is usually indirect. Entitlement to crucial resources is not always determined by ability to pay as it can be politically or socially influenced. Such social and cultural factors can make the impacts of environmental change difficult to predict (Kelly and Adger, 2004; Leary et al., 2008a). There are trends that make people more vulnerable than in the past: citizens are often out of touch with their environment; many have lost coping strategies and have few of the survival skills possessed by their ancestors. Modern crops are increasingly produced by agribusiness and, if there were to be a disaster, survivors could find the seeds they save are infertile or revert to something with undesirable traits.

Not all adaptation is conscious and deliberate as there are situations where people are unaware they are making a change. Adaptation can be anticipatory, concurrent or reactive, spontaneous or planned. Adaptation can be misjudged leading to greater vulnerability and/or less resilience. Adaptation can be at one or more level – by individuals or group (family, village, region or nation). The mechanism may involve change of habit or physiological or psychological adaptation. The timescale varies from short- to long-term. Institutions may play a key role in adaptation as a mechanism that formally or informally shapes expectations and behaviour of individuals and societies. Institutions include public organizations (e.g., government bodies), civic bodies (e.g., cooperatives) and private bodies (e.g., businesses or service organizations) (Berkhout et al., 2006). In order to make adaptations, it may be necessary to strengthen existing or establish new institutions, or weaken those that are a hindrance (Young, 2002). It is possible that efforts to adapt at different scales and/or in different places interact and damage or reinforce each other (Adger et al., 2004; Leary et al., 2008b).

Adaptation may not be planned or intentional. For example, male heads of household in some regions respond to drought or soil erosion by outmigration. Sometimes the impact is negative and sometimes it improves adaptation. Improvements have been seen, for example, in parts of Sudan, where the remaining women, in spite of a very patriarchal society, often prove to be more adaptable and innovative farmers. Future adaptation may be different to past because of things like the ageing of some societies. Genetically modified organisms (GMOs) could enable relatively rapid and flexible development of new crops and possibly effective carbon sequestration and methane emission reduction. However, citizen opposition to GMOs is widespread and tends to overlook benefits; so potential adaptation routes may be ignored. Adaptation often involves educating and winning over people.

Researchers and administrators may seek ways to build adaptive capacity and reduce vulnerability. An established modern socioeconomic response to future threat is insurance. Often societies in the past developed coping strategies that provide similar benefits. Many societies have established coping or adaptation strategies such as crop diversification, supplementary irrigation and so on but fresh strategies will be needed to suit novel environmental and socioeconomic challenges and for situations where existing strategies are breaking down (Adger et al., 2007: 719). One question, which should be asked by anyone involved in adaptation, is whose agenda is driving things? It could be one or more of the following: nations; signatories to the UNFCCC; big business; nongovernmental organizations (NGOs) or many other actors. With complex chains of causation, which are often unclear, vulnerability reduction might be an easier approach than adaptation.

International bodies had begun to react to the challenge of environmental change by the mid-1970s; for example, the UNEP and the IPCC launched a programme: Assessment of Impacts and Adaptations to Climate Change in Multiple Regions and Sectors[4]. This aims to improve the capacity of developing countries to assess vulnerability, identify possible adaptations and spread information on adaptation planning and action. When vulnerability is seen to be increasing, authorities may be able to strengthen existing coping strategies or provide other supports, such as infrastructure, capacity building, food-for-work programmes, emergency supplies, and so forth. Under the terms of the UNFCCC, developed countries have agreed to help developing countries, which are especially vulnerable to the adverse effects of global change (UNFCC, 2007). Holding conferences and getting nations to sign agreements is just an initial step. Getting signatories to actually act and pay can be a problem.

Challenges can offer opportunities. Cuba was dealt a blow when the Soviet Union cut trading in the 1990s. Having lost its market for sugar and other crops, and with a near embargo on oil imports and little hope of aid, Cuba modified its agriculture to make do with low energy inputs and few agrochemicals, making it more labour-intensive and 'organic'. It is now leading the world in some forms of sustainable agriculture (Fanelli, 2007).

3.2 Coping strategies

There is some overlap of adaptation and coping strategies. Coping strategies are the ways people develop to deal with adverse challenges. In the last half-century these have often become ineffective as a consequence of social development, political pressures, economic hardship and environmental problems. For example, nomadic pastoralists may no longer be able to migrate to try and avoid the impacts of drought because of recently introduced border restrictions. The appropriation and privatization of common resources is another reason people may suffer breakdown of their coping strategies. Environmental degradation and deterioration of coping strategies often increases people's vulnerability, reduces their resilience and makes sustainable development more of a challenge. Driven by harassment, or attracted by perceptions of better opportunities, people may adopt less

secure livelihoods and/or relocate to less advantageous locations. Such people are described as marginalized and they may need aid to help them improve their welfare.

4 THE WAY FORWARD

The start of the twenty-first century stimulated environment and development stock-taking and a number of these studies provide valuable information for environmental change adaptation and influence strategies. One of the most important is the Millennium Ecosystem Assessment (MEA), which was launched by the UN Secretary General in 2001 and published in 2005 with plans for updates every five to ten years (World Bank, 2005a). The MEA aimed to provide decision makers and public with information on ecosystem change and how that was affecting human wellbeing. It also sought to establish priorities for action, provide benchmarks, encourage development of assessment tools, identify best responses (especially to support sustainable development) and build institutional capacity. Another key millennium exercise was the publication of the Millennium Development Goals 1990–2025 (MDG) (UNDP, 2003b; World Bank, 2005b), which pledges UN Member States to seek by 2025 to eradicate extreme poverty and hunger, achieve universal primary education, combat HIV/AIDS, malaria and other diseases and ensure environmental sustainability. The MDGs are influencing aid, development planning and research. Other 'millennium assessments' include an atlas of environmental change (UNEP et al., 2005).

Change in the future may be faster and possibly different from that experienced in the past. Also, there is a much greater human population than ever before making huge demands on resources. The adaptive traits humans have relied upon for much of their evolution have been general intelligence, mobility and flexibility. Today many people are less flexible and probably less able to cope with change than their ancestors and opportunities for peaceful relocation are limited because virtually everywhere is occupied.

The choices that will be made in the coming years are likely to be part-based on assessments and modelling already underway. Are these reliable? Some researchers have questioned whether the widely accepted prediction of 0.6 m global sea-level rise by AD 2100 is too conservative and suggest 2.0 m or more might be more realistic (Anathaswamy, 2009). Environmental change modelling, no matter how impressive the computer and software, is a simplification of reality and may be insufficiently accurate. The wise strategy is to accept there is uncertainty and to plan for as much flexibility and diversity as possible to help adaptation and to do as much as possible to reduce vulnerability. Identifying reasonably accurate future scenarios is more realistic than seeking detailed and precise predictions. There are a number of possible response scenarios, including:

1 *business as usual* – there is no dramatic changes in the way problems are faced and in what people do;
2 *technological and ethical advance* – new technology and changed attitudes enable better adaptation and vulnerability reduction;
3 *disaster and/or human inertia* – hinders adaptation and vulnerability reduction (the disaster may be economic problems, unrest, disease, population growth);
4 *powerful groups buy adaptation and reduced vulnerability* – the rest might then be largely abandoned.

Increasingly, environmental change readiness should integrate with other activities, from energy supply to food security. Giddens (2009) provides an interesting study that treats climate change as a political, rather than technological or moral challenge. Some parts of the world lack capacity to adapt sufficiently if they are unaided. Africa is widely seen to be a problem continent that is vulnerable to environmental change because it already has food production difficulties, high rates of illness and poverty, frequent political instability and often-poor governance (Low, 2005).

4.1 What might be done?

As well as direct and obvious impacts, environmental change can cause covert indirect and cumulative impacts, some of these may have a sudden strong feedback effect (either a negative feedback that counters the trend, or a positive feedback that reinforces change, perhaps causing a 'runaway' situation). Caution is needed to ensure assessments and modelling of environmental change do not give a false sense of security. Decisions should be made with caution and based on multiple lines of evidence (Le Roy Ladurie, 1972). However, there is frequently a dilemma: to wait for proof means a problem develops, perhaps becoming costly or incurable. Yet without scientific evidence (which can be slow to gather) it is unwise to act and opponents could dismiss a call for action. Also, some who issue warnings work between disciplines or in other ways that allow critics to question whether they are qualified to advise. Nevertheless, it is important to have such messages; for example, Lomborg (2001) made enemies by criticizing scientists whilst not being one, but he prompted thought. Another example of this was report on *Limits to Growth* commissioned by the Club of Rome in 1972 which focused the world on limits and the need to rethink development ethics even though it was based on poor data and modelling (Meadows et al., 1972). Some of the threats humans face result from their own actions and solutions may require large numbers of people to alter their outlook and ethics and to prompt that social sciences are needed (Garvy, 2008). Valuable advances are sometimes made when someone synthesizes or reviews the studies of others who may be too narrow in focus to spot implications. These syntheses may be journalistic rather than academic. Lynas (2007) conducted one such synthesis, exploring possible impacts across the world for a 1°, 2°, 3°, 4°, 5° or 6°C rise in mean global temperature over the next 100 years. This sort of study can help develop frameworks and contingency plans before changes occur, and may encourage definitive research (UNDP, 2003a).

Infrastructure, which cannot be easily modified, should be built to cope with likely future conditions. Existing irrigation projects, bridges, dams and other large schemes should be checked against predictions and, if required, any 'fixes' should be carefully assessed. For example, embankments could protect a locality vulnerable to storm surges, but it may be cheaper and better in the long run to build elevated shelters or support conversion to appropriate land use. Cities should now be thinking about adapting to environmental change because they may be difficult to modify and make huge demands for inputs and disposal of outputs (Bicknell et al., 2009). Since 1960, the Aral Sea has shrunk by over 60 per cent of its area and has lost more than 80 per cent of its volume. Between 1960 and 1998, the surface level dropped about 18 m. Before the 1970s the Sea supported a rich fisheries industry; now there are few fish left in the three small lakes that remain. The impacts of Aral Sea shrinkage are clear: fisheries loss, contamination of surrounding land by polluted seabed sediment blown on the wind, considerable loss of biodiversity and regional environmental changes. Shrinkage has also concentrated salts, industrial pollutants and agricultural wastes, which have contaminated groundwater. The main cause of the disaster has been the extraction of water from the Syrdaria and Amudarya Rivers for irrigation. Much of that extracted water has been lost by evaporation and transpiration, so less water has fed the Sea. The puzzling thing is: how could the Soviet Union, a centrally planned economy with access to some of the world's best hydrologists, fail to note the serious changes and take remedial action? If such 'creeping change' is not addressed when it is under the jurisdiction of one country, the example should be taken as a warning by those concerned with complex and perhaps rapid transboundary global environmental changes. Streets and Glantz (2000) suggest the Soviet authorities failed to alter their perceptions fast enough despite warnings from experts since as far back as the 1920s. Clearly, perception studies are very important and should focus on those managing adaptation, not just on those potentially affected (Barrow, 2003: 116).

In the first half of the twenty-first century, economics has started to seriously address environmental change issues (Agrawala and Fankhauser, 2008; Ruth, 2008; Sinnott-Armstrong and Howarth, 2005). Economic responses to the threat of global warming, such as carbon taxes, tradable emission quotas and carbon offsetting, are underway (Jorgenson et al., 2004). Green insurance and pensions funds have also started to explore links with environmental change adaptation.

4.2 Sustainable development strategies and environmental change

How might different societies best sustain themselves in the face of environmental change? There is a move toward addressing climate change combined with seeking sustainable development – 'joint goals' or 'integrated adaptation' (Ågerup et al., 2004; Munasinghe and Swart, 2005; Salih, 2009). Article 3 of the UNFCCC calls for policies and measures to address climate change that foster sustainable economic growth and development, particularly in developing countries. Some argue that sustainable development is not a viable way forward until crucial survival problems are resolved (Lovelock, 2006). A better approach, Lovelock argues, is 'sustainable retreat' using unpopular nuclear power and GMOs to buy time until the environment settles and human numbers fall to a level at which sustainable development measures are worthwhile (see also Constanza et al., 2007). For sustainable development to have a chance of success, the following are desirable: (1) careful assessment of scenarios to determine the best approach and focus; (2) selection of the best scale at which to operate and (3) choice of the right timing of measures.

One problem with seeking sustainability in one locality or sector of an economy is that activities may threaten stewardship and adaptation elsewhere. For example, change of fashion, buying habits, or a decision to alter foreign aid can impact faraway places. Those causing impacts are often unaware of the problems they cause and those impacted may not fully grasp causation. Those managing development need adequate oversight and powers to enforce measures; often, however, that is not the case. Deforestation and land degradation in the Himalayas illustrates this. The impacts contribute significantly to flooding in distant Bangladesh, which has little chance of seriously influencing things (Hofer and Messerli, 2006). Similarly, acid deposition may be caused by industrial activities hundreds of kilometres from the biota or soils it degrades. This 'over-the-horizon' pattern of cumulative and indirect causation

frequently challenges policy makers and those trying to monitor or model (Lim and Spanger-Siegfried, 2004).

There are numerous proposals of how adaptation to environmental change could be undertaken in ways that support sustainable development and vice-versa and this is termed integrated adaptation (see Figures 44.2 and 44.3) (Beg et al., 2002; O'Brien and Leichenko, 2007; Swart and Mohan, 2005; Swart et al., 2003). Parry et al. (2007) and Metz et al. (2007) provide a comprehensive overview. The 'Gleneagles Plan of Action 2005' (Parliamentary Office of Science and Technology, 2006) undertook to assist developing countries to build capacity to respond to climate change in a sustainable manner. The UNEP has launched a 'Climate Change Outreach Programme' to advise developing countries. There are sources of funding to assist poorer nations to adapt to global change and enhance sustainability; for example, the 'Least Developed Countries Fund', the 'Special Climate Change Fund', the 'Kyoto Protocol Adaptation Fund', and the

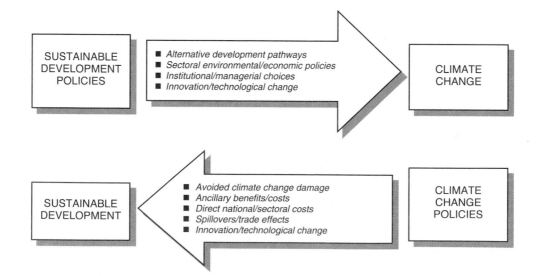

Figure 44.2   Linkages for climate change policies responding to climate change and seeking to support sustainable development; and for sustainable development policies seeking to support climate change adaptation/mitigation.

Source: Parry et al. (2007).

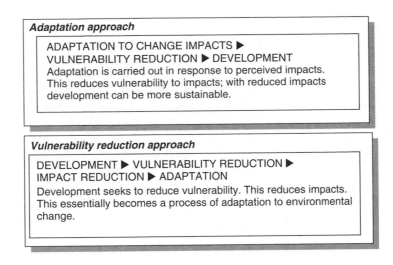

Figure 44.3 **Different approaches to linking adaptation and development.**

Source: Schipper (2007).

'Global Environmental Facility'. An approach that can be used to explore environmental change and sustainable development linkages is outlined in the WEHAB framework put forward by the World Summit on Sustainable Development in 2002.

The private sector (business, NGOs, and so forth) already plays a huge part in efforts to support sustainable development and environmental change adaptation. National Adaptation Plans of Action (NAPAs, established by the UNFCCC) offer a means for developing countries to recognize priorities to strengthen adaptation to environmental change (Adger et al., 2007: 732; Pelling, 2009). Ideally, there should be no risk of mitigation efforts leading to unexpected impacts. Lewis (1992: 147) explored 'Promethean' use of technology – urging bold adoption, but with caution and environmental sensitivity.

4.3 Environmental change and conflict

A number of historians and environmental determinists argue environmental change has triggered an historical event or shaped the outcome (see, e.g., Huntington, 1915). The problem is that insufficient evidence has too often been taken as proof. Also, humans can adapt or fail to adapt to any given challenge for a host of reasons. There has been a resurgence of environmental determinism since the mid-1980s. This neo-determinism has possibly been triggered by the growth of palaeoecological and archaeological studies of phenomena like ENSO events (Fagan, 1999). Conflict is a 'wild card' in any adaptation process. Modelling conflicts under conditions of environmental change is something that needs to be developed to assist avoidance and mitigation (Schnellnhuber, 2009). If environmental change results in the enforced movement of people (ecorefugees or environmental migrants) in problematic numbers, then serious disruption or conflict could result. Sometimes migration is an act of adaptation, but it can also be a resort after failure of other efforts to adapt. In the future, very large numbers could be displaced from low-lying areas by rising seas, altered disease transmission may cause peoples to relocate, and precipitation changes could also trigger movements. The problem is coping with large numbers of refugees if it occurs over a limited time span. Some see the current influx of migrants to Europe as in part caused by gradual environmental change or land degradation (i.e., mismanagement is involved), and argue it is a warning that problems in poor and distant countries can directly impact on developed nations.

Environmental change can impact on crucial resources, notably food and water (Dinar et al., 2008), and the result might be hunger and unrest (Ludwig et al., 2009). There is potential for conflict if Arctic and Antarctic ice recedes further and results in a rush to claim territory or seabed-drilling rights (Keskitalo, 2008). The study of conflicts related to environmental change and natural resource usage has generated growing interest from aid agencies, banks, large natural resources companies and those seeking 'secure' development (Schipper, 2007; WBGU, 2009).

5 CONCLUSION

Given the complex chains of causation involved in adaptation, mitigation, and vulnerability reduction, it is not easy to make reliable predictions. Environmental change may be unexpected and sudden, and even gradual shift might in time trigger sudden challenges (MacCracken, 2007). Planners and policy makers need to be ready for surprises and as much as possible be warned about critical thresholds and threats. Response to change is shaped by perceptions of a range of stakeholders: public and taxpayers, research community, special interest groups, business interests, and so forth. These should be monitored. One key group of stakeholders involved in sustainable development – future generations – is under-represented and effectively lacking in direct voice; however, their options must be safeguarded and someone should act for them now.

Adaptation, mitigation, vulnerability reduction and sustainable development have to work in the 'real world' and will never be precise and wholly reliable sciences (or arts). Nations and agencies are starting to 'mainstream' concern for environmental change (Kok and deConinck, 2007). However, policy making and law still have to evolve to address these challenges and there may be limited time available. A key thing is that planners and administrators are alert for sudden challenges and, as far as possible, seek flexibility and vulnerability reduction. Those concerned with environmental change must not focus only on human-induced global warming, sidelining vigilance for other environmental threats.

Currently, many feel adaptation and mitigation can be integrated with sustainable development efforts. Although the idea will be unpopular, focus on sustainable development may sometimes hinder adaptation. Sustainable development may have to wait while emergency responses try to cope with threatening environmental change problems. There is growing agreement that adaptation alone is not enough and should be accompanied by proactive vulnerability reduction and effective and safe mitigation measures. Much of what is currently being done in the name of 'development' is inflexible, lacking in diversity and is itself vulnerable.

NOTES

1 See http://www.eldis.org/go/topics/dossiers/climate-change-adaptation

2 See http://www.iisd.org

3 For a recent listing of books on the subject visit: http://www.economicsnetwork.ac.uk/books/EnvironmentalEconomicsand.htm

4 See http://www.aiaccproject.org/about/right_frame.html

REFERENCES

Adger W. N. 2003a. Adaptation to change in the developing world. *Progress in Development Studies* 3(5): 179–195.

Adger W. N. 2003b. Social capital, collective action, and adaptation to climate change. *Economic Geography* 79: 387–404.

Adger W. N. 2006. Vulnerability. *Global Environmental Change Part A* 16(3): 268–281.

Adger W. N., Agrawala M. M. Q., Mizra C., Conde K., O'Brien J., Pulhin R. et al. 2007. Assessments of adaptation practices, options, constraints and capacity, in Parry M. L., Canzani O. F., Palutikof J. P. van der Linden P. J. and Hanson C. E. (eds) *Climate Change 2007: Impacts, Adaptation and Vulnerability. Contribution of the Working Group II to the Fourth Assessment Report of the Intergovernmental Panel on Climate Change.* Cambridge: Cambridge University Press, pp. 717–743.

Adger W. N., Arnell N. W. and Tompkins E. L. 2004. Successful adaptation to climate change across scales *Global Environmental Change Part A* 15(2): 77–86.

Adger W. N., Kelly P. M. and Nguyen H. N. 2001. *Living with environmental Change: Social Vulnerability, Adaptation and Resilience in Vietnam.* London: Routledge.

Agrawal A. McSweeny C. and Perrin N. 2008. *The Social Dimensions of Climate Change.* Washington DC: Social Development Notes, World Bank.

Agrawala S. and Fankhauser S. (eds) 2008. *Economic Aspects of Adaptation to Climate Change: Costs, Benefits and Policy Instruments.* Paris: OECD.

Ågerup M., Ayodele T., Cordeiro J., Cudjoe F., Fernandez J. R., Hidalgo J. C. 2004. *Climate Change and Sustainable Development: A Blueprint from the Sustainable Development Network.* London: International Policy Network.

Applegate J. S. 2000. The precautionary principle: an American perspective on the precautionary principle. *Human and Ecological Risk Assessment* 6(3): 413–443.

Anathaswamy A. 2009. Waterworld awaits. *New Scientist* 203(2715): 28–33.

Barnett J. 2005. Titanic states? Impacts and responses to climate change in the Pacific Islands. *Journal of International Affairs* 59: 203–219.

Barnett J. and Campbell J. 2009. *Climate Change and Small Island States: Power, Knowledge and the South Pacific.* London: Earthscan.

Barrow C. J. 2003. *Environmental Change and Human Development: Controlling Nature.* London: Arnold.

Barrow C. J. 2006. *Environmental Management for Sustainable Development.* London: Routledge.

Beg N., Morlot J. C., Davidson O., Afrane-Okesse Y., Tyani L., Denton F. 2002. Linkages between climate change and sustainable development. *Climate Policy* 2(2/3): 129–144.

Berkhout F., Hertin J. and Gann D. M. 2006. Learning to adapt: organisational adaptation to climate change. *Climatic Change* 78: 135–156.

Bicknell J., Dodman D. and Satterthwaite D. (eds) 2009. *Adapting Cities to Climate Change: Understanding and Addressing the Development Challenges.* , London: Earthscan.

Birkmann J. (ed.) 2006. *Measuring Vulnerability to Natural Hazards: Towards Disaster Resilient Societies.* Tokyo: United Nations University Press.

Bouwer L. M. and Aerts J. C. J. H. 2006. Financing climate change adaptation. *Disasters* 30(1): 49–63.

Brooks M. 2009. Gone in 90 seconds. *New Scientist* 201(2700) 31–35.

Brooks N., Adger W. N. and Kelly P. M. 2005. The determinants of vulnerability and adaptive capacity at the national level and implications for adoption. *Global Environmental Change Part A* 15(2): 151–163.

Churchill R. R. and Freestone D. (eds.) 1991. *International Law and Global Climate Change.* London: Graham & Trotman (Martinus Nijhoff).

Constanza R., Graumlich L. J. and Steffen W. (eds) 2007. *Sustainability and Collapse? An Integrated History and Future of People on Earth.* Cambridge MA: MIT Press.

CSIRO. 2008. *Adaptation and Resilience in Rangeland Socio-economic Systems.* Available at: http: //www. csiro.au/science/RangelandSystemAdaptation.html

Davis D. K. 2007. *Resurrecting the Granary of Rome: Environmental History and French Colonial Expansion in North Africa.* Columbus: Ohio University Press and Swallow Press Ohio.

Davoudi S., Crawford J. and Mehmood A. (eds) 2009. *Planning for Climate Change: Strategies for Mitigation and Adaptation for Spatial Planners.* London: Earthscan.

De França Doria M., Boyd E., Tompkins E. L. and Adger W. N. (in press) Using expert elicitation to define successful adaptation to climate change. *Environmental Science & Policy.*

Diamond J. 2005. *Collapse: How Societies Choose to Fail or Survive.* London: Allen Lane.

Dinar A., Hassan R., Mendelsohn R. and Behin J. 2008. *Climate Change and Agriculture in Africa: Impact Assessment and Adaptation Strategies.* London: Earthscan.

Doyel D. E. and Dean J. S. (eds) 2004. *Environmental Change and Human Adaptation in the Ancient American Southwest.* Salt Lake City: University of Utah Press.

Fagan B. M. 1999. *Floods, Famines, and Emperors: El Niño and the Fate of Civilizations.* New York: Basic Books.

Fagan B. M. 2008. *The Great Warming: Climate Change and the Rise and Fall of Civilisations.* London: Bloomsbury Press.

Fairhead J. and Leach M. 1996. *Misreading the African Landscape: Society and Ecology in a Forest-Savanna Mosaic.* Cambridge: Cambridge University Press.

Fanelli D. 2007. Cuba flies lone flag for sustainability. *New Scientist* 196(2624): 10.

Füssel H-M. 2005. *Vulnerability in Climate Change Research: a Comprehensive Conceptual Framework.* University of California International and Area Studies Bresia Paper No. 6. Available at: http://repositories.cdlib.org/ucias/breslauer/6

Füssel H-M. and Klein R. J. T. 2006. Climate change vulnerability assessments: an evolution of conceptual thinking. *Climatic Change* 75(3): 301–329.

Galloway J. N. and Cowling E. B. 2002. Reactive nitrogen and the world: 200 years of change. *Ambio* 31(2): 64–71.

Gallopin G. C. 2006. Linkages between vulnerability, resilience, and adaptive capacity. *Global Environmental Change* 16(3): 293–303.

Garvey J. 2008. *The Ethics of Climate Change: Right and Wrong in a Warming World.* London: Continuum International Publishing.

Giddens A. 2009. *The Politics of Climate Change.* London: Polity Press.

Griffiths J., Rao M., Adshead F. and Thorpe A. (eds) 2009. T*he Health Practitioner's Guide to Climate Changes: Diagnosis and Cure.* London: Earthscan.

Handmer J. W., Dovers S. and Downing T. E. 2004. Societal vulnerability to climate change and variability mitigation and adaptation strategies. *Global Change* 4(3–4): 267–281.

Hofer T. and Messerli B. (eds) 2006. *Floods in Bangladesh: History, Dynamics and Rethinking the Role of the Himalayas.* Tokyo: UN University Press.

Holling C. S. 1973. Resilience and stability of ecological systems. *Annual Review of Ecology and Systematics* 4: 1–23.

Huntington E. 1915. *Civilization and Climate.* New Haven: Yale University Press.

IPCC. 2001. *Climate Change 2001: Impacts, Adaptation, and Vulnerability. Working Group II IPCC.* Cambridge: Cambridge University Press.

IPCC. 2007a. Implications for policy and sustainable development, in *IPCC Fourth Assessment Report: Working Group Report: Technical Paper VI: Climate Change And Water.* Geneva: IPCC, pp.121–131.

Jackson T. 2008. What politicians dare not say. *New Scientist* 201(2675); 40.

Janssen M. A., Schoon M. L., Weimao K. E. and Borner K. 2006. Scholarly networks on resilience, vulnerability and adaptation within the human dimensions of global environmental change. *Global Environmental Change* 16: 240–252.

Jorgenson D. W., Goettle R. J., Hurd B. H., Smith J. B. and Mills D. M. 2004. *US Market Consequences of Global Climate Change.* Cambridge, MA: Pew Center on Global Climate Change, Harvard University.

Kane S. M. and Yohe G. W. (eds) 2000. *Societal Adaptation to Climate Variability and Change.* London: Earthscan.

Kelly P. M. and Adger W. N. 2004. Theory and practice in assessing vulnerability to climate change and facilitating adaptation. *Climatic Change* 47(4): 325–352.

Keskitalo E. C. H. 2008. Climate Change and Globalization in the Arctic: an Integrated Approach to Vulnerability Assessment. London: Earthscan.

Kok M. T. J. and deConinck H. C. 2007. Widening the scope of policies to address climate change: directions for mainstreaming. *Environmental Science & Policy* 10(7–8): 587–599.

Leary N., Conde C., Kulkami J., Nyong A. and Pulhin J. (eds) 2008a. *Climate Change and Vulnerability.* London: Earthscan.

Leary N., Adejuwan J., Barros V., Burlon I., Kulkarni J. and Lasco R. (eds) 2008b. *Climate Change and Adaptation.* London: Earthscan.

Le Roy Ladurie E. 1972. *Times of Feast, Times of Famine: A History of Climate Since the Year 1000.* London: George Allen and Unwin.

Lewis M. 1992. *Green Delusions: An Environmentalist Critique of Radical Environmentalism.* Durham: Duke University Press.

Lim B. and Spanger-Siegfried E. (eds) 2004. *Adaptation Policy Frameworks for Climate Change: Developing Strategies, Policies and Measures.* Cambridge: Cambridge University Press.

Lomborg B. 2001. *The Sceptical Environmentalist: Measuring the Real State of the World.* Cambridge: Cambridge University Press,.

Lomborg B. 2009. *Cool it: the Sceptical Environmentalist's Guide to Global Warming.* New York: Alfred Knopf

Lovelock J. 2006. *The Revenge of Gaia: Why the Earth is Fighting Back - and How We Can Still Save Humanity.* London: Allen Lane.

Low P. S. (ed.) 2005. *Climate Change and Africa.* Cambridge: Cambridge University Press.

Ludwig F., Kabat P., von Schaik H. and van der Valk M. (eds) 2009. *Climate Change Adaptation in the Water Sector*. London: Earthscan.

Lynas M. 2007. *Six Degrees: Our Future on a Hotter Planet*. London: Fourth Estate.

MacCracken M. C. (ed.) 2007. *Sudden and Disruptive Climate Change: Exploring the Real Risks and How We Can Avoid Them*. London: Earthscan.

Mansfield B. 2009. Sustainability, in Castree N., Demeritt D., Liverman D. and Rhodes B. (eds) *A Companion to Environmental Geography*. Oxford: Wiley-Blackwell, pp. 37–49.

Markandya A. and Halsnaes K. (eds) 2002. *Climate Change and Sustainable Development: Prospects for Developing Countries*. London: Earthscan.

McCarthy J. J., Canziani O. F., Leary N. A., Dokken D. J. and White K. S. (eds) 2001. *Climate Change 2001: Impacts Adaptation, and Vulnerability Contribution of Working Group II to the Third Assessment Report of the IPCC*. Cambridge: Cambridge University Press.

McMichaels A. J., Haines A., Sloof R. and Kovats S. (eds) 1996. *Climate Change and Human Health*. Geneva: World Health Organization.

Meadows D. H., Meadows D. L., Randers J. and Begrens W. W. III. 1972. *The Limits to Growth (a Report for the Club of Rome's Project on the Predicament of Mankind)*. New York: Universal Books.

Metz B., Davidson Bosch P., Dave R. and Meyer L. (eds) 2007. *Climate Change 2007: Mitigation of Climate Change. Working Group III Contribution to the Fourth Assessment Report to the IPCC*. Cambridge: Cambridge University Press.

Munasinghe M. and Swart R. 2005. *Primer on Climate Change and Sustainable Development*. Cambridge: Cambridge University Press.

Niemeyer S., Pelts J. and Hobson K. 2005. Rapid climate change and society: assessing responses and thresholds. *Risk Analysis* 25: 1443–1455.

Nelson D. R., Adger W. N. and Brown K. 2007. Adaptation to environmental change: contributions to a resilience framework. *Annual Review of Environment and Resources* 32: 395–419.

O'Brien K. and Leichenko R. 2007. *UNDP Human Development Report 2007/2008: Fighting Climate Change: Human Solidarity in a Divided World (Human Security, Vulnerability and Sustainable Adaptation)*. New York: United Nations Development Programme.

Oxfam International. 2007. *Adapting to Climate Change: What's Needed in Poor Countries and Who Should Pay*. Oxfam Briefing Paper No. 104. Oxford: Oxfam International Secretariat.

Pandey N. 2005. Societal adaptation to abrupt climate change and monsoon variability: implications for sustainable livelihoods of rural communities. Washington DC: USAID. Available at: http://www. frameweb.org/ev02.php?ID=17053_201&ID2=DO_ TOPIC

Parliamentary Office of Science and Technology 2006. Adapting to climate change in developing countries *Postnote* October 2006. No. 269. Available at: http://www.parliament.uk/post

Parry M. L., Canziani O., Palutikof J., van der Linden P. and Hansen C. (eds) 2007. *Climate Change 2007: Impacts, Adaptation and Vulnerability Working Group II Contribution to the Fourth Assessment Report to the IPCC*. Cambridge: Cambridge University Press.

Parry M. L., Rosenzweig C., Iglesias A. and Fischer G. 2004. Effects of climate change on global food production under SRES emissions and socio-economic scenarios. *Global Environmental Change Part A* 14(1): 53–67.

Patt A. G., Schröter D., Klein R. J. T. and de la Vega-Leinert A. C. 2009. *Assessing Vulnerability to Global Environmental Change: Making Research Useful for Adaptation Decision Making and Policy*. London: Earthscan.

Pelling M. 2009. *Adaptation to Climate Change: A Progressive Vision of Human Security*. London: Routledge.

Pelling M. and Heigh C. 2005. Understanding adaptation: what can social capital offer assessments of adaptive capacity? *Global Environmental Change Part A*. 15(4): 308–319.

Philander S. G. (ed.) 2008. *Encyclopedia of Global Warming and Climate Change*. Los Angeles: Sage.

Reiter P. 2000. From Shakespear to Defoe: malaria in England in the Little Ice Age. *Emerging Infectious Diseases* 6: 1–11.

Rogers D. J., Wilson A. J., Hay S. I. and Graham A. J. 2006. Climate change and vector-borne diseases *Advances in Parasitology* 62: 345–381.

Ruth M. 2008. *Smart Growth and Climate Change*. London: Earthscan.

Salih, M. A. (ed.) 2009. *Climate Change and Sustainable Development: New Challenges for Poverty Reduction*. Cheltenham: Edward Elgar.

Schellnhuber H. J. 2009. *Climate Change as a Security Risk*. London: Earthscan.

Schipper E. L. F. 2007. *Climate Change Adaptation and Development: Exploring the Linkages*. Tyndall Centre Working Paper No. 107. University of East Anglia, Norwich.

Schipper E. L. F. and Burton I. (eds) 2008. *The Earthscan Reader on Adaptation to Environmental Change*. London: Earthscan.

Sinnott-Armstrong W. and Howarth R. B. (eds) 2005. *Perspectives on Climate Change: Science, Economics, Politics and Ethics*. Amsterdam: Elsevier.

Smith L. P. 1968. *Seasonable Weather*. London: George Allen and Unwin.

Solomon S., Quin D., Manning M., Marquis M., Averyt K., Tignor M. M. B. et al. (eds) 2007. *Climate Change 2007: The Physical Science Basis. Working Group I. contribution to the Fourth Assessment Report of the IPCC*. Cambridge: Cambridge University Press.

Stern N. H. 2007. *The Economics of Climate Change: The Stern Review on the Economics of Climate Change 2006*. (HM Treasury, UK). Cambridge: Cambridge University Press.

Stern N. H. and Patel I. G. 2009. *A Blueprint For A Safer Planet: How to Manage Climate Change and Create a New Era of Progress and Prosperity*. London: Bodley Head.

Swart R., Robinson J. and Cohen S. 2003. Climate change and sustainable development: expanding the options. *Climate Policy* 3(Suppl.1): S19–S40.

Swart R. and Mohan M. 2005. *Primer on Climate Change and Sustainable Development*. Cambridge: Cambridge University Press.

UNDP. 2003a. *User's Guidebook for the Adaptation Policy Framework (final draft)*. New York: United Nations Development Programme.

UNDP. 2003b. *Global Indicators (Millennium Development Goals)*. New York: United Nations Development Programme.

UNEP, NASA, USGS and University of Maryland 2005. *One Planet, Many People: Atlas of Our Changing Environment*. New York: UNEP.

UNEP. 2006. *Adaptation and Vulnerability to Climate Change: The Role of the Finance Sector*. UNEP FI Climate Change Working Group CEO briefing UNEP, Geneva.

UNEP. 2007. *Global Outlook for Ice and Snow*. New York: United Nations Environment Programme.

UNFCC. 2007. *Climate Change: Impacts, Vulnerabilities and Adaptation in Developing Countries*. Bonn: UN Framework Convention on Climate Change Secretariat.

USAID. 2007. *Adapting to Climate Variability and Change: A Guidance Manual for Development Planning*. Washington DC: USAID.

Watson R. T., Zinyowera M. C. and Moss R. H. (eds) 2001. *Climate Change 2001: Impacts, Adaptation, and Vulnerability Contribution of Working Group II to the Third Assessment Report of the IPCC*. Cambridge: Cambridge University Press.

World Bank 2005a. *Millennium Ecosystem Assessment: Current State and Trends Assessment*. Washington DC: Island Press.

World Bank 2005b. *Miniatlas of Millennium Development Goals*. Washington DC: World Bank.

World Bank 2007. *World Development Report 2008: Agriculture for Development*. Washington DC: World Bank.

WBGU. 2009. *Climate Change as a Security Risk*. London: Earthscan.

Yohe G., Malone E., Brenkert A., Schlesinger M. E., Meij H. and Xing X. 2006. Global distributions of vulnerability to climate change. *Integrated Assessment Journal* 6(1): 35–44.

Yohe G. 2007. Climate change, in B. Lomborg (ed.) *Solutions to the World's Biggest Problems: Costs and Benefits*. Cambridge: Cambridge University Press, pp. 103–124.

Young O. R. 2002. *The Institutional Dimensions of Environmental Change*. Cambridge MA: MIT Press.

Index

NOTE: When a page range next to an author's name appears in bold (eg. Bush, M. B. **113–40**), then this indicates the chapter in this Handbook that the author has written.

NOTE: All lakes, mountains and rivers are filed under their names – eg. Alexandrina (Lake).

Abbott, M. B. 121
Accelerator Mass Spectrometry (AMS) 220
acidification 48, 53–6, 72, 84–6, 89, 104–7, 200, 251–2, 269, 428, 440
  acid rain debate 53–4, 104–5, 406
  see also pH levels
acquired immune deficiency syndrome 387, 437
Adger, W. N. 380, 433
Adriatic Sea 82, 164
Advanced Along Track Scanning Radiometer (AATSR) 8, 10, 19
Aegean Sea 164, 170, 287–8
aerosols 101–8, 269, 328, 388, 390
Africa 10, 14, 16–17, 19, 25, 29, 56, 58–9, 105, 114–15, 117–19, 122–32, 154, 164–5, 170, 179, 215, 248, 430
  agriculture and domestication 335, 339
  arid and semi-arid regions 142–8, 156
  disease in 388, 392, 394–6
  hominid evolution and migration 301–9, 311–18
  mountain regions 263, 270, 276
  see also individual countries
African horse sickness (AHS) 389
agriculture 25, 34–5, 50, 209, 271, 292, 328–9, 344–50, 380, 432–3, 437
AIDS 387, 437
air pollution
  history of 95–7
  human impacts on the atmosphere 97–106
  see also carbon dioxide; greenhouse gases
Akkadian Empire 356, 361
Alaska 247, 253, 255
Albania 164
Alboran Sea 164, 166
Alexandrina (Lake) 60
Allen, J. A. 310
Alps 141, 266, 270, 274, 307
Alvarez-Filip, L. 88

Amazonia 115–19, 128–32, 288, 342, 344
Ambrose, S. H. 316
ammonia 48
Amudarya River 439
Anasazi civilization 356
Anatolia 164, 361
ancient DNA (aDNA) 329
Anderies, J. M. 381
Anderson, A. 339, 348
Anderson, L. 206
Andes 113–14, 117–22, 124, 127, 131, 236, 264, 266, 276, 342–4, 364
Angola 122
Antarctic 78, 98–101, 142, 153, 155, 245–61, 442
  geographic definitions 246–8
  polar environmental change 248–56
Antarctic Climate Change and the Environment (ACCE) 246
anthrax 388, 390
Anthropocene 65, 74, 82, 88, 95, 97, 99, 108, 245–6
ants 31–2
Appalachian Mountains 206
Arabian Desert 171
Arabian Sea 126–7
Aral Sea 59, 154
Arctic 245–61, 291, 442
  geographic definitions 246
  polar environmental change 248–56
Arctic Climate Impact Assessment (ACIA) 246, 256
Arctic Monitoring and Assessment Programme (AMAP) 252, 254
Arctic Ocean 84, 86, 125, 246–7, 251
Argentina 155, 163
arid and semi-arid regions 141–62
  environmental change in 143–53
  global impacts of environmental changes in 155–6
  land-use and desertification 153–5
  present-day distribution of 142–3

Arizona 37, 368
Asia 14, 17, 29, 103–7, 113–14, 116, 119, 123, 125–7,
    131–2, 165, 239, 285, 366, 430–1
    agriculture and domestication 331, 333, 335,
        339–40, 349
    arid and semi-arid regions 142, 148–50, 155, 157
    disease in 389, 396
    grasslands 218–21, 224–5, 229–38
    hominid evolution and migration 303, 306, 309
    mountain regions 263, 267, 270
    temperate forested regions 190, 192, 195, 199,
        201, 205
    see also individual countries
Asian Development Bank 428
Association for Small Island States (AOSIS) 431
Atacama Desert 152, 163
Atahan, P. 337
Athens, J. S. 121
Atlantic Meridional Overturning Circulation
    (AMOC) 115, 118, 124–5
Atlantic Ocean 81, 86–7, 103, 113, 115, 117–18,
    126–7, 148, 164–5, 167, 169, 171–3, 176, 179,
    188, 247–8, 285, 308, 317, 365
    North Atlantic Deep Water (NADW) 166
Australia 10, 19, 29–33, 58, 75, 77, 80–1, 86–7, 123,
    128–30, 132, 163, 248, 338, 396
    arid and semi-arid regions 142, 152–4, 156
    coastal regions 286, 289
    mountain regions 264, 267, 275
    temperate forested regions 190–2, 194–5,
        200–3, 206
Austria 264
avalanches 264
avian influenza 387, 390
Ayalon, A. 176–7
Azov Sea 229

Badain Jaran Desert 149
Bahamas 156
Baikal (Lake) 219
Balearic Islands 164
Balearic Sea 164
Baltic Sea 82
Bangladesh 270, 286, 440
Barbados 156
Barber, K. 356
Barents Sea 247
Bargagli, R. 245, 254
Bar-Matthews, M. **163–87**, 169, 173, 176–7
Barnes, D. K. A. 252
Barnosky, C. W. 203, 207
Barrow, C. J. **426–46**
Bartlein, P. J. **188–214**, 331
Barton, R. N. E. 313–14
Bar-Yosef, O. 178
Battarbee, R. W. **47–70**, 55, 59
Baylis, M. **387–405**
Behrensmeyer, K. 305

Belgium 400
Beniston, M. **262–81**
Bennion, H. **47–70**, 52, 59
Bergmann, C. 310
Bering Sea 247
Bettinger, R. L. 336
Beug, H.-J. 233
Beyer, L. 245
biomass 4, 11, 18–19, 48–9, 188, 195, 200–1, 203,
    208, 270, 328, 331, 421
    see also forests; vegetation
BIOME models 4, 223, 325–9
BIOPRESS project 16
Biosphere–Atmosphere Transfer Scheme (BATS) 4
Blaber, S. J. M. 79–80
Black Sea 82, 164, 219, 229–31, 234, 238
Blaut, J. 356
Blockley, S. M. **301–27**
Blockley, S. P. E. **301–27**, 321
Blue Mountains 275
bluetongue 396–8, 400
Bobe, R. 304–5
Böcher, J. 245
bogs 266
Bolivia 364
Bølling/Allerød warm event 114, 116, 173
Bolter, M. 245
Bond, G. 125–6
Bond Cycles 125–6, 177, 308
Bonneville (Lake) 151
Born, E. W. 245
Borneo 32
Boserup, E. 347, 357
Bosnia and Herzogovina 164
Botswana 145, 147
Bradley, R. S. 355, 360
Brahmaputra River 286
Braune, B. M. 253
Brazil 10, 18, 124–5, 380
Brimblecombe, P. 96
Britain *see* Great Britain
British Antarctic Survey 99–100
Bronze Age 49, 65, 178–9, 205
Brooke, B. 289
Brooks, N. 366, 368–9
Brunk, G. G. 359
Budain Jaran Desert 144
Burdekin River 81
burning *see* fire
Bush, M. B. **113–40**, 121
Butzer, K. 357, 368

Caddy, J. F. 79
Cadée, G. C. 84
cadmium (Cd) 61
Cairns Historical Society 80
Calaway, M. J. 364–5
California 150–1, 163, 218, 275, 435

Callaway, R. M. R. W. 274
Cambodia 123
Canada 52–3, 199, 234, 236, 245–7, 250, 253–5, 267, 286, 291
Canary Islands 286
Candy, I. **301–27**
cane toads 31
Cara (Lake) 123
carbon dioxide (CO_2) 3–4, 19, 53, 86, 97–103, 106, 108, 148, 156, 180, 191–3, 195, 208–9, 216, 222–3, 227, 236, 246, 251, 257, 270, 328, 338, 406, 419, 422, 428
Caribbean 88
Carpenter, K. E. 87–8
Carpenter, S. R. 36
Carrington Events 430
Carson, R. 406
Caspian Sea 229
Catto, G. 360
Catto, N. 360
causality see collapse scenarios
Cavallari, B. J. 178
caves 124, 128, 174, 176–9, 232, 305, 311, 320–1, 331, 343
Cenozoic 148
CENTURY model 4
CH_4 see methane
Chad Basin 147
Chalcolithic 178–9
Chapin, F. S. 378, 380
Chappell, J. 284
charcoal 127, 129, 131–2, 227, 229, 233, 321, 342
Charles, A. T. 79
Charlson, R. J. 105
Chepstow-Lusty, A. J. 121
Chesapeake Bay 78, 81–2
Chichancanab (Lake) 127
Chikungunya 387
Childe, V. G. 345
Chile 152, 163, 191, 193, 200, 204, 206
Chin, M. 104
China 30, 52, 61–2, 102, 124–6, 128, 141–2, 144, 148–50, 156–7, 188, 190–1, 193, 195, 200, 205, 219–22, 231–4, 236, 238, 268, 269, 287, 293, 305, 317, 335–8, 349, 366
    Loess Plateau 148–9
chlorofluorocarbons (CFCs) 101, 255
cholera 388
Chorreras (Lake) 341
Christmas Island 32
Clarke, A. 251
Clean Water Act (US) 47
CLIMAP 116
climate models 118, 123, 271–2, 379
    see also general circulation models (GCMs)
clouds 9, 86, 100
    Asian Brown Cloud 269
    polar stratospheric clouds (PSCs) 101–2

Clovis period 129–30, 205
Clow, D. W. 269
CO_2 see carbon dioxide (CO2)
coal 53, 95, 421, 432
coastal regions 71–94, 282–97, 380
    changes in coastlines 73, 88–9
    effect of climate change on ecosystems 84–8
    environmental histories of 288–91
    exploitation of ecosystems 73–81
    human impacts on 291–2
    non-human causes of environmental change in 283–8
    pollution of ecosystems 81–4
    prospects for 292–3
Cocklin, C. **406–25**
Cohen, A. J. 107
Colinvaux, P. A. 121
collapse scenarios 355–73
    lessons from the past 367–9
    trajectories of vulnerability 357–67
Colombia 343
Colorado 152
Columbus, C. 132
complexity see collapse scenarios
computers 5
Conservation of Arctic Flora and Fauna (CAFF) 252
Conseuelo (Lake) 120
Constantinople 96
Convention on Biological Diversity (CBD) 34
Cook, E. J. 206
Coombes, P. 356
Cooper, J. A. G. 289
Coordination of Information on the Environment (CORINE) 15–16
copper (Cu) 61
corals 72, 74, 79–80, 84–8, 106, 289–90, 319
Corcoran, P. L. 84
Corsica 164
Costa Rica 270
Costanza, R. 366
crabs 32, 252
Crete 335
Croatia 164
crocodiles 77
Cronon, W. 359
Crutzen, P. 97, 154, 246, 255
Cuba 432, 437
cyclones 151, 165, 219
Cyprus 168, 172, 335, 396

$\delta^{13}C$ 176–7, 227, 305
$\delta^{18}O$ 120, 124–7, 155, 164–71, 173–4, 176–9, 228, 265–6, 302, 305
Daley, B. **71–94**
Dang, H. H. 410
Dansgaard–Oeschger (DO) events 125, 151, 168, 172, 308, 317
Davies, M. 334–5

δD 124–5, 165
Dead Sea 173–5
dead zones 79, 81, 83
Defence Meteorological Satellite Program
 (DMSP) 18
deforestation 4, 17, 29, 205, 208, 229, 270, 276, 328,
 339, 348, 367, 395, 440
deMenocal, P. B. 303–4
dendrochronology *see* tree-ring dating
dengue fever 389
Denmark 49
Dennell, R. 306
Derraik, J. G. B. 84
D'Errico, F. 320, 321
desertification 17, 29–30, 153–6, 328
deserts 7–9, 155, 165, 189, 192, 195, 345
 see also arid and semi-arid regions; *individual*
  deserts
detergents 50
Devils Lake 61
Diamond, J. 130, 331, 355–6, 433
diaries 222
diatoms 115, 251
dichlorodiphenyltrichloroethane (DDT) 83–4
Dickinson, W. R. 291
dimethyl suphide (DMS) 86
dischlorodiphenyltrichloroethane (DDT) 252–4
diseases 33, 85, 87, 107, 128, 202, 209, 268, 380,
 387–405, 426, 433–4, 437–8
 climate and disease 388–9
 climate change and disease 389–99
 prospects for the future 399–400
dissolved organic carbon (DOC) 54, 64
DNA analysis 311–12, 318
 ancient DNA (aDNA) 329
 mitrochondrial DNA (mtDNA) 312, 335
Dnepr River 229, 231
dodos 80–1
Doebley, J. F. 330
dolphins 86
domestication 328–44, 346–8, 350
Don River 231
Douglas, M. S. V. **245–61**
droughts 17–18, 59, 126, 131, 219, 361–3, 380, 388,
 394–5, 409, 428
Drysdale, R. 179
dugongs 74, 76–7
dune fields 150, 201
Durden, L. 393
dust 155–6, 265
Dynesius, M. 207

E. coli 387
early warning systems 8
earthquakes 201, 286–7, 435
Easter Island 74, 356
ecotourism 36
Ecuador 343–4

Egypt 144, 179, 316, 335, 356, 361–2, 366
El Niño 19, 123, 387
 see also La Niña
El Niño southern oscillation (ENSO) 10, 115, 118,
 201, 206, 283, 286, 339, 341, 343, 389, 394,
 435, 441
Ellesmere Island 250–1
Endfield, G. **355–73**
Engelman, R. 78
England *see* Great Britain
Environmental Kuznets Curve (EKC) 27
Environmental Protection Agency (US) 269
Enzel, Y. 174
epidemics *see* diseases
Erickson, C. 364
Ethiopia 311
Europe 16, 53–5, 61–2, 64, 102–3, 105, 108, 164,
 173–4, 335, 355, 406, 430, 442
 disease in 388, 392, 396–7, 399–400, 433
 grasslands 218–21, 224–5, 229–39
 hominid evolution and migration 301, 303, 305–11,
  317–20, 322
 mountain regions 264, 269, 272, 274
 temperate forested regions 190–202, 204–5, 207–8
 see also individual countries
European Union 47, 52, 59, 61, 432
eutrophication 48–50, 52, 56, 72, 82, 292, 328
Evelyn, J. 96–7
evolution *see* human evolution
extinction 27–8, 35, 37, 72, 79–81, 87–8, 207–8, 252
 of archaic hominid species 301, 310–11, 320–2
 of megafauna 113, 129–32, 189, 202–5, 208–9
Eyre (Lake) 56, 152
Ezzati, M. 107

Fagan, B. M. 433
Farman, J. C. 99–100, 255
fasciolosis 389
Feary, D. A. 87
Fecke, J. 333, 336, 340
fertilizers 34, 50, 81–2, 328–9, 347, 406
Finland 255
fire 128–9, 131–2, 189, 201–4, 227, 229, 255, 275,
 328, 334, 339–40, 342, 348
fishing 72, 74–5, 77–80, 233, 251, 380, 439
Flannery, K. V. 346
Flannery, T. F. 203
flooding 116, 201, 264, 362, 388–9, 409, 435
Florida 156
flu *see* influenza
Fluin, J. 60
fluvial sediments 145, 148–9
Foley, J. A. 16–17
Foley, R. 306, 315–17
Food and Agriculture Organization (FAO) 16
Foody, G. M. 12
foot-and-mouth disease (FMD) 388, 390
foraminifera *see* microfossils

forests 17, 27–9, 32–3, 49, 130–1, 174–5, 188–214, 229, 233, 246, 255, 270, 274–5, 333–5, 337, 340, 342, 380
  anthropogenic impacts 205–7
  climatic and geomorphic change in temperate forested regions 192–5
  disease and insect attack 202
  distribution and environment of temperate forested regions 189–91
  fire 201–2
  loss of the megafauna 202–5
  soil changes and vegetation 200–1
  tertiary legacy 191–2
  vegetation responses to change 195–200
  see also deforestation
Forrester, J. W. 36
Forster, P. 315
fossil fuels 47, 53–4, 269, 328, 406, 409–10, 412–13, 429
fossils 28, 49, 127, 130, 152, 189, 199, 215, 239, 302–3, 306, 309, 311, 320
  macrofossils 122, 215, 267
  microfossils 266, 339
Foster, D. R. 202
fraction of photosynthetically active radiation (fAPAR) 4–5, 11–12
France 275, 319–20, 396, 400
French, H. M. 245
Frenzel, B. 224
Fritz, S. C. 61
Furgal, C. 246
Füssel, H.-M. 434

Galloway, J. N. 104
Ganges River 263, 270, 286
Gardiner, B. G. 99, 100
Garfinkel, Y. 178
Gell, P. 47–70
general circulation models (GCMs) 6, 222, 245, 249, 251, 271
genetically modified organisms (GMOs) 436, 439
geoengineering 422–3
Germany 306
Gibraltar 320–1
Giddens, A. 438
Gill, J. L. 130, 204
Giorgi, F. 271
glacial–interglacial cycles 148, 153, 157, 175–6, 192–3, 201, 206–7, 265, 302–4, 306, 308–9, 314, 323
glaciers 113–15, 192, 223, 245–6, 249, 252–3, 265–6, 269–70, 272–3
global climate models see general circulation models
Global Precipitation Climatology Centre (GPCC) 222
Goddard Institute of Space Studies (GISS) 6
Gogol, N. V. 215
Gordon, I. 25–46
Gordon, J. E. 245

Gosling, W. D. 113–40, 121
governments 80, 254, 428, 436
  see also mitigation
grasslands 7, 13–14, 215–44, 328, 333–4, 338, 340, 377
  environmental archives 219–22
  grassland modelling 222–3, 235–7
  during the Last Glacial Maximum 223–5
  during the Late Glacial and Holocene 225–34
  regional setting 218–19
  temperate grasslands during the last millennium 234–5
Gravina, B. 320
Great Barrier Reef 80–2
Great Britain 31, 49–50, 53, 63, 88, 96–7, 293, 306–8, 310, 346, 395–6, 399–400, 428, 432–3, 435
Great Oxidation Event 26
Great Plains 218, 224, 226–7, 237–8
Great Salt Lake 56
Greece 96, 164, 173, 178, 225, 229, 287, 289, 335, 361
greenhouse gases 72, 96, 99, 102, 105–6, 108, 154, 156, 209, 245, 249, 256, 272, 277, 407–9, 411–13, 416, 419–23
  see also carbon dioxide; methane
Greenland 126, 148, 150, 155, 166–7, 245, 247, 253, 308, 316–17, 320–1, 341, 361, 365–6
Grötzbach, E. F. 264
groundwater 57, 60, 88, 380
Guatemala 127, 130, 341–2, 363
Guinotte, J. M. 106
Gulf of Mexico 82–3
Gupta, A. K. 127

Hadley cells 118, 126
Hadley Centre 223
haemonchosis 389
Hanselman, J. A. 121
Hansen, B. C. 121
Hansen, M. C. 15
Hansom, J. D. 245
Harlan, J. R. 329
Harris, C. M. 251
Haug, G. H. 363
Hawaii 31, 98–9, 114–15, 287–8, 290
Hayes, A. 166
health see diseases
Heinrich Events 127, 151, 168, 172, 176, 308, 320–2
Helldèn, U. 154
Henderson-Sellers, A. 5–7
Henry, D. O. 345
herbicides 34
hexachlorocyclohexanes (HCHs) 252–4
Hg see mercury
Higham, T. 321
Hillman, G. 334–5
Hillyer, R. 121
Himalayas 107, 114, 263, 269–70, 276, 440
Hippocrates 96–7

historical documents 221–2
HIV 437
Hodell, D. A. 127, 341
Hodgson, D. A. 251
Hoegh-Guldberg, O. 85
Holdaway, R. N. 203
Holmgren, K. 147
Holocene 28, 55, 58, 74, 169, 171, 173–4, 176–9, 250,
      265–7, 302, 314, 323, 359
   agriculture and domestication 328–31, 333–46,
      348–50
   arid and semi-arid regions 146–51, 153, 156–7
   coastal regions 282–3, 285–6, 289, 291
   grasslands 218–19, 223, 225–34, 236–8
   temperate zone 192, 194–200, 205–6, 208
   tropical palaeoclimates 114–15, 117–28, 123–32
Hong Kong 88, 102
Hori, K. 286
housing 409
Hövermann, J. 147
Hudson Bay 247
Hughes, T. P. 87
human evolution and climate change 301–27
   anatomically modern humans 311–22
   early hominid evolution and migration 301–8
   hominid evolution during the middle and late
      Pleistocene 308–11
human–environment interactions 25–46, 50
   agriculture see agriculture
   on coastal regions and islands 282, 290–2
   domestication 328–44
   effects on health see diseases
   in grasslands 221–2, 232–3, 238
   human impacts on coastal and marine geo-
      ecosystems 71–94
   human impacts on the atmosphere 95–110
   human impacts on tropical palaeoclimates 128–32
   human-induced pressures on terrestrial
      ecosystems 29–33
   mitigation see mitigation
   in mountain regions 267–71
   in polar environments 246
   social discontinuity 355–73
   a social–ecological systems lens 26–7
   solutions 35–8
   in temperate forested regions 205–7
Hummel, J. 6
hunting 72, 74–7, 129, 233–4, 333, 337, 344, 358
Huntington, E. 363
Hurricane Katrina 432, 435
HYDE data 17

ice ages 130, 153, 209, 345
   Little Ice Age see Little Ice Age
ice cores 115, 126, 148, 150, 155, 166, 179, 265, 302,
      308–9, 316–17, 320–2, 341, 345, 361, 364
ice sheets 116, 124, 128, 192–5, 208, 223–4, 228
Iceland 365

India 52, 107, 117, 126–7, 150, 215, 219, 268–70, 317,
      338–9, 377
Indian Ocean 107, 117, 119, 126–7, 148, 248, 286–7,
      290, 317
Indiana 130
Indonesia 292, 305–6, 430
industrial pollution 32, 50, 75, 82
Industrial Revolution 25–6, 62
influenza 388, 391
insects 31–2, 189, 202, 389
   see also vector-borne diseases
Integrated Biosphere Simulator (IBIS) model 4
Integrated History and Future of the People on Earth
      (IHOPE) 368
Inter Tropical Convergence Zone (ITCZ) 114–15,
      117–19, 122, 124, 126–8, 143, 341–2
Intergovernmental Panel on Climate Change
      (IPCC) 33, 35, 108, 245, 272, 275, 377, 379, 418,
      427–8, 430, 436
International Arctic Science Committee (IASC) 256
International Institute for Sustainable Development 429
International Satellite Land-Surface Climatology
      Project (ISLSCP) 8
International Union for Conservation of Nature 77, 87
Ionian Sea 164
Iran 225, 229, 335
Iraq 268, 333, 335
Iron Age 6, 49
irrigation 47, 154, 274, 328, 347, 366, 411
islands
   environmental histories of 288–91
   human impacts on 291–2
   non-human causes of environmental change in
      283–8
   prospects for 292–3
Isreal 96, 175–6
Italy 164, 173, 175, 179, 195, 225, 321–2
IUCN Red Data Book 27
Iverson, J. 200

Jackson, J. B. C. 87
Jacquet, J. L. 80
Jäkel, D. 149
Jankaew, K. 286
Jansson, R. 207
Janzen, D. 203–4
Japan 75, 103, 188, 190–1, 193, 195, 199–200,
      264, 287
Java 131
Jennings, S. 79
Jensen, A. 77
Jeppesen, E. 52
Jimenez-Espejo, F. J. 321
Joint UK Land Environment Simulator (JULES) 4
Jones, G. 365
Jones, M. D. 173
Jordan 165
Jordan, P. 49

Joris, O. 320, 321, 322
Judean Mountains 174

Kaiser, M. J. 79
Kalahari Desert 145–7, 154
Kansas 224
Kaplan, J. O. 190, **215–44**
Karlèn, W. 266
Kasperson, J. X. 358
Kasperson, R. E. 358
Kazakhstan 231–2
Keeling, C. 98
Keller, F. 275
Kenya 264, 389
Kettle Lake 227
Khentey Mountains 233
Kilimanjaro 266
Kim, K. M. 107
Kinabalu 274
Klein, R. J. T. 434
Klein Goldewijk, K. 16
Kolata, A. L. 364
Korea 205
Kremenetski, K. V. 229
Kröpelin, S. 123
Kuznets, S. 27
  see also Environmental Kuznets Curve (EKC)
Kuznetsova, T. V. **215–44**
Kyoto Protocol 410, 440

La Niña 58, 123, 152, 339
  see also El Niño
La Yeguada (Lake) 342
Labrador Sea 247
lacustrine ecosystems 47–70, 149, 151, 245
  acidity 53–6
  nutrients 48–53
  salinity 56–61
  toxic substances 61–4
Lahr, M. M. 315–17
Lake 227 52
lakes 50–2, 56, 60, 119–20, 122–5, 127, 130, 141, 145,
    147, 149, 151–2, 157, 178–9, 189, 200, 218–21,
    229, 266, 314, 341–2, 364–5
  ecosystems see lacustrine ecosystems
  Mediterranean lakes 172–5
  polar lakes 250–1
Lancaster, N. 145
land cover 3–24
  biomass estimation from satellite data 11–12
  global land cover classifications 5–7, 12–16
  land-cover change 16–19
  monitoring from satellite 7–11
land use 221–2
  in arid and semi-arid regions 153–5
  land-use change 13, 16, 33, 53, 72, 270–1, 276
  see also agriculture
Landsat data 15–17

landslides 287–8, 409
Last Glacial Maximum 113–17, 144–5, 147, 150, 152,
    154–5, 157, 171–4, 176, 192–3, 195–201, 216,
    218, 220, 223–5, 235–9, 266–7, 283–5, 310, 323,
    333, 338, 342
Late Glacial
  agriculture and domestication 331, 337–8, 340, 350
  in grasslands 218, 225, 231
  in temperate forested regions 192, 194, 197,
    199, 204
Lau, K. M. 107
Laybourn-Parry, J. 250
lead (Pb) 61, 63, 81, 254
leaf area index (LAI) 4–5, 11, 223
Lebanon 174–5
Lejeusne, C. 85
Leroy, S. A. G. 367
Levant 164–5, 169–70, 172–5, 308, 311, 315, 331, 333
Li, Y. F. 252
Liao, H. 104
Libya 313–14, 316
Light Detection and Ranging (LiDAR) 11–12, 19
Ligurian Sea 164
Lisan (Lake) 173–5
Lisker, S. 174
Little Ice Age 58, 127–8, 266, 433
Liu, Z. 121
Loess Plateau 148–9
Los, S. O. **3–24**
Lovelock, J. 439
Lukes, S. 408
luminescence dating 141, 145, 153, 289, 311–13,
    315, 317
Lund–Potsdam–Jena (LPJ) model 4
Lyme disease 33
Lynas, M. 438
Lyons, W. B. 245

Macdonald, R. W. 252–3
MacNaghten, P. 358
macrofossils 122, 215, 267
Madagascar 270
Madsen, D. B. 336
malaria 389, 392, 398–9, 433, 437
Malawi (Lake) 122, 124–5
Malay Peninsula 116
Mallory, M. L. 84
Malthus, T. 346–7
manatees 74
marine geo-ecosystems 71–94
  effect of climate change on 84–8
  exploitation of 73–81
  polar marine ecosystems 251–2
  pollution of 81–4
Markgraf, V. 206
Marsh, H. 77
marshes 88
Martin, P. S. 203

mass extinction 27, 28
Matthews, E. 5–7, 14–16
Matthews, J. 369
Mauna Loa observatory 98–9
Maunder Minimum 128
Mauritania 144
Mauritius 32
Maxwell, A. L. 123
Maya civilization 356, 362–3
Mayewski, P. A. 177, 178
Mayle, F. E. 342
McCarthy, J. J. 427
McGlone, M. **188–214**
McLaughlin, C. J. 81
McNeill, J. R. 368
Meadow, R. H. 329
Meadows, D. H. 35
Mearns, L. O. 271
Medieval Climate Anomaly (MCA) 58, 127
Mediterranean region 163–87
    as bridge between Atlantic Ocean and Mediterranean
        Sea 164–5
    Mediterranean lakes 172–5
    palaeoclimate of the Mediterranean Sea 165–72
    present-day distribution of 163
    speleotherm record 175–80
Mediterranean Sea 82, 85, 164–72, 275, 314, 321
Medium Resolution Imaging Spectrometer (MERIS) 8,
    10, 15, 17, 19
Mega-Chad (Lake) 147, 157
megafauna 113, 129–32, 189, 202–5, 208–9, 236
Meghna River 286
Meinardus, W. 147
Mellars, P. 312, 314, 320
Menteith (Lake) 50–1
mercury (Hg) 61, 63–4, 81, 253
Mesolithic 25, 74, 205
Mesopotamia 29, 58, 178–9, 361, 366
Messerli, B. 358
Met Office (UK) 4
methane (CH$_4$) 97, 99, 102, 106, 154, 406
Mexico 115, 130, 151, 342, 344, 346, 363–4
Meyer, W. B. 361–2
mice 30–1
microfossils 266, 339
migration 207, 268, 301–8, 365–6, 436, 441–2
Milankovitch theory 302
Millennium Development Goals (MDGs) 437
Millennium Ecosystem Assessment (MEA) 26–8, 33,
    35, 437
Miller, C. **25–46**
Minnesota 224, 228
Miocene 148, 252, 301
Mississippi River 287
mitigation 406–25, 427–9, 433, 441
    plans 421–3
    policy 407–17
    programmes 417–21

mitrochondrial DNA (mtDNA) 312, 335
Moderate-resolution Imaging Spectroradiometer
    (MODIS) 8, 10, 15, 17–19
Mojave Desert 151, 157
Moldova 229–30
Molina, M. J. 101, 255
molluscs 83
Mongolia 150, 157, 219, 222, 231–4, 236, 238
Mono Lake 151
monsoons 60, 107, 114, 117–19, 122, 124, 126–8,
    147–8, 169–70, 179, 190, 193–5, 201, 208, 219,
    232, 238, 336–7
Montana 152
Montenegro 164
Moon, K. **406–25**
Moore, C. J. 84
Moreton Island 75
Morocco 164, 173, 264, 293, 313, 316, 431
Morrill, C. 126
Morse, A. P. **387–405**
mosquitoes 389, 394–5, 398
    see also vector-borne diseases
Mount St Helens 435
mountain regions 262–81
    anthropogenic environmental change in 267–71
    future environmental change in 271–7
    importance of 262–4
    past environmental change in 265–7
    see also individual mountain ranges
Moy, C. M. 121
Mueller, D. R. 251
Muir, D. C. G. 63, 254
Mullen, G. 393
Muller, R. A. 28
Murray, S. N. 80
Murray River 59–60
Myanmar 338

Namib Desert 163
Namibia 122, 145, 147
NASA 18
National Center for Atmospheric Research (NCAR) 6
National Oceanographic and Atmospheric Association
    (NOAA) 7, 9, 99, 100, 101
National Water Initiative (Australia) 47
Neanderthals 309–11, 315–22
Nebraska 151, 224
Negev Desert 165, 170, 176
Nelson, D. R. **374–86**
Neogene 207
Neolithic 49, 205, 230, 334–7, 340
Neotropics 117–22, 124, 130, 341–2, 350
Nepal 270
New Guinea 114, 123, 130, 270, 335, 338–9, 350
New Zealand 31, 74, 123, 190–5, 199–204, 206–7,
    248, 264, 266, 290
Newman, P. A. 255
Nichol, S. L. 290

nickel (Ni) 61
Nile 29, 144, 168–9, 170, 172, 314, 362
nitrogen 48, 50, 52–5, 81–3, 101, 102
Noone, K. J. **95–110**
Norin, E. 149
North America 19, 32, 53–4, 62, 85, 103, 105, 108,
    115, 126, 127, 129–32, 155, 216, 254, 290, 406,
    430, 432–3
  agriculture and domestication 340–4
  arid and semi-arid regions 142, 143, 150–2, 156–7
  disease in 396, 399
  grasslands 218, 220–2, 224–9, 234–9
  hominid evolution and migration 308, 323
  mountain regions 264, 267, 269, 275
  temperate forested regions 190–208
  see also individual countries/US states
North Atlantic Deep Water (NADW) 166
North Dakota 61
North Sea 82
Northern Contaminants Program (NCP) 254
Norton, D. A. 34
Norway 53, 255, 366
Norwegian Sea 232
nuclear power 439
Nunn, P. D. **282–97**

O'Brien, K. L. 377
Ocean Drilling Programme (ODP) 302, 304, 314
Oceania 17
Ogden, J. 290
oil 53, 83, 88, 95, 422
Okhotsk Sea 247
Oldfield, F. 85, 89
Olson, J. S. 5–6, 14–16
Oman 126–7, 179
optically stimulated luminescence (OSL) 141, 289,
    313, 315, 317
orang-utans 32
Oregon 12, 81
O'Riordan, T. 413
Orr, J. C. 106
Österle, H. **215–44**
Owens Lake 150
oxygen 26, 49
  see also $\delta^{18}O$
ozone depletion 101–2, 246, 254–6, 269, 328, 406,
    413, 426, 428

Pacific Ocean 86, 88, 115, 190, 218, 247–8, 285, 288,
    339, 341
Paduano, G. M. 121
Pakistan 264, 339
Palaeolithic 301, 320–2, 322, 347
Palaeotropics 122–3
Pallcachoca (Lake) 341
Panama 344
Pandis, S. N. 255
Pandolfi, J. M. 87

Pashennoe Lake 231–2
Patagonia 200, 204, 206
Pauly, D. 78, 80
Pavlopoulos, K. 289
Pb *see* lead
Pearsall, D. M. **328–54**, 332
Pedoja, K. 289
penguins 251
perfluorinated acids (PFAs) 64
permafrost 193, 245–6, 249–50, 252–3
Pershing, A. J. 76
persistent organic pollutants (POPs) 61–4, 252–3
Peru 124, 127, 152, 157, 264–5, 343–4, 366
peste des petits ruminants (PPR) 388, 390
pesticides 34, 83
Petén-Itzà (Lake) 130, 341
Petraglia, M. 317
pH levels 53–6, 58, 104–6, 269
  see also acidification
Phanerozoic 28
Philippines 29
phosphorus 48–52, 81–3
photosynthesis 3–4, 223
Pienitz, R. 251
Piperno, D. R. 338
plague 389
Pleistocene 117–18, 128, 130, 166, 173, 175, 191,
    204, 206
  agriculture and domestication 331, 334, 337,
    342–6, 349
  arid and semi-arid regions 142–53, 156
  grasslands 216, 221, 236
  hominid evolution and migration 302–4, 306–11,
    322–3
Pliocene 128, 130, 148, 191, 301, 303–4, 306
polar bears 252
polar environments *see* Antarctic; Arctic;
    Greenland
pollen analysis 114–15, 119–20, 122, 127, 150–2,
    170, 173–5, 189, 206, 215, 219–20, 227, 229–33,
    237–8, 341–2, 361
polychlorinated biphenyls (PCBs) 252–4
polycyclic aromatic hydrocarbons (PAHs) 252–3
Pongratz, J. 222, 234
population change 27–8, 34, 74, 95, 128–9, 154, 268,
    282, 292, 313, 315–18, 329, 339, 346–7, 358,
    363–4, 367, 417–18, 434
Portugal 171, 285, 318, 396
poverty 27, 429, 435–6, 438
Power, M. J. 342
Poyang Lake 337
precautionary principle (PP) 431
precipitation 32, 131, 147, 154, 163–4, 173–4, 178,
    193, 216, 233–5, 249, 268, 272, 275, 328, 341–2,
    349, 390, 441
Prober, S. M. 32
Prowse, T. D. 246, 251
Psenner, R. 55

Quaternary 116, 130, 141, 144–7, 149–50, 153, 155–6,
        188, 191–2, 202–3, 207, 218–21, 233, 282–5,
        289–90, 302, 311, 317, 323
    see also Holocene; Pleistocene

Rabalais, N. N. 82
rabies 390
radiocarbon dating 220, 231, 266–7, 318–19, 321
radiometric dating 303
rainfall 17–18, 28, 57, 59, 61, 64, 107, 117, 150, 155,
        163, 165, 177, 179, 189, 191, 333, 341, 346,
        378–80, 388–9, 394, 409
rainforests 6, 14, 31, 86, 131, 338
Ramanathan, V. 104
Ramankutty, N. 16–17
Ramsar 59–60
Rapa Nui 74, 356
Rassamakin, J. J. 230–1
rats 30–1
Raupach, M. R. 108
Reck, R. 6
Red Data Book 27
Red Sea 87, 176, 179, 311, 316
Reed, B. 15
regional climate models 271–2
Ren, G. 233
resilience studies 374–83, 427–9
    evolution of the concept 375–9
    key challenges 381–3
    overlap with vulnerability studies 379–81
    sustainability science 383–4
Richardson, J. B. 341
Richter, A. 102
Richthofen, F. 148
Ricklefs, R. E. 192
Rift Valley Fever (RVF) 389
River Euphrates 154
rivers 64, 115–16, 154, 164, 273–4, 349
    see also individual rivers
Roberts, N. 71, 173
Rocky Mountains 151, 218, 271
Rodhe, H. 104–5
Rodó, X. 341
Rodriguez-Arias, M.-A. 341
Roebroeks, W. 306
Rohde, R. A. 28
Romania 30
Rose, N. **47–70**, 63
Rosen, A. M. 331, 333–4
Rosenmeier, M. F. 178
Rothkin, K. 414
Rothman, H. 359–60
Round Loch of Glenhead 53–5
Rouse, W. R. 250
Rowland, F. S. 101
Rowland, S. 255
Rub Al Khali 150, 157
Ruddiman, W. 97, 107

runoff 48, 81, 153, 164, 262, 270, 273, 276
Russell, J. M. 122
Russia 247, 250, 321

Sahara 17, 105, 122, 126, 132, 142–7, 156, 163, 165,
        170–2, 176, 311, 313–14, 316–17
salinity 29–30, 34, 48, 56–61, 73, 88, 154, 172, 328
    sea surface salinity (SSS) 72–3, 84, 89, 164,
        170, 172
salmonellosis 388, 390
Sanchez-Goni, M. F. 320–1
Sandweiss, D. H. 341
SARS 387
Satellite Pour l'Obersvation de la Terre (SPOT)
        8, 10, 15
satellites 4, 19, 100, 102, 249, 257
    biomass estimation from 11–12
    land cover monitoring from 7–11
    land-cover change detection from 17–19
    land-cover classifications from 12–13
Sauer, C. O. 344
Scarborough, V. L. 363
Scavia, D. 85
Schindler, D. W. 250
Schmidt, R. 55
Scientific Committee on Antarctic Research
        (SCAR) 256
Scotland see Great Britain
Screen, J. A. 249
Scripps Institute of Oceanography 98
sea levels 72–3, 89, 115–16, 176, 246, 249, 257,
        282–6, 289–93, 316, 337, 379, 412, 431–2
Sea of Marmara 164
sea surface temperature 72–3, 84–9, 118, 164, 166,
        171–2, 175, 249, 302, 304
seals 72, 74, 76, 251
Sea-viewing Wide Field-of-view Sensor (SeaWiFS) 8,
        10, 19
Second World War 26–7, 62
sedimentology 48–9, 52, 57–8, 62, 64–5, 72, 81–2,
        88, 144–5, 151, 167–8, 215, 218, 224, 245, 251,
        265–6, 276, 289, 302, 305, 337, 341
    see also fluvial sediments
Seinfeld, J. H. 255
Seltzer, G. O. 121
semi-arid regions see arid regions
Senegal 144
Serreze, M. C. 249
severe acute respiratory syndrome (SARS) 387
sewage see wastewater
Shanklin, J. D. 99, 100
sharks 76, 253
shells 58, 74, 80, 251
Shi, N. 147
Shuman, B. 152
Siberia 106, 129, 219, 224–5, 231
Signy Island 250
Simmonds, I. 249

Simple Biosphere (SiB) model 4
Singapore 88
Slaymaker, O. 156, 245
Slovenia 164
Sluyter, A. 356
Smale, D. A. 252
Smart, K. 332
Smith, C. R. 83
Smith, V. H. 83
Smol, J. P. 250–1
social–ecological systems (SESs) 375, 378–82
societies 374–86
    social discontinuity 355–73
    sustainability science 383–4
    vulnerability and resilience studies 374–83
soils 5, 7, 31–2, 47, 53, 81, 104, 117, 154, 156, 189,
        193, 199–203, 208–9, 218, 223, 268–70, 275, 328,
        348, 364, 390, 409, 431–3
soluble reactive phosphorus (SRP) 48
Somalia 389
Sommaruga-Wögrath, S. 55
Sonak, S. 84
Sonoran Desert 163
South Africa 31, 145, 147–8, 163, 190, 216, 305, 389
South America 14, 17, 19, 114–19, 126, 129–30, 132,
        247–8, 254, 270, 323, 396
    agriculture and domestication 340–4
    arid and semi-arid regions 142–3, 150–2, 156
    grasslands 215–16, 235–6
    social discontinuity 363–4
    temperate forested regions 190–4, 199, 201, 204–7
    see also individual countries
Southern Ocean 251, 254
Spain 164, 173, 396
speleotherms 115, 170, 175–80, 317
Spörer Minimum 128
Sri Lanka 131
Srivastava, P. 145
Stahl, P. W. **328–54**
stalagmites 141, 147
Staubwasser, M. 178–9
Steel Lake 228
Sterken, M. 121
Sterman, J. D. 25, 36
Stern Review 428, 430
Stiros, S. C. 289
stock-and-flow models 36–7
Stockholm Convention 61
Stoermer, E. F. 246
Stohl, A. 103
Stone Age 312, 316–17
Stonehouse, B. 245, 247–8
storms 151, 155, 193, 246, 274, 287, 432
Street, M. 320–2
Strezlecki Desert 153
Sudan 144, 335
Sugden, D. 245
sulphur 53–5

Sumatra 32
sustainable development 426–46
    future of 437–42
    response to threats and opportunities 434–7
    sustainability science 383–4
Svenning, J. C. 204
Sweden 53
swine flu 387
Switzerland 264, 267, 273
Syrdaria River 439
Syria 165, 361
systems dynamics 36

Tahiti 287
Tainter, J. 362
Taklamakan Desert 149
Tanganyika (Lake) 122, 124–5
Tans, P. 98
Tarasov, P. 205, **215–44**
Tasmania 130, 190–1, 195, 206, 338
terrestrial biota and ecosystems 25–46
    agriculture and the future 34–5
    ecosystem services 33–4
    human-induced pressures on terrestrial
        ecosystems 29–33
    outlook 39
    status, conditions and trends 27–9
Tertiary 144, 191
Thailand 123, 338, 340
thermoluminescence (TL) dating 141, 311, 313
Thompson, L. G. 114
Tibesti Mountains 145
Tibetan Plateau 61, 107, 114, 148, 236, 263
Tierney, J. E. 122
Tigris River 154
Tirari Desert 153
Titicaca (Lake) 120, 364
Tiwanaku State 356
toads 31
Torres Strait 77
total nitrogen (TN) 48
total phosphorus (TP) 48–50, 52
Tottrup, C. 154
tourism 36, 50, 86, 252, 264, 271, 276
toxic substances 47–8, 61–4
tree-ring dating 151, 265, 267, 318–19, 363
tributyl tin (TBT) 61, 84
tropical palaeoclimates 113–40
    the Holocene 118–28
    human impacts 128–32
    the last deglaciation 113–18
tsetse flies 389, 393
Tsigaridis, K. 104
Tsodilo (Lake) 145
tsunamis 286–7, 290, 292, 426, 430
tuberculosis 387, 390
tuna 76
tundra 6–7, 28, 189, 195, 199, 235, 237, 246, 249, 267

Tunisia 316
Turkey 164, 174, 335
Turner, B. L. 376–7
turtles 74, 76–7
Tuttle, N. C. 31
Tzedakis, P. C. 321

Ukraine 229
ultraviolet radiation 73, 101, 246, 250, 254, 391
UNESCO 5–6
United Arab Emirates 311
United Kingdom *see* Great Britain
United Nations Conference on Environment and
    Development (UNCED) 153
United Nations Convention to Combat Desertification
    (UNCCD) 153–4
United Nations Economic Commission for Europe
    (UNECE) 108
United Nations Environment Programme (UNEP) 270,
    436, 440
United Nations Framework Convention on Climate
    Change (UNFCCC) 108, 412, 427, 436–7,
    439, 441
University of Maryland 14
urbanization 34
Urrego, D. H. 121
Urry, J. 358

Vaks, A. 174
Valencia, B. G. 121
Van der Leeuw, S. 359, 360
Vandenberghe, J. 148
variant Creutzfeldt–Jakob disease (vCJD) 387
vector-borne diseases 388–90, 391, 393–6, 399, 401
    bluetongue 396–8, 400
    malaria 389, 392, 398–9
Venezuela 127, 341
Verheyden, S. 178
Verstraete, M. 154–5
Vietnam 131, 433
Vincent, W. 250
Vogel, C. 358
volcanic activity 56, 117, 201, 204, 288, 290, 302, 306,
    311, 316–18, 321, 361, 426
Vrba, E. S. 304
vulnerability studies 374–83, 427–9, 434–5
    evolution of the concept 375–7
    key challenges 381–3
    overlap with resilience studies 379–81
    sustainability science 383–4

Wagner, M. **215–44**
Wake Island 81
Wales *see* Great Britain
Walker, B. 33
Wanner, H. 194
Ward, J. V. 121
Washington (Lake) 50
wastewater 48, 50, 52, 81–2
Water Framework Directive (EU) 47, 52, 59, 61
Watt-Cloutier, S. 256
Weiss, E. 334
Weiss, H. 178–9, 355, 360–1
Weng, C. 121
Weninger, B. 178
West Nile fever 387, 389
West Olaf Lake 228
Western Desert 144
wetlands 29
whales 72, 74–6, 86
Willcox, G. 334
Williams, J. 206, **215–44**, 228
Williams, M. A. J. 316–17
Wilson, H. 333, 336, 340
Wilson, M. F. 5–7, 14
Windermere (Lake) 50
Wolfe, B. B. R. A. 121
Wood, J. **188–214**
World Development Report 432
World War II 26–7, 62
Worthy, T. H. 203

Yang, X. **141–62**, 149
Yangtze River 205, 263, 335, 337–8
Yellow River 149, 335–8
Yellow Sea 219
Yellowstone National Park 275
Yohe, G. 430
Younger Dryas (YD) cold event 114–16, 127, 131, 172,
    175, 199, 333–4, 337, 339

Zagros mountains 335, 361
Zanchetta, G. 179
Zazari (Lake) 178
Zeribar (Lake) 229
zero-carbon *see* mitigation
Zhao, Z. 338
Zhou, W. 337
Zhu, Z. 149
Zilhao, J. 320
zinc (Zn) 61, 254